Jäger · Rannacher · Warnatz (Eds.)
Reactive Flows, Diffusion and Transport

Willi Jäger · Rolf Rannacher · Jürgen Warnatz
Editors

Reactive Flows, Diffusion and Transport

From Experiments via Mathematical Modeling
to Numerical Simulation and Optimization

Final Report of SFB
(Collaborative Research Center) 359

With 293 Figures, 200 in Color, and 39 Tables

 Springer

Editors

Willi Jäger
Rolf Rannacher
Jürgen Warnatz

Interdisziplinäres Zentrum
für Wissenschaftliches Rechnen (IWR)
Universität Heidelberg
Im Neuenheimer Feld 368
69120 Heidelberg, Germany

jaeger@iwr.uni-heidelberg.de
rannacher@iwr.uni-heidelberg.de
warnatz@iwr.uni-heidelberg.de

Library of Congress Control Number: 2006933613

Mathematics Subject Classification (2000): 92-XX, 76-XX, 85-XX, 35-XX, 49-XX, 65-XX

ISBN-10 3-540-28379-X Springer Berlin Heidelberg New York
ISBN-13 978-3-540-28379-9 Springer Berlin Heidelberg New York

This work is subject to copyright. All rights are reserved, whether the whole or part of the material is concerned, specifically the rights of translation, reprinting, reuse of illustrations, recitation, broadcasting, reproduction on microfilm or in any other way, and storage in data banks. Duplication of this publication or parts thereof is permitted only under the provisions of the German Copyright Law of September 9, 1965, in its current version, and permission for use must always be obtained from Springer. Violations are liable for prosecution under the German Copyright Law.

Springer is a part of Springer Science+Business Media

springer.com

© Springer-Verlag Berlin Heidelberg 2007

The use of general descriptive names, registered names, trademarks, etc. in this publication does not imply, even in the absence of a specific statement, that such names are exempt from the relevant protective laws and regulations and therefore free for general use.

Typesetting by the editors using a Springer TEX macro package
Production: LE-TEX Jelonek, Schmidt & Vöckler GbR, Leipzig
Cover design: WMXDesign GmbH, Heidelberg

Printed on acid-free paper   46/3100/YL - 5 4 3 2 1 0

Preface

This volume contains a series of articles which present new developments in the mathematical modeling, numerical simulation and optimization as well as the experimental characterization of complex processes in Physical Chemistry, Astrophysics and Environmental Physics.

The presented results have grown out of work done in the *Sonderforschungsbereich 359* (SFB 359) 'Reactive Flow, Diffusion and Transport' at the University of Heidelberg in which mathematicians, chemists and physicists collaborated on developing new methods for a better understanding of complex chemical and physical processes. The long-standing support by the German Research Foundation (DFG) is gratefully acknowledged.

The SFB 359 was established 1993 under the umbrella of the Interdisciplinary Center for Scientific Computing (IWR) and ended on December 31, 2004. It had about 100 members, i.e., senior scientists, post-docs and PhD students, who belonged to 10 of the 32 research groups of IWR. The research program of SFB 359 concentrated on mathematically based models for physical and chemical processes which share the difficulty of interacting diffusion, transport and reaction. Typical examples from Physical Chemistry are chemical reactions in flow reactors and catalytic combustion at surfaces, both requiring the consideration of multidimensional mass transport. In Astrophysics realistic modelling of star formation involves diffusive mass transport, energy radiation, and detailed chemistry. Processes of similar complexity occur in Environmental Physics in modeling pollutant transport in soil and waters. The central goal has been the development of solution methods for such problems which are theoretically supported and generally applicable.

According to the research program of SFB 359 the articles in this volume are organized in chapters as follows:

I. *Mathematical analysis for transport-reaction systems*
The articles in the first chapter deal with the analysis of fluid flows and free boundaries, and nonlinear evolution processes, including geometry aspects.

II. Navier-Stokes equations and chemical reactions
In the second chapter new numerical approaches are described for solving multidimensional reactive flow problems by adaptive finite element methods with emphasis on mesh and model adaptivity combined with multigrid techniques and parallel processing.

III. Optimization methods for reactive flows
The third chapter presents new developments in numerical optimal control, e.g., robust parameter estimation, optimal experimental design, partially reduced SQP methods, and adaptive finite element methods for PDE-constrained optimal control problems.

IV. Chemical reaction systems
The fourth chapter is concerned with numerical and experimental methods for the determination of kinetic parameters in laminar flow reactors, and the characterization and optimization of reactive flows in catalytic monoliths and on catalytically active surfaces.

V. Turbulent flow and combustion
The articles in the fifth chapter present results of the numerical simulation of turbulent flows by a multigrid LES method, of turbulent non-reacting and reacting spray flows, and of transport and diffusion processes in boundary layers of turbulent channel flow.

VI. Diffusion and transport in accretion discs
The sixth chapter deals with the evolution of proto-planetary disks including detailed chemistry and mineralogy and the efficient numerical solution of multidimensional radiative transfer problems.

VII. Flows in porous media
In the seventh chapter analytical and computational multiscale methods are presented for treating reactive flows in porous media and boundary and interface processes. An application is lake dynamics which involves the numerical simulation and experimental study of unsaturated water flow in heterogeneous systems.

VIII. Computer visualization
This last chapter contains descriptions of software tools for the computer visualization of numerical data resulting from computer simulations: the interactive VTK-based graphics packages HiVision and VisuSimple, and a tool for efficient volume rendering in scientific applications.

Heidelberg,
September 2006

Willi Jäger
Rolf Rannacher
Jürgen Warnatz

Contents

**Part I Mathematical Analysis
for Transport-Reaction Systems**

Preamble 3

Fluid Flows and Free Boundaries
B. Schweizer, S. Bodea, C. Surulescu, I. Surovtsova 5

Nonlinear Evolution Equations and Applications
W. Jäger, Th. Lorenz, A. Tambulea 25

**Part II Navier-Stokes Equations
and Chemical Reactions**

Preamble 45

Mesh and Model Adaptivity for Flow Problems
R. Becker, M. Braack, R. Rannacher, T. Richter 47

Parallel Multigrid on Locally Refined Meshes
R. Becker, M. Braack, T. Richter 77

**Solving Multidimensional Reactive Flow Problems with
Adaptive Finite Elements**
M. Braack, T. Richter 93

Part III Optimization Methods for Reactive Flows

Preamble 115

Robustness Aspects in Parameter Estimation, Optimal Design of Experiments and Optimal Control
H. G. Bock, S. Körkel, E. Kostina, J. P. Schlöder 117

Multiple Set Point Partially Reduced SQP Method for Optimal Control of PDE
H. G. Bock, E. Kostina, A. Schäfer, J. P. Schlöder, V. Schulz 147

Adaptive Finite Element Methods for PDE-Constrained Optimal Control Problems
R. Becker, M. Braack, D. Meidner, R. Rannacher, B. Vexler 177

Part IV Chemical Reaction Systems

Preamble 209

Determination of Kinetic Parameters in Laminar Flow Reactors. I. Theoretical Aspects
T. Carraro, V. Heuveline, R. Rannacher 211

Determination of Kinetic Parameters in Laminar Flow Reactors. II. Experimental Aspects
A. Hanf, H.-R. Volpp, J. Wolfrum 251

Optimization of Reactive Flows in a Single Channel of a Catalytic Monolith: Conversion of Ethane to Ethylene
H. G. Bock, O. Deutschmann, S. Körkel, L. Maier, H. D. Minh, J. P. Schlöder, S. Tischer, J. Warnatz 291

Reaction Processes on Catalytically Active Surfaces
O.R. Inderwildi, D. Starukhin, H.-R. Volpp, D. Lebiedz, O. Deutschmann, J. Warnatz 311

Stochastic Modeling and Deterministic Limit of Catalytic Surface Processes
J. Starke, C. Reichert, M. Eiswirth, K. Oelschläger 341

Part V Turbulent Flow and Combustion

Preamble 373

Multigrid Methods for Large-Eddy Simulation
A. Gordner, S. Nägele, G. Wittum 375

Modeling and Simulation of Turbulent Non-Reacting and Reacting Spray Flows
H.-W. Ge, D. Urzica, M. Vogelgesang, E. Gutheil 397

Transport and Diffusion in Boundary Layers of Turbulent Channel Flow
S. Sanwald, J. v. Saldern, U. Riedel, C. Schulz, J. Warnatz, J. Wolfrum 419

Part VI Diffusion and Transport in Accretion Discs

Preamble 435

Evolution of Protoplanetary Disks Including Detailed Chemistry and Mineralogy
H.-P. Gail, W. M. Tscharnuter 437

Numerical Methods for Multidimensional Radiative Transfer
E. Meinköhn, G. Kanschat, R. Rannacher, R. Wehrse 485

Part VII Flows in Porous Media

Preamble 529

Multiscale Analysis of Processes in Complex Media
W. Jäger, M. Neuss-Radu 531

Microscopic Interfaces in Porous Media
W. Jäger, B. Schweizer 555

High-Accuracy Approximation of Effective Coefficients
N. Neuss 567

Numerical Simulation and Experimental Studies of Unsaturated Water Flow in Heterogeneous Systems
P. Bastian, O. Ippisch, F. Rezanezhad, H. J. Vogel, K. Roth 579

Lake Dynamics: Observation and High-Resolution Numerical Simulation
C. von Rohden, A. Hauser, K. Wunderle, J. Ilmberger, G. Wittum, K. Roth 599

Part VIII Computer Visualization

Preamble 623

Advanced Flow Visualization with HiVision
S. Bönisch, V. Heuveline 625

VisuSimple: An Interactive Visualization Utility for Scientific Computing
Th. Dunne, R. Becker 635

Volume Rendering in Scientific Applications
S. Krömker, J. Rings 653

List of Contributors 667

Part I

Mathematical Analysis for Transport-Reaction Systems

Preamble

This chapter contains two articles which are concerned with the theoretical analysis of nonlinear evolution systems.

The first article *"Fluid flows and free boundaries"* gives an overview of results concerning viscous fluids with a free boundary modeling, for example, the motion of water with a water–air interface. The relevant tools and the major problems in the analysis of the corresponding equations are described. The focus is on results that concern existence, analysis of spectra, bifurcation, contact angles, coupling with elastic materials, and asymptotic behavior.

The second article *"Nonlinear evolution equations and applications"* deals with special kinds of evolution processes such as occurring in modeling the interaction of biological species in so-called "gradostats". The main issues of this study are persistence results for species competition, bifurcations, and coexistence equilibria. Another central aspect is the evolution of shapes if geometry plays a role. Set-valued maps are considered for describing evolution in metric spaces and Aubin's concept of mutational equations is extended.

Fluid Flows and Free Boundaries*

B. Schweizer[1], S. Bodea[2], C. Surulescu[2], and I. Surovtsova[2]

[1] Mathematisches Institut, Universität Basel
[2] Institut für Angewandte Mathematik, Universität Heidelberg

Summary. We give an overview over results concerning viscous fluids with a free boundary, modelling e.g. the motion of water with a water-air interface. We describe the relevant tools and the major problems in the analysis of the corresponding equations. Our focus are results that concern existence, analysis of spectra, bifurcation, contact angles, coupling with elastic materials, and asymptotic behavior.

1 Introduction

Due to their frequent appearance in physical or biological applications, systems involving phase boundaries or multiple-fluid flow are a fastly developing field in mathematical analysis. In this contribution we are concerned with a fluid that is described by the incompressible time-dependent Navier–Stokes equations and which exhibits a free boundary. The difficulty in these problems is that the domain which is occupied by fluid is not given in advance. Instead, the domain in which the equations are satisfied must be determined as a part of the problem.

We start this contribution by sketching the fundamental techniques for existence results in section 2. We will see why spaces of regular function are used and why the results are necessarily local in nature. All this is done for fluid flow in a domain with a free surface and with surface tension included. We will then go beyond the standard setting and describe some of the further results that were acchieved within this project. In section 3 we sketch results on the spectrum of the linearized operator and on bifurcations. In section 4 we indicate how the coupling with a second material can be analyzed. We see this contribution also as an opportunity to describe some older results from todays perspective.

---

*This work has been supported by the German Research Foundation (DFG) through SFB 359 (Project A1) at the University of Heidelberg.

2 Existence Theories

In this contribution we will always assume that the domain in which the fluid equations are posed is of a simple form. We denote the unknown domain by $\Omega = \Omega_t$ and assume that the free boundary can be written as a graph with a height function h, that is, we consider $\Omega_t = \{(\tilde{x}, x_3) | H < x_3 < h(\tilde{x}, t)\}$. We will describe three- as well as two-dimensional results. In order to do so, we let x denote either a real variable, $\tilde{x} = x_1 \in \mathbb{R}$, or a point in the plane, $\tilde{x} = (x_1, x_2) \in \mathbb{R}^2$. We will indicate the range of applicability in the results.

The range for x will be most of the time a bounded domain, that is $\tilde{x} \in (0,1)$ or $\tilde{x} \in (0,1)^2$, depending on the dimension. In the case of a bounded domain we assume periodicity conditions on the lateral boundaries, that is, the height extends to a periodic function on $\mathbb{R}$ (or $\mathbb{R}^2$), and bulk quantities like velocity and pressure also extend to periodic functions in x. The depth is assumed to be constant, either with a finite value, $H < 0$, or set to be $H = -\infty$. The Navier–Stokes equations with viscosity $\nu > 0$ are posed on $\Omega = \Omega_t$, in three dimensions

$$\partial_t v - \nu \Delta v + (v \cdot \nabla) v + \nabla p = 0, \tag{1}$$
$$\nabla \cdot v = 0, \tag{2}$$

and the boundary conditions on $\Gamma = \Gamma_t = \text{graph}(h(.,t))$ are

$$\partial_t h - v_3 + \partial_1 h\, v_1 + \partial_2 h\, v_2 = 0, \tag{3}$$
$$\partial_n v_\tau + \partial_\tau v_n = 0, \tag{4}$$
$$p - 2\nu \partial_n v_n + \beta \mathcal{H}(h) = g. \tag{5}$$

Equation (4) must hold for all tangential vectors τ, here and below n is the exterior normal to the domain. In addition, we are given initial values for v and h,

$$v(.,0) = v_0, \qquad h(.,0) = h_0. \tag{6}$$

In the boundary conditions, equation (3) is the kinematic condition. It expresses that the boundary moves with the fluid and is always parametrized by h. The equation can be derived by differentiating the relation $Y(t) = h(X(t), t)$ for Lagrangian coordinates (X, Y) with respect to time, and using $\partial_t(X, Y) = (v_1, v_2, v_3)$. Equations (4) and (5) are the balance of forces on the free boundary for the tangential and the normal components, respectively. The expression $\mathcal{H}(h)$ stands for the mean curvature of the curve or surface given by h,

$$\mathcal{H}(h)(x,t) = \nabla_x \cdot \left(\frac{\nabla_x h(x,t)}{\sqrt{1 + |\nabla_x h(x,t)|^2}} \right).$$

Note that, if applied to h, the gradient is understood as a gradient in the horizontal ($\tilde{x}$-) directions, only. The number $\beta > 0$ is the physical parameter for the surface tension. In the case of a finite depth, we impose a no-slip boundary condition on the bottom, $v = 0$ on the line $\{x_3 = H\}$.

Existence theories for the above equations are restricted to local results. They are either local in time or they are results for small initial values (initial domains close to a reference domain and small initial velocity). Judging from current methods, a non-local existence result seems to be out of sight. We will come back to this point in subsection 2.4. But first, we have to sketch the standard approach to existence theories, which is via linearized equations.

2.1 Linearizing the Equations

The above system is nonlinear: it contains the convective term, the mean curvature $\mathcal{H}$ is nonlinear, and the domain depends on the solution. The last point implies that we can not even define a sum of two solutions, since, in general, they live on different domains. The nonlinearity created by the changes in the domain turns out to be of higher order than the other two nonlinearities and poses the main difficulty in existence results. Typically, it is treated as follows. Given an initial domain Ω_0 which is close to a reference domain R (we assume in the following that R is a rectangle), we prescribe how to construct from the height function $h(.,t)$ a parametrization of the domain Ω_t by $X(.,t) : R \to \Omega_t$. Then, in the easiest case, we set

$$\hat{v} : R \times (0,T) \to \mathbb{R}^3, \quad \hat{v}(x,t) := v(X(t),t), \tag{7}$$

and similar for the pressure p to find $\hat{p} : R \to \mathbb{R}$. Transforming the differential operators into the X-coordinates, we find equations for $\hat{v}$ and $\hat{p}$ on the rectangle R, the boundary values are coupled to the unknown h. The system is now posed on the reference domain R. We can linearize all terms in the trivial solution and separate linear terms from the higher order terms, collecting the linear terms on the left hand side. Omitting the hat, the equations now read

$$\partial_t v - \nu \Delta v + \nabla p = f, \tag{8}$$
$$\nabla \cdot v = d, \tag{9}$$

with boundary condition on $\Sigma := \{(x,y) \in \partial R | y = 0\}$,

$$\partial_t h - v_3 = g_0, \tag{10}$$
$$\partial_3 v_i + \partial_i v_3 = g_i, \quad i = 1, 2, \tag{11}$$
$$p - 2\nu \partial_3 v_3 + \beta \Delta_x h = g_3. \tag{12}$$

Additionally we have an initial condition for v and h. The right hand side has to be understood as nonlinear terms, $f = f(v,p,h)$, $d = d(v,p,h)$ etc.

2.2 Treatment of the Nonlinear Equations

Once we have the nonlinear system in the form (8)–(12), there is a natural choice of an iteration scheme to treat the nonlinearity. One tries to find an existence result and regularity estimates for the linear system (8)–(12) where $f, ..., g_3$ are assumed to be given, time-dependent quantities. This defines a solution map $S : (f, ..., g_3) \mapsto (v, p, h)$. An iteration is then constructed by composing the evaluation of the nonlinear terms with the solution operator,

$$(v^k, p^k, h^k) \mapsto (f(v^k, p^k, h^k), ..., g_3(v^k, p^k, h^k)) \xrightarrow{S} (v^{k+1}, p^{k+1}, h^{k+1}). \quad (13)$$

It is shown for many situations that this iteration is contractive for small $T > 0$, i.e. a small time interval.

A transformation preserving the solinoidal structure

The transformation chosen in (7) introduces nonlinear terms in all the equations (8)–(12). We will see that it is desirable in the iteration to have vanishing right hand sides d and g_0. One can actually chose a transformation replacing (7) such that $d = 0$ and $g_0 = 0$. This transformation was exploited e.g. in [1], [5], [7], [8], [10]. One sets $J := \det DX$ and

$$\hat{v}(x) := J(x)(DX(x))^{-1} \cdot v(X(x)). \quad (14)$$

One may show with a direct calculation that $\hat{v}$ is again divergence-free. A calculation which provides more insight and which is much shorter exploits that $\hat{v}$ can be interpreted as the pull-back of a differential form. In three dimensions one identifies v with the closed 2-form $V = v_1 \, (dy_2 \wedge dy_3) + v_2 \, (dy_3 \wedge dy_1) + v_3 \, (dy_1 \wedge dy_2) =: v \cdot dS_y$. We can then define $\hat{v}$ as the coefficients of the form $\hat{V} := X^*V = \hat{v} \cdot dS_x$. As a pull-back, $\hat{V}$ is again a closed form and thus $\hat{v}$ is divergence-free. In order to calculate the coefficients, we use an arbitrary 1-form $\lambda = \lambda_1 \, dy_1 + \lambda_2 \, dy_2 + \lambda_3 \, dy_3$ to calculate two pull-backs under X,

$$V \wedge \lambda = \sum_i v_i \lambda_i dy_1 \wedge dy_2 \wedge dy_3,$$

$$X^*(V \wedge \lambda) = \sum_j v_j \lambda_j \, J \, dx_1 \wedge dx_2 \wedge dx_3,$$

$$X^*(V \wedge \lambda) = X^*V \wedge X^*\lambda = (\hat{v} \, dS_x) \wedge \sum_{j,k} \lambda_j \, \partial_k X_j \, dx_k = \sum_{j,k} \hat{v}_k \, \lambda_j \, \partial_k X_j.$$

Comparing the last two expressions we find (14) for $\hat{v}$.

As indicated, this transformation has actually a second advantage. If the parametrization X is chosen with $X_1(x_1, x_2, x_3) = x_1$ and $X_2(x_1, x_2, x_3) = x_2$, then, at the upper boundary of the rectangle, where $X_3(x_1, x_2, 0) = h(x_1, x_2)$,

$$DX = \begin{pmatrix} 1 & 0 & 0 \\ 0 & 1 & 0 \\ \partial_1 h & \partial_2 h & \partial_3 X_3 \end{pmatrix}, \quad DX^{-1} = \frac{1}{\partial_3 X_3} \begin{pmatrix} \partial_3 X_3 & 0 & 0 \\ 0 & \partial_3 X_3 & 0 \\ -\partial_1 h & -\partial_2 h & 1 \end{pmatrix}.$$

Hence the vertical component of $\hat{v}$ is

$$\hat{v}_3 = (-\partial_1 h, -\partial_2 h, 1) \cdot v,$$

which transforms (3) into $\partial_t h = \hat{v}_3$, and thus we have $g_0 = 0$ in (10). We will see in the next subsection that vanishing nonlinearities d and g_0 mean a great simplification if we want to make the iteration (13) well-defined and contractive.

2.3 Estimates for the Linearized System

The method described above restricts us in the choice of function spaces. In order to have well defined evaluations such as $(v, p, h) \mapsto f$ in (13), we must seek solutions in spaces of regular functions. In order to have energy estimates (or Fourier methods) available, we chose the scale of Sobolev spaces $H^k = W^{k,2}$ based on the Hilbert spaces L^2.

A typical choice of function spaces is the following. With $r \in \mathbb{N}$ we seek for $v(t) \in H^r(R)$, $p(t) \in H^{r-1}(R)$, and $h(t) \in H^{r+1/2}(\Sigma)$ for almost every $t \in (0, T)$. With this choice, we can associate to each small height function $h(., t)$ a parametrization $X(., t) \in H^{r+1}(R)$. By choosing the natural number r large enough, we achieve that the evaluation of all nonlinear terms is well defined. For large r ($r > 3$ in three space dimensions) we additionally achieve that the order of the nonlinearity coincides with the maximal number of derivatives. For example, if the nonlinearity contains second derivatives of v and only first derivatives of X, then the nonlinearity defines a map $H^r \times H^{r+1/2} \to H^{r-2}$.

Energy estimates

The estimates for the linear system are always based on the following observation which exploits the energy estimates for the linearized system. Let us assume that (v, p, h) solves the linear system on the bounded rectangle R. We assume that we have performed a Fourier-Transform in time such that the differential operator ∂_t is replaced by the multiplication with $\lambda \in \mathbb{R}$, which is the dual variable to time. To have transparent calculations, we assume to have a general f, but vanishing right hand sides in the other equations and $\nu = 1$.

Since the boundary conditions in (11) and (12) contain entries of the symmetrized gradient $Dv := [\nabla v + (\nabla v)^T]/2$, we use Dv also in integrations by parts. Multiplication of (11), that is, of $\lambda v - \Delta v + \nabla p = f$ with v and integrating over R yields

$$\lambda \int_R |v|^2 + 2 \int_R |Dv|^2 + \int_\Sigma n \cdot (p - 2Dv) \cdot v = \int_R f\, v. \tag{15}$$

Inserting the boundary conditions (11) and (12), which take the form $e_3 \cdot (p - 2Dv) = -\beta \Delta_x h\, e_3$, we find

$$\lambda \int_R |v|^2 + 2\int_R |Dv|^2 - \int_\Sigma \beta \Delta_x h\, v_3 = \int_R f\, v. \tag{16}$$

Using now (10) in the form $v_3 = \lambda h$ and with a multiplication by λ we have

$$\lambda^2 \|v\|_{L^2(R)}^2 + 2\lambda \int_R |Dv|^2 + \lambda^2 \beta \int_\Sigma |\nabla_x h|^2 \leq \|\lambda v\|_{L^2(R)} \|f\|_{L^2(R)}. \tag{17}$$

This provides the resolvent estimate

$$\|v\|_{L^2(R)} + \|h\|_{H^1(\Sigma)} \leq \frac{1}{\lambda} \|f\|_{L^2(R)}. \tag{18}$$

In order to have additionally regularity estimates, one first differentiates equation (11) in a horizontal ($\tilde{x}$-) direction, and then multiplies with v. Korn's inequality yields $L^2(R)$-estimates for all second derivatives of v if at least one derivative is in tangential direction. This provides the desired regularity of boundary values of v. We can therefore exploit the regularity results for the stationary Stokes system $-\Delta v + \nabla p = f - \lambda v$ which imply $\|v\|_{H^2} + \|p\|_{H^1} \leq C\|f - \lambda v\|_{L^2}$. At this stage one finally determines the regularity of the height function h. Since $\Delta_x h$ coincides with (a sum of) traces of p and first derivatives of v, we have $\Delta_x h \in H^{1/2}(\Sigma)$ and hence

$$\|v\|_{H^2} + \|p\|_{H^1} + \|h\|_{H^{2+1/2}(\Sigma)} \leq C\|f\|_{L^2}. \tag{19}$$

By differentiating with respect to horizontal directions more than once, one can raise the power of the Sobolev space by any natural number.

Interpretations of the Energy Estimates

Based on inequalities (18) and (19) there are different ways to construct solutions to the linear equations with maximal regularity. Beale carries out in [1] the concept of a Fourier-transform in time. In this setting, regularity estimates and resolvent estimates for the Fourier-transform $\tilde{v} : \mathbb{R} \to L^2(R, \mathbb{R}^n)$ yield corresponding estimates for the solution,

$$(\lambda \mapsto \tilde{v}(\lambda)) \in L^2(\mathbb{R}, H^2(R, \mathbb{R}^n)) \;\Rightarrow\; v \in L^2(\mathbb{R}, H^2(R, \mathbb{R}^n)),$$
$$(\lambda \mapsto \lambda\tilde{v}(\lambda)) \in L^2(\mathbb{R}, L^2(R, \mathbb{R}^n)) \;\Rightarrow\; \partial_t v \in L^2(\mathbb{R}, L^2(R, \mathbb{R}^n)).$$

As such, these implications hold for smooth maps that vanish on the negative time-axis. In order to deal with general initial values, extensions of the initial values with maximal regularity must exist.

Another approach is to phrase the problem in the language of semigroup theory. There are two difficulties to be circumvented. One problem is that no time derivative of the pressure appears in the equations. To deal with this problem one can use the operator of harmonic extensions. Let us denote by

$$\mathcal{P} : H^{r+1/2}(\Sigma) \to H^{r+1}(R),\; \varphi \mapsto u, \text{ where } \Delta u = 0,\; u|_\Sigma = \varphi,$$

such that u vanishes on the bottom and has a periodic extension. In the linear equations (we assume $\nabla \cdot f = 0$ for simplicity), we find that the pressure p is harmonic. Equation (12) with $g_3 = 0$ then allows to replace everywhere p by $\mathcal{P}(2\nu\partial_3 v_3 - \beta\Delta_x h)$. We set

$$\mathcal{L}\begin{pmatrix} v \\ h \end{pmatrix} := \begin{pmatrix} -\Delta v + \nabla \mathcal{P}(2\nu\partial_3 v_3 - \beta\Delta_x h) \\ -v_3|_\Sigma \end{pmatrix}.$$

By chosing a space of divergence-free functions (here we exploit $d = 0$) and incorporating (11) through the choice of the domain of $\mathcal{L}$, the problem takes the form

$$\frac{d}{dt}\begin{pmatrix} v \\ h \end{pmatrix} + \mathcal{L}\begin{pmatrix} v \\ h \end{pmatrix} = \begin{pmatrix} f \\ g_0 \end{pmatrix}. \tag{20}$$

After these formal manipulations there appears a problem concerning the orders of differentiability. If we demand the base-space to be of the form $H^r(R) \times H^{r+1/2}(\Sigma)$, then we can invert $\mathcal{L}$ only on subspaces of the form $H^{r-2}(R) \times H^{r-1/2}(\Sigma)$. A way to deal with this problem is to exploit that also the right hand side in the second row, g_0, vanishes. With semigroup methods one can show

Theorem 1. *The operator $\mathcal{L}$ generates an analytic semigroup on a subspace of $X^r := H^r(R) \times H^{r+1/2}(\Sigma)$. Solutions of (20) with $g_0 = 0$ and compatible initial values satisfy for $f \in C^\alpha([0,T], X^r)$ maximal regularity estimates,*

$$(v, h) \in C^\alpha([0,T], X^{r+2}) \cap C^{1+\alpha}([0,T], X^r).$$

For $r \geq 2$ there exists a solution to the nonlinear equations in the same function spaces.

We refer to [8] for a proof.

It may be of advantage to keep all proofs as elementary as possible. In fact, Renardy avoids in [7] both the Fourier-transform and the semigroup theory. He finds energy estimates in the primary variables without using a variable λ. Instead, estimates for λv are expressed as estimates for $\partial_t v$, and the multiplication of the equation with λ is replaced by a differentiation of the equation with respect to time. The existence of solutions is then shown with a time-discretized approximation.

2.4 Obstacles to Existence Theories

So far, we have outlined the *standard approach* to existence results for free boundary problems in fluid equations. We would like to emphasize that the quoted articles deal with additional topics and have goals beyond the sketched existence proof. In [1] an unbounded domain is treated. This introduces additional difficulties, e.g. Korn's inequality fails in this case. In [7] an inflow condition is considered. This reduces the maximal regularity and we can not, as assumed above, increase the order of differentiability to arbitrary order by differentiating in a tangential direction. We come back to this point below. In [8], the semigroup methods are a tool to prove a Hopf-bifurcation theorem.

On the existence of solutions far from equilibrium

We have claimed before that an existence result 'in the large' seems to be out of reach. We can now explain this statement concerning the standard approach sketched above. The standard approach needs function spaces of high regularity in order to have the nonlinearities well-defined as maps between the related function spaces. In [7] it was desirable to decrease the regularity as much as possible. Renardy achieved a result in which $h(t) \in H^{3/2}(\Sigma)$ is bounded for all t in the two-dimensional case (compare Theorem 9 of [7]). But this means, that $h(t)$ must still be Hölder continuous at all times.

Without smallness assumption on the initial data one can perform the following thought-experiment. Starting from a dumbell shaped drop with a very thin neck, surface tension will create a high pressure within the neck. The fluid is pushed out and a pinch-off is expected. Indeed, calculations in [4] confirm this believe by providing a self-similar solution with pinch-off in finite time. In the moment of the topology change, the surface loses its regularity and the standard method fails. We can not even solve the problem by performing surgery: The geometry at the moment of the pinch-off, re-parametrized over two topological balls and interpreted as new initial values, does not provide the regularity that is needed for initial values in the standard approach. In particular, we can not expect regular a priori estimates beyond pinch-off.

Regularity questions in the contact angle problem

Less severe are the regularity problems in the contact angle problem. Still, the standard approach does not work in this case. Let us first describe the questions concerning the dynamic contact angle. In the easiest setting we study a two-dimensional domain with independent spatial variables $(x, y) \in \mathbb{R}^2$. If the fluid is in contact with a solid material, say, at $x = 0$ and at $x = 1$, we have to pose conditions other than periodicity on the lateral walls $\Sigma_1 = \{(x,y) \in \bar{\Omega} | x = 0\}$ and $\Sigma_2 = \{(x,y) \in \bar{\Omega} | x = 1\}$. A good choice turns out to be (for some $\gamma_1 = -\gamma_2 \in \mathbb{R}$)

$$v_1 = 0 \quad \text{on } \Sigma_i, \quad i = 1, 2, \qquad (21)$$

$$\partial_x v_2 - \gamma_i v_2 = 0 \quad \text{on } \Sigma_i, \quad i = 1, 2. \qquad (22)$$

The first condition asserts that the fluid can not penetrate the wall, the second condition is the general Navier-slip condition in which the tangential velocity is proportional to the tangential stress across the wall. In order to complete the system it remains to pose boundary conditions for h, one choice is to consider a fixed angles

$$\partial_x h(0) = -\partial_x h(1) = \alpha, \qquad (23)$$

where $\alpha = \tan(90° - \Theta_0)$ and Θ_0 is the constant contact angle. Equation (1)–(5) together with (21)–(23) are a complete set of equations for the dynamic contact point problem.

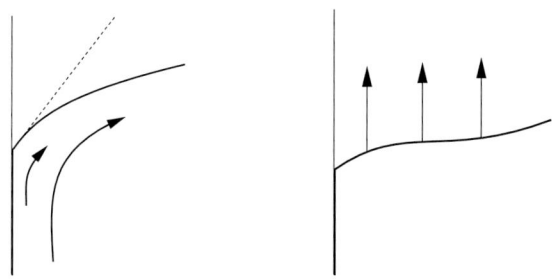

Fig. 1. Flow field near the contact point, stationary and instationary case

The reason to use (22) instead of the no-slip condition $v_2 = 0$ (formally, $\gamma_i = \infty$) is that there are no solutions of finite energy (finite H^1-norm) if we seek for $v_2 \neq 0$ on the upper surface and $v_2 = 0$ on Σ_i. The regularization with the Navier-slip condition is standard in the context of coating problems.

After the regularization we can expect a better regularity of the solutions. But let us perform a formal calculation that indicates that one must be careful in expecting too much. Let us assume that we can evaluate all functions below in the point of contact $(x, y) = (0, h(0, t))$, and that derivatives commute. We then find

$$0 = \partial_t \alpha = \partial_t \partial_x h(0, h(0)) = \partial_x \partial_t h(0, h(0))$$
$$= \partial_x v_2(0, h(0)) = \gamma_1 v_2(0, h(0)).$$

This means that the point of contact can not move, which is in contrast to physical experience (compare Figure 1; note that there is no problem in the stationary flow problem). On a mathematical level we find that we can not expect any estimates for $v(t) \in H^2(R)$.

In particular we find: even with the regularization of the Navier-slip condition, the standard approach to an existence result will fail. The regularity properties of solutions can not be increased by differentiating with respect to x, since then the boundary condition on Σ_i are lost. We can not hope for regularity estimates for $v(t)$ in $H^r(R)$ for $r \geq 2$.

Nevertheless, there is a positive statement for the dynamic contact angle problem, at least for a special choice of Θ_0.

Theorem 2. *For $\Theta_0 = 90°$, $\gamma_1 = -\gamma_2 > 0$, and compatible small initial values (v_0, h_0), there is a short time interval $(0, T)$ and a solution to the contact angle problem on $(0, T)$.*

For the precise statement and the choice of function spaces we refer to [10]. To our knowledge, until today, this is the only well-posedness result for the dynamic contact angle problem.

The proof of Theorem 2 consists in bringing together two facts.
1.) Renardy developed in [7] a method that allows to show existence results

for free boundary fluid problems exploiting only the regularity $v(t) \in H^s$, $s \in (1, 2)$.

2.) In the case $\Theta_0 = 90°$ a reflection principle can be used to find resolvent estimates for the linear equations. To be more precise, one decomposes solutions of the linearized equation into a solution of a free boundary problem with $\partial_x v_2|_{\Sigma_i} = 0$, and a solution of a Stokes problem for given $\partial_x v_2|_{\Sigma_i}$. The first contribution can be reflected across Σ_i to give a periodic solution, and the standard technique can be employed. An interpolation yields optimal estimates for the solution in H^s for $s \in (1, 2)$. Together with maximal regularity results for the second contribution, one finds the resolvent estimates for the linear equations. The technique of Renardy allows to treat the nonlinear problem.

3 Eigenvalues and Bifurcations

3.1 Eigenvalues of the Linearized Free Boundary System

With $\mathcal{L}$ being the generator of the linearized free boundary Stokes system, we now ask:

Can we characterize the spectrum of the operator $\mathcal{L}$?

In this context it is helpful to use the language of semigroup theorey and to write the linearized system in the form (20). But we want to emphasize that this is only a convenient way to state results. One can as well use the original (homogeneous) linearized system (8)–(12) and investigate the following question:

Can we characterize all values of $\lambda \in \mathbb{C}$ such that a solution of the form

$$(v(x, y, t), p(x, y, t), h(x, y, t)) = (v_0(x, y), p_0(x, y), h_0(x, y)) e^{\lambda t}.$$

exists for the linearized time-dependent problem?

An investigation of the operator $\mathcal{L}$ shows that, on a bounded spatial domain R, the operator $\mathcal{L}^{-1}$ is compact, hence the spectrum of $\mathcal{L}$ consists of eigenvalues. In particular, the two questions above are indeed equivalent. Furthermore, the operator $\mathcal{L}$ is a positive operator and every eigenvalue has a positive real part.

Let us first interpret the physical meaning of eigenvalues. A real eigenvalue $\lambda \in \mathbb{R}$ is necessarily positive and corresponds to an overdamped motion, i.e. to a solution that decays exponentially in a self-similar way without changing the sign of the similarity factor. Instead, a non-real eigenvalue corresponds to an oscillatory behavior. From physical reasoning we in fact expect such a behavior. Let us assume that we start with a sinusoidal height function h. The surface tension leads to a high pressure in the crest and a low pressure in the trough. The fluid is accelerated and developes a flow-profile which corresponds to a transport of mass from the crest to the trough. The height-profile now flattens. If the process is fast enough, then the wave reaches in finite time a state with flat height-profile and with a nontrivial flow profile. In this case, the sinusoidal shape of the height-function reappears with inverted sign, i.e. with the positions of crest and trough exchanged. The process repeats and leads to oscillations. We expect such oscillations if the surface tension is large enough in order to flatten out the wave in finite time.

The physical intuition is correct as can be shown with mathematical analysis. The following result was shown in two space dimensions in [8] and in three space dimensions in [2]. In both cases, periodicity conditions are used in one horizontal direction. In [2], on the lateral walls a slip condition was used and a vanishing normal derivative of the height function, i.e. equations (21)–(22) with $\gamma_1 = \gamma_2 = 0$ and (23) with $\Theta_0 = 0$.

Theorem 3. *For each wave number $k \in \mathbb{Z}$ (or $k \in \mathbb{Z}^2$ in three dimensions) there exists a critical value for the surface tension $\beta_0(k)$ such that the following is true: For $0 < \beta \leq \beta_0(k)$ all eigenvalues with wave number k (the x-dependence is e^{ikx}) are real and positive. For $\beta > \beta_0(k)$ there is a pair of conjugate complex nonreal eigenvalues with wave number k.*

We next indicate the main ideas in the proof which relies on a comparison of the free-boundary system with two auxiliary Stokes systems. For notational convenience we give formulas for the two-dimensional case.

1.) Decomposition into invariant subspaces. All eigenfunctions can be written in the form

$$v(x,y) = V(y)e^{ikx}, \quad p(x,y) = P(y)e^{ikx}, \quad h(x) = e^{ikx}, \quad \text{with } k \in \mathbb{Z}.$$

This can be shown as follows: For fixed k, functions of the above form constitute a subspace X_k of the chosen function space X, and the operator $\mathcal{L}$ maps the subspace X_k into itself. We have thus found a decomposition of the base-space into a family of closed invariant subspaces. Since the span of the $(X_k)_{k \in \mathbb{Z}}$ is all of X, the spectrum of $\mathcal{L}$ is the union of the eigenvalues of the operators $\mathcal{L}_k := \mathcal{L}|_{X_k} : X_k \to X_k$.

2.) Characterization of eigenvalues. We next characterize eigenvalues $\lambda \in \mathbb{C}$ of the operator $\mathcal{L}_k$ with the help of the Stokes system. We write down the eigenvalue problem for $\mathcal{L}$ by looking at equations (8)–(12) in two dimensions, leaving out equation (12) at this point. We thus consider

$$-\lambda v - \nu \Delta v + \nabla p = 0, \tag{24}$$

$$\nabla \cdot v = 0, \tag{25}$$

$$v_2|_\Sigma = -\lambda h, \tag{26}$$

$$\partial_2 v_1 + \partial_1 v_2 = 0. \tag{27}$$

This is a Stokes system and we assume periodicity conditions on the lateral walls and a no-slip condition on the bottom. The system is solvable for given

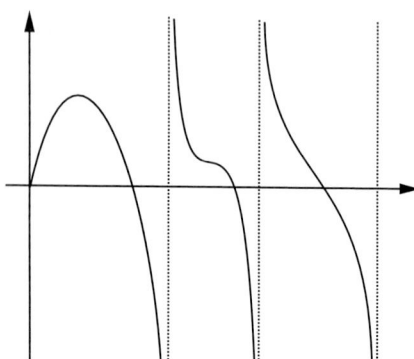

Fig. 2. The function g_k on the real axis

h except if λ is one of the positive real eigenvalues of the Stokes operator with Dirichlet boundary conditions on Σ. We denote the solution (v,p) of the system (24)–(27) with $h(x) = \Phi_0(x) := e^{ikx}$ by $(v,p) = S_k(\lambda)$. We can now give a characterization of eigenvalues of $\mathcal{L}$. A number $\lambda \in \mathbb{C}$ is an eigenvalue of $\mathcal{L}_k$, if the remaining equation (12) is satisfied, that is, if

$$(p - 2\nu\partial_2 v_2)|_\Sigma =: g_k(\lambda)\,\Phi_0 = \beta k^2\,\Phi_0 \qquad \text{holds for } (v,p) = S_k(\lambda). \qquad (28)$$

It is therefore enough to study the relation $g_k(\lambda) = \beta k^2$.

3.) *Properties of the function g_k.* It remains to observe that the function g_k can be described well, at least on the real axis. One can read off easily from equation (26), which contains the only right hand side in the Stokes system, that for $\mathbb{R}_+ \ni \lambda \to 0$ we find $g_k(\lambda) \to 0$. Furthermore, for each eigenvalue λ_N with wave number k of the (Neumann-) Stokes operator, we find $g_k(\lambda_N) = 0$, since there is a nontrivial solution with vanishing normal force in (28). One can furthermore calculate that for eigenvalues λ_D of the (Dirichlet-) Stokes operator, we find $|g_k(\lambda)| \to \infty$ for $\lambda \to \lambda_D$. More detailed investigations reveal the signs of g_k in the various limits and that g_k can have at most one turning point in each interval between two neighboring (Dirichlet-) Stokes eigenvalues. This provides a complete understanding of the function g_k on the real axis; it is illustrated in Figure 2.

We can now read off for which real values of λ we have an eigenvalue of $\mathcal{L}_k$. For vanishing surface tension, equation (28) is satisfied in the zeros of g_k, that is, in 0 and in the (Neumann-) Stokes eigenvalues. If $\beta \geq 0$ is increased, the first to eigenvalues move towards another, all the other eigenvalues move to the left. At a critical point of the surface tension, the first two eigenvalues meet and necessarily leave the real axis.

The analysis can actually be extended to a system with a forcing term. A model for the effect of wind on the free surface of water is obtained by introducing in (5) the right hand side $g = \gamma\partial_x h$. In this model, the effect of the wind is an additional normal force on the wind-facing sides of the wave. For every value of $\gamma \in \mathbb{R}$, this forced system is again tractable with the above method. The following result is shown in [8] and [2] for two and three space dimensions, respectively.

Theorem 4. *For every value of surface tension $\beta > 0$ and every wave number $k \in \mathbb{Z}$ (or $k \in \mathbb{Z}^2$ in three dimensions), there exists a critical value for the wind-force $\gamma_0(k)$ such that for $\gamma > \gamma_0$ there is a pair of conjugate complex purely imaginary eigenvalues. In dependence of γ, they cross the imaginary axis transversally.*

The essential part of the proof was done for Theorem 3. Let us assume that for the given $k \in \mathbb{Z}$ the surface tension β is above the critical values, that is, we have exactly two non-real eigenvalues. A continuation of the function g_k into the complex plane reveals that for $\gamma > 0$ the real eigenvalues (corresponding to $h(x) = e^{ikx}$) all move to the lower half of the complex plane (the conjugate complex eigenvalues correspond to the function $h(x) = e^{-ikx}$). For small $\gamma > 0$ we therefore have one complex eigenvalue in the upper half-plane, all other eigenvalues in the lower half-plane. The same situation appears if we start with a surface tension β below the critical value, since the derivative of g_k has the negative sign in the first eigenvalue and the positive sign in the other eigenvalues.

It now suffices to show that eigenvalues can not be real for $\gamma > 0$, which follows again by a study of the function g_k. For the isolated eigenvalue in the upper half plane one shows that it can not escape to infinity with non-negative real part. Since it can neither merge with other eigenvalues nor return to the real axis, this eigenvalue must cross the imaginary axis. We refer again to [8] and [2] for details and for the transversality result.

3.2 Bifurcation Analysis and Travelling Waves

We have already mentioned after Theorem 1 that a Hopf-bifurcation theorem can be shown with the help of semigroup theory. Theorem 4 provides the existence of imaginary eigenvalues for a critical wind-speed γ_0. The two results together allow to conclude the existence of a branch of non-trivial solutions to equations (1)–(5) for each $\beta > 0$. Parametrizing over a signed wave height $\varepsilon \in \mathbb{R}$, we find a branch of parameters γ_ε extending the critical wind-force γ_0 and non-trivial time-periodic solutions $u_\varepsilon = (v_\varepsilon, p_\varepsilon, h_\varepsilon)$ of the nonlinear system (1)–(5) with $g = \gamma_\varepsilon \partial_x h_\varepsilon$.

The invariance of the nonlinear equations imply that these solutions are in fact translations, i.e. we have found travelling wave solutions of the nonlinear equations. Therefore, there exists a wave-speed $c_\varepsilon \in \mathbb{R}$ such that $\partial_t u_\varepsilon = c_\varepsilon \partial_x u_\varepsilon$. A numerical analysis allows to determine the bifurcation diagram (ε versus γ_ε) and the diagram for wave-speeds (ε versus c_ε), see Figure 3. We refer to [11] for details and for an analysis of quantitative properties of the nonlinear solutions.

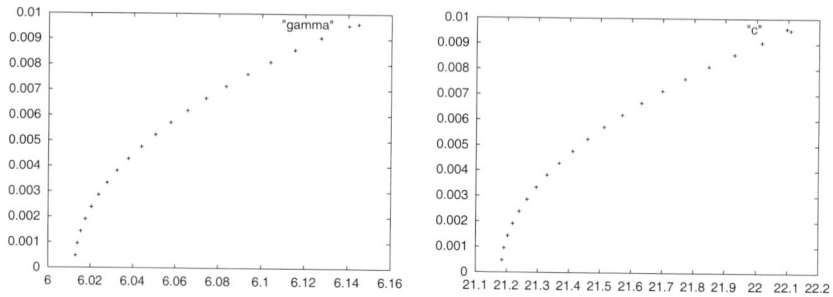

Fig. 3. The bifurcation diagram and values for the wave-speed

4 Coupling with Other Materials

So far, we studied the flow of a viscous fluid in a time-dependent domain. Deformations of the domain gave a feed-back to the fluid in two ways: Firstly, a change of the domain changes the flow geometry. Secondly, surface tension creates a normal force on the boundary which has the effect that curved boundaries try to straighten out. This is a model for the following physical situation: water is in contact with air, and, due to its low density, we can neglect the effect of the air.

In this section we describe some results that deal with the opposite case that we can not neglect the effect of the second material. This is the case for a water-air system in the case of high velocities (subsection 4.1), and for fluid flow coupled to an elastic material (subsections 4.2 – 4.4).

4.1 Coupling with an Inviscid Fluid

There are many contributions regarding the coupling with a second viscous fluid, and we do not intend to describe these results here. Instead, we wish to describe the situation where the second fluid (e.g. air) is a much less viscous fluid, such that the description with inviscid equations is desirable. In this case, the Navier–Stokes equations for the water as in (1)–(5) are coupled with the incompressible Euler equations in the domain $\tilde{\Omega}_t$ above the free surface. We consider the two-dimensional case, write (x, y) for the spatial co-ordinates and $\tilde{\Omega}_t = \{(x,y) | y > h(x,t)\}$. The equations in $\tilde{\Omega}_t$ are

$$\partial_t v + (v \cdot \nabla)v + \frac{1}{\rho}\nabla q = 0,$$

$$\nabla \cdot v = 0,$$

where $\rho > 0$ is the density of the inviscid fluid. In the boundary condition (5) one must set $g = q$, that is, the force felt by the water equals the pressure of the air near the free boundary. Additionally, initial conditions for v and h, and boundary conditions must be posed.

In [22] a local existence result for this system (with $\beta = 0$) is derived under the assumption that the density ρ is small. The result of [9] is a local existence result without the smallness assumption on ρ. In the case of a small density, one can utilize an iteration scheme of the following form:

1. For a given force g acting on the free boundary, solve the free boundary problem for the viscous fluid with force g.
2. The solution yields a time-evolution of the domain. Solve the Euler equations in this time-dependent domain to find anew force-field g.

An iteration of these steps yields the desired solution. The iteration is contractive if the density ρ is small.

For general ρ, we can not expect the above iteration to converge. The physical argument is that we assumed that the effect of the inviscid fluid is small and that it can be treated as a small perturbation of the system without air. The method works for small densities or small velocities. A generalization of the existence result must employ a new iteration scheme. We emphasize that this also gives a hint on how to construct a numerical method.

Every acceleration of the free boundary necessarily induces an acceleration of the inviscid fluid, and inertia leads to a pressure distribution on the free boundary that hinders the acceleration. If the velocities (or the density) are small, this is a small effect. Else, we have to incorporate this effect in the treatment of the single-fluid problem. The idea is to anticipate a simplyfied version of the inertia term in the iteration scheme. For a given boundary evolution $h(t)$ and a velocity field $v(t) : \Gamma(t) \to \mathbb{R}^2$ we define $\Phi : \tilde{\Omega}_t \to \mathbb{R}$ and $\tilde{v}$,

$$\tilde{v} := \nabla \Phi \quad \text{with} \quad \Delta \Phi(t) = 0,$$

with the boundary condition

$$\partial_n \Phi = v \cdot n \text{ on } \Gamma(t).$$

For a detailed description of other boundary conditions and the normalization of Φ we refer to [9]. We can now define a pressure field $\tilde{q}(t) := -\rho \partial_t \Phi$. With these definitions we find

$$\partial_t \tilde{v} + \frac{1}{\rho} \nabla \tilde{q} = 0,$$

in $\tilde{\Omega}_t$ and $\tilde{v} \cdot n = v \cdot n$ on $\Gamma(t)$. Hence $\tilde{v}$ and $\tilde{q}$ satisfy the Euler equations except for the convective term.

The iteration scheme to solve the nonlinear problem is the following.

1. Solve the single-fluid free boundary problem with the functional relation $g = \tilde{q}(h, v)$, with $\tilde{q}$ as defined above. Exploit in the resolvent estimates that $\tilde{q}$ is a positive term.
2. Solve the Euler equations in the given geometry to find the pressure q. Exploit that the correction $q - \tilde{q}$ is a compact perturbation compensating the lower order term of the convective derivative.

This scheme provides a fixed-point iteration with a compact iteration map. The Schauder fixed-point theorem yields the existence of a solution to the nonlinear problem.

4.2 Stationary Flow Coupled with an Elastic Material

We are interested in the flow of a viscous fluid in a domain with elastic walls. This physical problem appears for instance in the modeling of blood-flow, where the blood fills the arteries that are described as elastic tubes (see e.g. [6] for related model equations). In [19] such a three-dimensional fluid-elastic structure interaction problem is studied. The fluid is described by the incompressible Navier-Stokes equations and flows inside a smooth elastic cylinder with thickness.

As in the free boundary systems above, the fluid domain is unknown and the flow deformes the boundary. This deformation gives a two-fold feed-back to the fluid, the change of domain and via forces. In the case of the interaction with an elastic material, the excerted forces are of a complex nature. The changes of the fluid domain must be interpreted as deformation (Dirichlet-) conditions for the elastic material. The bulk equations for the material (modeled as a three-dimensional structure) determine the normal and tangential forces on the boundary which are the second feed-back to the fluid. The fluid equations (1)–(2) are coupled with the bulk equation for the deformation u of the elastic material,

$$-\nabla \cdot T_P(u) = g, \tag{29}$$

which holds in the domain occupied by the elastic material, T_P is the first Piola-Kirchhoff stress tensor and g are the exterior bulk forces. The coupling to the fluid is via the following replacement of conditions (3)–(5) on $\Gamma(t)$,

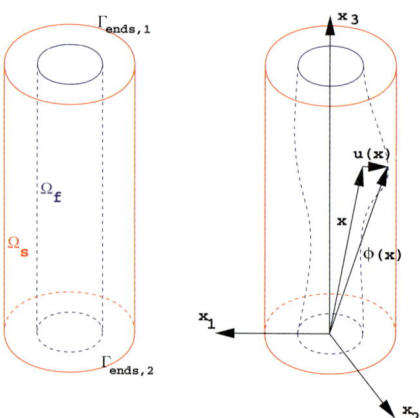

Fig. 4. Description of the flow-domain

$$v = 0 \tag{30}$$
$$T_P(u) \cdot n = pn - \nu(\nabla v + (\nabla v)^T) \cdot n. \tag{31}$$

Note that now both equations are vector valued, since the elastic material requires also tangential boundary conditions. For precise conditions at the other boundaries we refer to [21].

The problem in the description of this coupled system concerns coordinates. While for the elastic problem Lagrange coordinates are appropriate, velocity and stress of the fluid are described in Eulerian coordinates. In the balance of forces in equation (31) this means that the derivative on the right hand side is understood in Eulerian coordinates and must be converted into Lagrange coordinates with the cofactor matrix of the gradient of the transformation $\mathrm{Id} + u$ of the elastic material.

The results of [19]–[21] are existence results for the coupled system. As explained in section 2, these are necessarily results for small deformations and for smooth applied force-fields. The elastic structure is modeled as a St.Venant-Kirchhoff material and the geometry is as indicated in Figure 4. Different boundary conditions at the inflow and outflow sides were considered. In [20] a periodicity condition is assumed for the fluid and the elastic material. In [19] clamped elastic walls are considered and the proof exploits a reflection principle. The generality of the boundary conditions is improved in [21].

4.3 Instationary Flow Coupled with an Elastic Material

The next step was to study the time-dependent case. It turned out that in the instationary case the coupling of two three-dimensional media is not accessible with the above methods. Therefore, the elastic wall was described by a lower dimensional set of equations – in the case of elastic materials this means to describe the wall with plate equations.

In [18] a time-dependent 3D-2D fluid-elastic structure interaction problem was studied. It can be formulated in the following way: a viscous, incompressible fluid flows through a box having an elastic plate as cover, the bottom and two opposite lateral walls are rigid, the other two opposite lateral walls are the inflow and the outflow boundary, respectively. The fluid domain changes with time and must be determined as part of the problem. The fluid is now described by the time-dependent Navier-Stokes equations with prescribed pressures on the inflow and outflow parts of the boundary. These are nonstandard boundary conditions for a fluid flow, which constitutes the main difference with [3], where a similar problem has been considered, but where all lateral boundaries of the fluid domain were rigid. The elastic structure is viewed as a thin plate which is clamped at its boundary.

One of the results in [18] is a local existence theorem for the coupled system. The proof utilizes an auxiliary problem in which the convective term is regularized. This problem is linearized and the linearized equations are solved with Galerkin's method. A generalized Schauder fixed point theorem due to Zeidler yields the existence of a solution to the nonlinear auxiliary problem. The passage to the limit in the regularization parameter eventually yields a solution to the full coupled system.

4.4 A One-Dimensional Model for Fluid-Structure Interaction

In this section we sketch an alternative approach to describe the flow in a compliant vessel. This is the one-dimensional model which reduces the problem of the fluid-structure interaction to the equations similar to the Euler equations of gas dynamics ([14]–[17]).

The velocity profile of the fluid in an compliant vessel depends on the Womersley number $\alpha^2 = a^2 \frac{\omega}{\nu}$, where a is the tube radius and ω the circular frequency (see [6], for instance). At high Womersley numbers, the velocity profile is flat over most of the vessel and has a thin boundary layer at the wall. At very low α, the flow is essentially quasi static Poiseuille flow at each instant.

A one-dimensional model was derived in [14] by averaging the Navier–Stokes equations (1)–(2) for axialsymmetrical flow on the assumptions

$$\frac{a}{L} \ll 1, \qquad \frac{1}{\alpha^2} \ll 1,$$

so that the influence of the viscous boundary layer on the mean flow can be neglected (here L is the characterictic length of a tube). The conservation of mass and momentum requires:

$$\frac{\partial S}{\partial t} + \frac{\partial (S u)}{\partial x} = 0, \qquad \frac{\partial u}{\partial t} + u \frac{\partial u}{\partial x} + \frac{1}{\rho} \frac{\partial p}{\partial x} = 0,$$

where x is now a one-dimensional variable and denotes the longitudinal position within the tube, $u(t,x)$ and $p(t,x)$ denote the average axial velocity and average pressure over the cross-section, and $S(t,x)$ is the two-dimensional volume of the cross-section. The compliance of the wall has an effect corresponding to the compressibility in gas dynamics. For the details of model justification, we refer to [14],[15].

This model was applied to the blood flow in large arteries and studied both analytically (by the method of characteristics) and numerically. Surovtsova examined some features of solutions connected with the steeping of compression waves, leading to the possible formation of shock waves. They can occur in the human aorta, for instance, because of some cardiovascular illnesses (e.g. increase of vessel wall distensibility and some ventricles abnormalities). It was shown that the viscoelasticity of the vessel wall has a much larger influence on the damping of shock waves than the blood viscosity.

For studying the mechanical effects caused by the local stiffening of an artery (due to the vascular prosthesis or stent, for example), a modified one-dimensional model was derived in [16], [17]. The artery was supposed to be an orthotropical thin elastic shell composed of different homogeneous materials.

The solution was obtained by matched asymptotic expansions. The results proved the high flexure concentration close to the compliance jump. It can initiate an adaptive response in the vascular tissue (e.g. the restenosis). The use of orthotropical graft may reduce the peak value of these shear forces to a remarkable extent. Other features of the solution are the reflection of blood from the suture, as well as the pressure increase in the more rigid prosthesis (see [16], [17] for details).

References

1. J.T. Beale. Large time regularity of viscous surface waves. *Arch. Rat. Mech. Anal.*, 84:307–352, 1984.
2. S. Bodea. *Oscillations of a Fluid in a Channel.* Dissertationsschrift an der Universität Heidelberg, 2004.
3. A. Chambolle, B. Desjardins, M.J. Esteban, C. Grandmont, *Existence of Weak Solutions for an Unsteady Fluid-Plate Interaction Problem*, Cahier du CERE-MADE 0245 (2002).
4. J. Eggers. Theory of drop formation. *Phys. Fluids 7*, 5:941–953, 1995.
5. J.G. Heywood. On uniqueness questions in the theory of viscous flow. *Acta Math.*, 136:237–246, 1976.
6. A. Quarteroni, M. Tuveri, A. Veneziani, *Computational vascular fluid dynamics: Problems, models and methods*, Comput. Vis. Sci. 2, No.4 (2000), 163-197.
7. M. Renardy. An existence theorem for a free surface flow problem with open boundaries. *Comm. Part. Diff. Eq.*, 17:1387–1405, 1992.
8. B. Schweizer. Free boundary fluid systems in a semigroup approach and oscillatory behavior. *SIAM J. Math. Anal.*, 28:1135–1157, 1997.
9. B. Schweizer. A two-component flow with a viscous and an inviscid fluid. *Comm. PDE*, 25:887–901, 2000.
10. B. Schweizer. A well-posed model for dynamic contact angles. *Nonlinear Analysis TMA*, 43:109–125, 2001.
11. B. Schweizer. Bifurcation analysis for surface waves generated by wind. *SIAM J. Appl. Math.*, 62:407–423, 2001.
12. B. Schweizer. A stable time discretization of the Stefan problem with surface tension. *SIAM J. Numer. Anal.*, 40:1184–1205, 2002.
13. B. Schweizer. On the three-dimensional Euler equations with a free boundary driven by surface tension. *Annales de l'Institut Henri Poincaré (C) Analyse Non Linéaire*, 22(6):753–781, 2005.
14. I. Surovtsova. Blood flow in large vessels: A one-dimensional model. University Heidelberg, Preprint 2002-08 (SFB 359) (2002)
15. I. Surovtsova. Application of the one-dimensional model for blood flow to the vascular prosthesis. In: Capasso, V. (ed.) Mathematical Modelling & Computing in Biology and Medicine. The MIRIAM Project Series, Progetto Leonardo, ESCULAPIO Pub. Co., Bologna, Italy, pp: 242-248 (2003)
16. I. Surovtsova. Application of the one-dimensional model for blood flow through vascular prosthesis. University Heidelberg, Preprint 2003-04 (SFB 359) (2003)
17. I. Surovtsova. Effects of compliance mismatch on blood flow in an artery with endovascular prothesis. Journal of Biomechanics (accepted).

18. C. Surulescu, *Modeling Aspects and Mathematical Analysis of Some Fluid-Elastic Structure Interaction Problems*, Dissertationsschrift an der Universität Heidelberg, 2004.
19. C. Surulescu, *On the Stationary Motion of an Incompressible Fluid Flow Through an Elastic Tube in 3D*, IWR/SFB Preprint 06 (2003), Universität Heidelberg.
20. C. Surulescu, *On the Stationary Motion of a Stokes Fluid in a Thick Elastic Tube: A 3D/3D Interaction Problem* (submitted).
21. C. Surulescu, *The Stationary Motion of a Navier-Stokes Fluid Through an Elastic Tube Revisited*, IWR/SFB 359 preprint 39 (2004), Universität Heidelberg.
22. W.M. Zajaczkowski, *On the motion of a drop of a viscous incompressible fluid in an ideal incompressible fluid*, Arch. Mech. 42:307–325, 1990.

Nonlinear Evolution Equations and Applications*

W. Jäger[1], Th. Lorenz[2] and A. Tambulea[2]

[1] Institut für Angewandte Mathematik, Universität Heidelberg
[2] Interdisziplinäres Zentrum für Wissenschaftliches Rechnen, Universität Heidelberg

Summary. Mathematical analysis provides several powerful tools for describing dynamic processes in science and economics. Here the role of ordinary and partial differential equations within the SFB 359 is briefly sketched. Then two examples are presented in more detail: Firstly, the analytical description of gradostat demonstrates how sensitively a system of ODEs might depend on its parameters. Secondly, shape evolutions show how the tools of set-valued analysis provide an alternative to level-set methods. The key aim here is to extend evolution equations beyond vector spaces.

> *It is an error to imagine that evolution signifies a constant tendency to increased perfection.*
> *That process undoubtedly involves a constant remodelling of the organism in adaptation to new conditions; but it depends on the nature of those conditions whether the directions of the modifications effected shall be upward or downward.*
> Thomas H. Huxley   (English biologist, 1825-1895)

1 How to Approach Evolutions

Huxley's characterization of "evolution" holds to a more general extent than he might have been aware of. Indeed, it is not restricted to the historic aspect of life, but also describes the mathematical improvement in describing a dynamic process. Evolution equations, in particular, prove to be an excellent example in this context. Let us first sketch the situation in a more general way and then present two concrete examples in more details in §§ 2, 3.

---

*This work has been supported by the German Research Foundation (DFG) through SFB 359 (Project A1) at the University of Heidelberg.

The starting point is a system whose variation depends on its current state and (possibly) time. Analytically speaking, this situation is usually described by means of differential equations.

Their type and complexity depends very much on how many properties are taken into consideration explicitly. Many examples neglect spatial relations and thus deal with ordinary differential equations whereas additional dependence on space leads to partial differential equations.

At first glance, this conceptual sketch suggests to be the beginning of a straightforward field of analysis. Progress since Newton and Leibniz, however, has proved the contrary and, differential equations still provide many open questions today.

For systemizing the world of differential equations, the suggestions of Jacques Hadamard have been accepted as suitable criteria when coping with a new type of evolution problem. In a word, a problem is called *well-posed* if

1. a solution exists,
2. the solution is unique and
3. it depends continuously on the data.

(Otherwise it is called *ill-posed*.)   Uniqueness is usually closely related to the question in which set the solution is searched for and, this point arouses our interest in the regularity of existing solutions. So

- *existence*
- *regularity* and
- *stability*

summarize the traditional foci of investigation whenever a new type of differential equations occurs in modelling a dynamic process.

SFB 359 has been following this track for several years - and on various fields of applications. Here we briefly mention some of the topics and our former colleagues who elaborated interesting results:

- Existence and stability of traveling waves in porous media and combustion processes with complex chemical networks (Steffen Heinze)
- Marangoni convections in incompressible 3-d fluids (Alfred Wagner)
- Population dynamics of plankton in North Sea (Markus Kirkilionis)
- Aggregation induced by diffusing and nondiffusing media (Angela Stevens)
- Global classical solutions to a two-phase Stefan problem (Herbert Koch)
- Global solutions of quasilinear wave equations and stability of minimal surfaces (Ben Schweizer).

Of course, this list is far from being complete, but it gives a vague impression of how extended the field is. In the next section, a more recent example is presented. The gradostat demonstrates how sensitively a system might depend on its parameters. Moreover, the qualitative properties are usually impossible to justify intuitively whenever a new participant comes into play. In the special case of gradostat, the situation of three species does not result directly from the corresponding setting with two species.

Roughly speaking, evolution equations are based on the notion of extending ordinary differential equations from $\mathbb{R}^N$ to Banach spaces (such as function spaces). Although they have proved to be very useful in many applications, we cannot apply them beyond vector spaces. "Shapes (...) are basically sets [however], not even smooth" ([2]). So the classical tool of a single-valued function has to be replaced by sets if we want to describe shapes without any regularity restrictions. In § 3, two examples of shape evolutions provide an interesting alternative to level-set methods.

2 Competition and Persistence of Microorganisms in the Gradostat

Different species inhabiting a common environment are said to *compete purely* and *simply* if there is no other interaction among them and competition occurs for a single nutrient. For a homogeneous environment - spatially and temporally - the outcome of simple and pure competition is competitive exclusion.

Spatial heterogeneities can be created using interconnected reactors connected in both directions, known under the name of **gradostats**. Our main aim here has been the analytical derivation of persistence results for the competition of more than two species in the gradostat as well as the invadability of certain communities. For the competition of three species in n vessels we are able to give a quite detailed description of the dynamics, together with numerous numerical examples which agree with our analytical predictions.

The methods of lower and upper solutions provide the key tools for systems with quasimonotone reaction terms. Such a monotonicity property reduces the search for a solution of a nonlinear system of equations to finding iteratively a sequence of solutions of linear problems. This sequence will converge towards a solution of the nonlinear problem, with an initial iteration given by a pair of lower and upper solutions. The method is constructive and allows the possibility of obtaining comparison results and thus persistence without too much information on the ω-limit set of the solution.

Next, a bifurcation analysis is carried out and coexistence is obtained by varying a parameter. By continuing certain equilibria, numerical computations suggest that coexistence equilibria exist under the sufficient conditions we find analytically.

2.1 Persistence Conditions

As originally designed by Lovitt and Wimpenny, the gradostat is linearly connected, namely the contents of vessel i are transmitted only to vessels $i-1$ and $i+1$. We use the "reduced system" for the study of m species in n vessels ([7]):

for each $1 \leq j \leq m$, the vector $u^j(t) = (u_1^j(t), \ldots, u_n^j(t))$ denotes the concentration of species j at time t, with $u_i^j(t)$ abbreviating the concentration of species j in vessel i for $1 \leq i \leq n$ and, we have

$$\begin{cases} \frac{d}{dt} u^j(t) = [A + F_j(z - \sum_k u^k(t))] u^j(t), \\ u^j(t_0) \geq 0 \end{cases} \quad (j = 1 \ldots m) \quad (1)$$

on $\Omega = \{u \in \mathbb{R}_+^{nm} : \sum_j u^j \leq z\}$, with $z > 0$ the unique solution of $Az + e_0 = 0$,

$$A = \begin{bmatrix} -2 & 1 & 0 & \cdots & 0 & 0 \\ 1 & -2 & 1 & \cdots & 0 & 0 \\ \multicolumn{6}{c}{\dotfill} \\ 0 & 0 & 0 & \cdots & 1 & -2 \end{bmatrix},$$

$$F_j(x) = \begin{bmatrix} f_j(x_1) & 0 & \cdots & 0 \\ 0 & f_j(x_2) & \cdots & 0 \\ \multicolumn{4}{c}{\dotfill} \\ 0 & 0 & \cdots & f_j(x_n) \end{bmatrix}, \quad x \in \mathbb{R}^n$$

$e_0 = (1, 0, \ldots, 0) \in \mathbb{R}^n$, and $f_j(x) = \frac{m_j x}{a_j + x} =$ Michaelis-Menten kinetics.

Let $\underline{u}$ denote a lower solution for (1) and $\overline{u}$ an upper solution. It is easy to check that the reaction functions in (1) are quasimonotone and thus, following [13], if we can find a pair $(\underline{u}, \overline{u})$ of coupled lower and upper solutions, the sector $<\underline{u}, \overline{u}>$ will give us a positively invariant region for (1).

In fact, all matrices $A_j = A + F_j(z - x)$, $x \in \mathbb{R}_+^n$, are irreducible, with nonnegative off-diagonal elements, and if $s[A_j] = \max\{Re\,\lambda \,|\, \lambda \text{ eigenvalue of } A_j\}$ denotes the stability modulus of A_j, Perron-Frobenius theory implies the existence of a positive eigenvector $\phi_s^j[A_j]$ associated to $s[A_j]$.

We use the positivity of various such eigenvectors to build nonnegative lower solutions for (1), and if the lower solution is strictly positive, persistence follows.

In the case of only species j in the gradostat, $s[A + F_j(z)]$ will give us a measure of the "fitness" of species j: If

$$s[A + F_j(z)] > 0 \quad (2)$$

then species j is persistent, since $(\tilde{\gamma}\phi_s^j[A + F_j(z)], z)$, with $\tilde{\gamma} > 0$ a sufficiently small constant, is a pair of lower and upper solutions. As the reaction terms for the no competition case are quasimonotone nondecreasing, we have in addition that (1) has a unique positive equilibrium $\hat{E}^j = \hat{u}^j$ which is stable, and, following [16], $\hat{u}^j$ attracts all trajectories with initial positive values in Ω.

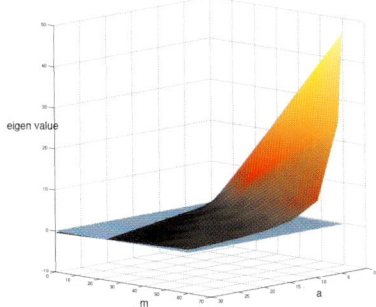

Fig. 1. The eigenvalue $s[A + F(z)]$ vs. the parameters m and a.

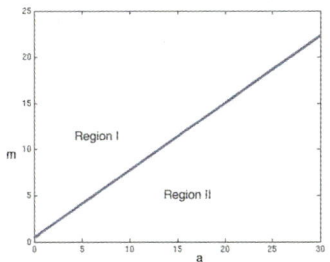

Fig. 2. An operating diagram for 1 species in 4 vessels: for parameters a and m in region I the species survives, and in region II, the species dies out.

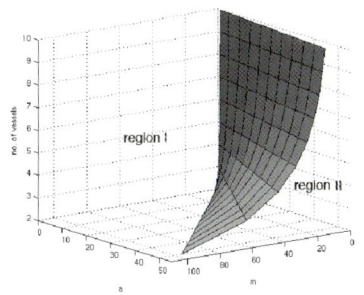

Fig. 3. An operating diagram for various number of vessels. Note that the size of region II decreases with the no. of vessels increasing.

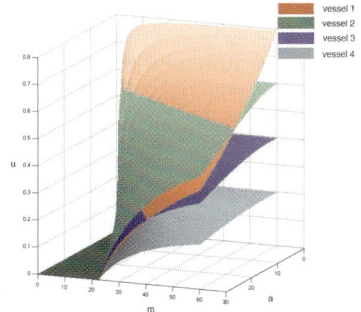

Fig. 4. Two parameter bifurcation diagram for 1 species in 4 vessels. We plot the equilibrium concentration vs. m and a.

In order to visualize the relation between (2) and the system's parameters, via a bifurcation analysis of the solution of $[A + F_j(z - u^j)] \cdot u^j = 0$, by varying parameters a_j and m_j, we obtain a region in the parameter space for which species j is persistent (Figure 2).

In order to survive in the gradostat in the competition case, a species must be at least "fit enough" to survive alone in the gradostat, thus we have (2) as a necessary condition for persistence and furthermore, $\hat{u}^j$ is an upper solution for u^j for all j.

In the next theorem, with the help of lower solutions of the type given for the no competition situation, we give sufficient conditions for persistence in the competition case.

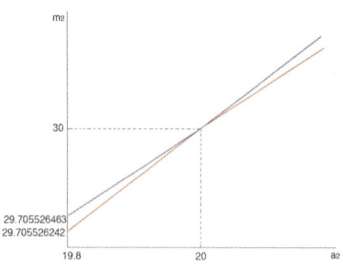

Fig. 5. 2 species in 4 vessels: an operating diagram for $m_1 = 30$, $a_1 = 20$. The size of the coexistence interval decreases as a_2 increases towards a_1 and increases as a_2 grows beyond a_1. In $a_2 = a_1$ the interval degenerates to one point.

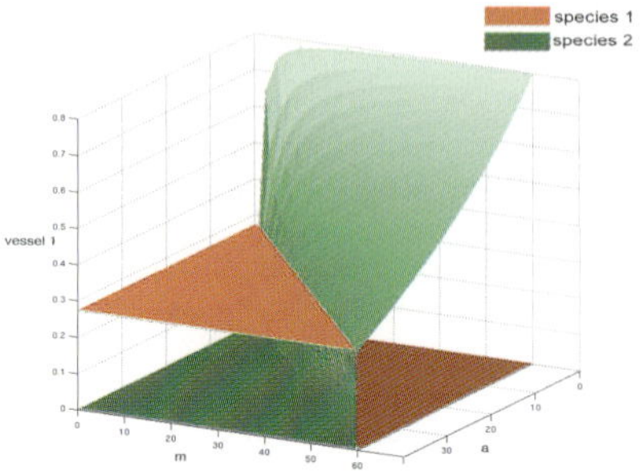

Fig. 6. Two parameter bifurcation diagram for 2 species in 4 vessels for $m_1 = 30$, $a_1 = 20$. We plot equilibrium concentrations of both species in vessel 1 vs. parameters a_2 and m_2.

Theorem 1 (Persistence of several species). *For all $1 \leq j \leq m$, let ξ^j be a time independent upper solution of the steady state problem*

$$[A + F_j(z - u^j)] \cdot u^j = 0, \tag{3}$$
$$u_0^j > 0.$$

Let $\theta^j \stackrel{def}{=} z - \sum_{k \neq j} \xi^k$ for all $1 \leq j \leq m$, where $z - \xi^j > 0$. Suppose that at time $t = t_0$ we have $0 < u_0^j \leq \xi^j$, and $\sum_{j=1}^m u_0^j \leq z$, for all $1 \leq j \leq m$. Moreover, assume

$$s[A + F_j(\theta^j)] > 0 \qquad \text{for } 1 \leq j \leq m. \tag{4}$$

Then, for all j, it is possible to find time independent lower solutions $\underline{u}^j > 0$, such that $\underline{u}^j \leq u^j(t) \leq \xi^j$, for all $t > t_0$, provided that this was true at $t = t_0$. This means that all m species are persistent.

For two vectors $x = (x^1, x^2)$ and $y = (y^1, y^2) \in \mathbb{R}^{2n}$, define the partial order relation $x \leq_k y \Leftrightarrow x^1 \leq y^1$ and $x^2 \geq y^2$. As noted in [7], (1) generates a strongly monotone, with regard to $\leq_k$, dynamical system in the interior of Ω.

For the case of two species, if (4) holds, we have the possible existence of two positive equilibria E^* and E^{**}, and, following [14], the sector $[E^*, E^{**}]_K$ attracting all trajectories with positive initial values in Ω. As for the case of one species, by a bifurcation analysis with parameters a_1, m_1 fixed, we obtain a region in the parameter space corresponding to persistence, as well as a bifurcation branch corresponding to positive equilibria of (1).

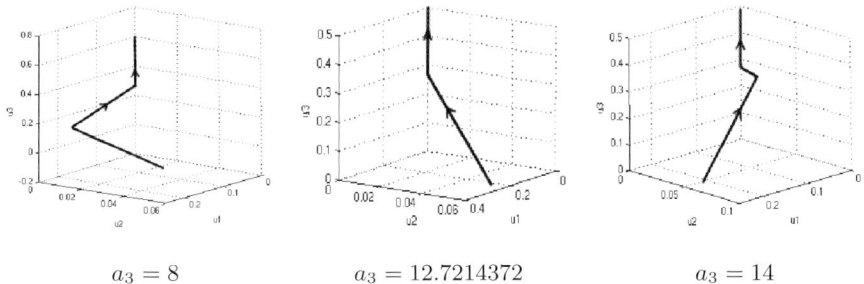

| $a_3 = 8$ | $a_3 = 12.7214372$ | $a_3 = 14$ |

Fig. 7. Bifurcation diagrams for (1) - corresponding to the first vessel - for fixed a_1, m_1, a_2, m_2, different values for a_3 and m_3 as bifurcation parameter.

2.2 Three Species

As a next step, we focus on improving the persistence conditions (4) for the case of three species: In all our numerical simulations, the positive two-species equilibria, if they existed, were unique, and a third species could successfully invade an established persistent two-species equilibrium $E_{1,2}^* = (u^{*1}, u^{*2}, 0)$ with $u^{*1}, u^{*2} > 0$, if parameters a_3, m_3 are such that $E_{1,2}^*$ becomes unstable:

$$s[A + F_3(z - u^{*1} - u^{*2})] > 0. \quad (5)$$

It is obvious that (4) implies (5), and thus, the region in the parameter space for which (4) holds, is included in the region corresponding to (5). Since $h^3 := [A + F_3(z - u^1 - u^2)]u^3$ is monotone decreasing in $u^1 + u^2$, for a fixed $u^3 \in [0, z]$, the solution $(u^1(u^3), u^2(u^3))$ of

$$\tfrac{d}{dt} u^j(u^3) = [A + F_j(z - u^1(u^3) - u^2(u^3) - u^3)]u^j(u^3), \qquad j = 1, 2$$

is such that $u^1(0) + u^2(0)$ is an upper solution for $u^1(u^3) + u^2(u^3)$. Thus species 3 can successfully invade a persistent two-species equilibrium (u^{*1}, u^{*2}) if $s[A + F_3(z - K)] > 0$ with $K = max\{u^1 + u^2 : (u^1, u^2) \in \Upsilon\}$ where Υ is an invariant region for (1) with $u_0^3 = 0$, and we have $K \geq u^{*1} + u^{*2} =: k$. Using the $\leq_K$-monotonicity of the system describing the competition of two species, we can construct various such invariant regions Υ for $(u^1, u^2, 0)$ and thus obtain various sufficient conditions for the persistence of species 3.

If we assume uniqueness for positive two-species equilibria, via a bifurcation analysis with fixed parameters for species 1 and 2, we look for persistence intervals for m_3 for various values of a_3. An interesting feature of the dynamics of the three-species competition system is that we can find (both numerically and analytically) bifurcation branches corresponding to the persistence of all three species even if not each combination of two species can coexist. Thus it can happen that a not persistent two-species system becomes persistent - due to introducing a third species.

In Figure 7 we present the three types of bifurcation diagrams we have found in our simulations. Let species 1 & 2 and species 2 & 3 coexist with the equilibria $(u^{*1}_{(1,2)}, u^{*2}_{(1,2)})$ and $(u^{*2}_{(2,3)}, u^{*3}_{(2,3)})$, respectively, and assume species 1 and 3 are not able to coexist, and denote with $k_1 = u^{*2}_{(2,3)} + u^{*3}_{(2,3)}$ and $k_2 = u^{*1}_{(1,3)} + u^{*3}_{(1,3)}$. Assume

$$s[A + F_1(z - k_{2,3} - \eta_1)] > 0, \tag{6}$$

ϕ_1 the positive eigenvector associated to $s[A + F_1(z - k_{2,3} - \eta_1)]$ and

$$s[A + F_3(z - k_{1,2} - \eta_3)] > 0, \tag{7}$$

ϕ_3 the positive eigenvector associated to $s[A + F_3(z - k_{1,2} - \eta_3)]$, and η_1 and η_3 nonnegative vectors such that we can build lower solutions $\underline{u}^1 := \gamma_1 \cdot \phi_1 \geq \eta_1$ and $\underline{u}^3 := \gamma_3 \cdot \phi_3 \geq \eta_3$, with γ_1 and γ_3 positive constants. Starting from the different invariant regions Υ, we can find (u^1, u^2) and choose the vectors η_1 and η_2 such that they both have some components equal to zero. Now the set $\Delta \subset \Omega$ is defined by:

$\eta_1 \leq \underline{u}_1 \leq u_1 \leq u^{*1}_{(1,2)}$,

$min\{u^{*2}_{(1,2),i}, u^{*2}_{(2,3),i}\} \leq u_i^2 \leq max\{u^{*2}_{(1,2),i}, u^{*2}_{(2,3),i}\}$, $i = 1, \ldots, n$,

$\eta_3 \leq \underline{u}_3 \leq u_3 \leq u^{*3}_{(2,3)}$,

$u^{*2}_{(2,3)} \leq u^1 + u^2 \leq k_{1,2}$,

$u^{*2}_{(1,2)} \leq u^2 + u^3 \leq k_{2,3}$,

$min\{u^{*1}_{(1,2),i}, u^{*3}_{(2,3),i}\} \leq u_i^1 + u_i^3 \leq max\{u^{*1}_{(1,2),i}, u^{*3}_{(2,3),i}\}$, $i = 1, \ldots, n$,

with $\underline{u}_1$ and $\underline{u}_3$ as above, which, if (6) and (7) hold, is an invariant region and thus system (1) is persistent for appropriate initial data. Moreover, if there exists an equilibrium $\tilde{E}$ corresponding to the persistence of all three species, then $\tilde{E} \in \Delta$.

Now assume that each combination of two species can coexist, and in addition to (6) and (7), assume

$$s[A + F_2(z - k_{1,3} - \eta_2)] > 0, \tag{8}$$

with $k_{1,3} = u^{*1}_{(1,3)} + u^{*1}_{(1,3)}$ and η_2 built as above. In this case, just as for species 1 and 3, we can find a lower solution $\underline{u}^2$ for u^2 with $\underline{u}^2 = \gamma_2 \cdot \phi_2 \geq \eta_2$ with $\gamma_2 > 0$ a sufficiently small constant, and ϕ_2 the positive eigenvector associated to $s[A + F_2(z - k_{1,3} - \eta_2)]$, such that for $u_0^2 \geq \underline{u}^2$, the solution of (1) satisfies $u^1 + u^3 \leq k_{1,3}$. Again we obtain that system (1) is persistent for appropriate initial data.

In all our numerical simulations, the region in the parameter space for which persistence holds is slightly larger than the region described by (6) and

(7), and, if each two species coexist, (8). Still, the intervals for parameters as m_3 in a bifurcation diagram, for which we find bifurcation branches corresponding to coexistence make a very small impression indeed. This leads to the interesting question whether coexistence could really occur in biology. As a first step towards an answer, we perturb the value for m_3 and solve the corresponding system (1) numerically. Consider, for example, $\bar{m}_3 := m_3(1-d-2dr)$ with d a disturbance factor and $r \in [0, 1]$ randomly generated at each time step. In order to make the simulation more realistic, we also introduced a "low-life" limit: Whenever a species reaches a concentration lower than the "low-life" limit, we consider it washed out of the gradostat, namely it no longer grows. Surprisingly, persistence of all three species still occurs. It is interesting to see that a species is able to recover even if it stays at a very low concentration for a long time.

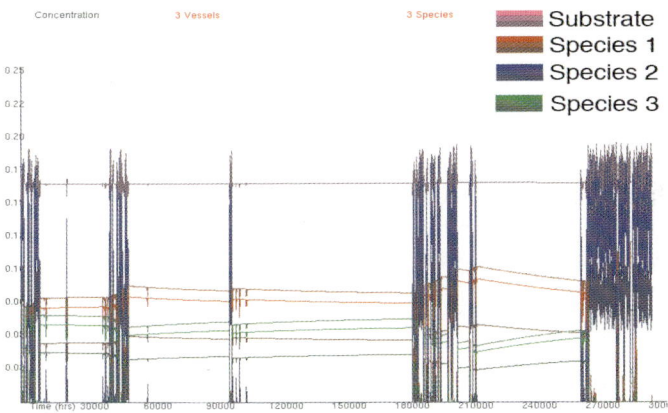

Fig. 8. The solution of (1) with parameters $a_1 = 20$, $m_1 = 30$, $a_2 = 5$, $a_3 = 30$, and perturbed m_3 with $d = 0.1$.

3 Evolution Equations in a more General Environment

In many applications, shapes play a key role and thus, the geometric aspect has to be taken into consideration. However, its mathematical description by means of functions often starts with restrictive preliminaries - like basic assumptions about regularity of boundaries.

In [4], [3], compact set-valued flows are suggested for applications in medical imaging and biological modeling, i.e. now compact subsets of $\mathbb{R}^N$ depending on time describe the evolution of a shape. Basically speaking, single-valued functions are replaced by set-valued maps, also called multivalued maps. In the first subsection, we follow [9] and apply this idea to image segmentation specifying the analytical background.

In comparison with traditional approaches using single-valued functions, set-valued maps lack any obvious structure of vector spaces. So in the second subsection, we consider evolutions in metric spaces and extend Jean-Pierre Aubin's concept of mutational equations ([1],[2]).

The main aim there is to unify the definition of "solution" for completely different types of evolutions, in particular for first-order geometric evolutions (i.e. time-dependent compact subsets $K(t) \subset \mathbb{R}^N$ whose deformation depend on nonlocal properties of $K(t)$ and its normal cones at the boundary) [10].

3.1 Set-Valued Maps for Image Segmentation

An important problem of computer vision is the detection of image segments which belong to one and the same object. Meanwhile many concepts have been developed to find their boundaries on grey-valued images. In particular, the so-called region growing methods seize the idea of improving the approximating segments in some sense while time is increasing.

So following that suggestion of [4], we developed such an algorithm combining 3 key properties: firstly, analytical tools close to graphical imagination, secondly, no a priori restrictions on the regularity of final contours and finally no parameterization of boundaries while expanding ([9]).

A 2-dimensional grey-valued image is given as a function $G : \mathbb{R}^2 \longrightarrow [0, \infty[$. Furthermore the "quality" of approximating segments has to be quantified. For example, the variance of grey-values $G|_X$ (restricted to a subset $X \subset \mathbb{R}^N$ with positive Lebesgue measure),

$$X \longmapsto \frac{1}{\mathcal{L}^N(X)} \int_X \left(G(x) - \frac{1}{\mathcal{L}^N(X)} \int_X G(y)\, dy \right)^2 dx$$

describes their oscillation within X. More generally, we consider a real-valued functional $\Phi : \mathcal{P}(\mathbb{R}^N) \longrightarrow \mathbb{R} \cup \{\infty\}$ of the form

$$\Phi(X) \stackrel{\text{Def.}}{=} \psi\left(\mathcal{L}^N(X), \int_X G\, dx, \int_X G^2\, dx \right)$$

for sets $X \subset \mathbb{R}^N$, $0 < \mathcal{L}^N(X) < \infty$ (and a fixed function $\psi \in C^2$). So the mathematical problem is: For a nonempty compact initial set $K_0 \subset \mathbb{R}^N$ given, construct a set-valued map $K(\cdot) : [0,T[\rightsquigarrow \mathbb{R}^N$ satisfying:

1. The values of $K(\cdot)$ are compact subsets of $\mathbb{R}^N$ with $K(0) = K_0$, $K(t_1) \subset K(t_2)$ for all $0 \leq t_1 \leq t_2 < T$,
2. $K(\cdot)$ is continuous (with respect to the Hausdorff distance)
3. $\Phi \circ K : [0,T[\longrightarrow \mathbb{R}$ is non-increasing,
4. $\bigcup_{t<T} K(t)$ is "critical" (in some sense - excluding simple continuation).

The ansatz for $K(t)$ is also based on set-valued maps. Indeed, we consider reachable sets of differential inclusions (generalizing flows along vector fields). For a set-valued map $F : \mathbb{R}^N \times [0,T[\rightsquigarrow \mathbb{R}^N$ and an initial set $K \subset \mathbb{R}^N$ given, the *reachable set* at time $t \in [0,T[$ is defined as

$$\vartheta_F(t,K) \stackrel{\text{Def.}}{=} \{ x(t) \mid x(\cdot) : [0,t] \longrightarrow \mathbb{R}^N \text{ absolutely continuous,} \\ \dot{x}(\cdot) \in F(x(\cdot),\cdot) \text{ almost everywhere, } x(0) \in K \}.$$

In comparison with flows along smooth vector fields, it has the advantage that topological changes may occur, e.g. "holes" might disappear while evolving.

The right-hand side F of the differential inclusion is now to be constructed such that $t \longmapsto \Phi(\vartheta_F(t,K))$ in non-increasing. This leads to the question how to differentiate the Lebesgue integral of G over the reachable set $\vartheta_F(t,K)$ with respect to time t. If the topological boundary of K is sufficiently smooth, then Reynold's transport theorem gives the answer for flows along smooth vector fields. However, dispensing with any regularity assumptions about the compact set K (and its boundary), the following theorem holds for reachable sets of differential inclusions:

Theorem 2. *Let $F : \mathbb{R}^N \times [0,T[\rightsquigarrow \mathbb{R}^N$ be a set-valued map satisfying*
1. the values of F are compact, convex and have nonempty interior,
2. F is continuous (w.r.t. Hausdorff distance) and has linear growth.
Then for any compact set $K \subset \mathbb{R}^N$ and $h \in L^1_{loc}(\mathbb{R}^N)$, the Lebesgue integral

$$[0,T[\longrightarrow \mathbb{R}, \quad t \longmapsto \int_{\vartheta_F(t,K)} h(x) \, dx$$

is absolutely continuous and has the weak derivative

$$\int_{\partial \vartheta_F(t,K)} h(x) \ \sup \left(F(x,t) \cdot \tilde{N}_{\vartheta_F(t,K)}(x) \right) \, d\mathcal{H}^{N-1}x.$$

with $\tilde{N}_{\vartheta_F(t,K)}(x)$ denoting all unit vectors of the Bouligand normal cone at x.

The detailed proof (using minimal time function of nonautonomous differential inclusions) is presented in [12]. Applying this result to grey-valued images and a "stabilized" form of variance

$$\Phi(X) \stackrel{\text{Def.}}{=} \tfrac{1}{\mathcal{L}^N(X)} \int_X \left(G(x) - \tfrac{1}{\mathcal{L}^N(X)} \int_X G(y) \, dy \right)^2 dx \ - \ \alpha \cdot \mathcal{L}^N(X)$$

(with a small parameter $\alpha > 0$) leads to the weak derivative $\frac{d}{dt} \Phi(\vartheta_F(t,K))$ being equal to

$$\int_{\partial \vartheta_F(t,K)} \varphi(x, \vartheta_F(t,K)) \ \sup \left(F(x,t) \cdot \tilde{N}_{\vartheta_F(t,K)}(x) \right) \, d\mathcal{H}^{N-1}x$$

with a quadratic polynomial of $G(x)$:

$$\varphi(x,K) \ = \ \tfrac{1}{\mathcal{L}^N(K)} G(x)^2 \ - \ \tfrac{2 \int_K G \, dy}{\mathcal{L}^N(K)^2} G(x) \ - \ \tfrac{\int_K G^2 \, dy}{\mathcal{L}^N(K)^2} \ + \ \tfrac{2 (\int_K G \, dy)^2}{\mathcal{L}^N(K)^3} \ - \ \alpha.$$

The ansatz for the set-valued map $F : \mathbb{R}^N \times [0, T[\rightsquigarrow \mathbb{R}^N$ is now motivated by the notion to prescribe the *speed* of expansion (and not the exact direction). This phenomenon is reflected by the choice $F(x,t) = \mathbb{B}_{r(x,t)}(0) \subset \mathbb{R}^N$, i.e. closed balls with center in 0 and a radius function $r : \mathbb{R}^N \times [0, T[\longrightarrow [0, \infty[$. As an essential advantage of this ansatz,

$$\sup (F(x,t) \cdot \tilde{N}_{\vartheta_F(t,K)}(x)) = r(x,t) > 0$$

and thus, the sign of the derivative $\frac{d}{dt} \Phi(\vartheta_F(t,K))$ is determined by the sign of $\varphi(\cdot, \vartheta_F(t,K))$ (restricted to the boundary $\partial\vartheta_F(t,K)$). This notion provides a sufficient condition on $F(\cdot)$ such that the variance of grey-values in $\vartheta_F(\cdot, K_0)$ is non-increasing.

For implementing this concept, we just have to decide whether a pixel belongs to the approximating segment or not. This leads to a quite simple implementation on an analytical basis and shows good results - for both 2-dimensional images (Fig. 9, 10) and image sequences (Fig. 11). Details are described in [9].

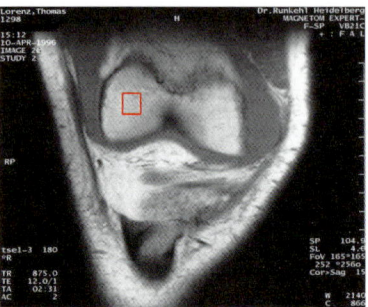 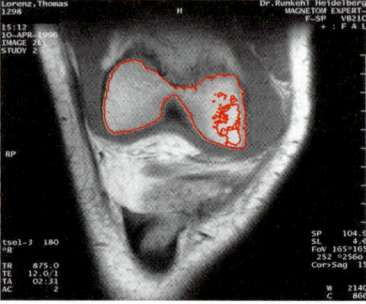

Fig. 9. MR of right human knee.   *Left:* the initial set   *Right:* the resulting set

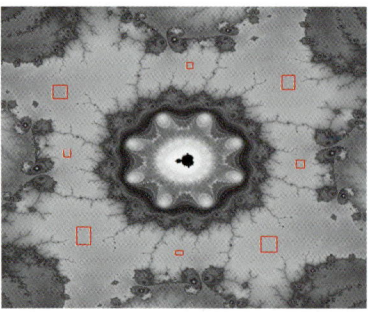 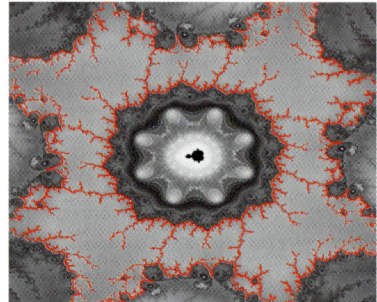

Fig. 10. Fractal sets give an impression of the precision at the boundary.

Fig. 11. Extract of a sequence of 50 images. Time is simply regarded as a third dimension. So marking a car in the first image, the algorithm follows its track while passing the crossroads.

3.2 Evolutions in Metric Spaces

Compact subsets of $\mathbb{R}^N$ are a typical example providing no obvious structure of vector spaces. So as a next generalizing step, we consider the Hausdorff metric d on the set $\mathcal{K}(\mathbb{R}^N)$ of all nonempty compact subsets of $\mathbb{R}^N$:

$$d(K_1, K_2) := \max\Big\{ \sup_{x \in K_1} \mathrm{dist}(x, K_2),\ \sup_{y \in K_2} \mathrm{dist}(y, K_1) \Big\}.$$

(As a topological advantage of $(\mathcal{K}(\mathbb{R}^N), d)$, every closed ball is compact.)

In the nineties, Jean-Pierre Aubin suggested a concept extending ordinary differential equations to a metric space like $(\mathcal{K}(\mathbb{R}^N), d)$, the so-called *mutational equations* ([1],[2]).

Lacking any form of addition or scalar multiplication, the typical term $x + h\,v$ of differential calculus in a vector space requires an adequate counterpart in a metric space (E, d). Geometrically speaking, a half-line in $\mathbb{R}^N$ (in a fixed direction $v \in \mathbb{R}^N$), i.e. $[0,1] \times \mathbb{R}^N \longrightarrow \mathbb{R}^N$, $(h, x) \longmapsto x + h v$ is replaced by a "half-curve" $\vartheta : [0,1] \times E \longrightarrow E$ such that any initial point $x \in E$ reaches the state $\vartheta(h, x) \in E$ at time $h \in [0,1]$.

In the metric space $(\mathcal{K}(\mathbb{R}^N), d)$ for example, such "half-curves" are induced by
- a vector $v \in \mathbb{R}^N$: $\vartheta_v(h, K) := \{ x + h v \mid x \in K \}$
- an ODE (with f) $\vartheta_f(h, K) := \{ x(h) \mid x(\cdot) \in C^1,\ \dot{x} = f(x),\ x(0) \in K \}$
- a diff. inclusion $\vartheta_F(h, K) := \{ x(h) \mid x(\cdot) \in AC,\ \dot{x} \in F(x)\ \text{a.e.},\ x(0) \in K \}$

Aubin showed that 4 properties are sufficient for following the track of ODEs: A map $\vartheta : [0,1] \times E \longrightarrow E$ is called *transition* on a metric space (E, d) if

1. $\vartheta(0, x) = x$ for all $x \in E$, i.e. the second argument is the initial point,
2. $\frac{1}{h} \cdot d(\vartheta(t+h, x), \vartheta(h, \vartheta(t, x))) \longrightarrow 0 \quad (h \downarrow 0) \quad$ for all x, t,
3. $\exists\, \alpha(\vartheta) < \infty: \ d(\vartheta(h, x), \vartheta(h, y)) \leq d(x, y) \cdot e^{\alpha(\vartheta)\,h} \quad$ for all x, y, h,
4. $\exists\, \beta(\vartheta) < \infty: \ d(\vartheta(s, x), \vartheta(t, x)) \leq \beta(\vartheta) \cdot |t - s| \quad$ for all x, s, t.

Transitions are the tools for introducing derivatives of a curve $x : [0,T] \longrightarrow E$ on the basis of first-order approximations. Indeed, a transition ϑ on E belongs to the so-called *mutation* of $x(\cdot)$ at time $t \in [0,T[$ (in the sense of Aubin) if

$$\vartheta \in \overset{\circ}{x}(t) \quad :\Longleftrightarrow \quad \limsup\nolimits_{h \downarrow 0} \tfrac{1}{h} \cdot d(x(t+h),\ \vartheta(h, x(t))) \;=\; 0.$$

Of course, ϑ need not be unique and thus, $\overset{\circ}{x}(t)$ is a set of transitions. In a *mutational equation*, one of its elements is prescribed by a function of the current state $x(t)$, i.e. $\overset{\circ}{x}(t) \ni f(x(t))$ with f given - corresponding to an ODE. Now the next step is to solve the initial value problem in a constructive way. Applying Euler method, for example, requires an estimate of the deformation along any 2 transitions ϑ, τ. A distance between ϑ, τ is induced by their effect on the same initial point: $D(\vartheta, \tau) := \sup_{x \in E} \limsup_{h \downarrow 0} \frac{d(\vartheta(h,x),\ \tau(h,x))}{h}$. Then, Gronwall's Lemma implies the global estimate for all $x, y \in E$, $h \in [0,1]$

$$d(\vartheta(h,x),\ \tau(h,y)) \;\leq\; \Big(d(x,y) + h \cdot D(\vartheta, \tau)\Big) \cdot e^{\alpha(\vartheta)\, h} \qquad (9)$$

and the Cauchy-Lipschitz Theorem is easy to extend to mutational equations:

Theorem 3. [2] Let (E,d) be a complete metric space whose closed balls are compact. Assume that f maps E to transitions on E such that $D(f(x),\ f(y)) \leq \text{const} \cdot d(x,y)$ for all $x, y \in E$, $\quad \sup_x \alpha(f(x)) < \infty$. Then for every $x_0 \in E$, there exists a unique solution $x(\cdot) : [0,T] \longrightarrow E$ of the initial value problem $\overset{\circ}{x}(t) \ni f(x(t))$ in $[0,T]$, $x(0) = x_0$, i.e. $\limsup_{h \downarrow 0} \tfrac{1}{h} \cdot d\big(x(t+h),\ f(x(t))(h, x(t))\big) = 0 \quad$ for all $t \in [0,T[$.

Applying this result to $(\mathcal{K}(\mathbb{R}^N), \boldsymbol{d})$ and reachable sets $\vartheta_F(h,K)$ of differential inclusions leads to the main examples of Aubin's interest (called morphological equations). They do not need regularity assumptions about the boundaries and can take *nonlocal* properties of compact subsets into consideration. Moreover they are not restricted to shape evolutions obeying the inclusion principle, i.e. if an initial set contains a second one, this inclusion need not hold forever.

On the other hand, the Hausdorff metric $\boldsymbol{d}$ on $\mathcal{K}(\mathbb{R}^N)$ has the structural weakness that it does not consider the topological boundary separately. So these mutational equations in $(\mathcal{K}(\mathbb{R}^N), \boldsymbol{d})$ cannot be directly applied to first-order geometric evolution (depending on the boundary and its normals).

Considering still deformations along differential inclusions, "holes" of the shapes might disappear in the course of time as indicated in Fig.12. Strictly speaking, connected components of the boundaries can evolve into interior points. This form of "losing boundary information" implies that the distance between boundaries of continuously evolving sets need not be continuous (with respect to time) any longer.

Then, estimate (9) does not result from Gronwall's Lemma any longer and furthermore, reachable sets $\vartheta_F(\cdot, K)$ of differential inclusions do not provide obvious bounds of the parameters $\alpha(\vartheta_F), \beta(\vartheta_F)$ when taking boundaries into consideration.

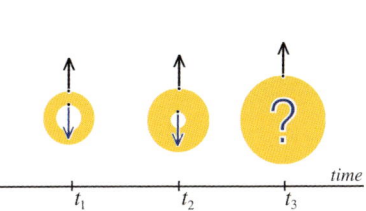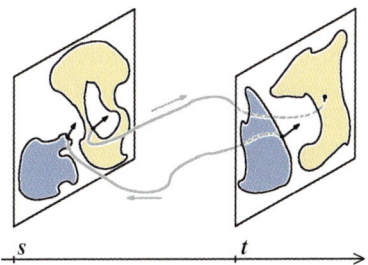

Fig. 12. Two obstacles for applying mutational equations due to the boundary: (a) lacking continuity in time   (b) lacking continuity w.r.t. initial set

In [10], two ideas are presented for coping with these difficulties:
Firstly, we compare only the boundaries of *later* sets with *earlier* ones since earlier sets usually contain "more information" about boundary components (see Fig.12 (a)). So the Hausdorff metric d is replaced by the nonsymmetric distance

$$q_{\mathcal{K}}(K_1, K_2) := d(K_1, K_2) + \mathrm{dist}\big(\mathrm{Graph}\ \tilde{N}^L_{K_2},\ \mathrm{Graph}\ \tilde{N}^L_{K_1}\big) \quad \text{for } K_j \in \mathcal{K}(\mathbb{R}^N)$$

with $\tilde{N}^L_{K_1}(x)$ containing all limiting normals of K_1 at x having norm ≤ 1.

In the general setting, the metric d on a given set E is replaced by a so-called *ostensible metric* $q : E \times E \longrightarrow [0, \infty[$ just satisfying $q(x, x) = 0$ for all x and the triangle inequality (in topology, q are also called *quasi-pseudo-metric*).
Then differential inclusions with sufficiently smooth right-hand side F satisfy

$$q_{\mathcal{K}}(\vartheta_F(s, K),\ \vartheta_F(t, K)) \leq \mathrm{const} \cdot (t - s) \qquad \text{for all } s < t.$$

The second idea concerns the continuity with respect to the initial state (quantified by the parameter α).
Consider the two initial compact sets $K_1, K_2 \subset \mathbb{R}^N$ evolving along the same differential inclusion as indicated in Fig.12 (b). Although each normal of K_2 might have a close counterpart among the normals of K_1, this relation need not be preserved. Indeed, any boundary point $x \in \partial \vartheta_F(s, K_1)$ can evolve into an interior point of $\vartheta_F(t, K_1)$ at time $t > s$.
This obstacle motivates to introduce weaker conditions on transitions and solutions - following the popular example of distributions. In an ostensible metric space, however, there are no obvious generalizations of linear forms or partial integration and so, distributions in their widespread sense cannot be introduced. So we suggest a more general interpretation: Their basic idea is to select a key property and demand it only for all elements of a given "test set". In fact, solving mutational equations has been based on estimate (9) and so we regard it as starting point for generalizing the definition of solutions.

Considering now $(\mathcal{K}(\mathbb{R}^N), q_{\mathcal{K}})$ and reachable sets $\vartheta_F(h, K)$ of differential inclusions, we choose compact sets $K_\circ \subset \mathbb{R}^N$ with $C^{1,1}$ boundary as test elements since their evolution is reversible for short times. Then differential inclusions with sufficiently smooth right-hand side F have a constant $\alpha(\vartheta_F)$ such that
$$q_{\mathcal{K}}(\vartheta_F(t, K_\circ), \vartheta_F(t, K)) \leq q_{\mathcal{K}}(K_\circ, K) \cdot e^{\alpha(\vartheta_F) \cdot t}$$
for all compact sets $K \subset \mathbb{R}^N$ and small $t > 0$.

Returning to the more general situation of an ostensible metric space (E, d), an essential distinction between Aubin's definition and our generalization is that the parameter $\alpha(\vartheta) < \infty$ of a transition ϑ has to satisfy
$$q(\vartheta(h, z), \vartheta(h, y)) \leq q(z, y) \cdot e^{\alpha(\vartheta) \cdot h}$$
for all $y \in E$, but "only" for all elements z of a fixed "test set" $D \subset E$ and for any small $h \geq 0$ (depending on z). Then, the distance $Q(\vartheta, \tau)$ between two (generalized) transitions $\vartheta, \tau : [0, 1] \times E \longrightarrow E$ can be defined in such a way that estimate (9) is essentially preserved, i.e.
$$q(\vartheta(h, z), \tau(h, y)) \leq \Big(q(z, y) + h \cdot Q(\vartheta, \tau)\Big) \cdot e^{\alpha(\tau) h} \qquad (10)$$
for all $y \in E$, $z \in D$ and small $h \geq 0$ (depending on z). This modified estimate (10) motivates how to specify a first-order approximation of a curve $x(\cdot) : [0, T] \longrightarrow (E, q)$ (in a generalized way): A generalized transition ϑ induces a first-order approximation of $x(\cdot)$ at time $t \in [0, T]$ if for any "test element" $z \in D$ and $h \searrow 0$,
$$q(\vartheta(h, z), x(t+h)) \leq \Big(q(z, x(t)) + h \cdot o(h)\Big) \cdot e^{\alpha h},$$
i.e. $\limsup_{h \downarrow 0} \frac{1}{h} \cdot \Big(q(\vartheta(h, z), x(t+h)) - q(z, x(t)) \cdot e^{\alpha \cdot h}\Big) \leq 0$.

Specifying the details about D and coining an adequate term of compactness guarantee existence and stability of solutions in ostensible metric spaces (see [10] for further details).

This concept of generalized mutational equations has been developed for the (rather general) environment of ostensible metric spaces and, first-order geometric evolutions in $(\mathcal{K}(\mathbb{R}^N), q_{\mathcal{K}})$ served as a motivating example.

In comparison with other geometric approaches (like level-set methods), several advantages of Aubin's (original) mutational equations are preserved:

- no regularity restrictions at the topological boundaries
- nonlocal properties of both the compact sets and their normal cones can be taken into consideration
- no restriction to the inclusion principle (whereas viscosity solutions of the level-set method are based on the maximum principle)

A key advantage of generalized mutational equations (over viscosity solutions, for example) is revealed by combining examples of different origin:
Existence of solutions holds for systems of generalized mutational equations.

In particular, we are free to combine geometric evolutions with other examples investigated meanwhile like

- semilinear evolution equations in reflexive Banach spaces ([10])
- nonlinear continuity equations of measures for BV vector fields ([11])

References

1. Aubin, J.-P. : Mutational equations in metric spaces. Set-Valued Anal. **1**, No.1, 3–46 (1993)
2. Aubin, J.-P. : Mutational and morphological analysis. Tools for shape evolution and morphogenesis. Birkhäuser, Boston (1999)
3. Demongeot, J., Kulesa, P., Murray, J.D.: Compact set valued flows. II: Applications in biological modelling. C. R. Acad. Sci., Paris, Sér. II, Fasc. b **324**, No.2, 107–115 (1997)
4. Demongeot, J., Leitner, F.: Compact set valued flows. I: Applications in medical imaging. C. R. Acad. Sci., Paris, Sér. II, Fasc. b **323**, No.11, 747–754 (1996)
5. Hofbauer, J., So, J. W.-H.: Competition in the Gradostat: the Global Stability Problem. Nonlinear Analysis, Theory, Methods and Applications **22**, No. 8, 1017-31 (1994)
6. Jäger, W., Smith, H. L., Tang, B.: Generic failure of persistence and equilibrium coexistence in a model of m-species competition in a n-vessel gradostat when m>n. In Busenberg, B., Martinelli,M. (ed) Lecture Notes in Biomathematics 92: Differential Equations Models in Biology, Epidemiology and Ecology. Springer, New York, 200-209 (1990)
7. Jäger, W., So, J. W.-H., Tang, B., Waltman, P.:Competition in the gradostat. J. Math. Biol., **25**, 23-42 (1987)
8. Lorenz, Th. : Mengenanalytischer Ansatz zur Bildsegmentierung. Diploma thesis, Ruprecht–Karls University, Heidelberg (1999)
9. Lorenz, Th. : Set-valued maps for image segmentation. Comput. Vis. Sci. **4**, No.1, 41–57 (2001)
10. Lorenz, Th. : First-order geometric evolutions and semilinear evolution equations : A common mutational approach. Doctor thesis, Ruprecht–Karls University, Heidelberg (2004)
11. Lorenz, Th. : Quasilinear continuity equations of measures for bounded BV vector fields, Preprint available at www.ub.uni-heidelberg.de/archiv/5993
12. Lorenz, Th. : Reynold's transport theorem for differential inclusions, Set–Valued Analysis, **14**, No.3, 209-247 (2006)
13. Pao, C.V.: Nonlinear Parabolic and Elliptic Equations. Plenum Press, New York (1992)
14. Smith, H. L., Tang, B., Waltman, P.: Competition in an n-vessel Gradostat. SIAM J. Appl. Math. Anal., **5**, 1451-71 (1991)
15. Tambulea, A.: Competition and persistence of microorganisms in the gradostat. Doctor thesis, Ruprecht-Karls University, Heidelberg (2005)
16. Tang, B.: Mathematical Investigations of Growth of Microorganisms in the Gradostat. J. Math. Biol., **23**, 319-339 (1986)

Part II

Navier-Stokes Equations and Chemical Reactions

Preamble

This chapter contains three articles which are concerned with the solution of multidimensional reactive flow problems by adaptive finite element methods. The approach to mesh as well as model adaptivity uses the dual weighted residual (DWR) method which employs residual-based a posteriori error estimation of physical quantities of interest and sensitivity information by numerically solving an extra dual problem.

The first article *"Mesh and model adaptivity for flow problems"* discusses the two main aspects of the DWR method: its practical realization for mesh adaptation in three-dimensional flow computations and its extension to controlling modeling errors. This provides the possibility to use on the one hand locally refined meshes, and on the other hand a hierarchy of models in order to reduce the numerical costs without sacrificing accuracy. The issue of modeling errors is of importance in many fields, especially in reactive flow computations, where sophisticated models are developed but not feasible to be used in multidimensional computations due to the high numerical costs.

The second article *"Parallel multigrid on locally refined meshes"* describes the practical use of adaptive finite element methods for solving three-dimensional flow problems to high accuracy at reasonable costs. This requires the combination of three modern numerical techniques: multigrid algorithms, local mesh refinement and parallelization. The following two topics of the development of a tool for the simulation of complex flows are discussed: (i) multigrid algorithms on locally refined meshes, and (ii) the modification of multigrid methods for use on parallel computers.

The third article *"Solving multidimensional reactive flow problems with adaptive finite elements"* presents the application of the DWR method to the compressible Navier-Stokes equations modeling chemically reactive flows. The emphasis is on the low-Mach number regime including the limit case of incompressible flow. The most important ingredients are appropriate finite element discretizations, residual driven a posteriori mesh refinement, fully coupled defect-correction iteration for linearization, and optimal multigrid preconditioning. The potential of automatic mesh adaptation together with approximate Newton iteration and multilevel techniques is illustrated by 2D and 3D simulations of laminar methane combustion including detailed reaction mechanisms.

Mesh and Model Adaptivity for Flow Problems[*]

R. Becker[1], M. Braack[2], R. Rannacher[2] and T. Richter[2]

[1] Laboratoire de Mathématiques Appliquées, Université de Pau et des Pays de l'Adour
[2] Institut für Angewandte Mathematik, Universität Heidelberg

Summary. This work addresses two aspects. The first one is the practical realization of the dual weighted residual (DWR) method of Becker & Rannacher [6] to three dimensional flow problems. The second aspect is its extension to modeling errors. This gives the possibility to use on the one hand locally refined meshes, and on the other hand, a hierarchy of models in order to reduce the numerical costs without sacrifying accuracy. The issue of modeling errors is of importance in many fields, especially in reactive flow computations where sophisticated models are developed but not feasable to be used in multidimensional computations due to the high numerical costs.

1 Introduction

We start with the abstract setting of a partial differential equation in variational form in a Hilbert space V. The solution $u \in V$ fulfills the equation

$$u \in V: \quad a(u)(\varphi) = \langle f, \varphi \rangle \quad \forall \varphi \in V, \tag{1}$$

with a semilinear form $a : V \times V \to \mathbb{R}$ beeing linear in the second argument, and a right hand side $f \in V'$. The approximate solution u_h in a finite dimensional subspace $V_h \subset V$ is given as the solution of the discrete system

$$u_h \in V_h: \quad a(u_h)(\varphi) = \langle f, \varphi \rangle \quad \forall \varphi \in V_h. \tag{2}$$

The space V_h results from finite elements on a triangulation $\mathcal{T}_h$ of the computational domain $\Omega \subset \mathbb{R}^d$ with mesh size h. Furthermore, we consider a functional

$$j : V \to \mathbb{R},$$

---

[*] This work has been supported by the German Research Foundation (DFG) through SFB 359 (Project A2) at the University of Heidelberg.

which gives us the quantity of interest $j(u)$. The a posteriori control aims at measuring the error with respect to this functional j and approximate it by a computable estimator η_h, i.e.

$$j(u) - j(u_h) \approx \eta_h \,.$$

We start in section 2 with the standard approach of dual weighted residuals, see [5, 6], where a dual solution is used for computing weights entering the estimator η_h. We focus on the approximation of the interpolation error for linear and quadratic elements and present an algorithm to adapt the mesh. This methodology is applied to a three-dimensional Navier-Stokes benchmark problem with tri-quadratic finite elements (Q_2).

In section 4 we extend this approach to modeling errors where the considered discrete nonlinear primal problem is given by

$$u_{hm} \in V_h : \ a_m(u_{hm})(\varphi) = (f, \varphi) \quad \forall \varphi \in V_h \,, \tag{3}$$

with a "simplified" semilinear form $a_m : V \times V \to \mathbb{R}$. In this case we do not have Galerkin orthogonality any more. The error estimator we are going to derive now splits into two parts:

$$j(u) - j(u_h) \approx \eta_h + \eta_m \,.$$

The part η_h of the estimator can be considered as contributions of the discretization, and the part η_m measures the influence of the model.

Finally, we apply the presented approach to different types of reactive flow problems where the diffusion model and the mesh size is locally adapted.

2 Mesh Adaptation Based on A Posteriori Control

2.1 Representation of Discretization Errors

As further notations, we need the directional derivatives of $a(u)(\cdot)$ denoted by $a'(u)(\cdot, \cdot)$, given by

$$a'(u)(v, \phi) = \lim_{\epsilon \to 0} \frac{1}{\epsilon} \{a(u + \epsilon v)(\phi) - a(u)(\phi)\} \,.$$

The estimator η_h makes use of the continuous dual solution $z \in V$ and the discrete dual solution $z_h \in V_h$:

$$a'(u_h)(z, \varphi) = j'(u_h)(\varphi) \quad \forall \varphi \in V \,, \tag{4}$$
$$a'(u_h)(z_h, \varphi) = j'(u_h)(\varphi) \quad \forall \varphi \in V_h \,. \tag{5}$$

We recall the result of [6] for a scalar output functional j. By $\mathcal{I}_h : V \to V_h$ we denote an arbitrary interpolation operator.

Theorem 1. *If $a(\cdot)(\cdot)$ and j are sufficiently often continuously differentiable it holds the error representation*

$$j(u) - j(u_h) = \frac{1}{2}\{\rho(u_h)(z - \mathcal{I}_h z) + \rho^*(u_h, z_h)(u - \mathcal{I}_h u)\} + R^*, \qquad (6)$$

with the primal and dual residuals

$$\rho(u_h)(\varphi) := (f, \varphi) - a(u_h)(\varphi),$$
$$\rho^*(u_h, z_h)(\varphi) := j'(u_h)(\varphi) - a'(u_h)(\varphi, z_h),$$

and a remainder term R^* formally of third order with respect to the error $\{u - u_h, z - z_h\}$.

In order to use this result as a mesh adaptation criterion we have to approximate the interpolation errors $z - \mathcal{I}_h z$ and $u - \mathcal{I}_h u$, and the resulting estimator has to be localized for giving information about the influence of the local mesh size h_K for each cell $K \in \mathcal{T}_h$.

A cheaper variant of (6) has a remainder term of only second order but only needs the evaluation of the primal residual and the interpolation of the dual solution, see [5]:

Theorem 2. *It holds*

$$j(u) - j(u_h) = \rho(u_h)(z - \mathcal{I}_h z) + R \qquad (7)$$

with a remainder term R formally of second order with respect to the error $\{u - u_h, z - z_h\}$.

In order to use these two error representations numerically, the interpolation errors $u - \mathcal{I}_h u$ and $z - \mathcal{I}_h z$ have to be approximated. We discuss this for Q_1 and for Q_2 elements in the following sections.

2.2 Linear Finite Elements

An efficient possibility for approximation of the interpolation errors is the recovery process of the computed quantities by higher-order polynomials, see [6]. For instance, in the case of triangles ($d = 2$) or tetrahedra ($d = 3$) and when V_h consists of piecewise linear elements, quadratic interpolation may be used. For quadrilaterals and piecewise d-linear elements, the interpolation can be done on d-quadratic elements. To this purpose, we assume that the triangulation $\mathcal{T}_h$ is organized patch-wise, i.e., $\mathcal{T}_h$ results from a global refinement of a mesh $\mathcal{T}_{2h}$.

By $i_{2h}^2 : V_h \to V_{2h}^{(2)}$ we denote the nodal interpolation of piecewise bilinears on $\mathcal{T}_h$ onto biquadratic finite elements on $\mathcal{T}_{2h}$. The interpolation errors will be numerically approximated by

$$z - \mathcal{I}_h z \approx i_{2h}^2 z_h - z_h, \qquad (8)$$

$$u - \mathcal{I}_h u \approx i_{2h}^2 u_h - u_h. \qquad (9)$$

This approximation is usually observed to be accurate enough. Taking into account that the residuals $\rho(u_h)(\varphi)$ and $\rho^*(u_h, z_h)(\varphi)$ with respect to a discrete test function $\varphi \in V_h$ vanish leads to the following estimator

$$j(u) - j(u_h) \approx \eta_h,$$

consisting of two additive terms:

$$\eta_h := \frac{1}{2}\left\{\rho(u_h)(i_{2h}^2 z_h) + \rho^*(u_h, z_h)(i_{2h}^2 u_h)\right\}. \qquad (10)$$

2.3 Localization to Nodal Quantities

We discuss the localization of the estimator for $(d-)$ linear finite elements. For each node $\mathcal{N}_i$ of the triangulation of $\mathcal{T}_h$, we have the $(d-)$ linear nodal function $\phi_i \in V_h$ and a quadratic nodal function $\phi_i^{(2)} := \mathcal{I}_{2h}^2 \phi_i \in V_{2h}^{(2)}$. We denote the coefficients of u_h and z_h with respect to the basis $\{\phi_i\}$ by the vectors $U, Z \in \mathbb{R}^n$, respectively. Let Ψ_i and Ψ_i^* be the primal and dual residual contributions with respect to the (scalar-valued) quadratic basis $\{\phi_i^{(2)}\}$:

$$\Psi_i = \rho(u_h)(\phi_i^{(2)}),$$
$$\Psi_i^* = \rho^*(u_h, z_h)(\phi_i^{(2)}).$$

Now, the estimator (10) can be expressed by the usual l^2-scalar products $\langle \cdot, \cdot \rangle$ in $\mathbb{R}^n$

$$\eta_h = \frac{1}{2}\{\langle \Psi, Z \rangle + \langle \Psi^*, U \rangle\}. \qquad (11)$$

The formulation for vector-valued functions is straightforward. Generally, direct localization of the terms $\langle \Psi, Z \rangle$ and $\langle \Psi^*, U \rangle$ results in a large overestimation of the error due to the oscillatory behavior of the residual terms; see Carstensen and Verfürth [13]. This can be reduced by a filter as described now.

We use the nodal interpolation operator $i_{2h}: V_h \to V_{2h}$ and the filtering operator $\kappa_{2h} := i_h - i_{2h}: V_h \to V_h$ giving the small-scale linear fluctuations. We denote the nodal vectors of the filtered primal solution $\kappa_{2h} u_h$ and dual solution $\kappa_{2h} z_h$ by U^π and Z^π, respectively:

$$(i_h - i_{2h})u_h = \sum_{i=1}^n \phi_i U_i^\pi, \qquad (i_h - i_{2h})z_h = \sum_{i=1}^n \phi_i Z_i^\pi.$$

This gives us the following localized estimator.

Lemma 1. *With the computable non-negative quantities on each node,*

$$\eta_{h,i} := \frac{1}{2}|\Psi_i Z_i^\pi + \Psi_i^* U_i^\pi|,$$

we have the following upper bound for the estimator:

$$|\eta_h| \leq \sum_{i=1}^{n} \eta_{h,i}.$$

Proof. The estimator defined in (10) can also be expressed by

$$\eta_h = \frac{1}{2}\left\{\langle \Psi, Z^\pi \rangle + \langle \Psi^*, U^\pi \rangle\right\},$$

because the quadratic interpolation operator $\mathcal{I}_{2h}^2$ is the identity on V_{2h}. For details we refer to [8].

2.4 Quadratic Finite Elements

The approximation of the interpolation errors $u - \mathcal{I}_h u$ and $z - \mathcal{I}_h z$ for Q_2 elements can be done by interpolation on Q_4 elements. However, this requires patches of Q_2 elements which becomes very expensive in 3-D. Therefore, we use another recovery process explained in the following for the variant (7) of the error representation. Note that the interpolation technique to be presented directly applies to representation (6).

As an example, we take the Navier-Stokes equations as the system under consideration described in the strong form by the nonlinear operator L. The viscosity will be denoted by ν and the velocity by v. After integration by parts the estimator reads:

$$\eta_h = \rho(u_h)(\omega_z)$$
$$= \sum_{K \in \mathcal{T}_h} (f - Lu_h, \omega_z)_K + \frac{\nu}{2} \sum_{K \in \mathcal{T}_h} ([\partial_n v_h], \omega_z)_{\partial K},$$

where the weight ω_z is an approximation of $z - \mathcal{I}_h z$. This estimator consists of cell residuals $(f - Lu_h, \omega_z)_K$ and jump terms across cell edges:

$$[\partial_n v_h](x) := (\partial_n v_h|_K - \partial_n v_h|_{K'})(x), \qquad (12)$$

for a point x on the edge $K \cap K'$. For quantitative error prediction, it is not necessary to compute both parts. It has been shown in Carstensen [14] that for linear finite elements the jump terms (12) dominate, whereas for quadratic elements the residual terms become dominant. Therefore, as cell error indicators we take the quantities

$$\eta_K := \|f - Lu_h\|_K \|\omega_z\|_K,$$

giving the estimator

$$|j(u) - j(u_h)| \approx \sum_{K \in \mathcal{T}_h} \eta_K .$$

For the definition of the weights ω_z, we approximate the interpolation error $z - \mathcal{I}_h z$ by discrete third-order difference quotients:

$$z - \mathcal{I}_h z \approx Ch^3 \nabla^3 z \approx Ch^3 \nabla_h^3 z_h ,$$

with an interpolation constant C. Therefore, the weights ω_z are chosen on each cell K as

$$\omega_z|_K := h_K^3 |\nabla_h^3 z_h| .$$

We still have to explain the definition of the discrete difference quotient ∇_h^3. For ease of presentation, we explain this procedure for a scalar quantity z. The extension to vector-valued quantities is straightforward. First, we approximate the Hessian of z by a matrix-valued discrete κ_z which is the L^2-projection of $\nabla^2 z_h$:

$$\kappa_z \in Q_h^{d \times d} : \sum_{K \in \mathcal{T}_h} (\nabla^2 z_h - \kappa_z, \phi)_K = 0 \quad \forall \phi \in Q_h^{d \times d} .$$

Note, that $\nabla^2 z_h$ is discontinuous across cell edges. Second, the expression $\nabla_h^3 z_h$ is defined cell-wise by the gradients of κ_z:

$$\nabla_h^3 z_h|_K := \sqrt{(\partial_x^2 + \partial_y^2 + \partial_z^2) \kappa_z|_K} \quad \text{for } d = 3 .$$

Remark 1. Instead of taking for κ_z a L^2-projection one may take the gradient of another projection of $\nabla^2 z_h$, for instance, by taking the average nodal values of $(\nabla^2 z_h)|_K(\mathcal{N})$, for the nodes $\mathcal{N}$ of $\mathcal{T}_h$. This procedure is much cheaper because it is a *local* projection.

2.5 The Mesh Refinement Strategy

In the previous section we have described the localization strategy for obtaining error indicators η_K (or η_i for nodal values). Next we have to choose a subset of cells of the triangulation $\mathcal{T}_h$ for refinement. There are several standard approaches, see for instance [7]. Without loss of generality we assume $\eta_{K_i} \geq \eta_{K_{i+1}}$ for all $i \in \{1, \ldots, n\}$, where n stands for the number of cells of $\mathcal{T}_h$. The question of adaptive refinement can be reduced to the determination of a non-negative number m with $1 \leq m \leq n$ and a refinement of all cells K_i with $i \leq m$.

We denote by $\mathcal{T}_m$ the locally refined mesh resulting from $\mathcal{T}_h$ by refining these m cells. The corresponding number of cells in a d-dimensional domain becomes for isotropic bisections

$$n_m = n + m(2^d - 1) .$$

For the mesh $\mathcal{T}_m$ we consider the "global mesh size" $h \sim n_m^{-1/d}$. Furthermore, we denote by $\alpha > 0$ the largest number such that $j(u)-j(u_h) = O(h^\alpha)$ assuming optimally fitted meshes. For Q_2 discretization in this work we have $\alpha = 4$. Using an error estimator η_m as an approximation of the discretization error on $\mathcal{T}_m$ we obtain $\eta_m \sim n_m^{-\alpha/3}$. Now, the adaptive process aims at minimizing the constant

$$c_m = \frac{\eta_m}{n_m^{-\alpha/3}},$$

where we compute η_m as an extrapolation of the already computed estimate η for the triangulation $\mathcal{T}_h$ to the finer mesh $\mathcal{T}_m$:

$$\eta_m := \eta - \beta \sum_{i=1}^{m} \eta_{K_i},$$

with $\beta = 1 - 0.5^\alpha$. Now, the optimal choice for m can be easily determined by taking

$$m = \arg \min_{1 \leq m \leq n} c_m.$$

Remark 2. For a mesh with equilibrated error indicators $\eta_K = \eta/N$ we have:

$$\eta_m = \eta(1 - \beta m/n),$$
$$c_m = \eta(1 - \beta m/n) n_m^{\alpha/3}$$
$$\sim (1 - \beta x)(1 + (2^d - 1)x)^{\alpha/3},$$

with $x := m/n$. For $d = 3$ and $\alpha = 4$, this number c_m has its minimum in $(0, 1]$ at $x = 1$. This results in $m = n$ and, therefore, to a global refinement. This is known to be optimal for equilibrated errors when the functional j is sufficiently regular.

3 Mesh Adaptation Applied to Navier-Stokes Benchmark Problems

Under the DFG Priority Research Program 'Flow Simulation on High Performance Computers' a set of benchmark problems has been defined by Schäfer & Turek [25]. One particular task is to calculate different coefficients of a three dimensional flow around obstacles at Reynolds number $Re = 20$. The flow is described by the Navier-Stokes equations for velocities v and pressure p:

$$-\nu \Delta v + (v \cdot \nabla)v + \nabla p = 0 \quad \text{in } \Omega,$$
$$\operatorname{div} v = 0 \quad \text{in } \Omega.$$

The domain Ω is a channel with height $H = 0.41\,m$ and length $L = 2.5\,m$. In the first configuration, the obstacle is a cylinder with diameter $D = 0.1\,m$. In the second one, the obstacle consists of a square cross-section with the

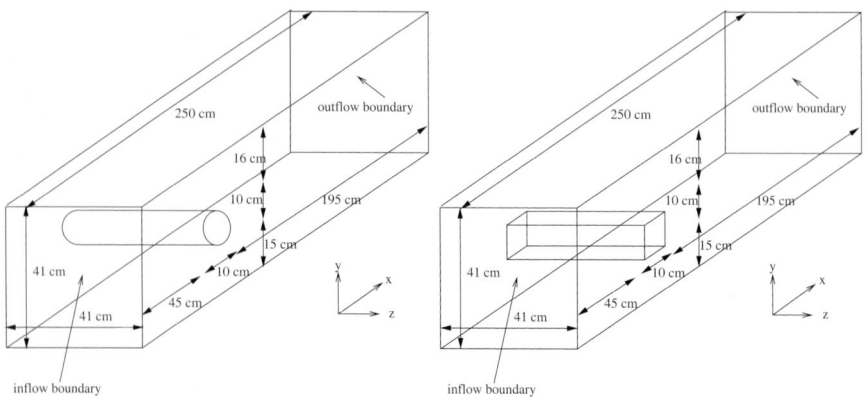

Fig. 1. Configurations of the benchmark problems. The obstacle is a cylinder with circular cross-section (left) and square cross-section (right).

same diameter D. The exact (non-symmetric) positions of these obstacles are displayed in Figure 1.

The boundary condition on the inflow boundary Γ_{in} is $v = \widehat{v}$, where the inflow velocity field $\widehat{v} = \{\widehat{v}_1, 0, 0\}$ is prescribed by the component in x-direction $\widehat{v}_1 = v^+ w(y) w(z)$ with the parabolic function $w(y) = 4y(H-y)/H^2$ and the peak velocity $v^+ = 0.45\,m/s$. Furthermore, we have $v = 0$ for solid walls Γ_{wall} and the so-called "do-nothing" condition on the outflow part Γ_{out} of the boundary:

$$\mu \partial_n v - pn = 0 \quad \text{on } \Gamma_{\text{out}}. \tag{13}$$

The kinematic viscosity is given by $\nu = 10^{-3}\frac{m^2}{s}$.

In this benchmark problem, certain local quantities are requested. Here we report on the results for the pressure difference Δp between the points $p_1 = (0.55, 0.2, 0.205)$ and $p_2 = (0.45, 0.2, 0.205)$:

$$\Delta p = p_2 - p_1,$$

and for the drag coefficient c_{drag} of the obstacle given by an integral over the surface $S \subset \Gamma_{\text{wall}}$ of the obstacle:

$$c_{\text{drag}} = C \int_S (\nu \partial_n v_t n_y - p n_x)\, ds, \quad C = \frac{2}{DH\overline{v}^2} = \frac{500}{0.41}.$$

Here, the vector $n = \{n_x, n_y, n_z\}$ denotes the unit normal vector on the surface S pointing into Ω, and v_t denotes the tangential velocity of v on tangent $t = \{n_y, -n_x, 0\}$. For the results of the lift coefficient, we refer to [11].

The discretization is done by Q_2 elements for pressure and velocity with local projection stabilization, see Becker & Braack [2, 3]. Computations on meshes with up to 2.5 million of degrees of freedom (dof) are performed on a single processor PC (AMD Opteron 246, 2 GHz) with 4 GB memory. On finer meshes, a parallel computer (PC cluster) was necessary. The used parallel algorithm is described in detail in the contribution [12] in this book.

3.1 Accuracy on Globally Refined Meshes

In the first step we use globally refined meshes, see Table 1. For the pressure drop Δp, the first two digits are reliable. With respect to the drag coefficient, also the fifth digit can now be considered as reliable $c_{\text{drag}} = 6.1853$. These values confirm the results of John [22] where piecewise triquadratic elements for the velocities (Q_2) and a discontinuous piecewise linear pressure (P_1^{disc}) are used.

Furthermore, we explore the far more difficult test case of the square cross-section. The solution can be found in a Sobolev space $H^{1+\alpha}(\Omega)$ with $0 < \alpha < 1$. Therefore, even for quadratic elements, the convergence order in L^2 (on globally refined meshes) can be expected only to be about $O(h)$. In Table 2 we also gather the values obtained with stabilized Q_2/Q_2 elements on structured meshes.

Table 1. Values obtained in this project for the circular cross-section on structured meshes.

#cells	dof	Δp	c_{drag}
480	18 720	0.188771	6.250365
3 840	136 000	0.178702	6.172750
30 720	983 040	0.173744	6.184551
245 760	7 864 320	0.171999	6.185323
1 966 080	62 914 560	0.171342	6.185331

Table 2. Values obtained in this project for the square cross-section on structured meshes.

#cells	dof	Δp	c_{drag}
78	3 696	0.183495	13.31491
624	24 544	0.208050	8.044496
4 992	177 600	0.177370	7.975926
39 936	1 348 480	0.176165	7.787849
319 488	10 787 840	0.175759	7.757928
2 555 904	86 302 720	0.175677	7.761191

3.2 Accuracy on Locally Refined Meshes

When local mesh refinement is carried out based on *insufficient* information about the discretization error, the quantities $j(u_h)$ may not converge or may converge to wrong values. Therefore, we perform one step of global refinement after the final adaptive step. This makes the values obtained much more reliable and guarantees the accuracy of the reference values. As expected, it turns out that the algorithm on locally refined meshes converges to the same solution as on globally refined meshes. In the following tables, we indicate this last global refinement by a straight line in order to separate it from the adaptive local refinement. In the plots given in the following subsections we do not display the result on the finest (final) mesh, because this defines the reference values.

Circular cross-section and locally refined meshes

We begin with the easier configuration with the circular cross-section. In the table of Figure 2, we list the values obtained of the pressure drop. The final step of a global refinement (increasing the number of cells by a factor of eight) shows that the local adaptation converges to the correct quantity. The first five leading digits of the pressure drop can be considered as correct: $j_1(u) = 0.17100$.

In Figure 2, we compare with the results shown before in Table 1 on globally refined meshes. On a mesh with $1\,383$ Q_2 cells, a relative error of 1% with respect to the pressure drop is achieved. This accuracy is similar to the one on a globally refined mesh with more than 2.5 million cells. For a relative accuracy of 10^{-4}, only $14\,375$ cells are necessary.

#cells	dof	Δp
480	18 720	0.1892506787
865	34 080	0.1798815121
1 138	45 520	0.1746605518
1 383	55 896	0.1729589594
6 556	243 224	0.1712704796
8 607	321 544	0.1710884110
10 742	404 376	0.1710308434
14 375	544 000	0.1709939769
924 704	32 260 736	0.1710070986

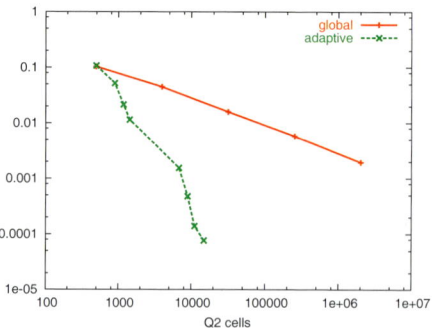

Fig. 2. Values of Δp for the circular cross-section (table on the left) and comparison of global and local mesh refinement with respect to the relative errors in Δp (plot on the left).

For the drag coefficient, the local refinement is not expected to give a gain that much since the primal and dual solution are smooth. On globally refined meshes, 4th order convergence is obtained, see the Table 1. However, as documented in the table of Figure 3 an enhancement of accuracy is observed. For instance, for a relative error of 10^{-6} a factor of 10 can be saved with local refinement. We can guarantee the first six leading digits of the drag: $c_{\text{drag}} = 6.18533$.

#cells	dof	c_{drag}
480	18 720	6.2503650
3 840	136 000	6.17275342
22 880	784 384	6.18471848
155 768	5 227 872	6.18533571
1 246 144	41 822 976	6.18533310
extrapolated		6.18533293

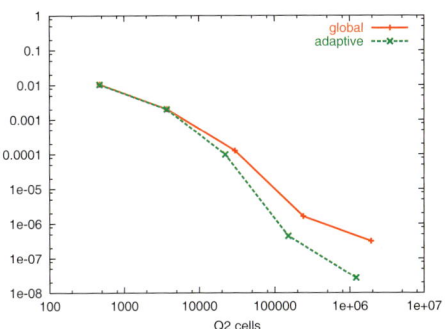

Fig. 3. Values of c_{drag} for the circular cross-section (table on the left) and a comparison of global and local mesh refinement with respect to the relative error in c_{drag} (plot on the left).

Square cross-section and locally refined meshes

For the geometry with corner singularities we expect an even better improvement by local mesh refinement, because neither the primal solution nor the dual one are smooth. The correct balance of resolving the singularities and the far field let us expect to be much more efficient than globally refined meshes.

#cells	dof	Δp
78	3 696	0.183495025
624	24 544	0.208049746
2 248	83 952	0.177348465
5 160	191 424	0.176449284
14 008	511 952	0.175792355
33 496	1 212 208	0.175713985
83 000	2 969 936	0.175676527
664 000	23 759 488	0.175686487

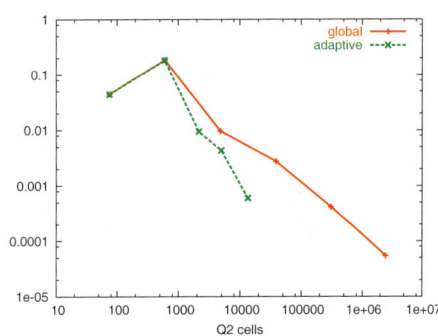

Fig. 4. Values of Δp for the square cross-section (table on the left) and comparison of global and local mesh refinement with respect to the relative errors in Δp (plot on the left).

Comparing the pressure drop on the penultimate mesh with 146 896 cells with the one on the last mesh (one global refinement) shows that the change is about $5 \cdot 10^{-6}$. Therefore, the reference solution can be considered as $\Delta p = 0.17569$. The comparison of the relative errors in Δp is shown in Figure 4.

#cells	dof	c_{drag}
78	3 696	13.31490984
372	15 296	8.045927135
1 849	69 816	7.973309702
5 363	200 840	7.787163358
8 639	331 944	7.796798708
554 744	20 314 304	7.766752694

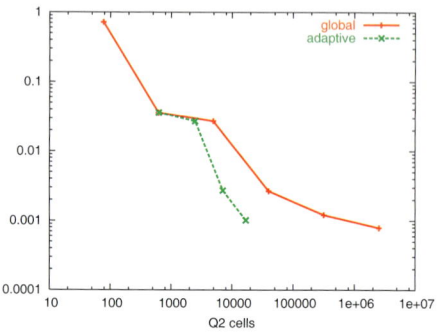

Fig. 5. Values of c_{drag} for the square cross-section (table on the left) and a comparison of global and local mesh refinement with respect to the relative error in c_{drag} (plot on the left).

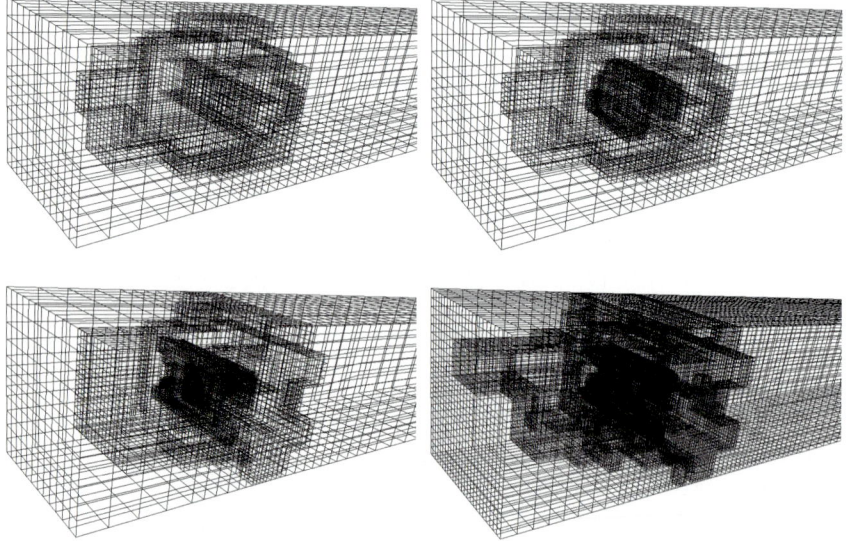

Fig. 6. Zoom of four meshes obtained on the basis of a posteriori error control of the pressure drop.

The drag was determined on globally refined meshes only up to the first two leading digits (cf. Table 2). With local refinement we obtain the first four digits: $c_{\text{drag}} = 7.768$. The plot in Figure 5 even shows a better asymptotic behaviour for the adaptive method. For 1% accuracy only 5 510 cells are needed instead of about 40 000 on a structured mesh (cf. Table 2).

Finally, we show in Figure 6 four meshes obtained within the adaptive mesh refinement for the pressure difference with increasing number of cells.

4 Model Adaptation with Finite Elements

Now, we extend the concept of a posteriori error estimation presented in the previous sections to measure model errors. More details on this topic can be found in Braack & Ern [8] for the general concept of model adaptivity, and in [9] for its application to combustion problems. For further work on adaptive modeling in computational mechanics we refer to the work of Oden & Prudhomme [24].

4.1 Error Representation for Non-Galerkin Methods

The nonlinear primal problem we are interested in is given as before in (1). The approximate discrete problem was already stated in (3). As the standard situation we consider the case where certain physical parts in the underlying equations are expressed by different mathematical models. As an example, one may think of a semilinear form split in two parts:

$$a(u)(\varphi) := b(u)(\varphi) + c_1(u)(\varphi),$$

where $c_1(u)(\varphi)$ stands for that part of the equation which is expensive to compute and is aimed to be replaced by a simpler model $c_2(u)(\varphi)$, at least in $\Omega \setminus \Omega_m$, whereas in $\Omega_m \subset \Omega$ the accurate model $c_1(u)(\varphi)$ should be used. In this case the simplified form reads

$$a_m(u)(\varphi) := b(u)(\varphi) + c_1(u)(\varphi|_{\Omega_m}) + c_2(u)(\varphi|_{\Omega \setminus \Omega_m}).$$

Here, we interpret $\varphi|_{\Omega_m}$ and $\varphi|_{\Omega \setminus \Omega_m}$ extended by zero to the whole of Ω. As a consequence, the two semilinear forms $a(\cdot)(\cdot)$ and $a_m(\cdot)(\cdot)$ coincide in Ω_m, i.e., the difference

$$\begin{aligned} d_m(u)(\varphi) &:= a(u)(\varphi) - a_m(u)(\varphi) \\ &= c_1(u)(\varphi|_{\Omega \setminus \Omega_m}) - c_2(u)(\varphi|_{\Omega \setminus \Omega_m}) \end{aligned}$$

vanishes if $\sup(\varphi) \subset \Omega_m$. The dual solution $z_{hm} \in V_h$ of the reduced problems is given by

$$z_{hm} \in V_h : \quad a'_m(u_{hm})(\phi, z_{hm}) = j'(u_{hm})(\phi) \quad \forall \phi \in V_h. \qquad (14)$$

In order to balance the model and discretization error, we have to include the discretization error in the a posteriori analysis. For the formulation of the error representation, we use the following notation for the primal and dual residual with respect to the reduced model and for test functions $\phi \in X$:

$$\rho(u_{hm})(\phi) := (f, \phi) - a(u_{hm})(\phi),$$
$$\rho^*(u_{hm}, z_{hm})(\phi) := j'(u_{hm})(\phi) - a'(u_{hm})(\phi, z_{hm}).$$

We have the following error representation, where similar to the previous sections, $\mathcal{I}_h : V \to V_h$ is an arbitrary interpolation operator:

Theorem 3. *If $a(u)(\cdot)$, $d(u)(\cdot)$ and the functional $j(u)$ are sufficiently differentiable with respect to u, then it holds that*

$$\begin{aligned}
j(u) - j(u_{hm}) = &-d_m(u_{hm})(z_{hm}) \\
&+ \frac{1}{2}\{\rho(u_{hm})(z - \mathcal{I}_h z) + \rho^*(u_{hm}, z_{hm})(u - \mathcal{I}_h u)\} \\
&- \frac{1}{2}\{d_m(u_{hm})(e_z) + d'_m(u_{hm})(e_u, z_{hm})\} + R,
\end{aligned}$$

where $e := \{e_u, e_z\} := \{u - u_{hm}, z - z_{hm}\}$, and R a remainder term cubic in the error e.

Proof. At first, we formulate the variational problems as optimization problems. The primal and dual solutions will be expressed by the variables $x = \{u, z\} \in X := V \times V$ and $x_{hm} = \{u_{hm}, z_{hm}\} \in X_h := V_h \times V_h$. In the variational space X we consider the functionals

$$L_m(x) = j(u) + (f, z) - a_m(u)(z), \tag{15}$$
$$\delta L(x) = -d_m(u)(z), \tag{16}$$
$$L(x) = L_m(x) + \delta L(x). \tag{17}$$

The derivative of L applied to a test function $y = \{\phi, \psi\} \in X$ is

$$L'(x)(y) = j'(u)(\phi) - a'(u)(\phi, z) + (f, \psi) - a(u)(\psi).$$

The original primal problem (1), the dual problem (4), the reduced discrete primal problem (3) and the reduced discrete dual problems (14) consist of finding the stationary points x and x_{hm} of L and L_m, respectively:

$$x \in X: \quad L'(x)(y) = 0 \quad \forall y \in X, \tag{18}$$
$$x_{hm} \in X_h: \quad L'_m(x_{hm})(y) = 0 \quad \forall y \in X_h. \tag{19}$$

Because the target quantities are given by evaluation of L and L_m at the stationary points x and x_{hm}:

$$j(u) = L(x),$$
$$j(u_{hm}) = L_m(x_{hm}),$$

we get the representation of the error in terms of L:

$$\begin{aligned}
j(u) - j(u_{hm}) &= L(x) - L(x_{hm}) + \delta L(x_{hm}) \\
&= \int_0^1 L'(x_{hm} + \lambda e)(e)\, d\lambda - d_m(u_{hm})(z_{hm}).
\end{aligned}$$

Applying the trapezoidal rule, we get

$$\int_0^1 L'(x_{hm} + \lambda e)(e)\, d\lambda = \frac{1}{2}\{L'(x)(e) + L'(x_{hm})(e)\} + R, \quad (20)$$

with remainder term R given by

$$R := \int_0^1 L'''(x_m + \lambda e)(e,e,e) \cdot \lambda(1-\lambda)\, d\lambda.$$

While the first term of the right hand side in (20) vanishes due to (18), the second term can be expressed by

$$L'(x_{hm})(e) = L'_m(x_{hm})(e) + \delta L'(x_{hm})(e)$$
$$= \rho(u_{hm})(z - z_{hm}) + \rho^*(u_{hm}, z_{hm})(u - u_{hm})$$
$$- d_m(u_{hm})(e_z) - d'_m(u_{hm})(e_u, z_{hm}).$$

Due to Galerkin orthogonality we may subtract arbitrary discrete test functions in the residual terms. Especially, we may use an arbitrary interpolation $\mathcal{I}_h u$ of u and $\mathcal{I}_h z$ of z:

$$0 = \rho(u_{hm})(z_{hm}) = \rho(u_{hm})(\mathcal{I}_h z),$$
$$0 = \rho^*(u_{hm}, z_{hm})(u_{hm}) = \rho^*(u_{hm}, z_{hm})(\mathcal{I}_h u).$$

This gives the assertion.

4.2 Numerical Evaluation of the Error Estimator

Now we split the error in contributions due to discretization and due to the models, respectively:

$$j(u) - j(u_{hm}) \approx \eta_h + \eta_m.$$

The error estimator η_h approximates the discretization error,

$$\eta_h \approx \frac{1}{2}\left\{\rho(u_{hm})(z - \mathcal{I}_h z) + \rho^*(u_{hm}, z_{hm})(u - \mathcal{I}_h u)\right\},$$

and the part η_m approximates the model error,

$$\eta_m \approx -d_m(u_{hm})(z_{hm}) - \frac{1}{2}\left\{d_m(u_{hm})(e_z) + d'_m(u_{hm})(e_u, z_{hm})\right\}.$$

Because the estimators should be computable numerically, we have to approximate various terms. The interpolation errors $u - \mathcal{I}_h u$ and $z - \mathcal{I}_h z$. This can be done as already presented in (8)–(9):

$$\eta_h := \frac{1}{2}\left\{\rho(u_{hm})(i^2_{2h} z_{hm}) + \rho^*(u_{hm}, z_{hm})(i^2_{2h} u_{hm})\right\}. \quad (21)$$

For the approximation η_m of the modeling error it is useful to obtain some a priori estimates of the various terms arising in the error representation. Assuming that the functional L' satisfies a stability property of the form

$$\|x_1 - x_2\|_X \leq \alpha \, \|L'(x_1) - L'(x_2)\|_{X'}$$

with $\alpha > 0$, we obtain

$$\|e\|_X = \|x - x_m\|_X \leq \alpha \, \|L'(x_m)\|_{X'} = \alpha \, \|\delta L'(x_m)\|_{X'},$$

thanks to relations (18) and (19). Assuming that the perturbation $d_m(\cdot)(\cdot)$ and its derivatives are sufficiently small, we may write $\|\delta L'(x_m)\|_{X'} \leq c(d) \, \|x_m\|_X$ with a constant $c(d) \ll 1$. Therefore, the error e is of first order in $c(d)$ and the contributions in R are thus of third order in $c(d)$. Furthermore, the terms $d_m(u_m)(e_z)$ and $d'_m(u_m)(e_u, z_m)$ are quadratic in $c(d)$, since they involve the nonlinear form $d_m(\cdot)(\cdot)$ and the error e.

If we neglect the higher-order terms in e with respect to the model $d(\cdot)(\cdot)$, we obtain the following computable quantity:

$$\eta_m := -d_m(u_{hm})(z_{hm}). \tag{22}$$

For complex models, the evaluation of η_m may be expensive. However, the gain of an adaptive algorithm with local model modification becomes substantial for nonlinear problems, since we do not need to include the (global) detailed model in each residual evaluation or in the Jacobian.

In the following section we show by numerical experiments that this estimator is reliable and efficient for several test problems. However, if more numerical effort in the error estimate is feasible, one may solve additional local dual problems involving the accurate model $d(\cdot, \cdot)$ in order to estimate the higher-order terms in (3).

4.3 Localization of the Estimator

In order to use the information of Theorem 3 for changing locally the model or the mesh size, we have to localize the estimator. Then we design an adaptive process in order to balance the two sources of error.

We look at the nodal contributions of $\eta_h + \eta_m$. For the part η_h of the estimator corresponding to the discretization error, we take advantage of the localization presented in Section 2.3. With the notation in that section, the estimator of the modeling error η_m can be expressed as a scalar product of Z and the vector $\Lambda = \{\Lambda_i\}$ built by the model residuals with respect to the Lagrangian nodal basis $\{\phi_i\} \subset V_h$:

$$\Lambda_i := d_m(u_{hm})(\phi_i),$$
$$\eta_m = \langle \Lambda, Z \rangle.$$

We localize the modeling error by the upper bound

$$|\eta_m| \leq \sum_{i=1}^{n} |\Lambda_i Z_i|.$$

Together with (12) we obtain the following localized estimator:

Lemma 2. *With the computable non-negative quantities on each node,*

$$\eta_{h,i} := \frac{1}{2}|\Psi_i Z_i^\pi + \Psi_i^* U_i^\pi|,$$
$$\eta_{m,i} := |\Lambda_i Z_i|,$$

we have the following upper bound for the estimator:

$$|\eta_h + \eta_m| \leq \sum_{i=1}^{n} (\eta_{h,i} + \eta_{m,i}).$$

To sum up, the computational extra work for evaluating the estimator and getting the nodal contributions consists of the following steps:

- Solving the (linear) dual problem for getting z_{hm}. There is no need to assemble a new Jacobian. Instead, the matrix corresponding to the primal problem can simply be transposed.
- Evaluation of the model with respect to the linear nodal basis, i.e. $d(u_{hm})(\phi_i)$, for getting Λ. This can be an expensive step for complex models. Taking the scalar product with the vector Z gives the model estimator η_m.
- Evaluation of the primal and dual residuals Ψ and Ψ^* with respect to the quadratic test functions and taking the scalar product with Z^π and U^π, respectively, for getting η_h.

4.4 Balancing Model and Mesh Size Adaptation

On the basis of the local indicators $\{\eta_{h,i}, \eta_{m,i}\}$, an adaptive process can be designed. Since the values are nodal-based, the information has to be shifted to the cells. This can be done in a straightforward way or by applying more sophisticated algorithms such that more smoothing is achieved.

The strategy for "refining the model" strongly depends on the problem. The subdomain $\Omega_m \subset \Omega$ may change in each adaptive step so that portions of $c_1(\cdot)(\cdot)$ are successively (and locally) shifted to $a_m(\cdot)(\cdot)$. In order to balance both sources of errors, it seems reasonable to work with modified error indicators for the refinement of the mesh and the model:

$$\tilde{\eta}_{m,i} = \begin{cases} \eta_{m,i} & \text{if } \eta_{m,i} \geq \alpha \eta_{h,i}, \\ 0 & \text{otherwise,} \end{cases}$$

$$\tilde{\eta}_{h,i} = \begin{cases} \eta_{h,i} & \text{if } \eta_{h,i} \geq \alpha \eta_{m,i}, \\ 0 & \text{otherwise.} \end{cases}$$

The parameter α is set to 0.2 in the numerical results presented in the next section. The mesh adaptation strategy we used is based on an optimization strategy; see [7, 10] for details. The change from the crude to the detailed model is done using an "error balancing" technique. Cells are marked whenever

$$\tilde{\eta}_{m,i} > tol,$$

with a tolerance tol given by the mean value of the cell contributions $\tilde{\eta}_{m,i}$ divided by 2. The detailed model is then used on marked cells in the next step.

5 Numerical Results for Model Adaptivity

We consider two examples. The first example consists of a system of convection-diffusion-reaction problems in order to validate the accuracy of the estimator and to illustrate the mechanism of error balancing. In the second example the model and mesh adaptation is applied to combustion problems where complex diffusion models are locally replaced by simpler ones.

5.1 System of Convection-Diffusion-Reaction Equations

The considered equations describe a reacting mixing layer of three species S_i, $i = 1, 2, 3$, diluted in an inert carrier gas

$$\beta \cdot \nabla u_i + \operatorname{div} \mathcal{F}_i = f_i(u), \quad i = 1, 2, 3,$$

where $u = (u_1, u_2, u_3)^T$ is the vector of concentrations and $\mathcal{F}_i$ is the diffusion flux of the i-th species given by

$$\mathcal{F}_i = -\sum_{j=1}^{3} D_{ij} \nabla u_j.$$

The flux diffusion matrix $D = (D_{ij})$ is taken in the form

$$D = \lambda \begin{pmatrix} d_1 & 0 & 0 \\ 0 & d_2 & 0 \\ 0 & 0 & d_3 \end{pmatrix} + \lambda \begin{pmatrix} d_{11}u_1 & d_{12}u_1 & d_{13}u_1 \\ d_{21}u_2 & d_{22}u_2 & d_{23}u_2 \\ d_{31}u_3 & d_{32}u_3 & d_{33}u_3 \end{pmatrix},$$

with $\lambda = (2 - u_1 - u_2 - u_3)^{-1}$ and fixed parameters $d_{ij} = \sqrt{d_i d_j}$, $d_1 = 10^{-4}$, and $d_2 = d_3 = 10^{-2}$. The diffusion matrix $D' = (D'_{ij})$ with $D'_{ij} = D_{ij}/u_i$ is symmetric positive definite. The advection velocity, $\beta = (1,0)^T$, is constant and divergence-free. The reaction terms read as $f_1(u) = 0$ and $f_3(u) = -f_2(u) = 10^3 u_1 u_2$ and describe the reaction

$$S_1 + S_2 \to S_1 + S_3.$$

Fig. 7. Solution of the reacting mixing layer problem; species S_1 (top), S_2 (middle), S_3 (bottom).

The species S_2 is consumed, S_3 is produced, and S_1 is a so-called third body. The computational domain is $\Omega = [0,1] \times [0, 0.2]$. The boundary conditions are homogeneous Neumann for all components on all parts of $\partial \Omega$, except for $x = 0$, where Dirichlet conditions are imposed: $u_1(0,y) = (1 + \tanh(10(y - 0.1)))/20$, $u_2(0,y) = 0.1 - u_1(0,y)$, and $u_3(0,y) = 0$.

This model problem describes the mixing of two diluted species S_1 and S_2 with equal molecular weight and with a realistic diffusion matrix. The diffusion coefficients d_i represent the diffusion of species i in the mixture. The species concentrations u_i are shown in Figure 7. The reference solution has been obtained as in the previous example by using biquadratic elements on a very fine mesh with the exact model. The third body S_1 diffuses very slowly, S_2 diffuses faster and is consumed, and S_3 is produced in the reaction layer and diffuses fast.

Non-diagonal diffusion introduces a coupling between gradients of the species. In practice, this yields additional couplings in the Jacobian matrix. For large reaction systems with many chemical species involved, this may become prohibitive. Moreover, multicomponent diffusion matrices are usually given only implicitly from the solution of a constrained singular system and their computation is numerically expensive; see [17, 20]. Therefore, instead of using the flux diffusion matrix D, one may prefer to use a simpler model consisting of a diagonal diffusion matrix. In this simpler model, the diffusion fluxes are given by
$$\widetilde{\mathcal{F}}_i = -\lambda d_i \nabla u_i.$$

Table 3. Error estimators and efficiencies for the simplified diffusion matrix.

#nodes	% exact m.	η_h	η_m	η	$j(e)$	I_{eff}
289	0	-7.54e-03	3.22e-04	-7.22e-03	-1.26e-03	5.712
537	0	-1.27e-03	3.01e-04	-9.66e-04	-1.68e-04	5.739
961	26.7	-3.15e-04	4.66e-05	-2.69e-04	-7.83e-05	3.429
1,959	36.6	-7.63e-05	5.99e-05	-1.64e-05	2.07e-05	-0.793
4,391	55.9	-2.17e-05	3.35e-05	1.17e-05	2.06e-05	0.569
10,195	75.3	-5.47e-06	1.78e-05	1.23e-05	1.18e-05	1.045
23,953	81.6	-1.33e-06	1.25e-05	1.12e-05	9.83e-06	1.136
64,175	88.1	-2.07e-07	6.93e-06	6.72e-06	5.75e-06	1.168
191,835	95.3	1.20e-08	4.07e-06	4.09e-06	3.81e-06	1.074

The simpler model is still slightly nonlinear because of the u-dependence of λ. At the first iterate of the adaptive algorithm, the nonlinear forms read as

$$a(u)(\phi) = \sum_{i=1}^{3} (\beta \cdot \nabla u_i, \phi_i) - (\mathcal{F}_i, \nabla \phi_i) - (f_i(u), \phi_i),$$

$$d_m(u)(\phi) = \sum_{i=1}^{3} (\widetilde{\mathcal{F}}_i - \mathcal{F}_i, \nabla \phi_i)_{\Omega \setminus \Omega_m} = \sum_{i,j=1}^{3} (\lambda d_{ij} u_i \nabla u_j, \nabla \phi_i)_{\Omega \setminus \Omega_m}$$

for test functions $\phi = \{\phi_1, \phi_2, \phi_3\}$. As the adaptive algorithm moves on, Ω_m becomes larger.

The convective terms are stabilized by the streamline diffusion method; see, for instance, Johnson [23]. For bilinear elements, the considered stabilizing form is given by a sum of element contributions:

$$s_h(u)(\phi) = \sum_{T \in \mathcal{T}_h} \sum_{i=1}^{3} \delta_{T,i} (\beta \cdot \nabla u_i - f_i(u), \beta \cdot \nabla \phi_i)_T,$$

with parameters $\delta_{T,i}$ evaluated by standard procedures using λd_i as the reference diffusion coefficient for the ith species. The discrete system to be solved reads as

$$u_{hm} \in V_h : \quad a(u_{hm})(\phi) + s_h(u_{hm})(\phi) = 0 \quad \forall \phi \in V_h.$$

Therefore, the model estimator η_m also includes the following stabilization terms:

$$\eta_m = -d(u_{hm})(z_{hm}) + s_h(u_{hm})(z_{hm}).$$

Table 3 presents the discretization and model errors, where the quantity of interest was chosen as a point value of the product, $j(u) = u_3(x_0)$ with $x_0 = (0.125, 0.1)$. Because on coarse meshes the discretization error dominates, the mesh is refined in the mixing layer, keeping the crude model in the whole domain. In the third iteration, both types of error are balanced, and the model

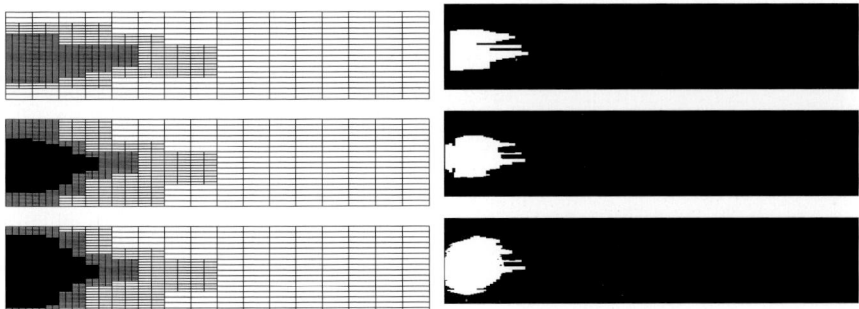

Fig. 8. Meshes (left) and corresponding model zones (right) for the reacting mixing-layer problem. Light areas indicate regions where the detailed model is used.

is adapted locally to the full flux diffusion matrix on 26.7% of the cells. The adaptive algorithm then refines simultaneously the model and the mesh. The error $j(u - u_{mh})$ is slightly overestimated on the first three meshes. On the fourth mesh, the error changes sign, and from the fifth mesh onward, the error is well represented by the estimator η with an asymptotic efficiency of 1. The discretization part η_h is of order $O(N^{-3/2})$ on the locally refined meshes, where N denotes the number of nodes. On tensor meshes, the order $O(N^{-1})$ would be expected for a second-order method. For very fine meshes, this would also be the case for locally refined meshes. However, local mesh refinement gives a better "asymptotic" behavior on relevant (coarse) meshes.

Some meshes and the areas where both types of flux diffusion matrices are used are presented in Figure 8. The detailed model (light area) is used only around the point x_0 where the functional is evaluated.

5.2 Model Adaptivity for Laminar Flames

The choice of diffusion models in gas mixtures is not straightforward. Although multicomponent diffusion models are accepted to be accurate [17], simpler and less accurate models, e.g. Fick's law, are often used in practice for two- and three-dimensional simulations. While simple diffusion models, as for instance Fick's law, may involve only diagonal diffusion, multicomponent diffusion models lead to couplings between all the chemical variables. For implicit solvers, these couplings yield substantial fill-ins in the (sparse) Jacobians. Therefore, it is desirable to apply detailed diffusion models only in those regions of the computational domain where necessary; for instance in the flame front where a complex balance of reaction, convection, and diffusion phenomena takes place. However, it is not a priori known where an accurate diffusion model is necessary.

In this section, we investigate the model adaptivity for such stationary combustion problems described by the system for the pressure p, velocity v, temperature T and chemical mass fractions $y_1, \ldots y_{n_s}$:

$$\operatorname{div}(\rho v) = 0,$$
$$\rho(v \cdot \nabla)v - \operatorname{div} \pi + \nabla p = 0,$$
$$(\rho c_p v + \alpha) \cdot \nabla T - \operatorname{div} \lambda \nabla T = f_T,$$
$$\rho v \cdot \nabla y_k + \operatorname{div} \mathcal{F}_k = f_k, \quad k = 1, \ldots, n_s,$$

where π denotes the stress tensor, c_p the heat capacity at constant pressure, λ the heat conductivity, f_T the heat source due to chemical reactions, f_k chemical production terms. The species mass diffusion fluxes $\mathcal{F}_k$ are dynamically chosen. The vector α accounts for heat flux due to diffusion of species with different specific enthalpies and depends on the diffusion fluxes:

$$\alpha = \sum_{i=1}^{n} c_{p,k} \mathcal{F}_k,$$

with the specific heat capacities $c_{p,k}$. The following two models for the fluxes $\mathcal{F}_k$ are considered:

Fick's law: The simple diffusion model is the diagonal diffusion driven by gradients of mole fractions, $x_k = \frac{\bar{m}}{m_k} y_k$, see [21],

$$\mathcal{F}_k^F = -\rho D_k^* \frac{y_k}{x_k} \nabla x_k, \quad k = 1, \ldots, s, \tag{23}$$

with diffusion coefficients D_k^* given by

$$D_k^* = \frac{1 - y_k}{\sum_{l \neq k} \frac{x_l}{D_{kl}^{\text{bin}}}},$$

and the binary diffusion coefficients D_{kl}^{bin} for the species pair (k, l).

Multicomponent diffusion: The accurate diffusion uses a full diffusion matrix acting on the gradients of species mole fractions and temperature

$$\mathcal{F}_k^M = -\rho y_k \sum_{l=1}^{n_s} D_{kl} \left(\nabla x_l + \chi_l \nabla (\log T) \right), \quad k = 1, \ldots, n_s, \tag{24}$$

where $(D_{kl})_{1 \leq k,l \leq n_s}$ are the species diffusion coefficients and $(\chi_l)_{1 \leq l \leq n_s}$ the thermal diffusion ratios. This form of the species mass diffusion fluxes results from the kinetic theory of gases in the first-order Enskog-Chapman expansion; see [17, 20]. The diffusion coefficients D_{kl} are compatible with the overall mass conservation constraint

$$\sum_{k=1}^{n_s} \mathcal{F}_k^M = 0.$$

To evaluate the coefficients $(D_{kl})_{1 \leq k,l \leq n_s}$ and $(\chi_l)_{1 \leq l \leq n_s}$, linear systems must be solved first. Although computational costs can be substantially reduced by using convergent iterative methods, they are still higher than those incurred with the previous diffusion model. For convenience, we express the multi-component diffusion fluxes (24) in terms of the gradients of the s first mass fractions

$$\mathcal{F}_k^M = -\rho y_k \sum_{l=1}^{s}(D_{kl} - D_{kn_s})\Big(\nabla x_l + \chi_l \nabla(\log T)\Big), \quad k=1,\ldots,s,$$

$$\nabla x_l = \frac{\bar{m}}{m_l}\left(\nabla y_l - y_l \sum_{i=1}^{s}(\frac{\bar{m}}{m_i} - \frac{\bar{m}}{m_{n_s}})\nabla y_i\right), \quad l=1,\ldots,s.$$

These two diffusion models are combined within the adaptive modeling approach and coupled with local mesh refinement. We consider two types of flames in two-dimensional geometries. The first example is an ozone flame in the configuration described in [4]. For this flame, the impact of the diffusion model is small. Moreover, the complexity of the problem size is moderate enough to obtain a reference solution on a very fine mesh. This reference solution is used for validation of the error estimators. The second example is an underventilated hydrogen diffusion flame. Since the impact of multicomponent diffusion and particularly of the Soret effect is often substantial in hydrogen flames (see for instance [15, 18, 19]), we expect a larger impact of the model error on the flame structure. In the computations presented below, multicomponent transport properties have been evaluated using the EGlib package [16].

Ozone flame:

The flame is a premixed flame with three chemical species, namely O_3, O_2, and O-atoms. The reaction mechanism consists of six reactions. The initial mesh is an equidistant mesh with 512 nodes. We start with Fick's law (23) for the species mass diffusion fluxes over the whole computational domain. After computing the stationary solution u_{hm} on this mesh with the crude diffusion model, we compute the dual solution z_{hm} associated with the functional

$$j(u) = c\int_\Omega y_O dx, \quad (25)$$

with a constant scaling factor c. This functional gives the mean value of mass fractions of O-atoms in the computational domain Ω. The error indicators η_h and η_m are obtained from (21) and (22). On the basis of η_h we change the mesh-size locally by bisection of element edges. According to η_m, we change the diffusion model locally from Fick's law to multicomponent diffusion.

The reference solution is computed on a very fine mesh and with the accurate diffusion model everywhere. The mass fractions of the O-atoms of the solution is shown in Figure 9. The adaptive algorithm balances both types of errors by adapting the mesh-size and the model. Figure 10 illustrates how the two sources of error are equilibrated. We plot the estimators η_h and η_m,

Fig. 9. Ozone flame: Mass fractions of O−atoms produced in the flame front.

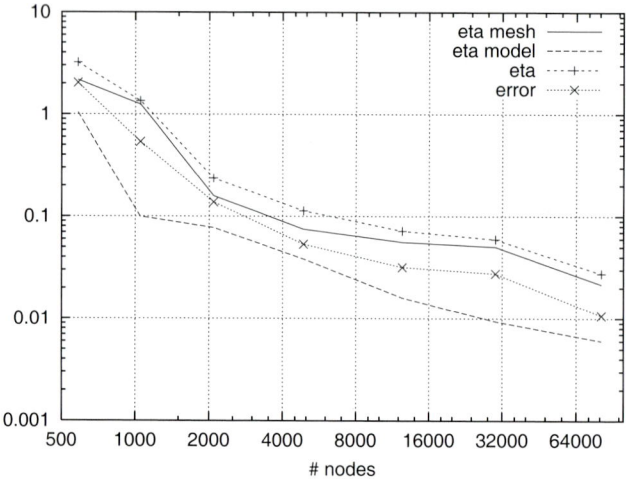

Fig. 10. Ozone flame: estimator η_m for the modeling error; estimator η_h for the discretization error; their sum η; true error $j(u - u_h)$ as a function of the number of nodes (mesh points).

their sum η, and the true error $j(u-u_{hm})$ as a function of the number of mesh nodes. The estimator and the true error clearly show the same asymptotic behavior. The efficiency index $I_{eff} = \eta/j(u - u_h)$ is in the range of $1.5 - 2.6$, which means that the estimator η slightly overestimates the error. However, the error estimator is reliable since it provides an upper bound for the actual error.

The sequence of locally refined meshes with 1 047, 2 085, and 4 871 nodes is shown in Figure 11. The darker areas indicate the part of the computational domain where the multicomponent diffusion model is used. In the remaining (light) part, Fick's law is used. We observe that the estimator quite well detects the reaction area where a difference in both diffusion models influences the accuracy of the output functional (25).

Fig. 11. The three figures show on the upper halfs the areas where the multicomponent diffusion model is used (dark areas) and where Fick's law is used (light areas); the lower half of each picture shows the corresponding locally refined mesh.

Hydrogen flame:

The final test case is an underventilated hydrogen diffusion flame stabilized over a series of rectangular and infinitely long slots. The flame configuration is modeled by taking only one slot and imposing periodic boundary conditions. A sketch of the geometry is displayed in Figure 12(a). The inflow boundary is located at the bottom of the figure. We consider a parabolic inflow with $1\,\mathrm{m/s}$ peak velocity and a hydrogen/nitrogen mixture ($y_{H2} = 0.1, y_{N2} = 1 - y_{H2}$). This flow is surrounded by a secondary flow of air ($y_{O2} = 0.1887, y_{N2} = 1 - y_{O2}$) that is also parabolic with $1\,\mathrm{m/s}$ peak velocity. The fuel and air flows are separated by burner lips which are $0.5\,\mathrm{mm}$ thick. At the left and right boundaries, periodicity is assumed for all variables. The upper boundary is a natural outflow boundary. The reaction mechanism contains 9 species participating in 19 reverse reactions.

To validate the adaptive method, we begin with two computations on a very fine structured mesh. In the first computation, we use Fick's law (23) as diffusion model in the whole domain, whereas in the second computation we use the accurate multicomponent diffusion model (24). Figures 12(b) and 12(c) present the resulting mass fractions of OH (the same color scale is used). In both computations, the flame front is above the inlet of air due to insufficient oxygen. However, Fick's law shifts the reaction zone downstream and over-

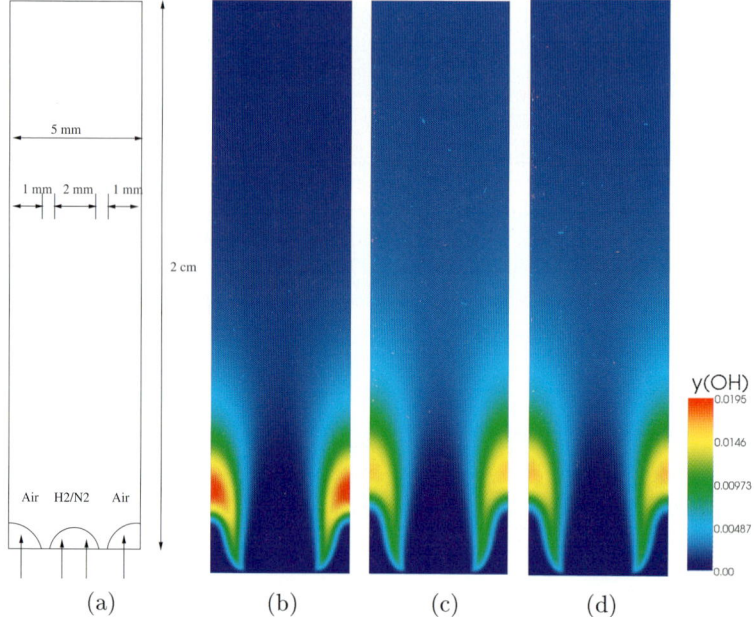

Fig. 12. Underventilated hydrogen diffusion flame: (a) Configuration sketch. (b)–(d) OH mass fraction. (b) Fick's law in the whole domain. (c) Multicomponent diffusion in the whole domain. (d) Second adaptive iteration with 15.5% cells flagged for multicomponent diffusion.

predicts the peak value of OH mass fractions by up to 20%. This is only a consequence of the diffusion model, since the reaction mechanism is the same in both computations.

The output functional used for mesh- and model adaptation is the mean value of mass fraction of OH along a straight horizontal line Γ:

$$j(u) = \int_{\Gamma} y_{OH} dx, \quad \Gamma = \{(x, 1.875\,mm) : 0 \leq x \leq 5\,mm\}. \qquad (26)$$

Referring to Figure 12(a), the x-axis is horizontal, the y-axis is vertical, and the origin is set at the lower left corner of the domain.

Figure 12(d) shows the OH mass fraction, y_{OH}, in the second adaptive iteration where only 15.5% of the cells are flagged for multicomponent diffusion. In the remaining part of the domain, Fick's law is used. A comparison with multicomponent diffusion in the whole domain (Figure 12(c)) reveals practically no difference.

A more precise comparison is performed by analyzing cross-sections of y_{OH}. The diagrams in Figures 13 display this quantity along the lines $x = 1.875\,mm$ and $y = 1.875\,mm$, respectively. We compare the profiles obtained using Fick's law over the whole domain (FL), the accurate diffusion model over the whole domain (MC), and the profiles obtained after two and three adaptive iterations (MC/FK). The first model adaptation (15.5% of the cells flagged for multicomponent diffusion) and mesh refinement (8901 nodes) already brings the numerical solution very close to the profile obtained with an accurate diffusion model over the whole domain. Note also that the profile along the line $y = 1.875\,mm$ shows that the OH mass fraction can be locally overpredicted by up to 80% when an inaccurate diffusion model is used.

The overall saving in CPU time due to model adaptivity is approximately a factor of 2. Finally, we show in Figure 14 a zoom of a locally refined mesh together with the isolines of OH radicals. Clearly visible is that the refinement is related to the region where the line integral $j(u)$ is evaluated.

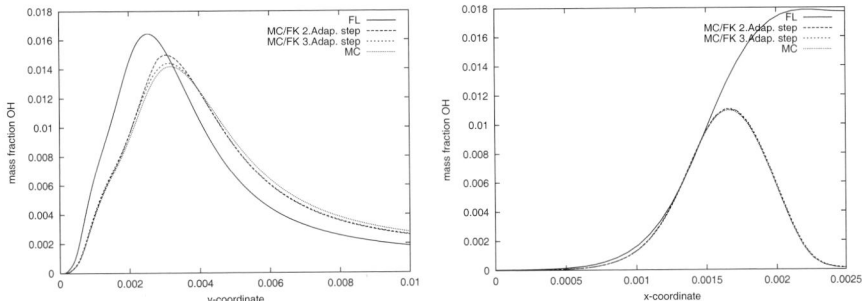

Fig. 13. Cross-section at $x = 1.875\,mm$ for the hydrogen flame. Comparison of y_{OH} for Fick's law (FL), the accurate model (MC), and two adaptive iterations (MC/FK). Left: cross-section at $x = 1.875\,mm$, right: cross-section at $y = 1.875\,mm$.

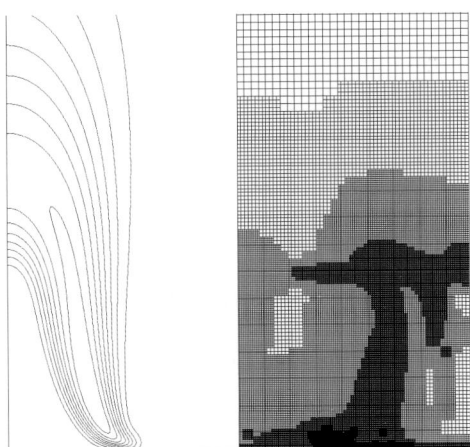

Fig. 14. Isolines of OH radicals (left) and locally refined mesh (right). The refinement is obviously related to the region where the line integral $j(u)$ is evaluated.

References

1. R. Becker and M. Braack. Multigrid techniques for finite elements on locally refined meshes. *Numerical Linear Algebra with Applications*, 7:363–379, 2000. Special Issue.
2. R. Becker and M. Braack. A finite element pressure gradient stabilization for the Stokes equations based on local projections. *Calcolo*, 38(4):173–199, 2001.
3. R. Becker and M. Braack. A two-level stabilization scheme for the Navier-Stokes equations. Numerical Mathematics and Advanced Applications, Enumath 2003, Prague. ed.: Feistauer et. al., Springer Verlag, 2004, p. 123-130.
4. R. Becker, M. Braack, and R. Rannacher. Numerical simulation of laminar flames at low Mach number with adaptive finite elements. *Combust. Theory Modelling*, 3:503–534, 1999.
5. R. Becker and R. Rannacher. A feed-back approach to error control in finite element methods: Basic analysis and examples. *East-West J. Numer. Math.*, 4:237–264, 1996.
6. R. Becker and R. Rannacher. An optimal control approach to a posteriori error estimation in finite element methods. In A. Iserles, editor, *Acta Numerica 2001*, volume 37, pages 1–225. Cambridge University Press, 2001.
7. M. Braack. An Adaptive Finite Element Method for Reactive Flow Problems. PhD Dissertation, Universität Heidelberg, 1998.
8. M. Braack and A. Ern. A posteriori control of modeling errors and discretization errors. *Multiscale Model. Simul.*, 1(2):221–238, 2003.
9. M. Braack and A. Ern. Coupling multimodeling with local mesh refinement for the numerical solution of laminar flames. *Combust. Theory Modelling*, 8(4):771–788, 2004.
10. M. Braack and R. Rannacher. Adaptive finite element methods for low-Mach-number flows with chemical reactions. In *30th Computational Fluid Dynamics*, VKI LS 1999-03, H. Deconinck, ed., pages 1–93. von Karman Institute for Fluid Dynamics, Brussels, Belgium, 1999.
11. M. Braack and T. Richter. Solutions of 3D Navier-Stokes benchmark problems with adaptive finite elements. *Computers and Fluids*, 2006, 35 (4), p. 372-392.
12. M. Braack and T. Richter. Parrallel multigrid on locally refined meshes. In R. Rannacher, editor, *Reactive Flows, Diffusion and Transport*. Springer, Berlin, 2005.
13. C. Carstensen and R. Verfürth. Edge residual dominate a posteriori error estimates for low order finite element methods. *SIAM J. Numer. Anal.*, 36(5):1571–1587, 1999.
14. C. Carstensen and R. Verfürth. Edge residuals dominate a posteriori error estimaters for low-order finite element methods. *SIAM J. Numer. Anal.*, 36:1571–1587, 1999.
15. J. de Charentenay and A. Ern. Multicomponent transport impact on turbulent premixed H_2/O_2 flames. *Combust. Theory Modelling*, 6:439–462, 2002.
16. A. Ern and V. Giovangigli. http://www.cmap.polytechnique.fr/www.eglib. EGlib server with user's manual.
17. A. Ern and V. Giovangigli. *Multicomponent Transport Algorithms*. Lecture Notes in Physics, m24, Springer, 1994.
18. A. Ern and V. Giovangigli. Thermal diffusion effects in hydrogen-air and methane-air flames. *Combust. Theory Modelling*, 2:349–372, 1998.

19. A. Ern and V. Giovangigli. Impact of detailed multicomponent transport on planar and. *Combust. Sci. Tech.*, 149:157–181, 1999.
20. V. Giovangigli. *Multicomponent Flow Modeling.* Birkhäuser, Boston, 1999.
21. J. O. Hirschfelder and C. F. Curtiss. *Flame and Explosion Phenomena.* Williams and Wilkins Cp., Baltimore, 1949.
22. V. John. Higher order finite element methods and multigrid solvers in a benchmark problem for 3D Navier-Stokes equations. *Int. J. Numer. Math. Fluids.*, 40:775–798, 2002.
23. C. Johnson. *Numerical Solution of Partial Differential Equations by the Finite Element Method.* Cambridge University Press, Cambridge, UK, 1987.
24. J. T. Oden and S. Prudhomme. Estimation of modeling error in computational mechanics. *J. Comput. Phys.*, 182:496–515, 2002.
25. M. Schäfer and S. Turek. Benchmark computations of laminar flow around a cylinder. (With support by F. Durst, E. Krause and R. Rannacher). In E. Hirschel, editor, *Flow Simulation with High-Performance Computers II. DFG priority research program results 1993-1995*, number 52 in Notes Numer. Fluid Mech., pages 547–566. Vieweg, Wiesbaden, 1996.

Parallel Multigrid on Locally Refined Meshes[*]

R. Becker[1], M. Braack[2], and T. Richter[2]

[1] Laboratoire de Mathematiques Appliquees, Universite de Pau et des Pays de l'Adour, France
[2] Institut für Angewandte Mathematik, Universität Heidelberg

Summary. The use of PC-clusters gives the opportunity to solve three-dimensional flow problems to high accuracy at reasonable costs. The use of adaptive mesh refinement can increase the efficiency of such computations even more. However, the combination of multigrid, local mesh refinement and parallelization is not trivial. Therefore, this work addresses the following two topics in order to develop a tool for the simulation of complex flows: (i) multigrid algorithms on locally refined meshes, and (ii) the extension of multigrid to be used on parallel machines.

1 Introduction

At first, we present a multigrid algorithm on locally refined meshes. This method is firstly documented in the work of Becker & Braack [4]. The basic idea is to perform "global coarsening" in order to obtain a nested hierarchy of meshes. This has certain advantages compared to the technique of *ghost* or *virtual nodes*, see Brandt [11], Rüde [19] and McCormick [14].

At second, we extend this multigrid algorithm to a parallel version. This is motivated by the fact that a parallelization of the smoother preserves an efficient multigrid algorithm even on parallel machines. For structured meshes, Oswald has shown in [16] that such a strategy may perform well for solving 3-D flow problems. Here, we combine the multigrid algorithm on locally refined meshes with the parallelization in order to use adaptive finite elements on parallel computers. Some of the developed concepts are extensions of the work of Bastian [2, 3], where parallel multigrid in the context of finite volumes for 2D problems are considered.

We begin with the presentation of the multigrid algorithm on nested locally refined meshes. Then, we present the developed parallel algorithm and analyze its parallel efficiency by help of an established 3D Navier-Stokes benchmark problem.

---
[*]This work has been supported by the German Research Foundation (DFG) through SFB 359 (Project A2) at the University of Heidelberg.

2 Multigrid on Locally Refined Meshes

Since the parallelization is independent of the specific partial differential equation, we firstly consider the linear equation

$$Ax = b, \qquad (1)$$

for $x, b \in \mathbb{R}^N$, which results from discretization by finite elements of a partial differential equation in 2 or 3 space dimensions. The discretization used in the numerical examples is based on Q_1 and Q_2 finite elements. However, the multigrid technique we are going to present can also be formulated for finite differences and finite volumes. Since we focus on the specific aspects of parallelization, we refer to existing literature for other numerical aspects: The discretization is explained in detail in Braack [8] and in Braack & Richter [9]. The evaluation of the estimator and the adaptation strategy used are described in [9] and [18].

The linear system (1) is solved by multigrid iterations in order to handle the bad conditioning of the matrices. A multigrid algorithm is based on solving linear systems

$$x_l \in \mathbb{R}^{n_l}: \quad A_l x_l = b_l,$$

on various levels $l = 1, \ldots L$. The domain $\Omega \subset \mathbb{R}^d$ is partitioned by a quadrilateral (or hexaedral) mesh $\mathcal{T}_h$ consisting of cells K. We consider hierarchical refinement with the initial coarse mesh $\mathcal{T}_1$. The refinement is done by bisection without introducing interface elements. This leads to irregular nodes ("hanging nodes") on the interface.

When using a multigrid algorithm for solving the linear systems, some additional difficulties arise compared to the case of globally refined meshes:

- There are several possibilities to define the spaces on each level of the multigrid algorithm. From a theoretical point of view, the main difference is the set of nodes on which the smoother acts. However, for practical implementation the difference is also in the computation of the residuals and in the data structure required to store both matrices and vectors.
- In contrast to standard multigrid, the unknowns on each level have to be numbered independently, because coarse-grid nodes may not be included in the set of fine-grid nodes.
- In order to be optimal, not only the convergence rate should be independent of the number of levels, but also the work should be linear in the number of unknowns. This is not as evident as in the standard case, since the number of unknowns per level is not multiplied by a constant factor (like for example by the factor of 4 in two dimensions) when refining the mesh.

In this section, we present the procedure how to construct a family of embedded finite element spaces $V_1 \subset \ldots \subset V_L \subset V$. The right-hand side and the solution of the algebraic system on a level l are denoted by f_l and u_l, respectively. The standard linear multigrid V-cycle is as follows:

Multigrid Iteration (V-cycle)
---
(0) If $l = 1$ solve exact; otherwise:
(1) Pre-smooth on level l.
(2) Compute the residual $r_l = b_l - A_l x_l$ on level l.
(3) Solve $A_{l-1} y_{l-1} = R_{l-1} r_l$ by the same algorithm, where R_{l-1} denotes the restriction from level l on $l-1$.
(4) Update $x_l = x_l + P_l y_{l-1}$ with the prolongation P_l.
(5) Post-smooth on level l.

Since we deal with embedded finite element spaces, the prolongation and restriction operators P_l and R_l are simply given by L^2-projection and inclusion.

In the literature, several approaches to multigrid on locally refined meshes are presented. The algorithms defined by Brandt in [11], Rüde [19] and Mc-Cormick [14] are based on constructing subdomains of locally refined areas. The first two authors use so-called *ghost* or *virtual* nodes in order to represent locally refined meshes. The technique of border cells are introduced by Mc-Cormick. A drawback is that the arising linear systems can not handle global restrictions, as for instance, a zero mean value for the pressure. This is the case for an incompressible fluid when the pressure is only determined up to a constant.

Another popular approach is the hierarchical-bases multigrid, see Yserentant [21] and Bank [1], which allows an easy implementation of local mesh refinement but does not lead to an $O(N)$ solver at least in three space dimensions.

Therefore, we propose an algorithm for constructing the V_l which cover the whole domain. Starting from the finest mesh, the lower-level meshes are constructed by recursive coarsening, called "global coarsening". This approach is slightly more expensive than the use of border cells, but is advantageous when a global restriction has to be fulfilled, see [4]. Note, that the usual multigrid theory applies directly.

The presented multigrid algorithms is applied to locally refined meshes with local condensation of the degrees of freedom at hanging nodes. However, it can be applied in a straightforward way on triangulations with interface elements. For numerical tests of this multigrid method we refer to [4].

2.1 The Technique of Global Coarsening

We construct a hierarchy of meshes $\widetilde{\mathcal{T}}_l$, always covering the whole domain Ω. We start with $\widetilde{\mathcal{T}}_L = \mathcal{T}_L$. The lower levels are recursively constructed. Such meshes on each level are shown in Figure 1. In the following we denote the number of cells of a triangulation $\widetilde{\mathcal{T}}_l$ by $\#\widetilde{\mathcal{T}}_l$.

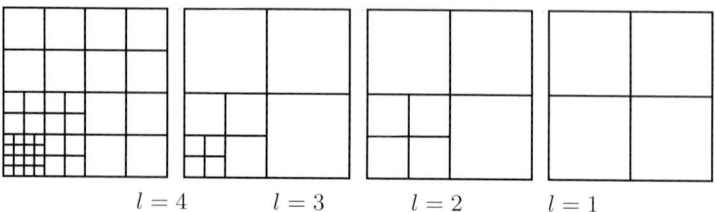

Fig. 1. Sequence of meshes $\widetilde{\mathcal{T}}_l$.

Construction of $\widetilde{S}_l$

(0) $M := \widetilde{\mathcal{T}}_{l+1}$,
(1) $M' := \{K \in \widetilde{\mathcal{T}}_{l+1} : brothers(K) \subset M, l_K > 1\}$,
(2) $P_1 := fathers(M')$, $P_2 := \widetilde{\mathcal{T}}_{l+1} \setminus M'$
  $\widetilde{\mathcal{T}}_l := P_1 \cup P_2$
(3) If $\widetilde{\mathcal{T}}_l$ is not a admissible mesh, delete at least one cell K in M with $l_K > l+1$. Go to (1).

Lemma 1. *For this construction of $\widetilde{\mathcal{T}}_l$ with $1 \leq l < L$ holds:*

(i) *The iteration converges, and* $max_{K \in \widetilde{\mathcal{T}}_l}\{l_K\} = max_{K \in \widetilde{\mathcal{T}}_{l+1}}\{l_K - 1, 1\}$.
(ii) *Each $K \in \widetilde{\mathcal{T}}_{l+1}$ itself or the father of K is element of $\widetilde{\mathcal{T}}_l$, but not both of them.*
(iii) *Each $K \in \widetilde{\mathcal{T}}_l$ itself or all children of K are in $\widetilde{\mathcal{T}}_{l+1}$.*
(iv) $\#\widetilde{\mathcal{T}}_l \leq \#\widetilde{\mathcal{T}}_{l+1}$.
(v) *If $\#\widetilde{\mathcal{T}}_l = \#\widetilde{\mathcal{T}}_{l+1}$, then the cells in $\widetilde{\mathcal{T}}_l$ are all of level 1.*

Proof. (i): On one hand, in each iteration (1)-(3), the set M becomes smaller by at least one cell. On the other hand, all cells of $\widetilde{\mathcal{T}}_{l+1}$ with highest level are always in M. If in step (3), no more cells are deleted in M, means that all cells in M are of the lowest level 1. In this case M is regular and the iteration stops.
(ii): Let $K \in \widetilde{\mathcal{T}}_{l+1}$ and suppose $K \notin \widetilde{\mathcal{T}}_l$. Then follows $K \notin P_2$, which concludes $K \in M'$ and the father of K is in $P_1 \subset \widetilde{\mathcal{T}}_l$.
(iii): Let $K \in \widetilde{\mathcal{T}}_l$; since $P_2 \subset \widetilde{\mathcal{T}}_{l+1}$, only the case $K \in P_1$ is interesting. But then exists a child of K in M' and, therefore, all children of K are in $M \subset \widetilde{\mathcal{T}}_{l+1}$
(iv): Follows immediately from (iii).
(v): $\#\widetilde{\mathcal{T}}_l = \#\widetilde{\mathcal{T}}_{l+1}$ holds only if all cells in M' are already of level 1. □

Each $\widetilde{\mathcal{T}}_l$ may include new hanging nodes, which have to be handled by the same technique as on the finest one $\mathcal{T}_L$. The equations for the smoother reads, on the lower levels $1 \leq l < L$, as follows:

$$v_l \in V^0(\widetilde{\mathcal{T}_l}): \quad a(v_l, \phi) = (r_l, \phi) \quad \forall \phi \in V^0(\widetilde{\mathcal{T}_l}), \tag{2}$$

where $r_l = R_l r_{l+1}$ is the restricted residual from level $l+1$ to l with boundary values $r_l|_{\widehat{\Omega}_l} = 0$. Here, $a(\cdot, \cdot)$ denotes the considered bilinear form.

A small drawback of this technique is that the size of V_l may not decrease fast enough to get an $O(N)$ algorithm. However, it is shown in [4] that this occurs only in very pathological situations and that the complexity is acceptable even on extremely locally refined meshes.

2.2 Numerical Tests

Relative work per mesh point

In order to test the complexity of the proposed multilevel technique we investigate the relative work per mesh point, given by

$$W = \frac{1}{N} \sum_{l=1}^{L} N_l,$$

where N_l denotes the degrees of freedom on level l and L the number of multigrid levels. Since this measure is independent of the underlying equation, we have to specify only the type of meshes. As an example, we sucessively refine a mesh along a circle, see Figure 2. These (artificial) refinement is chosen, to show the behaviour of the multigrid solvers for extreme local refinement.

In Figure 3 the value of W is plotted in dependence of the overall number of mesh points N. Obviously, the relative work is always lower than 2.4 and stabilizes for $N \to \infty$ at a value of about 2. Hence, the overall complexity is still linear in N.

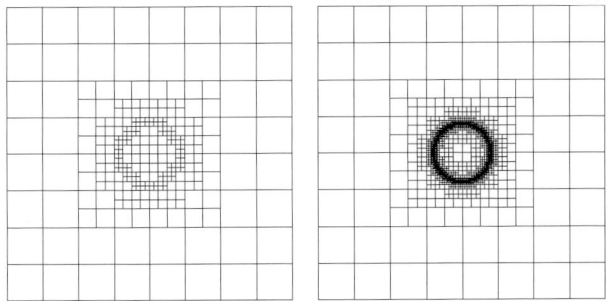

Fig. 2. Locally refined meshes along a circle for testing the multigrid work

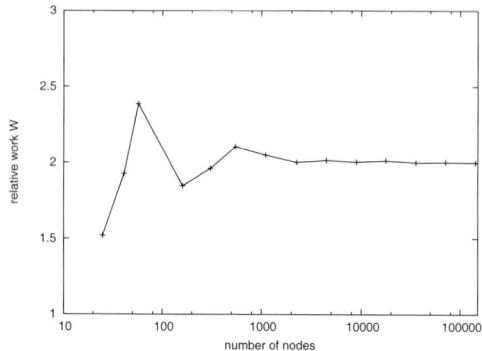

Fig. 3. Relative work W of the multigrid algorithm for the locally refined meshes along a circle.

Convergence of the multigrid solver for the Poisson problem

In order to test the convergence rate of the multigrid solver we consider the Poisson problem

$$-\Delta u = f \quad \text{in } \Omega \tag{3}$$
$$u = 0 \quad \text{on } \partial\Omega \tag{4}$$

on the unit square $\Omega = [0,1]^2$ with right hand side $f \equiv 1$. This equation is discretized by piece-wise bilinear finite elements using the locally refined meshes presented before.

As a smoother in this multigrid algorithm one Gauss-Seidel iteration in each pre- and post-smoothing step of the V-cycle is performed. factorization is used. For reducing the residual by a factor of 10^{-10} only 9 iterations are sufficient independent of the number of refinement levels. For comparison, on globally refined meshes 10 iterations are needed for the same tolerance. For further numerical results, especially for comparisons with other multilevel strategies and other types of problems, we refer to [4].

3 Parallel Multigrid on Locally Refined Meshes

In this section, we present a general concept of solving partial differential equations on locally refined meshes within a parallel multigrid algorithm. More implementational details and validation test cases can be found in [18]. We start with a parallel representation of linear systems.

3.1 Distribution of Data

We denote by P the number of processors (or the number of subdomains) and by n_p the number of mesh points corresponding to the processor number p,

$1 \leq p \leq P$. A solution vector $v \in \mathbb{R}^n$ is distributed over the processors, where the part $v_p = \mathcal{R}_p v$ is the restriction to processor p. The restriction matrix $\mathcal{R}_p \in \mathbb{R}^{n_p \times n}$ consists of zeros and ones according to the local numbering on processor p. The prolongation matrix $\mathcal{P}_p = \mathcal{R}_p^T$ is its transpose. Moreover, $\mathcal{R}_p \mathcal{P}_p$ is the identity in $\mathbb{R}^{n_p}$. The system matrix is partitioned as a sum

$$A = \sum_{q=1}^{P} \mathcal{P}_q A_q \mathcal{R}_q .$$

The restriction of a matrix-vector product Av to processor p can be obtained by local matrix-vector products $A_q v_q$ and balancing across the interfaces:

$$\mathcal{R}_p A v = \mathcal{R}_p \sum_{q=1}^{P} \mathcal{P}_q A_q \mathcal{R}_q v \qquad (5)$$

$$= \mathcal{R}_p \sum_{q=1}^{P} \mathcal{P}_q A_q v_q .$$

While the local matrix-vector products $A_q v_q$ can be done completely in parallel, the balancing across the interfaces requires communication.

3.2 Hierarchical Mesh Structure for Parallel Multigrid

For the use of multigrid we need a hierarchy of meshes

$$\mathcal{T}_1 \subset \ldots \subset \mathcal{T}_L = \mathcal{T}_h ,$$

where the mesh $\mathcal{T}_1$ should be as coarse as possible. Therefore, we need (different) partitionings for each grid $\mathcal{T}_l$. In a multigrid solver the biggest amount of work is related to the finest mesh. Hence, the partitioning of $\mathcal{T}_h$ is the most important, while the partitioning of the coarser meshes is of minor importance. For the prolongation and restriction operators between different levels it is advantageous to have the corresponding data on the same processor. However, this is not always possible, since children of one cell may belong to different subdomains on the finer mesh. Moreover, in order to reduce communication costs on coarser meshes, it may be more efficient to use less processors on coarser meshes than on finer meshes. An efficient parallel multigrid solver must find a compromise between well-balanced partitionings on each level and minimal communication costs due to the prolongation and restriction operators.

Partitioning of the finest mesh

We start with a description of the partitioning algorithm of the finest mesh $\mathcal{T}_h$ in such a way that further partitionings of the coarser meshes reduce communication of multigrid prolongation and restriction to a minimum.

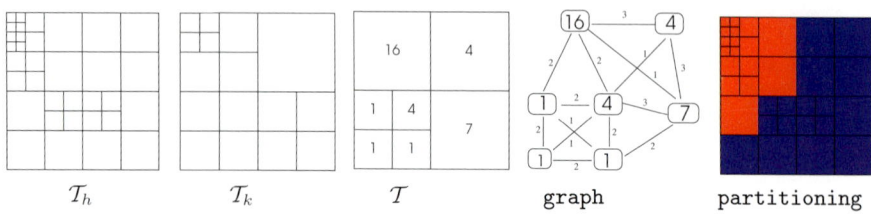

$\mathcal{T}_h$ $\mathcal{T}_k$ $\mathcal{T}$ graph partitioning

Fig. 4. Partitioning of the finest mesh: $\mathcal{T}_h$ denotes the mesh to be partitioned into two subdomains; $\mathcal{T}_k$ is a coarser mesh having a prescribed number of cells; $\mathcal{T}$ is a coarser mesh obtained by agglomeration of cells containing a hanging node with weights accounting the number of corresponding child cells on $\mathcal{T}_h$; the weighted graph is finally partitioned by METIS leading to a partitioning indicated by two colors.

By n_{sd} we denote the minimal number of cells in each subdomain (also on coarser meshes). In practice, this number is in the range $100 \sim 500$. We choose the minimal multigrid level k, so that the intermediate mesh $\mathcal{T}_k$ has at least $n_{sd}P$ cells. Next, we construct a new mesh $\mathcal{T} \subset \mathcal{T}_k$ which results by agglomeration of all cells of $\mathcal{T}_k$ containing a hanging node. The mesh $\mathcal{T}$ is not necessarily a mesh inside the multigrid hierarchy. The reason for this agglomeration is to avoid communication between processors due to the presence of hanging nodes.

On the basis of $\mathcal{T}$ we use the toolbox METIS [15] for the construction of the partitioning. This algorithm allows for a partitioning of weighted graphs. As weights for a cell $K \in \mathcal{T}$ we choose the number of descendant cells of the finest mesh $\mathcal{T}_h$. Since METIS balances these weights in each subgraph, we get an equilibrated partitioning of the finest mesh in terms of number of cells. We show this procedure for a relatively manageable mesh in Figure 4.

This partitioning of the intermediate mesh $\mathcal{T}$ can be easily transferred to all finer meshes $\mathcal{T}_k$ in the multigrid hierarchy. All these finer meshes are nested in a sense, that the multigrid transfer is performed without any communication.

The partitioning is done on one dedicated processor and the resulting partitioned meshes are handed out to the clients. If the finest mesh is large compared to the problem size, as for instance for a scalar equation like a Poisson problem on a mesh with several millions of mesh points, the partitioning can be more time consuming than the parallel computation of the solution. This bottleneck can only be overcome by performing also the partitioning in parallel. This is subject of future work. Complex flow problems can be handled by the algorithm developed so far.

However, we are able to perform global refinement of the locally adapted mesh $\mathcal{T}_h$ in parallel. In this case, the partitioning is performed complete locally on each processor.

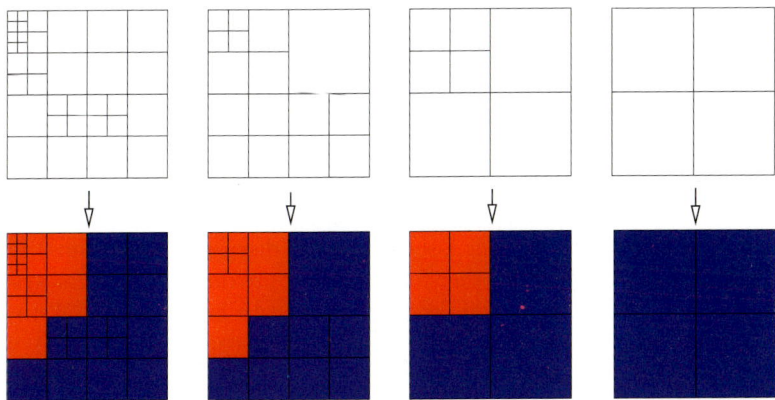

Fig. 5. Partitioning of the multigrid mesh hierarchy: The finest level (left) is partitioned according to the procedure explained in Figure 4. The coarsest mesh (right) has too few cells for distribution onto several processors.

Partitioning of the coarser meshes

In the following, we describe the partitioning $\mathcal{P}_l$ of $\mathcal{T}_l$ when the partitioning $\mathcal{P}_{l+1}$ of $\mathcal{T}_{l+1}$ is given:

- If all children $K_i \in \mathcal{T}_{l+1}$ of $K \in \mathcal{T}_l$ are located in the same subdomain of $\mathcal{P}_{l+1}$, K becomes cell of the subdomain on the same processor as the K_i.
- Otherwise, K is attached to the subdomain where most children K_i are located. If the K_i are equally distributed on different processors, K becomes cell of the subdomain with the fewest cells up to now. Of coarse, we consider only subdomains in coherence with the children cells.
- After distributing all cells $K \in \mathcal{T}_l$, those subdomains with less than n_d cells are combined with the smallest neighboring subdomain. This "coarse mesh agglomeration" limits the number of processors in accordance to the number of cells.

We show this procedure in Figure 5 for the example explained before in Figure 4.

3.3 Parallel Multigrid Solver

The parallel multigrid solver is identical to the sequential solver up to a minor modification in the smoother. The sequential smoother we mostly use consists of an incomplete block-LU factorization. While an iterative smoother like Jacobi iteration is naturally well-suited for parallelization, this is not the case for Gauss-Seidel iteration or methods of ILU type.

Before stating the difference between the sequential and the parallel version of the smoother, we analyze the parallel efficency of the mesh transfer operators: restriction and prolongation. Since the partitioning of the meshes is mainly nested (see Section 3.2), the mesh transfer operators perform locally on every processor. Only on coarser mesh levels, degrees of freedom must be fetched from neighboring processors. However, since the coarse grid agglomeration affects only coarse meshes of size $\mathcal{O}(1)$, this communication does not hurt the parallel efficiency. On intermediate meshes, we expect communication cost of order $\mathcal{O}(m)$ with $m := (N/P)^{(d-1)/d}$.

The evaluation of the defect employs a matrix-vector product and, therefore, needs communication. This issue is already discussed in Section 3.1.

In the sequential ILU smoother for solving approximately a linear system $Au = b$ with initial guess v the defect $d := b - Av$ is computed. With the approximate factorization $A \approx LU$, the solution becomes $u := v + w$, with update

$$w := U^{-1}L^{-1}d.$$

In the parallel version, this updated is replaced by local updates

$$y_p := \widetilde{U}_p^{-1}\widetilde{L}_p^{-1}d,$$

followed by balancing the values on the interfaces:

$$\widetilde{w} := \sum_p \mathcal{P}_p y_p.$$

A linear problem based on the sub-matrix A_p may correspond to a homogeneous Neumann problem, Since this is not solvable in general, the corresponding ILU decomposition would not be ensured. Therefore, we take the node-wise restriction of the global matrix A to the subdomains:

$$\widetilde{A}_p := \mathcal{R}_p A \mathcal{P}_p = \mathcal{R}_p \left(\sum_{q=1}^{P} \mathcal{P}_q A_q \mathcal{R}_q \right) \mathcal{P}_p.$$

The use of these local system matrices can be interpreted as local Dirichlet problems, with boundary values along the nodes situated just one layer of cells beyond the interfaces. The local matrices $\widetilde{L}_p, \widetilde{U}_p$ are ILU decompositions of $\widetilde{A}_p$.

If we would solve $\widetilde{A}_p y_p = d$ exactly, the parallel smoother would correspond to an additive Schwarz iteration with overlap h (one layer of cells) with homogeneous Dirichlet values.

The local ILU factorizations $\widetilde{L}_p \widetilde{U}_p \approx \widetilde{A}_p$ uses mainly the sparsity pattern of A restricted to the corresponding subdomain. However, it may happen that certain couplings of A corresponding to degrees of freedom on the boundary of the p-th subdomain are not considered in the local ILU factorization of $\widetilde{A}_p$. This is the case, when the coupling is only due to finite element test function with support outside of the p-th subdomain. Of course, such couplings remain in the ILU of another subdomain. This situation is sketched in Figure 6.

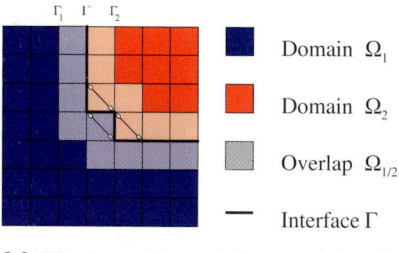

Fig. 6. Regions where the subdomain smoother act. The contribution of the deep red and the deep blue regions enter completely into the smoother on Ω_1 and Ω_2, respectively. The contributions of cells on the interface Γ enter only partially on each subdomain.

3.4 Labeling for Interface Communication

Once the partitioning is done, the communication between several processors for exchanging data at the interfaces can be done in a straightforward way. When for instance the sum above over the P processors in (5) is done sequentially one-by-one, the computation would be of complexity $\mathcal{O}(P)$. Therefore, we will present shortly an algorithm with a complexity independent of P.

The processors p are connected to n_e^p neighbor processors $\mathcal{N}_{p,i}$, $1 \leq i \leq n_e^p$. By $e_{p,i}$ we denote the edge inside a graph connecting processors p and $\mathcal{N}_{p,i}$. These edges are assumed to be ordered in such a way that $\mathcal{N}_{p,i-1} < \mathcal{N}_{p,i}$ for $2 \leq i \leq n_p$. The aim is to give each edge a label $l_{p,i}$ so that they are pair-wise different for every processor:

$$l_{p,i} \neq l_{p,j} \quad \forall 1 \leq i < j \leq n_p.$$

According to this labeling the sum in (5) can be done in parallel by parallel communication across all edges (interfaces) with the same label at a time.

We denote the maximum number of edges for one processor by n_e, i.e., $n_e^p \leq n_e$ for $1 \leq p \leq P$. A label l is called *unused on* p, if $l \neq l_{p,j}$ for all $j = 1, \ldots, n_e^p$. The following algorithm runs in parallel on each processor and labels the edges:

```
for i = 1,...,nᵖₑ
    set q = Nₚ,ᵢ
    set c = 1 and d = 0
    while (c ≠ d) do
        set c >=d minimal so that it is unused on p
        send c to processor q
        receive d from processor q
    set lₚ,ᵢ := c
```

Obviously, the result of this algorithm is an admissible labeling, i.e., with the property $l_{p,i} = l_{q,j}$ if $\mathcal{N}_{p,i} = q$ and $\mathcal{N}_{q,j} = p$. Moreover, the maximal label

is bounded by:
$$l_{p,i} \leq 2n_e \quad \forall p = 1, \ldots, P, \quad \forall i = 1, \ldots, n_e^p. \tag{6}$$

Once this algorithm has provided the labeling, each parallel communication uses this labeling. The following lemma is a direct consequence of (6) and states the complexity of such communications.

Lemma 2. *The exchange of data of size D_{pq} between processor p and q, with $1 \leq p, q \leq P$, according the labeling described above needs communication of complexity $\mathcal{O}(n_e D)$, with $D = \max\{D_{ij}\}$.*

The maximal number of degrees of freedom on one interfaces is of order $\mathcal{O}(m)$ with $m := (N/P)^{(d-1)/d}$. Hence, the overall communication cost is of complexity $\mathcal{O}(n_e m)$. However, this upper bound for the communication cost is still suboptimal, because the amount of data to be distributed is not distributed uniformly on all interfaces. In order to account for communication at interfaces with a smaller overlap, a minor modification of the algorithm described so far can reduce the costs significantly.

This variant reads for three-dimensional geometries ($d = 3$) as follows: At first, only communications across cell faces are taken into account. This corresponds to data of order $\mathcal{O}(m)$. Couplings across lines and single points are neglected at this step of communication. At second, the exchange of data corresponding to interface lines are carried out. Finally, data between processors sharing a single vertex are passed.

The gain of this variant can be illustrated by a uniform partitioning of the cube $\Omega = (0,1)^3$ into $P = Q^3$ cuboids. Each of the subdomains lying in the interior of Ω has 26 adjacent neighbor subdomains. Without taking into account that the sizes of the interfaces are of different sizes, the communication would be of order $\mathcal{O}(26m)$. In contrast, splitting the types of communications into communications across the 6 major interfaces of size $\mathcal{O}(m)$, and subsequently, communication across the 12 interfaces of size $\mathcal{O}((N/P)^{1/3})$, and finally across the 8 interfaces of size $\mathcal{O}(1)$, reduces the communication costs asymptotically by a factor of $26/6 \approx 4$.

4 Numerical Results for Testing the Parallel Efficiency

In this section, we perform some numerical studies on the efficiency of the parallel multigrid solver. The computations are done on the Linux Helix cluster situated in Heidelberg with 256 nodes. Each node consists of two 1.6 GHz Athlon MP processors and has 2 GByte memory. They are connected by a high-speed Myrinet 2000 network with 2 Gbits/s.

Table 1. Multigrid convergence rates for a Poisson problem with different number of cpu's on equidistant tensor grids and locally refined meshes.

number of cpu's	1	4	16	64	128	256
equidistant tensor grids	0.025	0.024	0.022	0.022	0.018	0.015
locally refined mesh	0.020	0.020	0.020	0.011	0.014	0.011

Table 2. CPU time in seconds for increasing problem sizes and numbers of cpu's. The increase of cpu time for 256 cpu's is due to the network topology of the cluster.

number of cpu's	1	4	16	64	256
cells	4^8	4^9	4^{10}	4^{11}	4^{12}
cpu time (s)	21.16	20.45	23.69	25.65	38.10

4.1 Parallel Efficiency for Solving a Poisson Problem

We begin with the Poisson problem on the two-dimensional unit-square and homogeneous Dirichlet conditions. The discretization is done with bilinear elements (Q1). In order to check the algorithm on locally refined meshes, we take two types of meshes. The first one is an equidistant tensor grid with 1024^2 cells. The second one consists of a stochastic locally refined mesh with about 500 000 cells. In the multigrid iteration we perform 2 ILU pre-smoothing and 2 post-smoothing steps. The convergence tolerance is 10^{-10}.

In Table 1, we report on the convergence rates for both types of meshes. Obviously, no negative influence is introduced by the parallel version. In contrary, the convergence rates slightly improves with increasing number of subdomains. This can be explained by the fact that the degrees of freedom lying on interfaces between the subdomains become smoothed by the ILU factorization from several subdomains.

In Table 2, the required cpu time is listed for increasing problem sizes. The number of cpu's also increases linearly with the problem size. The cpu time keeps nearly constant up to 128 cpu's. This states a very good parallel efficiency. On 256 nodes the cpu time increases significantly because in the Linux cluster only 64 nodes are linked with the high-speed Myrinet network while a much slower connection must be used between two groups of 64 nodes.

4.2 Parallel Efficiency for Solving the Navier-Stokes Equations

In this section we report on the parallel efficiency for a Navier-Stokes problem in a three-dimensional domain. The underlying equations for the pressure p and the velocity v are:

$$-\frac{1}{Re}\Delta v + (v \cdot \nabla)v + \nabla p = 0,$$
$$\operatorname{div} v = 0.$$

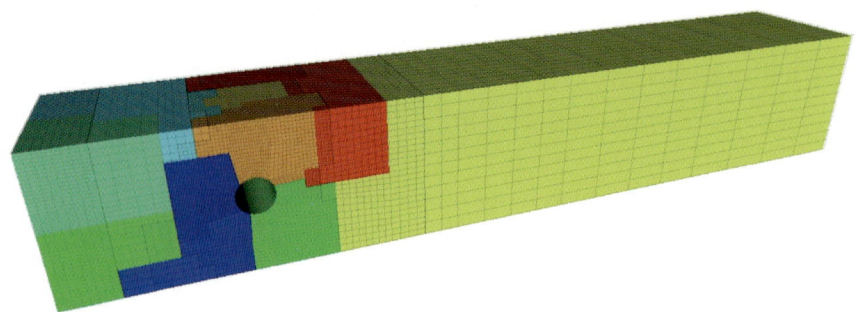

Fig. 7. Partitioning for the 3D Navier-Stokes benchmark problem into eight subdomains

The computational domain Ω consists of a channel with an obstacle of circular shape. The Reynolds number is chosen as $Re = 20$ in order to remain in the stationary regime. For a detailed description of the geometry and more details about the discretization we refer to the contribution [10] in this book. In Figure 7, we show a typical partitioning of Ω of locally refined meshes.

When $t_{seq}(n)$ denotes the cpu time of the sequential algorithm for problem size n and $t_{par}(n, P)$ the cpu time of the parallel algorithm for P cpu's and problem size n, the parallel efficiency is defined as:

$$E_{par}(n, P) = \frac{t_{seq}(n)}{P\, t_{par}(n, P)}.$$

A value of $E_{par}(n, P) = 1$ would be the optimal value (which is of course not possible in practice for arbitrary P). For the measurement of this parallel efficiency, Q_1 finite elements with local projection stabilization [5, 8].

In Figure 8 we plot this parallel efficiency in dependence of varying P for three different problem sizes. For a mesh with approximately 400 000 nodes, the efficiency is better than 0.9 for moderate number of cpu's (up to 15). For fixed numbers of cpu's and increasing problem sizes, $E_{par}(n, P)$ is listed in Figure 9. The parallel efficiency increases monotone for larger problem sizes, reaching a value of more than 0.8 for all numbers of cpu's considered.

The parallel efficiency $E_{par}(n, P)$ defined above is useful as long as the problem can be solved sequentially. Hence, on even finer meshes, where the problem cannot be solved on a single processor (of the same parallel computer), the parallel efficiency has to be measured differently. We refer to [18] for more details and numerical tests.

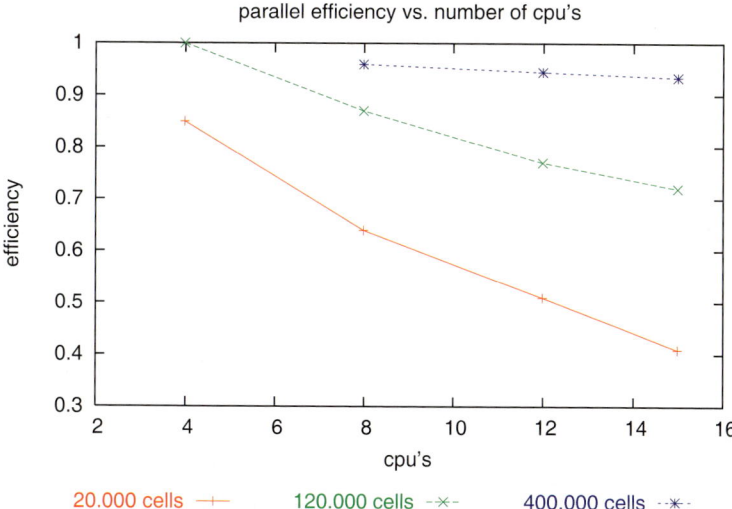

Fig. 8. Parallel efficiency for a 3D stationary incompressible flow with fixed problem size and variable numbers of cpu's.

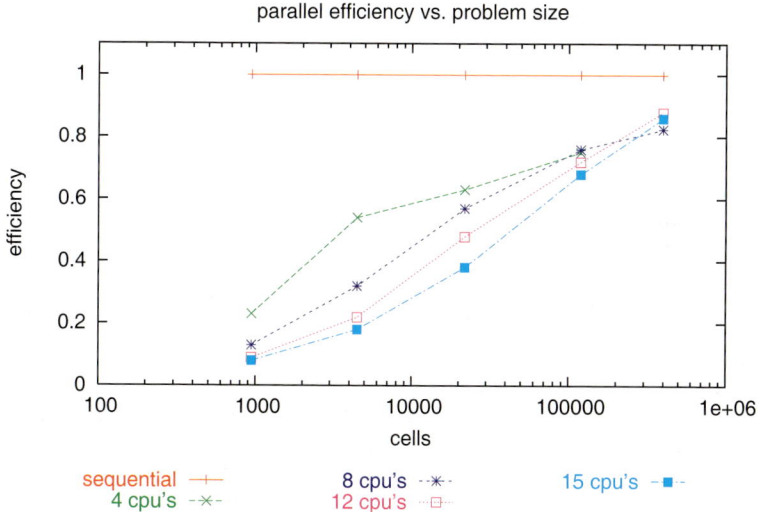

Fig. 9. Parallel efficiency for a 3D stationary incompressible flow with variable problem size and fixed numbers of cpu's.

References

1. R. E. Bank. *PLTMG: A Software Package for Solving Elliptic Partial Differential Equations, Users Guide 7.0*. Number 15 in Frontiers in Applied Mathematics. SIAM, 1994.
2. P. Bastian. *Parallele adaptive Mehrgitterverfahren*. Teubner Skripten zur Numerik, Stuttgart, 1996.
3. P. Bastian, K. Birken, K. Johannsen, S. Lang, V. Reichenberger, C. Wieners, G. Wittum, and C. Wrobel. Parallel solutions of partial differential equations with adaptive multigrid methods on unstructured grids. In *High performance computing in science and engineering II*, 1999.
4. R. Becker and M. Braack. Multigrid techniques for finite elements on locally refined meshes. *Numerical Linear Algebra with Applications*, 7:363–379, 2000. Special Issue.
5. R. Becker and M. Braack. A finite element pressure gradient stabilization for the Stokes equations based on local projections. *Calcolo*, 38(4):173–199, 2001.
6. R. Becker and R. Rannacher. Adaptive finite elements for optimal flow control. In W. Jäger et al., eds., *Reactive Flows, Diffusion and Transport*. Springer, Berlin, 2006.
7. M. Braack. An Adaptive Finite Element Method for Reactive Flow Problems. PhD Dissertation, Universität Heidelberg, 1998.
8. M. Braack. Solution of Complex Flow Problems: Mesh- and Model Adaptation with Stabilized Finite Elements. Habilitation thesis, Universität Heidelberg, 2005.
9. M. Braack and T. Richter. Solutions of 3D Navier-Stokes benchmark problems with adaptive finite elements. *Computers and Fluids*, 2006, 35 (4), p. 372-392.
10. M. Braack and T. Richter. Mesh and model adaptivity for flow problems. In W. Jäger et al., eds., *Reactive Flows, Diffusion and Transport*. Springer, Berlin, 2006.
11. A. Brandt. Multi-level adaptive solutions to boundary-value problems. *Math. Comp.*, 31:333–390, 1977.
12. A. Ern and V. Giovangigli. Thermal diffusion effects in hydrogen-air and methane-air flames. *Combust. Theory Modelling*, 2:349–372, 1998.
13. J. O. Hirschfelder and C. F. Curtiss. *Flame and Explosion Phenomena*. Williams and Wilkins Cp., Baltimore, 1949.
14. S. McCormick. *Multilevel Adaptive Methods for Partial Differential Equations*. SIAM, Frontiers in Appl. Math., Vol.3, 1989.
15. Metis. Serial graph partitioning, http://www-users.cs.umn.edu.
16. H. Oswald. *Parallele Lösung der instationären inkompressiblen Navier-Stokes-Gleichungen*. PhD thesis, Universität Heidelberg, 2001.
17. S. Parmentier, M. Braack, U. Riedel, and J. Warnatz. Modeling of combustion in a lamella burner. *Combustion Science and Technology*, 175(1):173–199, 2003.
18. T. Richter. Parallel multigrid for adaptive finite elements and its application to 3d flow problem. PhD Dissertation, Universität Heidelberg, to appear 2005.
19. U. Rüde. *Mathematical and Computational Techniques for Multilevel Adaptive Methods*. Number 13 in Frontiers in Applied Mathematics. SIAM, 1993.
20. M. D. Smooke. *Numerical Modeling of Laminar Diffusion Flames*. Progress in Astronautics and Aeronautics, Vol. 135, ed.: E. S. Oran and J. P. Boris, 1991.
21. H. Yserentant. On the multi-level splitting of finite element spaces. *Numer. Math.*, 49:379–412, 1986.

Solving Multidimensional Reactive Flow Problems with Adaptive Finite Elements*

M. Braack and T. Richter

Institut für Angewandte Mathematik, Universität Heidelberg

Summary. We describe recent developments in the design and implementation of finite element methods for the compressible Navier-Stokes equations modeling chemically reactive flows. The emphasize is on the low-Mach number regime including the limit case of incompressible flow. The most important ingredients are appropriate finite element discretizations, residual driven a posteriori mesh refinement, fully coupled defect-correction iteration for linearization, and optimal multigrid preconditioning. The potential of automatic mesh adaptation together with multilevel techniques is illustrated by 2D and 3D simulations of laminar methane combustion including detailed reaction mechanisms.

1 Governing Equations for Reactive Flow Problems

We denote the velocity by v, the pressure by p, the temperature by T and the density by ρ. Furthermore, we have n_s species mass fractions denoted by y_k, $k = 1, \ldots, n_s$. The basic equations for reactive viscous flow express the conservation of total mass, momentum, energy, and species mass in the form

$$\partial_t \rho + \operatorname{div}(\rho v) = 0, \tag{1}$$

$$\rho \partial_t v + \rho (v \cdot \nabla) v - \operatorname{div} \pi + \nabla p = g\rho, \tag{2}$$

$$\rho c_p \partial_t T + (\rho c_p v + \alpha) \cdot \nabla T + \operatorname{div} \mathcal{Q} = -\sum_{k=1}^{n_s} h_k m_k \dot{\omega}_k, \tag{3}$$

$$\rho \partial_t y_k + \rho v \cdot \nabla y_k + \operatorname{div} \mathcal{F}_k = m_k \dot{\omega}_k, \quad k = 1, \ldots, n_s, \tag{4}$$

where g is the gravitational force, c_p is the heat capacity of the mixture at constant pressure, and for each species k, m_k is its molar weight, h_k its specific enthalpy, $\dot{\omega}_k$ its molar production rate, and $\mathcal{F}_k$ its mass diffusion flux. The expression $\mathcal{Q}$ denotes the temperature flux. The viscous stress tensor π is given by

---

*This work has been supported by the German Research Foundation (DFG) through SFB 359 (Project A2) at the University of Heidelberg.

$$\pi = \mu \left(\nabla v + (\nabla v)^T - \frac{2}{3} \operatorname{div} v\, I \right).$$

The vector

$$\alpha = \sum_{k=1}^{n_s} c_{pk} \mathcal{F}_k \qquad (5)$$

is associated with an enthalpy flux due to diffusion fluxes of species having different specific heat capacities c_{pk}.

In equation (3), we have neglected heat release by pressure convection, volume viscosity effects, and viscous dissipation, because of their small impact on laminar flames.

The conservation equations (1)–(4) are completed by the ideal gas law for mixtures

$$\rho = \frac{p\overline{m}}{RT}, \qquad (6)$$

with the universal gas constant $R = 8.31451$, and mean molar weight $\overline{m}$ given by

$$\overline{m} = \left(\sum_{k=1}^{n_s} \frac{y_k}{m_k} \right)^{-1}.$$

We consider Fick's law for the species mass diffusion fluxes $\mathcal{F}_k$ driven by gradients of mass fractions:

$$\mathcal{F}_k^F = -\rho D_k^* \nabla y_k, \quad k = 1, \ldots, n_s. \qquad (7)$$

Following Hirschfelder & Curtiss[14], the diffusion coefficients D_k^* are given by

$$D_k^* = \frac{1 - y_k}{\sum_{l \neq k} \frac{x_l}{D_{kl}^{\text{bin}}}},$$

where D_{kl}^{bin} is the binary diffusion coefficient for the species pair (k, l). A more accurate diffusion model is the "extended Fick's law":

$$\mathcal{F}_k^F = -\rho D_k^* \frac{y_k}{x_k} \nabla x_k, \quad k = 1, \ldots, n_s. \qquad (8)$$

driven by gradients of mole fractions, $x_k = \frac{\overline{m}}{m_k} y_k$. However, this diffusion model is by far not the most accurate diffusion model. For hydrogen flames, for instance, multicomponent diffusion models are much more accurate. For more details on this topic we refer to Braack & Ern [6] and the contribution [8] in this book where diffusion models are locally adapted. In the numerical examples shown in this article we have used the simple diffusion law (7).

The temperature flux is modeled by Fourier's law

$$\mathcal{Q} = -\operatorname{div} \lambda \nabla T,$$

with the thermal conductivity λ. In this model the Dufour effect is not included, which is generally accepted to be of minor importance for the present flame configurations; see for instance Ern & Giovangigli [12].

The chemical production terms $\dot{\omega}_k$ are of the form

$$\dot{\omega}_k = \sum_{r \in \mathcal{R}} (\nu'_{rk} - \nu_{rk}) k_r \prod_{s \in \mathcal{S}} c_j^{\nu_{rj}},$$

with stoichiometric coefficients $\nu'_{rk}, \nu_{rk} \in \mathbb{N}$, and species concentrations $c_j = \rho x_j / \overline{m}$. The temperature dependent reaction rate is of Arrhenius type

$$k_r = A_r T^{\beta_r} \exp\left\{-\frac{E_r}{RT}\right\}.$$

The empirical coefficients A_r, β_r and E_r are given by the chemical mechanism. Furthermore, the coefficients λ and μ are modeled by the average of the arithmetic and harmonic mean of their specific quantities,

$$\lambda := \frac{1}{2} \sum_{k=1}^{n_s} x_k \lambda_k + \frac{1}{2} \left(\sum_{k=1}^{n_s} \frac{x_k}{\lambda_k}\right)^{-1},$$

$$\mu := \frac{1}{2} \sum_{k=1}^{n_s} x_k \mu_k + \frac{1}{2} \left(\sum_{k=1}^{n_s} \frac{x_k}{\mu_k}\right)^{-1},$$

with specific viscosities μ_k and heat conductivities λ_k, respectively. The heat capacity c_p is given by

$$c_p := \sum_{k=1}^{n_s} w_k c_{pk}.$$

The coefficients μ_k, λ_k and D_{kl}^{bin} are given in terms of exponential fits of experimental data,

$$\mu_k := \sum_{j=0}^{4} \alpha^{\mu}_{kj} T^j, \quad \lambda_k := \sum_{j=0}^{4} \alpha^{\lambda}_{kj} T^j, \quad D_{kl}^{bin} := \sum_{j=0}^{4} \alpha^{D}_{klj} T^j,$$

with coefficients $\alpha^{\mu}_{kj}, \alpha^{\lambda}_{kj}, \alpha^{D}_{klj} \in \mathbb{R}$. The specific enthalpies h_k are given by a similar exponential fit with 7 coefficients differently for the low and high temperature regime, $T \leq 1000\,K$ or $T > 1000\,K$, respectively. The specific heat capacity c_{pk} is obtained by taking the derivative of h_k with respect to the temperature:

$$c_{pk} = \frac{\partial h_{p,k}}{\partial T} = \sum_{j=1}^{6} j \alpha^{h}_{k,j} T^{j-1}.$$

Equations (1)–(4) are linearly dependent because the species mass fractions sum up to unity and

$$\sum_{k=1}^{n_s} \mathcal{F}_k = \sum_{k=1}^{n_s} m_k \dot\omega_k = 0\,.$$

Therefore, we omit one equation in (4), say that of the last species, and set with $s := n_s - 1$,

$$y_{n_s} := 1 - \sum_{i=1}^{s} y_i\,.$$

The system of equations is closed by suitable boundary conditions depending on the specific configuration to be considered: for temperature and species, we allow for non-homogeneous Dirichlet and homogeneous Neumann conditions. For the velocity v, we allow for non-homogeneous Dirichlet conditions at the inflow and rigid walls, and for the natural outflow boundary condition.

In order to account for compressible flows at low Mach number, the total pressure is split in two parts

$$p(x,t) = p_{th}(t) + p_{hyd}(x,t)\,.$$

While the so called thermodynamic pressure $p_{th}(t)$ is constant in space, the hydrodynamic pressure part $p_{hyd}(x,t)$ may vary in space and time. Hence, the pressure gradient in the momentum equation (2) can be replaced by ∇p_{hyd}. This is important for flows at low Mach number, because p_{hyd} is several magnitudes smaller than p_{th}.

2 Discretization by Stabilized Finite Elements

In this section, we describe the weak formulation and the discretization of the governing equations. This provides the framework to derive the a posteriori error estimator used in an adaptive procedure.

2.1 Variational Formulation

The vector u assembles the variables $u := \{p_{hyd}, v, T, y_1, \ldots, y_s\}$ while the density is considered as a coefficient determined by the ideal gas law

$$\rho = \frac{(p_{th} + p_{hyd})\overline{m}}{RT}\,. \tag{9}$$

By $\widehat{u}$ we denote an extension of non-homogeneous Dirichlet conditions to Ω. Hence, the solution is sought in the space $\widehat{u} + X$, where X is a function Hilbert space. The space X can be considered as a product of Hilbert spaces for each component, $L^2(\Omega) \times H^1(\Omega)^{d+1+s}$, with standard modifications to build in the boundary conditions and probably restrictions on the mean of pressure.

With the notation (u, φ) for the usual L^2–scalar product in Ω we define the semilinear form for stationary solutions of equations (1)–(6):

$$a(u)(\varphi) := (\text{div}\,(\rho v), \xi) + (\rho(v \cdot \nabla)v, \phi) + (\pi, \nabla\phi) - (p_{hyd}, \text{div}\,\phi) - (g\rho, \phi)$$

$$+ ((\rho c_p v + \alpha) \cdot \nabla T, \sigma) + (\lambda \nabla T, \nabla \sigma) + \sum_{k=1}^{n_s}(h_k m_k \dot{\omega}_k, \sigma)$$

$$+ \sum_{k=1}^{s}(\rho v \cdot \nabla y_k - m_k \dot{\omega}_k, \tau_k) - (\mathcal{F}_k, \nabla \tau_k)$$

for test functions $\varphi = \{\xi, \phi, \sigma, \tau_1, \ldots, \tau_s\}$. With this notations u is solution of the equation

$$u \in \hat{u} + X: \quad a(u)(\varphi) = 0 \quad \forall \varphi \in X.$$

2.2 Galerkin Formulation

For the discretization we use a conforming equal-order Galerkin finite element method defined on quadrilateral (hexaedrals in 3D) meshes $\mathcal{T}_h = \{K\}$ over Ω, with cells denoted by K. The mesh parameter h is defined as a cell-wise constant function by setting $h_{|K} = h_K$ and h_K is the diameter of K. In order to ease the mesh refinement we allow the cells to have nodes, which lie on midpoints of faces of neighboring cells. But at most one such *hanging node* is permitted for each face.

The discrete function space V_h consists of continuous, piecewise polynomial functions (so-called Q_1-elements) for all unknowns,

$$V_h = \{\varphi_h \in C(\overline{\Omega}); \varphi_h|_K \in Q_1(K)\},$$

where $Q_1(K)$ is the space of functions obtained by transformations of (isoparametric) bilinear polynomials from a fixed reference unit cell $\hat{K}$ to K. For a detailed description of this standard construction, see [11] or Johnson [15].

The case of hanging nodes requires some additional remarks. There are no degrees of freedom corresponding to these irregular nodes and the value of the finite element function is determined by pointwise interpolation. This implies continuity and therefore global conformity. For implementation details, see e.g. Carey & Oden [10].

Since we take for each component of the system the spaces V_h (with standard modifications for Dirichlet conditions), the discrete space X_h is a tensor product of the spaces V_h. The discrete Galerkin solution $u_h \in \hat{u} + X_h$ for a finite dimensional subspace $X_h \subset X$ reads:

$$u_h \in \hat{u} + X_h: \quad a(u_h)(\varphi) = 0 \quad \forall \varphi \in X_h. \tag{10}$$

The formulation (10) is not stable in general due to the following two reasons: (i) violation of the discrete inf-sup (or Babuska-Brezzi) condition for velocity and pressure approximation and (ii) dominating advection (and reaction). This problem will be addressed in more detail in the following.

2.3 Drawbacks of Standard Stabilization Techniques

The classical streamline diffusion (SUPG) stabilization for the incompressible Navier-Stokes problem, introduced by Hughes & Hughes [9], stabilizes the convective terms. Johnson & Saranen created in [16] an additional pressure stabilizing (PSPG) term in order to allow equal-order finite element approximations of velocity and pressure. Despite the success of the classical SUPG/PSPG stabilization technique for incompressible flow, one can find in recent papers a critical evaluation of this approach, see e.g. [22]. Drawbacks are basically due to the strong coupling between velocity and pressure in the stabilizing terms and the nontrivial construction of efficient algebraic solvers. The treatment of time-dependent problems requires the usage of space-time elements. Furthermore for reactive flows with a large number of species, the introduction of strong couplings between different chemical species becomes a severe problem. For illustration, we consider a single stationary convection-diffusion-reaction for species y_k equation of the form

$$\beta \cdot \nabla y_k - \text{div}\,(D_k \nabla y_k) = f_k \,, \qquad (11)$$

with a source term $f_k = f_k(T, y_1, \ldots y_{n_s})$ depending on temperature and other chemical species. For simplicity we assume homogeneous Dirichlet conditions (also assumed to be included in V_h). The weak formulation stabilized with SUPG reads:

$$y_k \in V_h : \quad (\beta \cdot \nabla y_k - f_k, \phi) + (D_k \nabla y_k, \nabla \phi)$$
$$+ \sum_{K \in \mathcal{T}_h} (\beta \cdot \nabla y_k - \text{div}\,(D_k \nabla y_k) - f_k, \delta_{kK} \beta \cdot \nabla \phi)_K = 0 \quad \forall \phi \in V_h,$$

with cell-dependent parameters δ_{kK}. In the corresponding stiffness matrix, the mass fractions of different chemical species are coupled due to the zero-order term (f_k, ϕ), because $f_k = f_k(T, y_1, \ldots y_{n_s})$, and due to the stabilization term $(f_k, \delta_k \beta \cdot \nabla \phi)$. The first coupling includes only the degrees of freedom corresponding to the same mesh points when the mass matrix is lumped. The latter coupling includes also degrees of freedom from different mesh points and different chemical species. Hence, the SUPG stabilization enlarges the number of couplings substantially.

Therefore, we present an alternative stabilization technique firstly introduced by Becker & Braack [1] which avoids this kind of additional couplings.

2.4 Stabilization by Local Projection

The nonlinear problem stabilized by local projection is of the form:

$$u_h \in \widehat{u} + X_h : \quad a(u_h)(\varphi) + s_h(u_h)(\varphi) = 0 \quad \forall \varphi \in X_h. \qquad (12)$$

The term $s_h(u_h)(\varphi)$ accounts for the saddle-point structure of the velocity and pressure coupling and for the convective terms.

For the definition of $s_h(\cdot)(\cdot)$ we suppose that the triangulation $\mathcal{T}_h$ is constructed in such a way that it results from a coarser quasi-regular mesh $\mathcal{T}_{2h}$ by one global refinement. By a "patch" of elements we denote a group of cells (four cells in 2D) which results from a common coarser cell in $\mathcal{T}_{2h}$. The corresponding discrete finite element spaces V_{2h} and V_h are nested: $V_{2h} \subset V_h$. By I_{2h}^h we denote the nodal interpolation operator $I_{2h}^h : V_h \to V_{2h}$. By

$$\kappa_h : V_h \to V_h, \quad \kappa_h \phi := \phi - I_{2h}^h \phi$$

we denote the difference between the identity and this interpolation. With this notation, the stabilization term added to the Galerkin formulation for an equation of type (11) reads

$$(\beta \cdot \nabla \kappa_h y_k, \delta_h \beta \cdot \nabla \kappa_h \phi),$$

where the parameter δ_h are chosen cell-wise constant depending on the local balance of convection and diffusion:

$$\delta_h|_K := \frac{\delta_0 h_K^2}{6 D_k + h_K \|\beta\|_{\infty,K}}.$$

Here, the quantity $\|\beta\|_{\infty,K}$ is the maximum of β on cell K. The parameter δ_0 is a fixed constant, usually chosen as $\delta_0 = 0.5$. Note, that κ_h vanishes on V_{2h}, and therefore, the stabilization vanishes for test functions of the coarse grid $\phi \in V_{2h}$.

The velocity-pressure coupling is stabilized for equal-order elements by a diffusive term acting on the pressure fluctuations:

$$(\nabla \kappa_h p_h, \alpha \nabla \kappa_h \xi).$$

For a stability proof and an error analysis for the Stokes equation we refer to [1].

For the full reactive flow system this stabilization technique is applied to all convective terms:

$$\begin{aligned}
s_h(u_h)(\varphi) := \quad & (\nabla \kappa_h p_h, \alpha_h \nabla \kappa_h \xi) \\
& + ((\rho_h v_h \cdot \nabla) \kappa_h v_h, \delta_{v,h} (\rho_h v_h \cdot \nabla) \kappa_h \phi) \\
& + ((\rho_h v_h \cdot \nabla) \kappa_h T_h, \delta_{T,h} (\rho_h v_h \cdot \nabla) \kappa_h \sigma) \\
& + \sum_{k=1}^{s} (\rho_h v \cdot \nabla \kappa_h y_{k,h}, \delta_k \rho_h v_h \cdot \nabla \kappa_h \tau_k).
\end{aligned} \qquad (13)$$

The proposed stabilization is consistent in the sense that the introduced terms vanish for $h \to 0$ with the correct order. Let us briefly compare the further couplings introduced by this technique. On the one hand, the stencil becomes larger due to the projection onto patches. On the other hand, and this is the crucial point for reactive flows, the stabilization does not act on the reactive source terms $m_k \dot{\omega}_k$. Hence no further couplings between different chemical species are introduced. We come back to this point when we discuss the matrix structure of the corresponding linear systems in a later section.

3 Local Mesh Refinement

Adaptive mesh refinement is well known to be an efficient method to increase the accuracy of the discrete solution with respect to the number of degrees of freedom, especially for an accurate resolution of a flame front. The efficiency of local mesh refinement strongly depends on the refinement strategy and the information about discretization errors. The *a posteriori error* estimates based on weighted residuals, introduced by Becker and Rannacher [3, 4] allows for a posteriori error control with respect to functional output $j(u)$. For a detailed description we refer to the contribution [8] in this book and the references cited therein. Here, we simply state the estimator used in the simulations presented below:

$$|j(u) - j(u_h)| \approx \sum_{i=1}^{n} \eta_i, \qquad \eta_i := |a(u_h)(\varphi_i^{(2)}) Z_i^{\pi}|,$$

where $\varphi_i^{(2)}$ denotes the Q_2 basis function corresponding to the node i of the mesh, and Z_i^{π} is a coefficient obtained by solving the linear dual problem

$$z_h \in X_h: \quad a_h^*(u_h)(\varphi, z_h) = j'(u_h)(\varphi) \quad \forall \varphi \in X_h.$$

with the bilinear form

$$a_h^*(u_h)(\varphi, z_h) := \lim_{\varepsilon \to 0} \frac{1}{\varepsilon} \Big\{ a(u_h + \varepsilon \varphi, z_h) + s_h(u_h + \varepsilon \varphi, z_h) - a(u_h)(z_h) - s_h(u_h)(z_h) \Big\}$$

Since this equation is linear, the computational cost of its solution relates to the cost of one Newton step in solving the primal problem. We note that the solution of the dual problem on the same mesh and by the same method as used for the primal problem is not mandatory. One may use a coarser mesh to save computational cost or even employ another (possibly higher-oder) discretization to get the weights with enhanced accuracy. In our computations, the dual solution is always computed by the same method as used for the primal problem.

4 Solution Procedure

The discrete equation system (12) is solved by quasi-Newton iteration with an approximate Jacobian $J = (J_{ij})$ of the stiffness matrix with block entries,

$$J_{ij} \approx a'(u^n)(\varphi_j, \varphi_i),$$

of size $n = s + d + 2$. The corresponding linear systems are solved with a multigrid algorithm. Due to the blocking of all the components of the system an incomplete LU factorization can be applied. This accounts for the strong local coupling of hydrodynamical variables as pressure, velocity and temperature with the chemical variables. This linear solver is described in detail in the contribution [2] of this book. Here we focus on the matrix structures because they may become extremely expensive for large chemical mechanisms.

Let us discuss the memory effort for storing a Jacobian with blocks J_{ij} of the form

$$\begin{bmatrix} A_{pp} & A_{pv} & A_{pT} & A_{py} \\ A_{vp} & A_{vv} & A_{vT} & A_{vy} \\ A_{Tp} & A_{Tv} & A_{TT} & A_{Ty} \\ A_{yp} & A_{yv} & A_{yT} & A_{yy} \end{bmatrix}. \tag{14}$$

Note, that the computational effort is proportional to the number of matrix entries. As already mentioned in Section 2, the biggest part in the blocks J_{ij} is due to the species couplings A_{yy}, at least for a large number of species $s \gg 1$. Considering trilinear finite elements (with the 27-point stencil on tensor grids), the matrix only containing matrix blocks of this type would have

$$27(5+s)^2$$

entries per grid point. In Table 1 we show the memory necessary for storing one system matrix in 2D and 3D when a reaction mechanism with $s = 20$ species is used.

The species couplings A_{yy} are due to various terms:

- For combustion problems, the chemical source terms $\dot{\omega}_k$ usually enforce extremely strong couplings between different species. However, if the mass matrix in the finite element discretization is lumped, these couplings do not appear in off-diagonal blocks J_{ij}, $i \neq j$.
- Some diffusion laws, as for instance the extended Fick's law (8), show off-diagonal couplings when mass fractions y_k are used as primary variables. Even more complex diffusion models, e.g., multicomponent diffusion, generate off-diagonal couplings even when mole fractions x_k are used. However, the off-diagonal couplings due to the extended Fick's law are usually of minor importance so that they may be neglected in the Jacobian. In this case, the contribution due to diffusion are diagonal in the blocks J_{ij}. For Fick's diffusion law (7) this is the case without modification of the Jacobian.

Table 1. Memory needed for storing one system matrix in single precision and $s = 20$ species in two and three dimensions.

	2D	3D
matrix couplings	9	27
block size	576	625
cells necessary for ≈ 5% error	10 000	250 000
memory (single prec.)	52 MB	4.2 GB

- SUPG stabilization generate off-diagonal couplings between species gradients due to the consistency terms, see discussion in Section 2.3.

In summary, due to the local projection stabilization the matrix blocks can be classified into two types: into dense diagonal blocks J_{ii} and into sparse off-diagonal blocks J_{ij}, $i \neq j$. The diagonal blocks remain of the form (14) while the off-diagonal blocks J_{ij}, $i \neq j$ of the system matrix are stored in the form

$$J_{ij} = \begin{bmatrix} A_{pp} & A_{pv} & A_{pT} & \\ A_{vp} & A_{vv} & A_{vT} & \\ A_{Tp} & A_{Tv} & A_{TT} & \\ & & & D_{yy} \end{bmatrix}$$

with a diagonal matrix D_{yy}. Such a block has only $(5^2 + s)$ entries for three velocity components (3D) compared to $(5 + s)^2$ entries of a full block J_{ii}. Using different matrix blocks for the off-diagonals, the storage usage reduces to

$$(5 + s)^2 + 26(5^2 + s).$$

If we use a reaction mechanism with $s = 20$ species, the memory needed to store a matrix is a factor

$$[27(5 + s)^2] : [(5 + s)^2 + 26(5^2 + s)] \approx 13.5$$

smaller than using the standard sparse matrix. Note, that the saving grows with larger reaction systems while the Newton residual is kept untouched. The Newton convergence may become slightly reduced. For details we refer to [5] and [7].

5 Simulation of a 2D Burner

The aim is the numerical simulation and its comparison with experimental data of a partially premixed methane flame at atmospheric pressure carried out at the Institute of Physical Chemistry (PCI) in Heidelberg. In particular, it was possible to measure absolute concentrations of formaldehyde (CH_2O) for the stationary laminar flame. A numerical simulation of such minor species poses high demands on the spatial accuracy of the discretization and on the reaction mechanism, as well.

Since the burner consists of seven identical slots it has a certain two dimensional structure, see the photo in Figure 1. Its capacity and energy density is comparable to a normal household burner. A numerical simulation of such a burner is a good test for the performance of the numerical method. Under the assumption of symmetry we restrict the simulation to *one* slot of the burner. Two simulations with different chemical mechanisms are performed:

- a so-called C1 mechanism with 15 chemical species and 42 reversible reactions of Smooke, see [21], and

Fig. 1. Methane burner at the PCI in Heidelberg.

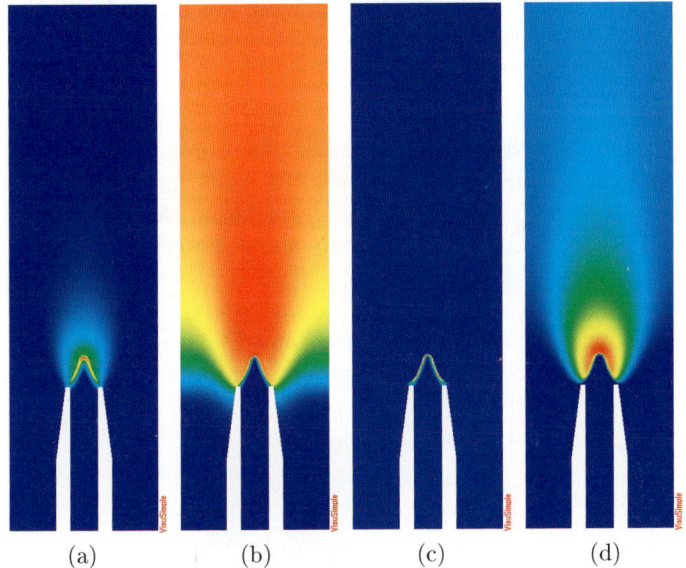

Fig. 2. Numerical simulation of the methane burner; mass fractions of (a) H_2, (b) H_2O, (c) CH_2O, (d) OH.

- a C2 mechanism with 39 chemical species and 151 reversible reactions of Warnatz [24]. This mechanism includes formation of several chemical species containing two carbon atoms.

The adaptation process is based on the mean value of formaldehyde mass fractions along the vertical line Γ at the center of the slot,

$$j(u) := \int_\Gamma y_{\mathrm{CH_2O}},$$

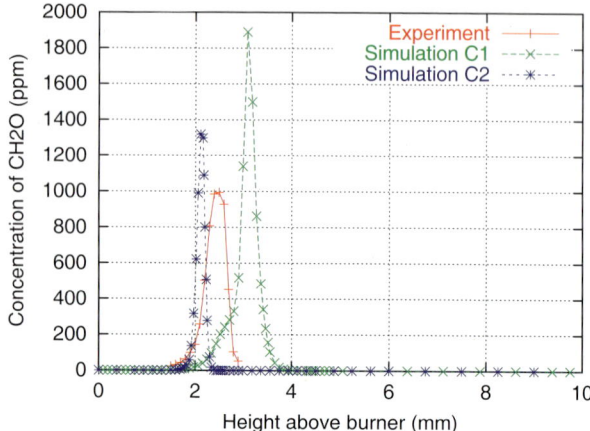

Fig. 3. Comparison of formaldehyde concentration: The $C2$ mechanism gives better agreement with experiments than the $C1$ mechanism in terms of absolute values and position of the flame front.

and produces a mesh with about 10^4 nodes. The corresponding CPU time on a single PC with 1 GHz was 4 hours for the small mechanism. In Figure 2 we show several species obtained by this simulation. A tensor grid with 10^6 nodes would give a similar accuracy, but about 1 week of CPU time would have been necessary.

Hence, the economy of the discretization gives the possibility to use the more exact C2 mechanism. The already obtained solution serves as a good initial guess for the Newton iteration. However, the mesh is adapted once more,

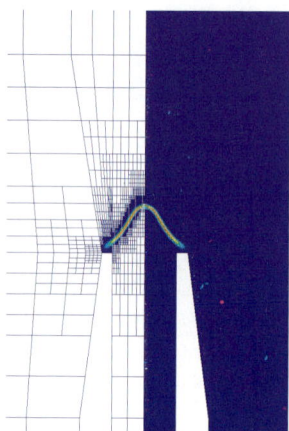

Fig. 4. Zoom into the flame front and mass fractions of ethylidyne C_2H_3. On the left hand side, the locally refined mesh is shown.

because the flame front moves upstream. This can also be seen in Figure 3 where a cross-section of formaldehyde is compared with the experiments. The C2 mechanism is obviously more accurate.

After 23 hours CPU time we get the stationary solution with 39 chemical species. Figure 4 shows a zoom of the obtained result for ethylidyne C_2H_3, a species with is not included in the smaller system. On the left hand side of the figure, the locally refined mesh is shown.

6 Simulation of a 3D Burner

Even more challenging is the computation of three-dimensional flames. The household burner constructed by BOSCH is an example of a burning facility with a three-dimensional laminar stationary flame, see Figure 5. Again this burner consists of several slots, but in contrast to the previous configuration, the symmetry is violated due to several cooling ducts transversal to the lamella. Within the PhD project [18], a two-dimensional approximation was arranged by neglecting the cooling ducts. With the software GASCOIGNE [13] a parameter study was carried out in order to obtain information concerning pollution formation under several loading situations of the burner, see [19].

As a first step towards three-dimensional combustion simulations we use the C1 and C2 reaction mechanism of the previous section with 15 and 39 chemical species, respectively. Due to the large number of chemical species, the nonlinearities due to chemical kinetics, the stiffness due to the differences in time scales for the fluid and the chemistry, this problem is much more complex than the corresponding flow problem without chemistry.

The sheer size of the problem with lots of chemical species leads to huge matrices which already in terms of memory usage make the use of parallel computers inevitable. In addition, adaptive mesh refinement is used to further reduce the problem dimension.

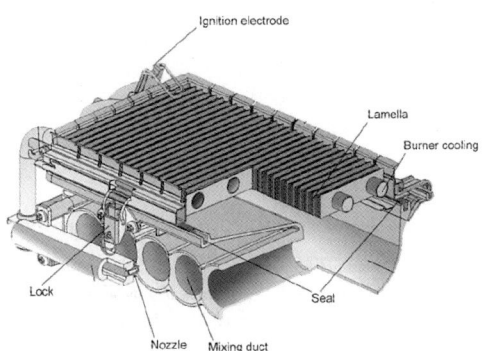

Fig. 5. Sketch of a household burner [17].

6.1 Two Dimensional Starting Guess with C1 Mechanism

The three dimensional simulation is started with a prolongation of the solution of a two dimensional (2D) simplification. In Figure 6, cut-outs of adaptively refined meshes from this 2D simplification are given. The mesh adaptation is driven by the adaptive process described before.

As quantity of interest we aim at detecting the concentration of formaldehyde CH_2O along a horizontal line through the three dimensional domain. In the two dimensional simplifications this quantity is represented by the evaluation at a single point. This evaluation point is highly resolved by the adapted meshes. Further, the edges of the lamellae are locally refined due to the produced singularities in the solution.

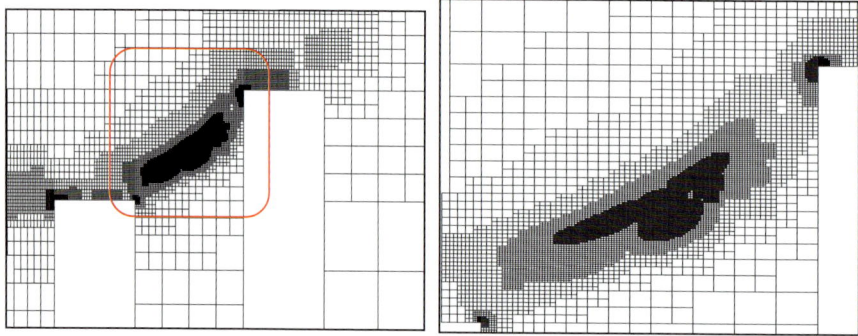

Fig. 6. Cut-outs of mesh used to generate the two dimensional simplification for the household burner.

6.2 Three Dimensional Computation with C1 Mechanism

In Figure 7 the computational domain including the cooling duct is described. The three dimensional simulations are performed on a PC cluster with the techniques described in this work and the parallelization technique presented in detail in the contribution [2] in this book. As mentioned before, mesh adaption is based on estimating the line functional. Although acting as an obstacle to the flow, the cooling duct does not require relevant mesh adaption, since its boundary is smooth and no chemical reaction takes place in this area. However, the influence of the cooling may not be neglected and will be analyzed later in this section.

In order to obtain a stationary solution the flame front has to be resolved sufficiently. Since the size of the flame front is determined by the species diffusion, an upscaling of the diffusion fluxes by a homotopy parameter $\tau \geq 1$ smears the flame front. For $\tau = 3$, a stationary solution is obtained even on the coarsest mesh with 26 008 cells only. On finer meshes the homotopy parameter is decreased down to 1 in order to recover the original diffusion model. The development of τ is listed in Table 2.

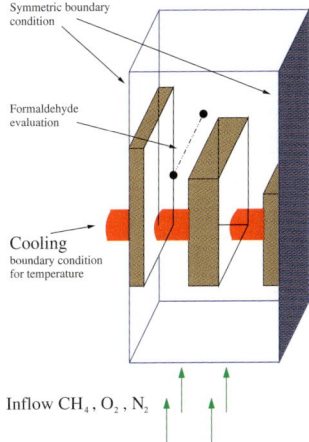

Fig. 7. Computational domain and measurement line for three dimensional simulation of the household burner.

In Table 3, the performance of the parallel solver is summarized. The finest mesh used corresponds to more than 11 million degrees of freedom. On 105 CPU's (AMD Athlon, 1.4 GHz) the solver needs about 22 minutes for on time step on this mesh. On the coarsest mesh, one time step last about 16 minutes on 32 processors. Since several time step were necessary, 6.5 hours are needed on this mesh in order to get the Newton iteration converged.

The computation is performed on a large Linux cluster with several tasks of different users at the same time. On those multi-user machines, it is in general not possible to have always the optimal number of CPU for this computation. Therefore, the number of CPU's is not necessarily increasing for larger problem sizes.

The problem is by far too large to be computed on a single processor, and hence the parallel efficiency can not be determined. However, an appropriate measure for the parallel efficency is

$$I_{eff}^{par} := 100 \frac{\#\text{CPU} \cdot \text{time}}{\text{dof}}$$

Table 2. Evolution of the homotopy parameter τ in dependence of the number of cells. For $\tau > 1$ the flame front is broader so that even on coarser meshes a stationary solver can be applied. For $\tau = 1$ the Fick's law is recovered. This 3D computation uses the C1 mechanism.

cells	dof's	τ
26 008	574 408	3.0
65 880	1 415 842	2.8
185 020	3 785 509	2.5
558 694	11 353 332	1.0

Table 3. Calculations for the C1 mechanism done on a AMD Athlon Linux clusters. On different meshes the number of cells, the number of degrees of freedom, the time necessary to solve one time step, the number of CPU's, the memory usage and the parallel efficiency index are given.

cells	dof's	#CPU	memory (GB)	time (sec)	I_{eff}^{par}
26 008	574 408	32	2.05	975	5.43
65 880	1 415 842	80	4.39	293	1.66
185 020	3 785 509	33	11.82	1930	1.68
558 694	11 353 332	105	40.62	1350	1.25

This quantity is listed in the last column. This efficency becomes better for larger problem size.

6.3 Comparison of the 2D and 3D Solutions

In Figure 8, the three-dimensional effect due to the cooling tubes are clearly visible. In particular, the flow velocity is reduced above the cooling tubes and the flame front becomes less pronounced. The visualization is done with the software VisuSimple [23].

For analyzing the three dimensional character of the burner – particularly the influence of the cooling ducts – we look at the course of the flame front along the z-axis of the burner. In Figure 9, we plot the process of the already mentioned functional evaluated along a line through the burner. This line is located in the flame front and crosses the domain atop of the cooling devices. In Figure 9, we show the velocity in main flow direction, the mass fraction of formaldehyde CH_2O and of HO_2 radicals. In addition, we plot the value of the corresponding solution component in the specific point obtained with the two dimensional simplification.

The plotted quantities feature a significant variation in z-direction, but also an overall discrepancy in comparison to the 2D simplification is observed. In Table 4 we list the maximal values of these four components identified in the whole domain for the two and the three dimensional setting. The difference in the temperature of about 100 Kelvin is quite remarkable and is caused by the cooling device. The lower temperature in the 3D-configuration leads to large differences in all other components. Taking in mind that the production of the pollutant NO_x is extremely sensitive to temperature it is questionable that a two-dimensional simulation is appropriate for an accurate prediction of NO_x.

Summarizing the results, the considered configuration yields real three-dimensional features. Two-dimensional simplifications of the geometry are not valid. However, detailed three-dimensional simulations are possible, if one combines efficient solvers with mesh adaptation and parallelization.

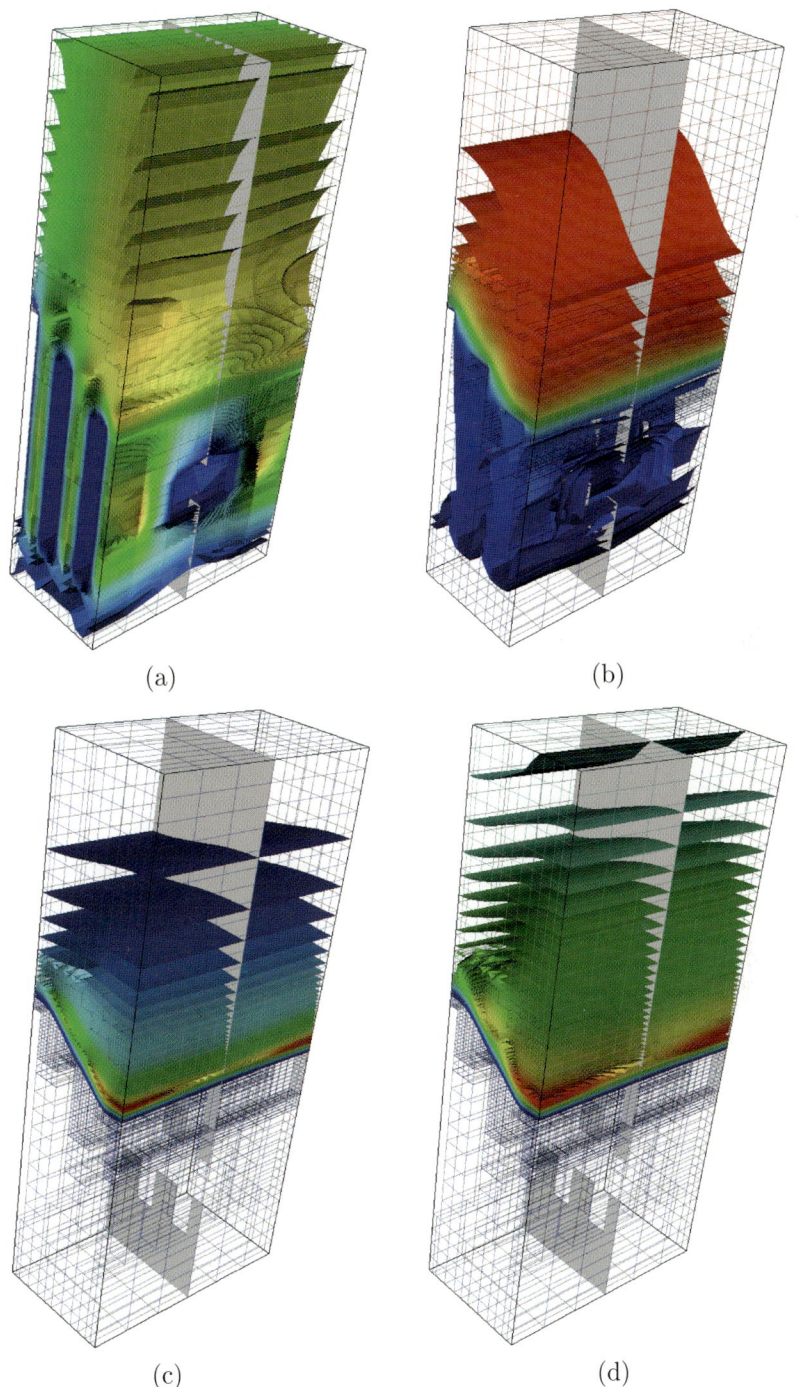

Fig. 8. Simulation of the three-dimensional methane burner: (a) Vertical velocity, (b) temperature, (c) H mass fractions, and (d) OH mass fractions.

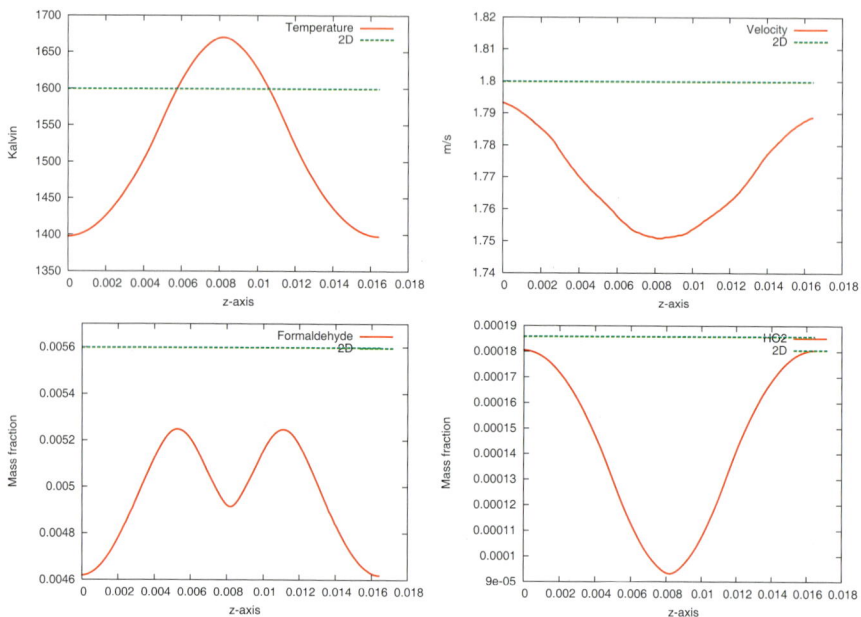

Fig. 9. Comparison of 2D and 3D simulations of the household burner. Cross sections of profiles of temperature, the velocity in main flow direction, the mass fractions of formaldehyde, and HO_2−radicals along the z-axis.

Table 4. Maximal values for the velocity, the temperature as well as the mass fraction of formaldehyde and HO_2 obtained in the 2D and 3D simulation.

	2D	3D
velocity (m/s)	3.39	2.61
temperature (K)	2156	2039
mass fraction $CH_2O \times 10^{-3}$	7.1	5.2
mass fraction $HO_2 \times 10^{-4}$	2.3	6.2

We refer to [20] for more details on the configuration and the comparison between the two- and three-dimensional simulations.

6.4 3D Simulation with 39 Chemical Species

The final computation breaks the dimensions of all the previous computations. We perform a three dimensional simulation of the burner with the C2 mechanism (39 chemical species). The solution looks quite similar to the C1 solution but looking on numbers, there is indeed a difference. In Table 5 some maximum values are listed in order to compare the effect of the dimension (2D or 3D) and the reaction mechanism used. While the mechanism has nearly no effect on the maximum temperature, the impact on CH_2O and HO_2 is quite large.

Table 5. Maximal values for the velocity, the temperature as well as the mass fraction of formaldehyde and HO_2 obtained in the 2D and 3D simulation for both reaction mechanisms.

	C1 mechanism		C2 mechanism	
	2D	3D	2D	3D
velocity (m/s)	3.39	2.61	3.51	3.31
temperature (K)	2156	2039	2099	2036
mass fraction $CH_2O \times 10^{-3}$	7.1	5.2	4.3	3.49
mass fraction $HO_2 \times 10^{-4}$	2.3	6.2	4.9	3.85

Table 6. Calculations for the C2 mechanism done on an Opteron Linux clusters. On different meshes the number of cells, the number of degrees of freedom, the time necessary to solve one time step, the number of CPU's, the memory usage and the parallel efficiency index are given. Further the used homotopy parameter τ is indicated.

cells	dof's	#CPU	memory (GB)	time (sec)	I_{eff}^{par}
15 872	825 084	3	1.10	1772	0.64
29 648	1 505 946	5	2.23	1583	0.53
60 784	3 136 248	10	5.26	2248	0.71
197 816	9 863 082	10	16.92	6096	0.61
291 102	14 299 478	18	25.25	4681	0.59

The homotopy parameter is taken as $\tau = 3$ on the coarsest mesh and reduces to $\tau = 1$ on the finest mesh. Hence, we recover the original Fick's law. The performance of the parallel solver is summarized in Table 6.

On the finest mesh we have more than 14 millions degrees of freedom. The solver needs about 78 minutes for one time step when 18 processors (Opteron CPU's) are used. Due to the sparse matrix structures designed for these type of problems only 25 GB of total memory is necessary.

References

1. R. Becker and M. Braack. A finite element pressure gradient stabilization for the Stokes equations based on local projections. *Calcolo*, 38(4):173–199, 2001.
2. R. Becker, M. Braack, and T. Richter. Parallel multigrid on locally refined meshes. In W. Jäger et. al., eds., *Reactive Flows, Diffusion and Transport*. Springer, Berlin, 2006.
3. R. Becker and R. Rannacher. A feed-back approach to error control in finite element methods: Basic analysis and examples. *East-West J. Numer. Math.*, 4:237–264, 1996.
4. R. Becker and R. Rannacher. An optimal control approach to a posteriori error estimation in finite element methods. In A. Iserles, editor, *Acta Numerica 2001*, volume 37, pages 1–225. Cambridge University Press, 2001.

5. M. Braack. An Adaptive Finite Element Method for Reactive Flow Problems. PhD Dissertation, Heidelberg University, 1998.
6. M. Braack and A. Ern. Coupling multimodeling with local mesh refinement for the numerical solution of laminar flames. *Combust. Theory Modelling*, 8(4):771–788, 2004.
7. M. Braack and R. Rannacher. Adaptive finite element methods for low-Mach-number flows with chemical reactions. In *30th Computational Fluid Dynamics*, VKI LS 1999-03, H. Deconinck, ed., pages 1–93. von Karman Institute for Fluid Dynamics, Brussels, Belgium, 1999.
8. M. Braack and T. Richter. Mesh and model adaptivity for flow problems. In W. Jäger et. al., eds., *Reactive Flows, Diffusion and Transport*. Springer, Berlin, 2005.
9. A. Brooks and T. Hughes. Streamline upwind Petrov-Galerkin formulation for convection dominated flows with particular emphasis on the incompressible Navier-Stokes equations. *Comput. Methods Appl. Mech. Engrg.*, 32:199–259, 1982.
10. G. Carey and J. Oden. *Finite Elements, Computational Aspects*, volume III. Prentice-Hall, 1984.
11. P. Ciarlet. *Finite Element Methods for Elliptic Problems*. North-Holland, Amsterdam, 1978.
12. A. Ern and V. Giovangigli. Thermal diffusion effects in hydrogen-air and methane-air flames. *Combust. Theory Modelling*, 2:349–372, 1998.
13. GASCOIGNE. http://www.gascoigne.de.
14. J. O. Hirschfelder and C. F. Curtiss. *Flame and Explosion Phenomena*. Williams and Wilkins Cp., Baltimore, 1949.
15. C. Johnson. *Numerical Solution of Partial Differential Equations by the Finite Element Method*. Cambridge University Press, Cambridge, UK, 1987.
16. C. Johnson and J. Saranen. Streamline diffusion methods for the incompressible euler and Navier-Stokes equations. *Math. Comp.*, 47:1–18, 1986.
17. Junkers Bosch Thermotechnik. Wandhängende Junkers Gas-Kesselthermen. Informationsbroschüre, Wernau, 1998.
18. S. Parmentier. *Modellierung und Simulation eines Lamellenbrenners und eines Diffusionsreaktors zur Reduzierung von Schadstoffen*. PhD thesis, Universität Heidelberg, 2002.
19. S. Parmentier, M. Braack, U. Riedel, and J. Warnatz. Modeling of combustion in a lamella burner. *Combustion Science and Technology*, 175(1):173–199, 2003.
20. T. Richter. Parallel multigrid for adaptive finite elements and its application to 3d flow problem. PhD Dissertation, Universität Heidelberg, to appear 2005.
21. M. D. Smooke. *Numerical Modeling of Laminar Diffusion Flames*. Progress in Astronautics and Aeronautics, Vol. 135, ed.: E. S. Oran and J. P. Boris, 1991.
22. G. T., G. Lube, M. Olshanskii, and J. Starcke. Stabilized finite element schmes with LBB-stable elements for incompressible flows. *J. Comp. Appl. Math.*, 177:243–267, 2005.
23. VisuSimple. http://visusimple.uni-hd.de.
24. J. Warnatz, U. Maas, and R. W. Dibble. *Combustion*. Springer, New York, 1996.

Part III

Optimization Methods for Reactive Flows

Preamble

This chapter contains three articles which present new computational approaches to the efficient numerical solution of optimization problems involving complex systems of ordinary as well as partial differential equations.

The first article *"Robustness aspects in parameter estimation, optimal design of experiments and optimal control"* presents new effective algorithms for robust parameter estimation and design of robust optimal experiments in dynamic systems. These methods overcome the difficulties due to the occurance of outliers in the measurements and ill-conditioning of the identification problem. Numerical results for a real-life application from biochemical engineering are presented.

The second article *"Multiple set point partially reduced SQP method for optimal control of PDE"* discusses the optimization of dynamic processes described by partial differential-algebraic equations (PDAE). Fast solutions methods are achieved by using the "all-at-once" approach, in which the optimization aspect of the overall algorithm is closely coupling with the solution method of the dynamic system, and partially reduced sequential quadratic programming (PRSQP) methods. Another important feature is the ability of efficient real-time optimization. Several industrial applications are presented such as shape optimization of turbine blades and the real-time optimization of a continuous distillation column.

The third article *"Adaptive finite element methods for PDE-constrained optimal control problems"* presents a systematic approach to error control and mesh adaptation in the numerical solution of optimal control problems governed by partial differential equations. By the Lagrangian formalism the optimization problem is reformulated as a saddle-point boundary value problem which is discretized by a finite element Galerkin method. The accuracy of the discretization is controlled by residual-based a posteriori error estimates. The main features of this method are illustrated by examples from optimal control of heat transfer, fluid flow and parameter estimation.

Robustness Aspects in Parameter Estimation, Optimal Design of Experiments and Optimal Control*

H. G. Bock, S. Körkel, E. Kostina, and J. P. Schlöder

Interdisziplinäres Zentrum für Wissenschaftliches Rechnen (IWR),
Universität Heidelberg

Summary. Estimating model parameters from experimental data is crucial to reliably simulate dynamic processes. In practical applications, however, it often appears that the data contains outliers. Thus, a reliable parameter estimation procedure is necessary that delivers parameter estimates insensitive (robust) to errors in measurements.

Another difficulty that occurs in practical applications is that the experiments performed to obtain measurements for parameter estimation are expensive, but nevertheless do not guarantee satisfactory parameter accuracy. The optimization of one or more dynamic experiments in order to maximize the accuracy of the results of a parameter estimation subject to cost and further technical inequality constraints leads to very complex non-standard optimal control problems. Newly developed successful methods and software for design of optimal experiments for nonlinear processes are based on the expansion of the problem at the nominal value of parameters which lie in a (possibly large) confidence region. Robust optimal experiments, that are insensitive against uncertainties in parameter values, should be obtained if we optimize the experiments in min-max fashion (worst-case design) over the whole range (confidence region) of an uncertainty set.

The paper presents new effective algorithms for robust parameter estimation and design of robust optimal experiments in dynamic systems. Numerical results for a real-life application from bio-chemical engineering are presented.

1 Introduction

Parameter estimation and optimal design of experiments are important steps in establishing models that reproduce a given process quantitatively correctly. The aim of parameter estimation is to reliably identify model parameters from sets of noisy experimental data. In practical applications, it often appears that the data contains outliers. Thus, a reliable parameter estimation procedure is

---

*This work has been supported by the German Research Foundation (DFG) through SFB 359 (Project A4) at the University of Heidelberg.

necessary that delivers parameter estimates insensitive (robust) to errors in measurements.

Another problem appearing in practical applications is obtaining "good accuracy" of the parameters. The accuracy of the parameters, i. e. their statistical distribution depending on data noise, can be estimated up to first order by means of a covariance matrix approximation and the corresponding confidence regions. In practice, however, one often finds that the experiments performed to obtain the required measurements are expensive, but nevertheless do not guarantee satisfactory parameter accuracy. In order to maximize the accuracy of the parameter estimates additional experiments can be designed with optimal experimental settings (e.g. initial conditions, measurement devices, sampling times, temperature profiles, feed streams etc) subject to constraints. As an objective functional a suitable function of the covariance matrix (e.g. trace, determinant, maximal eigenvalue, maximal diagonal element etc) can be used. The possible constraints in this problem describe costs, feasibility of experiments, model validity domains etc. Newly developed successful methods and software for design of optimal experiments for nonlinear processes are based on the expansion of the problem at the nominal value of parameters which lie in a (possibly large) confidence region. Robust optimal experiments, that are insensitive to uncertainties in parameter values, should be obtained if we optimize the experiments in min-max fashion (worst-case design) over the whole range (confidence region) of an uncertainty set.

The paper focuses on new effective algorithms for robust parameter estimation and design of robust optimal experiments in dynamic systems. It is organized as follows. Section 2 presents robust parameter estimation: problem statement, boundary value problem methods, theoretical background of generalized Gauss-Newton methods, convergence properties, special methods for solution of linearized problems. The problem of optimal experiment design is discussed in Section 3, including problem statement, properties of covariance matrix, methods for design of robust optimal experiments. Numerical results for a real-life application from bio-chemical engineering are presented at the end of each section.

2 Robust Parameter Estimation

2.1 Problem Statement

In this section we describe parameter estimation problems in dynamic processes. We assume that processes are modeled by systems of differential-algebraic equations (DAE)

$$\begin{aligned} \dot{y}(t) &= f(t, y(t), z(t), p, q, u(t)) \\ 0 &= g(t, y(t), z(t), p) \end{aligned} \quad t \in [t_0; t_f] \qquad (1)$$

where $y \in \mathbb{R}^{n_y}$ denotes the differential variables, $z \in \mathbb{R}^{n_z}$ denotes the algebraic variables, $u : [t_0; t_f] \to \mathbb{R}^{n_u}$ and $q \in \mathbb{R}^{n_q}$ are given control functions and control variables, and the right-hand sides f and g depend on a vector of parameters $p \in \mathbb{R}^{n_p}$ with unknown values. It is assumed that experiments have been carried out yielding at the given times t_j, $j = 1, ..., \mathcal{M}$, the measurements η_{ij}, $i = 1, ..., \mathcal{M}_j$, $j = 1, ..., \mathcal{M}$, of the observation functions b_{ij} of the state variables $x(t) = (y(t), z(t))$

$$\eta_{ij} = b_{ij}(t_j, x(t_j), p^{true}) + \varepsilon_{ij},$$

which are subject to measurement errors ε_{ij}. Here, p^{true} are the "true" values of the parameters. Note, that several model quantities can be measured at a time t_j.

Now we estimate parameters by minimizing the deviation between the model and the data:

$$\| r_1(x(t_1), ..., x(t_\mathcal{M}), p) \| := \left\| \begin{array}{c} \vdots \\ (\eta_{ij} - b_{ij}(t_j, x(t_j), p))/\sigma_{ij} \\ \vdots \end{array} \right\|$$

in an appropriate (weighted) *norm* with weights σ_{ij}. The type of the norm is motivated by the statistical distribution of the measurement errors. If the errors are independent, normally distributed with zero mean and known variances ($\mathcal{N}(0, \sigma_{ij}^2)$), minimizing a weighted least squares function

$$\min_p \sum_{i,j} (\eta_{ij} - b_{ij}(t_j, x(t_j), p))^2 / \sigma_{ij}^2$$

yields a *maximum likelihood estimate*, see e.g. [21, 22, 17]. A powerful software for l_2 based parameter estimation on dynamic systems is available, e.g. PARFIT [8, 9]. But in case of Laplace - distribution l_1 estimation

$$\min_p \sum_{i,j} |\eta_{ij} - b_{ij}(t_j, x(t_j), p)| / |\sigma_{ij}|$$

yields a maximum likelihood estimate, see e.g. [7]. Further, it is well-known that l_1 parameter estimation possesses a very remarkable property, namely, the optimal solution is typically *less sensitive to the effect of outliers* in data then e.g. an optimal solution of l_2 based parameter estimation and therefore l_1 estimation is used for *robust* parameter estimation, see [25]. l_1 optimization problems were first studied by Boscovic [16] in 1758.

Frequently, there is an additional knowledge about the state variables and the parameters in the model such as initial conditions, parameter bounds or boundary conditions, positivity etc. This information can be considered as equality or inequality constraints at times θ_i, $i = 1, ..., \mathcal{K}$,

$$r_2(x(\theta_1), ..., x(\theta_\mathcal{K}), p) = 0, \quad r_3(x(\theta_1), ..., x(\theta_\mathcal{K}), p) \geq 0. \tag{2}$$

Summing up, the parameter estimation problem can be formulated as

$$\min_{x,p} \ ||r_1(x(t_1),...,x(t_\mathcal{M}),p)||, \qquad (3)$$
$$\text{s.t.} \ (x,p) \text{ solves DAE (1)},$$
$$r_2(x(\theta_1),...,x(\theta_\mathcal{K}),p) = 0,$$
$$r_3(x(\theta_1),...,x(\theta_\mathcal{K}),p) \geq 0.$$

2.2 Boundary Value Problem Methods

A typical solution approach to parameter estimation which is found very often in practice is the initial value or single shooting approach: the DAE system is repeatedly solved as an initial value problem, and unknown parameters including possibly initial values are iteratively improved by some optimization procedure.

In contrast to that, our numerical solution of the parameter estimation problem is based on the Boundary Value Problem (BVP) approach going back to [8]. The basic idea consists in parameterizing the dynamic equations (initial or boundary value problem) like a boundary value problem (e.g., by multiple shooting) and then performing simultaneously (in one iteration loop) the minimization of the cost function subject to the constraints given by the parameterized boundary value problem. It has been shown [8, 9], that BVP methods (based on multiple shooting or collocation) are much more stable and efficient than the single shooting approach when solving parameter estimation problems.

Multiple Shooting

The scheme of multiple shooting is shown at Figure 1 and consists in the following. First one chooses a suitable grid of multiple shooting nodes τ_j

$$t_0 = \tau_0 < \tau_1 < \ldots < \tau_m = t_f,$$

covering the interval where measurements are given.

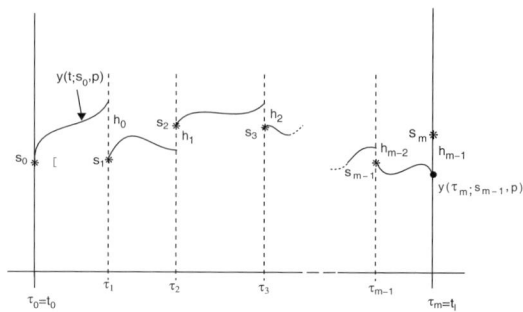

Fig. 1. Multiple shooting approach

At each grid point the values of the state variables s_j are chosen as additional unknowns and m relaxed DAE initial value problems [2]

$$\dot{y} = f(t, y, z, p),$$
$$0 = g(t, y, z, p) - \alpha(t)g(\tau_j, s_j^y, s_j^z, p),$$
$$y(\tau_j) = s_j^y, \quad z(\tau_j) = s_j^z,$$

are solved on each subinterval $[\tau_j, \tau_{j+1}]$ to yield a solution $x(t; s_j, p)$ for $t \in [\tau_j, \tau_{j+1}]$. The decreasing function $\alpha(t)$ ($\alpha(\tau_j) = 1, \alpha(t) \geq 0$) is chosen in order to improve consistency of the algebraic conditions in the course of the integration.

Solutions of dynamic systems, generated by this procedure, are usually neither continuous at τ_j, nor feasible for the original algebraic equation. This has to be enforced by additional matching conditions. Inserting the computed values $x(t, s_j, p)$, $\tau_j \leq t \leq \tau_{j+1}$, into problem (3) one obtains a constrained problem in the variables $(s, p) = (s_0, \ldots, s_m, p)$:

$$\min_{s,p} \|r_1(s, p)\|, \tag{4}$$
$$\text{s.t.} \quad r_2(s, p) = 0, \quad r_3(s, p) \geq 0,$$
$$h_j(s_j, s_{j+1}, p) := y(\tau_{j+1}; s_j, p) - s_{j+1}^y = 0, j = 0, \ldots, m-1,$$
$$g(\tau_j, s_j, p) = 0, \quad j = 0, \ldots, m.$$

Multiple shooting possesses several advantages:

1. It is possible to include a priori information about the state variables, e.g. from the measurements or from expert knowledge, by a proper choice of initial guesses for the additional variables s_j. Thus, it can be ensured that the initial solutions $y(t; s_j, p)$ remain close to the observed data. It can be shown that this damps the influence of poor parameter guesses.
2. The adequate choice of initial guesses for the state variables (and the application of a Gauss-Newton method for the solution of the constrained least squares problem) typically avoids convergence to local minima with large residuals.
3. The scheme is numerically stable. The splitting of the integration interval limits error propagation and allows to solve parameter estimation problems even for unstable or chaotic systems [2, 10].
4. The matching conditions induce a very specific BVP structure in the problem equations, which can be exploited in particular for the design of parallel algorithms.

---

[2] Since in the parameter estimation problem control functions u and control variables q are fixed, we omit them in this section from the problem formulation for the notation simplicity.

2.3 l_1 Gauss-Newton Method

Parameterization of the BVP constraint yields a finite dimensional, possibly large-scale, nonlinear constrained approximation problem (4) which can be formally written as

$$\min_X \sum_{i \in I_1} |F_i(X)|^\nu, \ \nu = 1 \text{ or } 2, \ I_1 := \{1, ..., m_1\} \tag{5}$$
$$\text{s.t. } F_i(X) = 0, \ i \in I_2 := \{m_1 + 1, ..., m_2\},$$
$$F_i(X) \geq 0, \ i \in I_3 := \{m_2 + 1, ..., m_3\}.$$

Note, that the equalities $F_i(X) = 0$, $i \in I_2$, include the matching conditions induced by multiple shooting. We assume that the functions $F_i : D \subset \mathbb{R}^n \to \mathbb{R}$, $i \in I = I_1 \cup I_2 \cup I_3$, are twice-continuously differentiable. The number of variables in problem (5) under consideration is equal to the number of differential and algebraic variables multiplied by the number of multiple shooting nodes plus the number of parameters.

To solve problem (5) we use a generalized Gauss-Newton method according to which a new iterate is (basically) generated by

$$X^{k+1} = X^k + t^k \Delta X^k, 0 < t^k \leq 1, \tag{6}$$

where the increment ΔX^k is the solution of the following linearized constrained problem at $X = X^k$

$$\min_{\Delta X \in \mathbb{R}^n} \sum_{i \in I_1} |F_i(X) + J_i(X)\Delta X|^\nu \ \nu = 1 \text{ or } 2, \tag{7}$$
$$\text{s.t. } F_i(X) + J_i(X)\Delta X = 0, i \in I_2, \ F_i(X) + J_i(X)\Delta X \geq 0, i \in I_3.$$

Here $J_i(X)$ denotes the gradient of $F_i(X)$, that is the row-vector $\nabla F_i(X) = (\frac{\partial F_i}{\partial X})^T$.

In the rest of this subsection we will consider the l_1 parameter estimation problem. An important question about the generalized Gauss-Newton method (6), (7) concerns about its convergence properties. Before their discussion we need to remind the optimality conditions for the l_1 problem (5). We define the active set $I(X)$ at the point X

$$I(X) = I_1(X) \cup I_2 \cup I_3(X), I_k(X) = \{i \in I_k : F_i(X) = 0\}, k = 1, 3,$$

and define a *regular* point X as a point for which *constraint qualification* holds, that is the Jacobian of all active constraints and zero components of the cost function at X has full row rank

$$\text{rank } \mathcal{J}_0(X) = m_0 := |I(X)|, \ \mathcal{J}_0(X) := \left[J_i(X)\right]_{i \in I(X)}. \tag{8}$$

Theorem 1. (Necessary conditions) *Let X^* be a regular solution of problem (5). Then there is a unique vector of Lagrange multipliers $\lambda^* \in \mathbb{R}^{m_3}$, such that the following conditions (the first-order necessary optimality conditions) are true for the pair (X^*, λ^*):*

$$\lambda^{*T} J(X^*) = 0, \tag{9}$$
$$|\lambda_i^*| \leq 1, i \in I_1(X^*), \lambda_i^* = -\text{sign} F_i(X^*), i \in I_1 \backslash I_1(X^*),$$
$$\lambda_i^* \geq 0, i \in I_3(X^*), \lambda_i^* = 0, i \in I_3 \backslash I_3(X^*).$$

Further, the point (X^*, λ^*) satisfies the second-order necessary optimality conditions:

$$d^T H d \geq 0, \ H = H(X^*, \lambda^*) := -\sum_{i \in I} \lambda_i^* \frac{\partial^2 F_i(X^*)}{\partial X^2}$$

for all directions d from the space $P(X^*, \lambda^*)$

$$P(X, \lambda) = \{d \neq 0 \mid J_i(X)d = 0, i \in I_2,$$
$$\begin{matrix} J_i(X)d = 0, & \text{if } |\lambda_i| < 1 \\ \lambda_i J_i(X)d \leq 0, \text{if } \lambda_i = 1 \text{ or } -1 \end{matrix} \ i \in I_1(X),$$
$$\begin{matrix} J_i(X)d = 0, \text{ if } \lambda_i > 0, \\ J_i(X)d \geq 0, \text{ if } \lambda_i = 0 \end{matrix} \ i \in I_3(X)\},$$

Theorem 2. (Sufficient optimality conditions) *Let (X^*, λ^*) satisfy the first-order necessary optimality conditions (9) and for all directions $d \in P(X^*, \lambda^*)$, the matrix H is positive definite, that is*

$$d^T H d > 0, \forall d \in P(X^*, \lambda^*). \tag{10}$$

Then X^* is a strict local minimum of (5).

Corollary 1. *Let (X^*, λ^*) satisfy the first-order necessary conditions (9), strict complementarity*

$$|\lambda_i| < 1, i \in I_1(X), \ \lambda_i > 0, i \in I_3(X). \tag{11}$$

and the Hessian H is positive definite for all $d \in Z(X^*)$,

$$Z(X) := \{d \neq 0 \mid J_i(X)d = 0, i \in I(X)\}.$$

Then X^* is a strict local minimum of (5).

Proofs follow from applying optimality conditions of nonlinear programming to the problem equivalent to problem (5)

$$\min_{X, \chi} \sum_{i \in I_1} \chi_i,$$
$$\text{s.t. } \chi_i + F_i(X) \geq 0, \ \chi_i - F_i(X) \geq 0, \ \chi_i \geq 0, i \in I_1,$$
$$F_i(X) = 0, i \in I_2, \ F_i(X) \geq 0, i \in I_3.$$

Applying optimality conditions to the linear l_1 problem (7), which read

$$\lambda_{LP}^{*T} J(X) = 0, \qquad (12)$$
$$|\lambda_{i,LP}^*| \leq 1, i \in I_1(\Delta X^*),$$
$$\lambda_{i,LP}^* = -\text{sign}(F_i(X) + J_i(X)\Delta X^*), i \in I_1 \setminus I_1(\Delta X^*),$$
$$\lambda_{i,LP}^* \geq 0, i \in I_3(\Delta X^*), \lambda_{i,LP}^* = 0, i \in I_3 \setminus I_3(\Delta X^*),$$

we may conclude that its solution is defined by a generalized inverse. Denote

$$\mathcal{F}(X) = \big[F_i(X)\big]_{i \in I}, \quad \mathcal{J}(X) = \big[J_i(X)\big]_{i \in I}.$$

Theorem 3. (Existence of a generalized inverse) *Let $\Delta X^\star$ minimize problem (7). Assume further that*

$$\text{rank } \mathcal{J}_0 = |I(\Delta X^*)|, \quad \mathcal{J}_0 = \mathcal{J}_0(\Delta X, X) := \big[J_i(X)\big]_{i \in I(\Delta X)} \qquad (13)$$
$$(14)$$

and that there exist Lagrange multipliers $\lambda_{LP}^\star$, satisfying the optimality conditions (12) for (7), such that $\text{rank}\mathcal{J}_0 = n$ and strict complementarity holds. Then

i) *$(\Delta X^\star, \lambda_{LP}^\star)$ is a unique Kuhn-Tucker point of the linear l_1 problem (7), and $\Delta X^\star$ is a strict minimum.*
ii) *There is a linear operator $\mathcal{J}^+ : R^{m_3} \to R^n$, such that*

$$\Delta X^\star = -\mathcal{J}(X)^+ \mathcal{F}(X).$$

iii) *The solution operator is a generalized inverse, that is, it possesses the property*

$$\mathcal{J}(X)^+ \mathcal{J}(X) \mathcal{J}(X)^+ = \mathcal{J}(X)^+,$$

and up to column permutation is explicitly given by

$$\mathcal{P}\mathcal{J}(X)^+ = [\mathbf{0}, \mathcal{J}_0^{-1}(\mathbf{X})].$$

Here $\mathcal{P}$ is a permutation matrix.

Analyzing further the optimality conditions, we may conclude that a solution X^* of the nonlinear l_1 problem (5) is characterized by an optimal "active set" which contains an information about active inequality constraints and zero components of the cost function. The following result shows that the Gauss-Newton method (6), (7) eventually identifies the optimal "active set" of the nonlinear l_1 problem (5).

Theorem 4. *Suppose that X^* is a solution of the problem (5) that satisfies constraint qualification (8) and $n = m_0$, that is exactly n components of the cost function and constraints in the nonlinear l_1 problem are zero. Assume further that strict complementarity holds. Then if (X^k, λ^k) is sufficiently close to (X^*, λ^*) the linear l_1 problem (7) has a unique solution $\Delta X^k, \lambda_{LP}^k$ whose active set $I(\Delta X^k)$ is the same as the active set $I(X^*)$ of the nonlinear problem (5) at X^*.*

Summing up, the iterations of the Gauss-Newton method (6), (7) are attracted by one of the local minima satisfying the regularity assumptions mentioned above. Neither the active sets in linear problems (7), constructed at X^k, will change, nor non-zero components of the cost function and the constraints will change their signs as the iterates X^k generated by the generalized Gauss-Newton method (6), (7) approach such a minimizer X^*. Moreover, these sets coincide with the corresponding optimal sets at X^*. Hence, the full-step ($t^k \equiv 1$) method becomes equivalent to the Newton method applied for solving a system of nonlinear equations

$$\mathcal{F}_{0,*}(X) = 0, \quad \mathcal{F}_{0,*}(X) := \big[F_i(X)\big]_{i \in I(X^*)} \tag{15}$$

thus obtaining quadratic rate of local convergence.

The equations (15) include the equality constraints, the active inequality constraints and the zero components of the cost function at the solution, so that we can ignore non-active constraints and non-zero components of the cost function. Consequently, the solution X^* interpolates $n - m_0$ "best" measurements which means that X^* is *less sensitive to outliers*.

Choosing the step length t^k by means of classical line search methods based on the exact penalty function

$$T_1(X) := \sum_{i \in I_1} |F_i(X)| + \sum_{i \in I_2} \alpha_i |F_i(X)| + \sum_{i \in I_3} \alpha_i \max\{0, -F_i(X)\}$$

with sufficiently large weights $\alpha_i > 0$, $i = 1, ..., m_2$, ensures global convergence. However, it is well known that already in mildly ill-conditioned problems such a step size strategy may be very inefficient since it may produce very small step sizes. Therefore we use the "restrictive monotonicity test", see [9], [11], that has proved to be very effective in practical applications. We interpret the Gauss-Newton method for l_1 parameter estimation problem as Newton's method applied to the nonlinear system of the active constraints and the zero components of the cost function (15) and use the restrictive monotonicity test for the "natural level function"

$$T_k(X^k + t\Delta X^k) = ||A\mathcal{F}_{0,k}(X^k + t\Delta X^k)||_2, \quad A = \mathcal{J}_{0,k}^{-1}(X^k).$$

The idea of the test is that at X^k we consider the relaxed system of nonlinear equations

$$\mathcal{F}_{0,k}(X) - (1-t)\mathcal{F}_{0,k}(X^k) = 0, \ 0 \le t \le 1, \tag{16}$$

and choose the maximal step size $t \in]0,1]$ for which the iterates of the simplified Newton method applied to (16) contract. The test may be also interpreted as following the Newton path. Note, however, that the active inequality constraints incorporated in $\mathcal{J}_{0,k}(X^k)$ as well as the zero components of the cost function are changing along the solution of the problem, so that the step-size strategy is combined with an additional monitoring of the changing active sets. The restrictive monotonicity test has shown very good performance in practice, for the theoretical justification of the test we refer the reader to [11].

2.4 "Statistically" Stable and Unstable Solutions

Now let us discuss some of the "disadvantages" of the Gauss-Newton method. We saw that the generalized Gauss-Newton method (6), (7) can be interpreted as a Sequential Linear Programming Method (SLP) since we ignore the information about second-order derivatives. As an SLP method the generalized Gauss-Newton method for (5) has several advantages. It does not use second order derivative information and the local linearized problems are linear. Under certain regularity assumptions about the solution the method shows quadratic rate of local convergence. Moreover, these solutions are strict local minima and according to the implicit function theorem applied to the optimality conditions (9) such a solution depends continuously and differentiable on (small) perturbations of the minimization problem, e.g. small perturbations of measurements. There are problems, however, for which the Gauss-Newton method may have rather slow local convergence rate or may even fail. The reason is that the linearized model (7), which forms the basis of the Gauss-Newton method, is an inadequate local representation of such nonlinear problems, since the second-order information cannot be ignored. Similar situations appear in applying the l_2 based Gauss-Newton method for parameter estimation problems with large residuals, see e.g. [9], for a thorough analysis (including constrained l_2 parameter estimation problems) why the Gauss-Newton method is appropriate for parameter estimation problem while quasi-Newton methods are not. In particular, it is shown there that if the full-step l_2 based Gauss-Newton method is not attracted by a minimum then this is a large residual minimum which does not satisfy a criterion for certain statistical stability. Using SQP-type methods for the nonlinear constrained l_1 optimization problem one could force convergence to such a solution even in such a case, see e.g. [33, 24] for unconstrained l_1 problems and [32] for constrained l_1 problems. However, such solutions are undesirable as parameter estimates in a certain sense. Recall, that we are interested not just in the solution of an optimization problem, but rather in getting reliable, trustworthy estimates for the parameters. We want to get estimates that are stable against some perturbations in measurement errors.

We will show that a solution that does not satisfy assumptions which guarantee uniqueness of primal and dual solutions in the linearized problems in its neighbourhood, even if it is a strict minimum of the nonlinear constrained l_1 problem (5), cannot be expected to be a continuous deformation of the "true" parameter values under perturbations caused by the measurement errors.

Theorem 5. *Let X^* be a local minimizer of the equality constrained l_1 nonlinear problem*

$$\min_{X \in \mathbb{R}^n} \sum_{i \in I_1} |F_i(X)|, \tag{17}$$
$$s.t. \ F_i(X, \tau) = 0, i \in I_2,$$

satisfying the following assumptions: the constraint qualification (8), $|I(X^*)| = n$ and the strict complementarity (11) of the corresponding λ^*. Then X^* is a strict local minimizer for any τ in problems

$$\min_{X \in \mathbb{R}^n} \sum_{i \in I_1} |F_i(X, \tau)|, \qquad (18)$$

$$\text{s.t. } F_i(X, \tau) = 0, i \in I_2,$$

where $F_i(X, \tau) = F_i(X) + (\tau - 1)F_i(X^*)$, $i \in I_1 \cup I_2$, $-1 \leq \tau \leq 1$, $\tau \neq 0$.

Proof. Under the assumptions of the theorem there is a unique vector of Lagrange multipliers λ^* such that

$$\sum_{i \in I_1 \cup I_2} \lambda_i^* J_i(X^*) = 0, \ |\lambda_i^*| \leq 1, \ \lambda_i^* = -\text{sign}(F_i(X^*)), \ i \in I_1 \setminus I_1(X^*). \quad (19)$$

Moreover, X^* is a strict local minimizer since $P(X^*, \lambda^*) = \emptyset$. Consider now problem (18). Obviously,

$$J_i(X, \tau) = J_i(X), \quad F_i(X^*, \tau) = \tau F_i(X^*), \ i \in I_1 \cup I_2,$$
$$F_i(X^*, \tau) = 0, i \in I(X^*), \forall \tau, \tau \neq 0,$$
$$\text{sign}(F_i(X^*, \tau)) = \text{sign}(F_i(X^*)), \ \tau > 0,$$
$$\text{sign}(F_i(X^*, \tau)) = -\text{sign}(F_i(X^*)), \ \tau < 0, \ i \in I_1 \setminus I_1(X^*).$$

Then the pair $(X^*, \lambda^*(\tau))$ satisfies the first-order necessary conditions (19) if we choose

$$\lambda^*(\tau) = \lambda^*, \text{ for } \tau > 0, \quad \lambda^*(\tau) = -\lambda^*, \text{ for } \tau < 0, \qquad (20)$$

Further, $P(X^*, \lambda^*, \tau) = P(X^*, \lambda^*) = \emptyset$, and, consequently, X^* is a strict local minimizer in problem (18) for any $\tau \neq 0$.

◇

Theorem 6. Let X^* be a regular strict local minimizer of the equality constrained l_1 problem (17). Assume further that $P(X^*, \lambda^*) \neq \emptyset$. Then

- X^* is a stationary point for any τ in problem (18);
- X^* is not a minimizer for any $\tau < 0$ in problem (18).

Proof. It follows from the proof of the previous theorem that $(X^*, \lambda^*(\tau))$ satisfies optimality conditions (19) with the special choice of the Lagrange multipliers (20). The Hessian $H(\tau)$ of the Lagrangian in problem (18) at the point $(X^*, \lambda^*(\tau))$ is given by

$$H(\tau) = H(X^*, \lambda^*(\tau)) = -\sum_{i \in I_1 \cup I_2} \lambda_i^*(\tau) \frac{\partial^2 F_i(X^*, \tau)}{\partial X^2}$$

$$= \begin{cases} H(X^*, \lambda^*) & \text{if } \tau > 0; \\ -H(X^*, \lambda^*) & \text{if } \tau < 0. \end{cases}$$

◇

What can we conclude from these theorems for parameter estimation problems? Consider just for a moment an unconstrained l_1 parameter estimation problem in the form

$$\min_p \sum_{i\in\mathcal{I}} |f_i(p)| := \sum_i |Q_i(p) - \eta_i| := |Q_i(p) - (Q_i(p^{true}) + \varepsilon_i)|, \quad (21)$$

where $Q_i(p)$ denotes a model response at some point i, p denotes unknown parameters, η_i denotes a measurement, which is assumed to be equal to the model response for the true values of the parameters subject to a measurement error ε_i, $i \in \mathcal{I}$. Let the strict local minimum $p^*(\varepsilon)$, $\varepsilon = (\varepsilon_i, i \in \mathcal{I})$, of problem (21) be a good estimate of the true parameter values $p^{true} = p^*(0) : p^*(\varepsilon) \approx p^{true}$. Then

$$f_i(p^*(\varepsilon)) = Q_i(p^*(\varepsilon)) - Q_i(p^{true}) - \varepsilon_i \approx -\varepsilon_i.$$

Assume further that $p^*(\varepsilon)$ is a solution which satisfies the conditions of Theorem 6, and consider the functions $\bar{f}_i(p, \tau) := f_i(p) + (\tau - 1)f_i(p^*(\varepsilon))$, $i \in \mathcal{I}$. Then for $\tau = 1$ we have the old problem (21) with $f_i(p) = Q_i(p) - \eta_i$, the measurements $\eta_i = Q_i(p^{true}) + \varepsilon_i$ and the measurement errors ε_i, $i \in \mathcal{I}$. If $\tau = -1$ we have new problem (21) with $\tilde{f}_i(p) = Q_i(p) - \tilde{\eta}_i$, the measurements $\tilde{\eta}_i = Q_i(p^*(\varepsilon)) + f_i(p^*(\varepsilon)) \approx Q_i(p^{true}) - \varepsilon_i$ and the measurement errors $-\varepsilon_i$, $i \in \mathcal{I}$, which means that we just inverted the signs of the measurement errors. According to Theorem 6, $p^*(-\varepsilon) \approx p^{true}$ is still a stationary point in the new parameter estimation problem, but not a solution anymore, although the measurement errors did not change their statistical properties. Thus, we can estimate the true values of parameters p^{true} for the errors ε, and cannot for $-\varepsilon$. We have a bifurcation: $p^*(-\varepsilon)$ jumps from the saddle point p^{true} to another point! This means that the point $p^*(-\varepsilon)$ is not a continuously differentiable estimate of $p^{true} = p^*(0)$.

Similarly to l_2 case [9] we have shown, that all minima requiring second-order information and which, for this reason, are not attractive for the Gauss-Newton method, are statistically unstable: they are not estimates, i.e. they cannot be considered to be continuous deformations of the true parameter values due to perturbations by measurement errors. From this point of view quasi-Newton methods have only limited use for parameter estimation since they do not deliver estimates for the true parameter values! Thus, slow local convergence or no convergence of the full-step Gauss-Newton method indicates deficiencies in the model or lack of data and can be considered as an *advantage* of the method.

2.5 Solution of Linear l_1 Problem

In this section we describe a method of solving a linear constrained l_1 problem (7), which has to be solved at each iteration of the Gauss-Newton method.

Condensing

The matrix $\mathcal{J}(X)$ of the l_1 problem under consideration shows the typical block structure of the BVP discretization which is induced by multiple shooting:

$$\mathcal{J} = \begin{bmatrix} D_0 & D_1 & \cdots & & D_m & D^p \\ G_0^l & G_0^r & & & & G_0^p \\ & \ddots & \ddots & & 0 & \vdots \\ & & \ddots & \ddots & & \vdots \\ & 0 & & \ddots & \ddots & \vdots \\ & & & G_{m-1}^l & G_{m-1}^r & G_{m-1}^p \end{bmatrix}.$$

Every block column corresponds to the derivatives with respect to the discretization variables and parameters in one subinterval. The block rows with G-matrices are the derivatives of the continuity conditions

$$G_j^l := \partial h_j / \partial s_j \quad G_j^r := \partial h_j / \partial s_{j+1} = -[\mathcal{I}, 0], \quad G_j^p := \partial h_j / \partial p.$$

Here $\mathcal{I}$ denotes an identity matrix. The block rows with D-matrices correspond to the derivatives of the functions F_i of the cost functional and the constraints of the nonlinear problem (5) excluding the continuity conditions.

We use a fast, stable and efficient structure exploiting decomposition of the matrix consisting of continuity and consistency conditions (see [8], [36]) to reduce a large linear l_1 problem to a linear constrained l_1 problem with smaller dimension. The number of variables in the resulting problem is in general case the number of parameters plus the number of differential variables. The number of constraints in the reduced problem is equal to the number of equality and inequality constraints in the original parameter estimation problem (3) $m_{r_2} + m_{r_3}$.

Solution of the condensed problem

In general the condensed problem is of the form

$$\min_{Y \in \mathbb{R}^{n_Y}} f_{l_1} = \sum_{i \in M_1} |A_i Y + c_i|, \tag{22}$$

$$\text{s.t.} \quad A_i Y + c_i = 0, \ i \in M_2, \ A_i Y + c_i \geq 0, \ i \in M_3,$$

where A_i denotes an n_Y-row-vector, $M_1 = \{1, ..., m_1\}$, $M_2 = \{m_1+1, ..., \bar{m}_2\}$, $M_3 = \{\bar{m}_2 + 1, ..., \bar{m}_3\}$, $\bar{m}_2 = m_1 + m_{r_2}$, $\bar{m}_3 = \bar{m}_2 + m_{r_3}$. In the problem under consideration (22) the number of variables $n_Y = n_y + n_p$ is moderate, but the number m_1 of the linear components of the cost function (=number of measurements in parameter estimation problem) may be rather large and the problem maybe dense.

Problem (22) is equivalent to the problem of linear programming with additional m_1 variables and $2 \times m_1$ inequality constraints

$$\min_{Y \in \mathbb{R}^{n_Y}, \xi \in \mathbb{R}^{m_1}} f_{LP}(Y) = \sum_{i \in M_1} \xi_i, \tag{23}$$

$$\text{s.t. } \xi_i - A_i Y \geq c_i, \ \xi_i + A_i Y \geq -c_i, \ \xi_i \geq 0, \ i \in M_1,$$
$$A_i Y + c_i = 0, \ i \in M_2, \ A_i Y + c_i \geq 0, \ i \in M_3,$$

and can in principle be solved by the simplex method of linear programming. However, it is more effective to solve the problem by the dual simplex method. Indeed, the dual problem for (23) and hence for (22) is an "ordinary" bounded-variable problem of linear programming

$$\min_{\lambda \in R^{M_1 + M_2 + M_3}} \varphi(\lambda) = c^T \lambda, \tag{24}$$

$$\text{s.t. } \sum_{i \in M_1 \cup M_2 \cup M_3} \lambda_i A_i^T = 0, \ |\lambda_i| \leq 1, i \in M_1, \lambda_i \geq 0, i \in M_3.$$

This kind of problems can be very effectively solved by the "long-step" dual simplex method [29], which takes into account the special structure of the inequality constraints in the linear programming problem (23). The details about this method can be found in [30].

It may happen that the matrix A in (22) does not have full rank and then the problem (22) may have an unbounded solution. This situation can be avoided by solving a regularized problem

$$\min f(Y) = \sum_{i \in M_1} |A_i Y + c_i|, \tag{25}$$

$$\text{s.t. } A_i Y + c_i = 0, \ i \in M_2, \ A_i Y + c_i \geq 0, \ i \in M_3,$$
$$L \leq Y \leq U,$$

which includes lower L and upper U bounds on Y. For the solution of the problem (25) a modification of the "adaptive" method [20] can be effectively applied.

2.6 Inexact l_1 Gauss-Newton Method

In each iteration of a generalized Gauss-Newton method (6), (7) we need to evaluate exact Jacobians, which can be computationally too expensive. In this section we present an inexact l_1 Gauss-Newton method, which does not need exact Jacobians. Instead, in each iteration of this method we need only to evaluate residuals and gradients of the Lagrange function, that can be effectively performed by automatic differentiation.

In each iteration of the inexact Gauss-Newton method (6) the increment ΔX^k solves the following problem at $X = X^k$, $\lambda = \lambda^k$

$$\min_{\Delta X \in \mathbb{R}^n} \sum_{i \in I_1} |F_i(X) + \mathcal{A}_i \Delta X| - \sum_{i \in I} \lambda_i (\mathcal{A}_i - \mathcal{J}_i(X)) \Delta X, \qquad (26)$$
$$\text{s.t.} \quad F_i(X) + \mathcal{A}_i \Delta X = 0, i \in I_2, \ F_i(X) + \mathcal{A}_i \Delta X \geq 0, i \in I_3.$$

Here the matrix $\mathcal{A} = \big[\mathcal{A}_i\big]_{i \in I}$ is an approximation of the Jacobian $\mathcal{J}(X)$; λ^k is a Lagrange vector from the previous iteration.

The following result shows that the inexact Gauss-Newton method (6), (26) eventually identifies the correct optimal "active set" of the nonlinear l_1 problem (5).

Theorem 7. *Suppose that X^* is a solution of the problem (5) that satisfies the constraint qualification (8) and $n = m_0$, that is exactly n components of the cost function and constraints in the nonlinear l_1 problem are zero. Assume further that strict complementarity holds. Then if (X^k, λ^k) is sufficiently close to (X^*, λ^*) the linear l_1-problem (26) has a unique solution $\Delta X^k, \lambda^k_{LP}$ for any fixed matrix $\mathcal{A}$ with*

$$\text{rank} \ \mathcal{A}_0 = |I(X^*)|, \mathcal{A}_0 = \big[\mathcal{A}_i\big]_{i \in I(X^*)}. \qquad (27)$$

Moreover, the active set $I(\Delta X^k)$ in the problem (26) is the same as the active set $I(X^*)$ of the nonlinear problem (5) at X^*.

Since the active set does not change near a solution satisfying regularity assumptions, the full-step ($t^k \equiv 1$) inexact Gauss-Newton method (6), (26) method becomes equivalent to a Newton-type method applied for solving a system of nonlinear equations (15). This allows us to establish conditions for local convergence of inexact Gauss-Newton method (6), (26), see [9].

Theorem 8. *Let D be the domain defined by Theorem 7 and the matrix $\mathcal{A}$ satisfying (27) is given. Assume further that*

$$\|\mathcal{A}_0^{-1}\big(\mathcal{J}_0(X + t\Delta X) + \mathcal{J}_0(X)\big)\Delta X\| \leq \omega t \|\Delta X\|^2, \omega < \infty, \qquad (28)$$
$$\|\mathcal{A}_0^{-1}\big(\mathcal{F}_0(X) + \mathcal{J}_0(X)\Delta X\big)\| \leq \kappa \|\Delta X\|, \kappa < 1, \qquad (29)$$

for all $X, X + \Delta X \in D$ with $\Delta X = -\mathcal{A}_0^{-1}\mathcal{F}_0(X) \neq 0$. Assume further that $\delta^0 := \frac{\omega}{2}\|\Delta X^0\| < 1$. Then the following holds:

- *if $D^0 := B\big(X^0, \|\Delta X^0\|/(1-\delta^0)\big) \subset D$, then the sequence of iterates defined by (6), (26) with $t^k = 1$ starting at X^0 remains in D^0,*
- *there exists $X^\star \in D^0$ with $\mathcal{A}_0^{-1}\mathcal{F}_0(X^\star) = 0$ and $X^k \to X^\star$ $(k \to \infty)$, and convergence is linear with*

$$\|\Delta X^{k+1}\| \leq \left(\frac{\omega}{2}\|\Delta X^k\| + \kappa\right) \|\Delta X^k\|.$$

Global convergence of the inexact Gauss-Newton method (6), (26) can be ensured by means of classical line search methods under the assumption that the matrix $\mathcal{A}$ is a sufficiently good approximation of the Jacobian.

2.7 Numerical Results: Biochemical Problem

Here we consider a biochemical problem from enzyme kinetics. The model involves four chemical species, one feed stream and a temperature control. The kinetics is described by Arrhenius mass action kinetics. Eight unknown parameters have to be estimated by only one indirect measurement function. Details about the model can be found in [14] and [12].

This real-life problem turned out to be very difficult since information from one experiment is insufficient for the identification of all parameters, and since the measurement error is large. To improve the information we first have applied new nonlinear optimum design methods (see Section 3) to design 5 supplementary temperature profiles for identifiability. Using the designed temperature profiles the measurements have been simulated without outliers (A), with outliers (B), with large outliers (C). Table 1 presents the "true" and the estimated values of the parameters using l_2 and l_1 based parameter estimation respectively. The number of iterations necessary for solving the estimation problems are given in Table 2. We may conclude that the l_1 based parameter estimation showed better performance than the traditional l_2 approach.

Table 1. Biochemical reaction: true and estimated parameter values

true values	20.31	32.19	0.591	208.9	-2.52	-24.06	-4.54	-12.91
A / l_2	20.29	32.14	0.587	207.5	-2.53	-23.89	-4.56	-12.88
A / l_1	20.28	32.13	0.585	206.9	-2.52	-23.85	-4.56	-12.88
B / l_2	22.08	36.92	0.588	207.3	-2.19	*-28.51*	-4.67	-11.52
B / l_1	20.17	31.82	0.587	207.8	-2.58	-22.81	-4.60	-12.73
C / l_2	20.11	31.69	*1.149*	*408.3*	*-4.95*	*4.63*	-6.07	-10.67
C / l_1	20.12	31.67	0.528	186.9	-2.51	-22.97	-4.75	-12.67

Table 2. Biochemical reaction: l_2 vs l_1

	l_2	l_1
A	2	2
B	6	4
C	7	3

3 Design of Robust Optimal Experiments

3.1 Covariance Matrix

It is important for parameter estimation problems to compute not only parameters but also a statistical assessment of the results. This can be done by means of the covariance matrix. A representation of the covariance matrix for *unconstrained* nonlinear parameter estimation problems is well-known, see e.g. [6]. In the following, this notion is generalized to *constrained* parameter estimation problems, see [9, 13].

In this section we consider the nonlinear constrained l_2 parameter estimation problem in the form

$$\min_{X \in \mathbb{R}^n} \frac{1}{2} \sum_{i \in I_1} F_i^2(X), \text{ s.t. } F_i(X) = 0, \ i \in I_2. \tag{30}$$

Here, as before $I_1 = \{1, ..., m_1\}$, $I_2 = \{m_1+1, ..., m_2\}$, the equality constraints $F_i(X) = 0, i \in I_2$, represent implicitly a dynamic model, e.g. a discretized boundary value problem, $F_i(X), i \in I_1$ present measurement errors.

In each iteration of the generalized Gauss-Newton method (6) applied to (30), we solve the linearized problem (7), which now reads

$$\min_{\Delta X \in \mathbb{R}^n} \frac{1}{2} \sum_{i \in I_1} (F_i(X) + J_i(X)\Delta X)^2, \tag{31}$$

$$\text{s.t. } F_i(X) + J_i(X)\Delta X = 0, i \in I_2.$$

to compute the increment ΔX^k.

If the Jacobians $\mathcal{J}_1(X) = \left[J_i(X)\right]_{i \in I_1}$ and $\mathcal{J}_2(X) = \left[J_i(X)\right]_{i \in I_2}$ satisfy two regularity assumptions on D

$$\text{rank } \mathcal{J}_2(X) = m_2 - m_1, \text{ rank } \mathcal{J}(X) = n, \tag{32}$$

then the linearized problem (31) has a unique solution ΔX^k and a unique Lagrange vector λ^k satisfying the following optimality conditions

$$\begin{aligned}\mathcal{J}_1^T(X)\mathcal{J}_1(X)\Delta X^k + \mathcal{J}_2^T(X)\lambda^k &= -\mathcal{J}_1^T(X)\mathcal{F}_1(X),\\ \mathcal{J}_2(X)\Delta X^k &= -\mathcal{F}_2(X).\end{aligned} \tag{33}$$

Here, $\mathcal{F}_1(X) = \left[F_i(X)\right]_{i \in I_1}$ and $\mathcal{F}_2(X) = \left[F_i(X)\right]_{i \in I_2}$. Using (33) one can easily show that under the regularity conditions (32) ΔX^k can be formally written with the help of a solution operator $\mathcal{J}^+$

$$\Delta X^k = -\mathcal{J}^+(X^k)\mathcal{F}(X^k).$$

The solution operator $\mathcal{J}^+$ is a generalized inverse, that is, it satisfies the defining conditions

$$\mathcal{J}^+\mathcal{J}\mathcal{J}^+ = \mathcal{J}^+,$$

and is explicitly given by

$$\mathcal{J}^+(X) = (\mathcal{I}\ 0) \begin{pmatrix} \mathcal{J}_1^T(X)\mathcal{J}_1(X) & \mathcal{J}_2(X)^T \\ \mathcal{J}_2(X) & 0 \end{pmatrix}^{-1} \begin{pmatrix} \mathcal{J}_1(X)^T & 0 \\ 0 & \mathcal{I} \end{pmatrix}. \tag{34}$$

Let $\mathcal{J}$ be the Jacobian at the solution X^* and $\mathcal{J}^+$ be the corresponding generalized inverse defined according to (34). Due to the statistical errors of the data as input of the parameter estimation problem, the estimate as the result of the solution procedure is a random variable. If the measurement errors are normally distributed with zero mean and variances σ^2, then up to the first order the estimated solution is normally distributed with expected value X^{true} and variances

$$\mathcal{E}\Big((X^* - X^{true})(X^* - X^{true})^T\Big). \tag{35}$$

We may approximate a variance-covariance matrix by the following matrix which we will call in the following, for the sake of brevity, a *covariance matrix*

$$C = \mathcal{J}^+ \begin{pmatrix} \mathcal{I}_{m_1 \times m_1} & 0 \\ 0 & 0_{m_2-m_1 \times m_2-m_1} \end{pmatrix} \mathcal{J}^{+T}. \tag{36}$$

Obviously, the matrix C is positive semi-definite matrix with $\mathrm{rank} C = \bar{m} := n - (m_2 - m_1)$.

Now let us show how to compute confidence regions for all variables in constrained parameter estimation problems. Generalizing the unconstrained case, we define a nonlinear confidence region for the solution X^* of the nonlinear *constrained* parameter estimation problem (30) by

$$G_N(\alpha) := \{X \mid \mathcal{F}_2(X) = 0, \|\mathcal{F}_1(X)\|_2^2 - \|\mathcal{F}_1(X^*)\|_2^2 \leq \gamma^2(\alpha)\}$$

where $\gamma^2(\alpha) := \chi^2_{\bar{m}}(1-\alpha)$ is the quantile of the χ^2 distribution for a value $\alpha \in [0,1]$ with $\bar{m}$ degrees of freedom. The nonlinear confidence region $G_N(\alpha)$ can be approximated by a linearized confidence region

$$G_L(\alpha) := \{X \mid \mathcal{F}_2(X^*) + \mathcal{J}_2(X^*)(X - X^*) = 0,$$
$$\|\mathcal{F}_1(X^*) + \mathcal{J}_1(X^*)(X - X^*)\|_2^2 - \|\mathcal{F}_1(X^*)\|_2^2 \leq \gamma^2(\alpha)\}.$$

The following lemma gives another, more illustrative, representation of the linear confidence region.

Lemma 1. *Let X^* be a solution of problem (30) satisfying the regularity assumptions (32). Then*

$$G_L(\alpha) = \bar{G}_L(\alpha) := \{X^* + \Delta X \mid \Delta X = -\mathcal{J}^+(X^*)\begin{pmatrix}\eta\\0\end{pmatrix}, \|\eta\|_2^2 \leq \gamma^2(\alpha)\} \tag{37}$$

The next result shows that the linearized confidence region $G_L(\alpha)$ is contained *exactly* in a minimal box which is the cross product of so-called confidence intervals.

Lemma 2. *Let X^* be a solution of parameter estimation problem (30) satisfying the regularity assumptions (32). Then*

$$G_L(\alpha) \subset \bigtimes_{i=1}^{n} [X_i^* - \theta_i, X_i^* + \theta_i],$$

where $\theta_i = C_{ii}\gamma(\alpha)$. Here C_{ii}^2 denotes the diagonal elements of the covariance matrix C. Further, the following exact bounds hold

$$\max_{X \in G_L(\alpha)} |X_i - X_i^*| = \theta_i, i = 1, ..., n.$$

The Lemma 2 shows that the diagonal elements of the covariance matrix play an important role in the statistical assessment of the estimates, namely they are the basis for joint confidence intervals. The proofs can be found in [9, 13].

Covariance matrix as a solution of a linear system

In this section we will show that the covariance matrix $C \in \mathbb{R}^{n \times n}$ for the constrained parameter estimation problem

$$C = \mathcal{J}^+ \begin{bmatrix} \mathcal{I} & 0 \\ 0 & 0 \end{bmatrix} (\mathcal{J}^+)^T \tag{38}$$

satisfies a linear system of equations. In this section for simplicity of notations we omit the dependence of C on the linearization point x. Throughout the section we assume that the matrices $\mathcal{J}_1$ and $\mathcal{J}_2$ satisfy the regularity assumption (32). Let us denote

$$\begin{pmatrix} \mathcal{J}_1^T \mathcal{J}_1 & \mathcal{J}_2^T \\ \mathcal{J}_2 & 0 \end{pmatrix}^{-1} := \begin{pmatrix} \mathcal{X} & \mathcal{Y} \\ \mathcal{Z} & \mathcal{T} \end{pmatrix} := \mathcal{M}^{-1}, \mathcal{X} \in \mathbb{R}^{n \times n}, \mathcal{Y} \in \mathbb{R}^{n \times (m_2 - m_1)},$$

$$\mathcal{Z} = \mathcal{Y}^T \in \mathbb{R}^{(m_2 - m_1) \times n}, \mathcal{T} \in \mathbb{R}^{(m_2 - m_1) \times (m_2 - m_1)}.$$

Lemma 3. *The covariance matrix C (38) is equal to the matrix $\mathcal{X}$ and satisfies the following linear equation system with respect to variables $C \in \mathbb{R}^{n \times n}$ and $\mathcal{Z} \in \mathbb{R}^{(m_2 - m_1) \times n}$*

$$\begin{aligned} \mathcal{J}_1^T \mathcal{J}_1 C + \mathcal{J}_2^T \mathcal{Z} &= \mathcal{I}, \\ \mathcal{J}_2 C &= 0. \end{aligned} \tag{39}$$

Proof. According (34) and (38) we have

$$C = (\mathcal{I}\ 0)\,\mathcal{M}^{-1} \begin{pmatrix} \mathcal{J}_1^T \mathcal{J}_1 & 0 \\ 0 & 0 \end{pmatrix} \mathcal{M}^{-1} \begin{pmatrix} \mathcal{I} \\ 0 \end{pmatrix}$$

$$= (\mathcal{X}\ \mathcal{Y}) \begin{pmatrix} \mathcal{J}_1^T \mathcal{J}_1 & 0 \\ 0 & 0 \end{pmatrix} \begin{pmatrix} \mathcal{X} \\ \mathcal{Z} \end{pmatrix} = \mathcal{X} \mathcal{J}_1^T \mathcal{J}_1 \mathcal{X}. \tag{40}$$

Since the blocks $\mathcal{X}$ and $\mathcal{Z}$ of the matrix $\mathcal{M}^{-1}$ satisfy the linear system

$$\begin{aligned} \mathcal{J}_1^T \mathcal{J}_1 \mathcal{X} + \mathcal{J}_2^T \mathcal{Z} &= \mathcal{I}, \\ \mathcal{J}_2 \mathcal{X} &= 0, \end{aligned} \tag{41}$$

relation (40) yields

$$C = \mathcal{X}(\mathcal{I} - \mathcal{J}_2^T \mathcal{Z}) = \mathcal{X}$$

which means that $C = \mathcal{X}$. ◇

Note, that according to Lemma 3 the covariance matrix C is a generalized inverse of the matrix $\mathcal{J}_1^T \mathcal{J}_1$, that is it satisfies $C(\mathcal{J}_1^T \mathcal{J}_1)C = C$. The covariance matrix can be computed by direct linear algebra methods. The following lemma shows that the columns of the covariance matrix solves special quadratic problems. This fact is used for derivation of a numerical procedure to compute the covariance matrix based on a conjugate gradient method for linearized parameter estimation problem (31), for details see [13].

Lemma 4. *The column $\mathcal{X}^{(i)}$ of the matrix $\mathcal{X}$ solves the following quadratic problem*

$$\min_{\kappa} f_i(\kappa) = \frac{1}{2}\kappa^T \mathcal{J}_1^T \mathcal{J}_1 \kappa - e_i^T \kappa, \tag{42}$$

$$\text{s.t.} \quad \mathcal{J}_2 \kappa = 0,$$

where the column $\mathcal{Z}^{(i)}$ of the matrix $\mathcal{Z}$ is the optimal Lagrange vector of the problem (42). The optimal value of the cost function in problem (42) is equal to

$$f_i(\mathcal{X}^{(i)}) = -\frac{1}{2}\mathcal{X}_i^{(i)}.$$

Here e_i denotes i-th unit vector and $\mathcal{X}_i^{(i)}$ denotes i-th component of the vector $\mathcal{X}^{(i)}$.

The following corollary of Lemma 4 shows how to compute the trace of the covariance matrix in the terms of the optimal values of the problems (42). The trace of the covariance matrix can be used as the cost functional for design of optimal experiments as we will see in what follows.

Corollary 2.

$$tr C = \sum_{i=1}^{n} e_i^T \mathcal{X}^{(i)} = -2 \sum_{i=1}^{n} f_i(\mathcal{X}^{(i)}). \tag{43}$$

Using the representation of the covariance matrix as a solution of the linear system we may derive the derivatives of the covariance matrix C and the matrix $\mathcal{Z}$, as the functions of the matrices $\mathcal{J}_1$ and $\mathcal{J}_2$

$$C = C(\mathcal{J}_1, \mathcal{J}_2) = \mathcal{X}(\mathcal{J}_1, \mathcal{J}_2),$$
$$\mathcal{Z} = \mathcal{Z}(\mathcal{J}_1, \mathcal{J}_2).$$

These derivatives are needed in numerical methods for design of optimal nonlinear experiments. Let $\mathcal{J}_1(t) = \mathcal{J}_1 + t\Delta\mathcal{J}_1$ and $\mathcal{J}_2(\mu) = \mathcal{J}_2 + \mu\Delta\mathcal{J}_2$, and compute the partial derivatives

$$\frac{\partial \mathcal{X}(\mathcal{J}_1, \mathcal{J}_2)}{\partial \mathcal{J}_1} \Delta\mathcal{J}_1 := \frac{\partial \mathcal{X}(\mathcal{J}_1(t), \mathcal{J}_2(\mu))}{\partial t}\Big|_{t=0,\mu=0} =: L_1,$$

$$\frac{\partial \mathcal{X}(\mathcal{J}_1, \mathcal{J}_2)}{\partial \mathcal{J}_2} \Delta\mathcal{J}_2 := \frac{\partial \mathcal{X}(\mathcal{J}_1(t), \mathcal{J}_2(\mu))}{\partial \mu}\Big|_{t=0,\mu=0} =: L_2,$$

$$\frac{\partial \mathcal{Z}(\mathcal{J}_1, \mathcal{J}_2)}{\partial \mathcal{J}_1} \Delta\mathcal{J}_1 := \frac{\partial \mathcal{Z}(\mathcal{J}_1(t), \mathcal{J}_2(\mu))}{\partial t}\Big|_{t=0,\mu=0} =: R_1,$$

$$\frac{\partial \mathcal{Z}(\mathcal{J}_1, \mathcal{J}_2)}{\partial \mathcal{J}_2} \Delta\mathcal{J}_2 := \frac{\partial \mathcal{Z}(\mathcal{J}_1(t), \mathcal{J}_2(\mu))}{\partial \mu}\Big|_{t=0,\mu=0} =: R_2.$$

Using the linear system (39) with $C = \mathcal{X}$ one may verify that the matrices L_1 and R_1 satisfy

$$L_1 = -\mathcal{X}(\Delta\mathcal{J}_1^T \mathcal{J}_1 + \mathcal{J}_1^T \Delta\mathcal{J}_1)\mathcal{X}, \tag{44}$$
$$R_1 = -\mathcal{Z}(\Delta\mathcal{J}_1^T \mathcal{J}_1 + \mathcal{J}_1^T \Delta\mathcal{J}_1)\mathcal{X}.$$

and the matrices L_2 and R_2 can be computed by

$$L_2 = -\mathcal{X}\Delta\mathcal{J}_2^T \mathcal{Z} - \mathcal{Z}^T \Delta\mathcal{J}_2 \mathcal{X}, \tag{45}$$
$$R_2 = -\mathcal{Z}\Delta\mathcal{J}_2^T \mathcal{Z} - T\Delta\mathcal{J}_2\mathcal{X},$$

where $T = -\mathcal{J}_2^+ \mathcal{J}_1^T \mathcal{J}_1 \mathcal{Z}^T$.

In case of the trace of the covariance matrix, which is one of the possible criteria for the design of the experiments, the computation of the derivatives is significantly simplified.

Lemma 5.

$$\frac{\partial trC(\mathcal{J}_1(t), \mathcal{J}_2)}{\partial t} = -\sum_{i=1}^{n} \mathcal{X}^{(i)T}(\Delta\mathcal{J}_1^T \mathcal{J}_1 + \mathcal{J}_1^T \Delta\mathcal{J}_1)\mathcal{X}^{(i)}, \tag{46}$$

$$\frac{\partial trC(\mathcal{J}_1, \mathcal{J}_2(\mu))}{\partial \mu} = -2\sum_{i=1}^{n} \mathcal{X}^{(i)T}\Delta\mathcal{J}_2\mathcal{Z}^{(i)}, \tag{47}$$

where $\mathcal{X}^{(i)}$ and $\mathcal{Z}^{(i)}$ denote i-th columns of matrices $\mathcal{X}$ and $\mathcal{Z}$ respectively.

3.2 Experimental Design for Nonlinear Dynamic Models

Since the experimental data is randomly distributed, the estimated parameters are also random variables. Data evaluated under different experimental conditions leads to estimations of parameters with good or poor confidence regions depending on the experimental conditions. We would like to find those experiments that result in the best statistical quality for the estimated parameters and at the same time satisfy additional constraints, e.g. experimental costs, safety, feasibility of experiments, validity of the model etc. The aim of optimum experimental design is to construct N_{ex} experiments by choosing appropriate experimental variables $q = (q_1, ..., q_{N_{ex}})$, and experimental controls $u = (u_1, ..., u_{N_{ex}})$ in order to maximize the statistical reliability of the unknown variables under estimation. For this purpose we solve a special optimization problem. Since the "size" of a confidence region is described by the covariance matrix C any suitable function of matrix C may be taken as a cost functional. Typical choices are

$$\Phi(C) = \text{trace}(C),$$
$$\Phi(C) = \lambda_{max}(C), \quad \text{where } \lambda_{max} \text{ denotes the largest eigenvalue of } C$$
$$\Phi(C) = \max_i c_{ii}.$$

The possible constraints in the problem may describe limitations on operability, safety, costs and model validity during the real experiments. Decision variables of the optimization problem are the control profiles $u_i(t)$ (e.g. temperature profiles of cooling/heating, inflow profiles) and the time-independent control variables q_i (e.g. initial concentrations, properties of the experimental device) for all experiments.

The optimum experimental design approach leads to an optimal control problem in dynamic systems, which can be stated in a fairly general form as follows

$$\min_{q,u} \Phi(C(x, p, q, u)), \tag{48}$$

C is the covariance matrix
of the underlying parameter estimation problem,
$x = (x_1, ..., x_{N_{ex}}), q = (q_1, ..., q_{N_{ex}}), u = (u_1, ..., u_{N_{ex}})$
s.t. $c_i(t, x_i(t), p, q_i, u_i(t)) \geq 0, \ t \in T_i = [t^i 0, t^i_f], \ i = 1, 2, ..., N_{ex}$,
$x_i(t) = (y_i(t), z_i(t)), q_i, u_i(t), p, \ t \in T_i$, satisfy model dynamics, e.g.
$\dot{y}_i(t) = f_i(t, y_i(t), z_i(t), p, q_i, u_i(t)), \ i = 1, 2, ..., N_{ex}$,
$0 \ \ = g_i(t, y_i(t), z_i(t), p), t \in T_i$,
and constraints of underlying parameter estimation problem
$r_{2,i}(y_i(\theta^i_1), ..., y_i(\theta^i_{K_i}), p, q_i, u_i) = 0$,
$\theta^i_s \in T_i, s = 1, ..., K_i, \ i = 1, 2, ..., N_{ex}$.

The experimental design optimization problem is a nonlinear constrained optimal control problem. The main difficulty lies in the non-standard objective function which is nonseparable and implicitly defined on the sensitivities of the underlying parameter estimation problem, i.e. on the derivatives of the solution of the differential equation system with respect to the parameters and initial values.

While experimental design for linear models is well established and discussed, e.g. in [1, 19, 34], numerical methods for design of experiments for nonlinear dynamic systems where first developed in SFB 359 "Reactive Flows, Transport and Diffusion", see [3, 4, 5, 26, 28].

The numerical methods are based on the direct approach, according to which the control functions are parameterized on an appropriate grid by local support functions, the solution of the DAE systems and the state constraints are discretized. As a result, we obtain a finite-dimensional constrained nonlinear optimization problem which can be formally written as a general nonlinear programming problem

$$\min_{\xi \in \mathbb{R}^{n_\xi}} \varphi(\xi, s, p) \quad \text{s.t.} \quad \psi_i(\xi, s, p) = 0, i \in \mathcal{E}, \psi_i(\xi, s, p) \leq 0, i \in \mathcal{I}. \quad (49)$$

Here we summarize all experimental design variables, the (parameterized) controls in the n_ξ vector ξ, $s \in \mathbb{R}^{n_s}$ denotes a parameterization of the state variables of the process models, p is an n_p-vector of parameters, the functions $\varphi : \mathbb{R}^{n_\xi + n_p} \to \mathbb{R}$, $\psi_i : \mathbb{R}^{n_\xi + n_p} \to \mathbb{R}$, $i \in \mathcal{E} \cup \mathcal{I}$, are twice-continuously differentiable. Note, that the equality constraints $\psi_i(\xi, s, p) = 0, i \in \mathcal{E}$, contain the discretized boundary value problem. The problem (49) is solved by an SQP method. The main effort for the solution of the optimization problem by the SQP method is spent on the calculation of the values of the objective function and the constraints as well as its gradients. Efficient methods for derivative computations combining internal numerical differentiation [9] of the solution of the DAE and automatic differentiation of the model functions [23] have been developed.

For more detailed discussion of the numerical methods for nonlinear optimum experimental design see [3, 4, 5, 26, 28].

3.3 Design of Robust Optimal Experiments

As we can see the experimental design optimization problem (48) is formulated for the assumed parameter values which are, however, only known to lie in a possibly large confidence region. In this section we discuss how to construct robust experiments that is experiments that are less sensitive to parameter uncertainty.

In [27], a sequential strategy is suggested which alternately designs experiments, collects experimental data and estimates the parameters. This approach allows to simultaneously improve the values and the confidence intervals of the parameters by maximizing the gain of information in every loop of

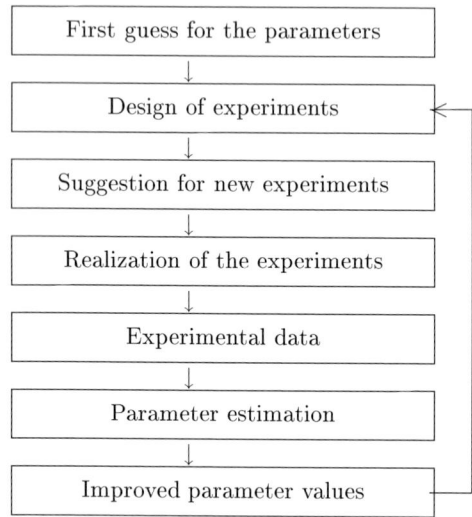

Fig. 2. Sequential approach to experimental design

the procedure. Case studies show that using this approach, we can get much more reliable estimates with a drastically less experimental effort compared to intuitively suggested experiments from an experienced chemical experimenter. The idea of the sequential approach is schematized in Figure 2. In this paper, our approach is to modify the experimental design optimization problem in order to treat the parameter uncertainty directly.

Robust nonlinear optimization

First, we discuss our approach for a general optimization problem (49). We assume that the parameters are only known to lie in a region defined by a "ball" around the nominal parameter values p_0

$$p \in \mathcal{P} := \{p : ||Q(p - p_0)|| \leq \gamma\},$$

where Q is a given nonsingular matrix. Our aim is to find a solution ξ^* which is robust, i.e. insensitive, to "small" perturbations in p. For this purpose, following one of the classical approaches since middle of the sixties, we may form and solve the worst-case design problem:

$$\begin{aligned}
\min_{\xi} \max_{p \in \mathcal{P}} \quad & \varphi(\xi, s, p), \\
\text{s.t.} \quad & \psi_i(\xi, s, p) = 0, i \in \mathcal{E}, \forall p \in \mathcal{P} \\
& \max_{p \in \mathcal{P}} \psi_i(\xi, s, p) \leq 0, i \in \mathcal{I}.
\end{aligned} \qquad (50)$$

The optimization problem (50) is a semi-infinite programming problem. The solution methods for such problems require the determination of global optima of nonlinear subproblems [35] which may be computationally extremely expensive. In order to compute robust solutions we suggest to approximate the worst-case problem (50) in the following way. First, we assume that we may reduce the problem to an inequality constrained one. For the sake of notation simplicity we assume that the number of constraints is equal to the number of the state variables s, the matrix

$$\frac{\partial \psi_\mathcal{E}(\xi, s_0, p_0)}{\partial s}, \psi_\mathcal{E}(\cdot) = \begin{pmatrix} \psi_i(\cdot) \\ i \in \mathcal{E} \end{pmatrix}$$

is nonsingular at the pair p_0, s_0, satisfying $\psi_\mathcal{E}(\xi, p_0, s_0) = 0$, and there exist a sufficiently smooth function $s = s(p)$, $s(p_0) := s_0$, $p \in \mathcal{P}$, such that $\psi_\mathcal{E}(\xi, s(p), p) \equiv 0, p \in \mathcal{P}$. Denote

$$\mathcal{R} := \frac{\partial \psi_\mathcal{E}(\xi, s_0, p_0)}{\partial s}^{-1} \frac{\partial \psi_\mathcal{E}(\xi, p_0, s_0)}{\partial p}.$$

Then we approximate the problem (50) by

$$\min_\xi \max_{p \in \mathcal{P}} \tilde{\varphi}(\xi, s_0, p_0) := \varphi(\xi, s_0, p_0) + \tag{51}$$

$$\left(-\frac{\partial \varphi(\xi, s_0, p_0)}{\partial s}^T \mathcal{R} + \frac{\partial \varphi(\xi, s_0, p_0)}{\partial p}^T\right)(p - p_0),$$

s.t. $\max_{p \in \mathcal{P}} \tilde{\psi}_i(\xi, s_0, p_0) := \psi_i(\xi, s_0, p_0) +$

$$\left(-\frac{\partial \psi_i(\xi, s_0, p_0)}{\partial s}^T \mathcal{R} + \frac{\partial \psi_i(\xi, s_0, p_0)}{\partial p}^T\right)(p - p_0) \leq 0, i \in \mathcal{I},$$

using Taylor expansions for the functions φ, ψ_i, $i \in \mathcal{I}$, with respect to p. The inner problems in (51) are the maximizations of linear functions subject to convex constraints which can be solved explicitly.

Lemma 6. *Consider a problem of the maximization of a linear function over a "ball"*

$$\max_{y \in \mathbb{R}^m} c^T y, \ s.\ t.\ , ||Qy||_\nu \leq \gamma, det Q \neq 0. \tag{52}$$

In case of the Euclidian norm ($\nu = 2$) the solution y^0 to (52) can be written in the form $y^0 = \frac{\gamma(Q^T Q)^{-1} c}{||Q^{-T} c||_2}$ and the value of the cost function in (52) is equal to $\gamma ||Q^{-T} c||_2$.

In case of the l_∞ norm solution y^0 to (52) can be written in the form

$$y_i^0 = \gamma Q^{-1} e, e = (sign(Q^{-T} c)_i, i = 1, ..., m)^T,$$

with the optimal value $\gamma ||Q^{-T}c||_1$.

In case of the l_1 norm ($\nu = 1$) solution y^0 to (52) can be written in the form

$$y^0 = \gamma Q^{-1}e, e = (e_{i_0} = 1, e_i = 0, i_0 \neq i, \ i = 1, ..., m),$$

$$i_0 = \text{argmax}\{|(Q^{-T}c)_i|, \ i = 1, ..., m\},$$

and the optimal value is equal to $\gamma ||Q^{-T}c||_\infty$.

Using Lemma 6, we may rewrite the approximate worst-case problem (51) as follows

$$\min_\xi \ \varphi(\xi, s_0, p_0) + \gamma ||Q^{-T} \left(-\mathcal{R}^T \frac{\partial \varphi(\xi, s_0, p_0)}{\partial s} + \frac{\partial \varphi(\xi, s_0, p_0)}{\partial p}\right)||_*, \qquad (53)$$

$$\text{s.t.} \quad \psi_i(\xi, p_0) + \gamma ||Q^{-T} \left(-\mathcal{R}^T \frac{\partial \psi_i(\xi, s_0, p_0)}{\partial s} + \frac{\partial \psi_i(\xi, s_0, p_0)}{\partial p}\right)||_* \leq 0, i \in \mathcal{I},$$

where $||\cdot||_*$ denotes a dual norm to $||\cdot||_\nu$. The second term in the cost function and the constraints can be interpreted as a penalty for uncertainty in the parameters.

Applying SQP-type method for solving (53), we need second-order derivative of functions ψ_i. However, in case of Euclidian norm we may compute the necessary derivatives very efficiently. Indeed, in this case we need directional derivatives of the form $\dfrac{\partial^2 \psi_i(\xi, s_0, p_0)}{\partial p \partial \xi} \dfrac{\partial \psi_i(\xi, s_0, p_0)}{\partial p}$ which can be computed by means of automatic differentiation. For computing the sensitivities $\mathcal{R}$ one may again apply methods of automatic differentiation.

Remark 1. Under assumption that the vector p is a vector of random independent normally distributed variables with mean value p_0 and variances $\Sigma = \text{diag}\{\sigma_i^2, i = 1, ..., n_p\}$, one can approximate the worst-case problem (50) with

$$\min_\xi \ \left(\mathcal{E}(\tilde{\varphi}(\xi, s_0, p_0)) + \sqrt{\mathcal{D}(\tilde{\varphi}(\xi, s_0, p_0))}\right) \qquad (54)$$

$$\text{s. t.} \ \mathcal{E}(\tilde{\psi}_i(\xi, s_0, p_0)) + \sqrt{\mathcal{D}(\tilde{\psi}_i(\xi, s_0, p_0))} \leq 0, i \in \mathcal{I},$$

where $\mathcal{E}(X)$ and $\mathcal{D}(X)$ denote an expected value and a variance of a random variable X respectively. One can show that the approximative problem (54) is equivalent to

$$\min_\xi \ \varphi(\xi, s_0, p_0) + ||\Sigma^{1/2} \left(-\mathcal{R}^T \frac{\partial \varphi(\xi, s_0, p_0)}{\partial s} + \frac{\partial \varphi(\xi, s_0, p_0)}{\partial p}\right)||_2$$

$$\text{s.t.} \ \psi_i(\xi, s_0, p_0) + ||\Sigma^{1/2} \left(-\mathcal{R}^T \frac{\partial \psi_i(\xi, s_0, p_0)}{\partial s} + \frac{\partial \psi_i(\xi, s_0, p_0)}{\partial p}\right)||_2 \leq 0,$$

$$i \in \mathcal{I}.$$

3.4 Numerical Results

In this section we present numerical results on comparing robust and non-robust designs for enzyme reaction kinetics, see Section 2.7 To illustrate the performance of the method experiments for identification of 4 out of 8 parameters were designed in the sequential mode. For the "true" values of the parameters $p_1^{true} = 27.77$, $p_2^{true} = 50.15$, $p_3^{true} = 0.55$, $p_4 = 185.25$ normally distributed data was simulated using an initial temperature profile consisting only of two phases: heating until some temperature and then keeping this temperature level. The results of parameter estimation based on this data

$$p_1 = 26.0455 \pm 0.371394$$
$$p_2 = 45.6499 \pm 0.973549$$
$$p_3 = 5.58527 \pm 4.4236$$
$$p_4 = 1885.01 \pm 1495.38$$

were used for the computation of a nonrobust and a robust designs. The diagonal matrix with the diagonal elements of the covariance matrix corresponding to the solution of parameter estimation problem was chosen as the matrix Σ in the robust design. Using the new optimized temperature profiles the data was simulated, parameter estimation was repeated with the new data and further new designs were constructed using the new estimates for the parameters. This procedure was repeated until we get trustworthy parameter estimates, see Table 3.

Table 3. Results of parameter estimation for initial design and 2 nonrobust experiments (second column), initial design and 4 nonrobust (third column) and for initial design and 2 robust experiments (fourth column)

"true" values	$p_i \pm c_{ii}$	$p_i \pm c_{ii}$	$p_i \pm c_{ii}$
$p_2 = 50.15$	49.31 ± 0.89	49.37 ± 0.69	49.65 ± 0.87
$p_3 = 0.55$	0.90 ± 0.20	0.69 ± 0.06	0.55 ± 0.04
$p_4 = 185.25$	304.50 ± 69.35	234.54 ± 15.40	186.34 ± 15.26

The sequential design of robust experiments after 2 designs provide us with reliable parameter estimates, while we need 4 nonrobust experiments to get similar quality of the estimates. Although a computational time for computing a robust experiment is three times greater than the time to compute a nonrobust experiment for this application, clearly application of the robust sequential design allows to reduce the number of real experiments and thus to reduce drastically experimental costs and the time necessary to identify the parameters.

4 Conclusions

We have presented newly developed methods for robust parameter estimation and design of optimal robust experiments. The methods have been applied successfully to real-world problems. The methods allow to estimate reliably unknown parameters and to reduce significantly experimental costs. The methods for robust nonlinear optimization presented were applied also for solving optimal control problems, see [18].

References

1. Atkinson, A. C., Donev, A. N.: Optimum Experimental Designs. Oxford University Press (1992)
2. Baake, E., Baake, M., Bock, H. G., Briggs, K.: Fitting ordinary differential equations to chaotic data. Physical Review A, **45** (1992)
3. Bauer, I.: Numerische Verfahren zur Lösung von Anfangswertaufgaben und zur Generierung von ersten und zweiten Ableitungen mit Anwendungen in Chemie und Verfahrenstechnik. Preprint, SFB 359, Universität Heidelberg (2001)
4. Bauer, I., Bock, H. G., Körkel, S., Schlöder, J. P.: Numerical methods for initial value problems and derivative generation for DAE models with application to optimum experimental design of chemical processes. In: Keil, F., Mackens, W., Voss, H., Werther, J. (eds) Scientific Computing in Chemical Engineering II. **2**, 282–289, Springer, Berlin-Heidelberg (1999)
5. Bauer, I., Bock, H. G., Körkel, S., Schlöder, J. P.: Numerical methods for optimum experimental design in DAE systems. Journal of Computational and Applied Mathematics, **120**, 1–25 (2000)
6. Beck, J. V., Arnold, K. J.:, Parameter estimation in engineering and science Wiley, New York (1977)
7. Birkes, D., Dodge, Y.: Alternative Methods of Regression. John Wiley and Sons (1993)
8. Bock, H. G.: Numerical treatment of inverse problems in chemical reaction kinetics. In: Ebert, K.-K., Deuflhard, P., Jäger, W. (eds) Modelling of Chemical Reaction Systems. Springer Series in Chemical Physics **18**, 102 – 125, Springer Verlag (1981)
9. Bock, H. G.: Randwertproblemmethoden zur Parameteridentifizierung in Systemen nichtlinearer Differentialgleichungen. Bonner Mathematische Schriften, **187**, Bonn (1987)
10. Bock, H. G., Kallrath, J., Schlöder, J. P.: Least squares parameter estimation in chaotic differential equations. Celestial Mechanics and Dynamical Astronomy, **56**, (1993)
11. Bock, H. G., Kostina, E. A., Schlöder, J. P.: On the role of natural level functions to achieve global convergence for damped newton methods. In: Powell, M.J.D., Scholtes, S. (eds) System Modelling and Optimization. Methods, Theory and Applications. Kluwer (2000)
12. Bock, H. G., Körkel, S., Kostina, E. A., Schlöder, J. P.: Methods for Design of Optimal Experiments with Application to Parameter Estimation in Enzyme Catalytic Processes. In: Hicks, M. G., Kettner C. (eds) Experimental

Standard Conditions of Enzyme Characterizations, Proccedings of the International Beilstein Workshop, Beilstein-Institut zur Förderung der Chemischen Wissenschaften, 45 –70 (2004)
13. Bock, H. G., Kostina, E. A., Kostyukova, O. I.: Covariance matrices for constrained parameter estimation problems. Submitted to SIAM Journal on Matrix Analysis and Applications (2004)
14. Bommarius, A., Estler, M., Kluge, A., Werner, H., Vollmer, H., Bock, H. G., Schlöder, J. P., Kostina E.: Method to determine the process stability of enzymes. Patent Application EP 1 067 198 A1, European Patent Office, Patentblatt, **2** (2001)
15. Bock, H. G., Körkel, S., Kostina, E. A., Schlöder, J. P.: Numerical Methods for Optimal Control Problems in Design of Robust Optimal Experiments for Nonlinear Dynamic Processes. Optimization Methods and Software, **19**, issue 3-4, 327 – 338 (2004)
16. Boscovic, R.: Theoria Philosophiae naturalis. Vienna (1758)
17. Box, G. E. P., Tiao, G. C.: Bayesian Inference in Statistical Analysis. Wiley Classics Library, John Wiley and Sons (1992)
18. Diehl, M., Bock, H. G., Kostina, E. A.: An Approximation Technique for Robust Nonlinear Optimization. To appear in Mathematical Programming, 2005.
19. Fedorov, V. V.: Theory of Optimal Experiments. Probability And Mathematical Statistics. Academic Press, London (1972)
20. Gabasov, R., Kirillova, F. M., Kostina, E. A.: An adaptive method of solving l_1 extremal problems. Zhurnal vychislitelnoy matematiki i matematicheskoy fiziki, **38**, 9, 1461 – 1472 (1998) (in Russian, trans. to English: Computational Mathematics and Mathematical Physics, **38**, 9, 1400 – 1411, 1998)
21. Gauss, C. F.: Theory of Combinations of Observations Least Subject to Errors. Original with translation. SIAM (1995)
22. Gauss, C. F.: Theoria Motus Corporum Coelestium in Sectionibus Conicis Solem Ambientium. F. Perthes and J. H. Besser, Hamburg (1809)
23. Griewank, A.: Evaluating Derivatives. Principles and Techniques of Algorithmic Differentiation. Frontiers in Applied Mathematics. SIAM (2000)
24. Hald, J, Madsen, K,: Combined linear programming and quasi-Newton methods for non-linear L_1 optimization. SIAM Journal on Numerical Analysis, **22**, 65–80 (1985)
25. Huber, P. J.: Robust Statistics. John Wiley and Sons (1981)
26. Körkel, S.: Numerische Methoden für Optimale Versuchsplanungsprobleme bei nichtlinearen DAE-Modellen. PhD thesis, Universität Heidelberg (2002)
27. Körkel, S., Bauer, I., Bock, H. G., Schlöder, J. P.: A sequential approach for nonlinear optimum experimental design in DAE systems. In: Keil, F., Mackens, W., Voss, H., Werther, J. (eds) Scientific Computing in Chemical Engineering II. **2**, 338–345, Springer, Berlin-Heidelberg (1999)
28. Körkel, S., Kostina, E. A.: Numerical methods for nonlinear experimental design. In: Bock, H. G., Kostina E. A., Phu H. X., Rannacher R. (eds) Modeling, Simulation and Optimization of Complex Processes, Proceedings of the International Conference on High Performance Scientific Computing, 2003, Hanoi, Vietnam, Springer (2004)
29. Kostina, E. A.: The long step rule in the bounded-variable dual simplex method: numerical experiments. Mathematical Methods of Operations Research, **55**, 3, 413 – 429 (2002)

30. Kostina, E. A.: Robust Parameter Estimation in dynamic systems. Preprint IWR/ SFB 359 (2001)
31. Kostina, E. A.: Robust Parameter estimation in dynamic dystems. Optimization and Engineering, **5**(4), 461 – 484 (2004)
32. Kostina, E. A., Prischepova, S. V.: A new algorithm for minimax and L_1-norm optimization. Optimization (Journal on Mathematical Programming and Operations Research), **44**, 263 – 289, (1998)
33. Murray, M., Overton, M. L.: A projected Lagrangian algorithm for nonlinear l_1 optimization. SIAM Journal on Scientific and Statistical Computing, **2**, 2, 207 – 224 (1981)
34. Pukelsheim, F.: Optimal Design of Experiments. John Wiley & Sons, Inc., New York (1993)
35. Reemtsen R., Rückmann, J.-J. (eds.): Semi-Infinite Programming. Nonconvex Optimization and its Applications. Kluwer, Boston (1998)
36. Schulz V. H., Bock H. G. and Steinbach M. C.: Exploiting invariants in the numerical solution of multipoint boundary value problems for DAE. SIAM Journal on Scientific Computing, **19**, 2, 440 - 467 (1998).

Multiple Set Point Partially Reduced SQP Method for Optimal Control of PDE*

H. G. Bock[1], E. Kostina[1], A. Schäfer[1], J. P. Schlöder[1], and V. Schulz[2]

[1] Interdisziplinäres Zentrum für Wissenschaftliches Rechnen (IWR), Universität Heidelberg
[2] Universität Trier, FB IV - Mathematik

Summary. Optimization of dynamic processes described by partial differential-algebraic equations (PDAE) is a challenging task due to dimension and complexity of the problems. Fast solutions methods are achieved by using a simultaneous approach for a close coupling of the optimization aspect of the overall algorithm with the solution method of the dynamic system. Especially using partially reduced sequential quadratic programming (PRSQP) approaches reduces the computational complexity while still being able to incorporate inequality constraints. An effective and straightforward generalization of the methods to treat optimization tasks modeled as multiple set point optimization problems is shown. Based on the simultaneous approach for optimization problems in NMPC an efficient real-time iteration technique is developed. As industrial applications we present shape optimization of turbine blades, operation optimization of a catalytic tube reactor and the real-time optimization of a continuous distillation column.

1 Introduction

Large-scale nonlinear optimization is an active research area of growing importance. A main source of large-scale optimization problems are the optimization of dynamic processes which are described by (time-dependent) partial differential-algebraic equations (PDAE). Today PDAE simulation is widespread in science and engineering applications. Often parameter identification, optimization and optimal control are the sequel to simulation and vice versa. Based on PDAE simulation software fast and reliable optimization methods were developed. The requirements of speed and reliability forbid the

---

*This work has been supported by the German Research Foundation (DFG) through SFB 359 (Project A4) at the University of Heidelberg. Furthermore, the authors thank the cooperation partners of the "Institut für Systemdynamik und Regelungstechnik" (ISR) and the "Institut für Systemtheorie Technischer Prozesse" (IST), University of Stuttgart, for the common work on NMPC of the distillation column, and Bayer Technology Services, Leverkusen, for providing a large-scale complex optimization task of the catalytic tube reactor.

use of so-called *black box* approaches, where an outer optimization loop iterates over the decision variables only and an inner simulation loop iteratively determines the state variables describing the dynamic system behaviour. As an alternative, we pursue the *simultaneous* approach, which typically requires a close coupling of the optimization aspect of the overall algorithm with the computation of the state variables of the dynamic system. It delivers the optimal solution at costs only a small factor higher than the costs for a simulation.

In this paper we assume that the dynamic system is discretized on a fixed spatial grid by the PDAE solver and the problem remains continuous in time. In the following we present two direct methods which both parameterize the controls on the time horizon by basis functions with local support such as piecewise constant, linear or cubic polynomials, resulting in a finite set of control parameters. The first method discretizes the state variables on the time horizon by *collocation*, the second method parameterizes the state variables by *multiple shooting*, where initial value problems on each multiple shooting interval have to be solved. The result is in both cases a finite-dimensional highly structured NLP problem.

Tailored partially reduced sequential quadratic programming (PRSQP) techniques have been developed to solve these NLP problems. With reduced SQP methods [17, 20] they share the property that they are simultaneous optimization approaches – or methods within the optimization boundary value problem framework [7, 8, 27, 11], i.e. iterating over all variables (state and decision) in one loop. With black box approaches, they share the property that only reduced Hessians are constructed and used in the algorithm – which results in computational gains. However, other reduced approaches (as mentioned above) are limited to optimization problems where the system equations are the only constraints and thus are not able to consider additional inequality constraints resulting, e.g., from geometrical design restrictions in shape optimization. Partially reduced methods, as established in [28, 29, 22, 26] overcome this limitation by incorporating the additional constraints into small quadratic subproblems to be solved in each iteration. Theoretical analysis and convergence proofs for this approach have been presented in [28].

In order to set up the reduced quadratic programming (RQP) problems and to compute the increments of the PRSQP algorithm two different approaches are presented. The *direct* variant computes the kernel-basis of the system sensitivities directly and in the sequel the reduced gradients of the RQP. This is advantageous in case of a low number of decision variables. The *adjoint* variant first computes the multipliers incorporating the transpose of the system sensitivities and then the reduced gradient avoiding the computation of the kernel-basis. In case of a low number of inequality constraints this variant is superior to the direct variant. In both cases an efficient recursive calculation of directional derivatives is important. For the multiple shooting approach a BDF solver [3] is used to compute the system states and directional state derivatives thereby intertwining constraint linearization and projection in the reduced approach. The result are efficient recursions avoiding the calculation of the full-space state derivatives. In case of collocation the reduced gradients and the multipliers are computed by efficient recursions from the sparse full-space derivatives. The reduced Hessians are approximated by (asymptotically correct) BFGS update strategies.

A striking feature of the methods presented in this paper is that they can be easily generalized to treat multiple set point optimization tasks. The obtained results are often much more important in practice than the single set point results. For the case of optimal shape design of turbine blades this will be demonstrated in detail in section 9. The approach provides a very natural parallelization possibility as well.

The simultaneous approaches mentioned in this paper are also extended towards Nonlinear Model Predictive Control (NMPC). Especially, real-time iteration schemes are developed for the direct multiple shooting approaches. It can be shown that the time for calculation of a feedback control after a disturbance can be drastically reduced compared to the solution of a complete optimal control problem.

The paper is organized in the following way. Section 2 describes three industrial applications in order to motivate the development of specially tailored simultaneous methods. Section 3 presents the general formulation of a multistage optimal control problem. Section 4 is devoted to simultaneous approaches which discretize and parameterize the before mentioned optimal control problems and lead to highly structured NLP problems. In section 5 PRSQP variants for the solution of the NLP problems are presented. The computation of directional derivatives is explained in section 6. The extension of the optimal control problem formulation to a multiple set point task is presented in section 7. In section 8 the extension towards NMPC is shown. Numerical results for the three industrial applications are given in section 9.

2 Motivation: Three Large-Scale Industrial Applications

In order to motivate the use of simultaneous methods we introduce three industrial applications: the shape optimization of turbine blades, the capacity maximization of a catalytic tube reactor, and the real-time optimization of a distillation column. The models are described by steady state Euler equations for the flow in axisymmetrical streamsurfaces of turbine blades, by time-dependent convection-diffusion equations for the reactive flow in a catalytic tube reactor and by stiff differential-algebraic equations of a continuous distillation column including hydrodynamic effects. They are used in the context of multiple set point optimization problem and in off- and on-line optimization problems. The difficulty in optimizing these problems are, e.g., the dimension of the coupled discretized PDEs, the nonlinearities of the model and the stability properties of the resulting DAEs. In the following we indicate the properties of the respective model equations and the consequences for optimization.

2.1 Shape Optimization of Turbine Blades

Shape optimization for airfoils has been investigated in many publications, see, e.g., [19, 25]. In contrast, the optimization of blades in turbines has received much less attention, e.g., [6]. Our numerical modeling of the flow is performed by the use of the solver MISES (Multiple blade Interacting Streamtube Euler Solver) [15, 31]. Considering the state of the art in computational methods, the most appropriate flow description would be presented by three-dimensional Navier–Stokes equations. However, their computational solution is still highly complex, and the results are generally considered as differing from practical flow measurements by approximately the same amount as the results obtained from the computations as described briefly below. Although the flow in axi-symmetrical turbomachinery is three-dimensional, a useful and often necessary simplification for design purposes is to approximate the flow through a stage as a set of two-dimensional blade-to-blade problems defined on axisymmetrical streamsurfaces. Examples are shown in figure 1. Axisymmetrical through-flow codes are used early in the preliminary design process to define circumferentially averaged conditions in one or more stages of the machine based on initial estimates of work and loss. These calculations define the flow in terms of axisymmetric streamsurfaces. At the next level of design refinement (pertaining to our situation), the streamsurface radius and spacing can be used to define a set of quasi-3D blade-to-blade design problems for each

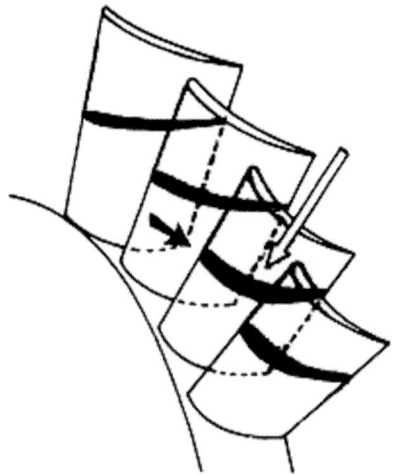

Fig. 1. Blade-to-blade flow on a streamsurface of revolution

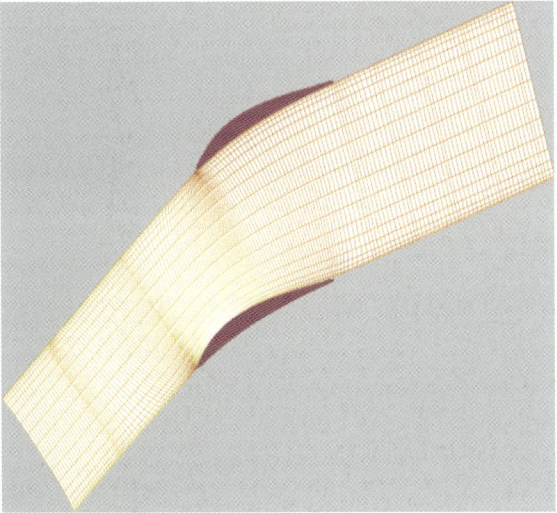

Fig. 2. Example for the computational grid for compressor blades.

stage. These allow the designer to select or design blade profiles at several radial stations, in order to define the complete three-dimensional rotor or stator blade. The blade-to-blade technique works very well for most design applications, limited in effectiveness largely by the estimates for boundary layer effects on the inner and outer walls and by three-dimensional effects not accounted for within the axisymmetric assumptions. Instead of solving the viscous flow directly, we use a zonal approach, where an equivalent inviscid flow in the interior of the computational domain is postulated by using a displacement surface in order to represent the viscous layer. The inner boundary is displaced outward from the wall by the boundary layer displacement thickness δ. The inviscid flow is modeled by the steady state Euler equations. These basic equations must be complemented by additional coupling equations and inflow/outflow boundary conditions. A complete and detailed description of the flow model and its discretization can be found in [31]. The discretization is performed on a streamline aligned structured grid as depicted in figure 2. The grid size needed for blade design is case dependent and ranges from about 150×20 to 300×40.

The blade profiles are defined in (m, θ) streamsurface coordinates. From cone coordinates (ρ, ξ, ϕ) for a given streamline $\rho(\xi)$ computed by an axisymmetrical through-flow code these can be obtained as

$$m = \int_{\xi_0}^{\xi} \frac{\sqrt{(dr(\zeta))^2 + (d\zeta)^2}}{r(\zeta)} d\zeta, \quad \theta = \phi - \theta_0.$$

The offset θ_0 is chosen in such a way that θ vanishes for the smallest vertical ξ-coordinate. The streamline is typically approximated by a straight line

$$\rho(\xi) = a + b\xi$$

with appropriately chosen parameters a and b. Based on previous practical experiences, the blade profiles themselves are represented by quintic B-Splines in both coordinates:

$$m(t) := \sum_{i=1}^{12} p_i B_i(t), \quad \theta(t) := \sum_{i=1}^{12} p_{i+12} B_i(t), \quad 0 \le t \le 1.$$

The discretized flow equations are summarized by $g(z,p) = 0$, $g : \mathbb{R}^{n_z+n_p} \to \mathbb{R}^{n_z}$, $G_z = \frac{\partial g}{\partial z}$ nonsingular with $z \in \mathbb{R}^{n_z}$ the discretized flow variables and $p \in \mathbb{R}^{n_p}$ the design variables (spline coefficients).

The spline profiles are subject to geometric constraints due to the actual contruction process of the blades and the necessity to cool the blades by the use of inner air pipes. On the other hand, geometric constraints are used in order to stabilize the optimization algorithm. Otherwise, intermediate blades may be the result which cannot be treated by MISES. For a detailed description of the geometric constraints see, e.g., [9]. The geometric constraints are summarized in the inequality $r(p) \ge 0$, $r \in C^2(\mathbb{R}^{n_p} \to \mathbb{R}^m)$.

One typical objective function is the minimization of the total pressure loss coefficient at a certain working set point, which is denoted by $\phi(z,p)$, $\phi : \mathbb{R}^{n_z+n_p} \to \mathbb{R}$. However, in general, turbomachine industry is much more interested in having blades for a whole range of working conditions (e.g., varying inflow angle), and also cross sectional shapes for various "heights" of the blade are searched for which are not undulating too much when put together – i.e. which possess a certain smoothness in the third dimension. In essence, this means a multiobjective or multicriteria optimization problem with a continuum of multiple objectives. One way to treat these problems – which we choose here – is to squeeze all objectives together into one single objective by integrating the objectives over the specific working range and using an apriori selected measure for this integration. This measure $\omega(\alpha)$ represents the importance of the working conditions parameterized by a vector (or a scalar) $\alpha \in \mathcal{A}$ within a range $\mathcal{A}$. After choosing an appropriate quadrature rule, the multiple set point problem can be formulated now with a weighted sum of the objectives at the discrete locations $\alpha_i, i = 1, \ldots, N$ using the abbreviations $z_i := z(\alpha_i), \omega_i := \omega(\alpha_i)$:

$$\min_{z_1,\ldots,z_N,p} \sum_{i=1}^{N} \omega_i \phi(z_i, p; \alpha_i)$$
$$\text{s.t.} \quad g(z_i, p; \alpha_i) = 0, \quad i = 1, \ldots, N,$$
$$r(p) \ge 0.$$

2.2 Optimal Operation of a Catalytic Tube Reactor

Industrial catalytic tube reactors for exothermal gas phase reactions typically consist of several thousand reaction tubes within a common heat exchanger shell. The reactant gas stream flows through a bed of porous catalyst particles inside the tubes, reacting to form the desired product. On the shell side, a stream of heat transfer fluid removes the heat generated by the reaction. Usually, the reaction takes place only in a relatively small section of the catalyst packing. This reaction zone slowly moves from the inlet side to the outlet side of the packing due to the deactivation of the catalyst, e.g., by coking. When the reaction zone reaches the end of the packing, i.e., just before there is a breakthrough of reactant, the reactor must be shut down for regeneration of the catalyst.

Based on a rigorous dynamic model of a single reaction tube, we have optimized the total capacity of an industrial tube reactor [26]. Apart from the time-dependent throughput, we have considered the (possibly staged) "dilution" of the catalyst by inert particles as a degree of freedom. While the catalyst regeneration time was assumed to be fixed, the optimal duration of the reaction cycle was implicitly determined by a reactant breakthrough condition. An upper limit had to be imposed on the temperature in the catalyst packing in order to avoid irreversible damage to the catalyst. The optimization results indicate that, at least for the specific reaction process examined here, catalyst dilution is a very interesting design parameter. A significant increase of total reactor capacity could be achieved using only about half the original amount of catalyst, although the duration of the reaction cycle was considerably reduced, i.e., the catalyst had to be regenerated more frequently.

For the optimization, we have employed a pseudo-homogeneous transient 2D model of the cylindrical reaction tube, formulated in the gPROMS modeling language [12], which treats the heterogeneous, two-phase system (connected fluid gas phase and dispers solid catalyst phase) as a quasi-continuum. Such models are common in chemical engineering for mathematical modeling of temperature and concentration flows in packed bed tube reactors [1, 2]. The heat and mass transfer transactions are described by Fourier's and Fick's laws [5]. For the gas phase velocity u_f superficial plug flow is assumed. Within the pseudo-homogeneous continuum inside the reaction tube, axial and radial dispersion were taken into account, and empirical correlations were used for the corresponding effective mass and heat transfer coefficients D_{eff} and λ_{eff} respectively. The unknown parameters of the reaction and catalyst deactivation kinetics were fitted to available experimental data. It was possible to make a quasi steady state assumption for the reaction kinetics, thus restricting the dynamic aspects of the model to the slow catalyst deactivation, the temperatures in the quasi-continuum and the cooling liquid. In our case we only calculate the concentrations of one educt c_A. Here r and z denotes the radial and axial direction, T the temperature in the quasi-continuum, a_{kat} the activity of the catalyst. The following coupled PDE describe the balance equations of the quasi-continuum:

mass balance:
$$-u_f \frac{\partial c_A}{\partial z} + D_{eff,z} \frac{\partial^2 c_A}{\partial z^2} + D_{eff,r}\left(\frac{\partial^2 c_A}{\partial r^2} + \frac{1}{r}\frac{\partial c_A}{\partial r}\right) = -\nu_A\, r_0\, \rho_{s,bed}\, a_{Kat},$$

energy balance:
$$(\varepsilon\, \rho_f\, c_{p,f} + \rho_{s,bed}\, c_{p,s}) \frac{\partial T}{\partial t} = -c_{p,f}\, u_f \frac{\partial(\rho_f T)}{\partial z} + \lambda_{eff,z} \frac{\partial^2 T}{\partial z^2}$$
$$+ \lambda_{eff,r}\left(\frac{\partial^2 T}{\partial r^2} + \frac{1}{r}\frac{\partial T}{\partial r}\right)$$
$$- \Delta H_R\, \nu_{An}\, r_0\, \rho_{s,bed}\, a_{Kat},$$

with the reaction rate for the main reaction:
$$r_0 = A_0 \exp\left(-\frac{E_a}{R_g T}\right) c_A\, a_{Kat},$$

and the catalyst deactivation:
$$\frac{\partial a_{Kat}}{\partial t} = -\left(k_{0,Des} + A_{0,Des} \exp\left(-\frac{E_{a,Des}}{R_g}\left(\frac{1}{T} - \frac{1}{T_{ref}}\right)\right)\right) c_A\, a_{Kat}^2$$
$$\text{with}\quad a_{Kat}|_{t=0} = 1.$$

At the inlet of tube the concentration of the reactands are prescribed, so that the time-dependent throughput is a boundary control. The PDE have nonlinearities due to convection, reaction rates and dependency of physical properties on the local states (e.g., the calculation of the density $\rho_{s,bed}$ from the ideal gas law). For a detailed description of the model see [26].

Spatial discretization of the resulting coupled partial differential and algebraic equations (PDAE) by the method of lines in gPROMS – upwind finite differences in axial direction and orthogonal collocation in radial direction – led to a differential-algebraic equation (DAE) model of index 1 with 548 differential states $x(t)$ and 1074 algebraic states $z(t)$. The scalar dilution parameter for the catalyst and the reaction duration are summarized in the design parameter p and the time-dependent throughput $c_{A,in}$ in the control vector $u(t)$.

Our aim is to optimize the total capacity of the tube reactor. We calculate the capacity by integrating the throughput $c_{A,in}$ over the time-horizon $[0, t_c]$ and multiplying with the number of cycles per year:

$$\int_{t=0}^{t_c} \Phi(u,p) = \int_{t=0}^{t_c} \frac{t_{prod}}{t_c + t_{reg}}\, c_{A,in}\, dt,$$

where t_{prod} is the production time per year and t_{reg} is the time for regeneration in each cycle. The upper limits on the temperatures and breakthrough conditions of the educts are summarized as the inequality constraints $h(x,z) \geq 0$ and $r_e(x,z) \geq 0$ respectively. The spatially discretized optimization problem of the catalytic tube reactor can now be formally stated as:

$$\min_{x,z,u,p,t_c} \int_{t=0}^{t_c} \Phi(u(t),p,t_c)$$

s.t. $\left. \begin{array}{l} B(\,\cdot\,)\dot{x}(t) = f(x(t),z(t),u(t),p,t) \\ 0 = g(x(t),z(t),u(t),p,t) \\ 0 \leq h(x(t),z(t)) \end{array} \right\}, \quad t \in [0,t_c]$

$$0 \leq r_e(x(t_c),z(t_c))$$
$$0 = r_0(x(0),z(0)).$$

2.3 Real-Time Operation of a Continuous Distillation Column

Distillation columns are used in chemical industry to separate mixtures of fluids with different boiling points. There are different objectives for this process: product purity and/or energy costs. In our case we treat a binary distillation with the components Methanol and n-Propanol, and the objective is to restore rapidly the purity of the products after a disturbance. This is achieved by controlling the temperature and thereby also the concentrations on specified reference trays.

The distillation column under consideration is located at ISR, University of Stuttgart, as a pilot plant and has a diameter of 0.1 m and a height of 7 m and consists of 40 bubble cap trays. The overhead vapor is totally condensed in a water cooled condenser which is open to atmosphere. The reboiler is heated electrically. A flowsheet of the distillation system is shown in figure 3. The preheated feed stream x_F enters the column at the feed tray 21 as saturated liquid. It can be switched automatically between two feed tanks in order to introduce well defined disturbances in the feed concentration. In the considered configuration, the process inputs that are available for control purposes are the heat input Q to the reboiler and the reflux flow rate L_{vol}. Control aim is to maintain high purity specifications for the distillate and bottom product streams D_{vol} and B_{vol}.

The mass and enthalpy balances and the hydrodynamics define a stiff DAE of index 1 and are used to model the concentrations and molar holdups of Methanol at each tray (summarized as 82 differential state variables $x(t)$) and the liquid and vapor (molar) fluxes out of the 40 trays and 42 temperatures of reboiler, trays and condenser (forming the 122 components of the algebraic state variables $z(t)$). For a complete description of the model see [14]. The two controls Q and L_{vol} are collected in $u(t)$. As usual in distillation control, the concentrations at the bottom and the reboiler are not controlled directly – instead, an inferential control scheme is used, which controls the deviation of the concentrations on tray 14 and 28 from a given set point. These two concentrations are much more sensitive to changes in the inputs of the system than the product concentrations: if they are kept constant, the product purities are safely maintained for a large range of process conditions. As concentrations are

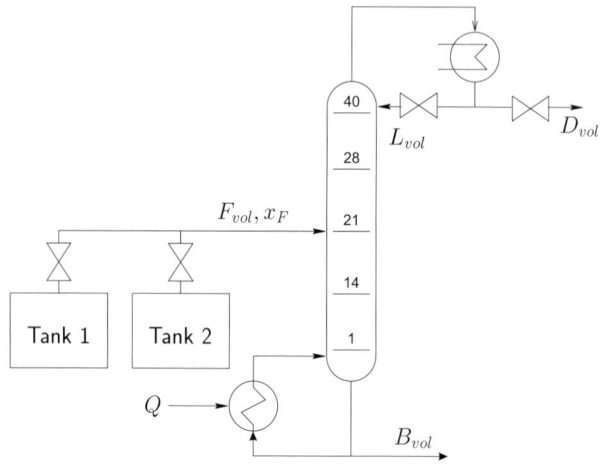

Fig. 3. Flowsheet of the distillation column

difficult to measure, we consider instead the tray *temperatures*, which correspond directly to the concentrations via the Antoine equation. In the following we will use the projection $\tilde{T}z := (T_{28}, T_{14})^T$ to extract the controlled temperatures from the vector z, and define $\tilde{T}_{ref} := \left(T_{28}^{ref}, T_{14}^{ref}\right)^T = (70\,°\mathrm{C},\,88\,°\mathrm{C})^T$ for the desired set points. The objective is formulated as the integral of a least squares term

$$\Phi(z, u, u_s) := \|\tilde{T}z - \tilde{T}_{\mathrm{ref}}\|_2^2 + \|R(u - u_s)\|_2^2,$$

where the second term is introduced for regularization, with a small diagonal weighting matrix R. Initially the system state x_0 is disturbed by, e.g., a breakdown of the reflux L_{vol} back into the column. The aim is to bring the system back into a steady state x_s with the corresponding control u_s which can be calculated in advance. In order to avoid the product streams flowing back into the column we have to pose path constraints: $h(x, z, u, p) := (D_{vol}, B_{vol})^T > 0 \;\; \forall t \in [t_0, t_0 + T_p]$. To ensure nominal stability of the closed-loop system, an additional prediction interval $[t_0 + T_c, t_0 + T_p]$ is appended to the control horizon $[t_0, t_0 + T_c]$, with the controls fixed to the set point values u_s. The off-line optimal control problem can now be formally summarized as follows:

$$\min_{x(t),z(t),u(t)} \int_{t_0}^{t_0+T_p} \Phi(z(t),u(t),p))$$

$$\text{s.t.} \quad \left.\begin{array}{l} \dot{x}(t) = f(x(t),z(t),u(t),p) \\ 0 = g(x(t),z(t),u(t),p) \\ 0 \leq h(x(t),z(t),u(t),p) \end{array}\right\}, \quad t \in [t_0, t_0+T_p]$$

$$x(t_0) = x_0.$$

The aim in on-line optimization is to deliver a feedback-control as soon as possible after the initial (disturbed) system state x_0 is known. To meet this goal fast on-line optimization methods have to be developed. Section 8 explains the features of the new on-line direct multiple shooting variants.

3 Multistage Optimal Control Problem Formulation

Many dynamic process optimization problems of practical relevance can be expressed as multistage optimal control problems in DAE. In the following we will consider a general class of M-stage optimal control problems, where the time horizon of interest $[t_0, t_M]$ is divided into M model stages corresponding to the subintervals $[t_i, t_{i+1}]$, $i = 0, 1, \ldots, M-1$:

$$\min_{x_i, z_i, u_i, p, t_i} \sum_{i=0}^{M-1} \phi_i(x_i(t_{i+1}), z_i(t_{i+1}), p, t_{i+1}) \tag{1a}$$

subject to the DAE model stages

$$\left.\begin{array}{l} B_i(\,\cdot\,)\,\dot{x}_i(t) = f_i(x_i(t), z_i(t), u_i(t), p, t) \\ 0 = g_i(x_i(t), z_i(t), u_i(t), p, t) \end{array}\right\}, \quad t \in [t_i, t_{i+1}], \quad i = 0, 1, \ldots, M-1, \tag{1b}$$

the control and path constraints

$$h_i(x_i(t), z_i(t), u_i(t), p, t) \geq 0, \quad t \in [t_i, t_{i+1}], \quad i = 0, 1, \ldots, M-1, \tag{1c}$$

the stage transition conditions

$$x_{i+1}(t_{i+1}) = c_i(x_i(t_{i+1}), z_i(t_{i+1}), p), \quad i = 0, 1, \ldots, M-2, \tag{1d}$$

and the multipoint boundary conditions

$$\sum_{i=0}^{M-1} \left(r_i^s(x_i(t_i), z_i(t_i), p, t_i) + r_i^e(x_i(t_{i+1}), z_i(t_{i+1}), p, t_{i+1}) \right) \begin{Bmatrix} = \\ \geq \end{Bmatrix} 0. \tag{1e}$$

The performance index (1a) of generalized Mayer type is minimized with respect to the differential and algebraic state profiles x_i and z_i, respectively, the control profiles u_i, a time-independent global parameter vector p, and the model stage grid points t_i. The vectors $x_i(t)$, $z_i(t)$, $u_i(t)$, and p are of dimensions n_i^x, n_i^z, n_i^u, and n^p. Note that initial conditions on the differential state can be included in the boundary conditions.

4 Simultaneous Approaches: Parameterization and Discretization

In order to solve the infinite-dimensional optimization problem (1) we parameterize the control space with a finite number of control parameters q. If a solution of the state dynamics exists for the parameterized controls and can be calculated the resulting finite-dimensional problem can be regarded as an NLP problem, which could be solved by, e.g., Sequential Quadratic Programming (SQP) methods. This approach, where the controls are treated as variables, is also known as the direct approach. For the numerical treatment of free end-times we perform a time transformation on each stage $i = 0, 1, \ldots, M-1$ by:

$$\theta_i(\tau, v) = t_i + \tau d_i, \quad t_i = t_0 + \sum_{k=0}^{i-1} d_k, \quad \tau \in [0, 1] \qquad (2)$$

with $v := (t_0, d_0, d_1, \ldots, d_{M-1})$ and a fixed (control) grid

$$0 = \tau_{i,0} < \tau_{i,1} < \ldots < \tau_{i,m_i} = 1, \qquad (3)$$

such that $\theta_i(\tau_{i,0}, v) = t_i$ and $\theta_i(\tau_{i,m_i}, v) = t_{i+1}$. Next we choose a piecewise representation u_i^τ of the control function u_i by basis functions φ_{ij} with *local* support (e.g., piecewise constant, linear or cubic polynomials):

$$u_i^\tau(\tau) := \varphi_{ij}(\tau, q_{ij}), \quad \tau \in I_{ij} = [\tau_{ij}, \tau_{i,j+1}], \quad j = 0, 1, \ldots, m_i - 1. \qquad (4)$$

In combination with the following simultaneous methods this choice of control parameterization is responsible for the structure of the NLP and is exploited by the methods mentioned in section 5, compare [11].

The simultaneous approaches are characterized by introducing additional intermediate variables of the state trajectory (and thereby higher-dimensional variable space with additional matching conditions) into the optimization problem. Thereby knowledge about the solution can be introduced by adequate choice of initial guesses for these variables. Here we describe the collocation discretization and the multiple shooting parameterization. In conjunction with SQP methods these simultaneous approaches exploit the intermediate information about the state trajectory to fasten convergence and to improve robustness of the algorithm compared to a sequential strategy, see, e.g., [7]. One difference between the two mentioned approaches is that in case of multiple shooting adaptivity in time discretization is realized within the integrator of the dynamic system where in case of collocation this has to be done on the optimization level. As additional information of the state trajectory we introduce startvalues s_{ij}^x, s_{ij}^z for the differential and algebraic states respectively on the grid (3).

The collocation approach discretizes the differential-algebraic systems (1b) by approximation of the solution with piecewise polynomials of degree k_{ij} on each interval of the grid (3) in each stage. The differential variables $x_i(\tau)$ are then approximated by:

$$x_{ij}^\pi(\tau) = s_{ij}^x + (\tau_{i,j+1} - \tau_{ij}) \sum_{l=1}^{k_{ij}} \dot{s}_{ij,l}^x \Psi_{ij,l}\left(\frac{\tau - \tau_{ij}}{\tau_{i,j+1} - \tau_{ij}}\right) \quad \forall\, \tau \in [\tau_{ij}, \tau_{i,j+1}], \tag{5}$$

with initial values $s_{ij}^x := x_{ij}^\pi(\tau_{ij})$ and collocation variables $\dot{s}_{ij,l}^x := \dot{x}_{ij}^\pi(\tau_{ij,l})$, $l = 1, \ldots, k_{ij}$ on a collocation grid $\{\tau_{ij,1}, \ldots, \tau_{ij,k_{ij}}\}$, $\tau_{ij,l} := \tau_{ij} + \rho_{ij,l}(\tau_{i,j+1} - \tau_{ij})$, $\rho_{ij,l} \in [0,1]$. The polynomials $\Psi_{ij,l}$ are a basis of the space of polynomials of degree k_{ij} on the interval $[0,1]$ with the properties $\Psi_{ij,l}(0) = 0$ and $\dot{\Psi}_{ij,l}(\rho_{ij,k}) = \delta_{lk}$ with δ the Kronecker symbol. The algebraic variables are approximated by the vectors $s_{ij,l}^z$, $l = 1, \ldots, k_{ij}$, representing the solution values $z_i(\tau_{ij,l})$ at the collocation points. The collocation discretization of the initial value problems (1b) with startvalues s_{ij}^x consists of the k_{ij} collocation conditions in each interval $[\tau_{ij}, \tau_{i,j+1}]$:

$$B_i(\,\cdot\,)\dot{s}_{ij,l}^x = f_i(x_{ij,l}^\pi, s_{ij,l}^z, \varphi_{ij}(\tau_{ij,l}, q_{ij}), p, \theta_i(\tau_{ij,l}, v))\, d_i \tag{6a}$$
$$0 = g_i(x_{ij,l}^\pi, s_{ij,l}^z, \varphi_{ij}(\tau_{ij,l}, q_{ij}), p, \theta_i(\tau_{ij,l}, v)) \tag{6b}$$

and the continuity conditions for each grid point:

$$x_{ij}^\pi(\tau_{i,j+1}) - s_{i,j+1}^x = 0. \tag{6c}$$

As alternative to collocation discretization multiple shooting parameterization is a basis for a flexible realization of a simultaneous approach. In the following we present a multiple shooting variant introduced in [10] for DAE of index 1 and extended for DAE of higher index in [30]. The main idea is to solve a relaxed initial value problem on each interval $I_{ij} := [\tau_{ij}, \tau_{i,j+1}]$:

$$\left.\begin{array}{l} B_i(\,\cdot\,)\, dx_i/d\tau = f_i(x_i, z_i, \varphi_{ij}(\tau, q_{ij}), p, \theta_i(\tau, v))\, d_i \\ 0 = g_i(x_i, z_i, \varphi_{ij}(\tau, q_{ij}), p, \theta_i(\tau, v)) \\ \quad - \alpha_{ij}(\tau)\, g_i(s_{ij}^x, s_{ij}^z, \varphi_{ij}(\tau_{ij}, q_{ij}), p, \theta_i(\tau_{ij}, v)) \end{array}\right\}, \quad \tau \in I_{ij}, \tag{7}$$

with the initial values $x_i(\tau_{ij}) = s_{ij}^x$ and $z_i(\tau_{ij}) = s_{ij}^z$. Following the idea of a homotopy strategy the algebraic conditions are modified to allow for inconsistent algebraic start values (as in the collocation approach). The scalar damping factor α_{ij} is defined as a scalar function, which is non-increasing and non-negative on I_{ij} and fulfills $\alpha_{ij}(\tau_{ij}) = 1$. One realization is the following variant:

$$\alpha_{ij}(\tau) = \exp\left(-\bar{\alpha}\,\frac{\tau - \tau_{ij}}{\tau_{i,j+1} - \tau_{ij}}\right), \quad \bar{\alpha} \geq 0. \tag{8}$$

In order to get a solution of the original problem (1b) we have to fulfill the continuity conditions on each grid point (i,j):

$$x_i(\tau_{i,j+1}; s^x_{ij}, s^z_{ij}, q_{ij}, p, v) - s^x_{i,j+1} = 0, \quad j = 0, 1, \ldots, m_i - 1 \quad (9a)$$

and the consistency conditions:

$$g_i(s^x_{ij}, s^z_{ij}, \varphi_{ij}(\tau_{ij}, q_{ij}), p, \theta_i(\tau_{ij}, v)) = 0, \quad j = 0, 1, \ldots, m_i. \quad (9b)$$

Summarizing the discretization and parameterization steps mentioned before we get the following structured NLP problem formulation:

$$\min_{\substack{s^x_{ij}, s^z_{ij}, q_{ij}, p, v \\ \text{collocation var.:} \ \dot{s}^x_{ij,l}, s^z_{ij,l}}} \sum_{i=0}^{M-1} \phi_i(s^x_{i,m_i}, s^z_{i,m_i}, p, \theta_i(\tau_{i,m_i}, v)) \quad (10a)$$

under the collocation conditions $l = 1, \ldots, k_{ij}, j = 1, \ldots, m_i$, $i = 0, \ldots, M-1$:

$$\begin{aligned} B_i(\cdot) \dot{s}^x_{ij,l} &= f_i(x^\pi_{ij,l}, s^z_{ij,l}, \varphi_{ij}(\tau_{ij,l}, q_{ij}), p, \theta_i(\tau_{ij,l}, v)) \, d_i \\ 0 &= g_i(x^\pi_{ij,l}, s^z_{ij,l}, \varphi_{ij}(\tau_{ij,l}, q_{ij}), p, \theta_i(\tau_{ij,l}, v)) \\ 0 &= x^\pi_{ij}(\tau_{i,j+1}) - s^x_{i,j+1} \end{aligned} \quad (10b)$$

or the multiple shooting conditions $i = 0, \ldots, M-1$:

$$\begin{aligned} x_i(\tau_{i,j+1}; s^x_{ij}, s^z_{ij}, q_{ij}, p, v) - s^x_{i,j+1} &= 0, \quad j = 0, 1, \ldots, m_i - 1, \\ g_i(s^x_{ij}, s^z_{ij}, \varphi_{ij}(\tau_{ij}, q_{ij}), p, \theta_i(\tau_{ij}, v)) &= 0, \quad j = 0, 1, \ldots, m_i, \end{aligned} \quad (10c)$$

the discretized control and path constraints

$$h_i(s^x_{ij}, s^z_{ij}, \varphi_{ij}(\tau_{ij}, q_{ij}), p, \theta_i(\tau_{ij}, v)) \geq 0, \quad \begin{cases} j = 0, 1, \ldots, m_i, \\ i = 0, 1, \ldots, M-1, \end{cases} \quad (10d)$$

the stage transition conditions

$$c_i(s^x_{i,m_i}, s^z_{i,m_i}, p) - s^x_{i+1,0} = 0, \quad i = 0, 1, \ldots, M-2, \quad (10e)$$

and linearly coupled boundary conditions

$$\sum_{i=0}^{M-1} \left(r^s_i(s^x_{i,0}, s^z_{i,0}, p, \theta_i(\tau_{i,0}, v)) + r^e_i(s^x_{i,m_i}, s^z_{i,m_i}, p, \theta_i(\tau_{i,m_i}, v)) \right) \begin{Bmatrix} = \\ \geq \end{Bmatrix} 0. \quad (10f)$$

Note that collocation grid can be much finer than the multiple shooting grid and the path constraints can be formulated independently of the multiple shooting grid.

5 Direct and Adjoint Partially Reduced SQP Methods

The NLP problem (10) resulting from discretization and parameterization of the original multistage optimal control problem (1) has a lot of internal structure which becomes visible in the KKT systems. SQP methods can be interpreted as Newton-type methods applied to the necessary KKT conditions for the NLP. Reduced SQP methods [17, 20] maintain the advantage of iterating over the whole set of variables (e.g., over the discretized or parameterized state variables) but reducing the costs for computation of the Newton-increment (to the order of the degrees of freedom) by solving only a reduced QP problem in each step. Partially reduced SQP methods [28, 29, 22, 26] allow to treat NLP problems with inequality constraints while maintaining the advantages of RSQP methods. In this section we present two approaches to compute an increment in partially reduced SQP methods. They differ in terms of explicit calculation of the kernel-basis and Lagrangian multipliers of the eliminated constraints. In order to present the concepts we formalize the NLP problem as follows:

[NLP]
$$\min\ F(w) \tag{11a}$$
$$\text{s.t.}\ G_1(w) = 0 \tag{11b}$$
$$G_2(w) = 0 \tag{11c}$$
$$H(w) \geq 0. \tag{11d}$$

The variables w consists of the discretized or parameterized states, controls and parameters. The equality constraints $G \equiv (G_1, G_2)$ are divided into two parts. G_1 is used for elimination of (some) state variables in w and therefore includes all (or parts) of the collocation or multiple shooting and initial conditions, in G_2 the remaining equality conditions such as multipoint boundary conditions are summarized. In the inequality conditions H discretized control and path constraints and multipoint boundary conditions are collected. The reduction methods result in different structure present in the KKT systems of the reduced QPs depending on which equality conditions are included in G_1. We start with the full-space SQP method with iterates:

$$w_{k+1} = w_k + \alpha_k \Delta w_k.$$

The increment Δw_k is defined as the solution of a quadratic program:

[QP]
$$\min_{\Delta w_k \in \Omega_k}\ \nabla F(w_k)^T \Delta w_k + \tfrac{1}{2}\Delta w_k^T B_k \Delta w_k \tag{12a}$$
$$\text{s.t.}\ G_1(w_k) + \nabla G_1(w_k)^T \Delta w_k = 0 \tag{12b}$$
$$G_2(w_k) + \nabla G_2(w_k)^T \Delta w_k = 0 \tag{12c}$$
$$H(w_k) + \nabla H(w_k)^T \Delta w_k \geq 0, \tag{12d}$$

where B_k is an approximation to the Hessian of the Lagrangian of the NLP problem (11)

$$L(w, \lambda, \mu) := F(w) - G_1(w)^T \lambda_1 - G_2(w)^T \lambda_2 - H(w)^T \mu$$

and the adjoint variables $\lambda \equiv (\lambda_1, \lambda_2), \mu$ are the Lagrangian multipliers of the constraints G_1, G_2, H. The step relaxation factor $\alpha_k \in (0, 1]$ can be used, e.g., in a line-search method described later. Bounds on the variables can be included in Ω_k and used for globalization by trust-region methods. In order to set up partially reduced methods for (11) we define a kernel-basis of the linearized equality constraints G_1. The variables w are divided into a first component w_1 so that the Jacobian of G_1 w.r.t. first component w_1 is regular over the whole variable domain. The other component is denoted by w_2. In the following we use a coordinate-basis since the sparsity structure of $\nabla_{w_1} G_1$ could be easily exploited in contrast to an orthogonal basis. As a result we have a partitioning of the step

$$\Delta w = (\Delta w_{1,k}^T, \Delta w_{2,k}^T)^T := S_{1,k}^{\mathcal{R}} y_k^{\mathcal{R}} + S_{1,k}^{\mathcal{N}} y_k^{\mathcal{N}}$$

with the range-space basis $S_{1,k}^{\mathcal{R}}{}^T = [\nabla_{w_1} G_{1,k}^{-1} : 0]$ and a null-space basis $S_{1,k}^{\mathcal{N}}{}^T = [-\nabla_{w_2} G_{1,k} \nabla_{w_1} G_{1,k}^{-1} : I]$, the range-space component $y_k^{\mathcal{R}} := -G_{1,k}$ and the null-space component $y_k^{\mathcal{N}}$ as the solution of the reduced QP:

[RQP]
$$\min_{y_k^{\mathcal{N}} \in \Omega_k^{\mathcal{N}}} \nabla F_k^T S_{1,k}^{\mathcal{N}} y_k^{\mathcal{N}} + C_k y_k^{\mathcal{N}} + \tfrac{1}{2} y_k^{\mathcal{N} T} S_{1,k}^{\mathcal{N}}{}^T B_{1,k} S_{1,k}^{\mathcal{N}} y_k^{\mathcal{N}} \quad (13a)$$

$$\text{s.t.} \quad G_{2,k} + \nabla G_{2,k}^T S_{1,k}^{\mathcal{R}} y_k^{\mathcal{R}} + \nabla G_{2,k}^T S_{1,k}^{\mathcal{N}} y_k^{\mathcal{N}} = 0 \quad (13b)$$

$$H_k + \nabla H_k^T S_{1,k}^{\mathcal{R}} y_k^{\mathcal{R}} + \nabla H_k^T S_{1,k}^{\mathcal{N}} y_k^{\mathcal{N}} \geq 0, \quad (13c)$$

where the second term $C_k y_k^{\mathcal{N}} := y_k^{\mathcal{R} T} S_{1,k}^{\mathcal{R}}{}^T B_{1,k} S_{1,k}^{\mathcal{N}} y_k^{\mathcal{N}}$ in the quadratic objective function is a crossterm involving both the range-space and null-space basis and in the manner of reduced SQP methods typically set to zero. This leaves the reduced Hessian $B_{1,k}^{\mathcal{N}} = S_{1,k}^{\mathcal{N}}{}^T B_{1,k} S_{1,k}^{\mathcal{N}}$ as the only place where second order information in the algorithm is considered. Formally the partitioning of the step can be seen as the elimination of Δw_1 of the QP problem (12). In the following we present two different methods to set up the reduced QP. One directly calculates the null-space basis $S_{1,k}^{\mathcal{N}}$ and computes the reduced gradients as products with the basis. This method is called the *direct* PRSQP variant. The other method first computes the reduced gradients as a solution of adjoint equations:

$$\nabla_{w_1} G_{1,k}^T \Lambda_k = \nabla_{w_1} (F_k, G_{2,k}, H_k),$$
$$S_{1,k}^{\mathcal{N}} \nabla (F_k, G_{2,k}, H_k) = -\nabla_{w_2} G_{2,k}^T \Lambda_k + \nabla_{w_2} (F_k, G_{2,k}, H_k).$$

Note that in case of a fully reduced method (no constraints in G_2, H) Λ is a vector and represents the Lagrange multiplier of G_1 w.r.t. the objective function F. This method is called *adjoint* PRSQP variant and is superior to the direct variant in case of a low number of constraints in G_2 and active constraints in H.

In general the strength of PRSQP methods arises from the fact that the reduced Hessians of the Lagrangian can be calculated by approximations $B_{1,k}^{\mathcal{N}}$ which avoid expensive applications of $S_{1,k}^{\mathcal{N}}$ or $S_{1,k}^{\mathcal{N}}{}^T$. For this approximation we employ, e.g., the BFGS update formula (cf. [16]):

$$B_{1,k+1}^{\mathcal{N}} := B_{1,k}^{\mathcal{N}} + U^{BFGS}(B_{1,k}^{\mathcal{N}}, \Delta y_k^{\mathcal{N}}, \Delta \gamma_k^{\mathcal{N}}),$$
$$\text{where} \quad U^{BFGS}(B, \Delta y, \Delta \gamma) := \frac{\Delta \gamma \Delta \gamma^T}{\Delta y^T \Delta \gamma} - \frac{(B \Delta y)(B \Delta y)^T}{\Delta y^T B \Delta y}.$$

The key property of this formula is the secant condition $B \Delta y = \Delta \gamma$. The intention for this kind of update is to collect second order information from first order magnitudes available in each iteration. Therefore Δy is formed by the null-space step, $\Delta y := y_k^{\mathcal{N}}$, and $\Delta \gamma$ by the resulting difference of reduced gradients of the Lagrangian

$$\Delta \gamma := \bar{S}_{1,k}^{\mathcal{N}\,T} \nabla_w L(\bar{w}_k, \lambda_{k+1}, \mu_{k+1}) - S_{1,k}^{\mathcal{N}\,T} \nabla_w L(w_k, \lambda_{k+1}, \mu_{k+1}),$$

where a bar over a symbol means evaluation at an intermediate point. Note that the reduced gradient does not depend on the Lagrange multipliers $\lambda_{1,k}$. The intermediate point may be chosen as w_{k+1}, which defines an update strategy in the spirit of [23] and is cheap to calculate, or

$$\bar{w}_k = w_k + S_{1,k}^{\mathcal{N}} y_k^{\mathcal{N}},$$

which defines an asymptotically correct update strategy. A proof for the resulting local 2-step superlinear convergence of the algorithm can be found in [28]. We implemented the first strategy together with a Han-Powell modification [24] to maintain positive definiteness of the Hessian. An option is a limited memory strategy to avoid blow-up of the condition number.

In order to give an overview of the simultaneous methods described in this article we state the different kinds of constraints again: the equality constraints are summarized in $G = (G_1^T, G_2^T)^T$ and include the collocation, consistency and continuity constraints as well as initial conditions and equality multipoint boundary constraints. The stage transition conditions are included in the continuity conditions since they have similar structures. The inequality constraints are summarized in H and include the control and path constraints and the inequality multipoint boundary constraints. The boundary, control and path constraints are abbreviated by BCP conditions. In figure 4 different PRSQP variants are sketched together with the Hessian structure of the corresponding KKT systems. Note that level 0 belongs only to the case of collocation discretization of the dynamic systems. The multiple shooting variants begin at level 1. The block structure in levels 0 to 2 can be exploited by high rank updates and leads to a faster convergence behaviour of the NLP solvers. In this article the collocation variant OCPRSQP [28] and the multiple shooting variants MUSCOD-II [21, 22] and MSOPT [26] are described and used to solve the industrial applications described in chapter 2. These three simultaneous methods explicitly calculate the coordinate-basis and therefore belong to the class of direct PRSQP methods. This has advantages in case of discretized path constraints on the state variables.

The globalization of the PRSQP variants can be done by line-search or trust-region methods where, e.g., an (exact) l_1-penalty function ϕ is used to measure (sufficient) decrease:

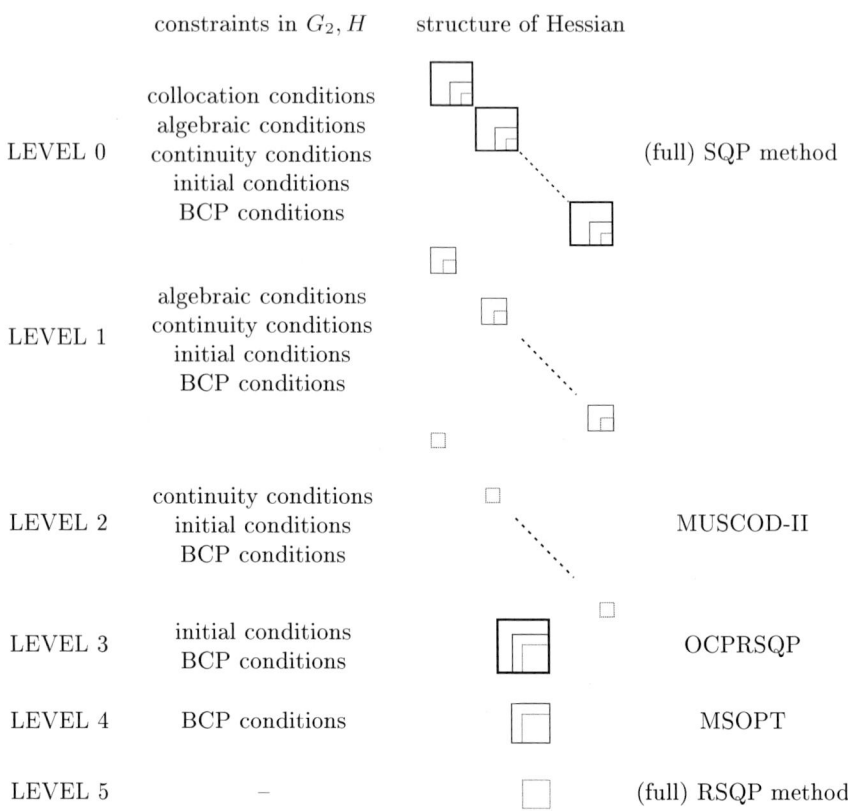

	constraints in G_2, H	structure of Hessian	
LEVEL 0	collocation conditions algebraic conditions continuity conditions initial conditions BCP conditions		(full) SQP method
LEVEL 1	algebraic conditions continuity conditions initial conditions BCP conditions		
LEVEL 2	continuity conditions initial conditions BCP conditions		MUSCOD-II
LEVEL 3	initial conditions BCP conditions		OCPRSQP
LEVEL 4	BCP conditions		MSOPT
LEVEL 5	–		(full) RSQP method

Fig. 4. Family of PRSQP methods for optimal control problems

$$\phi_{\rho,\sigma,\tau}(w) = F(w) + \sum_{i \in \mathcal{E}_1} \rho_i |G_{1,i}(w)| + \sum_{i \in \mathcal{E}_2} \sigma_i |G_{2,i}(w)| + \sum_{j \in \mathcal{I}} \tau_j H_j(w)^-$$

with $H^- := \max(-H, 0)$ and $\mathcal{E}_1, \mathcal{E}_2, \mathcal{I}$ the set of constraints indices in G_1, G_2, H respectively. In case of direct PRSQP methods where the Lagrangian multipliers are not calculated the compatibility between step and sufficient decrease of the penalty function can be achieved by partially multiplier free globalization method. Therefore we calculate lower bounds for the penalty parameters by:

$$\rho_i \|G_1\|_1 \geq |\lambda_1^T G_1|, \quad \sigma_i \geq |\lambda_{2,i}|, \quad \tau_j \geq \mu_j.$$

Note that $\lambda_1^T G_1$ can be computed cheaply by the following formula (motivated by the stationarity condition of the Lagrangian):

$$\lambda_1^T G_1 = -\nabla F S_1^{\mathcal{R}} y^{\mathcal{R}} + \lambda_2^T \nabla G_2^T S_1^{\mathcal{R}} y^{\mathcal{R}} + \mu^T \nabla H^T S_1^{\mathcal{R}} y^{\mathcal{R}}.$$

If the Lagrangian multipliers are calculated for all constraints (as is the case in adjoint PRSQP variants or if all Jacobians of G_1, G_2, H are calculated, e.g., in OCRPSQP) we can use the conventional strategy of penalizing the constraints individually: $\rho_i \geq |\lambda_{1,i}|$. Note that for ensuring convergence to a KKT point (w^*, λ^*, μ^*) the conditions for the penalty parameters have to be fulfilled also for the optimal multipliers (λ^*, μ^*). One way to achieve this is not to decrease the penalty parameters in a line-search step of the SQP method.

6 Efficient Computation of Directional Derivatives

In order to set up the reduced quadratic programs efficiently special care must be taken to the structure of the NLP problem. In case of collocation discretization in OCPRSQP all Jacobians (of objective and constraints) are calculated explicitly with, e.g., automatic differentiation tools like ADOL-C [18], combined with seed matrix compression as described in [13]. Then the reduced Jacobians and Lagrangian multipliers can be calculated recursively, see, e.g., [30].

In case of multiple shooting parameterization we have to compute the directional differential state derivatives in the null-space basis, see [22, 26]. This is done efficiently by the methods realized in the BDF integrator DAESOL [3]. The derivatives can be interpreted as the solution of the directional Variational Differential Algebraic Equations (VDAE). In case of MUSCOD-II the linearized BDF equations of the directional VDAE are *directly* solved at the solution of the nominal trajectory. Here the iteration matrix of the BDF equations has to be computed in every BDF step. For a high number of directions this approach is advantageous and shows good performance in practice. In case of MSOPT the nonlinear discretized BDF equations of the directional VDAE are solved *iteratively* together with discretized BDF equations of the nominal trajectory. Here the same iteration matrix is used. Both approaches conform to the principle of Internal Numerical Differentiation (IND), see [7]. The main idea is to differentiate the integrator scheme for calculation of the nominal trajectory. As a result the directional derivatives are computed efficiently and with high accuracy. For the BDF integrator the model Jacobians are evaluated. It is necessary to do this very fast with the required accuracy. This is done with the algorithmic differentiation tool Adol-C [18]. Since most model Jacobians are in general sparse a seedmatrix compression [13] is used to speed up computation time. Note that the sparsity structure of the model Jacobians has to be computed only once for each model stage and is then stored.

7 Multiple Set Point Optimization

Multiple set point problems as described in the application of shape optimization of compressor blades have a special structure. In our case we have an additive objective function together with global parameters connecting the different set points. Other formulations are min-max problems corresponding to an infinity norm of the objective components. In general each set point can be defined as a multistage optimal control problem described in chapter 3. The KKT system of the multiple set point problem has a block structure in the Hessian *and* in the constraint Jacobian which leads directly to a parallel computation approach. Note that the multiple set point structure of the shape optimization problem allows parallel computation of reduced gradients and multipliers for each set point independently. Only the solution of the QP is done sequentially.

8 Real-Time Iteration Scheme in On-Line Optimization

For steady state tracking problems in NMPC context we propose the following variant of the off-line optimization problem formulation (1) with the optimal cost function (1a) replaced a Bolza objective function to ensure nominal stability together with end point constraints in (1e):

$$\min_{u(\cdot),x(\cdot),z(\cdot)} \int_{t_0}^{t_0+T} L(x(t), z(t), u(t))\,dt \;+\; E(x(T), z(T)) \quad (14a)$$

subject to

$$\begin{align}
B(\cdot)\dot{x}(t) - f(x(t), z(t), u(t)) &= 0, \quad t \in [t_0, t_0+T], & (14b)\\
g(x(t), z(t), u(t)) &= 0, \quad t \in [t_0, t_0+T], & (14c)\\
x(t_0) - x_0 &= 0, & (14d)\\
r^{\mathrm{e}}(x(t_0+T), z(t_0+T)) &= 0, & (14e)\\
r^{\mathrm{i}}(x(t_0+T), z(t_0+T)) &\geq 0, & (14f)\\
h(x(t), z(t), u(t)) &\geq 0, \quad t \in [t_0, t_0+T]. & (14g)
\end{align}$$

Note that we additionally introduced fixed initial values x_0. Ideally this problem is solved instantaneously at every time t with initial value $x_0 \equiv x_0(t)$ and the optimal control solution of (14) is given to the real plant at time t. In the following we propose real-time algorithms to deliver an optimal feedback control, or, for moving horizons, a Receding Horizon Control (RHC) law. It basically consists of the direct multiple shooting algorithm for solution of (14) explained in section 5 where the number of solution iterations are restricted to one SQP iteration (real-time iteration) and the initial value x_0 is embedded as a homotopy parameter with the trivial equality constraint $s_0^x - x_0 = 0$ in the parameterization of problem (14). This allows us to divide the necessary computations during each real-time iteration into a long preparation phase where the QP problem is set up (except entries involving x_0) and a short feedback phase where the remaining entries are calculated, if x_0 is known and the QP problem is solved to deliver the feedback control. Note that the calculations for the preparation phase can be done in advance before the system state x_0 is known, thus minimizing the delay for a feedback control. A further advantage of the initial value embedding is that it delivers a second order prediction for the exact solution, thereby taking system nonlinearities into account. One main result is that the real-time iteration scheme on shrinking horizons is contracting under the *same* sufficient conditions as the corresponding off-line scheme with fully converged solutions, see [14]. Note that this result can be conceptually generalized to a shift strategy on infinite moving horizons and leads to a convergent closed-loop behaviour in this case.

For the two direct multiple shooting algorithms MUSCOD-II and MSOPT we present two real-time iteration schemes which differ in the PRSQP aspect and thereby in the necessary calculations in the feedback and preparation phase. Note that in case of integral least squares terms in the objective of (14) it's often appropriate to apply a Gauss-Newton approximation of the Lagrange-Hessian as a cheaper alternative to the exact Lagrange-Hessian. We abbreviate the NLP variable vector for the parameterization of problem (14) with $w := (s_0^x, s_0^z, q_0, \ldots, q_{n-1}, s_n^x, s_n^z)$, compare problem (10) and section 5.

The real-time iteration of MUSCOD-II consists of the following 5 steps:

1. Calculate coordinate-basis $S^{\mathcal{N}}$ and inhomogeneity $S^{\mathcal{R}} y^{\mathcal{R}}$ of the consistency constraints G_1.
2. Integrate and evaluate reduced constraints residuals of continuity, boundary and path constraints $G_2 + \nabla G_2^T S^{\mathcal{R}} y^{\mathcal{R}}$, $H + \nabla H^T S^{\mathcal{R}} y^{\mathcal{R}}$. Calculate reduced Jacobians of objective $\nabla F^T S^{\mathcal{N}}$ and constraints $\nabla (G_2, H)^T S^{\mathcal{N}}$ and reduced Hessian approximation.
3. Use linearized reduced continuity conditions of G_2 to eliminate differential state variables except first node (condensing).
4. At the moment that x_0 is known, solve the reduced QP in differential state at first node Δs_0^x and controls $\Delta q_0, \ldots, \Delta q_{N-1}$. The value $q_0 + \Delta q_0$ can immediately be given as a control to the real system.
5. Expand the fully reduced QP solution to yield the full QP solution $\Delta w = S^{\mathcal{R}} y^{\mathcal{R}} + S^{\mathcal{N}} y^{\mathcal{N}}$. Based on this QP solution, pass over to the next SQP iterate and go back to step 1.

Only step 4 has to done in the feedback phase, steps 5,1,2 and 3 in the preparation phase. Note that due to the derivative calculation the preparation phase takes several magnitudes longer than the feedback phase. In order to reduce the time in the preparation phase the number of directional derivatives can be reduced like in the extended PRSQP approach of MSOPT.

The real-time iteration of MSOPT is build of the following 5 steps:

1. Calculate coordinate-basis $S^{\mathcal{N}}$ of consistency, continuity and initial conditions G_1. Thereby integrate and solve relaxed directional variational DAE without inhomogeneity $S^{\mathcal{R}} y^{\mathcal{R}}$. Calculate residuals of the boundary and path constraints G_2, H.

2. Compute reduced Jacobians $\nabla(G_2, H)^T S^{\mathcal{N}}$ and reduced Hessian approximation.
3. At the moment that x_0 is known, calculate inhomogeneity $S^{\mathcal{R}} y^{\mathcal{R}}$ and the projected function values of the restrictions: $G_2 + \nabla G_2^T S^{\mathcal{R}} y^{\mathcal{R}}, H + \nabla H^T S^{\mathcal{R}} y^{\mathcal{R}}$.
4. Solve the reduced QP in $y^{\mathcal{N}} := (\Delta q_0, \ldots, \Delta q_{n-1})^T$ and give the feedback control $q_0 + \Delta q_0$ to the real system.
5. Expand the fully reduced QP solution to yield the full QP solution $\Delta w := S^{\mathcal{R}} y^{\mathcal{R}} + S^{\mathcal{N}} y^{\mathcal{N}}$. Based on this QP solution, pass over to the next SQP iterate and go back to step 1.

In this case step 3 and 4 have to be executed in the feedback phase and the remaining steps 5, 1 and 2 in the preparation phase. Since the differential *and* algebraic variables are eliminated in this PRSQP approach we have to calculate only directions derivatives w.r.t. controls which fasten the preparation phase for the burden of a longer feedback phase where the inhomogeneity has to be computed also. For the on-line optimization of the continuous distillation column (see section 2.3) we present numerical results of real-time iteration schemes for MUSCOD-II and MSOPT in section 9.3.

9 Numerical Results

9.1 Shape Optimization of Turbine Blades

Here we show examples for the efficacy of the adjoint PRSQP variant. In figure 5, the result of the optimization for a turbine blade at a specific working range set point – indicated by the downward arrow on the right hand side of this figure – is shown. Although the effect of the optimization is not much visible if the initial blade (industrial reference) is compared with the optimized one, the gain in the objective criterion (pressure loss) is approximately 14 % and thus considerable. Here, 13000 state variables are involved in the discretized flow description and 15 PRSQP iterations are needed for convergence. The overall computing time, however, was only 3 CPU minutes on an IBM workstation of type RS6000/900, which corresponds to 4 forward flow equation solutions (which takes 45 seconds each). This means that during the time of the simultaneous optimization run only 4 iterations of black box approaches can be done, which has to solve one forward flow equation in each iteration. Thus, the simultaneous optimization approach really pays off here, and modularity of the implementation is still maintained. The lower curve on the right hand side shows the objective functional over the whole working range (nevertheless, the blade is optimal only for the indicated set point). One can see that the optimized blade leads to good results all over the working range (figure 5).

However, the situation is completely different in the case of compressor blades, as it is shown in figure 6. On the left hand side two pairs of blades are plotted,

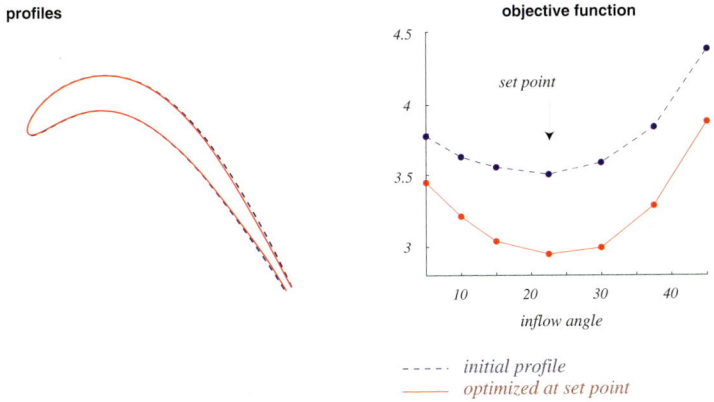

Fig. 5. Turbine blade, optimized at set point and over the working range, respectively.

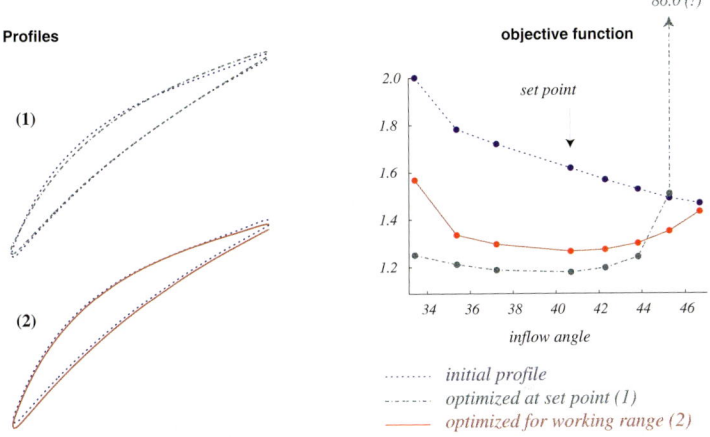

Fig. 6. Compressor blade, optimized at set point and over the working range, respectively.

where in each pair the initial blade before optimization and the results of the optimization are shown. The upper pair is for a single set point optimization for the set point (inflow angle) indicated on the right hand side by an arrow. This single set point optimization took 15 CPU minutes on a Pentium II/400-based Linux system, which again corresponds to 4 forward solution sweeps. Looking at the whole working range for the inflow angle, one recognizes that near the boundary of the working range the pressure loss is increasing dramatically for the blade optimized for a single set point. That is surely not what turbomachine engineers think of as an "optimal" blade. Therefore, for this case we employ also the working range formulation (2.1), where the eight set points $\{\alpha_1, \ldots, \alpha_8\}$ – i.e. discretizations of the working range $\mathcal{A}$ – are the inflow angles corresponding to the bullets in the right part of figure 6. The solution of this working range problem (in the form of a multiple set point problem) leads to blades which have better objective values than the initial blade profile at all set points. Furthermore, the pressure loss shows reasonable behavior all over the working range. However, as it can be expected, the blade optimized over the whole working range behaves slightly worse at the set point for which the single set point optimization was performed before (indicated by the downward arrow in figure 6).

9.2 Optimal Operation of a Catalytic Tube Reactor

For the capacity maximization of the tube reactor we use the simultaneous multiple shooting method MSOPT with a direct PRSQP approach. The discretized path constraints for the temperature makes an explicit calculation of the coordinate-basis advantageous. The optimization problem has over 25000 NLP variables but in each PRSQP iteration only 165 directional derivatives have to be computed. The CPU times for the optimization are in the range of 2 hours for 36 PRSQP iterations on a Intel Xeon 2.8 GHz Linux machine. Note that in case of path constraints (temperature bounds) the multiple shooting parameterization is particularly advantageous for defining startvalues for the states. In order to get an impression of the optimization results we compare the optimal operation with optimal catalyst dilution with a reference scenario of chemical industry with fixed catalyst dilution (100 %). In figure 7 both calculated throughputs are shown. The throughput with the optimized catalyst

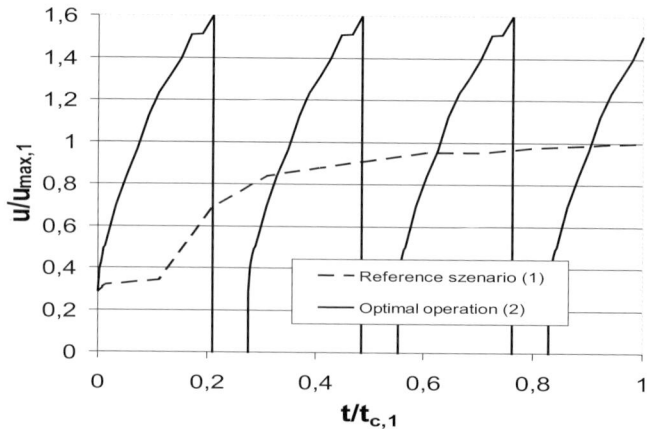

Fig. 7. Throughput of reference-szenario (with fixed catalyst dilution) and of optimal operation (with optimized catalyst dilution)

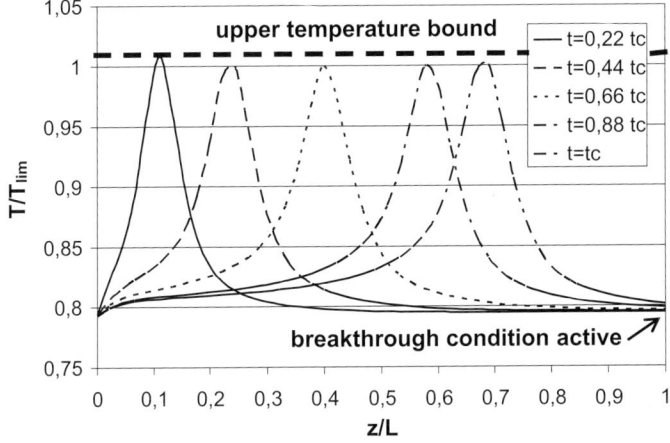

Fig. 8. Temperature of optimal operation in the center of tube

dilution is significantly higher than in the case of fixed dilution. Although the number of cycles (each cycle consists of a reaction phase and a regeneration phase) is increased, the yearly capacity is increased by about 12 %. The catalyst dilution also increased to 50 %, which additionally raises the profit. In figure 8 the temperature in the center of the tube is shown for consecutive time points. Note that the upper temperature bounds are not violated (so that the catalyst is not irreversibly deactivated) and the breakthrough condition of the reactands is active.

9.3 Real-Time Optimization of a Continuous Distillation Column

In figure 9 the results of the off-line optimization with a disturbed initial state are shown. We can see that at the end of the control horizon (1800 sec) the states are already in a steady state and the purity requirements on the products are fulfilled. The disturbed initial states are typically for on-line scenarios where the aim is to control the system such that the steady state is reached as fast as possible.

In table 1 the CPU times for one real-time iteration of MUSCOD-II and MSOPT are shown, where the computations were executed on a Intel Xeon 2.8 GHz Linux machine. The CPU times of the two real-time iteration schemes depend from the calculations which have to be done. In the extended PRSQP approach of MSOPT where the differential and algebraic variables are eliminated the residual of the initial state constraint is part of the range-space component (inhomogeneity). This is calculated by a forward evaluation of one directional derivative. After calculation of the range-space constraint residuals and the QP solution the feedback control is available. In the PRSQP approach of MUSCOD-II the elimination of the algebraic variables is local and the ini-

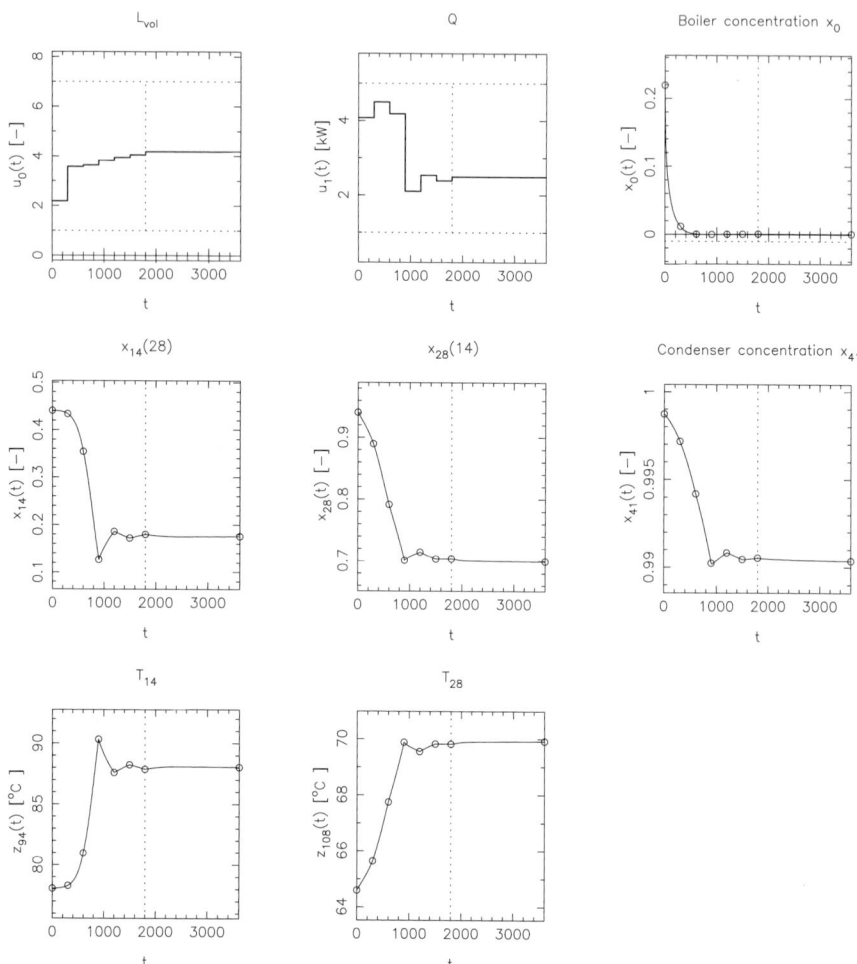

Fig. 9. Optimal operation of distillation column after a disturbance

tial constraints do not depend on the algebraic states. The only computation which has to be done after knowing the initial state is the solution of the QP. Directly afterwards the feedback-control is available. In contrast to the feedback phase the preparation phase is significantly shorter (factor 5) in case of MSOPT due to less directional derivatives which have to be computed in order to set up the reduced QP. Note that the times are given for the evaluation of directional VDAE and integration seperately in order to compare the evaluation time of the directional derivatives.

Table 1. Time comparison of real-time iteration schemes for MUSCOD-II and MSOPT

Phase	MUSCOD-II		MSOPT	
Feedback		< 0.01 sec		0.38 sec
			Inhomogeneity	0.36 sec
			Reduction	< 0.01 sec
	QP solution	< 0.01 sec	QP solution	< 0.01 sec
Preparation		17.44 sec		4.47 sec
	Reduction	0.22 sec	Reduction	0.11 sec
	Integration	1.21 sec	Integration	1.21 sec
	VDAE	15.98 sec	VDAE	3.15 sec
	Condensing	0.03 sec		

10 Conclusions

It has been shown that simultaneous methods for dynamic optimization problems can efficiently solve industrial applications whose dynamics are described by, e.g., a (time-dependent) PDE or stiff DAE. The close coupling of the simulation method with the optimization method is superior in terms of run-time and robustness compared to black box approaches (see the shape optimization of turbine blades). Special PRSQP variants have been developed which exploit the problem structure. For the multiple set point problems efficient extensions of the parallel simultaneous methods are available.

The shape optimization of the turbine blades (with stationary 2D-PDE dynamics) has been efficiently solved by an adjoint PRSQP method. The results of the multiple set point calculations are a low pressure loss over the whole working range (8 set points) compared to the single set point cases. A new multiple shooting variant with a direct PRSQP method has been developed (MSOPT). One application has been the capacity maximization of a catalytic tube reactor (with transient 2D-PDE dynamics). An optimal operation of a catalytic tube reactor has been calculated with increased profit through increased catalyst dilution *and* higher overall throughput. The real-time optimization of a continuous distillation column shows the potential of the direct multiple shooting variants MUSCOD-II and MSOPT for NMPC problems. It is shown that the time for one real-time iteration could be reduced by factor 5 (new developed MSOPT compared to MUSCOD-II).

References

1. R. Adler. Stand der Simulation von heterogen-gaskatalytischen Reaktionsabläufen in Festbettrohrreaktoren – Teil 1. *Chemie Ingenieur Technik*, 6:555–564, 2000.
2. R. Adler. Stand der Simulation von heterogen-gaskatalytischen Reaktionsabläufen in Festbettrohrreaktoren – Teil 2. *Chemie Ingenieur Technik*, 7:688–699, 2000.
3. I. Bauer. *Numerische Verfahren zur Lösung von Anfangswertaufgaben und zur Generierung von ersten und zweiten Ableitungen mit Anwendungen bei Optimierungsaufgaben in Chemie und Verfahrenstechnik*. PhD thesis, University of Heidelberg, 1999.
4. T. Binder, L. Blank, H. G. Bock, R. Bulirsch, W. Dahmen, M. Diehl, T. Kronseder, W. Marquardt, J. P. Schlöder, and O. v. Stryk. Introduction to model based optimization of chemical processes on moving horizons. In M. Grötschel, S.O. Krumke, and J. Rambau, editors, *Online Optimization of Large Scale Systems: State of the Art*, pages 295–340. Springer, 2001.
5. R. B. Bird, W. E. Stewart, and E. N. Lightfoot. *Transport Phenomena*. John Wiley & Sons, Inc., 2002.
6. J. Blazek. Aerodynamic shape optimization of turbomachinery cascades. Technical Report 97-2131 (A97-32475), AIAA, 1997.
7. H. G. Bock. Numerical treatment of inverse problems in chemical reaction kinetics. In K. H. Ebert, P. Deuflhard, and W. Jäger, editors, *Modelling of Chemical Reaction Systems (Springer Series in Chemical Physics 18)*, pages 102–125. Springer, Heidelberg, 1981.
8. H. G. Bock. *Randwertproblemmethoden zur Parameteridentifizierung in Systemen nichtlinearer Differentialgleichungen*, volume 183 of *Bonner Mathematische Schriften*. University of Bonn, Bonn, 1987.
9. H. G. Bock, W. Egartner, W. Kappis, and V. Schulz. Practical shape optimization for turbine and compressor blades by the use of PRSQP methods. *Journal on Optimization and Engineering*, 3:395–414, 2002.
10. H. G. Bock, E. Eich, and J. P. Schlöder. Numerical solution of constrained least squares boundary value problems in differential-algebraic equations. In K. Strehmel, editor, *Numerical Treatment of Differential Equations*. Teubner, Leipzig, 1988.
11. H. G. Bock and K.-J. Plitt. A multiple shooting algorithm for direct solution of optimal control problems. In *Proceedings of the 9th IFAC World Congress, Budapest*. Pergamon Press, 1984.
12. B. L. Braunschweig, C. C. Pantelides, H. I. Britt, and S. Sama. Process modeling: the promise of open software architectures. *Chemical Engineering Progress*, 96(9):65–76, 2000.
13. A. R. Curtis, M. J. D. Powell, and J. K. Reid. On the estimation of sparse Jacobian matrices. *J. Inst. Math. Appl.*, 13:117–119, 1974.
14. M. Diehl. *Real-Time Optimization for Large Scale Nonlinear Processes*, volume 920 of *Fortschr.-Ber. VDI Reihe 8, Meß, Steuerungs- und Regelungstechnik*. VDI Verlag, Düsseldorf, 2002.
15. M. Drela. *Two-dimensional transonic aerodynamic design and analysis using the Euler equations*. PhD thesis, MIT, 1985.
16. R. Fletcher. *Practical Methods of Optimization*. John Wiley & Sons, 1987.

17. D. Gabay. Reduced Quasi-Newton methods with feasibility improvement for nonlinear constrained optimization. *Mathematical Programming Study*, 16:18–44, 1982.
18. A. Griewank, D. Juedes, H. Mitev, J. Utke, O. Vogel, and A. Walther. ADOL-C: A package for the automatic differentiation of algorithms written in C/C++. *ACM TOMS*, 22(2):131–167, 1996.
19. A. Jameson, J. Reuther, L. Martinelli, and J. Vassberg. Aerodynamic shape optimization techniques based on control theory. Technical Report 98-2538 (A98-32805), AIAA, 1998.
20. F.-S. Kupfer. An infinite-dimensional convergence theory for reduced SQP methods in Hilbert space. *SIAM Journal on Optimization*, 6(1):126–163, 1996.
21. D. B. Leineweber. The theory of MUSCOD in a nutshell. IWR-Preprint 96-19, University of Heidelberg, 1996.
22. D. B. Leineweber. *Efficient reduced SQP methods for the optimization of chemical processes described by large sparse DAE models*, volume 613 of *Fortschr.-Ber. VDI Reihe 3, Verfahrenstechnik*. VDI Verlag, Düsseldorf, 1999.
23. J. Nocedal and M. L. Overton. Projected Hessian updating algorithms for nonlinearly constrained optimization. *SIAM Journal on Numerical Analysis*, 22(5):821–850, 1985.
24. M. J. D. Powell. A fast algorithm for nonlinearly constrained optimization calculations. In G. A. Watson, editor, *Numerical Analysis*, number 630 in Lect. Not. in Math., pages 144–157, 1978.
25. J. Reuther, A. Jameson, L. Martinelli, and D. Saunders. Aerodynamic shape optimization of complex aircraft configurations via an adjoint formulation. Technical Report 96-0094 (A96-18067), AIAA, 1996.
26. A. Schäfer. *Effiziente reduzierte Newton-ähnliche Verfahren zur Behandlung hochdimensionaler strukturierter Optimierungsprobleme mit Anwendung bei biologischen und chemischen Prozessen*. PhD thesis, University of Heidelberg, 2004.
27. J. P. Schlöder. *Numerische Methoden zur Behandlung hochdimensionaler Aufgaben der Parameteridentifizierung*, volume 187 of *Bonner Mathematische Schriften*. University of Bonn, Bonn, 1988.
28. V. H. Schulz. *Reduced SQP Methods for Large-Scale Optimal Control Problems in DAE with Application to Path Planning Problems for Satellite Mounted Robots*. PhD thesis, Ruprecht-Karls-Universität Heidelberg, 1996.
29. V. H. Schulz. Solving discretized optimization problems by partially reduced SQP methods. *Computing and Visualization in Science*, 1(2):83–96, 1998.
30. V. H. Schulz, H. G. Bock, and M. C. Steinbach. Exploiting invariants in the numerical solution of multipoint boundary value problems for DAEs. *SIAM J. Sci. Comp.*, 19:440–467, 1998.
31. H. Youngren. *Analysis and design of transonic cascades with splitter vanes*. PhD thesis, MIT, 1991.

Adaptive Finite Element Methods for PDE-Constrained Optimal Control Problems*

R. Becker[1], M. Braack[2], D. Meidner[2], R. Rannacher[2], and B. Vexler[3]

[1] Laboratoire de Mathématiques Appliquées, Université de Pau et des Pays de l'Adour, France
[2] Institut für Angewandte Mathematik, Universität Heidelberg
[3] Johann Radon Institute for Computational and Applied Mathematics, Austrian Academy of Sciences, Austria

Summary. We present a systematic approach to error control and mesh adaptation in the numerical solution of optimal control problems governed by partial differential equations. By the Lagrangian formalism the optimization problem is reformulated as a saddle-point boundary value problem which is discretized by a finite element Galerkin method. The accuracy of the discretization is controlled by residual-based a posteriori error estimates. The main features of this method are illustrated by examples from optimal control of heat transfer, fluid flow and parameter estimation. The contents of this article is as follows:

- Preliminary thoughts
- A general framework for a posteriori error estimation
- Solution process and mesh adaptation
- Examples of optimal control problems
- Conclusion and outlook
- References

1 Preliminary Thoughts

Let the goal of a numerical simulation be the computation of a quantity $J(u)$ from the solution of a continuous model $\mathcal{A}(u) = 0$ with accuracy TOL by using a discrete model $\mathcal{A}_h(u_h) = 0$ of dimension N. Accordingly, the goal of adaptivity is the optimal use of computing resources, i.e.,

$$\{TOL \text{ given}: \ N \to \min\} \quad \text{or} \quad \{N \text{ given}: \ TOL \to \min\}.$$

To achieve this goal, a posteriori information is used in terms of cell-wise refinement indicators $\eta_K := \rho_K(u_h)\omega_K$ based on 'smoothness' or 'residual'

---

*This work has been supported by the German Research Foundation (DFG) through SFB 359 (Project A2) at the University of Heidelberg.

information, where ω_K are certain weighting factors. These weights are obtained from the approximate solution of an associated global linear dual problem which is driven by the target functional $J(\cdot)$ as right-hand side. This approach, called the 'Dual Weighted Residual (DWR) Method', has been developed in Becker/Rannacher [15, 16] to facilitate systematic a posteriori error estimation and mesh adaptation in the finite element Galerkin approximation of general variational problems. For a comprehensive discussion of the DWR method and related references, we refer to the survey article Becker/Rannacher [17] and the book Bangerth/Rannacher [2]. Similar strategies of duality-based error control and mesh adaptation are described in Eriksson et al. [27] and Giles/Süli [30].

The application of mesh adaptation in the discretization of optimal control problems such as

$$J(u,q) \to \min, \qquad \mathcal{A}(u,q) = 0, \tag{1}$$

with states u and controls q, raises several fundamental questions:

- *What is the right notion of 'admissibility' of states $u = u(q)$?* Discretization inevitably introduces perturbation of the state equation. Achieving high accuracy in the discretization of PDEs is expensive. Hence, the extent to which 'admissibility' is relevant for the optimization process becomes a critical question.
- *How should admissibility be 'measured'?* In solving ODEs, one may require the error to be uniformly 'small', but in the context of PDEs the choice of an appropriate error measure is less clear.

In the following, we consider finite element discretization using continuous piecewise polynomial trial and test functions for all unknowns on meshes $\mathbb{T}_h = \{K\}$ which consist of non-degenerate quadrilaterals ('cells') K of width $h_K := \text{diam}(K)$. The 'global mesh size' is $h := \max_{K \in \mathbb{T}_h} h_K$. These meshes are allowed to have 'hanging nodes' for simplifying local mesh refinement. Figure 1 shows such a locally refined mesh and also indicates the interplay of local errors and local residuals (data perturbations) which is governed by a generalized global Green function.

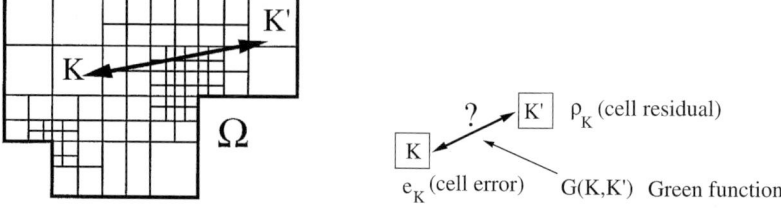

Fig. 1. Finite element mesh and scheme of error propagation.

2 A General Framework for A Posteriori Error Estimation

Let V be a Hilbert space of states and Q a space of controls. Then, with a three-times differentiable cost functional $J(\cdot,\cdot): V \times Q \to \mathbb{R}$ and semi-linear form $a(\cdot,\cdot)(\cdot)$ on $V \times Q \times V$, we consider the following optimization problem:

$$J(u,q) \to \min, \qquad a(u,q)(\psi) = 0 \quad \forall \psi \in V. \tag{2}$$

We assume that there is a locally unique minimum $\{u,q\} \in V \times Q$ which is characterized as corresponding to a saddle-point $\{u,q,\lambda\} \in V \times Q \times V$ of the Lagrangian functional

$$\mathcal{L}(u,q,\lambda) := J(u,q) - a(u,q)(\lambda),$$

where λ denotes the associated co-state ('adjoint' or 'dual' variable). Conditions under which this is true can be found in the standard literature on optimal control theory, e.g., Lions [36] and Fursikov [29]. The triplet $\{u,q,\lambda\} \in V \times Q \times V$ is determined by the saddle-point problem (so-called 'optimality system')

$$\begin{aligned}
a'_u(u,q)(\varphi,\lambda) &= J'_u(u,q)(\varphi) \quad \forall \varphi \in V, \\
a'_q(u,q)(\chi,\lambda) &= J'_q(u,q)(\chi) \quad \forall \chi \in Q, \\
a(u,q)(\psi) &= 0 \quad \forall \psi \in V.
\end{aligned} \tag{3}$$

Here, $a'_u(u,q)(\varphi,\cdot)$, $a'_q(u,q)(\chi,\cdot)$, $J'_u(u,q)(\varphi)$, and $J'_q(u,q)(\chi)$ denote the directional derivatives of the semi-linear form $a(u,q)(\cdot)$ and the functional $J(u,q)$ in the directions φ and χ, respectively.

The optimization problem is discretized by a standard Galerkin method using finite dimensional subspaces $V_h \times Q_h \subset V \times Q$:

$$J(u_h, q_h) \to \min, \qquad a(u_h, q_h)(\psi_h) = 0 \quad \forall \psi_h \in V_h. \tag{4}$$

Its local solutions $\{u_h, q_h\} \in V_h \times Q_h$ correspond to saddle points $x_h := \{u_h, q_h, \lambda_h\} \in X_h := V_h \times Q_h \times V_h$ of the Lagrangian $\mathcal{L}(\cdot,\cdot,\cdot)$, i.e., to solutions of the discrete saddle-point problems

$$\begin{aligned}
a'_u(u_h, q_h)(\varphi_h, \lambda_h) &= J'_u(u_h, q_h)(\varphi_h) \quad \forall \varphi_h \in V_h, \\
a'_q(u_h, q_h)(\chi_h, \lambda_h) &= J'_q(u_h, q_h)(\chi_h) \quad \forall \chi_h \in Q_h, \\
a(u_h, q_h)(\psi_h) &= 0 \quad \forall \psi_h \in V_h.
\end{aligned} \tag{5}$$

The idea is now to seek control of the discretization error with respect to the cost functional $J(\cdot,\cdot)$ of the optimal control problem in terms of the 'primal', 'dual', and 'control' residuals defined by

$$\begin{aligned}
\rho^*(\cdot) &:= J'_u(u_h, q_h)(\cdot) - a'_u(u_h, q_h)(\cdot, \lambda_h) \quad \text{(dual residual)} \\
\rho^q(\cdot) &:= J'_q(u_h, q_h)(\cdot) - a'_q(u_h, q_h)(\cdot, \lambda_h), \quad \text{(control residual)} \\
\rho(\cdot) &:= -a(u_h, q_h)(\cdot) \quad \text{(primal residual)}
\end{aligned}$$

From Becker/Rannacher [17] we recall the following general result.

Theorem 1. *There holds the a posteriori error representation*

$$J(u,q) - J(u_h, q_h) = \tfrac{1}{2}\rho^*(u - i_h u) + \tfrac{1}{2}\rho^q(q - i_h q) \\ + \tfrac{1}{2}\rho(\lambda - i_h \lambda) + \mathcal{R}_h, \qquad (6)$$

for arbitrary approximations $i_h u$, $i_h \lambda \in V_h$ *and* $i_h q \in Q_h$. *The remainder* $\mathcal{R}_h$ *is cubic in the errors* $e^u := u - u_h$, $e^q := q - q_h$, *and* $e^\lambda := \lambda - \lambda_h$.

Proof. On the product spaces $X := V \times Q \times V$ and $X_h := V_h \times Q_h \times V_h$ of triplets $x := \{u, q, \lambda\}$ and $x_h := \{u_h, q_h, \lambda_h\}$, respectively, we define the functional $L(x) := \mathcal{L}(u, q, \lambda)$. Then, solving the saddle-point problems (3) and (5) is equivalent to determining stationary points x and x_h of $L(\cdot)$ on X and X_h, respectively, i.e.,

$$L'(x)(y) = 0 \quad \forall y \in X, \qquad L'(x_h)(y_h) = 0 \quad \forall y_h \in X_h.$$

By elementary calculus and observing $L'(x)(e) = 0$, we have

$$L(x) - L(x_h) = \int_0^1 L'(x_h + se)(e)\,ds \\ - \tfrac{1}{2}\{L'(x_h)(e) + L'(x)(e)\} + \tfrac{1}{2}L'(x_h)(e).$$

Next, using the discrete equation $L'(x_h)(y_h) = 0$, $y_h \in X_h$, we obtain

$$L'(x_h)(e) = L'(x_h)(x - i_h x) + L'(x_h)(i_h x - x_h) = L'(x_h)(x - i_h x),$$

for arbitrary $i_h x \in X_h$. The well-known error formula of the trapezoidal rule,

$$\int_0^1 f(s)\,ds - \tfrac{1}{2}\{f(0) + f(1)\} = \tfrac{1}{2}\int_0^1 f''(s)s(s-1)\,ds,$$

yields the following a posteriori error representation, for arbitrary $i_h x \in X_h$,

$$L(x) - L(x_h) = \tfrac{1}{2}L'(x_h)(x - i_h x) + \mathcal{R}_h,$$

with a remainder term which is cubic in $e := x - x_h$,

$$\mathcal{R}_h := \tfrac{1}{2}\int_0^1 L'''(x_h + se)(e, e, e)\,s(s-1)\,ds.$$

Then, observing that

$$L(x) - L(x_h) = \mathcal{L}(u, q, \lambda) - \mathcal{L}(u_h, q_h, \lambda_h) \\ = J(u, q) - a(u, q)(\lambda) - J(u_h, q_h) + a(u_h, q_h)(\lambda_h) \\ = J(u, q) - J(u_h, q_h),$$

we conclude the identity (6).

The result of Theorem 1 requires some explanations for its application to practical problems:

- The derivation of the error representation (6) does not require the uniqueness of solutions. The a priori assumption

$$\{u_h, q_h, \lambda_h\} \to \{u, q, \lambda\} \quad (h \to 0)$$

  makes the result meaningful for non-unique solutions.
- The evaluation of the error identity requires guesses for the continuous solution $\{u, q, \lambda\}$ which are obtained from $\{u_h, q_h, \lambda_h\}$ by post-processing. The cubic remainder term $\mathcal{R}_h$ is usually neglected.
- The generally nonlinear saddle-point problem (5) can be solved by a Newton-like iteration with mesh adaptation in each step. Hence, the combined process may be viewed as a successive 'model enrichment'.
- The systematic mesh adaptation based on the a posteriori error representation (6) allows us to compute approximate optimal controls q_h^{opt} on rather coarse meshes such that the corresponding discrete states u_h^{opt} may lack admissibility, though the optimal cost is well achieved. In order to recover more admissible states, one may compute a better approximation u_H^{opt} to u^{opt} from the optimal control q_h by solving the state equation in a larger space $V_H \supset V_h$,

$$a(u_H^{opt}, q_h^{opt})(\psi_H) = 0 \quad \forall \psi_H \in V_H.$$

However, if accuracy in the optimal control itself is required, such as common in parameter identification problems, one needs to use a more refined approach as discussed below.

3 Solution Process and Mesh Adaptation

We briefly discuss the solution process of the optimal control problems discretized by Galerkin finite element methods and the essential steps of the corresponding mesh adaptation based on the a posteriori error representation (6). For more details and extensions we refer to Becker/Rannacher [16], Becker/Kapp/Rannacher [12], Bangerth/Rannacher [2], Becker [4], Vexler [43], and Becker/Meidner/Vexler [14]. For more details on the evaluation of a posteriori error estimates and their practical use in the mesh adaptation process, we refer to the article Braack/Richter [22] in this volume.

3.1 Solution Process

The constrained optimal control problem (1) can be reformulated as an unconstrained optimal control problem

$$j(q) := J(S(q), q) \to \min, \qquad (7)$$

where $S: Q \to V$ is the solution operator of the state equation

$$a(S(q), q)(\psi) = 0 \quad \forall \psi \in V.$$

The local existence and sufficient regularity of S is assumed. Then, the first and second-order necessary conditions for an optimal q are that

$$j'(q)(\delta q) = 0, \quad j''(q)(\delta q, \delta q) \geq 0 \quad \forall \delta q \in Q. \qquad (8)$$

The derivatives of the reduced functional can be computed using the Lagrangian $\mathcal{L}(u, q, \lambda) = J(u, q) - a(u, q)(\lambda)$ considered above. This is the basis of Newton-SQP methods for solving the optimality equation such as described in Tröltzsch [42] and Hinze/Kunisch [33]. For given $q \in Q$, let $u = S(q) \in V$ be the corresponding state and $\lambda \in V$ the corresponding co-state determined by the dual equation (see the optimality system (3))

$$\mathcal{L}'_u(u, q, \lambda)(\psi) = 0 \quad \forall \psi \in V.$$

Then, for $\delta q \in Q$, there holds

$$j'(q)(\delta q) = \mathcal{L}'_q(u, q, \lambda)(\delta q). \qquad (9)$$

Further for given $\delta q \in Q$, let $\delta u \in V$ and $\delta \lambda \in V$ fulfill the equation

$$\mathcal{L}''_{q\lambda}(u, q, \lambda)(\delta q, \varphi) + \mathcal{L}''_{u\lambda}(u, q, \lambda)(\delta u, \varphi) = 0 \quad \forall \varphi \in V,$$

and the equation

$$\mathcal{L}''_{qu}(u, q, \lambda)(\delta q, \varphi) + \mathcal{L}''_{uu}(u, q, \lambda)(\delta u, \varphi) + \mathcal{L}''_{u\lambda}(u, q, \lambda)(\delta \lambda, \varphi) = 0 \quad \forall \varphi \in V,$$

respectively. Then, for $\delta r \in Q$, there holds

$$\begin{aligned} j''(q)(\delta q, \delta r) &= \mathcal{L}''_{qq}(u, q, \lambda)(\delta q, \delta r) + \mathcal{L}''_{uq}(u, q, \lambda)(\delta u, \delta r) \\ &+ \mathcal{L}''_{\lambda q}(u, q, \lambda)(\delta \lambda, \delta r). \end{aligned} \qquad (10)$$

Now, with this notation the classical Newton method (expressed on the continuous level) for solving the optimality equation in (8) reads

$$j''(q^n)(\delta q^n, \chi) = -j'(q^n)(\chi) \quad \forall \chi \in Q, \qquad (11)$$

with the update step $q^{n+1} := q^n + \delta q^n$. If for the solution of this linear system the conjugate gradient method or one of its variants are used, only the evaluations $j''(q^n)(\delta q^n, \chi)$ and $j'(q^n)(\chi)$ are required for fixed χ. This can efficiently be achieved based on the formulas (9) and (10), respectively, after a suitable discretization (for details see Becker/Meidner/Vexler [14]).

3.2 Evaluation of A Posteriori Error Estimates

Neglecting the remainder in (6), we obtain the approximate error estimate

$$J(u,q) - J(u_h, q_h) \approx \eta := \tfrac{1}{2}\rho^*(u - i_h u) + \tfrac{1}{2}\rho^q(q - i_h q) + \tfrac{1}{2}\rho(\lambda - i_h \lambda). \quad (12)$$

Suppose that V_h is an intermediate finite element space obtained in the course of a successive mesh adaptation process. Let $\{u_h, q_h, \lambda_h\} \in V_h \times Q_h \times V_h$ be a corresponding solution of the saddle-point problem (5). In order to do the next adaptation step, we have to evaluate the cell-wise contributions to the residual terms in the error estimate (12). To this end, we first integrate cellwise by parts (as will be illustrated in the examples below) and then approximate the weights $u - i_h u$, $\lambda - i_h \lambda$, and $q - j_h q$ by locally post-processing the discrete solution $\{u_h, q_h, \lambda_h\}$. The simplest way is local higher-order interpolation, where for example, in the case of (isoparametric) bilinear finite elements on quadrilateral meshes $\mathbb{T}_h = \{K\}$ a patchwise biquadratic interpolation $i_{2h}^{(2)} u_h$ is constructed using the nodal values of u_h. Then it is set

$$(u - i_h u)_{|K} \approx (i_{2h}^{(2)} u_h - u_h)_{|K},$$

and analogously for λ and q. This results in local error indicators η_K associated to the cells of the mesh $\mathbb{T}_h$.

3.3 Mesh Adaptation Strategy

The cells K of the current mesh $\mathbb{T}_h$ are ordered with respect to the size of the associated error indicators, i.e., $\eta_{K_1} \leq \cdots \leq \eta_{K_{N_h}}$. The so-called 'fixed fraction strategy' splits the cells into two groups, according to the conditions

$$\sum_{i=1}^{N_*} \eta_{K_i} \approx a\eta, \quad \sum_{i=N^*}^{N_h} \eta_{K_i} \approx b\eta,$$

with pre-assigned fractions $0 < a < b < 1$ (say $a = b = 0.2$). Then, the cells $K_1, \ldots, K_{N_*}$ are refined and the cells $K_{N^*}, \ldots K_{N_h}$ are coarsened. This process is continued until the prescribed error tolerance TOL or the maximal available number $N_{\max}$ of cells are reached, i.e., $\eta \approx TOL$ or $N_h \approx N_{\max}$. We note that there are much more sophisticated strategies for mesh adaptation (see Becker/Rannacher [17]).

3.4 Nonlinear Nested Solution Process

We briefly describe the interplay of nonlinear Newton iteration and mesh adaptation used in the optimization cycle. For notational simplicity an abstract setting is used. Consider a nonlinear variational problem posed in variational form

$$a(u)(\varphi) = 0 \quad \forall \varphi \in V, \tag{13}$$

which represents the first-order optimality system (3). The goal of the computation is the evaluation of a functional $E(u)$.

Let a desired error tolerance TOL or a maximum mesh complexity $N_{\max}$ be given. Starting from a coarse initial mesh $\mathbb{T}_0$, a hierarchy of successively refined meshes $\mathbb{T}_i$, $i \geq 1$, and corresponding finite element spaces V_i, is generated by the following algorithm.

(0) *Initialization* $i = 0$: Compute an initial approximation $u_0 \in V_0$.
(1) *Defect correction iteration:* For $i \geq 1$, start with $u_i^{(0)} := u_{i-1} \in V_i$.
(2) *Iteration step:* For $j \geq 0$ evaluate the defect

$$(d_i^{(j)}, \varphi) := -a(u_i^{(j)})(\varphi), \quad \varphi \in V_i. \tag{14}$$

Select a suitable approximation $\tilde{a}'(u_i^{(j)})(\cdot, \cdot)$ to the derivative $a'(u_i^{(j)})(\cdot, \cdot)$ (with good stability and solvability properties) and compute a correction $\delta u_i^{(j)} \in V_i$ from the linear equation

$$\tilde{a}'(u_i^{(j)})(\delta u_i^{(j)}, \varphi) = (d_i^{(j)}, \varphi) \quad \forall \varphi \in V_i. \tag{15}$$

For that, Krylov space and multigrid methods are employed using the hierarchy $\{\mathbb{T}_i, ..., \mathbb{T}_0\}$ of already constructed meshes. Then, update $u_i^{(j+1)} = u_i^{(j)} + \lambda_i \delta u_i^{(j)}$, with some relaxation parameter $\lambda_i \in (0, 1]$, set $j := j+1$, and go back to (2). This process is repeated until a limit $\tilde{u}_i \in V_i$ is reached with a prescribed accuracy. This can be controlled by monitoring the algebraic residual $\|d_i^{(j)}\|$, which can additionally be weighted cell-wise by the current approximation z_i to the adjoint solution.

(3) *Error estimation:* Accept $u_i := \tilde{u}_i$ as the solution on mesh $\mathbb{T}_i$, solve the discrete linearized adjoint problem

$$z_i \in V_i: \quad a'(u_i)(\varphi, z_i) = E'(u_i; \varphi) \quad \forall \varphi \in V_i,$$

and evaluate the *a posteriori* error estimate

$$|E(u) - E(u_i)| \approx \eta_\omega(u_i). \tag{16}$$

For controlling the reliability of the bound (16), that is the accuracy of the linearization and the determination of the approximate dual solutions z_i, one may use the algorithm described below. If the error estimator $\eta_\omega(u_i)$ is detected to be reliable and $\eta_\omega(u_i) \leq \text{TOL}$ or $N_i \geq N_{\max}$, then stop. Otherwise mesh adaptation yields the new mesh $\mathbb{T}_{i+1}$. Set $i := i+1$ and go back to (1).

Remark 1. We emphasize that the evaluation of the *a posteriori* error estimate (16) involves only the solution of *linearized* problems. Hence, the whole error estimation may amount to only a relatively small fraction of the total cost for the solution process. This has to be compared to the usually much higher cost when working on non-optimized meshes.

4 Examples of Optimal Control Problems

4.1 A Stationary Model Problem

The material of this section is taken from Kapp [34] and Becker/Kapp/Rannacher [12]. We consider a stationary diffusion problem with 'boundary control',

$$-\Delta u + s(u) = 0 \quad \text{in } \Omega,$$
$$\partial_n u_{|\Gamma_N} = 0, \quad \partial_n u_{|\Gamma_C} = q, \tag{17}$$

in the two-dimensional domain shown in Fig. 2, with boundary $\partial \Omega = \Gamma_N \cup \Gamma_C$. The nonlinearity is taken as $s(u) = u^3$. The 'Neumann control' q on Γ_C, the 'control boundary', is to be determined such that

$$J(u, q) = \tfrac{1}{2}\|u - \bar{u}\|_{\Gamma_O}^2 + \tfrac{1}{2}\alpha\|q\|_{\Gamma_C}^2 \to \min,$$

for prescribed profile $\bar{u}$ on the 'observation boundary' $\Gamma_O \subset \partial \Omega$. The α-term, with $\alpha \geq 0$, may be viewed as a model for the control costs or simply as a mean of regularization.

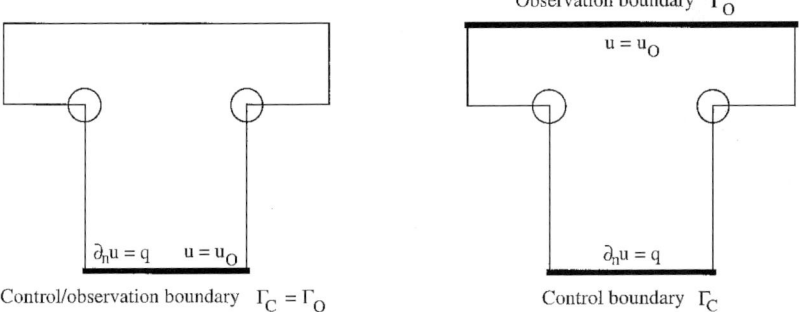

Fig. 2. Configuration 1 (left), Configuration 2 (right)

The variational formulation of the state equation reads

$$(\nabla u, \nabla \psi)_\Omega + (s(u), \psi)_\Omega - (q, \psi)_{\Gamma_C} = 0 \quad \forall \psi \in V,$$

with the state space $V := H^1(\Omega)$ and the control space $Q := L^2(\Gamma_C)$. Using the Lagrange approach described above, we obtain the following optimality system for stationary points $\{u, q, \lambda\} \in V \times Q \times V$ of the Lagrangian:

$$(\nabla \psi, \nabla \lambda)_\Omega + (s'(u)\psi, \lambda)_\Omega = (u - \bar{u}, \psi)_{\Gamma_O} \quad \forall \psi \in V,$$
$$(\lambda, \chi)_{\Gamma_C} = -\alpha(q, \chi)_{\Gamma_C} \quad \forall \chi \in Q,$$
$$(\nabla u, \nabla \varphi)_\Omega + (s(u), \varphi)_\Omega - (q, \varphi)_{\Gamma_C} = 0 \quad \forall \varphi \in V,$$

or written in strong form,

$$-\Delta\lambda + s'(u)\lambda = 0 \text{ in } \Omega,$$
$$\partial_n\lambda_{|\Gamma_N} = 0, \quad \partial_n\lambda_{|\Gamma_C\setminus\Gamma_O} = 0, \quad \partial_n\lambda_{|\Gamma_O} = u - \bar{u}, \tag{18}$$
$$\lambda_{|\Gamma_C} = -\alpha q, \tag{19}$$
$$-\Delta u + s(u) = 0 \text{ in } \Omega,$$
$$\partial_n u_{|\Gamma_N} = 0, \quad \partial_n u_{|\Gamma_C} = q. \tag{20}$$

The Galerkin approximation of this system uses spaces V_h of piecewise bilinear finite elements, on quadrilateral meshes, for the states and co-states and spaces Q_h of normal traces on Γ_C of elements in V_h for the controls. The corresponding discrete optimality system reads

$$(\nabla\psi_h, \nabla\lambda_h)_\Omega - (s'(u_h)\psi_h, \lambda_h)_\Omega = (u_h - \bar{u}, \psi_h)_{\Gamma_O} \quad \forall \psi_h \in V_h,$$
$$(\lambda_h, \chi_h)_{\Gamma_C} = -\alpha(q_h, \chi_h)_{\Gamma_C} \quad \forall \chi_h \in Q_h,$$
$$(\nabla u_h, \nabla\varphi_h)_\Omega + (s(u_h), \varphi_h)_\Omega - (q_h, \varphi_h)_{\Gamma_C} = 0 \quad \forall \varphi_h \in V_h.$$

For the solution of this weakly coupled saddle-point system, we employ the Newton-like iteration as described above within the abstract setting. For this situation Theorem 1 yields the a posteriori error estimate

$$|J(u,q) - J(u_h, q_h)| \approx \eta_\omega := \sum_{K \in \mathbb{T}_h} \{\rho^u_K \omega^\lambda_K + \rho^q_K \omega^q_K + \rho^\lambda_K \omega^u_K\}, \tag{21}$$

where the cell-residuals and weights of $\{u_h, q_h, \lambda_h\}$ are given by

$$\rho^\lambda_K := \|R^\lambda_h\|_K + h_K^{-1/2}\|r^\lambda_h\|_{\partial K}, \quad \omega^u_K := \|u - I_h u\|_K + h_K^{1/2}\|u - i_h u\|_{\partial K},$$
$$\rho^q_K := h_K^{-1/2}\|r^q_h\|_{\partial K}, \quad \omega^q_K := h_K^{1/2}\|q - i_h q\|_{\partial K},$$
$$\rho^u_K := \|R^u_h\|_K + h_K^{-1/2}\|r^u_h\|_{\partial K}, \quad \omega^\lambda_K := \|\lambda - i_h \lambda\|_K + h_K^{1/2}\|\lambda - I_h \lambda\|_{\partial K}.$$

with arbitrary approximations $\{i_h u, i_h q, i_h \lambda\} \in V_h \times Q_h \times V_h$. Here, $h_K := \mathrm{diam}(K)$ is the width of cell K and the subscripts $\|\cdot\|_K$ and $\|\cdot\|_{\partial K}$ indicate that the norms are taken over K and ∂K, respectively. The cell-wise residuals are defined by

$$R^u_{h|K} := \Delta u_h - s(u_h), \quad R^\lambda_{h|K} := \Delta\lambda_h - s'(u_h)\lambda_h,$$

$$r^u_{h|\Gamma} := \begin{cases} \frac{1}{2}[\partial_n u_h], & \text{if } \Gamma \not\subset \partial\Omega, \\ \partial_n u_h, & \text{if } \Gamma \subset \partial\Omega\setminus\Gamma_C, \\ \partial_n u_h - q_h, & \text{if } \Gamma \subset \Gamma_C, \end{cases} \quad r^q_{h|\Gamma} := \begin{cases} \lambda_h + \alpha q_h, & \text{if } \Gamma \subset \Gamma_C, \\ 0, & \text{if } \Gamma \not\subset \Gamma_C, \end{cases}$$

$$r^\lambda_{h|\Gamma} := \begin{cases} \frac{1}{2}[\partial_n \lambda_h], & \text{if } \Gamma \not\subset \partial\Omega, \\ \partial_n \lambda_h, & \text{if } \Gamma \subset \partial\Omega\setminus\Gamma_O, \\ \partial_n \lambda_h - u_h + \bar{u}, & \text{if } \Gamma \subset \Gamma_O. \end{cases}$$

We will compare the performance of mesh adaptation based on the weighted error estimator η_ω with that based on the following two heuristic 'energy-norm' error indicators:

$$\eta_E^* := \Big(\sum_{K \in \mathcal{T}_h} h_K^2 (\rho_K^u)^2 \Big)^{1/2}, \quad \eta_E := \Big(\sum_{K \in \mathcal{T}_h} h_K^2 \{(\rho_K^u)^2 + (\rho_K^\lambda)^2\} \Big)^{1/2}.$$

The accuracy of an error estimator η is measured by the 'effectivity index',

$$I_{\text{eff}} := \frac{|J(u,q) - J(u_h, q_h)|}{\eta}.$$

Configuration 1: This control problem is extreme since observation and control boundary coincide. Hence, the larger structure of the domain Ω, especially the reentrant corners, should not need to be resolved by the computational mesh. This is clearly confirmed by the results shown in Fig. 3 and 4 for the error estimators η_ω, η_E^*, and η_E.

N	E_{rel}	I_{eff}
596	2.5e-4	0.34
1616	2.3e-4	0.81
5084	8.2e-5	0.46
8648	4.2e-5	0.29
15512	3.9e-5	0.43

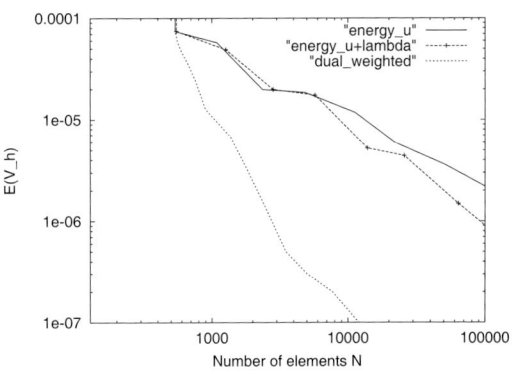

Fig. 3. Config. 1: Results for η_ω (left), and mesh efficiencies for η_E^*, η_E, and η_ω (right)

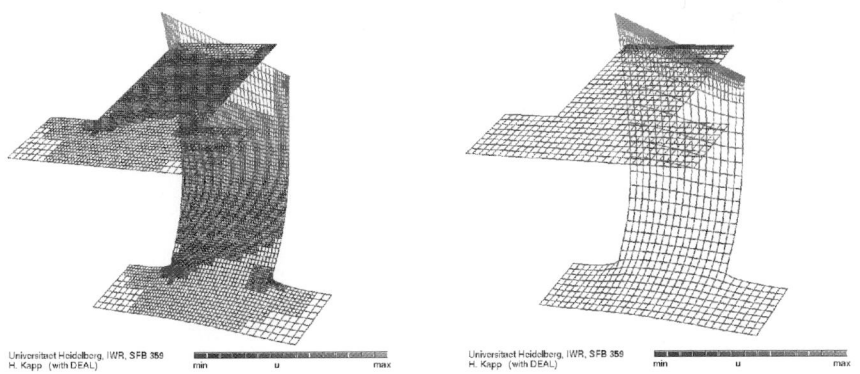

Fig. 4. Config. 1: Solutions and adapted meshes obtained by η_E (left) and η_ω (right)

Configuration 2: This control problem is more realistic since the effect of the control has to be passed through the whole domain Ω and, hence, the presence of the reentrant corners should have a stronger impact on the mesh refinement. This is confirmed by the results shown in Fig. 5 and 6, again for the error estimators η_ω, η_E^*, and η_E which show a comparable efficiency in this case.

N	E_{rel}	I_{eff}
512	9.3e-5	1.32
15368	8.1e-7	0.56
27800	4.9e-7	0.35
57632	2.3e-7	0.42
197408	4.6e-8	0.32

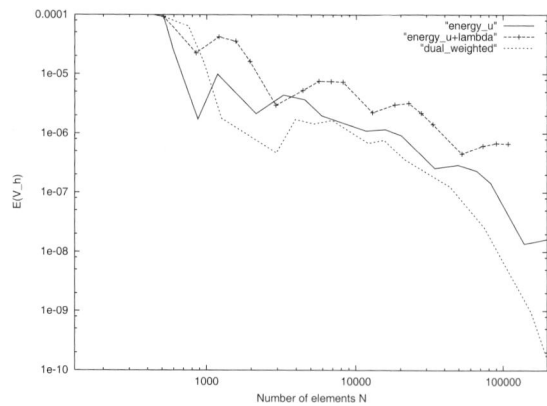

Fig. 5. Config. 2: Results for η_ω (left), and mesh efficiencies for η_E^*, η_E, and η_ω (right)

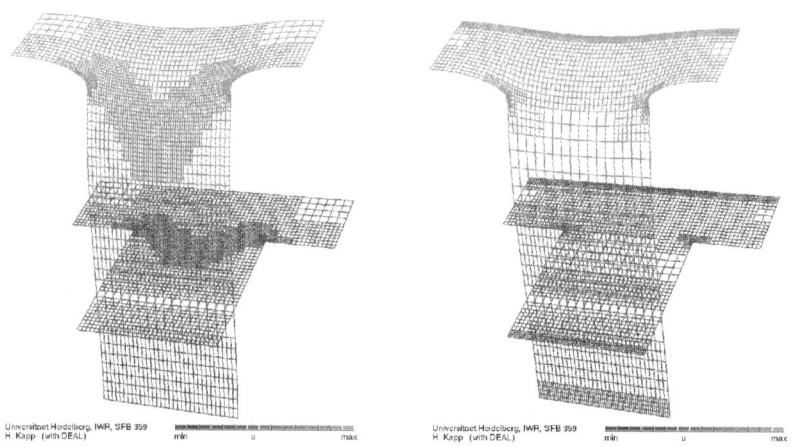

Fig. 6. Config. 2: Solutions and adapted meshes for η_E (left) and η_ω (right)

4.2 A Nonstationary Model Problem

The material of this section is taken from Becker/Meidner/Vexler [14]. In order to illustrate the difficulties arising in the adaptive solution of nonstationary control problems, we consider a nonstationary version of the control problem (17) with control in the initial condition:

$$\partial_t u - \Delta u + s(u) = f \quad \text{in } \Omega \times [0, T], \tag{22}$$
$$u_{|t=0} = q, \quad \partial_n u_{|\partial\Omega} = 0.$$

The corresponding variational formulation in the space-time region $Q_T = \Omega \times [0, T]$ reads

$$(\partial_t u, \psi)_{Q_T} + (\nabla u, \nabla \psi)_{Q_T} + (s(u), \psi)_{Q_T} = -(u(0) - q, \psi(0))_\Omega, \tag{23}$$

for all admissible test functions ψ. Here the initial condition is imposed in a variational sense. The 'initial control' q on Ω is to be determined such that

$$J(u, q) = \tfrac{1}{2}\|u(T) - \bar{u}\|_\Omega^2 + \tfrac{1}{2}\alpha\|q\|_\Omega^2 \to \min,$$

for a prescribed profile $\bar{u}$ on Ω at the final time $T > 0$. Now, following the Lagrange formalism, we seek to determine a stationary point $\{u, q, \lambda\}$ of the corresponding Lagrangian, i.e., a solution of the system

$$-(\varphi, \partial_t \lambda)_{Q_T} + (\nabla\varphi, \nabla\lambda)_{Q_T} + (\varphi, s'(u)\lambda)_{Q_T} = (\varphi(T), u(T) - \bar{u} - \lambda(T))_\Omega,$$
$$(\lambda(0), \chi)_\Omega = -\alpha(q, \chi)_\Omega,$$
$$(\partial_t u, \psi)_{Q_T} + (\nabla u, \nabla\psi)_{Q_T} + (s(u), \psi)_{Q_T} = -(u(0) - q, \psi(0))_\Omega,$$

for all admissible test triplets $\{\psi, \chi, \varphi\}$. In strong form,

$$\begin{aligned}
-\partial_t \lambda - \Delta\lambda + s'(u)\lambda &= 0 \quad \text{in } Q_T, \\
\lambda_{|t=T} &= u(T) - \bar{u}, \quad \partial_n \lambda_{|\partial\Omega} = 0, \\
\lambda_{|t=0} &= -\alpha q, \\
\partial_t u - \Delta u + s(u) &= 0 \quad \text{in } Q_T, \\
u_{|t=0} &= q, \quad \partial_n u_{|\partial\Omega} = 0.
\end{aligned} \tag{24}$$

For solving this saddle-point system, we consider a space-time finite element Galerkin discretization using the so-called 'dG(r)' ('discontinuous Galerkin') or 'cG(r)' ('continuous Galerkin') time discretization method (see, e.g., Eriksson/Johnson/Thomee [28]). For a time grid

$$0 = t_0 < \ldots < t_m < \ldots < t_M = T, \quad k_m := t_m - t_{m-1},$$

and a corresponding sequence of spatial meshes $\mathbb{T}_h^n$, we introduce finite element spaces V_h^k consisting of spatially continuous functions which are cellwise (isoparametric) bilinear in space and polynomial (constant or linear) in time.

Depending on the time discretization used, the dG or the cG method, the trial functions may be discontinuous or continuous in time, while the test functions are always discontinuous in time. Then, the two coupled subproblems for the primal state u_h and the adjoint state λ_h can be written in form of successive time-stepping schemes running from $t_0 = 0$ forward to $t_M = T$ and from T backward to t_0, respectively. Actually this space-time Galerkin finite element method is very similar to the backward Euler (cG(0) method) and the Crank-Nicolson (cG(1) method) scheme in time together with a time-varying (piecewise bilinear) spatial discretization. For more details, we refer to Becker [5] and Bangerth/Rannacher [2].

Efficient solution by windowing technique

As in the stationary case, for the solution of the coupled space-time system (24), we may employ a Newton-type iteration in the control variable q. However, due to the nonlinear coupling, the solution of the discretized adjoint equation (24) for λ_h requires storing of the discrete primal state u_h on the whole space-time domain Q_T. This difficulty can be overcome by trading storage for additional CPU-time using so-called 'windowing' or 'check-pointing' techniques (see Griewank [31], Berggren/Glowinski/Lions [20], Becker [5], and the detailed description in Becker/Meidner/Vexler [14]).

We briefly describe the central idea of 'windowing' in the solution of the above nonstationary saddle-point problem. Let the number of time steps M be factored like $M \sim PQ$ with $P, Q \in \mathbb{N}_+$. Accordingly, the set $\{t_0, \ldots, t_M\}$ of time points is decomposed into P sections each containing $Q - 1$ time steps and $P + 1$ subsets containing just one element:

$$\{t_0, \ldots, t_M\} = \{t_0\} \cup \{t_1, \ldots, t_{Q-1}\} \cup \ldots$$
$$\ldots \cup \{t_{(P-1)Q}\} \cup \{t_{(P-1)Q+1}, \ldots, t_{PQ-1}\} \cup \{t_{PQ}\}.$$

The 'windowing algorithm' now proceeds as follows. First, the primal state $\{u^m\}$, is computed over the whole time interval $[t_0, t_M]$ and the $P+1$ samples $\{u^0, u^Q, \ldots, u^{QP}\}$ are stored. Additionally, the $Q - 1$ values of u^m in the last section are stored such that now the adjoint state $\{\lambda^m\}$ can be computed backward in time in the last section. In the following process the $Q-1$ stored values of the primal state in the last section are no longer needed and their storage can be used for storing the primal state over other time sections. Next, starting from the stored value $u^{(P-2)Q}$ the primal state is recomputed in the $(P-1)$-st section and the resulting $Q-1$ of its values are stored. Then, in turn, the corresponding adjoint state can be calculated backward in time in the $(P-1)$-st section. This process is repeated until primal and adjoint state are computed on the first section. This 'one-level windowing' requires the memory $S_1(P,Q)$, measured in terms of multiples of the memory required for storing the primal state at one time level, and the total number $W_1(P,Q)$ of forward time steps needed to provide the primal state for computing the whole adjoint state is

$$S_1(P,Q) = (P+1) + (Q-1) = P+Q,$$
$$W_1(P,Q) = M + (P-1)(Q-1) = 2M - P - Q + 1.$$

This process can be extended to a 'multi-windowing'. Assuming that $M = M_0 \cdot \ldots \cdot M_L$, with certain $M_l \in \mathbb{N}_+$, the 'one-level windowing' can be applied for the factorization $M = PQ$ with $P = M_0$ and $Q = M_1 \cdot \ldots \cdot M_L$, and then recursively to each of the P sections. For the corresponding memory requirement and total number of forward time steps one finds by induction:

$$S_L = \sum_{l=0}^{L}(M_l - 1) + 2, \qquad W_L = (L+1)M - \sum_{l=0}^{L}\frac{M}{M_l} + 1.$$

Hence, for $L \sim \log_2(M)$ minimal storage S_{opt} and work count W_{opt} are achieved for $M_l \sim M^{1/(L+1)}, l = 0, \ldots, L$:

$$S_{\min} = O\bigl(\log_2(M)\bigr), \qquad W_{\min} = O\bigl(M \log_2(M)\bigr). \tag{25}$$

We see that multi-level windowing makes the simultaneous computation of primal and dual states feasible even for fine space-time meshes.

Numerical test: We consider the above model problem (22), with the nonlinearity $s(u) = u^2$, posed on the space-time region $Q_T := \Omega \times [0,T]$ where $\Omega = (0,1)^3 \subset \mathbb{R}^3$ and $[0,T] = [0,1]$. The initial control is assumed to be 'discrete' of the particular form

$$u_{|t=0} = g_0 + \sum_{i=1}^{8} q_i g_i,$$

where the shape functions g_i are given by $g_0 = (1 - 2\|x - \tilde{x}\|)^{30}$, with $\tilde{x} = (0.5, 0.5, 0.5)^T$, and $g_i = (1 - 0.5\|x - \tilde{x}_i\|)^{30}$, with $\tilde{x}_i \in \{0.2, 0.8\}^3$. Hence, the control space is $Q = \mathbb{R}^8$, while the space for states and co-states is $V = H^1(\Omega)$ as before. The given reference solution is $\bar{u}(x) = (3 + x_1 + x_2 + x_3)/6$, and the parameter in the regularization term is set to $\alpha = 10^{-4}$.

The discretization of this problem uses the cG(1) method in space (trilinear shape functions) and either the dG(0) or the cG(1) method in time with uniform mesh size $h = 0.0625$ in space and uniform time step $k = 0.01$, resulting in $N = 4096$ hexahedral cells and 100 time steps. The stationary point $q_{hk} \in Q$ of the discretized reduced functional $j_{hk}(\cdot)$ is obtained by the solution algorithm as described above with a Newton iteration (with and without building the entire Hessian matrix) containing n_{CG} inner CG steps. Table 1 and Fig. 7 show the obtained results. The difference in the limit values $j_{hk}(q_{hk}^{\mathrm{opt}})$ is due to the different accuracies of the dG(0) method (first order) and the cG(1) method (second order).

Table 1. Results of the optimization process with dG(0) and cG(1) time discretization and Newton iteration with n_{CG} inner CG steps, starting from $q^0 = 0$

Newton step	dG(0)			cG(1)		
	n_{CG}	$\|\nabla j_{hk}\|_2$	j_{hk}	n_{CG}	$\|\nabla j_{hk}\|_2$	j_{hk}
0	—	1.21e-01	2.76e-01	—	1.21e-01	2.76e-01
1	2	4.99e-02	1.34e-01	2	4.98e-02	1.34e-01
2	2	2.00e-02	6.28e-02	2	1.99e-02	6.33e-02
3	3	7.61e-03	2.94e-02	3	7.62e-03	3.00e-02
4	3	2.55e-03	1.64e-02	3	2.57e-03	1.70e-02
5	3	6.03e-04	1.32e-02	3	6.21e-04	1.37e-02
6	3	5.72e-05	1.29e-02	3	6.18e-05	1.34e-02
7	3	6.37e-07	1.29e-02	3	7.62e-07	1.34e-02
8	3	1.75e-10	1.29e-02	3	1.21e-10	1.34e-02

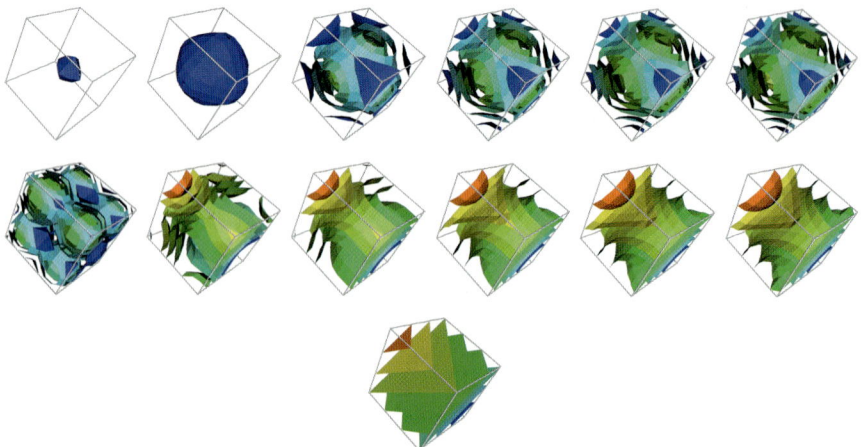

Fig. 7. Iso-surfaces of the state variable for time $t \in \{0.0, 0.2, 0.4, 0.6, 0.8, 1.0\}$ (from left to right) before (first row) and after (second row) optimization, compared to the reference solution $\bar{u}$ (third row)

In order to better illustrate the effect of multi-level windowing, in the next computation the spatial mesh is refined to 32768 cells and the dG(0) method uses 500 time steps. Table 2 confirms the expected reduction of the storage requirement. For the discretization used a storage reduction by about a factor of 1/30 can be achieved while the total number of time steps only grows by a factor of 3.2 for the Newton iteration with full Hessian matrix and by a factor of 4 for the Newton iteration without building the entire Hessian.

Table 2. Reduction of the storage requirement due to multi-level windowing with dG discretization and 32768 cells in each time step

	With Hessian		Without Hessian	
Factorization	Memory in MB	Time Steps	Memory in MB	Time Steps
500	1236	45000	274	35000
5 · 100	259	80640	58	87948
10 · 50	148	84690	32	90783
2 · 2 · 5 · 25	78	120582	17	118503
5 · 10 · 10	59	114174	13	113463
4 · 5 · 5 · 5	41	136512	9	130788
2 · 2 · 5 · 5 · 5	39	146646	9	138663

4.3 Boundary Control of Viscous Flow

The material of this section is taken from Becker [4, 5]. The next application concerns optimal flow control governed by the (stationary) incompressible Navier-Stokes system

$$-\nu \Delta v + v \cdot \nabla v + \nabla p = f, \quad \nabla \cdot v = 0, \quad \text{in } \Omega, \tag{26}$$

where the physical unknowns are the velocity and the pressure $u = \{v, p\}$ for prescribed viscosity ν, (constant) density $\rho \equiv 1$, and volume force $f = 0$, in the considered case. The flow configuration is depicted in Fig. 8.

Fig. 8. Configuration of the flow control problem

The prescribed boundary conditions are

$$v|_{\Gamma_{\text{rigid}}} = 0, \quad v|_{\Gamma_{\text{in}}} = v^{\text{in}}, \quad \nu \partial_n v - np|_{\Gamma_{\text{out}}} = 0,$$

with the 'rigid' boundary Γ_{rigid}, the 'inflow' boundary Γ_{in}, the 'outflow' boundary Γ_{out}, and the 'control' boundary Γ_Q. The Reynolds number is $\text{Re} = \bar{U}^2 D \nu^{-1} = 40$, defined in terms of viscosity ν, the characteristic length $D = \text{diam}(\text{cylinder})$, and the maximal inflow velocity $\bar{U}$, corresponding to stationary (uncontrolled) flow. The goal of the optimization is the minimization of the drag coefficient

$$J_{\mathrm{drag}}(u) := \frac{2}{\bar{U}D} \int_S n^T \sigma(v,p) e_1 \, ds,$$

taken over the surface S of the cylinder, with the stress tensor $\sigma(v,p) = -pI + \nu(\nabla v + \nabla v^T)$, and the unit vector e_1 of the main flow direction. The piecewise constant control $q = \{q_1, q_2\} \in \mathbb{R}^2$ acts at the two components of the control boundary Γ_Q as a Neumann control. The corresponding state equation posed in variational form seeks $u = \{v,p\} \in \{v^{\mathrm{in}} + H\} \times L^2(\Omega)$ and $q \in \mathbb{R}^2$, satisfying

$$\nu(\nabla v, \nabla \varphi^v)_\Omega + (v \cdot \nabla v, \varphi^v)_\Omega - (p, \nabla \cdot \varphi^v)_\Omega - (\varphi^p, \nabla \cdot v)_\Omega - (q, n \cdot \varphi^v)_{\Gamma_Q} = 0,$$

for test functions $\varphi = \{\varphi^v, \varphi^p\} \in V$, with the spaces $V = H \times L^2(\Omega)$ and $H = H_0^1(\Omega; \Gamma_{\mathrm{rigid}} \cup \Gamma_{\mathrm{in}})^2$. In this case the optimality system for the triplet $\{u, q, \lambda\}$, with $u = \{v, p\}$ and $\lambda = \{z, \pi\}$, in variational form reads as follows:

$$\nu(\nabla \psi^v, \nabla z)_\Omega + (v \cdot \nabla \psi^v, z)_\Omega + (\psi^v \cdot \nabla v, z)_\Omega - (\pi, \nabla \cdot \psi^v)_\Omega$$
$$-(\psi^p, \nabla \cdot z)_\Omega = J_{\mathrm{drag}}'(\psi) \quad \forall \psi \in V,$$
$$(r, z \cdot n)_{\Gamma_Q} = 0 \quad \forall r \in \mathbb{R}^2,$$
$$\nu(\nabla v, \nabla \varphi^v)_\Omega + (v \cdot \nabla v, \varphi^v)_\Omega - (p, \nabla \cdot \varphi^v)_\Omega$$
$$-(\varphi^p, \nabla \cdot v)_\Omega = (q, n \cdot \varphi^v)_{\Gamma_Q} \quad \forall \varphi \in V.$$

This nonlinear system is discretized by the finite element Galerkin method using (isoparametric) bilinear trial functions (equal-order Q_1-Stokes element) on quadrilateral meshes with consistent least-squares stabilization of pressure-velocity coupling and transport. We omit the explicit statement of the discrete Galerkin equations and the corresponding a posteriori error estimates. The resulting discrete equations are solved by a Newton-type method. For more details on this discretization see, e.g., Rannacher [39, 40].

By the theorem of Gauß the contour formula for the drag coefficient is transformed into a domain integral,

$$J_{\mathrm{drag}}(u) := \frac{2}{\bar{U}D} \int_\Omega \{(v \cdot \nabla v - f) \cdot \bar{e}^1 + (2\nu\tau - pI) : \nabla \bar{e}^1\} \, dx,$$

where $\bar{e}^1$ is a suitable extension of the directional unit vector e^1 from the contour S to Ω with small support along S. This formulation differs from the original one for the finite element solution and proves to be more robust and accurate. For more details on this 'old' trick see, e.g., Bangerth/Rannacher [2], and the literature cited therein.

Table 9 shows the advantages of using mesh adaptation in this optimal control model problem. The same minimal drag value can be reached on adapted meshes which have about 15-times less cells than the corresponding uniform

Fig. 9. Uniform refinement versus adaptive refinement for Re = 40

Uniform refinement		Adaptive refinement	
N	J_{drag}	N	J_{drag}
10512	3.31321	1572	3.28625
41504	3.21096	4264	3.16723
164928	3.11800	11146	3.11972

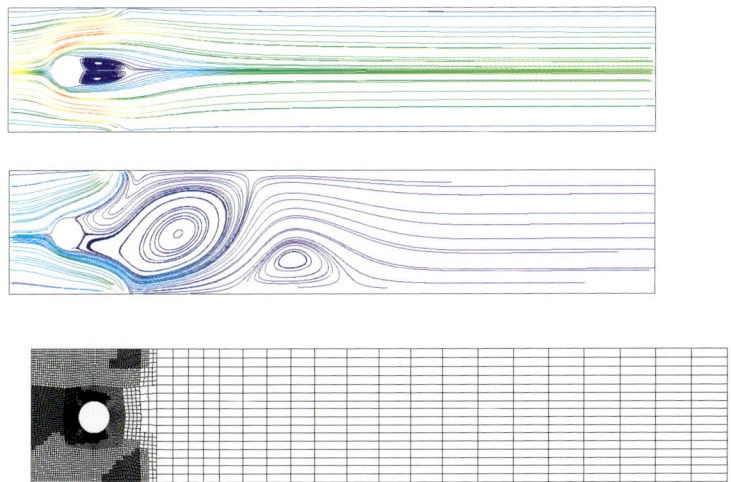

Fig. 10. Velocity of the uncontrolled flow (top), controlled flow (middle), corresponding adapted control mesh (bottom)

meshes. The nonlinear optimality system is solved by a Newton-like iteration. Fig. 10 shows the streamline plots for the uncontrolled flow ($q = 0$) and the optimally controlled flow together with the corresponding adapted 'control mesh'. The controlled flow has been computed by post-processing as described above on a globally refined mesh. The wild structure of this (stationary) flow pattern indicates that it my not be dynamically stable and hence not physically relevant. In fact, an accompanying linear stability analysis confirms this expectation (see Heuveline/Rannacher [32]).

4.4 Parameter Identification

The material of this section is taken from Becker/Vexler [18] and Vexler [43]. Our general approach to adaptive discretization of optimization problems can also be used in the context of parameter identification. Consider, for example the model problem

$$-\Delta u + qu = f \text{ in } \Omega, \quad u = 0 \text{ on } \partial\Omega. \tag{27}$$

The goal is to determine the a priori unknown coefficient q by comparing the corresponding state $u(q)$ with given measurements $\bar{u}$. The usual approach uses the least-squares minimization

$$J(u,q) := \tfrac{1}{2}\|u-\bar{u}\|^2 + \tfrac{1}{2}\alpha\|q\|^2 \to \min,$$

with a regularization parameter $0 \leq \alpha \ll 1$. The associated Lagrangian is

$$\mathcal{L}(u,q,\lambda) := J(u,q) - (\nabla u, \nabla \lambda) - (qu, \lambda) + (f, \lambda),$$

and the corresponding optimality system

$$\begin{aligned}(\nabla\varphi, \nabla\lambda) + (q\varphi, \lambda) &= (u-\bar{u}, \varphi) & \forall \varphi \in V, \\ (\chi u, \lambda) &= -\alpha(\chi, q) & \forall \chi \in Q, \\ (\nabla u, \nabla\psi) + (qu, \psi) &= (f, \psi) & \forall \psi \in V,\end{aligned} \tag{28}$$

with the spaces $V := H_0^1(\Omega)$ and $Q = \mathbb{R}^{n_p}$. This system may again be discretized by the finite element Galerkin method and the general Theorem 1 yields an a posteriori estimate for the error in the cost functional $J(\cdot,\cdot)$. However, this error is usually not of interest but rather the error in the control, i.e., in the unknown parameter q. In fact, assuming the parameter $q > 0$ to be identifiable, the adjoint equation

$$-\Delta\lambda + q\lambda = u-\bar{u} = 0 \text{ in } \Omega, \quad \lambda = 0 \text{ on } \partial\Omega,$$

implies that necessarily $\lambda \equiv 0$. Hence, in this case the a posteriori error estimate (6) is rendered useless. This difficulty has been treated in the following ways:

- An energy-norm-type a posteriori error estimate for $\alpha|q-q_h|^2$ can be derived based on a coercivity estimate for the saddle-point problem. However, the 'worst case' stability constant in this estimate is usually unknown and grows like α^{-1} (see Liu/Yan [37]).
- An a posteriori error estimate for a suitable norm of $q-q_h$, or more general for quantities of physical interest expressed in terms of functionals $E(u,q)$, can be derived by an 'outer' duality argument where the relevant stability property is captured in the dual variable (see Becker/Vexler [18, 19]).

A general approach to parameter identification

We present the following general approach to the numerical solution of (discrete) parameter identification problems based on the least-squares method. Let V denote the state space, $Q = \mathbb{R}^{n_p}$ the control space, $Z = \mathbb{R}^{n_m}$ the observation space with observation operator $C : V \to Z$, and observation $\bar{c} \in$

Z. Then, the corresponding optimization problem seeks $u \in V$, $q \in Q = \mathbb{R}^{n_p}$, such that

$$J(u) := \tfrac{1}{2}\|C(u) - \bar{c}\|_Z^2 \to \min, \qquad a(u,q)(\varphi) = 0 \quad \forall \varphi \in V, \qquad (29)$$

with an energy-semi-linear from $a(\cdot,\cdot)(\cdot)$ as described above. The corresponding necessary first-order optimality condition (analogous to (28)) reads

$$\begin{aligned} a'_u(u,q)(\varphi,\lambda) &= J'_u(u,q)(\varphi) \quad \forall \varphi \in V, \\ a'_q(u,q)(\chi,\lambda) &= J'_q(u,q)(\chi) \quad \forall \chi \in Q, \\ a(u,q)(\psi) &= 0 \quad \forall \psi \in V. \end{aligned} \qquad (30)$$

The discretization of (29) uses finite element spaces $V_h \subset V$ and determines $\{u_h, q_h\} \in V_h \times Q$, such that

$$J(u_h) = \tfrac{1}{2}\|C(u_h) - \bar{c}\|_Z^2 \to \min, \qquad a(u_h, q_h)(\varphi_h) = 0 \quad \forall \varphi_h \in V_h. \qquad (31)$$

If the form $a(\cdot, q)(\cdot)$ is regular for any $q \in Q$, we can define the solution operator $S : Q \to V$ and, setting $u = S(q)$, obtain the following unconstrained equivalent of (29) posed in the finite dimensional space $Q = \mathbb{R}^{n_p}$:

$$j(q) := \tfrac{1}{2}\|c(q) - \bar{c}\|_Z^2 \to \min, \qquad c(q) := C(S(q)), \ q \in Q.$$

The derivatives

$$G_{ij} := \partial_{q_j} c_i(q) = C'_i(u)(w_j), \qquad G = (G_{ij})_{i,j=1}^{n_p},$$

are determined by the solutions $w_j \in V$ of the tangent equations

$$a'_u(u,q)(w_j, \varphi) = -a'_{q_j}(u,q)(1,\varphi) \quad \forall \varphi \in V.$$

Using this notation the necessary first-order optimality condition is

$$j'(q) = 0 \quad \Leftrightarrow \quad G^*(c(q) - \bar{c}) = 0.$$

This equation may be solved by a fixed-point iteration of the form

$$q_{k+1} = q_k + \delta q, \qquad H_k \, \delta q = G_k^*(\bar{c} - c(q_k)), \qquad G_k = c'(q_k),$$

with a suitably chosen preconditioning matrix H_k. Popular choices are:

- The full 'Newton algorithm', $H_k := G_k^* G_k + \langle c(q_k) - \bar{c}, c''(q_k) \rangle_Z$.
- The 'Gauß-Newton algorithm', $H_k := G_k^* G_k$.
- The 'update method', $H_k := G_k^* G_k + M_k$ with certain corrections M_k.

However, the full Newton algorithm is rarely used in this context since it involves the evaluation of the term $\langle c(q_h^k) - \bar{c}, c''(q_h^k) \rangle_Z$, particularly the second derivative $c''_h(q_h^k)$. This is rather expensive since it requires the solution of several auxiliary problems, depending on the dimension of Q. Since in the limit $k \to \infty$ the deviation $c_h(q_h) - \bar{c}$ is expected to be small, the Gauß-Newton algorithm is justified and its convergence is sometimes even superlinear.

A posteriori error estimation

We have seen that controlling the error in the discretization of a parameter identification problem based on the control functional $J(\cdot)$ may be useless for guiding mesh adaptation. Hence, we have to follow another approach by choosing an error control functional $E(\cdot)$ which addresses the error in the controls more directly, e.g.,

$$E(u_h, q_h) := \tfrac{1}{2}\|q - q_h\|_Q^2.$$

Then, the systematic error control by the general approach described above applied to the Galerkin approximation of the saddle-point system (28) has to use an extra 'outer' (linearized) dual problem with solution $z = \{z^u, z^q, z^\lambda\} \in V \times Q \times V$,

$$A'(u, q, \lambda)(\varphi, \chi, \psi, z^u, z^q, z^\lambda) = E'(u, q, \lambda)(\varphi, \chi, \psi), \qquad (32)$$

for all $\{\varphi, \chi, \psi\} \in V \times Q \times V$, where

$$\begin{aligned}A(u, q, \lambda)(\varphi, \chi, \psi) &:= a'_u(u, q)(\varphi, \lambda) - J'_u(u, q)(\varphi) \\ &\quad + a'_q(u, q)(\chi, \lambda) - J'_q(u, q)(\chi) + a(u, q)(\psi).\end{aligned}$$

A careful analysis of this setting results in an a posteriori error representation of the following form (see Becker/Vexler [18] and Vexler [43]):

$$E(u, q) - E(u_h, q_h) = \eta + \mathcal{R} + \mathcal{P}. \qquad (33)$$

Here, the main part η of the estimator has the usual form

$$\eta = \tfrac{1}{2}\rho(u_h, q_h)(z - i_h z) + \tfrac{1}{2}\rho^*(u_h, q_h, z_h)(u - i_h u),$$

with the 'dual solution' $z \in V$ determined by the dual problem

$$a'_u(u, q)(\varphi, z) = -\langle G(G^*G)^{-1}\nabla E(q), C'(u)(\varphi)\rangle_Z \quad \forall \varphi \in V,$$

and the residuals

$$\rho(u_h, q_h)(\psi) := -a(u_h, q_h)(\psi),$$
$$\rho^*(u_h, q_h, z_h)(\varphi) := \langle G_h(G_h^*G_h)^{-1}\nabla E(q_h), C'(u_h)(\varphi)\rangle - a'_u(u_h, q_h)(\varphi, z_h).$$

The remainder $\mathcal{R}$ due to linearization is again cubic in the errors $u - u_h$, $q - q_h$, and $z - z_h$, and the additional error term $\mathcal{P}$ is bounded like

$$|\mathcal{P}| \leq \tilde{C}\,\|e\|_V\,\|C(u) - \bar{c}\|_Z. \qquad (34)$$

Due to the particular features of the parameter identification problem, we can expect that $\|C(u) - \bar{c}\|_Z \ll 1$ for the optimal state. Hence, this term is neglected compared to the leading term η. Based on the a posteriori error bound (33) the mesh adaptation is organized as described above. The performance of this method is illustrated by the following two examples.

Example 1: Fitting of reaction parameters

We consider the scalar reaction-diffusion problem

$$\beta \cdot \nabla u - \mu \Delta u + f(u) = 0 \quad \text{in } \Omega$$
$$u_{|\Gamma_{\text{in}}} = \hat{u}, \quad \partial_n u_{|\partial\Omega \setminus \Gamma_{\text{in}}} = 0, \tag{35}$$

on the domain depicted in Fig. 11. The reaction term $f(u)$ has the form of an Arrhenius-type reaction law

$$f(u) = A \exp\left(-\frac{E}{1-u}\right) u(1-u).$$

The goal is to identify the scalar parameters A and E from given integral averages of the solution along the broken lines shown in Fig. 11.

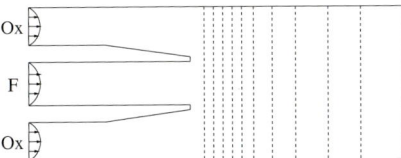

Measurements:
vertical line-integrals
of concentration

Fig. 11. Configuration of the reaction-diffusion problem

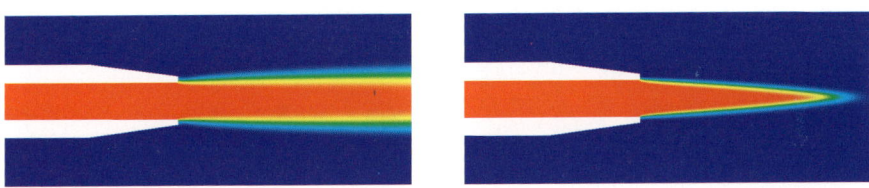

Fig. 12. Initial state (left: $A = 54.6$, $E = 0.15$) and final state corresponding to optimized parameters (right: $A = 992.3$, $E = 0.07$)

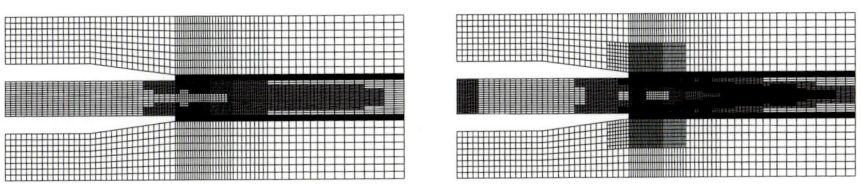

Fig. 13. Locally refined meshes

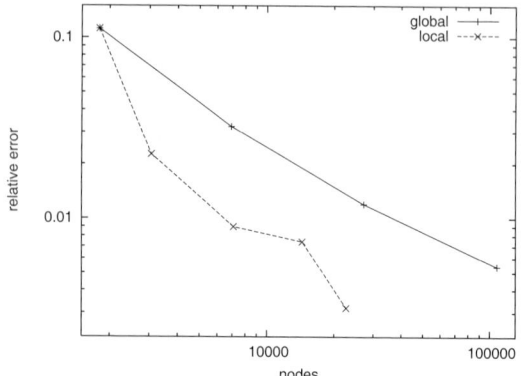

Fig. 14. Quality of generated meshes: error in estimated parameters versus number of nodes for uniform and adaptive mesh refinement

Example 2: Fitting of diffusion parameters

We consider a reactive flow problem governed by the full set of stationary conservation equations for mass, momentum, energy and species concentrations (the 'compressible' Navier-Stokes equations):

$$\text{div}\,(\rho v) = 0,$$
$$(\rho v \cdot \nabla) v + \text{div}\,\pi + \nabla p = 0,$$
$$\rho v \cdot \nabla T - c_p^{-1} \text{div}\, \mathcal{Q} = -\sum_{i \in \mathcal{S}} h_i f_i,$$
$$\rho v \cdot \nabla y_k + \text{div}\, \mathcal{F}_k = f_k, \quad k \in \mathcal{S},\ \#\mathcal{S} = 9,$$
$$\mathcal{F}_k = q_k D_k^* \nabla y_k, \quad D_k^* = (1 - y_k) \Big(\sum_{l \neq k} \frac{x_l}{D_{kl}^{\text{bin}}} \Big)^{-1}.$$

The set-up of this hydrogen diffusion flame is taken from Braack & Ern [23] and is shown schematically in Fig. 15. At the inflow boundary of the center tube, 10% mass fraction of hydrogen y_{O_2} and 90% of nitrogen y_{N_2} is prescribed. The inflow temperature is $T = 273\,\text{K}$. Along the upper and

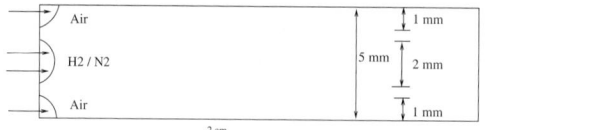

Fig. 15. Configuration of the diffusion-reactive problem

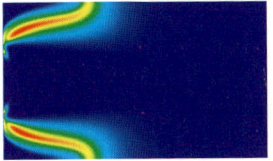  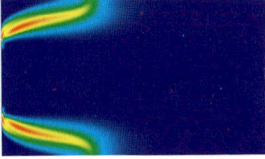

Fig. 16. Left: solution by multicomponent diffusion (reference solution), middle: solution by Fick's law (initial parameters), right: solution by optimized Fick's law

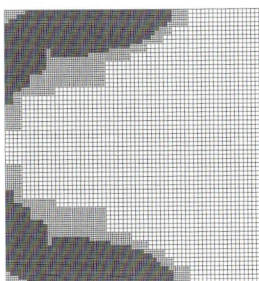 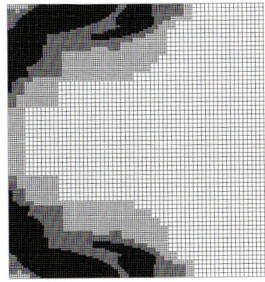

Fig. 17. Locally refined meshes (zooms)

lower boundary the constant values $y_{O_2} = 0.22$ and $y_{N_2} = 0.78$ are given. The peak velocity of the three parabolic velocity profiles is $1\,\mathrm{m/s}$. For a more detailed description of such a model, we refer to Braack/Rannacher [24] and Becker, Braack/Vexler [8, 9].

The numerical evaluation of the above so-called 'multi-component diffusion' model $\mathcal{F}_k = q_k D_k^* \nabla y_k$ is rather expensive since it requires the inversion of linear systems in all nodal points. Therefore, it is desirably to replace this complicated model by the much simpler 'Fick's model' $\mathcal{F}_k = q_k D_k \nabla y_k$, with diffusion coefficients independent of the other species. The goal is now to choose these coefficients D_k in such a way that the resulting solution is close to that obtained by using the full multi-component diffusion. This parameter identification is done on the basis of given measured point-values of species concentrations. Since the case of pointwise data is not directly covered by our Hilbert-space-based theory, we may replace *point-values* by certain local averages. An a priori error analysis of the discretization of point-value-based parameter estimation problems is given in Rannacher/Vexler [41].

Fig. 16 shows the result of a successful adaptation measured in terms of the position of the combustion front which is usually a very sensitive parameter. Fig. 17 shows zooms into automatically adapted meshes.

5 Conclusion and Outlook

Duality-based error estimation can be applied, in principle, for all problems posed in variational form. The approach described in this article for stationary and nonstationary model optimal control problems is currently extended into various directions:

- *Estimation of distributed parameters:* The identification of distributed parameters, for example in wave propagation problems in seismics, requires the proper discretization of the control space. Doing this adaptively may result in different meshes for states and control as suggested by the observations in Bangerth [1].
- *Adaptive nonstationary optimal control:* The multi-level windowing technique is to be realized in the context of duality based simultaneous mesh-size and time-step adaptation following ideas developed in Becker [5] and Becker/Meidner/Vexler [14].
- *Control- and state-constrained problems:* Most practical optimal control problems involve (inequality) constraints for the controls and, what is much more difficult to handle, also for the states. Though the algorithmic treatment of this complication is reasonably understood (see, e.g., Kunisch/Rösch [35]), its interaction with automatic mesh adaptation based on a posteriori error estimates is largely open. In this context techniques used for the adaptive discretization of variational inequalities corresponding to Signorini-type problems may be useful (see Blum/Suttmeier [21]).
- *Optimal experimental design:* Parameter estimation is usually troubled by inherent ill-conditioning due to insufficient data from measurements. This raises the question of how to find 'good' data which would be 'optimal', i.e., maximal sensitive, for the parameter identification process. First steps towards such 'optimal experimental design' in the context of PDEs and adaptive discretization are described in Carraro et al. [26] and [25] in this volume.

References

1. W. Bangerth: *Adaptive Finite Element Methods for the Identification of Distributed Parameters in Partial Differential Equations*, Dissertation, Univ. of Heidelberg, 2002.
2. W. Bangerth and R. Rannacher: *Adaptive Finite Element Methods for Differential Equations*, Lectures in Mathematics, ETH Zürich, Birkhäuser, Basel 2003.
3. R. Becker: *An optimal control approach to a posteriori error estimation for finite element discretizations of the Navier-Stokes equations*, East-West J. Numer. Math. 8, 257-274 (2000).
4. R. Becker: *Mesh adaptation for stationary flow control*, J. Math. Fluid Mech. 3, 317-341 (2001).
5. R. Becker: *Adaptive Finite Elements for Optimal Control Problems*, Habilitation thesis, Univ. of Heidelberg, 2001.

6. R. Becker, M. Braack: *The finite element toolkit GASCOIGNE*, Institute of Applied Mathematics, University of Heidelberg, URL http://www.gascoigne.de, 2005.
7. R. Becker, M. Braack, and R. Rannacher: *Adaptive finite element methods for flow problems*, in Foundations of Computational Mathematics (R.A. DeVore, A. Iserles, E. Süli, eds), pp. 21-44, Cambridge University Press, 2001.
8. R. Becker, M. Braack, and B. Vexler: *Numerical parameter estimation for chemical models in multidimensional reactive flows*, Combustion Theory and Modelling 8 (4), 661-682 (2004).
9. R. Becker, M. Braack, and B. Vexler: *Parameter identification for chemical models in combustion problems*, Appl. Numer. Math. 54 (3-4), 519-536, (2005).
10. R. Becker, T. Dunne, and D. Meidner: *VISUSIMPLE: An interactive visualization and graphics/mpeg-generation program*, Institute of Applied Mathematics, University of Heidelberg, URL http://www.visusimple.uni-hd.de, 2005.
11. R. Becker, V. Heuveline, and R. Rannacher: *An optimal control approach to adaptivity in computational fluid mechanics*, Int. J. Numer. Meth. Fluids. 40, 105-120 (2002).
12. R. Becker, H. Kapp, and R. Rannacher: *Adaptive finite element methods for optimal control of partial differential equations: basic concepts*, SIAM J. Optimization Control 39, 113-132 (2000).
13. R. Becker, D. Meidner, and B. Vexler: *RODOBO: A C++ library for optimization with stationary and nonstationary PDEs*, Institute of Applied Mathematics, University of Heidelberg, URL http://www.rodobo.uni-hd.de, 2005.
14. R. Becker, D. Meidner, and B. Vexler: *Efficient numerical solution of parabolic optimization problems by finite element methods*, submitted to Optimization Methods and Software, 2005.
15. R. Becker and R. Rannacher: *Weighted a-posteriori error estimates in FE methods*, Lecture ENUMATH-95, Paris, Sept. 18-22, 1995, in: Proc. ENUMATH-97, Heidelberg, Sept. 28 - Oct.3, 1997 (H.G. Bock, et al., eds), pp. 621-637, World Scientific Publ., Singapore, 1998.
16. R. Becker and R. Rannacher: *A feed-back approach to error control in finite element methods: Basic analysis and examples*, East-West J. Numer. Math., 4:237-264 (1996).
17. R. Becker and R. Rannacher: *An optimal control approach to error estimation and mesh adaptation in finite element methods*, Acta Numerica 2000 (A. Iserles, ed.), pp. 1-102, Cambridge University Press, 2001.
18. R. Becker and B. Vexler: *A Posteriori error estimation for finite element discretization of parameter identification problems*, Numer. Math. 96, 435-459 (2004).
19. R. Becker and B. Vexler: *Mesh refinement and numerical sensitivity analysis for parameter calibration of partial differential equations*, to appear in J. Comput. Phys. (2005).
20. M. Berggren, R. Glowinski, and J.-L. Lions: *A computational approach to controllability issues for flow-related models. (I): Pointwise control of the viscous Burgers equation*, Int. J. Comput. Fluid Dyn. 7, 237-253 (1996).
21. H. Blum and F.-T. Suttmeier. *Weighted error estimates for finite element solutions of variational inequalities*, Computing 65, 119-134 (2000).
22. M. Braack and T. Richter: *Solving multidimensional reactive flow problems with adaptive finite elements*, Reactive Flows, Diffusion and Transport, (W. Jäger et al., eds), Springer, Berlin-Heidelberg, 2006.

23. M. Braack and A. Ern: *Coupling multimodeling with local mesh refinement for the numerical solution of laminar flames*, Combust. Theory Modelling 8, 771-788 (2004).
24. M. Braack and R. Rannacher: *Adaptive finite element methods for low-mach-number flows with chemical reactions*, 30th Computational Fluid Dynamics (H. Deconinck, ed.), Volume 1999-03 of Lecture Series von Karman Institute for Fluid Dynamics, 1999.
25. T. Carraro, V. Heuveline, R. Rannacher, and C. Wague: *Determination of kinetic parameters in laminar flow reactors. I. Numerical aspects*, Reactive Flows, Diffusion and Transport, (W. Jäger et al., eds), Springer, Berlin-Heidelberg, 2006.
26. Th. Carraro, V. Heuvelinem, and R. Rannacher: *Parameter estimation and optimal experimental design in flow reactors*, Preprint, SFB 359, Univ. Heidelberg, March 2005.
27. K. Eriksson, D. Estep, P. Hansbo, and C. Johnson: *Introduction to adaptive methods for differential equations*. Acta Numerica 1995 (A. Iserles, ed.), pp. 105-158, Cambridge University Press, 1995.
28. K. Eriksson, C. Johnson, and V. Thomée: *Time discretization of parabolic problems by the discontinuous Galerkin method*, RAIRO Model. Math. Anal. Numer. 19, 611-643 (1985).
29. A. V. Fursikov: *Optimal Control of Distributed Systems: Theory and Applications*, Vol. 187, Trans l. Math. Monogr. AMS, Providence, 1999.
30. M. B. Giles and E. Süli: *Adjoint methods for PDEs: a posteriori error analysis and postprocessing by duality*, Acta Numerica 2002 (A. Iserles, ed.), pp. 145-236, Cambridge University Press, 2002.
31. A. Griewank: *Achieving logarithmic growth of temporal and spatial complexity in reverse automatic differentiation*, Optim. Methods Software 1, 35-54 (1992).
32. V. Heuveline and R. Rannacher: *Solution of stability eigenvalue problems by adaptive finite elements with application in hydrodynamic stability theory*, Preprint, SFB 359, Univ. of Heidelberg, March 2005.
33. M. Hinze and K. Kunisch: *Second order methods for optimal control of time-dependent fluid flow*, SIAM J. Optim. 13, 321-334 (2002).
34. H. Kapp: *Adaptive Finite Element Methods for Optimization in Partial Differential Equations*, Dissertation, Univ. of Heidelberg, 2000.
35. K. Kunisch and A. Rösch: *Primal-dual active set strategy for a general class of constrained optimal control problems*, SIAM J. Optim. 13, 321-334 (2002).
36. J.-L. Lions: *Optimal Control of Systems Governed by Partial Differential Equations*, Springer, Berlin-Heidelberg-New York, 1971.
37. W. Liu and N. Yan: *A posteriori error estimates for some model boundary control problems*, J. Comput. Appl. Math. 120, 159-173 (2000).
38. R. Rannacher: *Adaptive finite element discretization of optimization problems*, Proc. Conf. Numer. Analysis 1999 (D.F. Griffiths and G.A. Watson, eds), pp. 21-42, Chapman&Hall CRC, London, 2000.
39. R. Rannacher: *Finite element methods for the incompressible Navier-Stokes equations*, in Fundamental Directions in Mathematical Fluid Mechanics (P. Galdi, J. Heywood and R. Rannacher, eds), pp. 191-293, Birkhäuser, Basel, 2000.
40. R. Rannacher: *Incompressible Viscous Flow*, in Encyclopedia of Computational Mechanics (E. Stein, et al., eds), John Wiley, Chichester, 2004.

41. R. Ranncher and B. Vexler: *A priori error analysis for the finite element discretization of elliptic parameter identification problems*, SIAM J. Control Optim., 2005, to appear.
42. F. Tröltzsch: *On the Langreange-Newton-SQP method for the optimal control of semilinear parabolic equations*, SIAM J. Control Optim. 38, 294-312 (1999).
43. B. Vexler: *Adaptive Finite Element Methods for Parameter Identification Problems*, Dissertation, Univ. of Heidelberg, 2004.

Part IV

Chemical Reaction Systems

Preamble

The chapter contains five articles on modeling, simulation and optimization of chemically reacting systems and on supporting experiments.

The first two articles *Determination of kinetic parameters in laminar flow reactors. I. Theoretical aspects* and *Determination of kinetic parameters in laminar flow reactors. II. Experimental aspects* describe new techniques for the systematic determination of the kinetic parameters in benchmark elementary reaction steps in laminar flow reactors. The experimental data are obtained by laser spectroscopic measurement techniques such as Coherent Anti-Stokes Raman Spectroscopy (CARS) to detect vibrationally excited molecules and Laser Induced Fluorescence (LIF). The reactions studied are $H_2(\nu=1)+D_2(\nu=0) \to D_2(\nu=1)+H_2(\nu=0)$, $OH+H_2(\nu=1) \to H+H_2O$ at low temperature. A pulsed laser photolysis (LP)/laser induced fluorescence (LIF) pump-and-probe technique was used to investigate the temperature dependence of $O(^1D) + H_2 \to H + OH$.

The measurements are used in a systematic model-based parameter estimation process in which the mathematical model is formed by the full set of the compressible Navier-Stokes equations and the balance equations for the chemical species. This system is discretized by a finite element method with mesh adaptivity driven by duality-based a-posteriori error estimates ("DWR method"). The parameter estimation uses the Lagrangian formalism by which the problem is reformulated as a nonlinear saddle-point boundary-value problem which is solved on the discrete level by the Newton or Gauss-Newton method.

The third article *Optimization of reactive flows in a single channel of a catalytic monolith: conversion of ethane to ethylene* discusses the modeling, simulation, and, for the first time, optimization of the reactive flow in a channel of a catalytic monolith with detailed chemistry. Spatial semi-discretization of the boundary layer approximation of the problem leads to a high-dimensional stiff DAE system; the optimal control problem is solved by the SQP method. The optimization results show an improvement of a factor of two for the maximum ethane yield, achieved for temperatures around 1300 K.

The fourth article *Reaction processes on catalytically active surfaces* presents experimental studies on *in-situ* infrared-visible sum-frequency generation (IR-VIS SFG) surface vibrational spectroscopy. Investigated are the adsorption of carbon monoxide (CO) on rhodium (Rh) catalyst surfaces at elevated substrate temperatures ($T_s \geq 300K$) over a wide pressure range ($p_{CO} = 10^{-8} - 1000$ mbar). CO dissociation occurs via the Boudouard reaction: $2\,CO \rightarrow C(ad) + CO_2$. To rationalize the experimental findings, DFT studies were performed the results of which suggest the possibility of a novel low-temperature/high-pressure CO/Rh(111) dissociation pathway where a gas-phase CO(g) molecule reacts with a CO molecule chemisorbed on the catalyst surface, CO(ad), to yield gas-phase CO_2(g) and surface carbon C(ad) through an Eley-Rideal mechanism.

The topic of the fifth article *Stochastic modeling and deterministic limit of catalytic surface processes* is the modeling of CO oxidation on low-index platinum single-crystal surfaces on three levels of modeling microscopic, mesoscopic and macroscopic where the first two are stochastic while the model on the macroscopic level is deterministic and can be derived rigorously for low-pressure conditions from the microscopic model in the limit of infinite particle number. The mesoscopic model is also given by a many-particle system. However, the particles move on a lattice, such that, in contrast to the microscopic model, the spatial resolution is reduced. The models include a new approach to the platinum phase transition, which allows for a unification of existing models for Pt(100) and Pt(110). The rich nonlinear dynamical behavior of the macroscopic reaction kinetics is investigated and shows good agreement with low-pressure experiments.

Determination of Kinetic Parameters in Laminar Flow Reactors. I. Theoretical Aspects*

T. Carraro[1], V. Heuveline[1,2], and R. Rannacher[3]

[1] Institut für Angewandte Mathematik, Universität Karlsruhe (TH).
[2] Rechenzentrum Universität Karlsruhe (TH).
[3] Institut für Angewandte Mathematik, Universität Heidelberg.

Summary. This article describes the development of a numerical tool for the *simulation*, the *estimation of parameters* and the systematic *experimental design optimization* of chemical flow reactors. The goal is the reliable determination of unknown kinetic parameters of elementary reactions from measurements in a wide range of (laminar) flow conditions, from low to high temperature and from low to high pressure. The corresponding experiments have been set-up in the physical-chemistry group of J. Wolfrum at the PCI, Heidelberg; see the article Hanf/Volpp/Wolfrum [24] in this volume. The underlying mathematical model is the full set of the compressible Navier–Stokes equations accompanied by the balance equations for the chemical species. This system is discretized by a finite element method with mesh adaptivity driven by duality-based a posteriori error estimates ('DWR method'); see the article Becker et al. [12] in this volume. The parameter estimation uses the Lagrangian formalism by which the problem is reformulated as a nonlinear saddle-point boundary value problem which is solved on the discrete level by the Newton or Gauß-Newton method. The contents of this article is as follows:

- Introduction
- Mathematical model
- Numerical approach
- The low-temperature flow reactor
- The high-temperature flow reactor
- A step towards optimal experimental design
- Conclusion and outlook
- References
- Appendix

1 Introduction

Subject of this article is the numerical evaluation of kinetic data, obtained from controlled experiments in a flow reactor, and the modeling and optimization of the relevant complex reactive flow processes. The corresponding

---

*This work has been supported by the German Research Foundation (DFG) through SFB 359 (Project B1) at the University of Heidelberg.

experimental aspects are described in the article Hanf/Volpp/Wolfrum [24] in this volume.

The basic principle of flow-tube reactors is always the same: mixing of reactants takes place upstream in a 'mixing section' and their consumption or the build-up of products is followed along a measurement section by some detection methods for atoms, radicals, or molecules. A reaction rate constant is then deduced from measured axial concentration profiles. In order to favor diffusive processes, which minimize radial concentration gradients, a flow tube is traditionally operated at low pressure. An assumed mean flow velocity ('plug flow' assumption) allows to convert the axial coordinate (distance between the first point of mixing and the detection point) into reaction time. The reaction rate constants of interest can then be deduced by modeling the homogeneous reaction system (0-dimensional approximation). However, this approach is known to bear systematic errors, since it is based on the approximation of a perfect decoupling of chemical and hydrodynamical processes in the flow tube. Since the first complete description of the systematic errors in a 'plug flow' evaluation of reaction kinetic measurements in flow reactors by Kaufmann [29] much effort went into the incorporation of single aspects of real reactive flows in form of corrections to the approximately evaluated rate constants for idealized flow behavior. However, these approaches cannot take into account the mixing region with its complex hydrodynamics by which valuable reaction time for the observation is lost. Further, complex reaction mechanisms, such as occurring by including secondary reactions of the primary reaction products, cannot be implemented. This can result in a significant underestimation of reaction rates. Hence, in order to carry out a reliable evaluation of rate constants from experimental data, it is desirable to take into account all relevant physical and chemical processes occurring in the reactive flow.

This is done, on the modeling side, by a detailed numerical simulation of the reactive flow including convective and diffusive transport as well as heterogeneous wall effects and detailed gas-phase chemistry. On the experimental side, the flow reactor is specially designed to yield an almost undisturbed laminar, cylindrically symmetric and mostly stationary flow pattern which facilitates a reliable numerical simulation. The latter has to deal with the full system of conservation equations for mass, momentum, energy and species mass fractions ('compressible Navier–Stokes equations'). Due to the particular conditions of the experiments considered the flow is almost incompressible. This 'low-Mach number' condition with possibly strongly varying density due to large temperature gradients poses particular numerical difficulties which require a methodology oriented by the incompressible limit case.

At the start of the project, a decade ago, a simulation code was available for solving the two-dimensional laminar compressible Navier–Stokes equations at low-Mach number based on a finite-difference discretization on staggered tensor-product grids and semi-implicit operator splitting (see Segatz [38] and Segatz et al. [39]). Using this code systems with up to 18 elementary reactions in the flow reactor could be analyzed. The first system considered was the energy transfer between Hydrogen and Deuterium molecules (e.g., the wall deactivation reaction $H_2(v=1) \to H_2(v=0)$), which as chemically simplest substances can also serve as a model system accessible to a quantum mechanical treatment. However, for more complex simulations and systematic multi-parameter estimation the finite difference code was not flexible and efficient enough. This motivated a completely new code development which resulted in two PhD theses, Waguet [42] and Carraro [22]. This article gives an overview of the numerical approaches used in these new codes and of some of the results obtained for the laminar flow reactor. The implementation was based, in the first step, on the finite element tool box DEAL (Becker/Kanschat/Suttmeier [10]), and, in the second step, on the solver package HiFlow (Heuveline [26]).

The numerical approach used in the new code is based on a detailed two-dimensional modeling of all relevant chemical and physical processes in a specially designed flow system. The modeling is adapted to the special needs of this type of flow (laminar low-Mach number flow with variable density) by using a fully implicit adaptive finite element method with the pressure as a primal variable. The stationary solutions are computed by a Newton-type iteration on the finest mesh with pseudo-time-stepping only on the coarse meshes for generating appropriate starting values. This approach allows the accurate and efficient simulation of the whole chemically reactive flow process, by which, in turn, a systematic parameter estimation becomes possible. The input data for this parameter estimation are supplied by laser-spectroscopical methods, such as CARS (**C**oherent **A**nti-Stokes **R**aman **S**pectroscopy) and LIF (**L**aser **I**nduced **F**luorescence) Spectroscopy. The following prototypical situations will be described in some detail below:

- **Low-temperature reactor:** The investigation of the energy transfer from vibrationally excited hydrogen to deuterium molecules together with the corresponding wall-deactivation process, for room temperature and low pressure,

$$H_2(v=1) + D_2(v=0) \to H_2(v=0) + D_2(v=1),$$
$$H_2(v=1) \stackrel{\text{wall}}{\to} H_2(v=0),$$

in order to improve on the results of classical 'plug flow' methods which tend to strongly underestimate the velocity parameters. The goal is to determine the dependence of reaction velocities on vibrational excitation which is an important aspect in chemical kinetics.

- **High-temperature reactor:** Determination of the thermal reaction rate coefficients of the model reaction

$$O(^1D) + H_2 \to OH + H,$$

for high temperature in order to close the gap between earlier results obtained by the 0-dimensional evaluation of a measurement at 1200 K and standard results for room temperature around 300 K. This reaction is of basic importance for the analysis of chemical processes in the atmosphere.

Other reaction systems have also been analyzed, for example, the reaction of OH radicals with vibrationally excited hydrogen, $OH + H_2(v=1) \to H_2O + H$ and the CA-CVD (**C**ombustion **A**ssisted - **C**hemical **V**apour **D**eposition) of diamond in a low-pressure flow reactor (see Zumbach et al. [44]). For a discussion of the results and a detailed description of the experimental set-ups of the flow reactor experiments, we refer to the article Hanf/Volpp/Wolfrum [24] in this volume and the literature cited therein. More complex combustion processes (involving fast chemistry) such as the ozone recombination or a rather complex chemical model of methane combustion (39 species and 302 elementary reactions) have been treated by the same methods in Braack [18], see also Braack/Rannacher [20], Becker et al. [6], and the article Braack/Richter [19] in this volume.

2 Mathematical Model

We briefly describe the numerical methodology used in the simulation of the flow reactors. For more details, we refer to Braack/Rannacher [20], Becker et al. [6], Waguet [42], and the article Braack/Richter [19] in this volume. The governing system of equations consists of the equation of mass, momentum and energy conservation, written in terms of the non-conservative variables density ρ, pressure p, velocity v and temperature T, supplemented by the equations of conservation of species mass fractions $y = (y_i)_{i=1,\ldots,n_s}$. Since in all cases considered, the inflow velocity is small, a low-Mach approximation of the compressible Navier–Stokes equations is used, i.e., the pressure is split like $p(x,t) = P_{\text{th}}(t) + p(x,t)$ into a thermodynamical part $P_{\text{th}}(t)$ which is constant in space and used in the gas law, and a much smaller hydrodynamical part $p(x,t) \ll P_{\text{th}}(t)$ which occurs in the momentum equation.

2.1 Hydro- and Thermodynamical Model

The conservation of mass, momentum and energy implies the equations

$$\partial_t \rho + \nabla \cdot (\rho v) = 0, \tag{1}$$

$$\rho \partial_t v + \rho v \cdot \nabla v - \nabla \cdot \tau + \nabla p = \rho g, \tag{2}$$

$$c_p \rho \partial_t T + c_p \rho v \cdot \nabla T - \nabla \cdot (\lambda \nabla T) + p \nabla \cdot v - \tau : \nabla v = f_T(T, y), \tag{3}$$

where the shear stress tensor for a Newtonian fluid has the form $\tau = \mu(\nabla v + \nabla v^T - \frac{2}{3} \nabla \cdot v I)$. As usual, the diffusion term in the momentum equation is rewritten in the form $-\nabla \cdot (\mu \nabla v) + \nabla \bar{p}$ with the modified pressure $\bar{p} := p - \mu \frac{1}{3} \nabla \cdot v$. In the following, we will denote $\bar{p}$ again by p. In the case of multi-component flows the dynamic viscosity μ is a function of the partial viscosities and the mass fractions of the participating species. The coefficients c_p and λ are the specific heat capacity at constant pressure and the heat conductivity of the mixture, respectively. The terms $p\nabla \cdot v$ and $-\tau : \nabla v$ describing mechanical effects are neglected in the energy equation. The corresponding source term has the form

$$f_T(T, y) = -\sum_{i=1}^{n_s} h_i(T) f_i(T, y) \qquad (4)$$

with heat sources $f_i(T, y)$, which are described in more detail in Section 2, and the corresponding specific enthalpies $h_i(T)$ which are empirically modeled (for details see Braack/Richter [19]).

According to the low-mach number approximation, the law of an ideal gas is used in the simplified form

$$\rho = \frac{P_{\text{th}} \bar{m}}{RT}, \qquad \frac{1}{\bar{m}} = \sum_{i=1}^{n} \frac{y_i}{m_i}, \qquad (5)$$

with the thermodynamical pressure $P_{\text{th}} = P_{\text{th}}(t)$, the molar masses m_i of the i-th species, the mean molar mass $\bar{m}$, and the ideal gas constant R. In the present configuration of an 'open' system, P_{th} is given as the spatial mean value over the outflow boundary,

$$P_{\text{th}}(t) = |\Gamma_{\text{out}}|^{-1} \int_{\Gamma_{\text{out}}} p(x, t)\, do,$$

and is to be prescribed.

2.2 Chemical Model

Gas-Phase Reactions

For the description of the chemical conversions in the gas phase the chemical mechanisms are composed of elementary reactions. The n_r elementary reactions of n_s species can be generally described by

$$\sum_{i=1}^{n_s} a_{ir} \chi_i \xrightarrow{k_r} \sum_{i=1}^{n_s} \tilde{a}_{ir} \chi_i, \qquad r = 1, \ldots, n_r,$$

where the χ_i represent the i-th species and k_r the reaction rate of the r-th reaction. The $\tilde{a}_{ir}$ and a_{ir} are the corresponding stoichiometric coefficients of species i as educt and product in the reaction r. In order to conserve mass, these coefficients must satisfy the balance equation

$$\sum_{i=1}^{n_s} m_i(\tilde{a}_{ir} - a_{ir}) = 0.$$

Then, the species mass conservation equations have the form

$$\rho \partial_t y_i + \rho v \cdot \nabla y_i + \nabla \cdot F_i = f_i(T, y), \quad i = 1, ..., n_s, \tag{6}$$

with the source terms

$$f_i(T, y) = m_i \dot{\omega}_i(T, y), \quad i = 1, \ldots, n_s.$$

The production rate $\dot{\omega}_i$ for species i is obtained by adding the participation of all the reactions considered to the reaction or destruction of species i,

$$\dot{\omega}_i(T, y) = \sum_{r=1}^{n_r} \left\{ (\tilde{a}_{ir} - a_{ir}) k_r(T) \prod_{j=1}^{n_s} c_j^{a_{jr}} \right\},$$

with $c_j := \rho y_j / m_j$ being the concentration of species j. The dependence on temperature of the reaction rate is given in form of an Arrhenius law

$$k_j(T) = A_j \left(\frac{T}{300K} \right)^{\beta_j} \exp\left(-\frac{E_{aj}}{RT} \right), \tag{7}$$

with constants A_j, β_j and E_{aj} (activation energy) which have to be determined by evaluating experimental data.

Surface Reactions

The model for surface reactions introduces a reaction probability γ (so-called 'sticking coefficient'; see Warnatz/Maas/Dibble [43]) for particles in the gas phase which hit a wall surface. These particles can react (recombination, decomposition) or diffuse further unchanged in the gas phase. Here, the case of surface reactions is considered in which there is only one gas-phase reactant. These reactions are described by the scheme

$$a_{jr} \chi_j \xrightarrow{k_r} \sum_{i=1}^{n_s} \tilde{a}_{ir} \chi_i, \quad j = 1, \ldots, n_s, \; r = 1, \ldots, n_r^0.$$

The corresponding reaction rate per surface unit for species i over all the n_r^0 surface reactions is given by

$$\dot{\omega}_i^0(T, y) = \sum_{r=1}^{n_r^0} \left\{ \gamma_r \frac{1}{4} \sqrt{\frac{8RT}{\pi m_j}} c_j (\tilde{a}_{ir} - \delta_{ij} a_{jr}) \right\},$$

where j refers to the single educt species of the reaction r. In this wall reaction model there is exactly one educt species for each surface reaction. The deactivation probabilities are assumed to be of the form

$$\gamma_r = a_r \left(\frac{T}{300K} \right)^{b_r} \exp\left(-\frac{c_r}{RT} \right), \quad r = 1, \ldots, n_r^0, \tag{8}$$

where again the constants a_r, b_r, and c_r are to be determined by evaluating experimental data.

Diffusion models

In the considered flow reactor for the viscosity and heat conductivity the empirical laws from Warnatz/Maas/Dibble [43] are used. The viscosity μ of a mixture can be modeled with an accuracy of about 10% by the partial viscosities μ_i and the mole fractions $x_i := y_i \bar{m}/m_i$ of the species:

$$\mu(T,y) = \frac{1}{2} \left(\sum_{i=1}^{n_s} x_i \mu_i + \left(\sum_{i=1}^{n_s} \frac{x_i}{\mu_i} \right)^{-1} \right),$$

where the $\mu_i = \mu_i(T)$ are nearly proportional to $\sqrt{T}$. For this polynomial fits with experimentally determined coefficients are used (see Kee/Rupley/Miller [30]). The heat conductivity λ has a similar representation:

$$\lambda(T,y) = \frac{1}{2} \left(\sum_{i=1}^{n_s} x_i \lambda_i + \left(\sum_{i=1}^{n_s} \frac{x_i}{\lambda_i} \right)^{-1} \right),$$

with λ_i the partial heat conductivities for which also a polynomial dependence on the temperature is assumed.

The diffusive fluxes $F_i(T, y)$ in the species conservation equations consist of three components, the 'mass diffusion' due to gradients in molar fractions (Fick's law), 'thermo-diffusion' due to temperature gradients (Soret effect), and 'pressure diffusion' due to pressure gradients. Here, the simple Fick's law is used, in which the diffusive fluxes are expressed in terms of the mole fractions $x_i := y_i \bar{m} m_i^{-1}$:

$$F_i := -\rho D_i \frac{m_i}{\bar{m}} \nabla x_i = -\rho D_i \nabla y_i - \rho D_i \frac{y_i}{\bar{m}} \nabla \bar{m}. \tag{9}$$

with the specific diffusion coefficients D_i of the different species which are obtained from chemical data banks. A further simplification is achieved by neglecting variations of the mean molar mass in the species diffusion,

$$F_i := -\rho D_i \nabla y_i. \tag{10}$$

This simplest model is used in the following discussion. A more detailed description of this setting is given in the article Braack/Richter [19] in this volume within the context of more general methods for simulating reactive flows.

Physical constraints

By definition, the mass fractions satisfy $0 \leq y_i \leq 1$, their sum must be one, and the sum over the diffusive fluxes must vanish:

$$\sum_{i=1}^{n_s} y_i = 1, \quad \sum_{i=1}^{n_s} F_i = 0. \tag{11}$$

However, the approximations used in the diffusion model are not fully consistent and may lead to mass fractions which do not some up to one, i.e., do not obey mass conservation. In the computations described in this article, the following ad-hoc correction is used:

$$\tilde{y}_i := \begin{cases} 10^{-12}, & \text{if } y_i \leq 10^{-12}, \\ y_i, & \text{otherwise,} \end{cases} \qquad y_i := \frac{\tilde{y}_i}{\sum_{j=1}^{n_s} \tilde{y}_j}.$$

In all computations the order of the corrections are locally at most 10% of the species mass fractions.

Geometry simplification

In view of the particular geometry of the flow reactors considered, the system (1) - (6) is written in cylinder coordinates and full rotational symmetry of all variables is assumed. This assumption has to be checked a priori by computations for realistic flow parameters. We summarize the set of conservation equations written in cylinder coordinates for the variables $\{\rho, v_r, v_z, \bar{p}, T, y\}$:

$$\bar{m}^{-1}(\partial_t \bar{m} + v \cdot \nabla \bar{m}) - T^{-1}(\partial_t T + v \cdot \nabla T) + \nabla \cdot v = -P_{\text{th}}^{-1} \partial_t P_{\text{th}}, \quad (12)$$

$$\rho \partial_t v_r + \rho v \cdot \nabla v_r - \nabla \cdot (\mu \nabla v_r) - \nabla \mu \partial_r v + \mu v_r r^{-2} + \partial_r p = 0, \quad (13)$$

$$\rho \partial_t v_z + \rho (v \cdot \nabla) v_z - \nabla \cdot (\mu \nabla v_z) - \nabla \mu \partial_z v + \partial_z p = \rho g, \quad (14)$$

$$\rho \partial_t T + \rho (v \cdot \nabla) T - c_p^{-1} \nabla \cdot (\lambda \nabla T) = c_p^{-1} f_T(T, y), \quad (15)$$

$$\rho \partial_t y + \rho c_p (v \cdot \nabla) y - \nabla \cdot (D \nabla y) = f_y(T, y), \quad (16)$$

with the abbreviations $y := (y_1, \ldots, y_{n_s})^T$, $f_y := (f_1, \ldots, f_{n_s})^T$, and $D := \text{diag}(D_i)$. The mean molar mass $\bar{m}$ and the density ρ are linked to the other variables by the law of an ideal gas in the form (5).

Boundary conditions

The system (12) - (16) has to be supplemented by appropriate initial and boundary conditions. If the simulation of the reactor flow is started from rest, the corresponding initial conditions are

$$\rho_{|t=0} = \text{const.}, \quad v_{|t=0} = 0, \quad T_{|t=0} = 0, \quad y_{i|t=0} = 0.$$

The boundary of the flow domain Ω is decomposed like $\partial \Omega = \Gamma_{\text{in}} \cup \Gamma_{\text{rigid}} \cup \Gamma_{\text{out}} \cup \Gamma_{\text{sym}}$ into the 'inflow part', the 'rigid part', the 'outflow part', and the axis of symmetry. Along the axis of symmetry the gradients of all variable vanish and the radial velocity is set to zero. At the inflow boundary Dirichlet conditions are prescribed for all variables with values to be experimentally determined. At the reactor walls 'no-slip' is assumed for the velocity, and for the temperature adiabatic or isothermal boundary conditions are used. For the species mass fractions a mixed boundary condition is used in which an equilibrium condition for diffusive transport and surface reaction rate (describing the heterogeneous processes at the wall) determines the concentration gradient. At the outflow boundary the usual 'free stream' condition is imposed, e.g., $n \cdot \tau - pn|_{\Gamma_{out}} = 0$ for the velocity. The total pressure is adjusted by prescribing the mean pressure P_{out} at the outflow boundary.

2.3 Goals of the Numerical Computation

The goals of the numerical computation for this model are as follows:

- *Direct simulation:* The computation of temperature and species densities for known model parameters and comparison with measured data, for validating the code and calibrating the experimental set-up.
- *Flow-model calibration:* The computation of velocity and temperature without chemistry and comparison with experimental data for determining unknown temperature boundary conditions.
- *Parameter estimation:* The determination of unknown kinetic parameters from measured concentrations of some of the chemical species.
- *Optimal experimental design:* The determination of the sensitivities in the parameter estimation process and optimizing them by re-designing the experimental set-up

Though most flow processes considered in this article are essentially stationary, the nonstationary version of the model is used in cases of possible nonstationary effects, such as in the high-temperature reactor under gravity, and for generating starting values for the Newton iteration.

Due to exponential dependence on temperature (Arrhenius law) and polynomial dependence on y, the source terms $f_i(T, y)$ are highly nonlinear. In general, these zero-order terms lead to a coupling between all chemical species mass fractions. For robustness the resulting system of equations is to be solved implicitly in a fully coupled manner on locally refined meshes.

3 Numerical Approach

The finite element (FE) Galerkin discretization of the coupled system (12) - (16) is based on its variational formulation written in compact form as

$$A(u)(\varphi) = 0 \quad \forall \varphi \in V, \qquad (17)$$

where V is a Hilbert space representing simultaneously the spatial as well as the temporal dependence of functions. The solution u and the test function φ are vector-valued according to the underlying model, i.e., $u = \{p, v, T, y\}$ and $\varphi = \{\varphi^p, \varphi^v, \varphi^T, \varphi^y\}$. The semi-linear form $A(\cdot)(\cdot)$ is (in cartesian coordinates) given by

$$A(u)(\varphi) := (\bar{m}^{-1}\partial_t\bar{m} + \bar{m}^{-1}v \cdot \nabla\bar{m} - T^{-1}\partial_t T - T^{-1}v \cdot \nabla T + \nabla \cdot v, \varphi^p)$$
$$+ (\rho\partial_t v + \rho v \cdot \nabla v, \varphi^v) + (\mu\nabla v, \nabla\varphi^v) - (\bar{p}, \nabla \cdot \varphi^v) - (\rho g, \varphi^v)$$
$$+ (\rho\partial_t T + \rho v \cdot \nabla T, \varphi^T) + (c_p^{-1}\lambda\nabla T, \nabla\varphi^T) - (c_p^{-1}f_T(T,y), \varphi^T)$$
$$+ (\rho\partial_t y + \rho v \cdot \nabla y, \varphi^y) + (\rho D\nabla y, \nabla\varphi^y) - (f_y(T,y), \varphi^y),$$

where $(\cdot, \cdot)$ denotes the usual L^2-inner product of scalar or vector-valued functions on the space-time domain $\Omega \times (0,T)$.

The spatial discretization by an FE method uses continuous piecewise polynomial trial and test functions for all unknowns on general meshes $\mathbb{T}_h = \{K\}$ consisting of non-degenerate quadrilaterals ('cells') K of width $h_K := \mathrm{diam}(K)$. The 'global mesh size' is $h := \max_{K \in \mathbb{T}_h} h_K$. These meshes are allowed to have 'hanging nodes' for local mesh refinement which is crucial for an accurate simulation of the models considered (see Fig. 1).

For pressure and transport stabilization standard least-squares techniques are used. We do not state the corresponding discrete equations since they have the same abstract structure as the continuous variational equations and can be found in the literature mentioned above. The basic ingredients of this numerical approach are listed below and will be described in the following subsections in more detail:

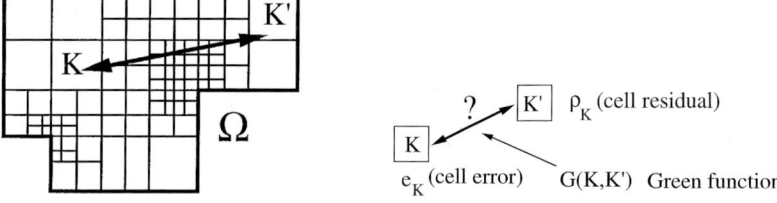

Fig. 1. Finite element mesh and scheme of error propagation.

- Second-order and third-order discretization in space by the finite element Galerkin method using Q_1/Q_1 or Q_2/Q_1 (biquadratic/bilinear Taylor-Hood) "Stokes-elements" for the flow variables, Q_2 elements for the temperature, and Q_1 elements for the chemical species. This discretizaton is on locally refined but hierarchically structured quadrilateral meshes utilizing "hanging nodes" for flexible mesh refinement or coarsening.
- Fully consistent 'least-squares' stabilization of pressure-velocity coupling (for the Q_1/Q_1 Stokes-element) and advective transport.
- Implicit first- or third-order discretization in time by the backward Euler ("dG(0)" method) embedded in a "dG(2)" interation in time (block strategy in time).
- Automatic mesh adaptation based on a posteriori error estimates following the DWR (Dual Weighted Residual) approach. In the nonstationary case,

the (linearized) dual problems are solved using a "check-pointing" strategy for storage saving, as described in Heuveline/Walther [27].
- Parameter identification by a quasi-Newton method called "limited-memory" BFGS (**B**royden, **F**letcher, **G**oldfarb, and **S**hanno method).
- Fast computation of stationary solutions by Newton-type iterations avoiding pseudo-time stepping on fine meshes.
- Multigrid solution of linear subproblems on the hierarchies of locally refined meshes.

3.1 Duality-Based Mesh Adaptation

The mesh adaptation aims at an efficient use of computer resources in terms of CPU time and memory requirements for computing the desired quantities in the reactive flow model. To this end, a posteriori estimates for the error in these quantities are derived by duality arguments similar to the representation of the solution of an elliptic boundary value problem in terms of the associated Green function. In such an estimate the cell-residuals of the approximate solution are multiplied by certain weights obtained from an associated linearized 'dual problem' in which the given error functional acts as right-hand side. The solution of this 'dual' problem (generalized Green function) is usually computed by the same method as used for solving the 'primal' problem. The cell-weights are the 'influence factors' of the cell-residuals with respect to the error in the target quantity. This principle is indicated in Fig. 1.

In the course of the adaptation process a hierarchy of meshes is generated which are particularly tailored according to the prescribed goal of the computation. The use of locally adapted meshes yields an improved robustness of the computation compared to ad-hoc designed tensor-product meshes since the sensitivities of the single cells on the accuracy in the approximation of the target quantity are observed. This allows the computation of stationary solutions by 'stationary methods', such as the Newton method, which are more economical than the common pseudo-time stepping procedures.

We present a unified description of this approach, called the 'DWR (**D**ual **W**eighted **R**esidual) method', which can be applied to the direct simulation for computing certain quantities of physical interest as well as in the context of parameter optimization and identification. To this end, we use a rather abstract set-up which contains both situations, purely stationary or fully nonstationary, and refer to the articles Becker et al. [12] and Braack/Richter [19] in this volume for more details.

Let V be a Hilbert space and $A(\cdot)(\cdot)$ a differentiable semi-linear form on $V \times V$. We consider the variational equation

$$A(u)(\psi) = 0 \quad \forall \psi \in V, \tag{18}$$

which is assumed to posses a locally unique solution. We seek approximations $u_h \in V_h$ in finite element subspaces $V_h \subset V$ by solving the Galerkin equations

$$A(u_h)(\psi_h) = 0 \quad \forall \psi_h \in V_h. \tag{19}$$

Let $J(u)$ be a quantity of physical interest which is to be computed, where $J(\cdot)$ is a differentiable functional on V. Then, it makes sense to estimate the error in this approximation with respect to this functional, i.e., in terms of $J(u) - J(u_h)$. To the Galerkin equation (19), we associate the residual functional

$$\rho(\cdot) := -A(u_h)(\cdot).$$

Further, we introduce the 'dual variable' $z \in V$ associated to the functional $J(\cdot)$ as the solution of the 'dual problem'

$$A'_u(u)(\varphi, z) = J'_u(\varphi) \quad \forall \varphi \in V, \tag{20}$$

which is a linear evolution problem running from $t = T$ backward in time to $t = 0$. >From Becker/Rannacher [14], we recall the following abstract a posteriori error representation

$$J(u) - J(u_h) = \rho(z - i_h z) + \mathcal{R}_h, \tag{21}$$

for arbitrary approximations $i_h z \in V_h$, where the remainder $\mathcal{R}_h$ is quadratic in the error $e^u := u - u_h$.

For illustrating this abstract result, we state the strong form of the dual problem in the *stationary* case (suppressing terms related to the least-squares stabilization and those containing the mean molar mass $\bar{m}$) used in our computations. For economical reasons, we do not use the full Jacobian of the coupled system in setting up the dual problem, but only include its dominant parts. The same simplification is used in the nonlinear iteration process. Accordingly, an (approximate) dual solution $z = (z^p, z^v, z^T, z^y)$ is determined by the system (written in cartesian coordinates):

$$\begin{aligned}
-\nabla \cdot z^v &= J^p, \\
\nabla(\rho v z^v) + \nabla \cdot (\mu \nabla z^v) - \rho \nabla v \cdot z^v - T^{-1} z^p \nabla T - \nabla z^p & \\
+ \rho z^T \nabla T + \rho \nabla y \cdot z^y &= J^v, \\
T^{-2} v \cdot \nabla T z^p - \nabla \cdot (T^{-1} v z^p) - \nabla \cdot (\rho v z^T) - \nabla \cdot (c_p^{-1} \lambda \nabla z^T) & \\
- c_p^{-1} D_T f_T(T, y) z^T - D_T f_y(T, y) z^y &= J^T, \\
-\nabla \cdot (\rho v z^y) - \nabla \cdot (\rho D \nabla z^y) - D_y f_y(T, y) z^y - c_p^{-1} D_y f_T(T, y) z^T &= J^{y_i},
\end{aligned} \tag{22}$$

where $J = (J^p, J^v, J^T, J^y)^T$ represents the prescribed output functional. In setting up the dual problem the nonlinear dependence of diffusion coefficients as well as the stabilization terms are neglected. With this notation, and neglecting the remainder term $\mathcal{R}_h$, the abstract a posteriori error representation (21) after cell-wise integration by parts and using Hölder's inequality results in the following concrete error estimate:

$$|J(u) - J(u_h)| \approx \sum_{K \in \mathbb{T}_h} \sum_{\alpha \in \{p,v,T,y_i\}} h_K^4 \{\rho_K^\alpha + \sigma_K^\alpha\} \omega_K^\alpha. \tag{23}$$

Here, the so-called cell residuals

$$\rho_K^p := h_K^{-1} \|R^p(u_h)\|_K, \qquad \rho_K^v := h_K^{-1} \|R^v(u_h)\|_K + j_{\partial K}(v_h)$$
$$\rho_K^T := h_K^{-1} \|R^T(u_h)\|_K + j_{\partial K}(T_h), \qquad \rho_K^{y_i} := h_K^{-1} \|R^{w_i}(u_h)\|_K + j_{\partial K}(y_{h,i}),$$

are expressed in terms of the cell-wise equation residuals $R^\alpha(u_h)$ and certain boundary terms $j_{\partial K}(\cdot)$ involving normal-jumps of the discrete solution $u_h = (p_h, v_h, T_h, y_h)^T$ across inter-element boundaries. Further ingredients are the contributions from the least-squares stabilization

$$\sigma_K^p := \delta_K^p h_K^{-1} \|R^v(u_h)\|_K, \qquad \sigma_K^v := \delta_K^v \|\rho v\|_{\infty;K} h_K^{-1} \|R^v(u_h)\|_K,$$
$$\sigma_K^T := \delta_K^T \|\rho v\|_{\infty;K} h_K^{-1} \|R^T(u_h)\|_K, \qquad \sigma_K^{y_i} := \delta_K^{y_i} \|\rho v\|_{\infty;K} h_K^{-1} \|R^{y_i}(u_h)\|_K.$$

and the weights

$$\omega_K^p := h_K^{-3} \|z^p - z_h^p\|_K, \qquad \omega_K^v := h_K^{-5/2} \|z^v - z_h^v\|_{\partial K},$$
$$\omega_K^T := h_K^{-5/2} \|z^T - z_h^T\|_{\partial K}, \qquad \omega_K^{y_i} := h_K^{-5/2} \|z^{y_i} - z_h^{y_i}\|_{\partial K}.$$

For a more detailed description of such a posteriori error estimates see the article Becker et al. [12] in this volume.

In the case of bilinear trial functions as used in these computations, the jump terms $j_{\partial K}(\cdot)$ in the cell residuals ρ_K^α are dominant and determine the relevant size of the local error indicator. The weights in the a posteriori error bound (23) are evaluated by solving the dual problem numerically on the current mesh and approximating the exact dual solution z by patchwise higher-order interpolation of the computed dual solution z_h,

$$(z - i_h z)_{|K} \approx (i_{2h}^{(2)} z_h - z_h)_{|K},$$

This technique shows sufficient robustness and does not require the determination of any interpolation constant; see the articles Becker et al. [12] and Braack/Richter [19] in this volume.

Remark 1. The most important feature of the a posteriori error estimate (23) is that the local cell residuals related to the various physical effects governing flow and transfer of temperature and chemical species are systematically weighted according to their impact on the error quantity to be controlled. To illustrate this, let us consider control of the mean radial concentration of any of the chemical species

$$J(p, v, T, y) = |\Omega|^{-1} \int_\Omega y_i \, dx \,.$$

Then, in the dual problem the right-hand-sides J^p, J^v, J^T and J^{y_j} ($j \neq i$) vanish, but because of the coupling of the variables all components of the dual solution z will be non-zero. Consequently, although the error term to be controlled only enters the i-th species balance equation it is also affected by the cell-residuals of the other equations. This sensitivity is quantitatively represented by the weights involving the dual components z^p, z^T, z^v and z^{y_j} ($j \neq i$). This frees us from heuristic guessing in balancing the various residual terms in the error estimator.

3.2 Mesh Adaptation and Solution Process

The strategy of mesh adaptation on the basis of the a posteriori error estimate (23) aims at the equilibration of the local 'error indicators' η_K corresponding to the mesh cells $K \in \mathbb{T}_h$. A heuristic but practically useful strategy proceeds as follows: The cells K of the current mesh $\mathbb{T}_h$ are ordered with respect to the size of the associated error indicators, i.e., $\eta_{K_1} \leq \cdots \leq \eta_{K_{N_h}}$. The so-called 'fixed fraction strategy' splits the cells into two groups, according to the conditions

$$\sum_{i=1}^{N_*} \eta_{K_i} \approx a\eta, \quad \sum_{i=N^*}^{N_h} \eta_{K_i} \approx b\eta,$$

with pre-assigned fractions $0 < a < b < 1$ (say $a = b = 0.2$). Then, the cells $K_1, \ldots, K_{N_*}$ are refined and the cells $K_{N^*}, \ldots K_{N_h}$ are coarsened. This process is continued until the prescribed error tolerance TOL or the maximal available number $N_{\max}$ of cells are reached, i.e., $\eta \approx TOL$ or $N_h \approx N_{\max}$.

This mesh adaptation step is successively used within an outer solution process, e.g., an approximate Newton iteration, which starts from an initial coarse mesh and converges on a sequence of locally adapted meshes to the solution on the final 'optimized' fine mesh. For a more detailed description of this process see Becker et al. [12] in this volume.

3.3 Parameter Identification

Next, we describe the extension of the adaptive solution approach presented so far for the 'forward' simulation of an reactive flow model to the estimation of kinetic parameters used in these models. Such a parameter estimation procedure relies on the definition of a 'merit-function' which measures the agreement between the experimental data and the model with a particular choice of parameters. This merit-function is usually chosen such that small values represent close agreement. The parameters of the model are then chosen to achieve a minimum in the merit-function, yielding 'best-fit parameters'. In our approach, we restrict ourselves to the case of least-squares type functionals for the merit-function. This choice offers a convenient set-up for the numerical solution of parameter identification problems. However it relies on specific statistical assumptions for the measurement errors which are of great importance in the context of optimum experimental design which will be discussed below.

In the following, we outline the solution approach for (finite) parameter identification problems based on the least-squares method. For more details and extensions we refer to the article Becker at al [12] in this volume; see also Becker/Vexler [15], Vexler [41].

Following the abstract setting from above, let V denote the state space, $Q = \mathbb{R}^{n_p}$ the control space, $Z = \mathbb{R}^{n_m}$ the observation space with observation operator $C : V \to Z$, and observation $\bar{c} \in Z$. Then, the corresponding optimization problem seeks $u \in V$, $q \in Q = \mathbb{R}^{n_p}$, such that

$$J(u, q) := \tfrac{1}{2}\|C(u) - \bar{c}\|_Z^2 \to \min. \tag{24}$$

The connection between control and state variable is given by the equation

$$A(u, q)(\varphi) = 0 \quad \forall \varphi \in V, \tag{25}$$

with a parameter-dependent semi-linear form $A(\cdot, \cdot)(\cdot)$ defined on $(V \times Q) \times V$ describing the underlying model. The discrete problems are posed in finite element spaces $V_h \subset V$ and $Q_h = Q$ and read

$$J(u_h) = \tfrac{1}{2}\|C(u_h) - \bar{c}\|_Z^2 \to \min, \tag{26}$$

with the constraint

$$A(u_h, q_h)(\varphi_h) = 0 \quad \forall \varphi_h \in V_h. \tag{27}$$

If the form $A(\cdot, q)(\cdot)$ is regular for any $q \in Q$, for sufficiently good approximations $V_h \subset V$, we can define the discrete solution operator $S_h : Q \to V_h$ and, setting $u_h := S_h(q)$, obtain the following unconstrained equivalent of the discretized optimization problem (26, 27) posed in the space $Q = \mathbb{R}^{n_p}$:

$$j_h(q) := \tfrac{1}{2}\|c_h(q) - \bar{c}\|_Z^2 \to \min, \tag{28}$$

where $c_h(q) := C(S_h(q))$, $q \in Q$. The partial derivatives

$$G(q)_{ij} := \partial_{q_j} c_{h,i}(q) = C'_{h,i}(u_h)(w_h^j), \qquad G(q) = (G(q)_{ij})_{i,j=1}^{n_p},$$

are determined by the solutions $w_h^j \in V$ of the tangent equations

$$A'_u(u_h, q)(w_h^j, \varphi_h) = -A'_{q_j}(u_h, q)(1, \varphi_h) \quad \forall \varphi_h \in V_h. \tag{29}$$

Using this notation the necessary first-order optimality condition is

$$j'_h(q_h) = 0 \quad \Leftrightarrow \quad G(q_h)(c_h(q_h) - \bar{c}) = 0. \tag{30}$$

This equation is solved by the standard Gauß-Newton algorithm (see, e.g., Nocedal/Wright [31]),

$$q_h^{k+1} = q_h^k + \delta q_h^k, \tag{31}$$

where δq_h^k is determined by the least-squares property

$$\tfrac{1}{2}\|c_h(q_h^k) + G(g_h^k)\delta q_h^k\|^2 \to \min. \tag{32}$$

This is equivalent to the normal equation

$$G(g_h^k)^* G(g_h^k)\delta q_h^k = -G(g_h^k)^* c_h(q_h^k). \tag{33}$$

A posteriori error estimation

For the finite element Galerkin approximation of the optimal control problem (24) a posteriori error estimates can be derived following very much the approach described above for the case of a variational problem. For more details on this we refer to the article Becker et al. [12] in this volume. The argument starts from the corresponding Lagrangian functional

$$\mathcal{L}(u, q, \lambda) := J(u, q) - A(u, q)(\lambda),$$

where λ denotes the associated co-state ('adjoint' or 'dual' variable). The triplet $\{u, q, \lambda\} \in V \times Q \times V$ is determined by the saddle-point system (first-order optimality condition)

$$\begin{aligned} A'_u(u, q)(\varphi, \lambda) &= J'_u(u, q)(\varphi) \quad \forall \varphi \in V, \\ A'_q(u, q)(\chi, \lambda) &= J'_q(u, q)(\chi) \quad \forall \chi \in Q, \\ A(u, q)(\psi) &= 0 \quad \forall \psi \in V. \end{aligned} \quad (34)$$

This is discretized by a standard Galerkin method using finite dimensional subspaces $V_h \times Q_h \subset V \times Q$ which results in the discrete saddle-point system

$$\begin{aligned} A'_u(u_h, q_h)(\varphi_h, \lambda_h) &= J'_u(u_h, q_h)(\varphi_h) \quad \forall \varphi_h \in V_h, \\ A'_q(u_h, q_h)(\chi_h, \lambda_h) &= J'_q(u_h, q_h)(\chi_h) \quad \forall \chi_h \in Q_h, \\ A(u_h, q_h)(\psi_h) &= 0 \quad \forall \psi_h \in V_h, \end{aligned} \quad (35)$$

for $\{u_h, q_h, \lambda_h\} \in V_h \times Q \times V_h$. The idea is now to seek control of the error in this discretization with respect to some functional $E(\cdot)$ defined on $V \times Q \times V$. The simplest choice, namely the cost functional of the optimal control problem itself,

$$E(u_h, q_h) := J(u_h, q_h) = \tfrac{1}{2} \|C(u_h) - \bar{c}\|_Z^2.$$

makes sense in the case of a real optimization problem. For example in seeking the stabilization of a nonstationary flow by boundary control, one is interested in a control q for which $J(u, q)$ is close to its minimum, but the 'optimal' control itself is only of secondary interest. In this case the resulting a posteriori error estimate is particularly simple since the corresponding 'dual solution' turns out to coincide with the adjoint variable λ. However, in the case of pure parameter identification this approach is not appropriate since here the error in the control parameter q itself is of primary interest. In fact, for almost identifiable parameters the cost functional $E(\cdot, \cdot) = J(\cdot, \cdot)$ may vanish even on coarse meshes, while the discrete control parameter q_h is still far away from the exact one, q. Hence, the error control functional should address the error in the control more directly, for instance by setting

$$E(u_h, q_h) := \tfrac{1}{2} \|q - q_h\|_Q^2.$$

Then, the systematic error control by the general approach described above applied to the Galerkin approximation of the saddle-point system (34) employs an extra 'outer' (linearized) dual problem resulting in an a posteriori error representation of the form (see Becker/Vexler [15] and Vexler [41])

$$E(u,q) - E(u_h, q_h) = \eta_h + \mathcal{R}_h + \mathcal{P}_h. \tag{36}$$

Here, the main part η_h of the estimator has the usual form

$$\eta_h = \tfrac{1}{2}\rho(u_h, q_h)(z - i_h z) + \tfrac{1}{2}\rho^*(u_h, q_h, z_h)(u - i_h u),$$

with the 'dual solution' $z \in V$ determined by the dual problem

$$A'_u(u,q)(\varphi, z) = -\langle G(G^*G)^{-1}\nabla E(q), C'(u)(\varphi)\rangle_Z \quad \forall \varphi \in V, \tag{37}$$

and the residuals

$$\rho(u_h, q_h)(\psi) := -A(u_h, q_h)(\psi),$$
$$\rho^*(u_h, q_h, z_h)(\varphi) := \langle G_h(G_h^*G_h)^{-1}\nabla E(q_h), C'(u_h)(\varphi)\rangle - A'_u(u_h, q_h)(\varphi, z_h).$$

The remainder $\mathcal{R}_h$ due to linearization is again cubic in the errors $u-u_h$, $q-q_h$, and $\lambda-\lambda_h$, and the additional error term $\mathcal{P}_h$ is bounded like

$$|\mathcal{P}_h| \leq \tilde{C} \|e\|_V \|C(u) - \bar{c}\|_Z. \tag{38}$$

Due to the particular features of the parameter identification problem, we can expect that $\|C(u) - \bar{c}\|_Z \ll 1$ for the optimal state. Hence, this term is neglected compared to the leading term η. Based on the a posteriori error representation (36) the mesh adaptation is organized as described above.

4 The Low-Temperature Flow Reactor (CARS experiment)

We consider a laminar flow reactor operated at room temperature around $300\,\text{K}$ and low pressure around $4\,\text{Torr}$, for determining the vibration energy transfer velocity between vibrationally excited hydrogen and deuterium. The configuration of the flow reactor is shown in Fig. 2.

(I) In the first experiment the wall-deactivation process is analyzed:

$$H_2(v{=}1) \stackrel{\text{wall}}{\to} H_2(v{=}0). \tag{39}$$

The chemical model consists of a set of 7 reactions involving 4 species. Helium He is used as inert gas.

(II) In the second experiment the exchange of vibrational energy of H_2 molecules with D_2 molecules together with the wall-deactivation process is analyzed:

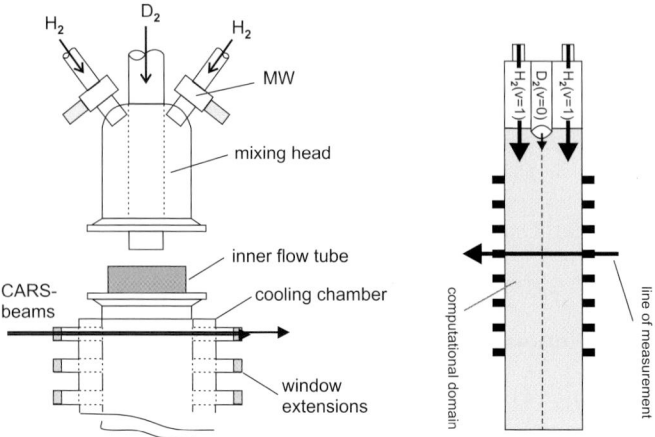

Fig. 2. Configuration of the low-temperature flow reactor (CARS experiment): experimental setting (left) and computational domain (right)

$$H_2(v=1) \stackrel{\text{wall}}{\to} H_2(v=0),$$
$$H_2(v=1) + D_2(v=0) \to H_2(v=0) + D_2(v=1). \tag{40}$$

Vibrationally excited species are treated 'state-selective', i.e., each species with a different vibrational level represents a new species with specific thermodynamic and physical properties for which enthalpies and transport coefficients are changing correspondingly. Initially, close to 200 reaction steps were considered for the present system. However, a sensitivity analysis using a zero-dimensional kinetic algorithm revealed that many of them were of marginal importance. Therefore, to handle the present full reactive flow problem with reasonable computational work, including the optimization procedure, the reaction mechanism was reduced to the sequence listed in Tables 4 and 5 in the Appendix. The complete chemical model used consists of 29 gas-phase and 5 wall reactions involving 8 species. Helium (He) is used as inert gas.

The concentrations of hydrogen and deuterium have been determined by laser spectroscopical means, particularly a two-color CARS system (**C**oherent **A**nti-Stokes **R**aman **S**pectroscopy) at several fixed measurement points along the reactor axis; for details see the article Hanf/Volpp/Wolfrum [24] in this volume. The proportion in mole of the vibrationally excited H_2 molecules at the inlet is 0.5%, the proportion of H atoms is 0.3%, and the rest 99.2% are not vibrationally excited H_2 molecules. With an inflow velocity of 50 m/s the Mach number is $Ma = 0.018$, such that the low-Mach number approximation is justified. The thermodynamical pressure is set constant to $P_{\text{th}} = 5.33\,\text{mbar}$. The assumption about the structure (rotationally symmetric and stationary) of the flow has been checked by an a priori calculation using the flow data relevant for the reactive flow simulation to be done.

The goal is the determination of the kinetic constants A_j, β_j and E_{aj} in the temperature range between $180K$ and $300K$ from measured CARS signals which are of the form

$$J(u) := I_{\text{as}}(p) = \int n(r)^2 f(r-p) \, dr,$$

where $n(r)$ is the radial concentration profile of the measured species and p is the location of the beam focus along the beam axis. The accuracy of these measurements is about 20%. The weighting function $f(r-p)$ is equipment dependent and has to be determined by experimental calibration.

4.1 Results

The wall-deactivation of vibrationally excited hydrogen is reflected by an decrease of the $H_2(v=1)$-CARS signal. An optimal adjustment of the simulated CARS signals to the experimental data is reached for a value of $\gamma = \mathbf{1.49 \cdot 10^{-3}}$ for the wall deactivation probabilty for $H_2(v=1)$. The vibrational energy transfer of hydrogen to deuterium is reflected by an decrease of the $H_2(v=1)$-CARS signal and the increase of the $D_2(v=1)$-CARS signal. An optimal adjustment of the simulated CARS signals to the experimental data is reached for a value of $\mathbf{k = 1.0 \cdot 10^{-14}}$ [$\mathbf{cm^3}$ molecule^{-1} s^{-1}] for the room-temperature rate constant for vibrational energy transfer of hydrogen to deuterium. This value coincides well with the value $k = 1.0 \cdot 10^{-14}$ [cm^3 molecule^{-1} s^{-1}] given in Bott et al. [17], and the value $k = 9.8 \cdot 10^{-15}$ [cm^3 molecule^{-1} s^{-1}] given by Pirkle et al. [33], which were obtained by other experimental methods. The 'plug flow' evaluation yields the systematically too small value $k = 7,1 \cdot 10^{-15}$ [cm^3 molecule^{-1} s^{-1}]. This is explained by the strong hydrodynamical effects on the distribution of the reactants which are not taken into account by the 0-dimensional analysis.

The mass fractions of $H_2(v=0/1)$ and $D_2(v=0/1)$ obtained by the simulation of the full two-dimensional reactive-flow model are shown in Tables 1. The values obtained on the automatically adapted meshes have a significantly better accuracy compared to those on hand-adapted tensor-product meshes and show a monotonic behavior which allows for extrapolation to the limit $h = 0$. Corresponding species concentrations and adapted meshes are shown in Fig. 3. The increase of performance of the new adaptive finite element code compared to the earlier (tensor-product) finite difference code is demonstrated in Table 2.

Table 1. Numerical results for the H_2 and D_2 mass fraction on hand-adapted (left) and on automatically adapted (right) meshes (Waguet [42])

Heuristic refinement				Adaptive refinement			
L	N	$H_2(v=0)$	$H_2(v=1)$	L	N	$H_2(v=0)$	$H_2(v=1)$
1	137	0.6556	0.005294	1	137	0.6556	0.005294
2	481	0.7373	0.006610	2	282	0.7382	0.006063
3	1793	0.7962	0.007096	3	619	0.7958	0.007132
4	6913	0.8172	0.007434	4	1368	0.8149	0.007323
5	7042	0.8197	0.007419	5	3077	0.8257	0.007457
6	7494	0.8240	0.007473	6	6800	0.8295	0.007534
7	8492	0.8269	0.007504	7	15100	0.8317	0.007564
8	10482	0.8286	0.007521	8	33462	0.8328	0.007587
9	15993	0.82853	0.007545	9	-	-	-

Heuristic refinement				Adaptive refinement			
L	N	$D_2(v=0)$	$D_2(v=1)$	L	N	$D_2(v=0)$	$D_2(v=1)$
1	137	0.7761	0.000000	1	137	0.7761	0.000000
2	481	0.7422	0.002541	2	244	0.7380	0.004020
3	1793	0.7801	0.002531	3	446	0.7450	0.002600
4	1923	0.7829	0.002729	4	860	0.7567	0.002010
5	2378	0.7851	0.001713	5	1723	0.7806	0.001390
6	3380	0.7917	0.001162	6	3427	0.7859	0.001130
7	5374	0.7916	0.001436	7	7053	0.7997	0.001090

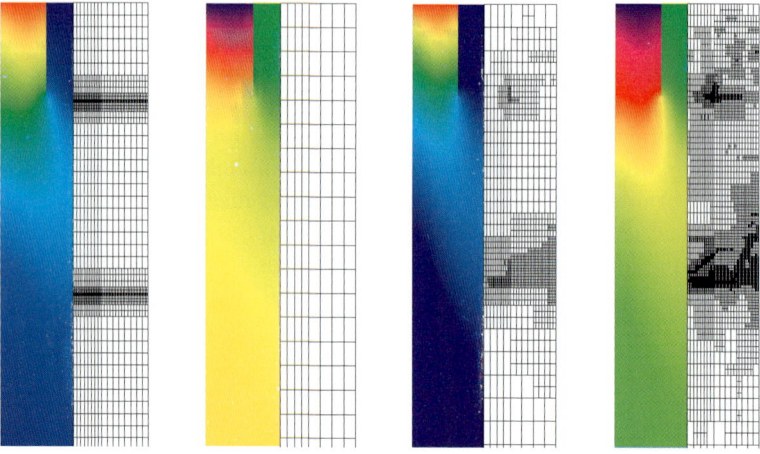

Fig. 3. Mass fraction of $D_2(v=1)$ in the flow reactor on a heuristically (left) and a sequence of adaptively (middle and right) refined meshes (Waguet [42])

Table 2. Comparison of CPU time and memory requirements of the tensor-product finite difference (FD) and the adaptive finite element method (FE) for simulating the wall deactivation reactions (Waguet [42])

Code	# of nodes	CPU time (sec.) total	per node	memory total	per node
FE	26884	18720 ($\sim$ 5 h)	0.70	127 MB	4,7 KB
FD	16000	85750 ($\sim$ 24 h)	5.35	153 MB	9.5 KB

4.2 Extensions to Other Reactions

After the demonstration that the direct PDE-based approach to parameter estimation is feasible for a first simple reaction system, now the rate constants of other important elementary reactions can be investigated. With the extension of the considered reaction processes beyond the hydrogen system the use of other more direct diagnosis tools such as LIF (**L**aser **I**nduced **F**luorescence) spectroscopy for OH and H atom detection becomes possible by which the design of the reactor can be largely simplified. For example, the optical windows can be avoided which act as sinks for the excited reaction partners and disturb the rotational symmetry. Further, the information about the concentrations are directly contained in the fluorescence signal and not, as in the CARS signal, in integral form with a resolution which can only be experimentally determined. For the reaction

$$OH + H_2(v=1) \rightarrow H_2O + H$$

the parameter estimation on the basis of the two-dimensional simulation yielded a value of $1.2 \cdot 10^{-13}$ [cm^3 molecule^{-1} s^{-1}] for the room-temperature rate constant, which is in good agreement with the results of full dimensional quantum scattering calculations performed by Meyer and co-workers [40] (see also Hanf/Volpp/Wolfrum [24] in this volume).

5 The High-Temperature Flow Reactor (LIF measurement)

This reactor is specially designed for the study of the reaction kinetics at elevated and high temperatures. Its configuration is shown in Fig. 4. Details on the experimental set-up are given in Hanf/Volpp/Wolfrum [24] in this volume. As prototypical example the mechanism of the following reaction has been analyzed in detail,

$$O(^1D) + H_2 \rightarrow OH + H, \tag{41}$$

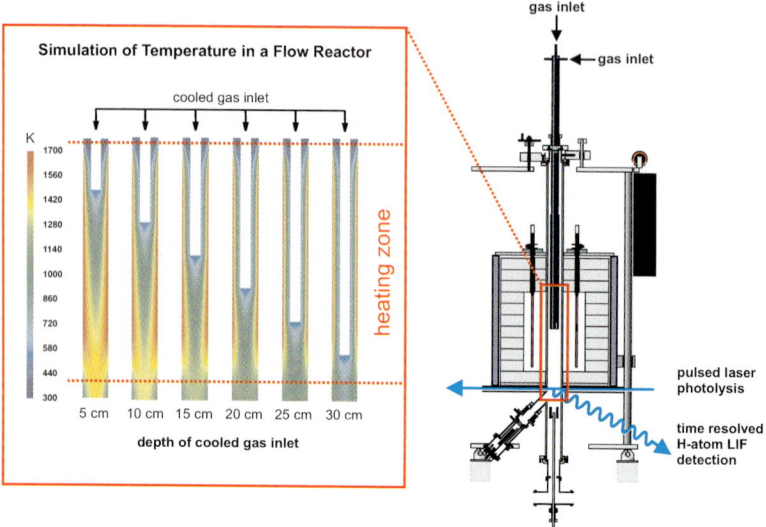

Fig. 4. Configuration of the high-temperature reactor

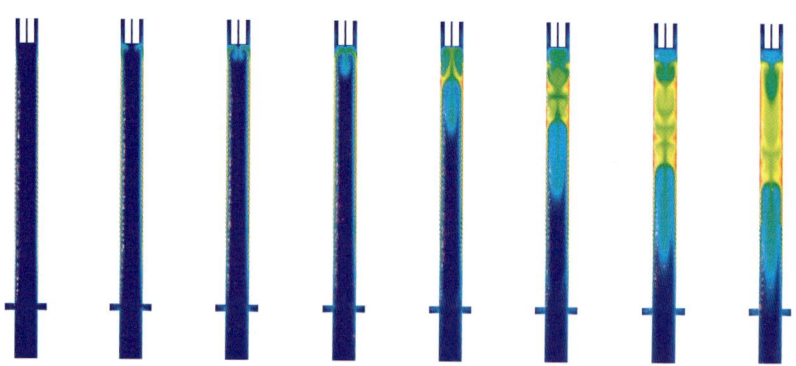

Fig. 5. A time series of temperature plots in the case of non-zero gravity indicating nonstationary behavior (Carraro [22])

where argon Ar has been used as inert gas. This system involves 7 chemical species and 6 elementary reactions. The goal is the determination of the reaction rates k_j or the complete set of kinetic parameters A_j, β_j and E_{aj} from measured H atom LIF signals (**L**aser **I**nduced **F**luorescence), i.e. mean values of the hydrogen H concentration over a small cylinder with a diameter of $\epsilon = 1$ mm located at the center of the main tube:

$$J(y) = \frac{1}{4\epsilon^2} \int_{a_z-\epsilon}^{a_z+\epsilon} \int_{a_r-\epsilon}^{a_r+\epsilon} y_H(r,z)\,dr dz.$$

A detailed description of this setting can be found in Carraro [22].

The design of the reactor is largely based on a priori numerical simulation. Especially the question of the achievable temperature in the measurement area is crucial and has been systematically investigated. These computations have shown that at high temperature and 'high' pressure (atmospheric pressure) the inclusion of gravity may result in persisting nonstationary flow behavior; see Fig. 5. This has enforced the restriction of the experiment to low pressure conditions (15 – 50 Torr), such as used in the CARS experiment discussed above. In order to do controlled measurements in the reactor at high temperature *and* atmospheric pressure the flow needs to be stabilized. An approach to achieve this will be briefly described below.

5.1 Model Calibration

The first step in preparing for the simulation of the reactive flow in the reactor is the determination of a complete set of initial and boundary conditions for which the model can be expected to be well-posed. For the initial concentration of the species, their partial pressures are calculated from the values of their fluxes and the total pressure, and from the partial pressures the initial number of molecules per m^3 is obtained. The in-flow profiles are calculated from the geometry of the inflow and the fluxes measured experimentally by the flux controllers. The out-flow boundary conditions, the no-slip condition at rigid walls and the symmetry condition along the cylinder axis are imposed as usual. Only the temperature boundary condition at the outer wall is not so clearly defined. This is a typical difficulty for the simulation of real-life experiments. The reactor is heated at the exterior wall in order to achieve the desired high temperature in the interior. However, due to constructional constraints the exact temperature distribution along the outer wall cannot be directly measured and is actually nonuniform due to the cold inflow at the upper inlets. The only information available is the temperature profile along the inner reactor axis which is obtained from thermo-sensor measurements. This is taken as input data for the implicit determination of the unknown temperature boundary data by a parameter estimation process (see the geometry description in Fig. 6).

- The temperature profile along the wall is assumed as a piecewise linear function described by 6 parameters, the temperature values at the positions $y_2, \ldots, y_7$, to be determined.
- Input data for the parameter estimation are measured profiles T_{meas} of the temperature along the symmetry axis.
- The parameters are determined by minimizing the least squares functional

$$F(T) = \frac{1}{2} \int_{\Gamma_{\text{sym}}} |T_{\text{sim}} - T_{\text{meas}}|^2 \, dz$$

for the measured and computed temperature profile along the symmetry axis subject to the flow equations.

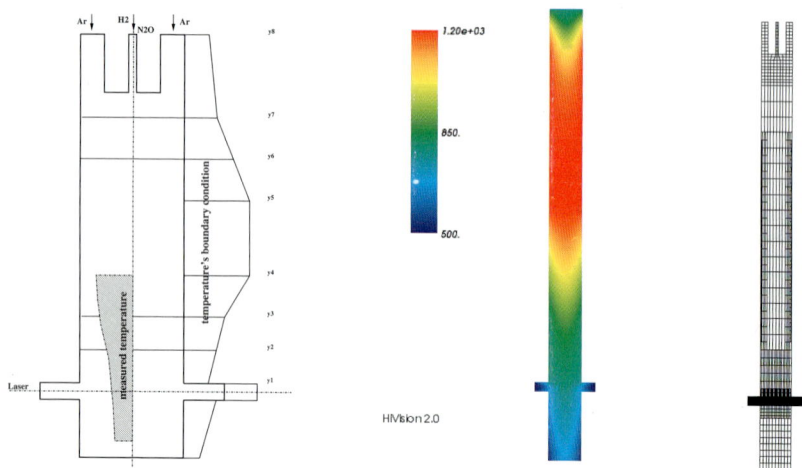

Fig. 6. Geometry sketch of the experiment for the calibration of the temperature boundary condition (left), computed temperature with $780\,K$ in the measurement area (middle), and an adapted mesh for the flow simulation (right) (Carraro [22])

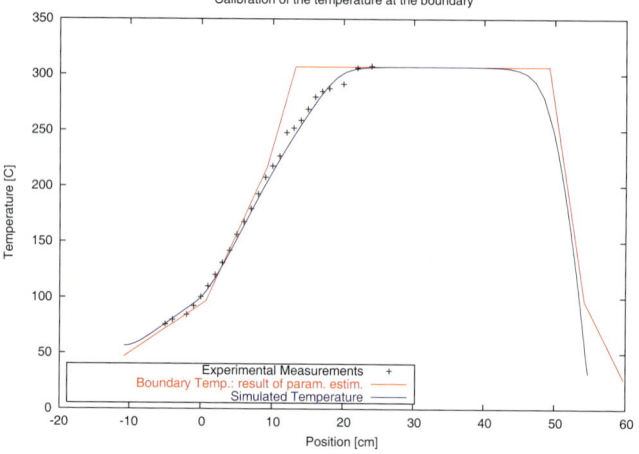

Fig. 7. Temperature profiles along the symmetry axis in the calibration of boundary conditions (Carraro [22])

Such a 'model calibration' has to be done within each measurement cycle in the reactor. The result of this calibration of temperature boundary conditions for an experiment with $580\,K$ in the heating zone is shown in Fig. 7.

Experimentally the reaction starts with the photodissociation of N_2O in N_2 and $O(^1D)$ produced by an ArF (193 nm) excimer laser. The detection

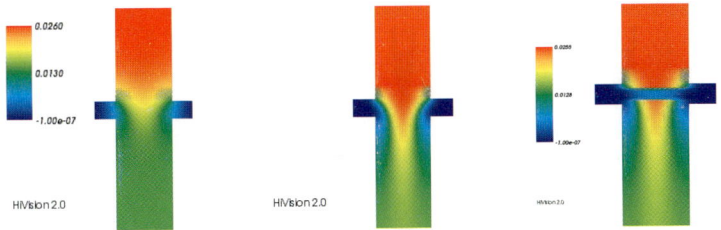

Fig. 8. H_2 (left) and N_2O distribution before ignition (left/middle) and during ignition (right) (Carraro [22])

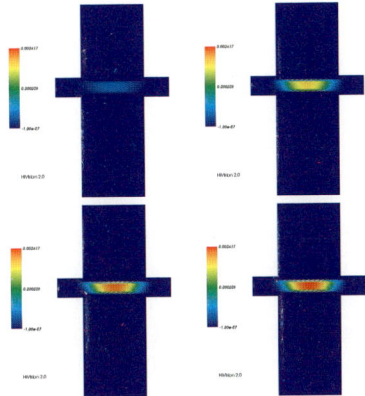

Fig. 9. Computed H mass fractions at different times (time step 100 ns) during the reaction process (Carraro [22])

of H atoms is done by the laser-induced fluorescence (LIF) technique. In the numerical case the photodissociation has been simulated by an ignition mechanism and the detection is an integration in the measurement volume of the mass fraction of the H atoms. The LIF signal is proportional to the number of H atoms that are in the measurement volume, so to compare the two signals (the experimental and the numerical one), we have to consider an additional parameter, that is the unknown scaling factor of the LIF signal.

Fig. 8 shows zooms into the measurement area before and after ignition. The depicted distributions of H_2 and N_2O indicate that the inflow from the side tubes at the optical zone does not much affect the concentration at the measurement point. Fig. 9 shows H mass fractions at different times (time step 100 ns) during the reaction process. Computed distributions of several of the relevant quantities are shown in Fig. 10.

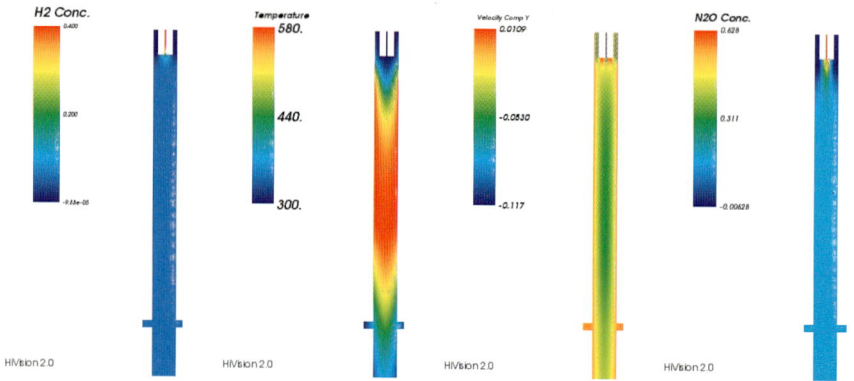

Fig. 10. Computed H_2, temperature, axial-velocity component, and N_2O (from left to right) in the stationary solution under zero gravity (Carraro [22])

5.2 Flow Stabilization

At high temperature and atmospheric pressure the flow in the flow reactor shows fully nonstationary behavior caused by temperature-gradient driven convection. But for a reliable LIF measurement we need stationary flow conditions in the measurement zone. This can be achieved by artificial stabilization of the flow by optimal control techniques. The idea is to modulate the boundary heating in such a way that at least in the measurement zone Ω_{meas} the nonstationary behavior is suppressed. To this end, we try to minimize the cost functional

$$J(u) = \frac{K}{2} \int_0^T \left\{ \|\partial_t v(t)\|^2_{\Omega_{\text{meas}}} + \|\partial_t T(t)\|^2_{\Omega_{\text{meas}}} \right\} dt + \text{'Regularization'},$$

where $u = \{v, p, T, y\}$ is determined by the flow model (12) - (16), and $q(t)$ is the prescribed time-dependent exterior heat distribution which is used to control the dynamics of the flow behavior, $T(\cdot,t)_{|\partial\Omega} = q(\cdot,t)$, $t \in [0,T]$. The application of this approach to the high-temperature flow reactor is subject of current work. It is expected that in this way the LIF measurements can be extended to atmospheric pressure.

5.3 Parameter Estimation

Due to the limitations imposed by the flow conditions discussed above, the LIF measurements in the high-temperature flow reactor had to be restricted to low pressure, $15 - 50$ Torr, and intermediate temperature, $300 - 800$ K. The quantity to be determined is the reaction rate

$$k(T) = A \left(\frac{T}{300K} \right)^\beta \exp\left(-\frac{E_a}{RT} \right), \tag{42}$$

of the reaction of interest, $O(^1D) + H_2 \to OH + H$, while the rates of all the other participating elementary reactions are supposed to be known with sufficient accuracy over the full temperature range. Since the experiment is carried out for fixed known temperature and the activation energy E_a is rather small the reaction rate $k(T)$ can directly be determined from measured values of the concentration of hydrogen H at different temperature.

The experiments have been repeated with different initial conditions, changing the ratio of the concentration of the species at the inflow: H_2, N_2O and Ar. The number of molecules of each species is determined by the flows in the cooled gas inlet and in the bathgas inlet, determined by flow regulators. In all cases the concentration of H_2 can be considered in excess so that during the reaction it can be regarded as almost constant for the calculation of the reaction rate. For the determination of the reaction rate 6 different setups have been used, obtained by varying the inflow regulator of H_2 from 5% to 50%. From these 6 data sets a pseudo first order reaction rate can be estimated and from this the 'true' reaction rate, for a given temperature.

With this technique the reaction rate has been determined 'experimentally'; for more details, we refer to the article Hanf/Volpp/Wolfrum [24] in this volume. For the numerical estimation of the reaction rates, we just need one experimental curve for one of the values of the initial concentration. We demonstrate this in the case of $T = 300\,K$ comparing the numerical results in two extreme cases: the first with the flow regulator for H_2 at 50% of the maximum value and the second at 5%. The measured values shortly after the ignition time are most significant for the reaction parameter, while later the measured values are rather scattered.

Experiment 1: $T = 300\,K$ and 50% H_2-inflow concentration

- Pressure $p = 1866\,Pa$ (14 Torr).
- Inflow mass fraction $y_{H_2} = 0.016$, $y_{N_2O} = 0.0236$.
- Inflow velocity in the main tube $v = 0.86\,m/s$ and in the optical cooling system $v = 0.55\,m/s$.

The parameter identification process determines the desired reaction rate as

$$k = 1.0 \cdot 10^{-10}\,[\text{cm}^3\,\text{molecule}^{-1}\,\text{s}^{-1}].$$

Fig. 11 gives the comparison between the corresponding computed and measured H concentrations. The two curves for the parameter values $k = 0.9 \cdot 10^{-10}\,[\text{cm}^3\,\text{molecule}^{-1}\,\text{s}^{-1}]$ and $k = 1.1 \cdot 10^{-10}\,[\text{cm}^3\,\text{molecule}^{-1}\,\text{s}^{-1}]$ are shown to demonstrate the sensitivity of the result with respect the value of the reaction rate.

Experiment 2: $T = 300\,K$ and 5% H_2-inflow concentration

- Pressure $p = 1866\,Pa$ (14 Torr).
- Inflow mass fraction $y_{H_2} = 0.0022$, $y_{N_2O} = 0.0239$.

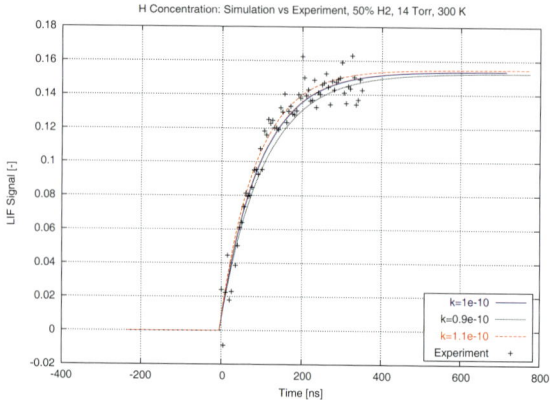

Fig. 11. Computed H concentrations using estimated parameters versus experimental data (H atom LIF signal) at room temperature $300\,\text{K}$ (Carraro [22])

- Inflow velocity in the main tube $v = 0.86\,\text{m/s}$ and in the optical cooling system $v = 0.55\,\text{m/s}$.

The parameter identification process determines the desired reaction rate as

$$k = 1.0 \cdot 10^{-10}\,[\text{cm}^3\,\text{molecule}^{-1}\,\text{s}^{-1}].$$

Fig. 12 gives the comparison between the corresponding computed and measured H concentrations.

Experiment 3: $T = 780\,\text{K}$ and 50% H_2-inflow concentration

- Pressure $p = 5866\,\text{Pa}\,(43.8\,\text{Torr})$.
- Inflow mass fraction $y_{H_2} = 0.02$, $y_{N_2O} = 0.0249$.
- Inflow velocity in the main tube $v = 0.86\,\text{m/s}$ and in the optical cooling system $v = 0.55\,\text{m/s}$.

The parameter identification process determines the desired reaction rate as

$$k = 1.5 \cdot 10^{-10}\,[\text{cm}^3\,\text{molecule}^{-1}\,\text{s}^{-1}].$$

Fig. 13 shows the comparison between the corresponding computed and measured H concentrations.

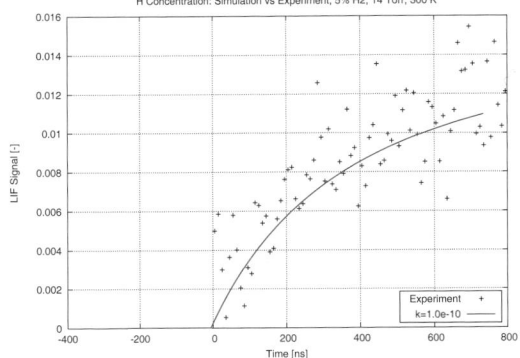

Fig. 12. Computed H concentrations using estimated parameters versus experimental data (H atom LIF signal) at room temperature $300\,\text{K}$ (Carraro [22])

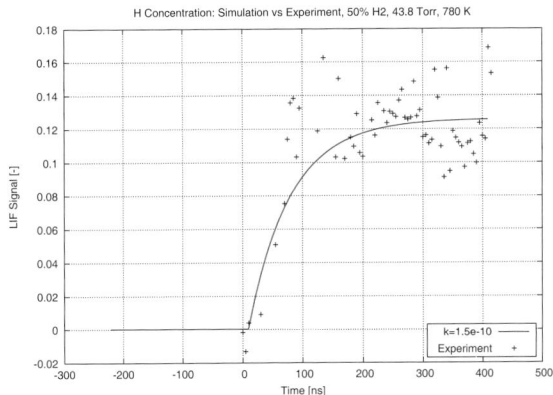

Fig. 13. Computed H concentrations using estimated parameters versus experimental data (H atom LIF signal) at higher temperature $780\,\text{K}$ (Carraro [22])

6 A Step Towards Optimal Experimental Design

For the high-temperature reactor also the question of systematic experiment optimization ('optimal experimental design') has been considered. The variances of parameter estimates and prediction depend upon this design. Unnecessarily large variances and imprecise predictions resulting from a poorly designed experiment largely wastes resources. Hence, the goal is to choose the model parameters, e.g., the inflow velocity and the position of the inner pipe, in such a way that the sensitivity of the reaction parameter to be determined is maximal. The proposed approach takes also into account the statistical properties of a random error in the measurements.

The parameter estimation procedure uses a merit-function of least-squares type which is particularly convenient for its numerical solution. We assume a physical model which is described by means of partial differential equations depending on a finite number of parameters $q := (q_1, \cdots, q_M) \in Q \subset \mathbb{R}^M$. In their weak form the corresponding *state equations* can be formulated as

$$A(u, q)(\varphi) = 0 \quad \forall \varphi \in V, \tag{43}$$

where u is the *state variable* defined in an appropriate Hilbert space V. As before, the semi-linear form $A(\cdot, \cdot)(\cdot)$ which is defined on the Hilbert space $(V \times Q) \times V$ may be nonlinear only with regard to its first two arguments, the state variable u and the control parameter q. We assume that each $q \in Q$ defines a unique solution $u_q \in V$ of the problem (43). Further, let the experimental set-up be characterized by a set $D \subset \mathbb{R}^{\tilde{N}}$ of so-called *experiment explanatory variables* $\xi = (\xi_1, \ldots, \xi_{\tilde{N}})$. Then, in order to determine the correct parameters q^* describing the considered physical problem, we suppose that for each parameter set $\xi \in D$ and any given state u, we have measurements of N observable quantities denoted by $c(\xi, u) \in \mathbb{R}^N$. The measurements obtained by means of the experiment explanatory variable ξ which describes the experimental design are denoted by $\bar{c}(\xi)$, where $\bar{c} : D \to \mathbb{R}^N$. As depicted in Fig. 14, the measurements are subject to random errors $\epsilon = (\epsilon_1, \cdots, \epsilon_N)$ and $\bar{c}(\xi)$ should therefore be understood as an N-dimensional random variable. For simplifying notation, in the following we assume that $\tilde{N} = N$.

One crucial assumption is that the underlying model is an exact description of the physical process considered, i.e., modelling errors are negligibly small compared to measurement errors, and we have

$$\bar{c}(\xi) = c(\xi, u) + \epsilon. \tag{44}$$

Further, the statistical assumptions are made that the random errors ϵ have a Gauß distribution,

$$\mathrm{var}(\epsilon_i) = \sigma^2, \tag{45}$$

and posses covariance and expectation value zero,

$$\mathrm{cov}(\epsilon_i, \epsilon_j) = 0, \quad E(\epsilon_i) = 0. \tag{46}$$

The formulation of the parameter identification problem as a least-squares problem is consistent with these statistical assumptions.

As an example of this setting, we consider the case of point measurements of the state variable u at certain points x_i in the region $\bar{\Omega}$. Accordingly, we set $D := \Omega_{\mathrm{obs}} = \{x_i, i = 1, \ldots, x_N\} \in \mathbb{R}^N$ and

$$c(x_i, u) := \frac{1}{|B_\epsilon(x_i)|} \int_{B_\epsilon(x_i)} \omega(|x_i - x|) u_q(x) \, dx,$$

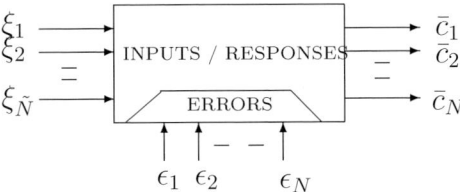

Fig. 14. Schematic representation of an experiment: The *experiment explanatory variables* $(\xi_1, \cdots, \xi_{\tilde{N}})$ describe the current experimental design. The measurement responses $(\bar{c}_1, \cdots, \bar{c}_N)$ are random variables. Their dependence on the experiment explanatory variables $(\xi_1, \cdots, \xi_{\tilde{N}})$ is blurred by the uncontrollable errors $(\epsilon_1, \cdots, \epsilon_N)$.

where $B_\epsilon(x_i) \subset \mathbb{R}^d$ denotes the ball centered in x_i with radius $\epsilon > 0$, and $\omega(\cdot)$ is an adequate weighting function. If well defined in the considered functional analytical set-up, the mapping $c(\cdot, \cdot)$ could even reduce to sharp point evaluations (Dirac functionals) $c(x_i, u) := u_q(x_i)$, $i = 1, \ldots, N$.

The parameter identification problem aims at solving the following problem for a given $\xi \in H$:

$$\begin{cases} \frac{1}{2}\|C(u,\xi) - \bar{c}(\xi)\|_Z^2 \to_{q \in Q} \min, \\ A(u,q)(\varphi) = 0 \quad \forall \varphi \in V. \end{cases} \quad (47)$$

Now, the question is to determine 'optimal' explanatory variables ξ in order to improve statistically the results of the parameter identification problem (47). This means that the covariance matrix $coV := \sigma^{-2}(G^*G)^{-1}$ of the parameters should be optimized. Common optimization criteria are the maximization of

$$\Phi(coV) = \begin{cases} \det(coV) & \text{(determinant: Criterion D)}, \\ \operatorname{tr}(coV) & \text{(trace: Criterion A)}, \\ \sigma(coV) & \text{(spectral radius: Criterion E)}, \end{cases}$$

with respect to the design parameter $\xi \in D$ of the model, the mentioned position of the inner pipe or the inflow velocity in the present case,

$$\Phi(coV(\xi)) \to_{\xi \in D} \min. \quad (48)$$

6.1 Numerical Experiment

To illustrate the principles of our approach to optimal experimental design in the context of the chemical flow reactor, as a model case, we consider the reaction between two chemical species, where the balance equation for the mass fraction u of one species is our model,

$$\mu(\nabla u, \nabla \varphi) + (b \cdot \nabla u, \varphi) = (m\omega, \varphi) \quad \text{in } \Omega, \quad u = 0 \quad \text{on } \partial\Omega. \quad (49)$$

Table 3. Optimization results and 95%-confidence region (Carraro [22])

	no OED	D	A	E	True Value
Meas 3	(0.9,0.5)	(0.12,0.27)	(0.11,0.2)	(0.11,0.2)	-
Param 1: μ	0.236819	0.194105	0.184985	0.184985	0.2
Param 2: k	8817.56	11654.1	12215.3	12215.3	15000
Area Ellipse	1.0220e+04	2586.0	2655.6	2655.6	-
Min Axis	0.068075	0.051487	0.054365	0.054365	-
Max Axis	4.7788e+04	1.5988e+04	1.5549e+04	1.5549e+04	-
Sum Axes	4.7788e+04	1.5988e+04	1.5549e+04	1.5549e+04	-

The domain is the unit square $\Omega = (0,1) \times (0,1)$, and the mass fraction v of the other species is obtained by $v := 1 - u$. Here, μ is the corresponding diffusion coefficient, b the velocity of the base flow, m the molecular mass, and ω the molar production rate of the species. The latter has the form $\omega = kc$, with k the reaction rate and c the species concentration. An Arrhenius law is assumed for k,

$$k = A\left(\frac{T}{300K}\right)^{\beta} \exp\left(-\frac{E_a}{RT}\right), \tag{50}$$

where the factor A, the temperature T, the activation energy E_a, and the gas constant R are given parameters. We want to identify the rate constant k and the diffusion coefficient μ from given 'measurement' data of the solution. To this end, we assume 3 point measurements of u at 3 positions in the flow domain. The first two measurement points are the same for all considered cases, while the third measurement is determined as the result of the optimal experimental design using the three different optimality criteria (D), (A) and (E), described above. In one case the measurement is taken at a position which is not an optimal design point. To each measurement we have added the same error, that is given by the 10% of the maximum of the value u in the domain. The fixed system parameters are

$$b = (1.5, 0.3)^T, \quad x_1^{\text{meas}} = (0.4, 0.3)^T, \quad x_2^{\text{meas}} = (0.6, 0.5)^T,$$

and the measurement error is $\epsilon = 10\% \, u_{\max} = 0.016$.

The results of the design optimization are shown in Table 3: the values of the estimated parameters for all criteria D, A, and E, the values of the error added and the dimensions of the 95% confidence regions, that are the area of the ellipses (minimized by the D criterion), the length of the axes (minimized by the E criterion) and the sum of the lengths of the axes (minimized by the A criterion). The least squares functional is the distance between the measured values of u_{meas} and the approximated ones by the finite element model,

$$F = \tfrac{1}{2}\|u_{\text{meas}} - u\|^2.$$

Fig. 17 shows corresponding confidence regions (ellipses). Obviously the confidence regions in the "non OED" case are much larger than in the other cases.

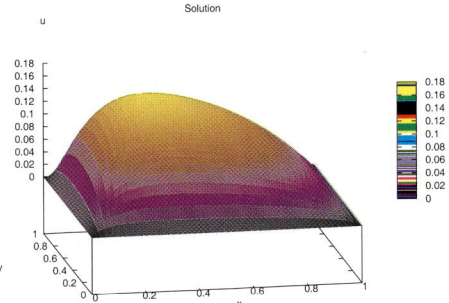

Fig. 15. Solution u for the nominal parameters (Carraro [22])

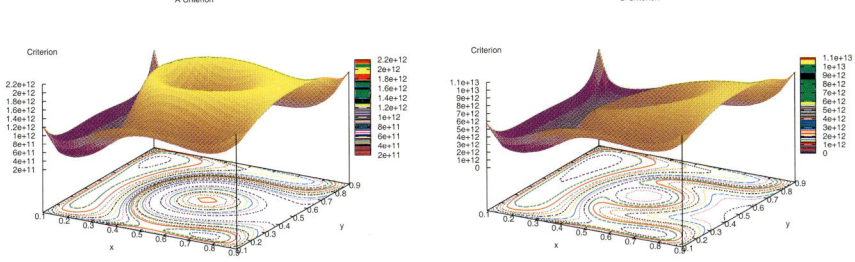

Fig. 16. Determinant (left) and trace (right) of the covariance matrix (Carraro [22])

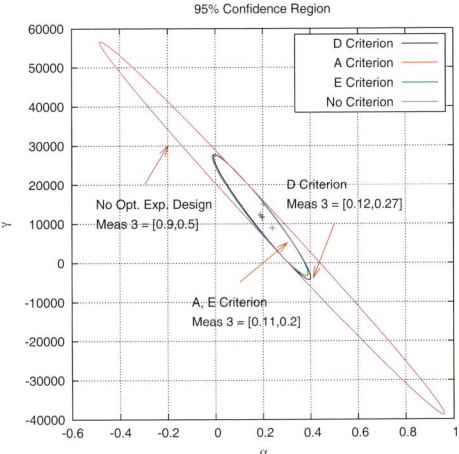

Fig. 17. Confidence regions for criteria (A), (D) and "non OED" (Carraro [22])

7 Conclusion and Outlook

In this project it has been demonstrated that experiments combined with numerical evaluation can be an effective tool for determining kinetic parameters of elementary reactions in flow reactors in a wide temperature range. The approach is currently extended into various directions:

- *Other reaction mechanisms:* After the setting up of the high-temperature flow-reactor system and its first successful application in kinetics studies of the $O(^1D)+H_2$ reaction(41), more complex systems, such as $NH+NO$, $NH + O_2$, $NH_2 + NO$, $CH_2 + NO$, and $CH + N_2$ which have multiple reaction product channels can be studied over an extended range of temperature.
- *Optimal experimental design:* The approach to optimal experimental design described in this article has to be developed further in order to be applied to the high-temperature reactor as well as to other types of flow reactors which contain more control parameters.
- *Optimal experimental control:* In the case of higher, i.e., atmospheric pressure, the flow in the reactor is nonstationary due to temperature-driven convection. It should be possible to stabilize this flow by nonstationary modulation of the wall heating such that it becomes (almost) stationary at least in the measurement area. This would allow the extension of the kinetics studies to atmospheric pressure conditions.

8 Appendix

In the following, we list the detailed reaction systems used in the simulations of the flow reactors described in this article.

Table 4. Mechanism of the $H_2(v=0/1)$ wall relaxation process at room temperature used in the CARS experiment with He as inert gas. Species: $H_2(0) := H_2(v = 0)$, $H_2(1) := H_2(v = 1)$, H, He denote neutral hydrogen and He atoms, respectively; in the notation for units $n = 1$ refers to bimolecular and $n = 2$ to trimolecular reactions; the kinetic parameters determined in this work are indicated by boldface type.

	Participating Species						$k(300\,\text{K})$
	H_2 – He Mechanism						$[\text{m}^{3n}/(\text{mol}^n\cdot\text{s})]$
1	$H_2(1)$	$+H$		$>$	$H_2(0)$	$+H$	$3.73E+04$
2	$H_2(1)$	$+H_2(0)$		$>$	$H_2(0)$	$+H_2(0)$	$7.80E+01$
3	$H_2(1)$	$+He$		$>$	$H_2(0)$	$+He$	$1.56E+01$
4	H	$+H$	$+He$	$>$	$H_2(0)$	$+He$	$2.50E+03$
5	H	$+H$	$+H_2(0)$	$>$	$H_2(0)$	$+H_2$	$2.90E+03$
	Wall reactions						$\gamma(300\,\text{K})$
6	$H_2(1)$			$>$	$H_2(0)$		**1.49E − 03**
7	H	$+H$		$>$	$H_2(0)$		$1.00E-04$

Table 5. Mechanism of the H_2/D_2 wall relaxation and vibrational energy exchange process at room temperature used in the CARS experiment. Species: $H_2(0)$:= $H_2(v = 0)$, $H_2(1)$:= $H_2(v = 1)$, $D_2(0)$:= $D_2(v = 0)$, $D_2(1)$:= $D_2(v = 1)$, $HD_2(0)$:= $HD_2(v = 0)$, $HD_2(1)$:= $HD_2(v = 1)$, H, D denote neutral hydrogen and He atoms, respectively; in the notation for units $n = 1$ refers to bimolecular and $n = 2$ to trimolecular reactions; the kinetic parameters determined in this work are indicated by boldface type.

		Participating Species				$k(300\,\text{K})$
		$H_2 + D_2-$ Mechanism				$[\text{m}^{3n}/(\text{mol}^n\cdot\text{s})]$
1	D	$+H_2(0)$		>H	$+HD(0)$	$1.28E+02$
2	D	$+HD(0)$		>H	$+D_2(0)$	$2.00E+01$
3	H	$+D_2(0)$		>D	$+HD(0)$	$1.02E+01$
4	H	$+HD(0)$		>D	$+H_2(0)$	$6.02E+01$
5	D	$+H_2(1)$		>H	$+HD(1)$	$7.77E+04$
6	D	$+HD(1)$		>H	$+D_2(1)$	$1.00E+04$
7	H	$+D_2(1)$		>D	$+HD(1)$	$7.00E+03$
8	H	$+HD(1)$		>D	$+H_2(1)$	$1.00E+04$
9	D	$+H_2(1)$		>H	$+HD(0)$	$2.00E+04$
10	D	$+HD(1)$		>H	$+D_2(0)$	$1.00E+04$
11	H	$+D_2(1)$		>D	$+HD(0)$	$2.00E+03$
12	H	$+HD(1)$		>D	$+H_2(0)$	$1.10E+04$
13	H	$+H_2(1)$		>H	$+H_2(0)$	$3.73E+04$
14	D	$+D_2(1)$		>D	$+D_2(0)$	$6.02E+03$
15	H	$+D_2(1)$		>H	$+D_2(0)$	$1.10E+05$
16	$D_2(1)$	$+D_2(0)$		$>D_2(0)$	$+D_2(0)$	$8.40E+00$
17	$HD(0)$	$+D_2(1)$		$>HD(0)$	$+D_2(0)$	$6.30E+01$
18	$H_2(1)$	$+H_2(0)$		$>H_2(0)$	$+H_2(0)$	$7.80E+01$
19	$D_2(0)$	$+H_2(1)$		$>D_2(0)$	$+H_2(0)$	$1.30E+02$
20	D	$+H_2(1)$		$>D$	$+H_2(0)$	$5.42E+04$
21	$D_2(0)$	$+H_2(1)$		$>D_2(1)$	$+H_2(0)$	$\mathbf{6.00E+03}$
22	$HD(0)$	$+H_2(1)$		$>HD(1)$	$+H_2(0)$	$1.13E+05$
23	$D_2(1)$	$H_2(0)$		$>D_2(0)$	$+H_2(1)$	$6.02E+01$
24	$HD(0)$	$+D_2(1)$		$>HD(1)$	$D_2(0)$	$7.00E+02$
25	$HD(1)$	$+H_2(0)$		$>HD(0)$	$+H_2(1)$	$9.00E+03$
26	$HD(1)$	$+D_2(0)$		$>HD(0)$	$+D_2(1)$	$1.00E+04$
27	H	$+H$	$+H_2(0)$	$>H_2(0)$	$+H_2$	$2.90E+03$
28	D	$+D$	$+H_2(0)$	$>D_2(0)$	$+H_2(0)$	$2.90E+03$
29	D	$+H$	$+H_2(0)$	$>HD(0)$	$+H_2(0)$	$2.90E+03$
		Wall reactions				$\gamma(300\,\text{K})$
30	$H_2(1)$			$>H_2(0)$		$\mathbf{1.49E-03}$
31	$D_2(1)$			$>D_2(0)$		$\mathbf{1.49E-03}$
32	$HD(1)$			$>HD(0)$		$\mathbf{1.49E-03}$
33	H	$+H$		$>H_2(0)$		$1.00E-04$
34	D	$+D$		$>D_2(0)$		$1.00E-04$

Table 6. Mechanism of the $O(^1D) + H_2$ reaction with Ar as inert gas used in the high-temperature LIF experiment: primary, secondary and tertiary reactions; the abbreviations $O := O(^3P)$ and $O^* := O(^1D)$ are used; the = sign means that the backward reaction is included; in the notation for units $n = 1$ refers to bimolecular and $n = 2$ to trimolecular reactions; the kinetic parameters determined in this work are indicated by boldface type.

						A	β	E_a
		Participating Species				$[m^{3n}/(mol^n \cdot s)]$		[kJ/mol]
1	O^*	+Ar		>O	+Ar	$3.011E+05$	0.0	0.000
		Primary and secondary reactions (48 species)						
2	O^*	$+H_2$		>OH	+H	**6.020E+06**	**1.96**	**−5.750**
3	O^*	$+N_2O$		>O	$+N_2O$	$6.022E+05$	0.0	0.000
4	O^*	$+N_2O$		>NO	+NO	$4.336E+07$	0.0	0.000
5	O^*	$+N_2O$		$>N_2$	$+O_2$	$2.650E+07$	0.0	0.000
6	N_2O	$+H_2$		$>H_2O$	$+N_2$	$3.451E+06$	0.5	0.000
7	OH	$+N_2O$		$>N_2$	$+HO_2$	$8.431E+06$	0.0	41.572
8	OH	$+N_2O$		=HNO	+NO	$6.082E+00$	4.3	104.762
9	OH	$+H_2$		>H	$+H_2O$	$9.334E+05$	1.6	13.802
10	O	$+N_2O$		=NO	+NO	$6.624E+07$	0.0	111.414
11	O	$+N_2O$		$>N_2$	$+O_2$	$1.018E+08$	0.0	117.234
12	O	$+H_2$		>H	+OH	$2.072E+05$	2.7	26.274
13	NO	$+N_2O$		$>NO_2$	$+N_2$	$1.728E+05$	2.2	193.727
14	NO	$+O^*$		>O	+NO	$2.409E+07$	0.0	0.000
15	NO	$+O^*$		>N	$+O_2$	$5.119E+07$	0.0	0.000
16	NO	$+H_2$		=HNO	+H	$1.078E+06$	2.3	212.019
17	H_2O	$+O^*$		>OH	+OH	$1.319E+08$	0.0	0.000
18	H	$+N_2O$		$>N_2$	+OH	$9.635E+07$	0.0	63.190
19	H	$+N_2O$		=NH	+NO	$3.029E+11$	−2.2	155.481
		Tertiary reactions (16 species)						
20	OH	+OH		$>H_2O$	+O	$3.101E+04$	2.4	−8.813
21	O	+OH		=H	$+O_2$	$2.608E+07$	−0.5	0.249
22	O	+O	+Ar	$>O_2$		+Ar	$4.098E+01$	0.0	−4.407
23	NO	+OH		$=HNO_2$		$1.499E+06$	−0.1	3.018
24	NO	+OH	+Ar	$=HNO_2$		+Ar	$3.130E+05$	−2.5	0.283
25	NO	+O		=N	$+O_2$	$5.378E+05$	1.0	162.132
26	NO	+O		$=NO_2$		$1.813E+07$	0.3	0.000
27	NO	+O	+Ar	$>NO_2$		+Ar	$2.430E+04$	−1.4	0.000
28	NO	+NO		$>N_2$	$+O_2$	$3.071E+06$	0.5	253.591
29	H_2O	+OH		$=H_2O$	+OH	$1.391E+05$	0.0	17.460
30	H_2O	+O		>OH	+OH	$7.528E+06$	1.3	71.504
31	H	+OH		$>H_2O$		$1.620E+08$	0.0	0.624
32	H	+OH	+Ar	$>H_2O$		+Ar	$9.393E+04$	−2.0	0.000
33	H	+OH		>O	$+H_2$	$4.131E+04$	2.8	16.213
34	H	+O		>O	+H	$1.581E+04$	−1.0	0.000
35	H	+NO		=N	+OH	$2.168E+08$	0.0	207.030
36	H	+NO		=O	+NH	$5.600E+08$	−0.1	292.030
37	H	+NO		=HNO		$1.469E+08$	−0.4	0.000
38	H	+NO	+Ar	=HNO		+Ar	$4.859E+04$	−1.3	3.076
39	H	$+H_2O$		>OH	$+H_2$	$4.107E+06$	1.6	80.817
40	H	+H	+Ar	$>H_2$		+Ar	$2.190E+04$	−1.0	0.000

References

1. J. P. D. Abbatt, K. L. Demerjian, and J. G. Anderson: *A new approach to free-radical kinetics: radially and axially resolved high-pressure discharge flow with results for $OH+(C_2H_6, C_3H_8, n\text{-}C_4H_{10}, n\text{-}C_5H_{12}) \rightarrow$ products at 297 K*, J. Chem. Phys. 94, 4566-4575 (1990).
2. J. Arnold, T. Bouché, T. Dreier, J. Wichmann, J. Wolfrum: *CARS studies on the heterogeneous relaxation of vibrationally excited hydrogen and deuterium*, Chem. Phys. Lett. 203, 283 (1993).
3. W. Bangerth and R. Rannacher: *Adaptive Finite Element Methods for Differential Equations*, Lectures in Mathematics, ETH Zürich, Birkhäuser, Basel 2003.
4. R. Becker: *An optimal control approach to a posteriori error estimation for finite element discretizations of the Navier-Stokes equations*, East-West J. Numer. Math. 8, 257-274 (2000).
5. R. Becker: *Adaptive Finite Elements for Optimal Control Problems*, Habilitationsschrift, Institut für Angewandte Mathematik, Universität Heidelberg, 2001.
6. R. Becker, M. Braack, R .Rannacher, and C. Waguet: *Fast and reliable solutions of the Navier-Stokes equations including chemistry*, Comput. Visual. Sci. 2, 107–122 (1999).
7. R. Becker, M. Braack, and B. Vexler: *Numerical parameter estimation for chemical models in multidimensional reactive flows*, to appear in Combustion Theory and Modelling 8, 661–682 (2004).
8. R. Becker, M. Braack, and B. Vexler: *Parameter identification for chemical models in combustion problems*, to appear in Appl. Numer. Math. (2004).
9. R. Becker, V. Heuveline, and R. Rannacher: *An optimal control approach to adaptivity in computational fluid mechanics*, Int. J. Numer. Meth. Fluids. 40, 105-120 (2002).
10. R. Becker, G. Kanschat, F.-T. Suttmeier: *DEAL, Differential Equations Analysis Library*, Release 1.0, Technical Report, University of Heidelberg 1995.
11. R. Becker, H. Kapp, and R. Rannacher: *Adaptive finite element methods for optimal control of partial differential equations: basic concepts*, SIAM J. Optimization Control 39, 113-132 (2000).
12. R. Becker, H. Kapp, D. Meidner, R. Rannacher, and B. Vexler: *Adaptive finite element methods for PDE-constrained optimal control problems*, Reactive Flows, Diffusion and Transport, (R. Rannacher et al., eds), Springer, Berlin-Heidelberg, 2005.
13. R. Becker and R. Rannacher: *Weighted a-posteriori error estimates in FE methods*, Lecture ENUMATH-95, Paris, Sept. 18-22, 1995, in: Proc. ENUMATH-97, Heidelberg, Sept. 28 - Oct.3, 1997 (H.G. Bock, et al., eds), pp. 621-637, World Scientific Publ., Singapore, 1998.
14. R. Becker and R. Rannacher: *An optimal control approach to error estimation and mesh adaptation in finite element methods*, Acta Numerica 2000 (A. Iserles, ed.), pp. 1-102, Cambridge University Press, 2001.
15. R. Becker and B. Vexler: *A Posteriori error estimation for finite element discretization of parameter identification problems*, Numer. Math. 96, 435–459 (2004).
16. G. D. Billing, R. E. Kolesnik: Chem. Phys. Lett. 215, 571 (1993).
17. J. F. Bott: *Vibrational energy exchange between $H_2(v=1)$ and D_2, N_2, HCl, and CO_2*, J. Chem. Phys. 65, 3921 (1976).

18. M. Braack: *An Adaptive Finite Element Method for Combustion Problems*, Dissertation, Institut für Angewandte Mathematik, Universität Heidelberg, 1998.
19. M. Braack and T. Richter: *Solving multidimensional reactive flow problems with adaptive finite elements*, Reactive Flows, Diffusion and Transport, (R. Rannacher et al., eds), Springer, Berlin-Heidelberg, 2005.
20. M. Braack and R. Rannacher: *Adaptive finite element methods for low-mach-number flows with chemical reactions*, 30th Computational Fluid Dynamics (H. Deconinck, ed.), Volume 1999-03 of Lecture Series von Karman Institute for Fluid Dynamics, 1999.
21. M. Cacciatore and G. D. Billing: *State-to-state vibration-translation and vibration-vibration rate constants in H_2-H_2 and HD-HD collisions*, J. Phys. Chem. 96, 217 (1992).
22. T. Carraro: *Parameter estimation and experimental design in flow reactors*, Dissertation, Institut für Angewandte Mathematik, Universität Heidelberg, 2005.
23. A. Hanf: *Konstruktion eines Hochtemperaturstrmungsreaktors und Untersuchung der Reaktion $O(^1D) + H_2 \rightarrow OH + H$ bei hohen Temperaturen*, Dissertation, Physikalisch-Chemisches Institut, Universität Heidelberg, 2005.
24. A. Hanf, H.-R. Volpp, and J. Wolfrum: *Determination of kinetic parameters in laminar flow reactors. II. Experimental aspects*, Reactive Flows, Diffusion and Transport, (R. Rannacher et al., eds), Springer, Berlin-Heidelberg, 2005.
25. F.-K. Hebeker and R. Rannacher: *An adaptive finite element method for unsteady convection-dominated flows with stiff source terms*, SIAM J. Sci. Comput. 21, 799–818 (1999).
26. V. Heuveline: *HiFlow: A finite element solver package*, http://www.hiflow.de/.
27. V. Heuveline and A. Walther: *Online checkpointing for adjoint computation in PDEs: Application to goal-oriented adaptivity and flow control*, Preprint, Computer Center, University of Karlsruhe, accepted for publication.
28. P. Houston, R. Rannacher, and E. Süli: *A posteriori error analysis for stabilized finite element approximations of transport problems*, Comput. Methods Appl. Mech. Engrg 190, 1483–1508 (2000).
29. F. Kaufman: *Progress in Reaction Kinetics*, Vol. 1, Pergamon Press, 1961.
30. R. J. Kee, F. M. Rupley, and J. A. Miller: *The chemkin thermodynamical data base*, Report SAND87-8215.UC4, Sandia National Laboratories, USA, 1987
31. J. Nocedal and S. J. Wright: *Numerical Optimization*, Springer Series in Operations Research, Springer, New York, 1999.
32. R. J. Pirkle and T. A. Cool: *Vibrational energy transfer for H_2-D_2 and H_2-HCl mixtures from 220-450 K*, Chem. Phys. Lett. 42, 58 (1976).
33. R. V. Poirier and R. W. Carr, Jr.: *The Use of tubular flow reactors for kinetic studies over extended pressure ranges*, J. Phys. Chem. 75, 1593-1601 (1971).
34. R. Rannacher: *Adaptive finite element discretization of optimization problems*, Proc. Conf. Numer. Analysis 1999 (D.F. Griffiths and G.A. Watson, eds), pp. 21-42, Chapman&Hall CRC, London, 2000.
35. R. Rannacher: *Finite element methods for the incompressible Navier-Stokes equations*, in Fundamental Directions in Mathematical Fluid Mechanics (P. Galdi, J. Heywood and R. Rannacher, eds), pp. 191-293, Birkhäuser, Basel, 2000.
36. R. Rannacher: *Incompressible Viscous Flow*, in Encyclopedia of Computational Mechanics (E. Stein, et al., eds), John Wiley, Chichester, 2004.

37. Y. Schneider-Kühnle, T. Dreier, and J. Wolfrum: *Vibrational relaxation and energy transfer in the Hydrogen system at temperatures between 110 and 300 K*, Chem. Phys. Lett. 294, 191–196 (1998).
38. J. Segatz: *Simulation und Optimierung von reaktiven Innenströmungen*, Dissertation, Physikalisch-Chemisches Institut, Universität Heidelberg, 1995.
39. J. Segatz, J. Wichmann, K. Orlemann, T. Dreier, R. Rannacher, J. Wolfrum: *Detailed numerical simulations in flow reactors - a new approach in measuring absolute rate constants*, J. Phys. Chem. 100, 9323-9333 (1996).
40. S. Sukiasyan: *Investigation of three- and four-atomic reactive scattering problems with the help of the multiconfiguration time-dependent Hartree method*, PhD thesis, Physikalisch-Chemisches Institut, University of Heidelberg, 2005.
41. B. Vexler: *Adaptive Finite Element Methods for Parameter Identification Problems*, Dissertation, Institut für Angewandte Mathematik, Univ. Heidelberg, 2004.
42. C. Waguet: *Adaptive Finite Element Computation of Chemical Flow Reactors*, Dissertation, Institut für Angewandte Mathematik, Univ. Heidelberg, 2001.
43. J. Warnatz, U. Maas, R. W. Dibble: *Combustion: Physical and Chemical Fundamentals, Modeling and Simulation, Experiments, Pollutant Formation*, Springer (3rd edition): Berlin-Heidelberg-New York, 2005.
44. V. Zumbach, J. Schäfer, J. Tobai, M. Ridder, Th. Dreier, Th. Schaich, and J. Wolfrum: *Experimental investigation and computationalk modeling of hot filament diamond chemical vapour deposition*, J. Chem. Phys. 107, 5918–5928 (1997).

Determination of Kinetic Parameters in Laminar Flow Reactors. II. Experimental Aspects*

A. Hanf, H.-R. Volpp, and J. Wolfrum

Institut für Physikalische Chemie, Universität Heidelberg

Summary. In the present contribution laser spectroscopic studies are described in which the chemical kinetics of benchmark elementary reaction steps in different laminar flow reactors were experimentally investigated along with detailed numerical modeling calculations (see the article Carraro/Heuveline/Rannacher [5] in this volume). Coherent anti-Stokes Raman spectroscopy (CARS) was utilized to study the collisional relaxation and vibrational energy transfer of vibrationally excited molecular hydrogen $H_2(v=1)$ in a low-temperature discharge flow reactor ($T = 110 - 300\,K$). In theses studies wall deactivation probabilities and thermal rate constants for the vibrational energy transfer gas-phase reaction $H_2(v=1) + D_2(v=0) \rightarrow D_2(v=1) + H_2(v=0)$ could be derived from a direct comparison of measured concentration profiles with results from a detailed numerical modeling. Further experiments were performed, in which CARS for molecular hydrogen detection along with OH laser-induced fluorescence (LIF) spectroscopy was utilized to determine the rate constant for the gas-phase reaction $OH + H_2(v=1) \rightarrow H + H_2O$. Finally a high-temperature flow reactor setup will be described, which allows for kinetics studies of elementary reactions using the pulsed laser photolysis (LP)/laser induced fluorescence (LIF) pump-and-probe technique. The latter technique was employed in the present work to investigate the temperature dependence of the reaction of electronically excited oxygen atoms with molecular hydrogen, $O(^1D) + H_2 \rightarrow H + OH$.

The contents of the present article is organized as follows:

- Introduction
- Experimental Section
- Results and Discussion
- Summary

1 Introduction

Since the early days of gas-phase reaction rate measurements in discharge flow systems [1, 2] the flow tube technique kept its importance in modern chemical

---

*This work has been supported by the German Research Foundation (DFG) through SFB 359 (Project B1) at the University of Heidelberg.

kinetics experiments as one of the most powerful tools for the determination of thermal rate constants of chemical elementary reactions. While new technologies, especially in the field of spectroscopic detection methods, have extended its range of applicability, the basic principle of the flow tube itself remains the same: mixing of reactants takes place upstream in a mixing section and consumption of reactants or build-up of products is monitored along a measurement section of the flow tube using suitable methods for atom, radical, or stable molecule detection in order to deduce axial concentration profiles.

In the evaluation of such measurements a mean flow velocity has to be assumed in order to convert the axial coordinate (the distance between the point of mixing and the point of detection) into a reaction time from which the reaction rate constant of interest can be deduced by modeling the homogeneous chemical reaction system. In case of carefully designed kinetic measurements this "plug flow" assumption is a useful approximation and has produced a wealth of trustworthy kinetic data. However, the method is known to bear systematic errors, since it is based on the approximation of a perfect decoupling of chemical and hydrodynamic processes in the flow tube. Especially in the mixing section of the reactor this assumption is not valid.

Concentration profiles of reactants and products in a flow tube usually show spatial variations in radial direction caused by the characteristics of the underlying flow field. Generally, the flow velocity is higher in the central region of the reactor than near the wall, giving rise to different residence times of the reactant species depending on their radial position. Those traveling in low speed domains near the wall experience a longer reaction time than those traveling near the center at the same axial position. In addition, for unstable species such as radicals or internally excited atoms and molecules, despite diffusional transport heterogeneous reactions and relaxation on reactive surfaces also entail radial concentration gradients depending on geometric constraints of the flow tube and wall activity. Often concentrations are deduced from spatially integrating optical detection methods, i.e., absorption measurements, which disregard these spatial variations.

In order to favour diffusive processes, which minimize radial concentration gradients, a flow tube traditionally is operated at low pressure (less than 9.75 Torr), in the so-called "plug flow limit" [3]. On the other hand a rapid radial diffusional transport will increase the relative importance of wall reactions with respect to the gas phase reactions of interest. Besides viscous pressure drop, back diffusion will become important: a diffusional transport process in the axial direction governed by axial concentration gradients that are caused by the chemical reactions. In the past, for the evaluation of kinetic rate constants many studies have been conducted to consider the various mentioned aspects by approximations to the ideal reactive flow. Kaufmann was one of the first who addressed the basic problems arising from viscous flow and diffusion [4]. In Ref. [4] approximate expressions for axial diffusion as well as for the treatment of wall reactions in terms of a pseudo homogeneous rate constant and an estimation of the radial concentration gradient due to homogeneous and heterogeneous reactions in a Poiseuille flow were derived.

One purpose of the combined experimental and theoretical work to be presented in the following is to show that the detailed modeling of reactive flow fields within a suitable laminar discharge flow reactor is a powerful tool for the determination of elementary reaction rate constants (see also the article Carraro/Heuveline/Rannacher [5] in this volume). Comprehensive analytical and experimental validations guarantee that the simulations properly describe transport and chemical processes in the reactive flow. It will be shown that, even for low pressures, the approximations made within the simplified plug flow evaluation can lead to an underestimation of rate constants. As a model system for a simple kinetic process the heterogeneous relaxation of vibrationally excited molecular hydrogen, $H_2(v = 1)$, and its energy transfer in collisions with deuterium molecules in the vibrational ground state, $D_2(v = 0)$, was considered:

$$H_2(v = 1) \xrightarrow{\text{wall}} H_2(v = 0) \tag{1}$$

$$H_2(v = 1) + D_2(v = 0) \longrightarrow D_2(v = 1) + H_2(v = 0) \tag{2}$$

Reaction (2) is of interest because of its model-like character as a simple example for the theoretical understanding of vibrational energy transfer (VV) processes in molecular collisions. For reaction (2) there is still a discrepancy between results of theoretical approaches and experimentally determined rate constants [6].

Furthermore, the reactions (1) and (2) are of importance in the context of experimental rate measurements for other reactions involving $H_2(v = 1)$ reagents such as the $OH + H_2(v = 1)$ reaction.

$$OH + H_2(v = 1) \rightarrow H + H_2O \tag{3}$$

In kinetics measurements of this reaction the presence of $H_2(v = 1)$ reactants requires the knowledge of the molecular hydrogen deactivation pathways, such as heterogeneous relaxation and vibrational energy transfer, which can compete with the actual reaction pathway (3). Besides its practical importance in combustion [7], in atmospheric [8], and in interstellar chemistry [9], the reaction $OH + H_2 \rightarrow H + H_2O$ along with its isotopic variants such as $OH + D_2$, and $OH + HD$ and the back reaction $H + H_2O$ - as one of the most "simplest" four-atom reactions – have become benchmark systems towards the development of rigorous quantum mechanical diatom-diatom and atom-triatom reactive scattering methods [10, 11].

However, while the $OH + H_2 \rightarrow H + H_2O$ reaction takes place adiabatically on the $H_3O(1^2A')$-ground-state potential energy surface (PES), reactions which involve electronically excited species such as e.g. the reaction of metastable oxygen atoms, $O(^1D)$, with H_2, which plays an important role in atmospheric chemistry [8], can proceed via a non-adiabatic mechanism involving both the ground-state PES and one or more electronically excited

PESs [10, 12].

$$O(^1D) + H_2 \rightarrow H + OH \qquad (4)$$

As outlined in Ref. [10], in case of reaction (4) five adiabatic surfaces correlate with the $O(^1D) + H_2$ reagents, two of which can be neglected in the theoretical treatment if the electronic fine structure of the reagents is not considered. The influence of the three remaining PESs, denoted as $1^1A'$, $2^1A'$, and $1^1A''$, on the dynamics and thermal kinetics of reaction (4) has been investigated in recent years in a variety of quasi-classical "surface hopping" [13] and quantum scattering calculations (see e.g. [10, 14]). These calculations revealed that the contribution of the ground-state PES $1^1A'$, which has a deep well corresponding to the electronic ground state of the water molecule, dominates at low collision energies. These calculations further showed that on the ground-state PES $1^1A'$ reaction (4) proceeds *via* insertion of the oxygen atom into the H_2 bond leading to a highly vibrationally excited H_2O^* transient reaction intermediate, which rapidly (within a few molecular vibrations) decomposes into H + OH products. Because the reaction on the ground-state PES $1^1A'$ proceeds without a barrier the rate coefficient of reaction (4) should be almost temperature independent, which is in agreement with the results of thermal kinetics measurements performed at low temperatures ($250\,\text{K} < T < 350\,\text{K}$) [15]. Based on the latter measurement and an extensive evaluation of other experimental data obtained at low-temperatures a temperature-independent value of $1.1 \times 10^{-10}\,\text{cm}^3\text{molecule}^{-1}\text{s}^{-1}$ is recommended in the temperature range ($200\,\text{K} < T < 350\,\text{K}$) for the reaction rate coefficient of reaction (4) [16].

However, reaction dynamics studies of reaction (4), in which "hot" $O(^1D)$ atoms with a translational temperature of about 1200 K were generated via pulsed laser flash photolysis, yielded a considerably high rate constant of $(2.7 \pm 0.6) \times 10^{-10}\,\text{cm}^3\text{molecule}^{-1}\text{s}^{-1}$ [12], indicating a pronounced increase of the reaction rate constant with reagent temperature. The latter observation suggests that at higher reagent energies the chemical kinetics of reaction (4) is significantly influenced by the participation of at least one (or more) excited-state reaction pathways. This proposition is supported by results from OH rovibrational product distributions and angular momentum polarization measurements [17], the results of which could only be reproduced if two PESs, the ground-state PES($1^1A'$) and the electronically excited PES($1^1A''$), were considered in the respective quantum scattering calculations [18]. Because the electronically excited PES($1^1A''$) has a collinear transition state, the participation of this PES opens an additional abstraction pathway which exhibits a reaction barrier. This in turn could results in a thermal rate constant, which increases with temperature. In order to provide further experimental evidence for the proposed mechanism the focus of the present work was on the extension of the thermal rate constant measurements of reaction (4) towards higher temperatures to bridge the gap between the available low-temperature thermal rate constant measurements [15, 16] and the "hot" $O(^1D)$ atom studies [12].

2 Experimental Section

2.1 Low-Temperature Flow Reactor

The principal outline of the low-temperature flow reactor employed in the experimental investigations of the $H_2(v=1)$ deactivation and the $H_2(v=1) + D_2(v=0) \rightarrow D_2(v=1) + H_2(v=0)$ vibrational energy transfer reaction is depicted in Figure 1 and 2. The reagent gas flows (H_2, purity 99,999 %, D_2, 99, 7 %) as well as the inert buffer gas (He, 99,999 %) flow was metered by calibrated flow controllers. The reagent gases were passed through liquid nitrogen cold traps in order to remove water impurities before entering the flow reactor. The latter consists of a reactor head where the gases enter the actual flow tube. The reactor head and the flow tube were attached to one another by an O-ring sealed flange system, to allow for an easy change of all design parameters of the flow reactor system.

The flow reactor walls were thermostated with a regulated flow of cold nitrogen from a liquid-nitrogen reservoir in the temperature range 300 – 110 K. The flow was contained in a stainless-steel chamber surrounding the main flow reactor. The liquid nitrogen in the storage tank was resistively heated at the end section of the extractor tubing, whose amount in turn was determined by the temperature difference between a preset value in the electronics of the cold gas system and the temperature reading of a thermocouple located in the cooling chamber near the flow tube wall. For better insulation the cooling chamber was packed on the outside with styrofoam of 30 mm wall thickness. To avoid water condensation the optical windows were flushed with dry nitrogen. Two different kinds of reactor heads were used in the experiments. A conventional one consisting of a T-shaped Pyrex glass tubing (Figure 1a) where the gas flows impinge on each other directly causing a fast, turbulent mixing. In the literature, in most kinetic studies using flow tubes measurements are made further downstream from the inlet ports where mixing is complete, and close to plug flow conditions are established. Due to the complex flow conditions, concentration measurements in this mixing section cannot be used for our data evaluation. Therefore, a second reactor head was designed (Figure 1b) in order to circumvent these difficulties in the data evaluation and simultaneously providing a swirl-free, two-dimensional rotationally symmetric flow field, which is accessible to a detailed numerical treatment from the beginning of the reaction. One reactant is admitted through an inner axial tube (diameter 10 mm, wall thickness 2 mm), the other through four equally spaced side arms. Reactants get in contact at the outlet of the central inlet tubing, the length of which is large enough to guarantee fully developed laminar flow fields in both the inner and outer gas flow. Mixing is mainly governed by diffusive transport. For comparison all kinetics measurements have been performed under the same conditions in both the laminar and the conventional flow system, and have been evaluated either by numerical simulations or under the plug flow assumption, respectively. The main flow tube (see Figure

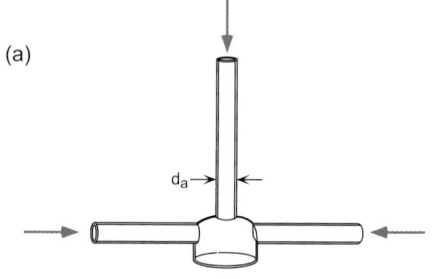

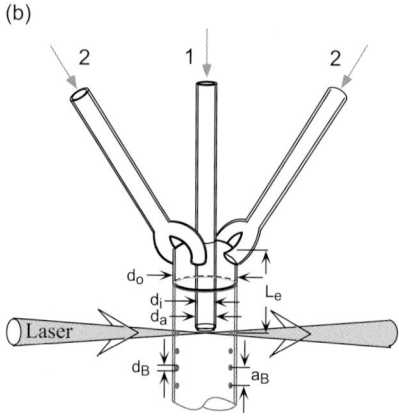

Fig. 1. Sketch of the two different flow reactor mixing heads used in the present work: (a) head for turbulent mixing, (b) mixing head for generation of a laminar flow through a fourfold gas inlet system symmetrically arranged around the central tube. The geometrical parameters are the following: $L_e = 45$ mm, $d_o = 32$ mm, $d_i = 10$ mm, $d_a = 14$ mm, $d_B = 2$ mm, $a_B = 12$ mm (for further details see text).

2) consists of a straight, 32 mm internal diameter section equipped with an array of diametrically opposed 2 mm diameter holes in the wall to allow for optical CARS diagnostics with focused laser beams. In order to avoid damage of window material located too close to the beam foci, small diameter 70 mm long tubes fitted with quartz windows at one end are glued to the central flow tube so that beam intensities are low enough on the window surfaces. This precaution also reduces the generation of spurious nonresonant CARS signals within the window material. Various materials and coatings of the flow tube wall have been tested to minimize heterogeneous relaxation of vibrationally excited species. From CARS measurements it was found that stainless steel coated with Teflon (Du Pont FEP 856200) leads to the lowest wall deactivation probability for $H_2(v = 1)$, two times lower than the value obtained with

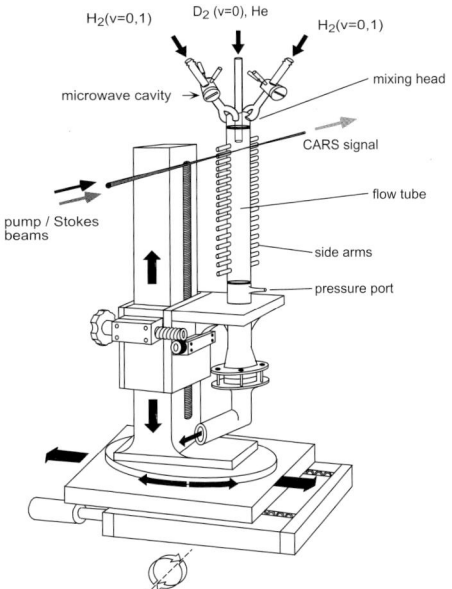

Fig. 2. Three-dimensional representation of the flow reactor system used for the kinetics studies of the $H_2(v=1)$ deactivation and $H_2(v=1) + D_2(v=0) \rightarrow D_2(v=1) + H_2(v=0)$ vibrational energy transfer reactions. For generation of $H_2(v=1)$ molecules microwave discharge cavities are attached to the quartz inlet tubing (denoted as (2) in Figure 1). The reactor is attached to translation and rotation stages to allow for proper spatial alignment.

HF-cleaned Pyrex glass [19, 20]. As is shown in Figure 2, the flow reactor was positioned into the beam path and aligned at each axial measurement location by a translation/rotation stage. Low pressure conditions in the reactor were maintained by a mechanical pump. The pressure in the reactor was measured with a precision capacitance manometer (MKS Baratron, 290H). Vibrationally excited hydrogen molecules were generated in two microwave cavities assembled to the side arms of the respective reactor head (cf. Figure 2), and driven by a magnetron microwave generator (2450 GHz) at powers between 20 and 100 W.

2D acetone flow tracer LIF setup: Rotational symmetry and the absence of vortices and swirls in the flow were checked experimentally by adding small amounts of acetone as a tracer to the outer or inner gas flow and visualizing their spatial distribution by planar laser induced fluorescence (2D-LIF) techniques [21] taken in vertical and horizontal sections at different axial locations along the flow tube (see Figure 3). For acetone LIF excitation the fourth harmonic of a pulsed Nd:YAG laser at 266 nm was formed into a light manifold by a cylindrical lens (f = 700 mm). Care was taken to place the focal line behind the flow reactor, which ensured a slowly converging light sheet thickness

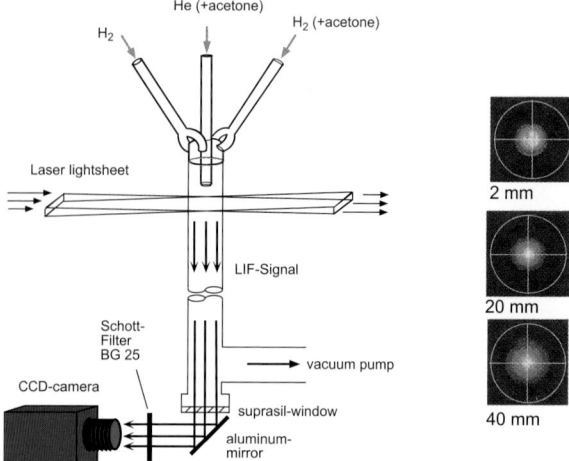

Fig. 3. (Left side) Schematic of the experimental setup used for two-dimensional (2D) laser-induced fluorescence (LIF) monitoring of the spatial distribution of acetone tracer species inside the flow tube. As depicted, the laser light sheet excites a horizontal cross section of the flow downstream the inlet manifold, the LIF-signal is detected through a quartz window at the bottom next to the pumping port by a CCD-camera. (Right side) 2D-LIF images taken after the acetone tracer was added through the central tubing in the laminar mixing head shown in Figure 1b. The 2D-LIF images represent horizontal cross sections at three distances z from the exit of the center tubing (1) in Figure 1b. In the images the LIF intensity varies from high (white pixels) to low (dark pixels).

in the measurement section. For these experiments the regular flow tube was replaced by a transparent Suprasil tube of equal dimensions. 2D acetone LIF concentration maps (se e.g. right side of Figure 3) were imaged perpendicular to the plane of laser excitation either through the side walls (vertical cuts) or, as is shown in Figure 3, through a quartz window at the bottom of the flow tube (horizontal cuts). The broadband acetone fluorescence signal was detected through a blue glass filter (transmission between 350 and 550 nm) by means of a CCD camera (Proxitronic Nanoscan PC 1811) with adjustable image intensifier.

Room-temperature flow reactor: The modified flow reactor used in the present work for room-temperature kinetics studies of the $OH + H_2(v = 1)$ reaction is depicted in in Figure 4. In these experiments NO_2 was added to the flow in order to convert the H atoms generated in the microwave cavity assembled to the movable inner flow tube into OH radicals via the fast reaction $H + NO_2 \rightarrow OH + NO$ [22]. By moving the inner flow tube inside the main flow tube the time for reaction of the generated OH radicals with $H_2(v = 1)$ reagent molecules could be varied. At the end of the main flow tube laser spectroscopic molecular hydrogen (using CARS as described above) and OH

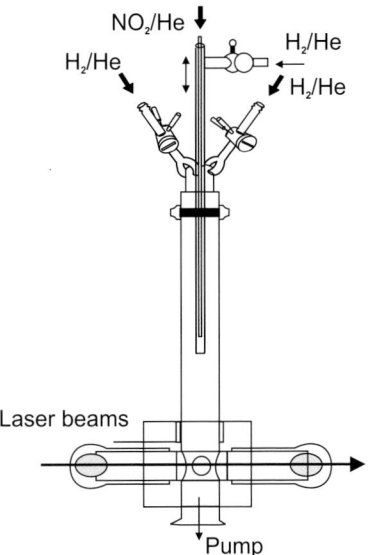

Fig. 4. Schematic of the flow reactor system employed in the room-temperature kinetics studies of the OH + $H_2(v = 1) \rightarrow H + H_2O$ reaction.

radical concentration measurements (*via* LIF) were performed for different reaction times in order to obtain OH concentration versus reaction time plots for the rate constant evaluation for a given $H_2(v = 0)/H_2(v = 1)$ ratio.

CARS spectrometer and LIF spectrometer for $H_2(v = 0, 1)$ and OH radical detection: The experimental CARS setup employed for molecular hydrogen detection is schematically depicted in Figure 5. A pulsed (8 ns) single mode Nd:YAG-laser (Spectra Physics, GCR-3) and a narrowband ($\Delta\omega = 0.1\,\text{cm}^{-1}$) dye laser (LUMONIX, HD500) pumped by the same laser provide the pump and Stokes laser beams for the generation of the CARS signals in hydrogen and other gases in a collinear beam geometry. Pyridin I dye (660-740 nm) and DCM (610-675 nm) in methanol provide the appropriate Stokes wavelengths for hydrogen and deuterium, respectively. Telescopes in each beam path enable the separate adjustment of the beam waists for proper focusing and recollimation of both beams to a common focal spot using lenses with a focal length of 300 mm.

For the measurement of radial concentration profiles inside the reactor the lenses could be translated along the optical axis in order to move the focal spot along the beam direction. The CARS signal emerging from the focal volume was separated from the pump and Stokes laser beams by three longwave pass dichroic mirrors, interference filters and a 0.25 m monochromator (McPherson, 218) before it was detected by a photomultiplier (EMI 9635B). To eliminate systematic errors in concentration measurements due to interference of non-resonant signals from buffer gases and windows with the resonant part of the

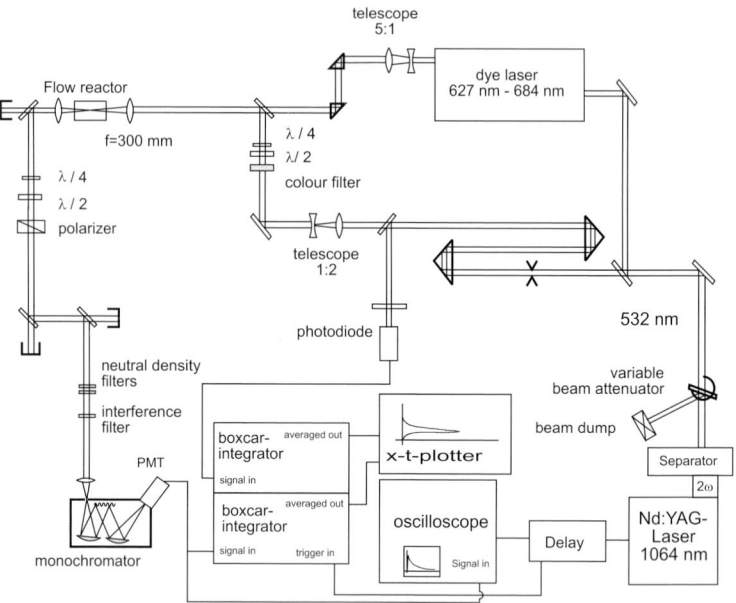

Fig. 5. Laser spectroscopic arrangement for molecular hydrogen and deuterium detection *via* coherent anti-Stokes Raman spectroscopy (CARS) in the $H_2(v=1)$ deactivation and $H_2(v=1) + D_2(v=0) \rightarrow D_2(v=1) + H_2(v=0)$ vibrational energy transfer reaction kinetics studies ($\lambda/4$ and $\lambda/2$: quarter- and halfwave plates, respectively; PMT: photomultiplier tubes).

third-order susceptibility of the detected hydrogen species, the nonresonant contribution to the CARS signal was rejected by polarization selective excitation and detection [23]. This technique reduced the nonresonant background by 3 orders of magnitude, giving at 3.98 Torr total pressure a detection sensitivity of 0.06% corresponding to a quantum state specific detection limit of 7×10^{12} molecules cm^{-3}.

The combined laser spectroscopic setup employed in the OH + $H_2(v=1)$ room-temperature kinetics studies, which allows CARS measurements for molecular hydrogen concentration determination to be performed along with OH radical detection *via* pulsed $OH(A^2\Sigma^+ - X^2\Pi)$-LIF in the ultraviolet spectral region, is schematically depicted in Figure 6. Relative number densities of vibrational ground-state OH radicals in the absorbing fine-structure states could be derived from $OH(A^2\Sigma^+ - X^2\Pi)$-LIF-spectra [27] recorded under low-excitation-laser-beam-energy (linear) conditions taking into consideration the respective Einstein coefficients of absorption [28, 29].

Measurement Procedure and Data Evaluation: Vibrationally excited hydrogen molecules $H_2(v=1)$ were generated by microwave discharges in the sidearms of the mixing head (cf. Figure 2). Populations in higher vibrational states were below the current CARS detection limit. Dilution ratios in rare gas helium

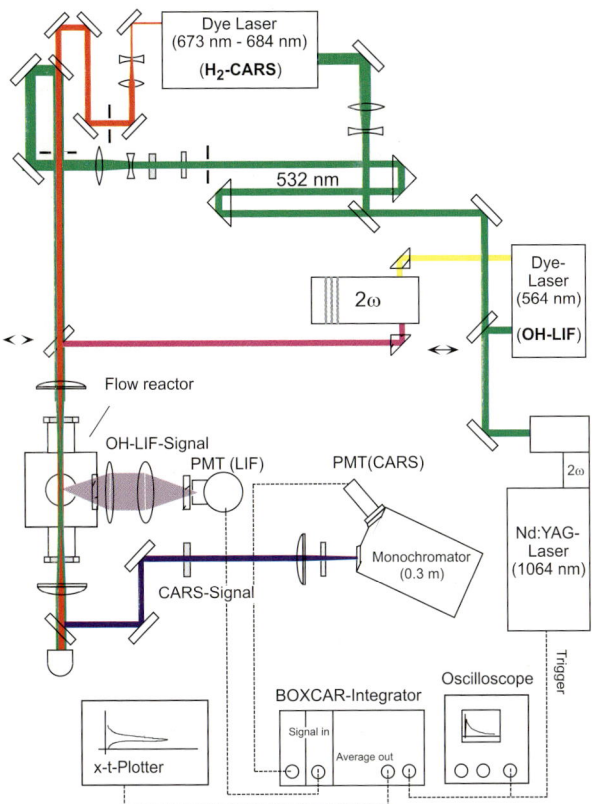

Fig. 6. Laser spectroscopic arrangement for combined molecular hydrogen and deuterium detection *via* coherent anti-Stokes Raman spectroscopy (CARS) and OH radical detection *via* laser-induced fluorescence (LIF) as it was employed in the OH + $H_2(v = 1) \rightarrow H + H_2O$ reaction kinetics studies (PMT: photomultiplier tubes).

and discharge conditions were optimized for minimum atom yield and maximum vibrational excitation. Hydrogen atom concentrations were determined by quantitative gas phase titration with ethylene using small quantities of NO to generate a detectable chemiluminescence from electronically excited HNO [20, 24, 25].

Generally, for highest $H_2(v = 1)$ yield with simultaneous low H atom yield in the discharge a high hydrogen partial pressure and a low microwave power (20 W) in the two side arms were preferred. Typical H atom partial pressures were in the range of 15 mTorr, which corresponds to a dissociation yield of 0.5%. Hydrogen molecules, on the other hand, can be excited vibrationally with an efficiency of 0.8%, which gives a concentration of

3.5×10^{14} molecule $\cdot$ cm^{-3} at the present conditions. The CARS lines of individual Q-branch transitions of hydrogen are well separated and their full width at halfmaximum (fwhm) at room temperature and below and at low pressure are entirely determined by Doppler broadening. Due to efficient suppression of the non-resonant background, the measured CARS signal intensity is proportional to the square of the population difference of the Raman coupled states. For determination of relative population densities from the measured population differences the assumption is made that for the highest vibrational level probed with detectable signal intensity the population in the upper Raman level is zero. Thus, the relative population densities of all lower levels can be calculated successively, if account is taken of the different Raman cross sections of the respective transitions. Conversion to absolute concentrations was possible from the known volumetric flow rates and total pressure (4 Torr). Generally, the most intense Q(1) and Q(2) CARS lines for hydrogen and deuterium, respectively, are used for concentration measurements. When vibrational population ratios are determined, care was taken to employ low laser beam energies to exclude errors from vibrational quantum number dependent saturation [26].

2.2 High-Temperature Flow Reactor Setup

A schematic of the photochemistry high-temperature laminar flow reactor setup, which was designed and constructed in the framework of the present work [30], is reproduced in Figure 7. As outlined in detail in part I [5], the design of the reactor setup is largely based on detailed a priori numerical simulations [31, 32]. The actual reactor consists of a vertically oriented main flow tube with a inner diameter of 50 mm made out of alumina ceramics (with a heated length of 250 mm), which could be heated up to a temperature of 1700 K by means of MoSi$_2$ radiation heating elements (KS 1900) purchased from KANTHAL SUPER (Sweden). To achieve a homogeneous temperature distribution six heating elements were positioned concentrically around the flow tube at a distance of 46 mm [33]. To minimize heat losses the heating element arrangement was surrounded by insulation. The whole system was then included into a stainless steal double-wall water-cooling system. The temperature profile along the inner reactor axis could be measured by a movable thermocouple.

In the upper part the reactor was equipped with a gas inlet system, which allows introducing buffer gas, as well as reagent and photolytic precursor compounds. The gas inlet system consists of two different types of inlets: one through which inert buffer gas is introduced directly into the heated region of the main flow tube and a second inner one, which was cooled by means of a separate water circulation system, and which could be used to introduce reagent and photolytic precursor mixtures. By changing the position of the inner gas inlet tube the contact time of the reagent and photolytic precursor mixtures with the hot exterior wall of the main flow tube could be varied

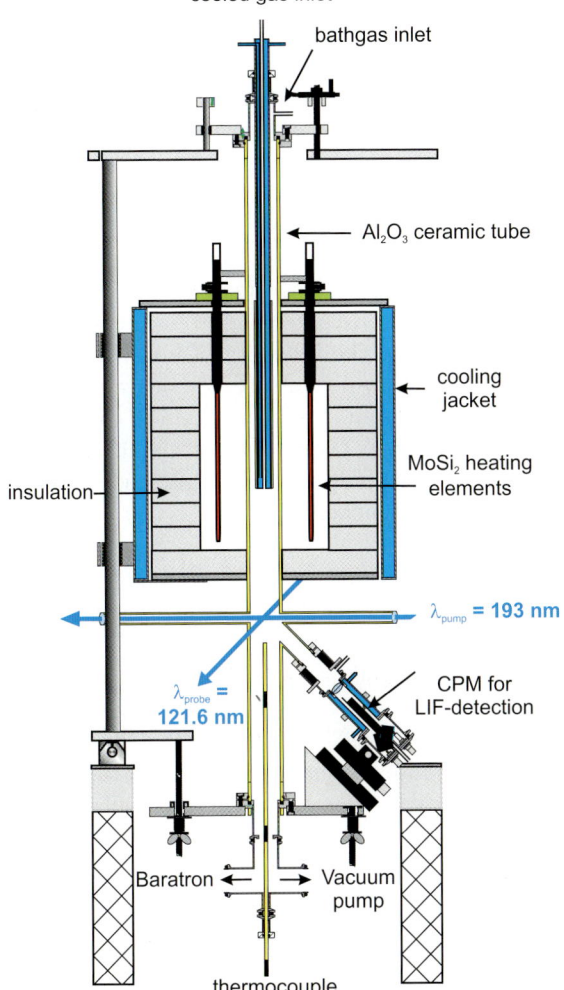

Fig. 7. Sketch of the photochemistry high-temperature laminar flow reactor employed in the investigations of the temperature dependence of the $O(^1D) + H_2 \rightarrow H + OH$ rate constant. (CPM: channel photomultiplier).

in order to minimize heterogeneous thermal decomposition of the respective compounds under the constrain of maintaining a mixing time still long enough to reach a homogeneous mixture with the desired temperature at the point of the photochemical "pump" and "probe" reaction kinetics measurements (*vide infra*).

The flows of the gases introduced into the reactor were controlled by calibrated flow meters and the pressure in the reactor was measured with a precision capacitance manometer. The reactor was pumped by a combination of a rotary pump and an oil diffusion pump. The gases had a flow velocity of around 40 cm/min to ensured a complete renewal of the gas mixture in the reaction volume (which is defined by the intersection region of the pump and the probe laser beams) between two successive laser shots. In the present work all measurements were performed at a laser repetition rate of 6 Hz.

The lower part the reactor was designed in a way to allow for introducing a photodissociation "pump" laser beam (for pulsed flash photolytic generation of reactive species) as well a "probe" laser beam for pulsed time-resolved LIF detection of atomic products. Therefore the reactor was equipped with four ceramic alumina side arms (1.5 cm i.d.), installed perpendicular to the reaction tube, which provide pair wise opposing access for pump- and probe-laser and the possibility to monitor the intensity of both. At their end this arms are sealed by MgF_2 windows, which ensure a high transmittance for vacuum ultraviolet (VUV) laser radiation. A fifth sidearm (2 cm i.d.) for fluorescence detection is positioned between the two sidearms at an angle of 45 ° with respect to the reaction tube, whose centreline intersects the crossing point of the pump and the probe laser beams.

Experimental arrangement for pulsed laser photolytic generation of $O(^1D)$ atoms and time-resolved H atom product detection: The spectroscopic arrangement used in the present laser photolysis "pump" and "probe" rate constant measurements for the $O(^1D) + H_2 \rightarrow H + OH$ reaction is reproduced in Figure 8. In these studies the reaction is photochemically initiated in a homogeneous flow of N_2O(purity 99.998%)/H_2(99.995%)/Ar buffer gas (99.998%) by a pulsed photolysis "pump" laser beam. In the present work, an ArF excimer laser (emission wavelength λ_{pump}= 193 nm, pulse duration: 15-20 ns) provided the photolysis laser beam with a typically intensity of $2 \, mJ/cm^2$. The photolysis laser beam was then slightly focused into the reactor where it generates $O(^1D)$ reagents *via* the selective photodissociation of the N_2O precursor molecules [34]: $N_2O + h\nu(193 \, nm) \rightarrow O(^1D) + N_2$. As in the latter photochemical reaction the nascent $O(^1D)$ species are formed with a non-thermal velocity distribution [12], appropriate partial pressures of Ar were maintained in the present kinetics measurements in order to achieve rapid thermalization of the $O(^1D)$ reagent atoms by collisions with Ar buffer gas atoms. All experiments were performed under pseudo-first order conditions: $[O(^1D)] << [H_2]$.

H atom reaction product formation was monitored using time resolved pulsed vacuum ultra-violet (VUV) LIF at the hydrogen atom $(2p^2P - 1s^2S)$ Lyman-α transition (121.567 nm) [35]. Tuneable pulsed VUV "probe" laser radiation (band width $\Delta\omega_{VUV(probe)} = 0.35 \, cm^{-1}$, pulse duration: 15-20 ns), was generated using the Wallenstein method [36] for resonant third-order sum-difference frequency conversion of pulsed dye laser radiation (pulse duration 15-20 ns) in a phase-matched Kr/Ar mixture. In the Kr mixing scheme *via* which the VUV radiation ($\omega_{VUV(probe)} = 2\omega_1 - \omega_2$) was generated, the laser radiation of $\omega_1(\lambda_1 = 212.55 \, nm)$ is two-photon resonant with the Kr 4p 5p

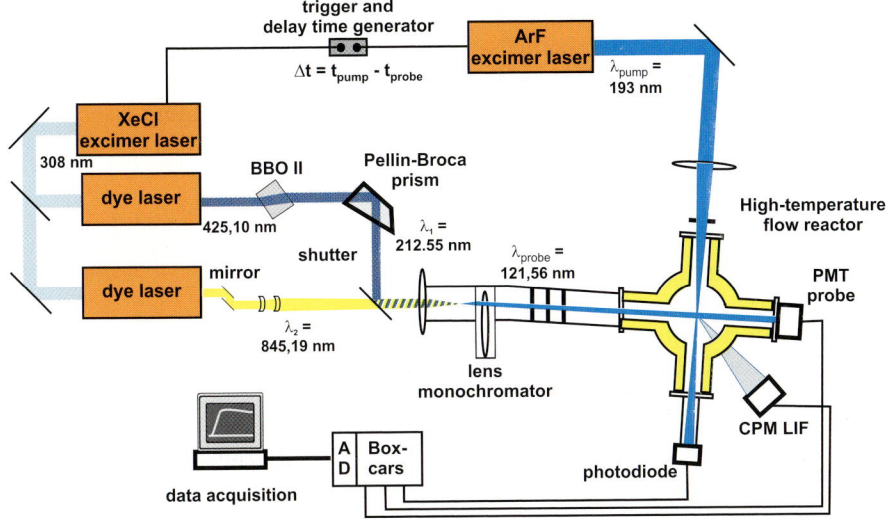

Fig. 8. Laser spectroscopic arrangement used in the laser photolysis (LP)/laser induced florescence (LIF) "pump" and "probe" kinetics studies of the $O(^1D) + H_2 \rightarrow H + OH$ reaction (AD: analogue to digital converter, PMT: photomultiplier tube, CPM: channel photomultiplier).

(1/2, 0) transition and was held fixed during the experiments, while ω_2 could be tuned from 844 nm to 846 nm to generate VUV laser radiation covering the H atom Lyman-α transition [37]. The laser radiation was obtained from two dye lasers, simultaneously pumped by a XeCl excimer laser (Lambda Physik, LPX 205i). In the first dye laser (Lambda Physik FL 2002), Coumarin 120 dye was used to generate the initial 425.10 nm radiation, which was subsequently frequency doubled in a BBO II crystal to obtain $\lambda_1 = 212.55$ nm. $\lambda_2 = 844 - 846$ nm was generated by operating the second dye laser (Lambda Physik Scanmate 2E) with Styryl 9 dye. The Lyman-α laser light was separated from the fundamental laser light (λ_1, λ_2) by a lens monochromator followed by a light baffle system.

The probe laser beam was aligned to overlap the pump laser beam at right angles in the viewing region of the LIF detector (denoted as CPM LIF Figure. 8). The H atom LIF signal was measured through a band pass filter (ARC, model 130B-1D, $\lambda_c = 122$ nm, fwhm = 20 nm) by a channel photomultiplier (CPM, Perkin Elmer C921P solar blind). The VUV probe laser intensity was monitored by another photomultiplier (Hamamatsu R1459 solar blind, denoted as PMT VUV in Figure 8). The signals of both photomultipliers and the signal from the photodiode, which was used to measure the photolysis laser intensity behind the reaction cell, were recorded by a three-channel boxcar system (SRS 250) and transferred *via* an analogue to digital converter (SRS 235) to a personal computer, where the H atom LIF signal was normalized to the VUV probe and the photolysis laser intensity. To improve the S/N ratio, the H atom signal recorded at a given delay time was averaged over 40 laser shots. A digital pulse generator (SRS DG 535) was used to control the delay time setting ($\Delta t = t_{\text{pump}} - t_{\text{probe}}$) between probe and pump laser pulse. Typically, the delay time was varied from -100 ns up to 1000 ns to record transient H atom concentration profiles.

3 Results and Discussion

3.1 Room-Temperature (T = 300 K) Kinetics Studies of H_2(v = 1) Collisional Relaxation and Vibrational Energy Transfer

Experimental spatial resolution of the CARS measurements: In the CARS measurements, due to finite spatial resolution along the beam path a radial CARS signal profile contains spatially averaged concentration information through a convolution of the square of the radial concentration profile $n(r)^2$ with a spatial instrument function $f(r)$, which depends on the characteristics of the focused laser beam geometry:

$$I_{as}(p) = \int n(r)^2 f(r-p) \, dr \tag{5}$$

In this expression r is the radial coordinate and p the common location of the beam foci along the beam axis. The function $f(r-p)$ quantifies the contribution to the CARS signal created at location r while the CARS focus is situated at position p. For comparison with simulated concentration profiles, one can either deconvolve the experimental CARS data or otherwise perform a numerical convolution for the simulated concentration profiles to arrive at calculated CARS signal intensities. Because the first method is more demanding as it needs entirely recorded radial CARS profiles, which are not available for all measurement points, in the present study the second method was used.

The CARS spatial resolution function $f(r-p)$ (instrument function) can be determined by translating a thin (100 μm) glass plate along the focal region of the beams and measuring the intensity of the generated nonresonant CARS signal [38]. It has been argued, however, that in gas-phase measurements this procedure will not reproduce the true spatial variation of the CARS signal generation due to different growth mechanisms of the macroscopic CARS signal in solid and gaseous samples. In case of a plate the signal mainly originates from the sample boundaries whereas the contribution generated inside the thin plate is negligible [39]. On the contrary, in the gas phase the variation of the refractive index is very small and the signal generated inside the medium dominates. Quantitative estimates are difficult since the effect strongly depends on the geometrical setup and beam characteristics. Druet and Taran pointed out that the spatial extension of $f(r)$ if measured with a glass plate will be strongly underestimated [40].

As a consequence, in order to provide reliable input data for the modeling two approaches were investigated for the measurement of the CARS instrument function: the translation of a glass plate as well as of a thin hydrogen gas jet emerging from a 1×5 mm rectangular nozzle with an air coflow. Probing was performed right above the nozzle exit (as depicted in the lower part of Figure 9). The results from both measurements are reproduced in the upper part of Figure 9, where the CARS intensities of both the nonresonant signal from the glass plate and the Q(1) transition in molecular hydrogen are plotted as a function of distance from the point of maximum intensity. The longitudinal spatial resolution, defined as the distance along the common focus of the beams where 75% of the emerging signal is generated [40], turns out to be on the order of 15 and 9 mm if measured with the nozzle and glass slide, respectively. For reasons mentioned above, the instrument function measured by means of the hydrogen jet was considered the more reliable representation of the "true" spatial resolution function.

Hydrodynamics in the Laminar Flow Tube: For well-defined initial conditions in kinetic data evaluations it is desirable that a swirl free transition from inner and surrounding laminar flow profiles into a Poiseuille type flow takes place at the exit plane of the inner tube (see Figure 1). In gas dynamic terms this means streamlined flow in the mixing region so that the mixing will be dominated by diffusive processes. Experimentally the ratio of outer (Q_o) to inner (Q_i) gas volume flow rates can be adjusted to meet these requirements. From extensive test calculations and comparisons with experimental CARS radial concentration profiles an optimum volumetric flow ratio of Q_o/Q_i =4.5 (total flow rate 3.96×10^3 cm^{3}min^{-1} at STP, total pressure 4 Torr) was found by changing this parameter in the numeric modeling. Figure 10 illustrates in a three-dimensional plot the calculated axial flow velocity field in the mixing section under these conditions. For a detailed experimental investigation of convective processes especially at the end of the inner inlet tube, part of the inner helium flow was substituted by acetone, and its spatial distribution was monitored by two-dimensional laser-induced fluorescence (LIF) measurements. Because diffusional transport of acetone is significantly slower than hydrogen, the former will mainly follow the streamlines and therefore will be a tracer of convective processes. The tracer is added through the inner flow tube (the total flow is kept fixed in all experiments), while hydrogen is added through the outer side arms in the laminar mixing head (see Figure 1b); The right side of Figure 3 shows its distribution in horizontal cross sections at three locations downstream from the lower end of the tube. These images demonstrate the uniform diffusive mixing and rotational symmetry of the two merging gas flows. Further downstream (corresponding to reaction times larger than 4.7 ms) signal levels, relative to dark noise and scattered light, are too low for meaningful interpretations.

The vertical development of the flow was investigated by rotating the light sheet 90° from horizontal and exciting in a plane of the flow axis. Figure 11 shows experimental acetone LIF intensity traces (dots) through the image

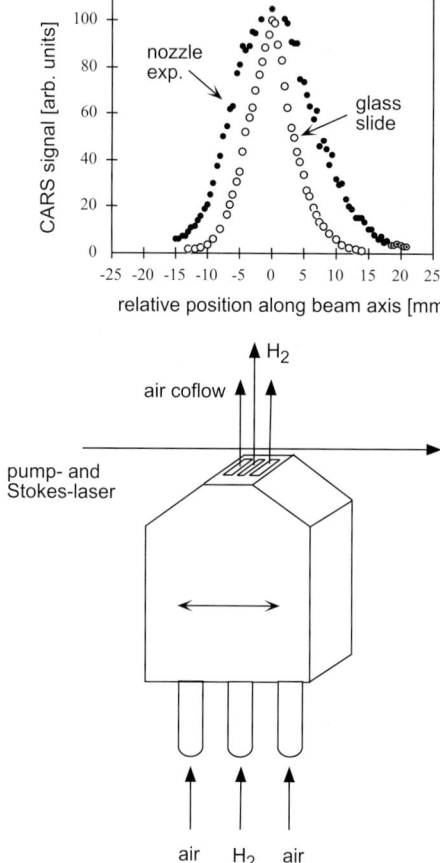

Fig. 9. (Upper part) Results of the CARS spatial resolution function measurements obtained by translating a 0.1 mm thick glass sheet (open circles) and a thin molecular hydrogen flow from a nozzle (filled circles) through the common foci of the pump and Stokes laser beams, respectively. (Lower part) Schematic of the nozzle used for the generation of a spatially thin molecular hydrogen gas flow.

plane and perpendicular to the main flow axis. It again clearly shows the gradual diffusive mixing of the tracer gas with the surrounding gas at a location starting directly below the exit of the inner tube (upper part of Figure 11) and 42 mm downstream. The excellent agreement between measured radial concentration profiles at different axial locations and results obtained from numerical simulations (solid lines through the data points) demonstrates that hydrodynamics and transport within the flow reactor is well described by the model.

Optimization Procedure: Figure 12 shows in a projection onto the two-dimensional parameter space the path of the optimization procedure. Both the

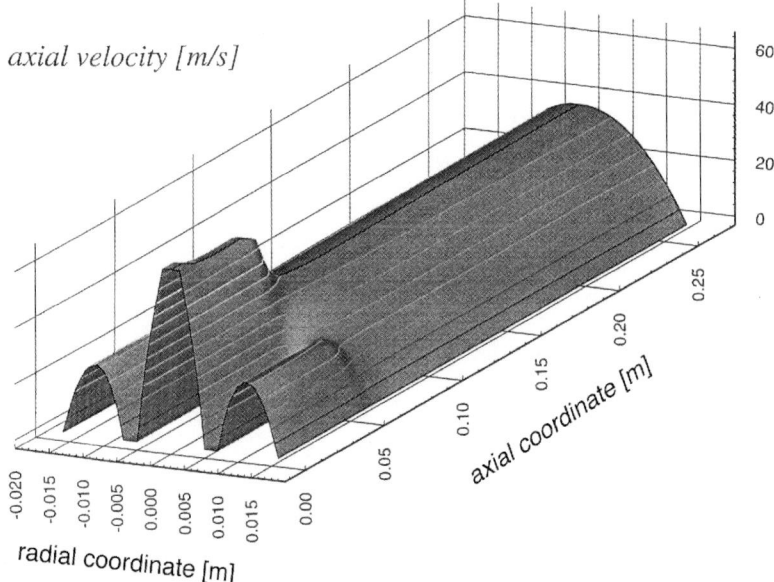

Fig. 10. Calculated axial velocity profile of the total gas flow in the mixing section of the laminar flow tube reactor depicted in Figure 2. The gases enter from the lower left (further details are given in the text).

rate constant of heterogeneous relaxation of vibrationally excited hydrogen (x axis) and of vibrational energy transfer (y axis) are optimized simultaneously by determination of the slope of the quality function QF in both variables. The elongated valley reflects different singularities of the Hesse matrix. Perpendicular to this valley the minimum is found within only a few iterations. After that, optimization is continued along the valley. Finally a test step is made to test for a possible saddle point of the approached minimum. Using this optimization procedure, the rate constants given in the following paragraphs were derived.

Vibrational Relaxation of $H_2(v=1)$: A visualization of the three-dimensional spatial distribution of the mass fraction of vibrationally excited hydrogen in the mixing section (with helium supplied through the central tube) is given in Figure 13. The experimental conditions were: Total gas pressure: 4 Torr; p(He)= 2.8 Torr; p(H_2)= 1.2 Torr; total gas flow $3.96 \times 10^3 cm^3 min^{-1}$ ($6.67 \times 10^{-5} m^3 s^{-1}$) at STP; mean flow velocity 17 m/s; microwave power 2×20 W; temperature 300 K. The almost instantaneous mixing after both gas flows merge, as well as the significant back-diffusion of helium from the in-

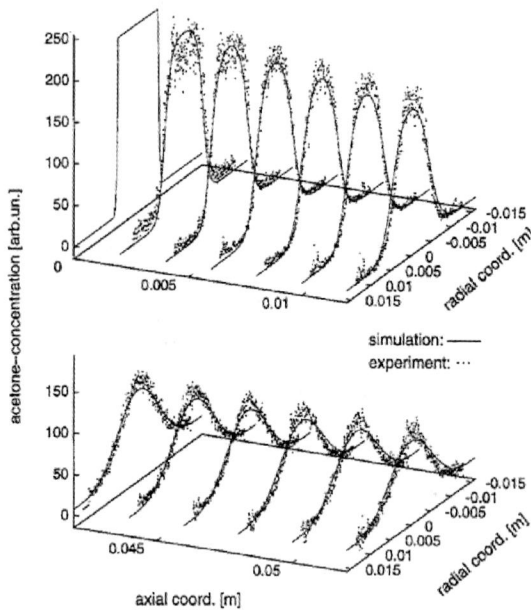

Fig. 11. Three-dimensional representation of line cuts of the acetone LIF intensity distributions (depicted as points) measured at various distances from the inner inlet port (denoted as (1) in Figure 1). The intensity distributions were obtained with a vertically oriented laser light sheet. The acetone LIF signal was detection through the sidewall of the quartz reactor. (Upper part) Section from 2 to 17 mm. (Lower part) Section from 42 to 51 mm. The solid represent the results of a numerical simulation.

ner tube into the supply tubes for hydrogen is clearly visible. Vibrationally excited hydrogen then is further diluted by diffusive and convective mixing and diminishes by homogeneous and heterogeneous collisional relaxation. It has to be noted from Figure 13 that to check for the effect of backdiffusion of reactive species into the center tubing the inflow border (and corresponding start of simulation) was shifted 5 mm upstream. This caused part of the inner flow tube to be "visible" in the concentration plot (sharp step at axial coordinate equal to zero). For grid points that were located in this spatial region of the inner flow tube concentration was arbitrarily set to the average value of the incoming $H_2(v = 1)$ mass fraction (0.0065). It turned out, however, that this effect does not influence the CARS data, since, because diffusion is small in comparison with the high axial convection velocity, this changed the $H_2(v = 1)$ concentrations by not more than 1%. With the typical flow conditions and vibrationally excited hydrogen present (microwave discharges switched on), concentrations were measured by tuning the Stokes dye laser to the Q(1) transition in the first vibrationally excited state of molecular hydrogen. Hydrogen and helium are added through the outer sidearms and the central

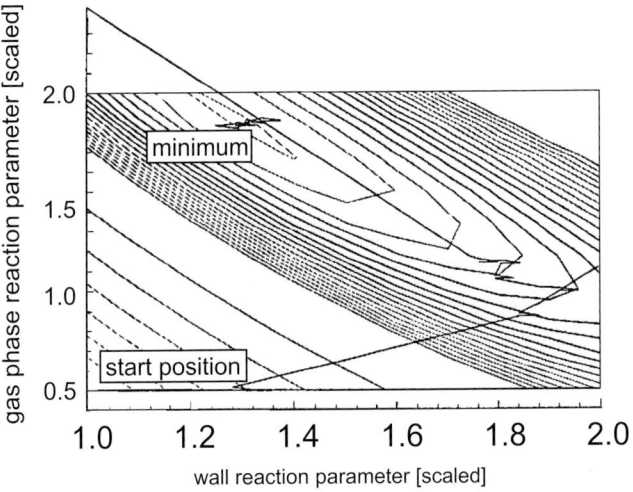

Fig. 12. Quality function (QF) for the two-parameter optimization procedure applied to the reactive flow of vibrationally excited hydrogen in deuterium in the flow tube with helium as carrier gas. Two-dimensional projection of QF on the optimization parameters for $H_2(v = 1)$ wall deactivation (x axis) and the $H_2(v = 1) + D_2(v = 0) \rightarrow D_2(v = 1) + H_2(v = 0)$ gas-phase vibrational energy transfer rate constant (y axis). Solid line through the valley traces optimization path.

tube into the mixing head, respectively. Figure 14 shows the comparison between the experimentally optained CARS intensity profiles (points with error bars) and the results of the numerical simulation (solid lines through the data points). In the upper right part of Figure 14 a radial CARS intensity profile at z = 48 mm downstream from the inner flow tube exit. In the lower left and right part of Figure 14 the axial relative concentration profile with the center of the probe volume located on the axis of the flow tube are presented. Along this symmetry line the initial increase of the CARS signal is due to diffusion of $H_2(v = 1)$ molecules from the outer regions into the center of the CARS probe volume. The subsequent decay of the signal intensity can be attributed to loss of vibrationally excited species due to heterogeneous relaxation by collisions of with the reactor wall as well as homogeneous vibration-translational (VT) relaxation in collisions with hydrogen atoms, helium and molecular hydrogen (see Table 5 in the appendix of the article Carraro/Heuveline/Rannacher [5] in this volume). The experimentally obtained axial and radial concentration profiles were best fitted by a simulation (solid lines in Figure 14), in which a wall deactivation probability for hydrogen equal to $\gamma_w = (1.5 \pm 0.3) \times 10^{-3}$ was used.

The axial CARS intensity profile measured in the flow system with turbulent mixing (cf. Figure 1a) resembles data taken in the laminar system except for the mixing region (the first few centimeters). In terms of a simplified plug

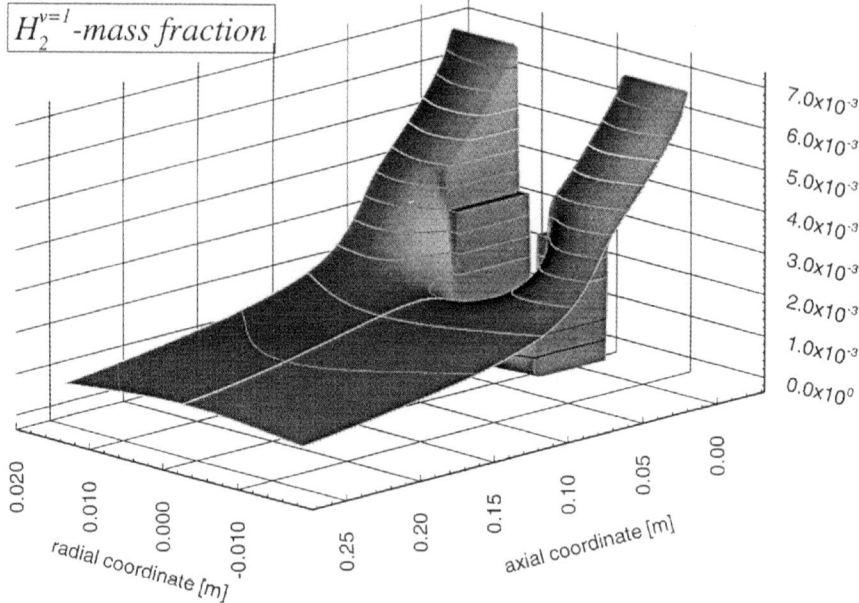

Fig. 13. Three-dimensional visualization of calculated mass fraction of $H_2(v = 1)$ in the mixing section of the flow reactor (for details see text).

flow evaluation the axial coordinate z can be converted into a reaction time $t = z/v^{plug}$, with the mean volumetric flow velocity v^{plug} determined by the volume flow rate. Since the time zero of the reaction is not well-defined in the turbulent mixing head, it can only be roughly estimated with the help of the plug flow velocity and an approximate distance between the end of the gas inlet tubes and the first downstream measurement location. The unsaturated CARS signal is assumed to directly correspond to the $H_2(v = 1)$-concentration squared and it is further assumed to be constant along the diameter of the flow tube. From half-logarithmic plots of relative $H_2(v = 1)$ concentrations *versus* reaction time (neglecting data taken in the mixing region), a total first-order rate constant for vibrational deactivation of hydrogen molecules in the flow system of $k^{plug} = 82 \pm 10$ s^{-1} can be obtained.

The contribution to the pseudo-first-order rate constant from homogeneous relaxation pathways (see Table 6 in the appendix of the article Carraro/Heuveline/Rannacher [5] in this volume) *via* VT transfer reactions $H_2(v = 1) + H \rightarrow H_2(v = 0) + H$, $H_2(v = 1) + H_2(v = 0) \rightarrow H_2(v = 0) + H_2(v = 0)$ and $H_2(v = 1) + D_2(v = 0) \rightarrow H_2(v = 0) + D_2(v = 0)$ is calculated to be $k_H = 42$ s^{-1}. Using the latter value a wall deactivation rate constant $k_w = k^{plug} - k_H = 40 \pm 10$ s^{-1} can be obtained. Within the zero-dimensional approximation this value for k_w translates into a wall deactivation probability γ_w^{plug} *via* the Kaufmann formula [4]:

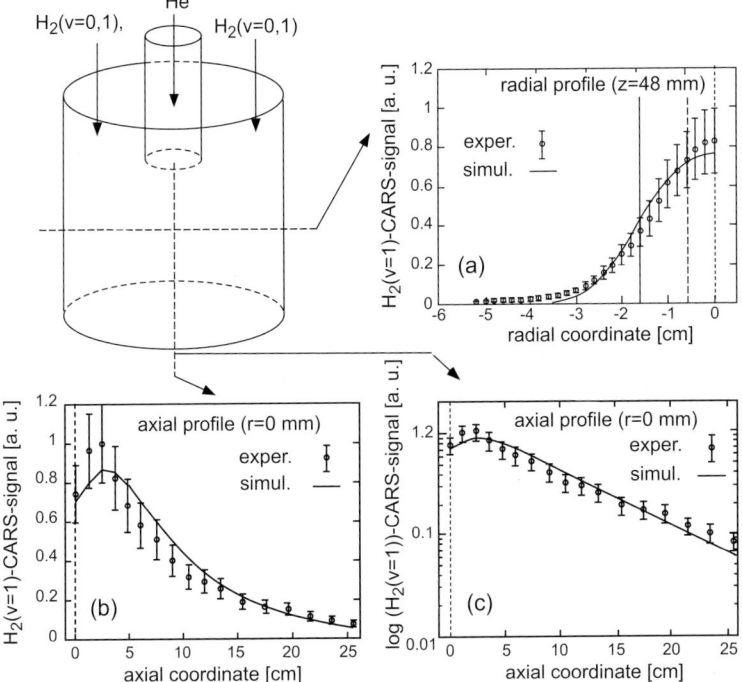

Fig. 14. Collisional relaxation of $H_2(v=1)$: comparison of experimental $H_2(v=1)$ CARS spatial concentration profiles (represented as open circles with error bars) with detailed numerical simulation results (solid lines). In the upper left part a schematic of the flow tube section is shown where the experimental profiles have been obtained. In the upper right part a $H_2(v=1)$ radial concentration profile obtained 48 mm downstream from the exit of the inner gas supply tubing (denoted as (1) in Figure 1) in the laminar mixing head is shown. The dashed and solid vertical lines mark the location of inner and outer tube walls of the reactor, respectively. In the lower left part a experimental $H_2(v=1)$ axial concentration profile measured along the centreline of the flow tube is depicted along with the corresponding simulated profile. The graphics in the lower right part depicts the same data on a logarithmic scale.

$$\gamma_w^{plug} = k_w^{plug} 2r_0/v_m = (7 \pm 2) \times 10^{-4} \tag{6}$$

Here r_0 is the radius of the flow tube and v_m represents the mean thermal velocity of $H_2(v=1)$. Hence, the so obtained plug flow value γ_w^{plug} is about two times smaller than the value of $\gamma_w = (1.5 \pm 0.3) \times 10^{-3}$ obtained by the detailed numerical simulation.

Vibrational Energy Transfer of $H_2(v=1)$ in Collisions with $D_2(v=0)$: In this experiment deuterium is added through the central tube while hydrogen enters through the outer side arms (2) with the two microwave discharges switched on (see Figure 1b). Both hydrogen and deuterium were monitored in their first excited vibrational state by tuning the Stokes laser to their respective Q(1) and Q(2) transition. In the lower left part of Figure 15 the axial decay of $H_2(v = 1)$, which is mainly caused by VV transfer collisions with $D_2(v = 0)$ molecules and by deactivating collisions with the wall, is depicted. In the lower right part of Figure 15 the corresponding build-up of vibrationally excited deuterium, $D_2(v = 1)$ products, due to the VV processes and its subsequent decay, which is mainly due to wall loss, is shown. The $D_2(v = 1)$ radial concentration distribution (measured 48 mm downstream from the mixing section) is given in the upper right part of the figure. The relevant reaction mechanism used in the simulations is given in the appendix of part I of this contribution [5]. The previously obtained value for the wall deactivation probability $\gamma_w = (1.5 \pm 0.3) \times 10^{-3}$ is taken for both isotopic hydrogen species [20], since within the accuracy of the measurements reported there, an isotope effect could not be observed. This can partly be explained by the far removed optical phonon frequencies of the fully fluorinated carbon bonds used in the present type of Teflon wall coating [41] from the hydrogen vibrational frequencies. The experimental data points were best fitted by the results of the detailed numerical simulation if a rate constant value for the vibrational energy transfer from hydrogen to deuterium of $k_{vv} = (1.0 \pm 0.1) \times 10^{-14}$ cm^3 molecule^{-1} s^{-1} was employed.

For comparison with the plug flow evaluation, again the axial coordinate was replaced by the reaction time $t = z/v^{plug}$. In the present reactor data from the turbulent mixing region are not available, so the initial build-up of $D_2(v = 1)$ product concentration could not be observed experimentally. Therefore an attempt was made to fit concentration-time histories by results from a numerical integration of the homogeneous reaction mechanism, by varying the rate constant for the vibrational energy transfer k_{vv}. Heterogeneous relaxation was taken into account in terms of the pseudo homogeneous rate constant k_w^{plug} determined previously. The nearly exponential $H_2(v = 1)$ decay is best fitted by a rate constant $k_{vv} = 7.1 \times 10^{-15}$ cm^3 molecule^{-1} s^{-1}, a value which is almost 40% smaller than the value obtained by the detailed numerical simulation. The $D_2(v = 1)$ concentration profile could only be fitted after the data points were shifted by 2 ms on the time axis. This means that the mean residence time of reactants in the mixing region is longer than the time estimated by the plug flow velocity and the distance between the end of the inlet tubes (Figure 1b) and the first measurement location. This mismatch can be reasonably explained by the increased residence times in the mixing region caused by the turbulent motion, and again shows the importance of a detailed knowledge of the mutual dependence between reaction and flow field characteristics.

From the comparison of the plug flow results obtained using the Kaufmann equation (6) with those obtained by a detailed numerical simulation within a suitable reactor, it is evident that, even at low pressures within the plug flow limit, the sum of neglected aspects can lead to significant underestimations of evaluated rate constants, approaching the 40%-range [42]. The

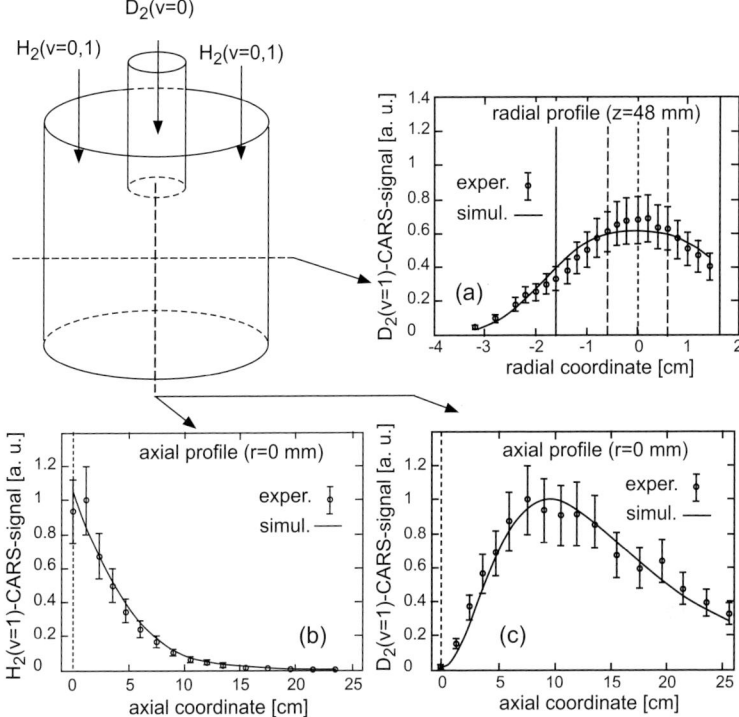

Fig. 15. Collisional relaxation and vibrational energy transfer between $H_2(v = 1)$ and $D_2(v = 0)$: comparison of experimental $H_2(v = 1)$ and $D_2(v = 1)$ CARS spatial concentration profiles (represented as open circles with error bars) with detailed numerical simulation results (solid lines) in the flow tube. In the upper left part a schematic of the flow tube section is shown where the experimental profiles have been obtained. In the upper right part a $D_2(v = 1)$ radial concentration profile 48 mm away from the exit of the inner gas supply tubing in the laminar mixing head shown in Figure 1b. Dashed and solid vertical lines are the geometric dimensions of the inner and outer tube walls of the reactor, respectively. In the lower left part a experimental $H_2(v = 1)$ axial concentration profile measured along the centreline of the flow tube is depicted along with the corresponding simulated profile. The graphics in the lower right part shows a experimental $D_2(v = 1)$ axial concentration profile measured along the centreline of the flow tube.

Kaufmann equation (6) is derived under the assumption that the number of molecules, ΔN, deactivated at the surface $S = 2\pi r_0 \Delta z$ in a flow tube with radius r_0 and length element Δz within the time Δt, $\Delta N = (1/4)v_m nS\gamma_w \Delta t$, is distributed homogeneously within the volume $V = \pi r_0^2 \Delta z$. This results in an uniform decrease in concentration $\Delta n = \Delta N/V = v_m/(2r_0)\gamma_w n\Delta t$ without radial concentration gradients. This holds for small γ_w and fast diffusive transport. However, as it is evident from Figure 11 or 13 the actual concentration profiles do show radial variations giving rise to a systematic error in derived rate constants, if the deactivation probability is evaluated in terms of a homogeneous process. Due to the presence of a nonuniform radial concentration distribution (concentrations being higher in the high-speed domains in the center of the reactor and lower in the low-speed domains near the wall) the average axial transport of excited species takes place with an effective average transport velocity larger than the plug flow velocity. This effect is further enhanced in our measurements of axial concentration profiles because CARS detection is more weighted towards the location of the beam foci (i.e., on the centreline of the reactor), so that molecules in high-speed domains mainly contribute to the total CARS signal. In an assumed "worst case" of a parabolic concentration profile as a result of less effective diffusion, and detection limited to one point located on the reactor axis, the measured effective average transport velocity will equal the maximum of the Poiseuille flow velocity which, in this case, is twice the plug flow velocity. Since velocity directly enters into the evaluated time scale, rate constants determined by plug flow evaluations are systematically lower.

In summary, the room-temperature rate constant obtained in the present work for the $H_2(v = 1) + D_2(v = 0) \rightarrow D_2(v = 1) + H_2(v = 0)$ VV transfer reaction is in good agreement with the experimental results of Pirkle and Cool [43] and Bott [44], who obtained values of 9.8×10^{-15} and 1.0×10^{-14} cm^3 molecule^{-1} s^{-1}, respectively. In the latter studies vibrational excitation of hydrogen was achieved *via* VV transfer in collisions with vibrationally excited HF molecules generated by resonant absorption of infrared laser light and the subsequent relaxation of $H_2(v = 1)$ due to VV transfer to $D_2(v = 0)$ was monitored by time resolved HF infrared fluorescence measurements.

3.2 Low-Temperature Kinetics (T = 110 − 300 K) study of $H_2(v = 1)$ collisional relaxation and vibrational energy transfer

In order to extend the temperature range of the kinetics studies of the $H_2(v = 1)$ collisional relaxation and vibrational energy transfer reactions with $D_2(v = 0)$ to lower temperatures were quantum effects might occur, CARS experiments were performed in a thermostated version (see Section 3.1) of the flow reactor, which was characterized in detail in the course of the room-temperature kinetics studies described above [45]. Experiments were performed under low-pressure conditions (4 Torr) in the temperature range T = 110-300 K along with a detailed numerical modeling to study the temperature dependence of the $H_2(v = 1)$ wall deactivation probability $\gamma_w(T)$ and the rate constant for the VV transfer reaction $H_2(v = 1) + D_2(v = 0) \rightarrow D_2(v = 1) + H_2(v = 0)$. Values for the $H_2(v = 1)$ wall deactivation probabilities (at fixed reactor temperatures) were determined by fitting calculated $H_2(v = 1)$ concentration profiles obtained in the simulation to the experimental $H_2(v = 1)$ concentrations measured along the flow tube axis with $\gamma_w(T)$ as an adjustable parameter. In these measurements the deuterium flow through the inner tube was replaced with an equal amount of helium. The thermal rate coefficient

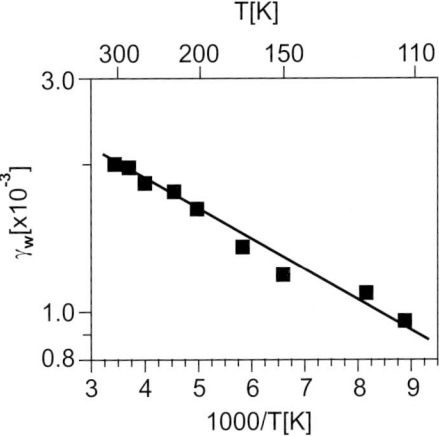

Fig. 16. Temperature dependence of the measured wall deactivation probability $\gamma_w(T)$ for $H_2(v = 1)$ collisional removal in the Teflon-coated flow reactor (solid squares). The solid line represents the result of a fit of the experimental data points (for details see text).

$k_{vv}(T)$ for vibrational energy transfer between hydrogen and deuterium at low temperatures was determined in a similar way by a direct comparison between measured concentration profiles of vibrationally excited species along the flow tube centerline with corresponding results from the detailed modeling of the reactive flow. Further CARS measurements were performed to measure temperature profiles on the centerline of the flow tube between the first and last CARS observation port in the flow reactor for different temperature settings of the cooling system under normal operating conditions. These measurements revealed small but noticeable temperature gradients in the mixing section of the flow tube, which increased with decreasing wall temperature. By fitting the measured temperature profiles with polynomials the appropriate temperature distribution input data sets for the numerical modeling could be generated.

Figure 16 depicts the measured temperature dependence of the wall deactivation probability $\gamma_w(T)$ for $H_2(v = 1)$ plotted in Arrhenius form. For comparison an empirical Arrhenius expression of the form $\gamma_w(T) = A_d \exp\{-E_d/RT\}$ is depicted as a solid line, which was used to fit the experimental wall relaxation data. Here the E_d denotes the apparent activation energy for the deactivation process of $H_2(v = 1)$ in collisions with the reactor wall [41]. The fit of the experimental results yielded a temperature independent preexponential factor of $A_d = (5.9 \pm 0.9) \times 10^{-3}$ and a value of $E_d = 1.3 \pm 0.4$ kJ/mol.

Collisional relaxation and vibrational energy transfer between $H_2(v = 1)$ and $D_2(v = 0)$: comparison of experimental $H_2(v = 1)$ and $D_2(v = 1)$ CARS spatial concentration profiles (represented as open circles with error bars)

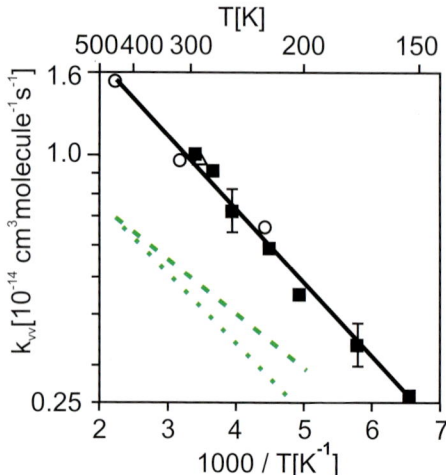

Fig. 17. Temperature dependence of the rate constant k_{vv} for the $H_2(v = 1) + D_2(v = 0) \rightarrow D_2(v = 1) + H_2(v = 0)$ vibrational energy transfer reaction. Results of the present study are depicted as solid squares. Open circles and the open triangle are experimental results reported in Ref. [43] and [44], respectively. The solid black line represents the result of a fit of all experimental data points. The two dashed green lines represent results of semi-classical collision model calculations in which two different *ab inito* potential energy surfaces were employed [6].

with detailed numerical simulation results (solid lines) in the flow tube. In the upper left part a schematic of the flow tube section is shown where the experimental profiles have been obtained. In the upper right part a $D_2(v = 1)$ radial concentration profile 48 mm away from the exit of the inner gas supply tubing in the laminar mixing head shown in Figure 1b. Dashed and solid vertical lines are the geometric dimensions of the inner and outer tube walls of the reactor, respectively. In the lower left part a experimental $H_2(v = 1)$ axial concentration profile measured along the centreline of the flow tube is depicted along with the corresponding simulated profile. The graphics in the lower right part shows a experimental $D_2(v = 1)$ axial concentration profile measured along the centreline of the flow tube.

The results of the $k_{vv}(T)$ thermal rate constant measurements for the vibrational energy transfer reaction $H_2(v = 1) + D_2(v = 0) \rightarrow D_2(v = 1) + H_2(v = 0)$ are summarized in Figure 17 and compared with available experimental and theoretical data. If all available experimental data is taken into account values of $A = (4.3 \pm 0.3) \times 10^{-14}$ cm^3 molecule^{-1} s^{-1} and $E_a = (3.7 \pm 0.6)$ kJ/mol can be derived for the preexponential factor and the reaction activation energy, respectively, from a nonlinear list square fit of the experimental data set employing a Arrhenius expression of the kind $k_{vv}(T) = A \exp\{-E_a/RT\}$. However, as can be seen in Figure 17, although there is mutual agreement between the results of different independent experimental studies, the available theoretical results (dashed green lines Figure 17) obtained in semi-classical collision model calculations employing two different *ab initio* PESs do not agree with the experimental ones (for details about the PESs employed in the calculations see Ref. [6]). The theoretical results systematically underestimate the VV energy transfer reaction rate over the whole range of temperatures indicating that even for one of the simplest diatom-diatom energy transfer processes a quantitative modeling based on *ab initio* potential energy surface data is still lacking.

3.3 Room-Temperature Kinetics (T = 300 K) measurements of the OH + H_2(v = 1) reaction rate constant

Due to is great practical importance in combustion and atmospheric chemistry the thermal rate coefficient of the reaction OH + H_2 → H + H_2O has been measured over a wide temperature range using various experimental approaches [7]. On the theoretical side numerous quasi-classical and quantum mechanical calculations were performed for this four-atom diatom-diatom benchmark reaction which provided thermal rate constant values which were found to agree well with the experimental ones (see e.g. [10] and references therein). In contrast to the large number of experimental and theoretical thermal kinetics studies, which provided detailed information about the temperature dependence of the OH + H_2 reaction rate constant, similar detailed information about the influence of the selective excitation of the OH and H_2 vibrational degree of freedom on reactivity is not yet available.

On the experimental side the influence of OH(v = 1) vibrational excitation was investigated by Spencer *et al.*, who observed that at room temperature the reaction rate is enhanced by less than a factor of two [46]. The latter observation can be rationalized based on the results of an analysis of the H-O-H-H transition state (TS) geometry, which indicate that the OH bond acts as a spectator bond as it is essentially unchanged during the course of the reaction (see e.g. [47, 48, 49] and references therein). The same TS analysis, however, revealed that it is the H_2 bond, which is considerably extended in the TS-geometry. As a consequence one can expect selective vibrational excitation of H_2 to be much more effective in promoting reactivity than vibrational excitation of the OH spectator bond.

In the first study, in which the influence of H_2 vibrational excitation on the on the room-temperature rate constant was investigated only an upper limit of 5.5×10^{-12} cm^3 molecule^{-1} s^{-1} could be derived for the OH + H_2(v = 1) reaction rate constant [50]. Subsequent flow tube studies by Zellner and Steinert [51], and Glass and Chaturvedi [52], yielded absolute values of $(7.5 \pm 3) \times 10^{-13}$ cm^3 molecule^{-1} s^{-1} and $(9.9 \pm 2.4) \times 10^{-13}$ cm^3 molecule^{-1} s^{-1}, respectively, which are in agreement with the upper limit of Ref. [50]. The latter values indicate that the room temperature rate constant is enhanced by a factor of ca. 120 and 155, respectively, if H_2 is selectively excited into its first vibrational excited state.

On the theoretical side full dimensional quantum scattering calculations, in which the influence of H_2 vibrational excitation on the reactivity of the $OH + H_2$ reaction was investigated, were performed only recently [53]. In this study the Heidelberg multiconfiguration time-dependent Hartree (MCTDH) package developed by Meyer and co-workers [54, 55] was utilized. In these calculations, in which the YZCL2-PES [56] was employed as an interaction potential for the $OH + H_2$ diatom-diatom reactive scattering system, initial vibrational state selected reaction cross sections were calculated for a wide range of reagent translational energies. The latter quantities were then used to obtain, by averaging over the respective thermal translational energy distribution, a value of 9.0×10^{-14} cm^3 molecule^{-1} s^{-1} for the $OH + H_2(v = 1)$ room-temperature rate. This value is considerably lower that the available experimental values [51, 52], indicating the presence of a significantly smaller rate enhancement upon H_2 vibrational excitation.

In Figure 18, experimental results of the present room-temperature kinetics investigation of the $OH + H_2(v = 1)$ reaction are reproduced; the experimentally observed decrease of the OH concentration as a function of reaction time in the presence of vibrationally excited H_2 is shown as open circles. As described in Section 3.1 relative OH concentrations were derived from LIF measurements while CARS detection of molecular hydrogen allowed the determination of the actual $H_2(v = 1)$ excitation efficiency and hence the $H_2(v = 1)$ concentration (*vide supra*) at different reaction times. Measurements under various experimental conditions were performed to quantify the non reactive deactivation pathways of $H_2(v = 1)$, such as heterogeneous relaxation and vibrational energy transfer, and the OH removal rate originating from wall reactions.

The actual value of the $OH + H_2(v = 1)$ rate constant was determined by means of the parameter estimation procedure described in the article by Carraro/Heuveline/Rannacher [5] in this volume on the basis of a two-dimensional simulation of the flow reactor system using the HiFlow packet [57]. The solid line depicted in Figure 18 represents the result of such a two-dimensional simulation, which yielded a optimum value of $(1.2 \pm 0.1) \times 10^{-13}$ cm^3 molecule^{-1} s^{-1} for the $OH + H_2(v = 1)$ reaction rate constant at room temperature. The latter value is in much better agreement with the theoretical value derived in the quantum scattering calculations than the previously measured values [51, 52].

In order to access the influence of the dimensionality of the simulation code on the value of the rate constant, additional simulations of the experimental data were performed in which the zero-dimensional HOMREACT simulation code was employed [58]. The latter zero-dimensional simulation resulted in a value of 7.5×10^{-13} cm^3 molecule^{-1} s^{-1}. The fact that the latter value is considerably higher, and therefore much closer to the previously measured values [51, 52] quoted above, which were also derived from a zero-dimensional analysis of the respective experimental data, emphasizes the importance of

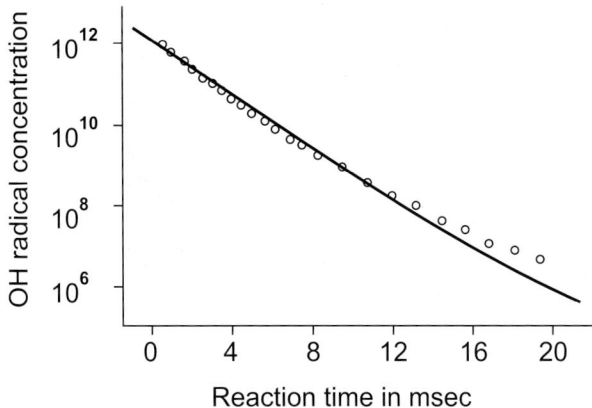

Fig. 18. Reaction time dependence of the OH radical concentration (in molecules/cm^3) obtained in the room-temperature OH + H$_2$(v = 1) → H + H$_2$O kinetics study. Experimental results are shown as open circles. The solid line represents the result of a two-dimensional simulation of the reaction system [57], which was used for the parameter estimation of the reaction rate constant. The experiments were performed at a total pressure of 4 Torr (p$_{H2(v=0)}$ = 1.03 Torr, p$_{H2(v=1)}$ = 0.01 Torr), p$_{He}$ = 2.96 Torr) at a flow rate of 16.4 m/sec.

the detailed numerical approach developed in the framework of the present work [5] in the evaluation of accurate kinetics data from flow tube studies.

3.4 High-Temperature Measurements of the O(^{1}D) + H$_2$ reaction rate constant

The OH + H$_2$ reaction together with the reaction of electronically excited oxygen atoms, O(^{1}D), with H$_2$ molecules represent the two major sinks of molecular hydrogen in the eraths´ stratosphere [8]. As a consequence, the thermal kinetics of the O(^{1}D) + H$_2$ reaction has so far been studied mainly at low temperatures (250 K < T < 350 K) [15]. In this temperature range the rate constant was found to be almost independent of temperature. Reaction dynamics studies, in which "hot" O(^{1}D) atoms with a translational temperature of 1200 K were photolytically generated, however, indicated that the rate constant increases with increasing translational excitation of the reagents [12]. The latter finding initiated a large number of theoretical studies both quasiclassical and quantum mechanical ones on different PESs, the results of which cover a wide range of temperature (see e.g. [10, 59, 60] and references therein). On the experimental side, however, kinetics measurements for elevated temperatures are still missing. Therefore studies at higher temperatures were performed in the course of the present work, the results of which allow for a more detailed comparison with the available theoretical results.

In Figure 19 the time profile of H atom product concentration (symbols) following the photolysis "pump" laser pulse, which generated the O(^{1}D)

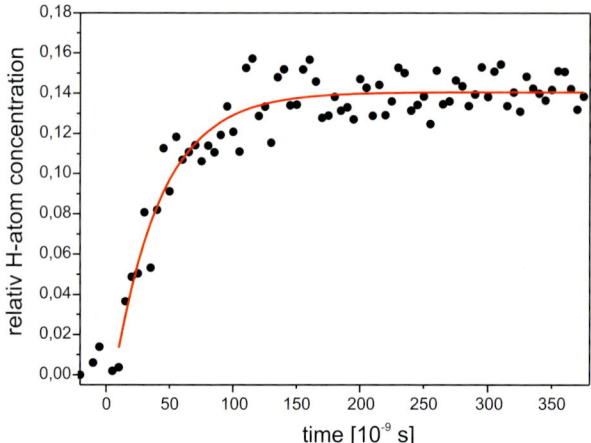

Fig. 19. Transient concentration profiles (symbols) of H atoms formed in the O(^{1}D) + H$_2$ → H + OH reaction obtained *via* VUV LIF measurements at the H atom Lyman-α transition. The solid red line represents the result of a numerical simulation [5], which was used to derive the H atom formation pseudo-first-order rate constant k′ = (24.9 ± 0.25) × 10^6 s^{-1}. The measurements were performed in a flowing mixture of N$_2$O, H$_2$, Ar at a total pressure of 50 Torr (p_{N2O} = 891 mTorr, p_{H2} = 12.997 Torr, p_{Ar} = 36.112 Torr) at a temperature of 580 K.

reagent atoms *via* the photodissociation of N$_2$O at Δt = 0, is depicted. The experimental H atom product concentration *versus* time profiles could be analyzed using the parameter estimation procedure described in the article by Carraro/Heuveline/Rannacher [5]. The solid line in Figure 19 shows the result of a numerical simulation, which was performed using the reaction mechanism given in Table 7 in the appendix of Ref. [5], to determine the H atom formation pseudo-first-order rate constant (k′).

In Figure 20 the k′-values obtained at a temperature of T = 580 K are plotted against the H$_2$ concentration. As can be seen in Figure 20 the measured H atom formation rates show a linear dependence on the H$_2$ reagent concentration. Hence the value of the bimolecular rate constant of the O(^{1}D) + H$_2$ → H + OH reaction could be derived from the slope of the k′ *versus* H$_2$ reagent concentration plot. The latter one was obtained from the straight line depicted in the figure, which represents the result of a non-linear least squares fit to the measured k′ values. The values of the rate constant of the O(^{1}D) + H$_2$ → H + OH obtained in the present study are shown as an Arrhenius plot in Figure 21 along with the results of previous experimental studies [12, 16]. The dashed curve in Figure 21 represents the result of a non-linear least squares fit of the experimental values obtained in the present study (solid circles) and the "high-temperature" value reported in our previous work [12] (open circle) using a modified Arrhenius expression: k(T) = A (T/300 K)$^\beta$ exp {-E$_a$/RT}. The non-linear least squares fit analysis resulted in values of A

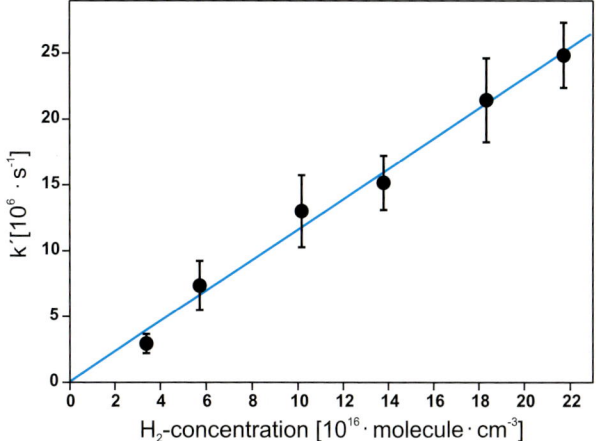

Fig. 20. Plot of the pseudo-first-order rate constants (k') *versus* H_2 reagent concentration for the $O(^1D) + H_2 \rightarrow H + OH$ reaction at a temperature of 580 K.

$= 1.0 \times 10^{-11}$ cm^3 molecule^{-1} s^{-1}, $\beta = 1.96$, and $E_a = -5.75$ kJ/mol, respectively. The latter results indicate pronounced non-Arrhenius behavior with a significant upward shift in the slope of the Arrhenius curve of the thermal rate constant as a function of temperature for T > 370 K.

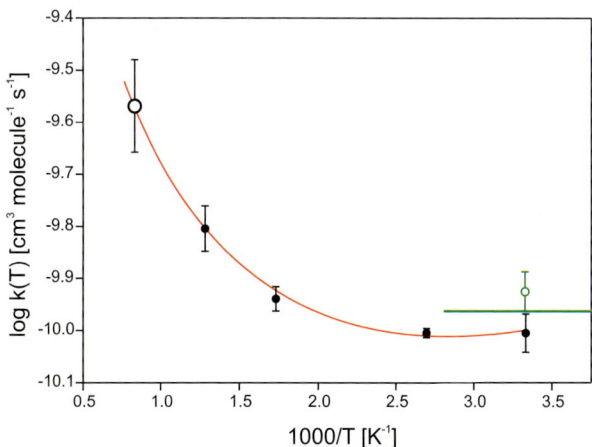

Fig. 21. Arrhenius plot of the rate constant k(T) for the $O(^1D) + H_2 \rightarrow H + OH$ reaction. Solid circles (this work), open black circle (from Ref. [12]), green line (data recommended in Ref. [16]), open green circle (experimental value from Ref. [65]). The red curve represents the result of a non-linear least squares fit analysis of the present experimental results (including the value determined in Ref. [12]) using a modified Arrhenius expression (for details see text).

In Figure 22 finally a comparison between the experimental data (black circles depict the results of the present work) and results of quasi-classical trajectory (QCT) [59, 60] and quantum scattering (QS) calculations [61] is presented. In the QS calculations of Lin and Guo thermal rate constants (black curve in Figure 22) were calculated over a wide range of temperatures using an improved version of the $H_2O(1^1A')$ ground-state PES, which includes a proper representation of the long-range reagent interactions [62, 63]. Excited-state PESs, however, were not taken into account in these calculations. In the QCT studies of Brandão and Rio [60], who used the same version of the $H_2O(1^1A')$ ground-state PES, contributions to the thermal rate constant originating from the $H_2O(1^1A'')$ exited-state PES and the non-adiabatic electrostatic coupling between the $^1\Sigma^+$ ($1^1A'$) and $^1\Pi$ ($2^1A'$ and $1^1A''$) exited states of H_2O, which take place for collinear collision geometries, were taken into account in an approximative fashion. The results of the latter calculations are shown as a red curve in Figure 22. In QCT studies performed by Schatz and co-workers, where the influence of the $H_2O(1^1A'')$ exited-state PES on the thermal rate constant was investigated for the first time [59], an earlier version of the $H_2O(1^1A')$ ground-state PES was employed [64], which yields - as can be seen in Figure 22 - a somewhat higher value for the room-temperature rate constant than the more recent and more accurate $H_2O(1^1A')$ ground-state PES of Brandão, and Rio [62, 63].

In the studies of Schatz and co-workers non-adiabatic coupling between the ground and exited-state PESs were not included in the calculations [59]. Schatz and co-workers were also the first to note that the inclusion of the $H_2O(1^1A'')$ exited-state PES can results in a notable upward shift in the slope of the Arrhenius curve of the thermal rate constant at higher temperatures, as it is observed in the present study (see Figure 21 and Figure 6 in Ref. [59]). The QCT studies of Schatz and co-workers further showed that the room-temperature (T = 300 K) rate constant is due exclusively to reaction on the $H_2O(1^1A')$ ground-state PES.

A detailed inspection of Figure 22 reveals that at low temperatures (T = 300 K and T = 370 K) the present results agree reasonably well with the most recent experimental result of Talukdar and Ravishankara [65], who obtained a value of k(T = 300 K) = $(1.2 \pm 0.1) \times 10^{-10}$ cm^3 molecule^{-1} s^{-1}, and with the QS results performed on the $H_2O(1^1A')$ ground-state PES. With increasing temperature, however, the rate constants obtained in the present study are significantly higher than the QS results, indicating the onset of H_2O exited-states contributions, which leads to the significant increase of the rate constants as observed in the QCT calculations where H_2O exited-states contributions were included [59, 60]. Within the experimental uncertainty the rate constant value obtained in the present study at T = 780 K agrees with the results of the QCT calculations.

As can be seen in Figure 22, the present experimental results (along with the experimental results obtained previously [12, 65]) clearly demonstrate that the rate constant of the $O(^1D) + H_2 \rightarrow H + OH$ reaction exhibits a pro-

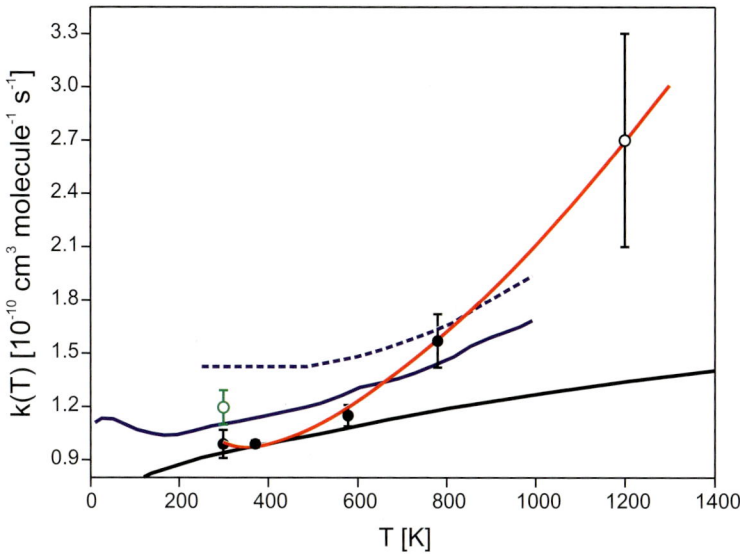

Fig. 22. Comparison of the temperature dependence of the experimental rate constant (symbols with error bars) of the $O(^1D) + H_2 \rightarrow H + OH$ reaction with results of quasi-classical trajectory (QCT) [59, 60] and quantum scattering (QS) calculations [61]. The latter QS calculations the results of which are shown as a black curve were performed on the $H_2O(1^1A')$ ground-state PES while the QCT results of Ref. [59] (dashed blue curve) and Ref. [60] (solid blue curve) account for contributions from the H_2O exited-state PESs to the thermal rate constant. The solid red curve represents the results of the non-linear least squares fit analysis using a modified Arrhenius expression of the experimental values (see Fig. 21). Further details about the theoretical calculations are given in the text and in Refs. [59,60,61].

nounced increase with temperature, which is considerably higher than the increase predicted by QS calculations on the $H_2O(1^1A')$ ground-state PES, indicating that a proper theoretical description of the high-temperature reaction kinetics requires the inclusion of electronically exited H_2O-PESs. As outlined in Ref. [59], a more accurate treatment would require QS calculations employing the most accurate available PESs along with an appropriate inclusion of the possible non-adiabatic transitions between them.

4 Summary

In the framework of the present project three prototype elementary reaction systems in chemical kinetics were investigated in different laminar flow reactor systems by means of modern laserspectroscopic techniques along with detailed numerical two-dimensional simulations which were applied to determine accurate rate constants. In cases of the $H_2(v = 1) + D_2(v = 0) \rightarrow D_2(v = 1) +$

$H_2(v = 0)$ vibrational energy transfer reaction comparison between the results obtained using a detailed numerical simulation and the results obtained using the plug flow approximation revealed that, even at low pressures within the plug flow limit, the sum of aspects neglected in that approximation can lead to significant underestimations of evaluated reaction rate constants, approaching the 40% range. Similar results were observed in case of the room-temperature kinetics investigation of the four-atom benchmark reaction $OH + H_2(v = 1) \rightarrow H + H_2O$ where a considerable difference between the rate constants derived using a zero-dimensional and a detailed two-dimensional numerical simulation approach was observed. Finally results of reaction kinetics studies of the $O(^1D) + H_2 \rightarrow H + OH$ reaction obtained in a high-temperature photochemical flow reactor, which was designed using extensive *a priori* simulations (see the article Carraro/Heuveline/Rannacher [5] in this volume), were presented. In the latter case a parameter estimation procedure was applied to derive rate constants in the temperature range $T = 300$-760 K, which were found to exhibit a pronounced increase with temperature, indicating that at temperatures above ca. 600 K exited-states-chemistry contributes significantly to the thermal rate constant.

References

1. Wood, R.W.: *Spontaneous incandescence of substances in atomic hydrogen gas*, Proc. R. Soc. (London), **A102**, 1-9, (1922)
2. Bonhoeffer, K.F.: *The behavior of active hydrogen*. Z. Phys. Chem., **113**, 199-219, (1924)
3. Schlichting, H. In: Braun, G. (ed) *Boundary layer theory*. Karlsruhe, 1982.
4. Kaufmann, F.: *Progress reaction kinetics*. Pergamon Press, New York, 1961; Vol. **1**.
5. Carraro, T., Heuveline, V., Rannacher, R.: *Determination of kinetic parameters in laminar flow reactors. I. numerical aspects, reactive flows, diffusion and transport*, (R. Rannacher et al., eds), Springer, Berlin-Heidelberg (2005)
6. Billing, G.D., Kolesnick, R.E.: *Semi-classical calculations of rate constants for vibrational transitions in hydrogen*, Chem. Phys. Lett., **215**, 571-575 (1993)
7. Warnatz, J., Maas, U., and Dibble, R.W.: *Combustion: physical and chemical fundamentals, modeling and simulation, experiments, pollutant formation*, 3d ed. Springer, Berlin (2005)
8. Wayne, R.P.: *Chemistry of atmospheres*, 2nd ed. Oxford University Press, Oxford, (1991)
9. Millar, T.J. and Williams D.A. (eds.): *Rate coefficients in astrochemistry*. Kluwer, Dordrecht (1988)
10. Althorpe, S.C., Clary, D.C.: *Quantum scattering calculations on chemical reactions*, Annu. Rev. Phys. Chem. **54**, 493-529 (2003)
11. Volpp, H.-R.: *The dynamics of the $H + X_2O (X = H, D)$ gas-phase isotope reactions: comparison between experiment and quantum theory*. Proceedings of the Indian National Science Academy (PINSA), Part A, Vol. **66/1** "Special Issue on Chemical Dynamics and Photochemistry I" pp. 1–28 (2000)

12. Koppe, S., Laurent, T., Naik, P.D., Volpp, H.-R., Wolfrum, J., Arusi-Parpar, T., Bar, I.,Rosenwaks, S.: *Absolute rate constants and reactive cross sections for the reactions of $O(^1D)$ with molecular hydrogen and deuterium,* Chem. Phys. Lett., **214**, 546-552 (1993)
13. Berg, P.A., Sloan, J.J., Kuntz, P.J., *The effect of reagent excitation on the dynamics of the reaction $O(^1D_2) + H_2 \rightarrow OH(X^2\Pi) + H$,* J. Chem. Phys., **95**, 8038-8047 (1991)
14. K. Drukker, G. C. Schatz: *Quantum scattering study of electronic Coriolis and nonadiabatic coupling effects in $O(^1D) + H_2 \rightarrow OH + H$* J. Chem. Phys. 111, 2451-2463 (1999)
15. Davidson, J.A., Schiff, H.I., Streit, G.E., McAfee, J.R. Schmeltekopf, A.L., Howard, C.J.: *Temperature dependence of $O(^1D)$ rate constants for reactions with $N_2O, H_2, CH_4, HCl,$ and NH_3,* J. Chem. Phys., **67**, 5021-5025 (1977)
16. Atkinson, R., Baulch, D.L., Cox, R.A., Hampson, R.F., Kerr, J.A., Rossi, M.J., Troe J.: *Evaluated kinetic and photochemical data for atmospheric chemistry: supplement VI. IUPAC subcommittee on gas kinetic data evaluation for atmospheric chemistry,* J. Phys. Chem. Ref. Data, **26**, 1329-1499 (1997)
17. Aoiz, F.J., Bañares, L., Castillo, J.F., Brouard, M., Denzer, W., *Insertion and abstraction pathways in the reaction $O(^1D_2) + H_2 \rightarrow OH + H$,* Phys. Rev. Lett., **86**, 1729-1732 (2001)
18. Honvault, P., Launay, J.-M., *A quantum-mechanical study of the dynamics of the $O(^1D) + H_2 \rightarrow OH + H$ insertion reaction* J. Chem. Phys., **114**, 1057-1059 (2001)
19. Buchenau, H., Toennies, J.P., Arnold, J., Wolfrum, J., $H + H_2$*: The current status,* Ber. Bunsenges. Phys. Chem., **94**, 1231-1248 (1990)
20. Arnold, J., Bouché, T., Dreier, T., Wichmann, J., Wolfrum, *CARS studies on the heterogeneous relaxation of vibrationally excited hydrogen and deuterium,* J. Chem. Phys. Lett., **203**, 283-288 (1993)
21. Seitzman, J.M., Hanson, R.K., in: Taylor, A.M.K.P. (ed.) *Instrumentation for flows with combustion.* Academic Press, London (1993)
22. Tsang, W., Herron, J.T.: *Chemical kinetic data base for propellant combustion. I. Reactions involving $NO, NO_2, HNO, HNO_2, HCN$ and N_2O,* J. Phys. Chem. Ref. Data., **20**, 609-663 (1991)
23. Rahn, L.A., Zych, L.J., Mattern, P.L., *Background-free CARS studies of carbon monoxide in a flame,* Opt. Commun., **39**, 249-253 (1979)
24. Clyne, M.A.A., Thrush, B.A., *Rates of elementary processes in the chain reaction between hydrogen and oxygen. I. Reactions of oxygen atoms,* Proc. R. Soc., **A275**, 544-558 (1963)
25. Hartley, D.B., Thrush, B.A., *The rates of elementary processes in the chain reaction between hydrogen and oxygen. III. kinetics of the combination of hydrogen atoms with nitric oxide,* Proc. R. Soc., **A297**, 520-533 (1967)
26. Péalat, M., Lefebvre, M., Taran, J-P.E., Kelley, P.L., *Sensitivity of quantitative vibrational coherent anti-Stokes Raman spectroscopy to saturation and Stark shifts,* Phys. Rev. A, **38**, 1948-1965 (1988)
27. Dieke, G.H. , Crosswhite, M.H., *The ultraviolet bands of OH Fundamental data,* J. Quantum Spectrosc. Radiative Transfer, **2**, 97-199 (1962)
28. Smyth, K.C., Crosley, D.R., in: Jeffries, J., Kohse-Höinghaus, K. (eds.) *Applied combustion diagnostics.* Taylor & Francis (2002)

29. Chidsey, I.L., Crosley, D.R., *Calculated rotational transition probabilities for the A-X system of OH*, J. Quantum Spectrosc. Radiative Transfer, **23**, 187-199 (1980)
30. Hanf, A.,*Konstruktion eines Hochtemperaturströmungsreaktors und Untersuchung der Reaktion $O(^1D) + H_2 \to H + OH$ bei hohen Temperaturen*, PhD thesis, University of Heidelberg, (2005)
31. Carraro, T., *Parameter estimation and experimental design in flow reactors*, PhD thesis, University of Heidelberg, (2005)
32. Carraro, T., Heuveline, V., *Parameter estimation and experimental design in flow reactors*, Preprint, SFB 359, University of Heidelberg, (March 2005)
33. KANTHAL SUPER, *Electric heating Element Handbook*, Hallstahammar, Sweden (1994)
34. Felder, P., Haas, B.-M., Huber J. R., *The photoreaction $N_2O \to O(^1D) + N_2(^1\Sigma)$ at 193 nm studied by photofragment translational spectroscopy*, Chem. Phys. Letters **186**, 177-182 (1991)
35. Brownsword, R.A., Hillenkamp, M., Laurent, T., Volpp, H.-R., Wolfrum, J., Vatsa, R.K., Yoo, H.-S., *Excitation function and reaction threshold studies of isotope exchange reactions: $H + D_2 \to D + HD$ and $H + D_2O \to D + HOD$*, J. Phys. Chem. **101**, 6448-6454 (1997) "Y.T. Lee Special Issue".
36. Hilber, G., Lago, A., Wallenstein, R.,*Broadly tunable vacuum-ultraviolet/extreme-ultraviolet radiation generated by resonant third-order frequency conversion in krypton.* , J. Opt. Soc. Am. B., **4**, 1753-1764 (1987)
37. Brownsword, R.A., Hillenkamp, M., Laurent, T., Vatsa, R.K., Volpp, H.-R., *Photodissociation dynamics in the UV laser photolysis of DNCO: comparison with HNCO*, J. Chem. Phys. **106**, 4436-4447 (1997)
38. Boquillon, J. P., Pealat, M., Bouchardy, P., Collin, G., Magre, P.,*Spatial averaging and multiplex coherent anti-Stokes Raman scattering temperature-measurement error*, Opt. Lett., **13**, 722-725 (1988)
39. Bloembergen, N., *Nonlinear optics*; W. A. Benjamin, Inc.: New York, (1965)
40. Druet, S. A. J., Taran, J. P. E., *Cars spectroscopy*, Prog. Quant. Electron., **7**, 1-72 (1981)
41. Gershenzon, Yu. M., Ivanov, A. V., Kucheryavi, S. I., Lyapunov, A. Ya., Rozenshtein, V. B.,*Reactions of vibrationlly excited hydrogen and deuterium molecules with atoms and radicals: Heterogeneous relaxation of the vibrational energy of H_2 and D_2 on quartz and Teflon surfaces*, Kinet. Catal., **27**, 928-932, (1986)
42. Segatz, J., Rannacher, R., Wichmann, J., Orlemann, C., Dreier, T., Wolfrum J., *Detailed numerical simulations in flow reactors: A new approach in measuring absolute rate constants*, J. Phys. Chem. **100**, 9323-9333 (1996)
43. Pirkle, R. J., Cool, T. A., *Vibrational energy transfer for $H_2 - D_2$ and $H_2 - HCl$ mixtures from 220 - 450 K*, Chem. Phys. Lett., **42**, 58-63, (1976)
44. Bott, J. F., *Vibrational energy exchange between $H_2(v=1)$ and D_2, N_2, HCl, and CO_2*, J. Chem. Phys., **65**, 3921-3928, (1976)
45. Schneider-Kühnle, Y., Dreier, T., Wolfrum, J., *Vibrational relaxation and energy transfer in the hydrogen system at temperatures between 110 and 300 K*, Chem. Phys. Letters **294**, 191-196, (1998)
46. Spencer, J. E., Endo, H., Glass, G. P., *16th Symp. (Intern.) on Combustion*, The Combustion Institute, Pittsburgh, p. 829. (1976)
47. Schatz, G. C., Elgersma, H.,*A quasi-classical trajectory study of product vibrational distributions in the $OH + H_2 \to H_2O + H$ reaction* , Chem. Phys. Lett. **73**, 21-25 (1980)

48. Walch, S. P., Dunning, T. H., *A theoretical study of the potential energy surface for OH + H_2* , J. Chem. Phys. **72**, 1303-1311 (1980)
49. Koppe, S., Laurent, T., Naik, P.D., Volpp, H.-R., Wolfrum, J., *Reaction dynamics studies of a simple tetraatomic system: $AA + BA \rightarrow A + ABA$,* Can. J. Chem. **72**, 615-624 (1994): "Paper dedicated to Prof. John C. Polanyi on the occasion of his 65th birthday".
50. Light, G. C., Matsumoto, J. H., *The effect of vibrational excitation in the reactions of OH with H_2,* Chem. Phys. Lett. **58**, 578-581 (1978)
51. Zellner, R., Steinert, W., *Vibrational rate enhancement in the reaction $OH + H_2(v = 1) \rightarrow H_2O + H$,* Chem. Phys. Lett., **81**, 568-572 (1981)
52. Glass, G. P., Chaturvedi, B. K., *The effect of vibrational excitation of H_2 and OH on the rate of the reaction $H_2 + OH \rightarrow H_2O + H$,* J. Chem. Phys. **75**, 2749-2752 (1981)
53. Sukiasyan, S., *Investigation of three- and four-atomic reactive scattering problems with the help of the multiconfiguration time-dependent Hartree method,* PhD thesis, University of Heidelberg, (2005)
54. Beck, M. H., Jäckle, A., Worth, G. A., Meyer, H.-D., *The multiconfiguration time-dependent Hartree method: A highly efficient algorithm for propagating wavepackets,* Phys. Rep. **324**, 1-105 (2000)
55. Meyer, H.-D., *The Heidelberg MCTDH Package,* http://www.pci.uni-heidelberg.de/tc/usr/mctdh/index.html#package/.
56. Yang, M., Zhang, D. H., Collins, M. A., Lee, S.-Y., *Ab initio potential energy surface for the reactions $H_2 + OH \rightarrow H_2O + H$,* J. Chem. Phys. **115**, 174 (2001)
57. Heuveline, V., *HiFlow: A finite element solver package,* http://www.hiflow.de/
58. Warnatz. J., *HOMREACT: code for the simulation of HOMogeneous REACTion Systems,* http://reaflow.iwr.uni-heidelberg.de/homreact.php/
59. Schatz, G. C., Papioannou, A., Pederson, L. A., Harding, L. S., Hollebeek, T., Ho, T.-S., Rabitz, H., *A global A-state potential surface for H_2O: Influence of excited state on the $O(^1D) + H_2$ reaction,* J. Chem. Phys., **107**, 2340-2350 (1997)
60. Brandão, J., Rio, C. M. A., *Quasiclassical and capture studies on the $O(^1D) + H_2$ reaction using a new potential energy surface for H_2O,* Chem. Phys. Lett., **377**, 523-529 (2003)
61. Lin, S. Y., Guo, H.,: *Quantum integral cross-section and rate constant of the $O(^1D) + H_2 \rightarrow OH + H$ reaction on a new potential energy surface,* Chem. Phys. Lett., **385**, 193-197 (2004)
62. Brandão, J., Rio, C. M. A.,: *Long-range interactions within the H_2O molecule,* Chem. Phys. Lett., **372**, 866-872 (2003)
63. Brandão, J., Rio, C. M. A.,: *Quasiclassical and capture studies on the $O(^1D) + H_2 \rightarrow OH + H$ reaction using a new potential energy surface for H_2O,* Chem. Phys. Lett. **377**, 523-529 (2003)
64. Ho, T.-S., Hollebeek, T., Rabitz, H., Harding, L. B., G. C.Schatz,: *A global H_2O potential energy surface for the reaction $O(^1D) + H_2 \rightarrow OH + H$,*J. Chem. Phys., **105**, 10472-10486 (1996)
65. Talukdar, R. K., Ravishankara, A. R.,: *Rate coefficients for $O(^1D) + H_2, D_2,$ HD reactions and H atom yield in $O(^1D) + HD$,* Chem. Phys. Lett., **253**, 177-183 (1996)

Optimization of Reactive Flows in a Single Channel of a Catalytic Monolith: Conversion of Ethane to Ethylene*

H. G. Bock[1], O. Deutschmann[2], S. Körkel[1], L. Maier[2], H. D. Minh[1], J. P. Schlöder[1], S. Tischer[2], and J. Warnatz[1]

[1] Interdisziplinäres Zentrum für Wissenschaftliches Rechnen (IWR), Universität Heidelberg
[2] Institut für Chemische Technologie und Polymerchemie, Universität Karlsruhe

Summary. We discuss the modeling, simulation, and, for the first time, optimization of the reactive flow in a channel of a catalytic monolith with detailed chemistry. We use boundary layer approximation to model the process and obtain a high dimensional PDE. We discuss numerical methods based on the efficient solution of high dimensional stiff DAEs arising from spatial semi-discretization and SQP method for the optimal control problem parameterized by the direct approach. We have investigated the application of conversion of ethane to ethylene which involves a complex reaction scheme for gas phase and surface chemistry. Our optimization results show that the maximum yield, an improvement of a factor of two, is achieved for temperatures around 1300 K.

1 Introduction

The application of so-called short-contact-time reactors to the autothermal production of ethylene from ethane has led to a promising technology. In these devices, mixtures of ethane, oxygen, and nitrogen (and possibly hydrogen) flow through a ceramic monolith coated with a catalytic metal such as platinum. Mild heating of the reactor initiates an autocatalytic reaction that yields a mixture of ethylene, carbon monoxide, hydrogen, water, and smaller amounts of other hydrocarbons. The residence time in the reactor is typically a few milliseconds and the reactor temperatures appear to be in the range 900 – 1000 °C. Experiments with platinum-catalyzed systems show that ethane conversions and ethylene selectivity comparable to conventional steam cracking can be achieved [17].

One of the questions arising is the role that homogeneous reactions play in the oxidative dehydrogenation at high temperatures and short contact times.

---
*This work has been supported by the German Research Foundation (DFG) through SFB 359 (Project B1) at the University of Heidelberg.

While Huff and Schmidt [10] proposed a purely heterogeneous mechanism, experiments by Beretta et al. provided evidence for a major influence of gas-phase reactions [3]. The catalyst may serve as a heat supply by supporting complete combustion of hydrocarbons, and thus initiating endothermic gas-phase reactions.

In a first step, a single channel of a monolithic reactor is modeled. We assume that the cross-section of the channel can be approximated by a cylindrical shape. The inlet conditions and the wall temperature profile are given parameters, which in a second step become subject to the optimization. Detailed models for transport processes as well as for gas-phase and surface reactions are applied to the numerical simulation.

2 Simulation

2.1 General Mathematical Formulation

Modeling of the fluid dynamical process: boundary layer equations

To model flows in a channel, we employ the boundary layer equations which are a simplification of the Navier–Stokes equations. Since our considered channel is an axisymmetrical cylinder, we assume that the flow in it is also axisymmetrical, which can be described by two spatial coordinates, namely the axial one z and the radial one r. By applying von Mises transformation

$$\psi = \int_0^r \rho u r' dr',$$

where ψ is the stream variable, u is the axial velocity and ρ is the mass density, we obtain the following equation system, which we use for modeling the reactive flow in channel of monoliths. Here, we assume that the system is in a steady state. Figure 1 shows a typical catalytic monolith with flow conditions and the model assumption.

Momentum equation:

$$\rho u \frac{\partial u}{\partial z} + \frac{\partial p}{\partial z} = \rho u \frac{\partial}{\partial \psi}\left(\rho u \mu r^2 \frac{\partial u}{\partial \psi}\right), \qquad (1)$$

$$\frac{\partial p}{\partial \psi} = 0. \qquad (2)$$

Energy equation:

$$\rho u c_p \frac{\partial T}{\partial z} = \rho u \frac{\partial}{\partial \psi}\left(\rho u \lambda r^2 \frac{\partial T}{\partial \psi}\right) - \sum_{k=1}^{N_g} \dot{\omega}_k W_k h_k - \rho u r \sum_{k=1}^{N_g} J_{kr} c_{pk} \frac{\partial T}{\partial \psi}. \qquad (3)$$

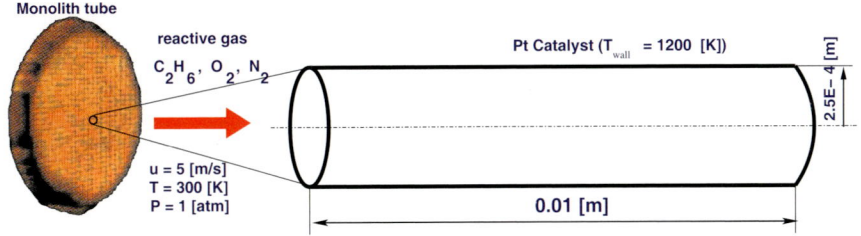

Fig. 1. Catalytic Monolith

Species equation:
$$\rho u \frac{\partial Y_k}{\partial z} = \dot{\omega}_k W_k - \rho u \frac{\partial}{\partial \psi}(r J_{kr}), \quad (k = 1, \ldots, N_g). \tag{4}$$

State equation:
$$\rho = \frac{P\overline{W}}{RT}. \tag{5}$$

This equation is used for calculating ρ. The relation between the stream variable ψ and the radial variable r is

$$\frac{\partial r^2}{\partial \psi} - \frac{2}{\rho u} = 0. \tag{6}$$

The meaning of the notations used above and in the following is as follows.

- z and r are the cylindrical coordinates.
- u and v are the axial and radial components of the velocity vector.
- p is the pressure.
- T is the temperature.
- Y_k is the mass fraction of the kth species.
- μ is the viscosity.
- ρ is the mass density.
- c_p is the heat capacity of the mixture.
- λ is the thermal conductivity of the mixture.
- c_{pk} is the specific heat capacity of the kth species.
- J_{kr} is the radial component of mass flux vector of the kth species.
- $\dot{\omega}_k$ is the rate of production of the kth species by gas phase reactions.
- h_k is the specific heat enthalpy of the kth species.
- W_k is the molecular weight of the kth species.
- N_g is the total number of gas phase species.
- K_g is the total number of elementary reactions.
- $\overline{W}$ is the mixture mean molecular weight.
- R is the universal gas constant.

Note that $\mu = \mu(Y,T)$, $\lambda = \lambda(Y,T)$, $c_p = c_p(Y,T)$, $c_{pk} = c_{pk}(Y,T)$, $h_k = h_k(Y,T)$, $\dot{\omega}_k = \dot{\omega}_k(Y,T,p)$, $Y = (Y_1, Y_2, \cdots, Y_{Ng})$, and the diffusion flux J_{kr} is given by

$$J_{kr} = -D_k^m \frac{W_k}{\overline{W}} \rho \frac{\partial X_k}{\partial r} - \frac{D_k^T}{T} \frac{\partial T}{\partial r}, \tag{7}$$

where D_k^m and D_k^T are diffusion coefficients, X_k is the mole fraction of the kth species. More details can be found in, e.g., [6], [12] and [13].

Modeling of the chemical process in the gas phase

When moving in a channel, the gas mixture takes part in several chemical reactions. To model this process, we use the detailed chemistry describing elementary reactions on molecular level, the description and derivation of which can be found, e.g., in [16, 8]. A chemical reaction involving N_g species can be represented in the general form

$$\sum_{k=1}^{N_g} v'_{ki} \chi_k \to \sum_{k=1}^{N_g} v''_{ki} \chi_k, \tag{8}$$

where v'_k and v''_k are the stoichiometric coefficients of the kth species and χ_k is the chemical symbol for the kth species and i denotes the reaction number. The rate of production $\dot{\omega}_k$ of the kth species, which appears in equations (4) and (3), is determined by

$$\dot{\omega}_k = \sum_{i=1}^{K_g} v_{ki} k_{fi} \prod_{j=1}^{N_g} [X_j]^{v'_{ji}} \quad (k = 1, \ldots, N_g), \tag{9}$$

where
$v_{ki} := v''_{ki} - v'_{ki}$,
$[X_j]$: the concentration of the jth species,
k_{fi} : the forward rate coefficient of the ith reaction.
The forward rate coefficient k_{fi} is calculated by the Arrhenius expression:

$$k_{fi} = A_i T^{\beta_i} \exp\left(-\frac{E_{ai}}{RT}\right) \quad (i = 1, \ldots, K_g), \tag{10}$$

where
A_i : the pre-exponential factor of the ith reaction
[the units are given in terms of m, mol, and s],
β_i : the temperature exponent of the ith reaction,
E_{ai} : the activation energy of the ith reaction [J/mol],
R : the universal gas constant being 8.314 [J/(mol · K)],
T : the gas temperature [K].

The backward (reverse) rate coefficient k_r is determined based on the chemical equilibrium.

The mass fractions Y_k of the species are computed from the concentrations

$$Y_k = \frac{W_k[X_k]}{\sum_{j=1}^{N_g} W_j[X_j]}. \tag{11}$$

They must satisfy

$$0 \leq Y_k \leq 1 \quad (k = 1, \ldots, N_g), \quad \sum_{k=1}^{N_g} Y_k = 1. \tag{12}$$

Modeling of the surface chemistry

The chemistry source term $\dot{s}_i$ appearing later in the boundary conditions (17) and (19) states the creation or depletion rate of the ith species due to the adsorption/desorption process and is given by

$$\dot{s}_i = \sum_{k=1}^{K_s} v_{ik} k_{f_k} \prod_{j=1}^{N_g+N_s} [X_j]^{v'_{jk}}, \tag{13}$$

where K_s and N_s are the numbers of elementary surface reactions and of adsorbed species, respectively. If the jth species is a gas-phase species then $[X_j]$ is its concentration, otherwise, if it is a surface species then $[X_j] = \theta_i \Gamma$, where θ_i is the surface coverage and Γ the surface site density.

Depending on the reaction mechanism, in some reactions, the rate coefficient k_{f_k} is calculated by the standard Arrhenius formula

$$k_{f_k} = A_k T^{\beta_k} \exp\left(-\frac{E_{ak}}{RT}\right), \tag{14}$$

and in some other reactions, it is described by the modified Arrhenius formula

$$k_{f_k} = A_k T^{\beta_k} \exp\left(-\frac{E_{ak}}{RT}\right) \prod_{i=1}^{N_s} \theta_i^{\mu_{ik}} \exp\left(\frac{\epsilon_{ik} \theta_i}{RT}\right) \tag{15}$$

where μ_{ik} and ϵ_{ik} are surface parameters, and θ_i is surface coverage, which must satisfy

$$0 \leq \theta_i \leq 1 \quad (i = 1, \ldots, N_s), \quad \sum_{i=1}^{N_s} \theta_i = 1. \tag{16}$$

2.2 Initial and Boundary Conditions

As initial conditions, the values of u, p, T, and Y_k are specified at the inlet of the channel:

$$u = u_0, \ p = p_0, \ T = T_0, \ Y_k = Y_{k0} \ (k = 1, \ldots, N_g) \text{ at } z = 0.$$

At $\psi = 0$, which corresponds to the centerline of the cylindrical channel, we can deduce the following boundary conditions from the assumed axisymmetry

$$r = 0, \ \partial u/\partial \psi = 0, \ \partial p/\partial \psi = 0, \ \partial T/\partial \psi = 0, \ \partial Y_k/\partial \psi = 0.$$

At $\psi = \psi_{\max}$ which is definied as

$$\psi_{\max} = \int_0^{r_{\max}} \rho u r' dr' \bigg|_{z=0},$$

and corresponds to the channel wall, it holds

$$r = r_{\max}, \ u = 0, \ T = T_{\text{wall}}.$$

In addition, since an essential part of chemical reactions takes place at the catalytic wall and because of the state steady assumption, the gas species mass flux produced by heterogeneous chemical reactions and the mass flux of that species in the gas must be balanced there, i.e.

$$\dot{s}_k W_k = -J_{kr} \quad (k = 1, \ldots, N_g), \tag{17}$$

where $\dot{s}_k$ is the rate of creation/depletion of the kth gas-phase species by surface reactions. Note that J_{kr} represent the radial components of the diffusive flux vector pointing from the center to the wall. In general we additionally have to take the convective flux $\rho Y_k v_{\text{Stef}}$ into account, where v_{Stef} is the Stefan velocity calculated by

$$v_{\text{Stef}} = \frac{1}{\rho} \sum_{k=1}^{N_g} \dot{s}_k W_k.$$

At the steady state, the Stefan velocity vanishes. Also due to the steady state assumption, the surface coverage θ_i does not depend on time, i.e. $\partial \theta_i/\partial t = 0$. By definition,

$$\frac{\partial \theta_i}{\partial t} = \frac{\dot{s}_i}{\Gamma} \quad (i = N_g + 1, \ldots, N_g + N_s). \tag{18}$$

Hence, we have

$$\dot{s}_i = 0 \quad (i = N_g + 1, \ldots, N_g + N_s). \tag{19}$$

Equations (17)–(19) stand for the reactions at the catalytic surface, which play a crucial role in the whole physical-chemical process.

The solution of the boundary equation is crucial since it strongly influences the solution of the whole PDE system. Later in this paper we discuss how we compute consistent boundary values.

By (7), (13), (15), and (17), these equations are highly nonlinear with respect to the unknowns Y_k and θ_i. This fact causes major difficulties in the numerical treatment and makes an essential difference between our problem and the one without catalytic surface.

Note that the boundary conditions of our PDEs are not standard ones, such as *Dirichlet conditions* where the dependent variables are explicitly specified at the boundary or *Neumann conditions* where the first-order derivatives of dependent variables are known, but they are given as the algebraic equations (17) and (19).

Using subscript for denoting partial derivatives and the abbreviations

$$\mathcal{E} = \begin{bmatrix} \rho u u_z + p_z \\ 0 \\ \rho u c_p T_z \\ 0 \\ \rho u Y_{1z} \\ \vdots \\ \rho u Y_{N_g z} \end{bmatrix}, \quad \mathcal{F} = \begin{bmatrix} \rho u (\rho u \mu r^2 u_\psi)_\psi \\ p_\psi \\ \rho u (\rho u \lambda r^2 T_\psi)_\psi - \sum_{k=1}^{N_g} \dot{\omega}_k W_k h_k - \rho u r \sum_{k=1}^{N_g} J_{kr} c_{pk} T_\psi \\ \dfrac{\partial r^2}{\partial \psi} - \dfrac{2}{\rho u} \\ \dot{\omega}_1 W_1 - \rho u (r J_{1,r})_\psi \\ \vdots \\ \dot{\omega}_{N_g} W_{N_g} - \rho u (r J_{N_g,r})_\psi \end{bmatrix}$$

equations (1), (2), (3), (4), and (6) can be summarized to the system

$$\mathcal{E} = \mathcal{F}, \tag{20}$$

which forms, along with (5) and (17)–(19), our entire mathematical model. Next, we discuss how to solve this problem numerically.

2.3 Simulation Method

Semi-discretization

We choose the approach of semi-discretization of the PDE system in the direction ψ by the *method of lines* [14]. Here, we have two independent spatial variables z and ψ, but do not have the independent variable "time". The axial direction z is now treated as the time-like direction.

The considered interval of ψ is discretized by ψ_i, $i = 1, ..., N$. Let us denote the function section corresponding to $\psi = \psi_i$ by the subscript i. For instance,

$$u_i := u_i(z) := u(z, \psi_i).$$

This rule is also applied to partial derivatives, e.g.,

$$u_{\psi_i} := u_{\psi_i}(z) := \left.\frac{\partial u(z,\psi)}{\partial \psi}\right|_{\psi=\psi_i}$$

and other quantities, such as temperature T, pressure p, radial coordinate r, and mass fraction Y_k.

Let $\mathcal{A} = (A_{j,k})$ be the matrix defined by

$$A_{j,k} = \begin{cases} \rho u, & \text{if } j = k = 1 \text{ and } 5 \leq j = k \leq N_g + 4 \\ 1, & \text{if } j = 1, k = 2 \\ \rho u c_p, & \text{if } j = k = 3 \\ 0, & \text{otherwise.} \end{cases}$$

and let

$$\mathcal{Q} = \left[u, p, T, r, Y_1, Y_2, \ldots, Y_{N_g}\right].$$

Then we have

$$\mathcal{E} = \mathcal{A}\, \mathcal{Q}_z^T. \tag{21}$$

By our convention,

$$\mathcal{E}_i = \mathcal{E}|_{\psi=\psi_i},\ \mathcal{A}_i = \mathcal{A}|_{\psi=\psi_i},\ \mathcal{Q}_i = \mathcal{Q}|_{\psi=\psi_i},\ \mathcal{Q}_{zi} = \mathcal{Q}_z|_{\psi=\psi_i}.$$

With

$$\begin{aligned} E &= \left[\mathcal{E}_1^T, \mathcal{E}_2^T, \ldots, \mathcal{E}_{N-1}^T\right]^T, \\ A &= \operatorname{diag}(\mathcal{A}_i), \\ Q &= [\mathcal{Q}_1, \mathcal{Q}_2, \ldots, \mathcal{Q}_{N-1}], \\ Q_z &= [\mathcal{Q}_{z1}, \mathcal{Q}_{z2}, \ldots, \mathcal{Q}_{zN-1}], \end{aligned}$$

(21) implies

$$E = AQ_z^T,$$

which is the discretization result of the left-hand side of equation (20).

$\mathcal{A}_i$ depends on ρ_i, u_i, and c_{p_i}, which in turn depend only on $\mathcal{Q}_i$. Note that $\mathcal{A}_i$, $i = 1, \ldots, N-1$, are band matrices with upper bandwidth equal to 1 and lower bandwidth equal to 0. Therefore, A inherits this property, too.

We use the forward finite difference to approximate p_ψ:

$$p_{\psi_i} = \frac{p_{i+1} - p_i}{\psi_{i+1} - \psi_i}. \tag{22}$$

Central finite differences are applied to the following partial derivatives with respect to ψ:

$$u_\psi,\ T_\psi,\ (\rho u \mu r^2 u_\psi)_\psi,\ (\rho u \lambda r^2 T_\psi)_\psi,\ (r J_{kr})_\psi,\ \text{and } X_{k\psi}. \tag{23}$$

(Note that $X_{k\psi}$ does appear in J_{kr} as given in (7).)

The fourth component of $\mathcal{F}$ is discretized by the trapezoidal rule:

$$\left[\frac{\partial r^2}{\partial \psi} - \frac{2}{\rho u}\right]_i = \frac{r_i^2 - r_{i-1}^2}{\psi_i - \psi_{i-1}} - \frac{4}{\rho_i u_i + \rho_{i-1} u_{i-1}}. \qquad (24)$$

In addition, we have the boundary condition $r_1 = 0$.

Let F_i denote the semi-discretized form of $\mathcal{F}_i = \mathcal{F}|_{\psi=\psi_i}$ by using the approximation scheme described in (22), (23), and (24). Then

$$F = [F_1^T, F_2^T, \ldots, F_{N-1}^T]^T$$

is the discretization result of the right-hand side of equation (20). Note that $F = F(Q)$. Due to the central finite difference scheme, F_i depends on the values at three points ψ_{i-1}, ψ_i, and ψ_{i+1}, i.e., $F_i = F_i(Q_{i-1}, Q_i, Q_{i+1})$.

Hence, the PDE system (20) corresponds to

$$A(Q)Q_z^T = F(Q). \qquad (25)$$

With

$$P = \begin{cases} \dot{s}_k W_k + J_{kr}|_{\psi=\psi_N}, & \text{if } 1 \leq k \leq N_g \\ \dot{s}_k, & \text{if } N_g + 1 \leq k \leq N_g + N_s \end{cases}$$

the boundary conditions (17)–(19) can be written as

$$0 = P. \qquad (26)$$

Initial conditions are

$$u = u_0, \ p = p_0, \ T = T_0, \ Y_k = Y_{k0} \ (k = 1, \ldots, N_g) \text{ at } z = 0. \qquad (27)$$

At the channel wall, u, T, p, and r must fulfill

$$0 = \begin{bmatrix} u \\ p \\ T \\ r \end{bmatrix}_{\psi=\psi_N} - \begin{bmatrix} 0 \\ p_{N-1} \\ T_{\text{wall}} \\ r_{\max} \end{bmatrix}. \qquad (28)$$

Finally, the equations (25)-(28) together form the Differential Algebraic Equation (DAE) system with the unknowns

$$[Q_1, Q_2, \ldots, Q_N, \theta_1, \ldots, \theta_{N_s}],$$

which satisfy, in addition, conditions (12) and (16).

Note that $\theta_1, \ldots, \theta_{N_s}$ appear in the formulas to compute $\dot{s}_k$.

It is worth to say that the partial derivatives of the left- and right-hand side of the DAE with respect to the unknowns and the iteration matrix are of band structure, with total bandwidth $3 \times \dim(Q_i)$. The mentioned band structure arises from choosing suitable indices for Q and F. It is used for efficient computation and storage of derivatives and iteration matrix.

Solution of the DAE system

Since the DAE system is derived from the discretization of a PDE and it contains the model of chemical reaction kinetics, it is stiff. The DAE is of index 1. Therefore we choose an implicit integration method, based on Backward Differentiation Formulas (BDF) for the solution of the initial value problems. For the practial computation, based on the code DAESOL [1, 2], we develop a new code that allows us to solve this problem. Features of this code are variable step size and variable order controlled by error estimation, modified Newton's method for the solution of the implicit nonlinear problems, a monitor strategy to control the computation and decomposition of the Jacobian and Internal Numerical Differentiation for the computation of derivatives of the solution w.r.t. initial values and parameters.

In our problems the linear systems arising in Newton's method are very ill-conditioned. We have developed an appropriate scaling. The variables are scaled with the same weighting vector that is used in the BDF error estimation and then we perform row equilibration. The scaling factors are chosen to be integer powers of the machine base in order to avoid scaling roundoff errors. Using this technique the condition number of the linear system is reduced from more than 10^{18} to around 10^7.

The BDF method needs derivatives of the DAE model functions. Here, we exploit the band structure. Instead of computation of the full Jacobian, we apply a compression technique which only requires few directional derivatives. We use Automatic Differentiation (implemented in the tool ADIFOR [4]) which allows us to compute derivatives with accuracy up to machine precision. This is crucial for a fast performance of the overall solution method.

To solve the DAE, we supply consistent initial values for the algebraic variables. During the integration the consistency is preserved because the algebraic constraints and the equation from the implicit integration scheme are solved simultaneously.

Computation of consistent initial values of the DAE

To integrate the DAE system (25)–(28), a set of consistent initial values is needed. Some of them are explicitely given, as stated in section 2.2. But the mass fractions Y_k ($k = 1, \ldots, N_g$) at the catalytic wall $\psi = \psi_N$ and the surface coverage fractions θ_i ($i = 1, \ldots, N_s$) are only implicitly determined by the nonlinear equations at the boundary (26) and the constraints (12) and (16).

These equations are highly nonlinear due to the Arrhenius kinetics. The solution is the steady state of a dynamic system of the surface process. The steady state is an asymptotic limit of the corresponding transient system. We only know initial values of the transient system, which can change very drastically until the system goes to steady state. It is very difficult to find values that are sufficiently close to the consistent values in order to have convergence of a Newton type method. Techniques of globalization of the convergence often fail because non-singularity of the Jacobian cannot be guaranteed. Therefore, we use a time-stepping method for solving the corresponding transient system to find an initial guess close to the solution and then apply Newton's method to converge to the solution.

With the variables $(Y_{1N}, \ldots, Y_{N_gN}, \theta_1, \ldots, \theta_{N_s}) \in \mathbb{R}^{N_g+N_s}$ the nonlinear equation system for the boundary is

$$\dot{s}_k W_k + J_{kr}|_{\psi=\psi_N} = 0 \quad (k = 1, \ldots, N_g) \tag{29}$$

$$\dot{s}_k = 0 \quad (k = N_g + 1, \ldots, N_g + N_s). \tag{30}$$

The left-hand sides of equation (30) $\dot{s}_k$, $(k = N_g+1, \ldots, N_g+N_s)$ are the rates of creation/depletion of the surface coverage of the surface species multiplied by the site density Γ:

$$\frac{\partial \theta_i}{\partial t} = \frac{\dot{s}_k}{\Gamma} \quad (k = N_g + 1, \ldots, N_g + N_s). \tag{31}$$

Similarly, the left-hand side of equation (29) can be considered as the mass rate of creation/depletion of the kth gas species by surface reactions and diffusion process multiplied by a some length dr, i.e.,

$$\rho dr \frac{\partial Y_k}{\partial t} = \dot{s}_k W_k + J_{kr}|_{\psi=\psi_N} \quad (k = 1, \ldots, N_g). \tag{32}$$

The differential equations (31) and (32) describe the corresponding transient state model for the nonlinear equations (29) and (30).

Starting from initial values for mass fractions and surface coverage at the beginning $(Y_{1N}, \ldots, Y_{N_gN}, \theta_1, \ldots, \theta_{N_s})(t_0)$ we integrate the ODE (31)-(32) until it nearly reaches steady state. In our implementation, we monitor the value of $\|P\|$, when it decreases below a certain value then we switch to Newton's method. >From our practical experience, this method is quite stable even for ill-conditioned problems.

The system (31)–(32) describes a chemical process modeled using detailed chemistry and therefore is very stiff. For solution, we also use the BDF method implemented in the new DAESOL.

To speed up the computation, we only integrate until we are near steady state and then use Newton's method for fast convergence. Therefore, we do not need high tolerance for the ODE integration which makes the integration procedure fast.

It is interesting to note that the conditions (12) and (16) are satisfied during the integration of ODE if they are fulfilled at the initial and the corresponding ODE model has a physical meaning.

2.4 Simulation Results

The presented methods have been implemented in the software package BLAYER. The software has been developed for general reaction schemes and can be applied for any arbitrary process in a catalytic monolith with detailed gas-phase and surface chemistry.

The user of the software has to provide the reaction mechanism and thermodynamic data, the initial and boundary values, i.e., velocity, gas temperature, pressure and mass or mole fractions of the gas species at the inlet and the surface temperature at the wall. Output of the software are the trajectories of the state variables in Tecplot format.

In the following we present simulation results for the process of conversion of ethane to ethylene. The chemical reactions are modeled by detailed elementary-step reaction mechanisms featuring 261 gas-phase [11] and 82 surface reactions [17] involving 25 gas-phase and 20 surface species. This leads to 29 PDEs.

The channel has the length $z_{max} = 0.01$ [m] and the radius $r_{max} = 2.5 \times 10^{-4}$ [m].

At inlet the mass fractions are $X_{C_2H_6} = 0.44$, $X_{O_2} = 0.26$ and $X_{N_2} = 0.30$. All other gases do not occur at the inlet. The inlet gas temperature is $T_{gas} = 650$ [K] and surface temperature is $T_{wall} = 930$ [K]. The inlet velocity has the value $u_0 = 0.5$ [m/s].

For the simulation, we use a spatial grid with 20 grid points. Figure 2 shows the trajectories of selected species.

The computational time for one simulation run is 30 seconds on a 2.5 GHz Pentium 4 Linux PC.

2.5 Chemical Interpretation of the Results

Figure 2 illustrates typical trajectories for the species concentrations during the oxidative dehydrogenation of ethane to ethylene. The oxygen is consumed completely within the first millimeter of the reactor. Ethane is combusted completely, giving mainly CO_2 and water. The gradients in the concentration profiles for these species indicate that catalytic surface reactions dominate the oxidation.

The dehydrogenation of ethane is a much slower process. The gradients in the species concentrations vanish. Therefore the process is kinetically controlled. This can be achieved by either gas-phase reactions or by slow surface reactions. The continuous formation of CO is due to the water-gas-shift reaction.

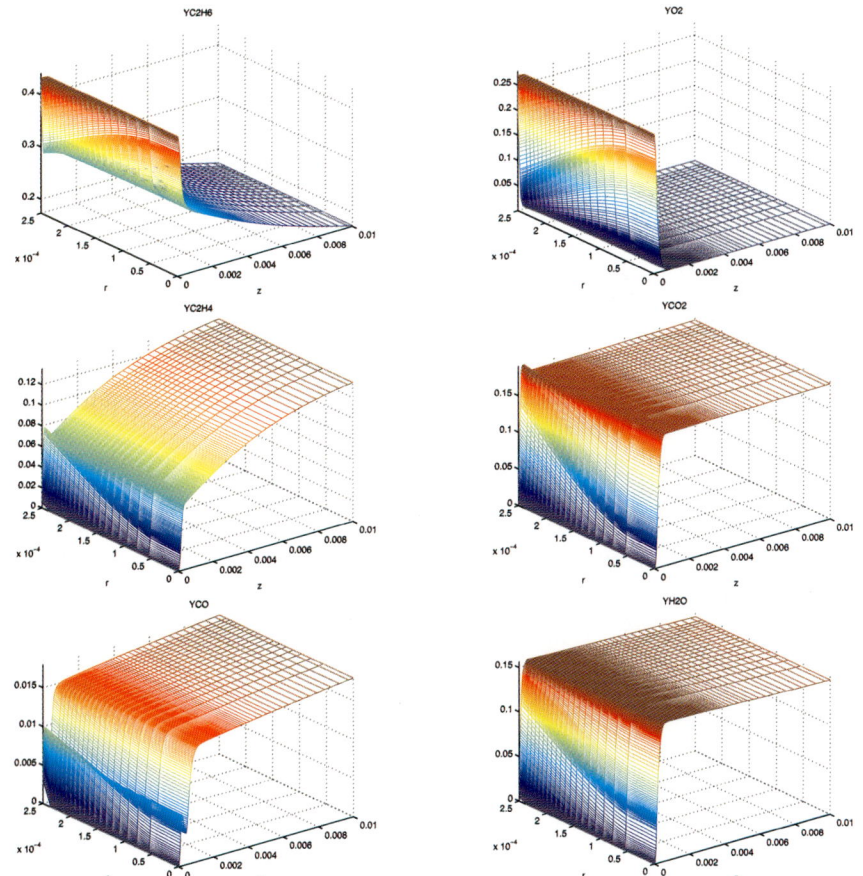

Fig. 2. Simulation results for the conversion of ethane to ethylene, trajectories of ethane, oxygen, ethylene, carbon dioxide, carbon monoxide and water.

3 Optimization

3.1 Formulation of an Optimal Control Problem

In the catalytic combustion process, the initial and boundary conditions can be used to optimize the performance of the reactor, e.g., maximize the gas conversion or maximize the selectivity. In particular, at the inlet of the catalytic monolith, the mass or molar fractions of the species, or the initial velocity $u(r,z)|_{z=0}$, or the initial temperature $T(r,z)|_{z=0}$ can be changed resp. be optimized, and at the catalytic wall, the temperature profile $T_{\text{wall}}(z)$ can be controlled. Moreover, the length of the catalytic tube $z_{\max}$ can be optimized.

In a general formulation, this optimization problem can be stated as

$$\min_{w,q} \varphi(w) \tag{33}$$

subject to PDE Model Equations(w, q) (34)

Initial and Boundary Conditions(w, q) (35)

State and Control Constraints(w, q) (36)

where the PDE equations describe the fluid dynamical process (1)-(5) and the gas-phase (9)-(10) chemistry. The initial and boundary conditions are described in (17)–(19) and contain in particular the surface chemistry (13)-(16). w denotes the state vector $w = (u, p, T, r, Y_1, Y_2, \ldots, Y_{N_g}, \theta_1, \ldots, \theta_{N_s})$ and q are the controls.

For practical reasons, there are often equality and inequality constraints on the control and state variables, such as an upper and lower bounds for the wall temperature, or sum of all mass fractions must be one.

Again, as described above, we semi-discretize the PDE using the method of lines on the grid ψ_i, $i = 1, \ldots, N$. This transforms the optimal control problem in a PDE (33)-(36) to an optimal control problem in a DAE which, using the notation of section 2.3, can be written as

$$\min_{x,q} \Phi(x, q) \tag{37}$$

subject to $A(Q)Q_z^T = F(Q)$ (38)

$0 = s_k W_k - J_{kr}|_{\psi=\psi_N}$, if $1 \leq k \leq N_g$ (39)

$0 = \dot{s}_k$, if $N_g + 1 \leq k \leq N_g + N_s$ (40)

$u = u_0$, $p = p_0$, $T = T_0$ at $z = 0$ (41)

$Y_k = Y_{k0}$ ($k = 1, \ldots, N_g$) at $z = 0$ (42)

$$0 = \begin{bmatrix} u \\ p \\ T \\ r \end{bmatrix}_{\psi=\psi_N} - \begin{bmatrix} 0 \\ p_{N-1} \\ T_{\text{wall}} \\ r_{\max} \end{bmatrix} \tag{43}$$

State and Control Constraints(x, q) (44)

The vector of state variables is

$$x = [Q_1, Q_2, \ldots, Q_N, \theta_1, \ldots, \theta_{N_s}],$$

and the control variable is

$$q = T_{\text{wall}}(z).$$

3.2 Direct Method

To transform the optimal control problem (37)-(44) to a finite dimensional optimization problem, we apply the direct approach, this means we parameterize the control functions by a finite number of degrees of freedom.

Parameterization of the control functions

The temperature profile at the wall $T_{\text{wall}}(z)$ is treated as control function in the optimal control problem.

Control functions are discretized on an appropriate user-defined grid using a suitable finite functional basis. Usually, the controls are approximated by piecewise continuous functions, e.g., piecewise constant or piecewise linear but also other schemes are applicable. The coefficients in these schemes will be control parameters replacing the control functions. By this way, the control function in infinite-dimensional space is approximated by its piecewise representation using in a finite-dimensional space. If the piecewise linear approximation is applied, then, e.g.,

$$T_{\text{wall}}(z) = T_{\text{wall}}(z_j) + (T_{\text{wall}}(z_{j+1}) - T_{\text{wall}}(z_j))\frac{z - z_j}{z_{j+1} - z_j}.$$

Note that in this case the bounds on the controls are transformed to bounds on the parameterization coefficients.

Representation of the DAE solution

For given initial values and control parameters, we solve the DAE initial value problem (38)-(43). This yields a representation of the DAE solution which we use for the evaluation of the objective function and the constraints.

3.3 Nonlinear Optimization Problem

The parameterization of the controls and the states turns the dynamic optimization problem into a finite dimensional optimization problem. It is a nonlinear constrained optimization problem which can be written in abbreviated form as

$$\begin{aligned}\min_{v} \quad & F(v) \\ \text{subject to} \quad & E(v) = 0 \\ & G(v) \geq 0\end{aligned} \qquad (45)$$

The time-independent control variables q and the control parameters introduced by the parameterization of the control functions are the optimization variables v in the NLP.

3.4 Optimization methods

To solve constrained nonlinear optimization problems, the method of Sequential Quadratic Programming (SQP) is the most efficient available method. It consists of the solution of a sequence of quadratic optimization problems and can be regarded as a Newton-like method for the optimality conditions of the problem (45). We use the implementation SNOPT [9] which employs BFGS updates for the approximation of the Hessian and an Active-Set strategy for the treatment of the inequalities.

As discussed above, a solution of the semi-discretized PDE only makes sense if the algebraic equations (the boundary condition of the PDE) are consistent. As a consequence, our optimization follows the so-called sequential approach solving the algebraic constraints in every iteration. Fortunately, in our case this is not time consuming and the computing time for consistenency calculations is negligible compared to the solution time for the whole discretized PDE.

Computation of derivatives

For the application of the SQP method, derivatives of the objective function and the constraints have to be provided. In our case, this is some what intricate because these functions are implicitly defined on the solution of the DAE system derived from the semi-discretization of the PDE.

The derivatives of the DAE solution w.r.t. the control parameters are solution of a variational DAE which, in principle, can be derived by differentiation of the DAE w.r.t. the control parameters. Because discretization of the DAE by a BDF scheme and differentiation commute, we can solve the variational DAE by differentiating the BDF scheme where we freeze all adaptive components as step size and order control and monitor strategy. The step size control is computed from an error estimator for the system consisting of the nominal and variational DAE. Nominal and variational DAE use the same iteration matrix to compute the BDF step. This approach of Internal Numerical Differentiation was introduced by Bock [5] and is implemented in our new BDF code DAESOL.

We apply the automatic differentiation tool ADIFOR (see [4] for more details) to generate Fortran codes for the required derivatives of the model functions of the DAE system. As we described in the previous sections, for our problems, the Jacobian matrices of the model functions are banded-structured. We exploit this structure by seed matrix compression according Curtis, Powell and Reid [7]. This reduces the number of directional derivatives to the total bandwidth of the matrix.

3.5 Optimization Results

For our optimization case study we keep the initial values at the inlet fixed: $X_{C_2H_6} = 0.44$, $X_{O_2} = 0.26$, $X_{N_2} = 0.30$, the inlet gas temperature $T_{gas} = 650$ [K], the inlet velocity $u_0 = 0.5$ [m/s].

The wall temperature profile is optimized. We use a piecewise linear parameterization with 8 intervals. Objective is to maximize the mass fraction of

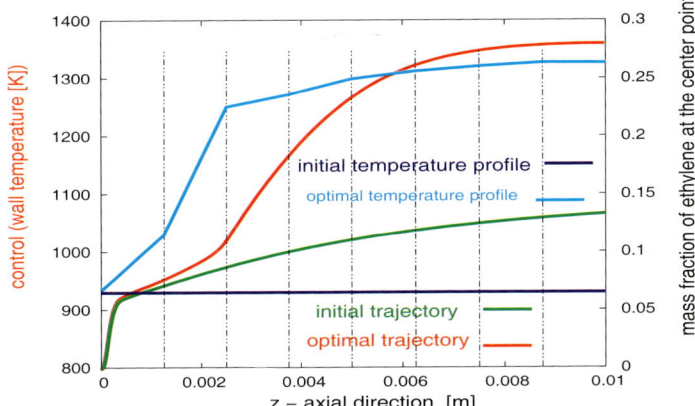

Fig. 3. Initial and optimal temperature profile and initial and optimal trajectory of ethylene at the centerline.

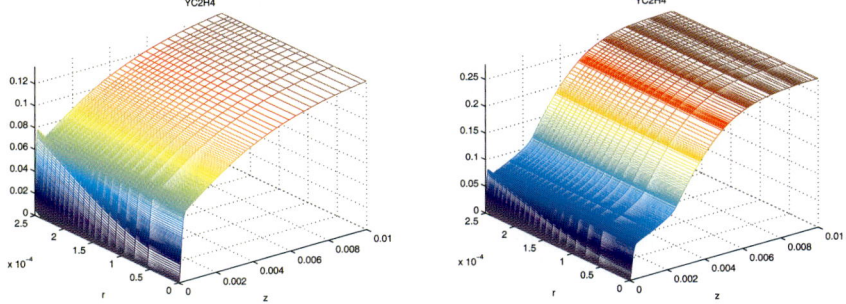

Fig. 4. Initial and optimal profile of ethylene.

ethylene at the outlet. As constraint the temperature is required to be between 600 [K] and 1500 [K].

The optimization was started with a constant temperature profile of 930 [K] leading to an objective value of 0.132. The optimization run took 30 min computational time on a 2.5 GHz Pentium 4 Linux PC. In the optimal solution the objective value is 0.280. Figure 3 shows the temperature profile and figure 4 the mass fraction of ethylene before and after optimization.

The work presented in this paper is the first realization of optimization for this problem.

3.6 Chemical Interpretation of the Results

The results show that temperatures around 1300 K give maximum yield in the ethylene production. At inlet the temperatures only need to be sufficiently high enough for ignition of the combustion to occur. An autothermal reactor – where the temperature is only controlled by the exothermic reaction – should therefore maintain a temperature around 1300 K. This is nearly the same temperature as observed in experiments [10]. The optimal oxygen content can be determined by the amount of heat necessary to maintain this temperature.

4 Conclusions

This project continuous work on conversion of natural gas done previously in the SFB: In [15] the homogeneous oxidative coupling of methane to achive ethylene was investigated. By optimization of temperature and residence time the concentration of ethylene could be more than doubled [15].

In this project, we have developed simulation and optimization software packages for investigating the chemically reacting flow in a catalytic channel. Using the boundary-layer approximation, we model the process by a system of parabolic partial differential equations (PDEs), with nonlinear boundary conditions for coupling gas-phase and surface chemistry. The PDEs are semi-discretized by using the method of lines which leads to a large-scale stiff differential algebraic equations (DAEs). The solution of the DAEs requires a set of consistent initial values. A combination of a time-stepping and of Newton's method is employed for finding consistent initial values at the boundary. Based on the BDF code DAESOL, we develop a new code which allows us to solve the DAEs efficiently. Given the inlet conditions and wall temperature profile as parameters, the reactive flow field can be computed very fast, below one minute, even for very complex reaction mechanisms.

The fast simulation of single channel flow fields lets optimization procedures become applicable. Based on the simulation code, we develop the optimization program using the direct approach with an SQP method. The SQP method requires derivatives, which are efficiently computed by the Internal Numerical Differentiation technique.

The program has been applied to the problem of catalyst-supported dehydrogenation of ethane to ethylene. The results are in good agreement with experimental observations. The numerical simulation of the single channel flow field gives detailed insight into the catalytic and gas-phase processes occurring. By this, the program provides a useful tool for the validation of reaction mechanisms.

By the extension to optimization problems the simulation tool has gained a new qualitative level. For the first time, systematic parameter optimization for a reactive tubular flow with detailed chemistry becomes possible. In our case of ethane to ethylene conversion, the temperature profile of the reactor wall was subject to optimization. The ethylene yield could be more than doubled by optimization. The maximum yield was achieved for temperatures around 1300 [K].

Besides the wall temperature, other parameters – such as inlet composition or catalyst distribution along the channel – will be of interest. The method described offers excellent opportunities for further studies of optimization problems.

References

1. I. Bauer, H. G. Bock, and J. P. Schlöder. DAESOL – a BDF-code for the numerical solution of differential algebraic equations. Technical report, IWR, University of Heidelberg, 1999. SFB 359.
2. I. Bauer, F. Finocchi, W. Duschl, H. Gail, and J. Schlöder. Simulation of chemical reactions and dust destruction in protoplanetary accretion disks. *Astronomy & Astrophys.*, 317:273–289, 1997.
3. A. Beretta, E. Ranzi, and P. Forzatti. Oxidative dehydrogenation of light paraffins in novel short contact time reactors. Experimental and theoretical investigation. *Chemical Engineering Science*, 56(3):779–787, 2001.
4. C. Bischof, A. Carle, P. Hovland, P. Khademi, and A. Mauer. *ADIFOR 2.0 User's Guide*, 1995.
5. H. G. Bock. Numerical treatment of inverse problems in chemical reaction kinetics. In K. H. Ebert, P. Deuflhard, and W. Jäger, editors, *Modelling of Chemical Reaction System*, volume 18 of *Chemical Physics 18*, pages 102–125. Springer, Heidelberg, 1981.
6. M. E. Coltrin, R. J. Kee, and J. A. Miller. A mathematical model of the coupled fluid mechanics and chemical kinetics in a chemical vapor deposition reactor. *J. Electronchem. Soc.*, 131(2):425–434, 1984.
7. A. R. Curtis, M. J. D. Powell, and J. K. Reid. On the estimation of sparse Jacobian matrices. *J. Inst. Math. Appl.*, 13:117–119, 1974.
8. O. Deutschmann. *DETCHEM - User manual, version 1.4*. IWR, University of Heidelberg, 2000.
9. P. E. Gill, W. Murray, and M. A. Saunders. SNOPT: An SQP algorithm for large-scale constrained optimization. *SIAM J. Opt.*, 12:979–1006, 2002.
10. M. Huff and L. D. Schmidt. Ethylene formation by oxidative dehydrogenation of ethane over monoliths at very short contact times. *Journal of Physical Chemistry*, 97(45):11815, 1993.
11. V. Karbach. *Validierung eines detaillierten Reaktionsmechanismus zur Oxidation von Kohlenwasserstoffen bei hohen Temperaturen*. Diplomarbeit, Universität Heidelberg, 1997.
12. R. J. Kee, M. E. Coltrin, and P. Glarborg. *Chemically Reacting Flow: Theory and Practice*. Willey, 2003.
13. L. L. Raja, R. J. Kee, O. Deutschmann, J. Warnatz, and L. D. Schmidt. A critical evaluation of Navier-Stokes, boundary-layer, and plug-flow models of the flow and chemistry in a catalytic-combustion monolith. *Catalysis Today*, 59:47–60, 2000.
14. W. E. Schiesser. *The Numerical Method of Lines: Integration of Partial Differential Equations*. Academic Press, San Diego, CA, 1991.
15. M. . v. Schwerin, O. Deutschmann, and V. Schulz. Process optimization of reactives systems by partial reduced SQP methods. *Computers and Chemical Engineering*, 24:89–97, 2000.
16. J. Warnatz, R. Dibble, and U. Maas. *Combustion, Physical and Chemcial Fundamentals, Modeling and Simulation, Experiments, Pollutant Formation*. Springer-Verlag, New York, 1996.

17. D. K. Zerkle, M. D. Allendorf, M. Wolf, and O. Deutschman. Understanding homogeneous and heterogeneous contributions to the partial oxidation of ethane in a short contact time reactor. *J. Catal.*, 196:18–39, 2000.

Reaction Processes on Catalytically Active Surfaces*

O.R. Inderwildi[1], D. Starukhin[2], H.-R. Volpp[2], D. Lebiedz[3], O. Deutschmann[4], and J. Warnatz[3]

[1] Department of Chemistry, University of Cambridge.
[2] Physikalisch-Chemisches Institut (PCI), Universität Heidelberg.
[3] Interdisziplinäres Zentrum für Wissenschaftliches Rechnen (IWR), Universität Heidelberg.
[4] Institut für Technische Chemie und Polymerchemie, Universität Karlsruhe (TH).

Summary. In this article results of studies are reported in which *in-situ* infrared-visible sum-frequency generation (IR-VIS SFG) surface vibrational spectroscopy was combined with density functional theory (DFT) ab initio computations to investigate the adsorption of carbon monoxide (CO) on rhodium (Rh) catalyst surfaces at elevated substrate temperatures ($T_s \geq 300K$) over a wide pressure-range ($p_{CO} = 10^{-8} - 1000$ mbar). The experimental studies demonstrated the reversible molecular adsorption of CO up to a pressure of ca. 10 mbar. For higher CO pressures, however, the onset of a new irreversible dissociative CO adsorption pathway could be observed already at a substrate temperature of $T_s = 300$ K. CO dissociation was found to result in the formation of carbon on the surface as the only detectable dissociation product, indicating that CO dissociation occurs via the Boudouard reaction: $2CO \rightarrow C(ad) + CO_2$. To rationalize the experimental findings DFT studies were performed the results of which suggest the possibility of a novel low-temperature/high-pressure CO/Rh(111) dissociation pathway where a gas-phase CO(g) molecule reacts with a CO molecule chemisorbed on the catalyst surface, CO(ad), to yield gas-phase $CO_2(g)$ and surface carbon C(ad) through an Eley-Rideal mechanism.

The contents of the article is as follows:
- Introduction
- Experimental and theoretical methods
- Pressure and temperature dependence of CO adsorption on rhodium
- Summary

1 Introduction

Heterogeneously catalyzed reaction steps contribute to approximately 90% of the production processes of industrial chemicals, fuels, and pharmaceuticals

---

*This work has been supported by the German Research Foundation (DFG) through SFB 359 (Project B3) at the University of Heidelberg.

(see e.g. [1]). Furthermore, heterogeneous catalysts play a key role in pollutant emission control processes such as the three-way catalyst (TWC) used for the after-treatment of automobile exhaust gases (see e.g. [2]). In the TWC, a catalyst containing Pt/Pd/Rh converts the two reducing pollutants, CO and unburned hydrocarbons (HC), as well as the oxidizing pollutant, NO, to H_2O, CO_2, and N_2 products [3]. However, because heterogeneous reactions sensitively depend on the surface concentrations of reactants and products that are connected with adsorption and desorption equilibria and with gas-phase transport processes, depending on the actual operating conditions, different elementary reaction steps can become rate-determining. As a consequence, the development of appropriate numerical models for the simulation of complex surface reaction systems is essential for the detailed understanding of heterogeneous catalysis under technical relevant conditions.

Simulation tools for the description of heterogeneous reaction systems have been developed in recent years (see e.g. [4]), which include detailed surface chemistry (by extending the concept of elementary reactions originally introduced by Bodenstein for gas-phase reaction systems [5] to heterogeneous catalysis) as well as detailed models for molecular multi-species transport. In contrast to gas-phase combustion systems (see e.g. [6]), however, only a few complete surface reaction mechanisms have been derived, which so far are mainly based on experimental studies of elementary surface reaction steps carried out under ultra-high vacuum (UHV) conditions and on well-defined single crystal surfaces (see e.g. [4] and references therein). The use of this kind of surface kinetics data in the modeling of technical processes, which – as schematically illustrated in Figure 1 – take place at high pressure (pressure gap) and on polycrystalline catalyst materials (materials gap), emphasizes the importance of the application of *in-situ* diagnostics techniques, such as the optical infrared-visible sum-frequency generation (IR-VIS SFG) method that allow for molecular level studies of adsorbed species under technically relevant operating conditions [7, 8].

In [9, 10], for example, IR-VIS SFG surface vibrational spectroscopy of adsorbed CO species was successfully combined with detailed numerical simulations and a sensitivity analysis to investigate the heterogeneous CO oxidation on polycrystalline Pt under laminar flow conditions at mbar reactant pressures. A detailed surface reaction mechanism was developed which takes into account the surface heterogeneity of the polycrystalline catalyst. With this model the experimental results could be successfully described indicating that for the CO/O_2/Pt-foil system UHV surface kinetics data obtained on well-defined Pt single crystals can reproduce the catalytic behavior of the more realistic polycrystalline catalyst for elevated CO reactant pressures as typically present in the exhaust gas of spark ignited engines. IR-VIS SFG surface vibrational spectroscopy could also be applied along with kinetic modeling calculations to investigate the pressure- and temperature-dependence of the NO adsorption on a Pt(111) single-crystal catalyst [11] as well as the heterogeneous oxidation of CO on Rh(111) [12]. In addition, as demonstrated by the

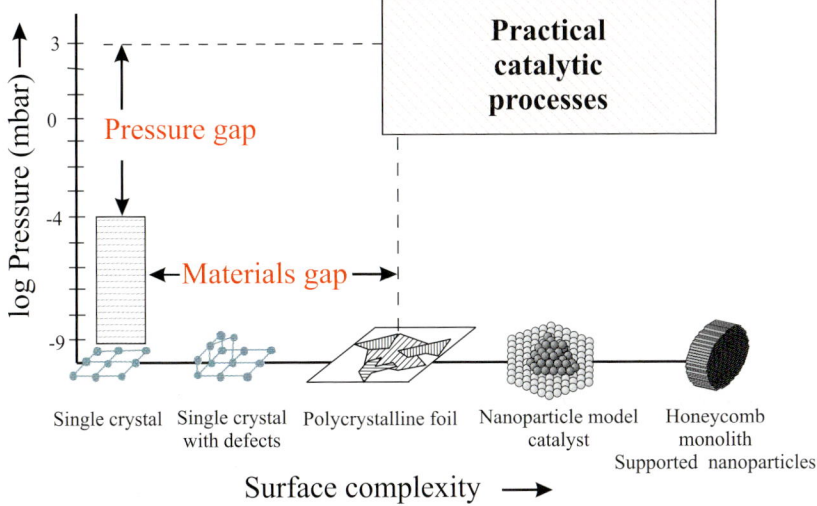

Fig. 1. Schematic illustration of the so-called pressure and materials gaps that separate ultra-high-vacuum (UHV) single-crystal model studies from technical catalytic reaction investigations.

pioneering studies of Freund and co-workers, IR-VIS SFG can also be utilized to investigate CO adsorption on alumina-supported Pd nanoparticle-model-catalysts [13].

On the theoretical side, due to recent computational advances, Density Functional Theory (DFT) has become the primary mathematical tool for the *ab initio* description of elementary surface reactions on the molecular level (see e.g. [14, 15, 16]). Today DFT calculations are on the way to become a quite useful mathematical tool for performing realistic simulations of complex systems like solid surfaces and adsorbates at surfaces as well as for the determination of surface reaction kinetics parameters such as coverage dependencies of adsorption and activation energies [17, 18], which are in some cases difficult to determine experimentally.

In the following results of studies will be presented where IR-VIS SFG surface vibrational spectroscopy was employed for *in situ* monitoring of CO adsorbed on Rh(111) single-crystal surfaces along with DFT *ab initio* calculations to investigate CO adsorbate structures formed under adsorption/desorption equilibrium conditions, as well as the possibility for a new dissociative CO adsorption mechanism at elevated pressures and temperatures. In addition, results of IR-VIS SFG studies of CO adsorption on a polycrystalline Rh-foil will be presented where CO adsorption was investigated over 12 orders of magnitude in CO pressure.

2 Experimental and Theoretical Methods

2.1 Infrared-Visible (IR-VIS) SFG Surface Vibrational Spectroscopy

Nonlinear optical infrared-visible (IR-VIS) sum frequency generation (SFG) allows one to measure vibrational spectra of molecular adsorbates from sub-monolayer coverages typically present under UHV conditions as well as vibrational spectra of dense adsorbate layers formed under adsorption/desorption equilibrium conditions at ambient pressure. As an optical method – unlike conventional surface spectroscopic methods in which beams of charged particles are used as a probe [19] – the application of SFG spectroscopy is not restricted to UHV conditions (see e.g. [20, 21]).

The present experiments were carried out in a reaction chamber, which allows surface spectroscopic studies over a wide pressure range from UHV conditions (base pressure 3×10^{-10} mbar) up to atmospheric pressure. A detailed description of the experimental setup can be found in Ref. [22].

The principles of SFG surface vibrational spectroscopy have been described in detail elsewhere (see e.g. [23] and references therein). As schematically depicted in Figure 2, SFG is a second order nonlinear optical process where a tunable infrared laser beam (ω_{IR}) is mixed with a visible (ω_{VIS}) laser beam to generate a sum frequency output ($\omega_{SFG} = \omega_{IR} + \omega_{VIS}$), which is reflected from the substrate, according to the phase-matching condition, $\omega_{SFG} \sin\theta_{SFG} = \omega_{VIS} \sin\theta_{VIS} + \omega_{IR} \sin\theta_{IR}$, at an angle θ_{SFG} with respect to the surface normal. Because in the electric dipole approximation this process is only allowed in a medium without centro-symmetry for the CO gas-phase/Rh-substrate system the SFG signal is highly specific to the interface region bounded by the two centro-symmetric media [24]. Hence, the SFG signal originates predominately from the gas-phase/surface interface region with no contribution from the isotropic gas phase.

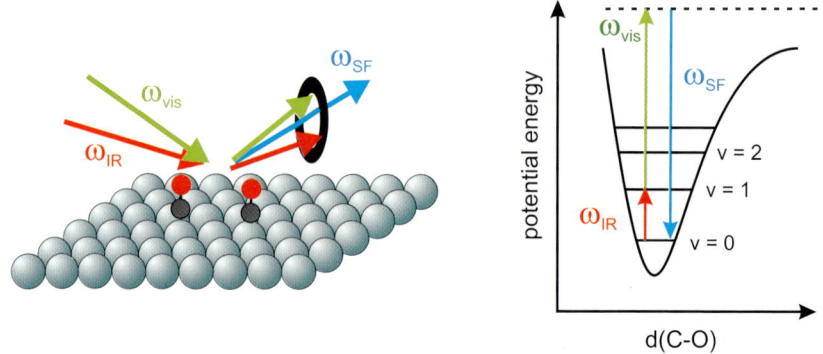

Fig. 2. Schematic illustration of the IR-VIS sum-frequency generation (SFG) process for surface vibrational spectroscopy of adsorbed molecules.

In the SFG experiment the visible beam is held at a fixed frequency while the IR beam is tuned over the vibrational frequency range of interest. When the IR beam is tuned through a surface species vibrational resonance the effective surface non-linear susceptibility $\chi_S^{(2)}$ is resonantly enhanced. Because the SFG intensity $I_{SFG}(\omega_{IR})$ is proportional to the absolute square of the effective surface nonlinear susceptibility [23],

$$I_{SFG}(\omega_{IR}) \propto |\chi_S^{(2)}|^2 = |\chi_R^{(2)} + \chi_{NR}^{(2)}|^2 \quad (1)$$

a vibrational spectrum of the adsorbed molecules can be measured. In the present study, SFG spectra were recorded by tuning the IR laser over the frequency region $\omega_{IR} = 1700\text{–}2100$ cm^{-1} in which C-O stretching vibrations of adsorbed CO molecules can be excited.

SFG spectra can be simulated using equation 1, together with the following additive expression for the vibrationally resonant contributions and the nonresonant contribution of $\chi_S^{(2)}$ [23, 25]:

$$\chi_S^{(2)} = \sum_n \frac{A_{R(n)}}{(\omega_{IR} - \omega_n + i\Gamma_n)} + A_{NR} \cdot e^{i\Phi} \quad (2)$$

A_{NR} represents the amplitude of the vibrationally nonresonant susceptibility $\chi_{NR}^{(2)}$, and Φ is its phase relative to the vibrational resonances that are characterized by their resonance frequencies ω_n and their homogeneous Lorentzian linewidths $2\Gamma_n$. $A_{R(n)}$ are the vibrationally resonant amplitudes which are proportional to the number density of adsorbed molecules and the infrared and Raman transition moment of the vibrational mode. As a consequence, if a vibrational mode is both Raman and infrared active, the vibrational resonant contribution becomes significant as the infrared laser is tuned through the vibrational resonance. Hence, a SFG spectrum provides similar information as can be obtained from conventional reflection absorption infrared spectroscopy (RAIRS) with the major difference that the SFG signal originates predominantly from the surface, with no contribution from the gas phase [23, 24]. In addition, the SFG signal output is coherent and highly directional.

In principle, SFG experiments can be carried out in different polarization combinations. However, because on metal surfaces the surface light field of a s-polarized IR laser beam is efficiently screened by the conduction electrons in IR-VIS SFG studies over metal surfaces, the IR laser beam is always p-polarized. Because of the lower dielectric constants of the metals in the VIS spectral region, in which the surface light fields are not as efficiently screened by the conduction electrons, the VIS beam can be both s- or p-polarized. Although the SFG signal obtained in *ssp* configuration (s-polarized SFG signal, s-polarized VIS and p-polarized IR laser beam) is in general considerably weaker than the SFG signal obtained in *ppp* configuration[11], the analysis of SFG spectra recorded for these two polarization configurations allows us to gain information about the surface orientation of the adsorbed molecules (*vide infra*).

2.2 Density Functional Theory Calculations

In density functional theory (DFT) the energy of a system of interacting electrons in an external potential is described as a functional of its electron density. The foundation of DFT is the Kohn-Hohenberg theorem, which states that the ground state energy of such a system is a functional of its ground state electron density and that this functional is unique. Unfortunately, this unique functional is not known, while, fortunately, approximations to this functional have been developed, which describe several physical as well as chemical systems extremely accurately (see [16, 17, 18] and references therein).

In the present work on CO adsorption on Rh(111) DFT calculations were performed using CASTEP (Cambridge Sequential Total Energy Package) [26] to determine CO adsorbate structures and to investigate the possibility for a low-temperature high-pressure CO dissociation mechanism on the Rh(111) surface. In the calculations the generalized gradient approximation (GGA) as proposed by Perdew and Wang [27] was applied in combination with Vanderbilt ultrasoft pseudopotentials [28]. Where necessary, spin effects were taken into account by using the spin polarized (dependent) GGA functional (GGSA). The plane wave basis set was truncated at a kinetic energy of 300 eV. Computations were performed over a range of k-points within the Brillouin zone as generated by the Monkhorst-Pack scheme [29]. The preciseness of the Monkhorst-Pack scheme was chosen so that the k-point spacing is similar for all investigated unit cells.

The catalyst surface was modeled as a rhodium slab with the thickness of 3 atomic layers. Periodic-boundary conditions extrapolate from a metal cluster to an extended surface. Various studies demonstrated that the thickness of 3 layers is sufficient to generate a model of a transition metal surface (see e.g. [14]. Location and extent of the elementary cell for the DFT calculations were chosen in a way to obtain the desired surface coverage. A 10 vacuum was placed in between the periodic slabs to ensure that adsorbate and subsequent slab do not interact for geometry optimization routine and 14 vacuum slab have been used for reaction transition state search procedure. The positions of the metal atoms were fixed in (111) surface configuration, while the positions of the CO surface adsorbates were fully mobile.

To determine the adsorption energies ΔE_{ad}, the Rh(111) surface was geometry optimized with the CO added on the one hand and without the CO molecule on the other hand; the energies of the optimized surfaces were calculated subsequent to the geometry optimization. The geometry of the CO was optimized within a supercell identical to the cell of the surface and the energy of this optimized CO (E_{CO}) was calculated subsequently. Finally the adsorption energies were determined according to equation (3).

$$\Delta E_{ad} = E_{slab+CO} - (E_{slab} + E_{CO}) \tag{3}$$

In cases in which co-adsorbates are present on the surface, the energy of the slab (E_{slab}) in equation (3) refers to the energy of the slab including the co-adsorbates.

Prior to calculating the surface reaction, the structures of the reactants and the products were relaxed. The transition state (TS) of the reaction was then determined on the potential energy surface (PES) by performing a linear synchronous, combined with a quadratic synchronous transit calculation and conjugate gradient refinements [30]. The total energies for the reactants, the transition state and the products were computed. Heats of reaction $\Delta_r H$ and activation energies E_{diss} were calculated via equation (4) and (5), respectively, where ZPE stand for the respective harmonic zero-point energy corrections.

$$\Delta_r H_{0K} = E_{reactants} + ZPE_{reactants} - E_{products} - ZPE_{products} \quad (4)$$

$$E_{diss} = E_{TS} - E_{reactants} \quad (5)$$

3 Pressure and Temperature Dependence of CO Adsorption on Rhodium

Besides its importance as an essential component of the TWC, rhodium represents one of the most important catalyst constituents in many industrial catalytic processes such as, for example, the CO methanation and the Fischer-Tropsch synthesis (see e.g. [31]). As a consequence, the adsorption of CO on Rh single crystal "model" catalysts has been investigated quite extensively since the first structural results of low energy electron diffraction (LEED) study of the CO adsorption on Rh(111) was published in 1970 [32]. The results of these studies, which were mainly carried out under UHV conditions, have provided detailed information about the variety of CO adsorbate structures which are formed as a function of CO coverage (see e.g. [33, 34, 35] and references therein). However, only a few studies were reported for the CO/Rh(111) system in which CO adsorption was investigated under high reactant pressure conditions [12, 21, 36, 37].

In the group of Somorjai, IR-VIS SFG [21, 36] and scanning tunneling microscopy (STM) [37] was employed for in situ studies of CO adsorption on Rh(111) at room temperature up to a CO pressure of 1000 mbar. The "high pressure" STM measurements revealed new CO adsorbate structures and the IR-VIS SFG studies indicated new C–O stretching vibrational features at CO pressures around 900 mbar. In [21], one of the new vibrational which could be identified at a frequency of about 2040 cm^{-1} was assigned to CO molecules terminally (ontop) adsorbed on Rh(111) defect sides which are formed by a pressure-induced displacive reconstruction (surface roughening) processes of the catalyst surface. It was further noted that the high-pressure vibrational feature disappeared after the CO pressure was reduced and it was suggested that the surface roughening is reversible [21]. In [36], however, it was reported that after exposing the sample to a CO pressure of about 900 mbar the intensity of the CO stretching vibrational spectra (recorded after pumping down the CO pressure to about 1 mbar) is considerably smaller than that of the spectrum recorded at the same pressure in the increasing pressure cycle. It was therefore proposed that the high CO pressure may cause some irreversible changes in the Rh(111) surfaces, which is stabilized by the adsorbed CO.

To shed more light on the actual mechanism responsible for the irreversibility observed in these high-pressure CO/Rh(111) adsorption studies, experiments were carried out in the present work in which IR-VIS SFG for *in situ* CO detection on Rh(111) and on a polycrystalline Rh-foil in the pressure range $p_{CO} = 10^{-8}$–1000 mbar was combined with "postreactive" Auger electron spectroscopy (AES) studies. Further experiments in which the temperature dependence of the CO adsorption was studied in the substrate temperature range $T_s = 300$–800 K were also performed.

3.1 CO Adsorption Studies on Rhodium from UHV to 1000 mbar

SFG spectra of CO adsorbed on the Rh(111) single-crystal substrate and on a polycrystalline Rh-foil sample were obtained in the CO pressure range $p_{CO} = 10^{-8}$–1000 mbar at a substrate temperature of 300 K under adsorption/desorption equilibrium conditions. Spectra were recorded in the *ppp* and *ssp* configuration (see e.g. [22]) by tuning the IR laser over the frequency region $\omega_{IR} = 1850$–2150 cm^{-1} in which stretching vibrations of adsorbed CO species can be excited.

Figure 3 depicts SFG spectra of CO adsorbed on a polycrystalline Rh-foil surface, which were recorded in the *ppp* configuration in an increasing CO pressure cycle in the range $p_{CO} = 10^{-8}$–1000 mbar. All spectra shown in Figure 3 were normalized to the actual intensity of the IR laser beam at the Rh substrate in order to account for the different absorption of the IR laser radiation by the CO gas phase. The solid lines in Figure 3 represent results of numerical simulations of the spectral line shapes.

A SFG spectrum, which was recorded after the CO pressure was pumped down again to $p_{CO} = 10^{-6}$ mbar, is also shown (red spectrum in Figure 3). Comparison of the SFG intensity of the latter spectrum with the intensity of the same spectrum obtained at the same CO pressure in the increasing pressure cycle clearly indicates that the high-pressure adsorption process is not reversible. After further evacuation of the cell down to about 10^{-9} mbar, Auger electron (AE) spectra were recorded which revealed the presence of carbon on the surface indicating that CO must have dissociated during the adsorption study (*vide infra*).

SFG spectra of CO adsorbed on Rh(111), which were recorded under the same "high pressure" conditions, have been reported previously in [12]. For

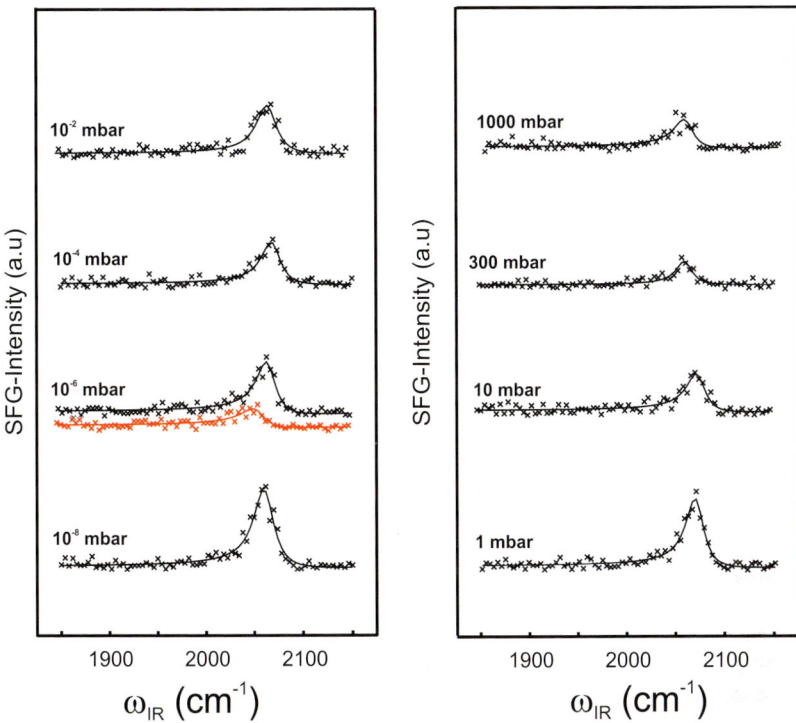

Fig. 3. SFG vibrational spectra of CO adsorbed on a polycrystalline Rh-foil catalyst at a substrate temperature $T_s = 300$ K in the pressure range of $p_{CO} = 10^{-8} - 1000$ mbar. The SFG signal intensity I_{SFG} is plotted versus the frequency ω_{IR} of the tunable IR laser. The depicted SFG spectra are on the same vertical scale. Crosses represent experimental data points, solid lines are results of a numerical simulation of the measured SFG line-shapes (details are given in the text). The spectra were recorded under thermal equilibrium conditions for the *ppp* polarization combination (*p*-polarized sum frequency signal, *p*-polarized visible and *p*-polarized IR laser beam). The SFG spectra shown in black were recorded in one increasing pressure cycle. The SFG spectrum shown in red was obtained after the pressure was reduced back to $p_{CO} = 10^{-6}$ mbar.

both adsorption systems the series of spectra are dominated by a single vibrational feature originating from ontop adsorbed CO molecules. In case of the CO/Rh(111) adsorption system additional vibrational features around a frequency of 1900 cm^{-1} were observed in the high-pressure region, which could be assigned to CO molecules bound to 3-fold coordinated hollow sites [33, 34]. The latter observation of both ontop and hollow-site CO on Rh(111) is consistent with the high-coverage Rh(111)+(2 × 2)-3CO ($\theta_{CO} = 0.75$ ML) adsorbate structure observed in previous "high pressure" STM studies [37]. One monolayer (ML) equals the number of rhodium atoms in the Rh(111) surface plane (1.60 × 10^{15} atoms/cm^2). However, as discussed in [12], due to the reduced sensitivity of SFG for the detection of multiple coordinated CO surface species the relative SFG signal intensities did not reflect the actual ontop *versus* hollow site CO coverage ratio of 1:2 of the Rh(111)+(2 × 2)-3CO (θ_{CO} = 0.75 ML) adsorbate structure. The reduced sensitivity of SFG for detecting multiple coordinated surface species is usually attributed to their smaller Raman polarizability compared to the ontop adsorbed species [38]. In addition, dynamic dipole-dipole coupling can result in signal intensity transfer from the low-frequency band of the 3-fold coordinated hollow CO to the higher-frequency band of the ontop bound CO (see e.g. [22] and references therein). In addition, as demonstrated in [34], dynamic dipole-dipole coupling is also responsible for the relatively strong ontop CO vibrational frequency blue shift observed for CO coverages on Rh(111) up to 0.5 ML, while the somewhat weaker blue shift of the onto CO frequency as observed for higher coverages (0.5 ML < θ_{CO} ≤ 0.75 ML) is due to chemical effects. The latter findings are in agreement with the results obtained in [12], where SFG measurements were combined with thermal desorption spectroscopic (TDS) studies to investigate the coverage dependence of the ontop CO stretching frequency.

The ontop CO stretching frequency values obtained from a line shape analysis of the ontop CO vibrational bands recorded in the CO/Rh(111) and CO/Rh-foil studies are plotted against the CO pressure in Figure 4a. The relative ontop CO adsorbate number densities $n_{CO-ontop}$ derived in the SFG experiments are depicted in Figure 4b. For comparison the pressure dependence of the CO equilibrium surface coverage θ_{CO} for the CO/Rh(111) system [12] is depicted in Figure 4c.

In case of the CO/Rh(111) system, Figure 4a shows that when the pressure is increased the ontop CO stretching vibrational frequency is blue shifted from a value of 2053 ± 2 cm^{-1} at p_{CO} = 10^{-8} mbar to value of 2075 ± 2 cm^{-1} for p_{CO} = 10^{-2} mbar. If the CO pressure is further increased, however, the frequency remains almost constant in the pressure range p_{CO} = 10^{-3}– 100 mbar indicating that due to the strong repulsive interaction between the adsorbate molecules the maximum CO coverage cannot be significantly altered after the saturation coverage of 0.75 ML has been reached.

In case of the CO/Rh-foil system the ontop CO stretching vibrational frequency is blue shifted from an initial value of 2062 ± 3 cm^{-1} at p_{CO} = 10^{-8} mbar to an average value of about 2072 cm^{-1} at pressures between p_{CO} = 10^{-2} mbar and p_{CO} = 10 mbar. The fact that at a pressure of 10^{-8} mbar the ontop CO frequency in the CO/Rh-foil system is substantially higher than in the CO/Rh(111) system indicates that the equilibrium CO surface coverage is higher on the Rh-foil. This could be confirmed by TPD studies, which showed that at a CO pressure of 10^{-8} mbar the CO surface coverage of the Rh-foil is actually about 2.5 times higher than the CO coverage of the Rh(111) single crystal. In addition, the CO/Rh-foil TPD spectra revealed considerably higher CO desorption temperatures indicating that CO molecules strongly adsorbed at defect sides of the polycrystalline Rh-foil surface are responsible for the

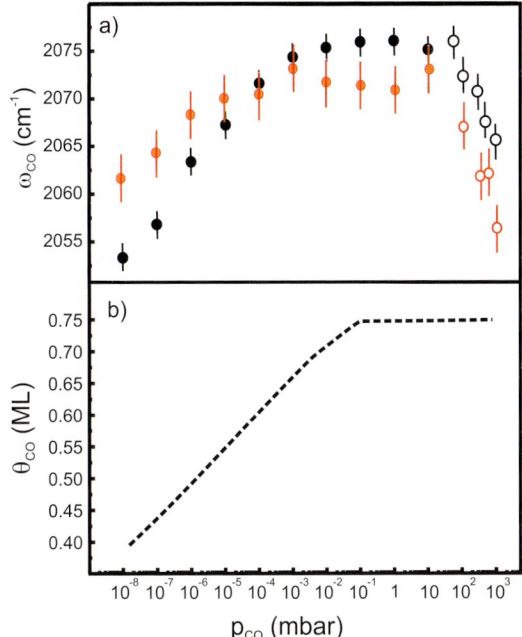

Fig. 4. (a) C-O stretching vibrational frequency ω_{CO} of CO molecules ontop adsorbed on a Rh(111) single-crystal (black circles) and on a polycrystalline Rh foil (red circles) catalyst surface. (b) Calculated equilibrium coverage of CO on the Rh(111) single-crystal surface (for details see Ref. [12]). All values are plotted against the CO gas pressure.

higher CO equilibrium coverages observed in the SFG study for CO pressures below $p_{CO} = 10^{-2}$ mbar.

In the high pressures regime (100 mbar < p_{CO} < 1000 mbar), however, for both the CO/Rh(111) and the CO/Rh-foil adsorption system the CO frequencies decrease, finally reaching a value of $\omega_{CO} = 2065 \pm 3$ cm^{-1} and $\omega_{CO} = 2056 \pm 3$ cm^{-1}, respectively. The observed decrease of the ontop CO frequency above $p_{CO} = 100$ mbar, which is accompanied by an irreversible decrease of the ontop CO adsorbate density, along with the presence of a considerable carbon signal observed in the "postreactive" Auger spectrum indicated the appearance of a new high-pressure dissociative CO adsorption mechanism, which will be further discussed below.

3.2 Polarization Dependence of the CO SFG Vibrational Spectra on Rh(111)

Although the vibrational resonant SFG signal obtained in the *ssp* configuration was considerable lower than in the *ppp* configuration, it was possible to record vibrational spectra of CO adsorbed on Rh(111) in the *ssp* configuration in the CO pressure range $p_{CO} = 10^{-8}$–10^{-4} mbar. In Figure 5, SFG spectra of ontop adsorbed CO obtained in the *ppp* and *ssp* configuration at a substrate temperature of $T_s = 300$ K for different CO pressures in the range $p_{CO} = 10^{-8}$–10^{-5} mbar are shown. In this pressure range for both polarization combinations no vibrational resonant SFG signal of CO bound to 3-fold coordinated hollow sites could be observed. As described in [39, 40] the orientation of the CO adsorbates with respect to the surface normal can be derived from the SFG vibrational resonant signal intensities measured for these two polarization combinations. In the *ssp* configuration no vibrational resonant SFG signal could be discerned for CO pressures above 10^{-4} mbar.

In Figure 6, the average CO tilt angle φ (solid triangles) derived from the *ssp* and *ppp* polarization SFG spectra is plotted as a function of the CO equilibrium surface coverage calculated using the adsorption/desorption kinetics data given in [41]. For comparison the tilt angle derived in a LEED study [42] by Somorjai, Van Hove and co-workers at a CO coverage of $\theta_{CO} = 1/3$ is included as a open triangle. Figure 6 clearly indicates that with increasing CO surface coverage the on-top adsorbed CO changes from a tilted species at coverages in the range $\theta_{CO} = 0.33$–0.4 ML into one which is adsorbed upright at higher coverages ($0.55 \leq \theta_{CO} \leq 0.75$ ML).

In [43], it was found that for the Rh(111)+($\sqrt{3} \times \sqrt{3}$)R30°-CO adsorbate structure which is formed at $\theta_{CO} = 1/3$ ML all CO molecules are terminally bound on top of one rhodium surface atom, hence, $\theta_{CO} = \theta_{CO-ontop} \approx 0.33$ ML. In addition, the presence of a buckling of the topmost Rh(111) substrate layer was noted with the Rh atoms directly underneath the adsorbed CO molecules being pulled outwards relative to the other Rh atoms in that layer [42] and it was further suggested that the tilt of the ontop CO adsorbate molecules is due to an anisotropic bending vibration [44]. In the present DFT calculations of the Rh(111)+($2\sqrt{3} \times \sqrt{3}$)R30°-2CO adsorbate structure a static tilt angle of $\varphi = 5.5°$ could be obtained, which supports the proposed presence of an anisotropic CO-bending potential for that low-coverage adsorbate structure.

Detailed studies of the temperature behavior of the Rh(111)+($\sqrt{3} \times \sqrt{3}$)R30°-CO phase were reported by Ertl and co-workers who employed temperature-dependent LEED in combination with helium scattering measurements [45]. In the latter experiments, it was found that the Rh(111)+($\sqrt{3} \times \sqrt{3}$)R30°-CO structure undergoes an order-disorder transition at a substrate temperature of $T_s = 330 \pm 5$ K. Further high-resolution core-level photoemission studies of the temperature and coverage dependence of the CO adsorption site occupation on Rh(111) under UHV conditions were reported in [46, 47]. These studies confirmed that in the CO/Rh(111) adsorption system for low coverages ($0 \leq \theta_{CO} \leq 0.33$ ML) only ontop sites are occupied [43]. For higher coverages it was found that both ontop and 3-fold hollow sites are occupied [47]. The controversy between the latter result and the previous structure model of [48], in which in the high-coverage Rh(111)+(2×2)-3CO adsorbate structure two

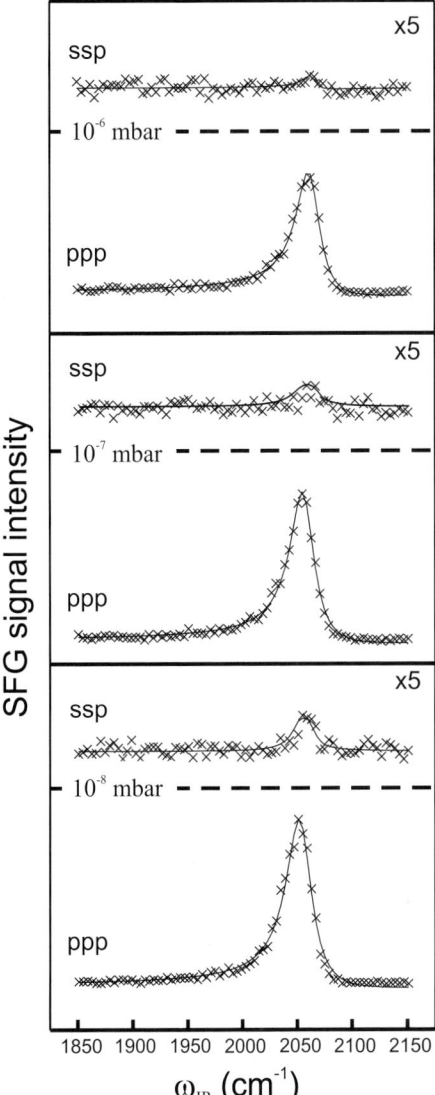

Fig. 5. SFG spectra of ontop adsorbed CO on Rh(111) at room temperature ($T_s = 300$ K) for different CO pressures as indicated in the figure. Spectra were recorded for two different polarization combinations: *ssp* (*s*-polarized sum frequency signal, *s*-polarized visible and *p*-polarized IR laser beam) and *ppp* (*p*-polarized sum frequency signal, *p*-polarized visible and *p*-polarized IR laser beam).

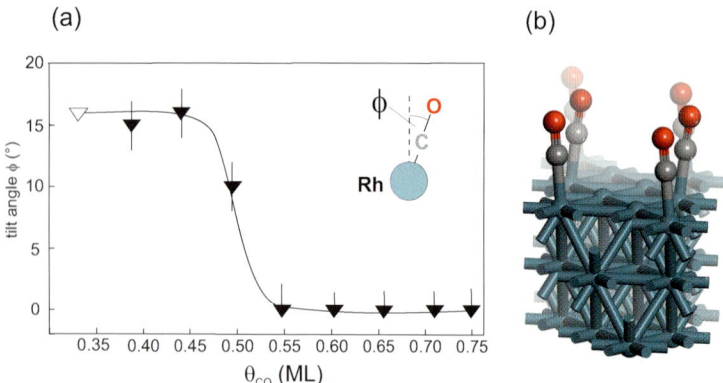

Fig. 6. Left side: Coverage dependence measurement of the average tilt angle of CO ontop adsorbed on Rh(111). Solid symbols represent the results obtained in the present SFG measurements with different polarization combinations (*ssp* and *ppp*). The open symbol represents a value obtained in LEED studies [42]. The solid line is just to guide the eye. The results indicate the presence of a tilted CO species at low CO coverages ($1/3 \leq \theta_{CO} \leq 0.55$ ML), which converts into an upright adsorbed species at higher coverage ($0.55 < \theta_{CO} \leq 3/4$ ML). Right sides: Rh(111)+($2\sqrt{3} \times \sqrt{3}$)R30°-2CO adsorbate structures obtained in DFT calculations (PW91), which indicate the possibility for a tilted CO adsorption geometry for $\theta_{CO} = 1/3$ ML. For details see text.

CO molecules occupy "near-on-top" positions and the third CO molecule is adsorbed in a 2-fold coordinated bridge position, was finally resolved in a reanalysis of the high-coverage LEED structures [42]. The results of this reanalysis finally confirmed the adsorption site assignment given in [47], where the Rh(111)+(2×2)-3CO adsorbate structure formed at a coverage of $\theta_{CO} = 3/4$ ML $= 0.75$ ML contains one ontop CO molecule adsorbed in an upright position, and two 3-fold hollow site CO adsorbate molecules in the (2×2) unit cell.

Subsequent "high-pressure" STM studies, in which the formation of dense CO adsorbate structures on Rh(111) at room temperature for $\theta_{CO} \geq 0.5$ ML was studied, showed that the same Rh(111)+(2×2)-3CO adsorbate structure is also formed in equilibrium with gas-phase CO at $\theta_{CO} = 0.75$ ML [37]. These STM measurements further revealed in the coverage range 0.43 ML $\leq \theta_{CO} \leq 0.57$ ML the presence of a Rh(111)+($\sqrt{7} \times \sqrt{7}$)R19°-CO phase. This adsorbate structure could not be observed in the previous UHV studies in which high initial CO coverages were achieved by cooling the substrate to low (liquid nitrogen and helium) temperatures [47]. The STM experiments also indicated a coverage dependent shift in the CO adsorption site population from one where all CO molecules are adsorbed at ontop sites (for coverages of $\theta_{CO} = 1/3$ ML and below) to one in which CO is adsorbed at ontop and 3-fold hollow sites with a ratio of 1:3 (for intermediate CO coverages around 0.57

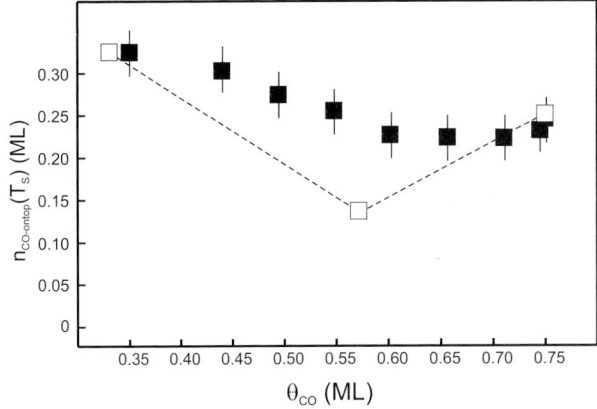

Fig. 7. Plot of the ontop CO coverage derived from the SFG measurements on Rh(111) (solid squares) against the total CO coverage calculated using the adsorption/desorption kinetics data for molecular adsorption of CO given in [41]. Open squares represent results of STM studies reported in [47]. The dashed line is drawn just to guide the eye. The experimental ontop CO coverages were normalized with respect to the total CO coverage of $\theta_{CO} = 1/3$ ML were only ontop CO is present at the surface [42].

ML) and finally to one where CO occupies ontop and 3-fold hollow sites with a ratio of 1:2 (at the CO saturation coverage of 0.75 ML). From the latter CO adsorption site ratios, the ontop CO coverage, $\theta_{CO-ontop}$, at a given total CO coverage, θ_{CO}, can be derived. In Figure 7, the so obtained values for $\theta_{CO-ontop}$ are plotted (open squares) against the total CO coverage. For comparison ontop CO coverages derived from the integrated ontop CO vibrational band intensities of the *ppp* SFG spectra recorded in the pressure range 10^{-8}–10 mbar are included in Figure 7 as solid squares. The experimental ontop CO coverages were normalized with respect to the coverage of $\theta_{CO} = 1/3$ ML associated with the Rh(111)+($\sqrt{3}\times\sqrt{3}$)R30°-CO adsorbate structure where all CO molecules are adsorbed at ontop sites. As Figure 7 shows, the ontop CO coverage derived from the SFG measurements exhibits the same overall variation as the ontop CO coverage derived from the STM studies.

However, at this point it is worth to note that the present as well as previous SFG spectroscopy results [36] clearly indicate that under elevated and high-pressure adsorption/desorption equilibrium conditions the ontop CO species represents a major constituent of the dense adsorbate structures, which are formed. In both SFG studies no indication was found for a global surface adsorbate structure in which CO molecules are solely adsorbed on 3-fold hollow sites. The experimental data reproduced in Figure 7 indicates that with increasing pressure the ontop CO coverage changes from a value of $\theta_{CO-ontop} = 1/3$ ML at 10^{-8} mbar to a value of about $\theta_{CO-ontop} = 0.22$ ML in the high-pressure limit. Hence, it is suggested that the new 3-fold hollow site Rh(111)+($\sqrt{7}\times\sqrt{7}$)R19°-CO-adsorbate structure ($\theta_{CO}=$ 3/7 ML) observed in the STM studies [37] must coexists with ontop CO containing structures such as the ($\theta_{CO}=$ 4/7 ML) Rh(111)+($\sqrt{7}\times\sqrt{7}$)R19°-CO-adsorbate structure. The latter suggestion is supported by the DFT results of the present work in which comparable adsorption energies were obtained for these two adsorbate structures. In Figure 8, the sequence of CO/Rh(111) adsorbate structures derived in the present DFT computations for coverages between $\theta_{CO}=$ 1/4 ML and $\theta_{CO}=$ 3/4 ML are reproduced, and the corresponding adsorption energies E_{ad} are given in the figure caption.

3.3 Temperature Dependence of the CO SFG Vibrational Spectra on Rh(111)

To study the temperature dependence of the CO adsorption on Rh(111) various sets of SFG spectra were recorded in the substrate temperature range T_s = 300–800 K at different CO pressures in the range 10^{-8}–100 mbar. All measurements were performed under adsorption/desorption equilibrium conditions.

Figure 9 depicts SFG spectra of ontop adsorbed CO recorded in the *ppp* configuration in one single increasing substrate temperature cycle at a constant CO pressure of 10^{-8} mbar. For substrate temperatures above 300 K no vibrational resonant SFG signal could be obtained in the *ssp* configuration. The numerical analysis of the SFG spectral line shapes (solid lines) revealed that the ontop adsorbed CO frequency decreases with decreasing equilibrium coverage from ω_{CO} = 2053 ± 2 cm^{-1} at T_s = 300 K to ω_{CO} = 2000 ± 2 cm^{-1} at T_s = 480 K. For $T_s \geq$ 500 K no vibrational resonant SFG signal could be observed anymore.

As can bee seen in Figure 9, even at the highest temperature where CO could be detected, the SFG spectrum exhibits a single vibrational resonance feature. The measured SFG spectra were completely reversible and reproducible with variation of the substrate temperature. All SFG spectral line shapes could be well described by a single vibrational resonance origination from ontop adsorbed CO species. AE spectra recorded after pumping down the cell pressure to 10^{-9} mbar and cooling down the substrate to T_s = 300 K showed no indication for carbon deposition on the surface. The AE spectra were virtually identical with those recorded for the clean Rh surface before the adsorption study. Similar results were obtained for CO pressures of 10^{-6} and 10^{-4} mbar. Under the latter pressure conditions, ontop CO could be detected up to substrate temperatures of T_s = 520 K and T_s = 600 K, respectively, without any indication for dissociation [49]. The latter results are in agreement with previous TPD studies, which demonstrated that for $p_{CO} < 1.33 \times 10^{-5}$ mbar and $T_s <$ 600 K CO dissociation does not occur [50, 51]. The experimental results can be attributed to the high barrier for dissociation which

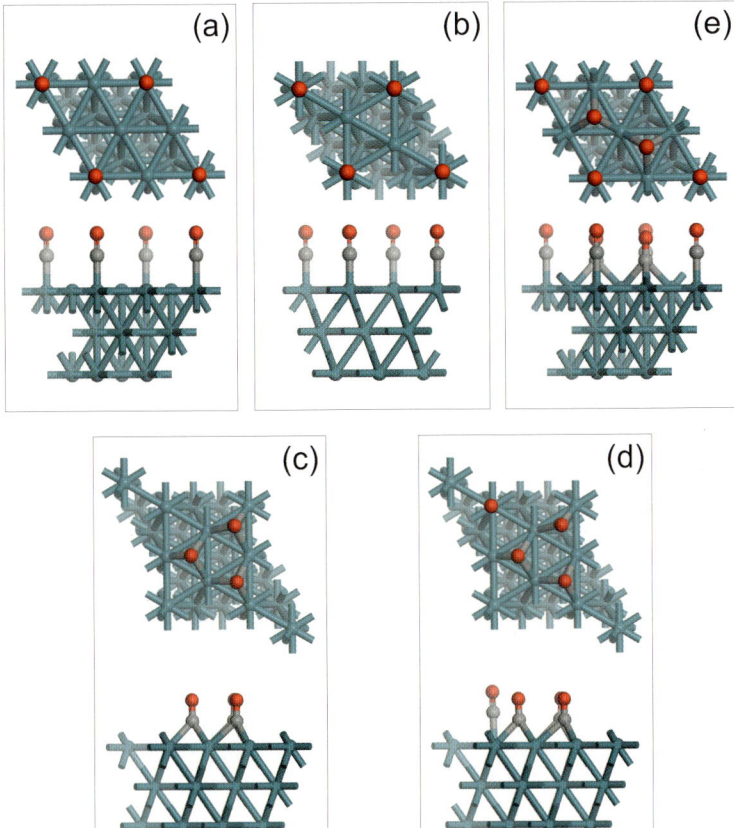

Fig. 8. CO/Rh(111) adsorbate structures and adsorption energies E_{ad} obtained in DFT calculations using the PW91 (RPBE) functionals:
(a) Rh(111)+(2× 2)-CO, $\theta_{CO} = 1/4$ ML, $E_{ad} = -1.79$ eV (-1.49 eV),
(b) Rh(111)+($\sqrt{3}\times\sqrt{3}$)R30°-CO, $\theta_{CO} = 1/3$ ML, $E_{ad} = -1.73$ eV (-1.44 eV), **(c)** Rh(111)+($\sqrt{7}\times\sqrt{7}$)R19°-CO, $\theta_{CO} = 3/7$ ML, $E_{ad} = -1.76$ eV (-1.48 eV),
(d) Rh(111)+($\sqrt{7}\times\sqrt{7}$)R19°-CO, $\theta_{CO} = 4/7$ ML, $E_{ad} = -1.74$ eV (-1.50 eV),
(e) Rh(111)+(2×2)-3CO, $\theta_{CO} = 3/4$ ML, $E_{ad} = -1.77$ eV (-1.36 eV).

was obtained both in previous [14, 15] and in the present DFT calculations for CO molecules adsorbed at the Rh(111) surface (see Section 3.4).

At higher pressures and substrate temperatures, however, in the present experiments an additional low-frequency vibrational feature became observable, as can bee seen in Figure 10, which shows CO SFG spectra recorded in one single increasing substrate temperature cycle at a constant CO pressure of 1 mbar. SFG spectra recorded after the substrate temperature was reduced back to $T_s = 300$ K yielded considerably lower signal intensities than the corresponding SFG spectrum recorded in the increasing substrate temperature cycle and AE spectra (recorded after evacuation of the cell down

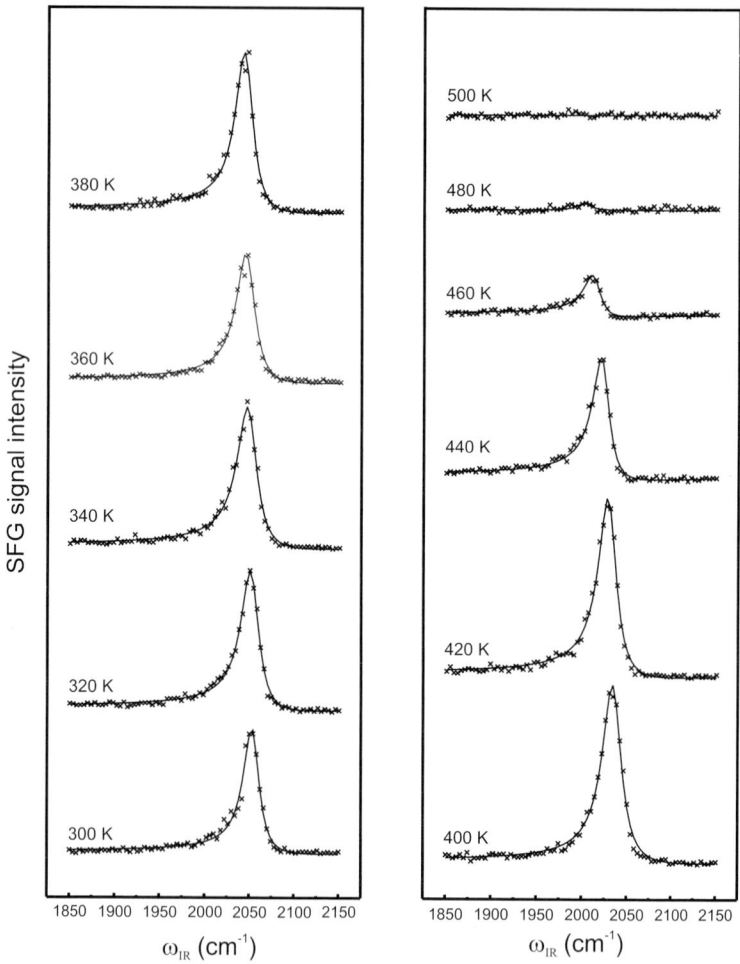

Fig. 9. SFG vibrational spectra of ontop CO adsorbed on Rh(111) recorded for the ppp polarization combination at different substrate temperatures T_s at a fixed CO pressure of 10^{-8} mbar. Crosses represent experimental data points; solid lines are results of a numerical simulation of the measured SFG line shapes.

to about 10^{-9} mbar) revealed the presence of carbon at the surface indicating the occurrence of irreversible dissociative CO adsorption for $T_s \geq 600$ K. The dissociative adsorption of CO at high substrate temperatures is further demonstrated by the SFG and AE data reproduced in Figure 11.

The SFG spectrum shown in Figure 11a was recorded at a CO pressure of 1 mbar at a substrate temperature of 300 K. The AE spectrum of the clean Rh surface measured before the substrate was exposed to CO is depicted as an inset in Figure 11a. Figure 11c finally depicts a "SFG spectrum" recorded for $T_s = 680$ K after the CO pressure was raised to 100 mbar in which no

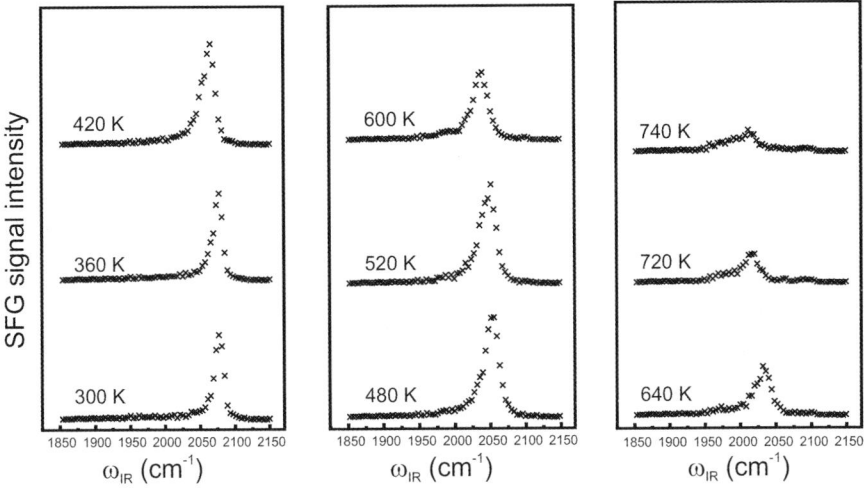

Fig. 10. A series of SFG vibrational spectra of ontop CO adsorbed on Rh(111) recorded for the *ppp* polarization combination at different substrate temperatures T_s (as indicated in the figure) at a fixed CO pressure of 1 mbar. For $T_s > 520$ K the appearance of a new low-frequency vibrational feature indicates the onset CO dissociation. For further details see text.

vibrational resonant signal could be discerned anymore. The fact that also after reducing the substrate temperature no noticeable vibrational resonant signal in the SFG spectra reappeared demonstrates the irreversibility of the CO adsorption process under these pressure and temperature conditions. The AE spectrum of the Rh surface measured after the substrate temperature was reduced down to 300 K and after evacuation of the cell down to about 10^{-9} mbar is depicted as an inset in Figure 11c. Comparison of the latter "postreactive" AE spectrum with the one of the clean Rh surface shown in Figure 11a clearly demonstrates that a considerable amount of surface carbon, C(ad), is deposited (see below) during the high-temperature and high-pressure adsorption of CO. The fact that in contrast to the CO dissociation studies on Pt(111) reported in [22] no oxygen signal was observed in the "postreactive" AE spectrum of the Rh(111) surface suggests that on Rh(111) the dissociation of CO proceeds predominately *via* the exothermic Boudouard reaction $2CO \rightarrow C(ad) + CO_2$.

The data reproduced in Figure 12 illustrate the considerable efforts, which had to be made to remove the surface carbon layer formed during the high-temperature/high-pressure CO adsorption study described above. Figure 12a shows in the lower part the AE spectrum of the Rh surface after 10 hours of Ar^+ ion sputtering followed by annealing in the UHV, in the upper part a SFG spectrum is depicted, which was recorded on this Rh surface at a

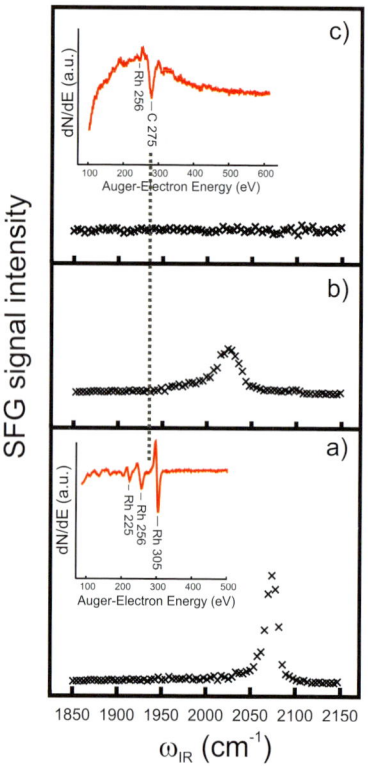

Fig. 11. (a) Ontop CO SFG spectrum recorded for $T_s = 300$ K at a CO pressure of $p_{CO} = 1$ mbar on Rh(111). **(b)** The same spectrum after the substrate temperature was raised to $T_s = 680$ K. **(c)** The same "SFG spectrum" after the pressure was increased to $p_{CO} = 100$ mbar. Shown as insets: **(a)** AE spectrum (red curve) of the clean surface recorded before the adsorption study. **(c)** "Postreactive" AE spectrum (red curve) recorded after the adsorption measurement.

substrate temperature of 300 K at a CO pressure of 10^{-8} mbar. Figure 12b depicts an AE spectrum of the same Rh surface after another 10 hours of Ar^+ ion sputtering in the UHV along with the SFG spectrum recorded under the same pressure condition as the one reproduced in Figure 12a. Figure 12c finally shows the AE spectrum of the clean surface, which could be obtained after another 10 hours of Ar^+ ion sputtering followed by extensive cleaning and annealing cycles in oxygen. The SFG spectrum recorded on the clean surface under the same substrate temperature and CO pressure conditions as the SFG spectra shown in Figure 12a and 12b is depicted in the upper part of Figure 12c.

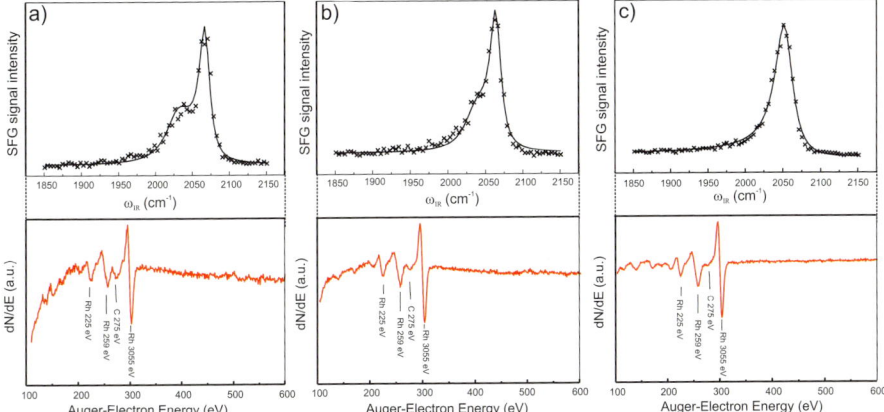

Fig. 12. CO SFG spectra recorded at $p_{CO} = 10^{-8}$ mbar for $T_s = 300$ K (**a**) after ca. 10 h of cleaning (following the adsorption study illustrated in Figure 11. (**b**) and (**c**), each SFG spectrum recorded after another 10 h of cleaning. The corresponding AE spectra (red curves) indicate a correlation between the amount of surface carbon and the low frequency shoulder of the ontop CO SFG vibrational spectra.

An analysis of spectral line-shapes of the SFG spectra shown in Figure 12a and 12b, in which a two-surface species expression for the nonlinear surface susceptibility was employed, revealed that the decrease of the intensity of the low frequency component with a center-frequency of about 2035 cm^{-1} correlates with the decrease of carbon at the surface. This correlation suggests that the down shift in the CO ontop stretching frequency as observed in the present and in the previous [36] room temperature CO adsorption studies for CO pressures ≥ 100 mbar (see Figure 4) is due to the formation of ontop CO co-adsorbed with carbon on the Rh(111) surface rather than due to a irreversible adsorbate induced surface roughening as it has been tentatively proposed in [21].

3.4 A Novel Mechanism for Low-Energy High-Pressure CO Dissociation on Rh(111)

Due to its importance as a first step in the methanation reaction and in the Fischer-Tropsch synthesis the possibility for CO dissociation has been investigated on a variety of transition metal surfaces [31]. In case of rhodium, a series of DFT studies were performed [14, 15] in order to elucidate the mechanism underlying the pronounced particle size dependence of the CO dissociation activity observed in previous experiments with size-selected Rh particles supported on thin alumina films [52]. These comparative DFT computations, in which the thermochemistry for CO dissociation on flat (111) and stepped (211) Rh single-crystal facets was examined [14, 15], revealed that the dissociation of CO out of the chemisorbed state *via* reaction (6)

$$CO(ad) \to C(ad) + O(ad) \qquad (6)$$

to yield chemisorbed carbon C(ad) and oxygen O(ad) atoms is an extremely structure sensitive process. The calculations demonstrated that for CO molecules adsorbed ontop of steps sides the dissociation energy is lowered by ca. 32% compared to that of CO ontop adsorbed on Rh(111) terrace sides (see Figure 1 of Ref. [15]), which might explain the enhanced CO dissociation activity observed for supported Rh nanoparticles [52]. For CO molecules ontop adsorbed on the Rh(111) surface, the calculations indicated a negligible small dissociation rate, due to the considerable energy barrier of $E_{diss} = 3.16$ eV in the reaction pathway of reaction (6) [14]. The latter value was derived in DFT computations with the DACAPO program package [53] employing the RPBE functional [54].

In the present work, we used the CASTEP [26] employing RPBE [54] and Perdew-Wang-91 (PW91) functionals [55] to study CO adsorption on Rh(111) for coverages between $\theta_{CO} = 1/4$ ML and $\theta_{CO} = 3/4$ ML, results of which were presented in Figure 8. The CO adsorption energies of $E_{ad} = -1.49$ eV (RPBE) and -1.79 eV (PW91), which were obtained in the latter computations for the lowest CO coverage of $\theta_{CO} = 1/4$ ML are consistent with results of TPD measurements, which yielded values between 1.57 eV and 1.71 eV for the CO desorption energy in the low-coverage limit [34, 56].

To study the energetics of the CO dissociation reaction pathway (6) further DFT calculations for an initial CO coverage of 1/6 ML were performed. As indicated by Figure 13, the overall process of CO adsorption followed by CO dissociation *via* the transition state TS1, which leads to carbon and oxygen atom products adsorbed at hexagonal-close-packed (hcp) hollow sides, is endothermic by 0.08 eV (PW91). Only if the possibility for subsequent diffusion of the oxygen atom dissociation product to face-centered-cubic (fcc) hollow side (*via* the transition state TS2 shown in Figure 13) is considered in the DFT calculations, the overall process becomes exothermic by -0.27 eV (PW91). Furthermore, as can be seen in Figure 13, the present results confirm, in concordance with the previous DFT studies [14, 15], that a high barrier to CO dissociation *via* reaction (6) exists, $E_{diss} = 3.33$ eV (PW91). Additional calculations, performed for a CO coverage of 1/4 ML, yielded values of $E_{diss} = 3.45$ eV (PW91) and 3.63 eV (RPBE), indicating that the barrier to CO dissociation *via* reaction (6) increases with CO coverage. In summary, the DFT results for reaction (6) are in general agreement with the previous [50, 51] and the present low-pressure CO adsorption studies on Rh(111), in which no indications for CO dissociation could be discerned for $p_{CO} \leq 10^{-4}$ mbar.

However, the results of the high-pressure CO adsorption experiments, which revealed for CO pressures above ca. 10 mbar the onset of irreversible dissociative CO adsorption on Rh(111) already at a substrate temperature of $T_s = 300$ K, cannot be explained if (6) is considered as the only reaction pathway. In addition, as described in the previous section, surface carbon but

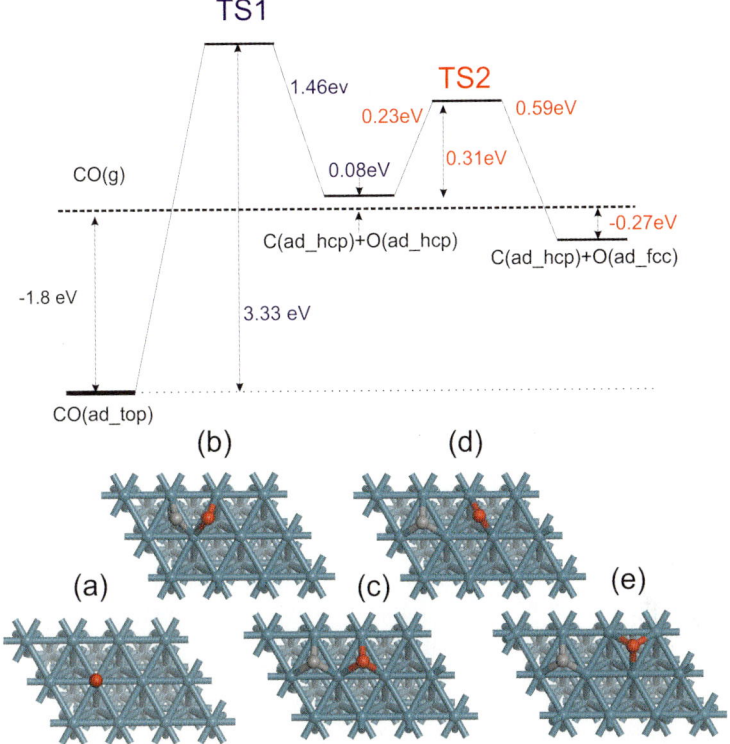

Fig. 13. Upper part: Schematic one-dimensional potential energy diagram for the $CO(ad) \rightarrow C(ad) + O(ad)$ high-barrier dissociation reaction pathway on Rh(111) as obtained in the present DFG calculations using the PW91 functional for a CO coverage of 1/6 ML. Lower part (from left to right): Sequence of snapshots of geometries obtained along the minimum energy path: **(a)** CO reactant state, **(b)** transition state (TS1) for C-O bond dissociation, **(c)** intermediate product state with the C and O atom dissociation products adsorbed at hcp-hollow sites, **(d)** transitions state (TS2) for the surface O atom diffusion process, $O(ad\_hcp) \rightarrow O(ad\_fcc)$, **(e)** final product state configuration.

no surface oxygen signals could be observed in the AE spectra of the Rh(111) surface after the high-pressure CO dissociation studies. As a consequence, DFT computational studies were performed in the course of the present work, where – as an alternative to reaction (6) – the chemical interaction between a gas-phase carbon monoxide molecule, $CO(g)$, and carbon monoxide molecule adsorbed on the Rh(111) surface, $CO(ad)$, was investigated, which can result in surface carbon $C(ad)$ and gas-phase $CO_2(g)$ products through an Eley-Rideal (ER) mechanism (7):

$$CO(ad) + CO(g) \quad C(ad) + CO_2(g) \tag{7}$$

In Figure 14, the energetic and structural results of DFT computations of reaction (7) are reproduced. In the latter computations, the adsorbed CO reactant molecules were located at a fcc-hollow side of the Rh surface, CO(ad_hcp), and the final product configuration was a CO_2 gas-phase molecule and a carbon atom adsorbed at a hcp-hollow side. The present results indicated that the overall reaction $2CO(g) \rightarrow CO(ad\_hcp) + CO_2(g)$ is exothermic with a heat of reaction of $\Delta_r H_{0K}$ = -1.83 eV. Furthermore, as indicated by the plot of the potential energy against the internal reaction coordinate, as shown in the upper part of Figure 14, reaction (7) does indeed provide a low-energy pathway along which the molecular bond of the adsorbed CO reactant is broken during a direct abstraction reaction with the incoming gas-phase CO reagent leading to the formation of a gas-phase $CO_2(g)$ molecule and a strongly adsorbed carbon surface atom. The latter of which could be observed in the AE spectra of the Rh(111) surface after the high-pressure CO adsorption studies described in the previous section.

A more detailed analysis of the structural results of the DFT computations (see lower part of Figure 14) further revealed that reaction (7) exhibits a "late" transition state, the structure of which more closely resembles that of the final reaction products than the reactants. Therefore, based on the general principles derived by Polanyi and co-workers in a series of quasi-classical trajectory studies of the dynamics of gas-phase model reactions, where different "diagnostic" potential energy surfaces (PES) were employed (see e.g. Refs. [57, 58, 59], one might anticipate that reactant vibrational excitation is more efficient in promoting reactivity than relative reagent translational excitation. Furthermore, in the particular case of an abstraction reaction such as reaction (7), it is not unreasonable to assume that vibrational excitation of the C-O bond of the adsorbed reactant molecule – the bond to be broken during the reaction – will be more efficient than vibrational excitation of the incoming gas-phase CO molecule, the bond of which might play the role of a spectator during reaction. In order to verify the latter propositions, however, molecular reaction dynamics simulations on full dimensional *ab initio* based PESs, which can describe the interaction potential between gas-phase CO molecules and chemisorbed CO molecules for different CO/Rh(111) adsorbate structures with a high global accuracy, would be needed. However, in the light of the considerable efforts, which are necessary to construct global PESs for the molecular reaction dynamics approach, an alternative computationally less demanding surface reaction kinetics approach might be taken, in which modern DFT for the computation of activation energies and transition state properties is combined with transition state theory (TST) for the calculation of reaction rate constants (see e.g. Ref. [60]). The so obtained kinetics data could then be used in numerical simulations of more complex heterogeneous reaction systems, such as the e.g. the Fischer-Tropsch synthesis process [61], in order to access the importance of the newly proposed low-energy CO dissociation pathway (7) under technically relevant operation conditions.

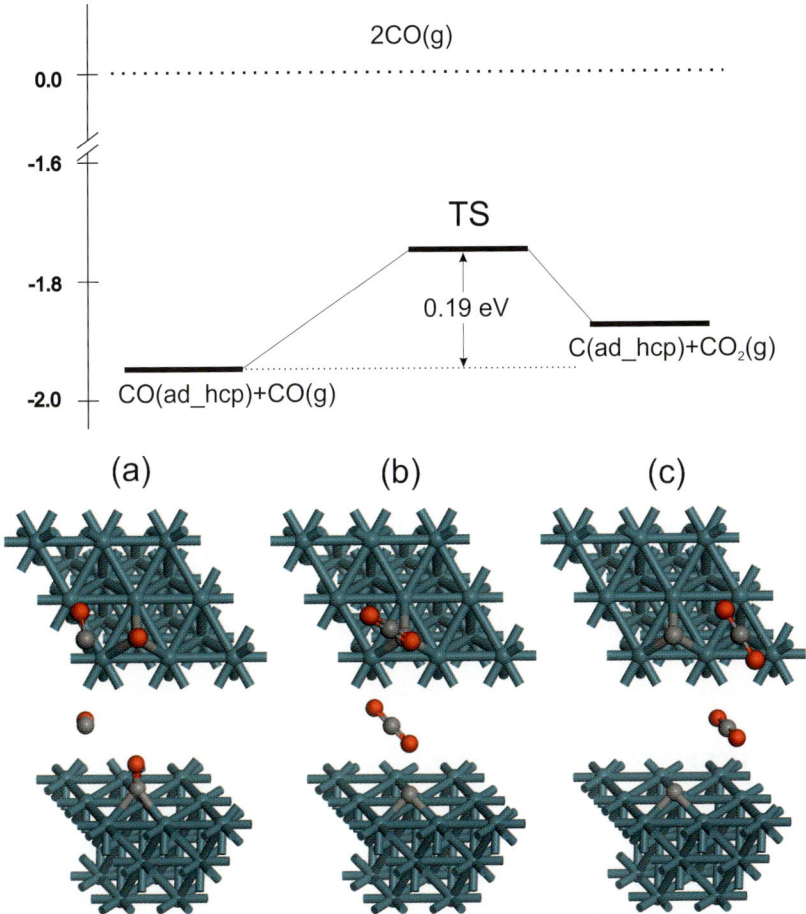

Fig. 14. Upper part: Schematic one-dimensional potential energy diagram for the CO(ad_hcp) + CO(g) → C(ad) + CO_2(g) low-barrier direct abstraction reaction pathway on Rh(111) obtained in the present DFT calculations, in which a incoming gas-phase molecule CO(g) reacts with an adsorbed molecule CO(ad_hcp) *via* an Eley-Rideal mechanism to yield atomic carbon adsorbed at a hcp-hollow sides of the Rh(111) surface and a gas-phase molecule CO_2(g) as reaction products. Lower part: Sequence of snapshot of geometries obtained along the minimum energy path: **(a)** Reactant state, **(b)** transition state (TS), **(c)** product state. The calculations were performed for a (2×2) unit cell with a CO coverage of 1/4 ML employing a PW91 functional.

4 Summary

The present SFG studies of CO adsorption on Rh(111) and on a polycrystalline Rh-foil, which were performed at a substrate temperature of 300 K over 12 orders of magnitude in CO pressure, demonstrated the reversible molecular adsorption of CO up to a CO pressure of ca. 10 mbar. For the CO/Rh(111) adsorbate system SFG measurement performed under different excitation/detection polarization combinations carried out at room-temperature in the pressure range $p_{CO} = 10^{-8}$–10 mbar revealed the presence of tilted CO ontop species at intermediate CO surface coverages ($\theta_{CO} = 0.33$–0.4 ML). In agreement with the results of DFT calculations with increasing coverage the ontop adsorbed CO was found to change from a tilted species to one, which is adsorbed in an upright geometry at higher coverages ($0.4 < \theta_{CO} \leq 0.75$ ML).

For CO pressures above 10 mbar, the onset of irreversible dissociative CO adsorption could be observed already at a substrate temperature of $T_s = 300$ K both on Rh(111) and on the polycrystalline Rh-foil. At this temperature, as well as at substrate temperatures up to 800 K, CO dissociation was found to result in the formation of carbon on the surface as the only detectable dissociation product, and it was concluded that CO dissociation on rhodium occurs *via* the Boudouard reaction: $2CO \rightarrow C(ad) + CO_2$. Results of DFT studies of the CO/Rh(111) gas-phase/adsorbate system, which were performed to rationalize the experimental findings, suggest the occurrence of a "low-temperature/high-pressure CO/Rh(111) dissociation pathway" where a gas-phase CO molecule reacts with a CO molecule chemisorbed on the catalyst surface through an Eley-Rideal mechanism to yield gas-phase CO_2 and surface carbon C(ad).

References

1. Thomas, J.M.: *Principles and practice of heterogeneous catalysis*, VCH-Verlag, Weinheim (1996)
2. Heck, R.M., and Farrauto, R.J.: *Catalytic air pollution control*, Wiley, New York (1995)
3. Ertl, G., Knözinger, H., and Weitkamp J.: *Handbook of heterogeneous catalysis*, vol. 4, VCH-Verlag, Weinheim (1997)
4. Deutschmann, O, Warnatz, J., Chap. 20:*Diagnostics for catalytic combustion* in: *Applied combustion diagnostics* (Eds. K. Kohse-Höinghaus, J.B. Jeffries) Taylor & Francis (2002)
5. Wolfrum, J., Volpp, H.-R., Rannacher, R., Warnatz, J. (Eds.): *Gas-phase chemical reaction systems: 100 Years after Max Bodenstein*, Springer Series in Chemical Physics, Volume 61, Springer, Heidelberg (1996)
6. Warnatz, J., Maas, U., and Dibble, R.W.: *Combustion: physical and chemical fundamentals, modeling and simulation, experiments, pollutant formation*, 3d ed. Springer, Berlin (2005)
7. Härle, H., Metka, U., Volpp, H.-R., Wolfrum, J.: *Pressure dependence (10^{-8}–1000 mbar) of the vibrational vpectra of CO chemisorbed on polycrystalline platinum studied by infrared-visible sum frequency generation*, Phys. Chem. Chem. Phys. **1**, 5059-5064 (1999)
8. Somorjai, G.A., and Rupprechter, G.: *Molecular studies of catalytic reactions on crystal curfaces at high pressures and high temperatures by infrared-visible sum*

frequency generation (SFG) surface vibrational spectroscopy, J. Phys. Chem. B, **103**, 1623-1638, (1999)
9. Härle, H., Lehnert, A., Metka, U., Volpp, H.-R., Willms, L., and Wolfrum, J.: *In situ detection and surface coverage measurements of CO during CO oxidation on polycrystalline platinum using sum frequency generation*, Appl. Phys. B., **68**, 567-572 (1999)
10. Kissel-Osterrieder, R., Behrendt, F., Warnatz, J., Metka, U., Volpp, H.-R., Wolfrum, J.: *Experimental and theoretical investigation of CO-oxidation on platinum: Bridging the pressure and the materials gap*, Proc. Combust. Inst., **28**, 1341-1348 (2000)
11. Metka, U., Schweitzer, M. G., Volpp, H.-R., Wolfrum, J., Warnatz, J.: *In-situ detection of NO chemisorbed on platinum using infrared-visible sum-frequency generation (SFG)*, Zeit. Phys. Chem., **214**, 865-888 (2000)
12. Pery, T., Schweitzer, M. G., Volpp, H.-R., Wolfrum, J., Ciossu, L., Deutschmann, O., Warnatz J.: *Sum-frequency generation in situ study of CO adsorption and catalytic CO oxidation on rhodium at elevated Pressures*, Proc. Combust. Inst., **29**, 973-980 (2002)
13. Dellwig, T., Rupprechter, G., Unterhalt, H., and Freund, H.-J.: *Bridging the pressure and materials gaps: High pressure sum frequency generation study on supported Pd nanoparticles*, Phys. Rev. Letters, **85**, 776-779 (2000)
14. Mavrikakis, M., Rempel, J., Greeley, J., Hansen, L. B., Nørskov, J. K.: *Atomic and molecular adsorption on Rh(111)*, J. Chem. Phys. **117**, 6737-6744 (2002)
15. Mavrikakis, M., Bäumer, M. Freund, H.-J., Nørskov, J. K.: *Structure sensitivity of CO dissociation on Rh Surfaces*, Catal. Lett. **81**, 153-156 (2002)
16. Honkala, K., Hellman, A., Remediakis, I. N., Logadottir, A., Carlsson, A., Dahl, S., Christensen, C. H., Nørskov, J. K.: *Ammonia synthesis from first-principles calculations*, Science **307**, 555-558 (2005)
17. Inderwildi, O., Lebiedz, D. Deutschmann, O., Warnatz, J.: *Coverage dependence of oxygen decomposition and surface diffusion on Rhodium (111): A DFT study*, J. Chem. Phys. **122**, 034710(8 pages) (2005)
18. Inderwildi, O., Lebiedz, D. Deutschmann, O., Warnatz, J., *Influence of initial oxygen coverage and magnetic moment on the NO decomposition on Rhodium (111)*, J. Chem. Phys. **122**, 154702(7 pages) (2005)
19. Ertl, G., and Küppers J.: *Low energy electrons and surface chemistry*, VCH-Verlag, Weinheim (1985)
20. Cremer, P.S., McIntyre, B.J., Salmeron, M., Shen, Y.-R., Somorjai, G.A.: *Monitoring surfaces on the molecular level during catalytic reactions at high pressure by sum frequency generation vibrational spectroscopy and scanning tunneling microscopy*, Catalysis Letters, **34**, 11-18 (1995)
21. Somorjai, G.A., Rupprechter, G.: *Molecular studies of catalytic reactions on crystal surfaces at high pressures and high temperatures by infrared-visible sum frequency generation (SFG) surface vibrational spectroscopy*, J. Phys. Chem. B., **103**, 1623-1638 (1999)
22. Volpp, H.-R., Wolfrum, J.: Chap. 12 *Sum-frequency generation (SFG) vibrational spectroscopy as a means for the investigation of catalytic combustion* in: *Applied combustion diagnostics* (Eds. K. Kohse-Höinghaus, J.B. Jeffries) Taylor & Francis (2002)
23. Shen, Y.R.: *A Few selected applications of surface nonlinear optical spectroscopy*, Proc. Natl. Acad. Sci. USA, **93**, 12104-12111 (1996)

24. Shen, Y.R.: *The principles of nonlinear optics*, Wiley, New York (1984)
25. Miragliotta, J., Polizzotti, R.S., Rabinowitz, P., Cameron, S.D., Hall, R.B., *IR-Visible sum-frequency generation study of methanol adsorption and reaction on Ni(100)*, Chem. Phys., **143**, 123-130 (1990)
26. Segall, M.D.,Lindan, P.L.D., Probert, M.J., Pickard, C.J., Hasnip, P.J., Clark, S.J., Payne, M.C., *First-principles simulation: ideas, illustrations and the CASTEP code* J. Phys.: Condens. Matter **14**, 2717-2743 (2002)
27. Perdew, J. P., Chevary, J. A., Vosko, S.H., Jackson, K. A., Pederson, M.R. and Fiolhais C., *Atoms, molecules, solids, and surfaces: Applications of the generalized gradient approximation for exchange and correlation*, Phys. Rev. B **46**, 6671-6687 (1992)
28. Vanderbilt, D., *Soft self-consistent pseudopotentials in a generalized eigenvalue formalism*, Phys. Rev. B **41**, 7892-7895 (1990)
29. Monkhorst, H.J. and Pack J.D., *Special points for Brillouin-zone integrations*, Phys. Rev. B **13**, 5188-5192 (1976)
30. Govind, N., Petersen, M., Fitzgerald, G., King-Smith, D., Andzelm, J., *A generalized synchronous transit method for transition state location*, Comp. Mater. Sci. **28**, 250-258 (2003)
31. Somorjai, G. A.: *Introduction to surface chemistry and catalysis*, Wiley, New York (1994)
32. Grant, J.T., Haas, T.W., *A study of Ru(0001) and Rh(111) surfaces using LEED and Auger electron spectroscopy*, Surf. Sci. **21** 76-85, (1970)
33. Van Hove, M.A., Isr. J. Chem. *Special issue in honour of the 1998 Wolf prize recipients G. Ertl and G. A. Somorjai*, **38** 349-352 (1998)
34. Linke, R., Curulla, D., Hopstaken, M.J.P. and Niemantsverdriet, J.W., *CO/Rh(111): Vibrational frequency shifts and lateral interactions in adsorbate layers*, J. Chem. Phys. **115**, 8209-8216 (2001)
35. Birgersson, M., Almbladh, C.-O., Borg, M., Andersen, J. N.: *Density-functional theory applied to Rh(111) and CO/Rh(111) systems: Geometries, energies, and chemical shifts*, Phys. Rev. B. **67**, 045402(14 pages) (2003)
36. Somorjai, G.A.,Su, X., McCrea, K.R., Rider, K.B., *Molecular surface studies of adsorption and catalytic reaction on crystal (Pt, Rh) surfaces under high pressure conditions (atmospheres) using sum frequency generation (SFG) - surface vibrational spectroscopy and scanning tunneling microscopy (STM)*, Topics in Catalysis **8**, 23-34 (1999)
37. Cernota, P., Rider, K., Yoon, H. A., Salmeron, M., Somorjai, G.A., *Dense structures formed by CO on Rh(111) studied by scanning tunneling microscopy*, Surf. Sci. **445**, 249-255 (2000)
38. Bandara, A., Dobashi, S., Kubota, J., Onda, K., Wada, A., Domen, K., Hirose, C., Kano, S.S.: *Adsorption of CO and NO on NiO(111)/Ni(111) surface studied by infrared-visible sum frequency generation spectroscopy*, Surface Science, **387**, 312-319 (1997)
39. Heinz, T. F., H. W. K. Tom, Y. R. Shen, *Determination of molecular orientation of monolayer adsorbates by optical second-harmonic generation*, Phys. Rev. A **28**, 1883-1885 (1983)
40. Chen, Z., Gracias, D.H., Somorjai, G.A., *Sum frequency generation (SFG)-surface vibrational spectroscopy studies of buried interfaces: Catalytic reaction intermediates on transition metal crystal surfaces at high reactant pressures; polymer surface structures at the solid-Gas and solid-liquid interfaces*, Appl. Phys. B. **68,** 549-557 (1999)

41. Zhdanov, V.P., Kasemo, B., *Mechanism and kinetics of the NO—CO reaction on Rh*, Surf. Sci. Rep. **29**, 31-90 (1997)
42. Gierer, M., Barbieri, A., Van Hove, M.A., Somorjai, G.A., *Structural reanalysis of the Rh(111)+($\sqrt{3}\times\sqrt{3}$)R30°-CO and Rh(111)+(2×2)-3CO phases using automated tensor LEED*, Surf. Sci. **391**, 176-182 (1997)
43. Koestner, R.J., Van Hove, M.A., Somorjai, G.A., *A surface crystallography study by dynamical LEED of the ($\sqrt{3}\times\sqrt{3}$)R30° CO structure on the Rh(111) crystal surface*, Surf. Sci. **107**, 439-458 (1981)
44. Witte, G., Range, H., Toennies, J. P., Wöll, C., *External vibrations of molecules of surfaces: The case of CO and benzene*, J. Electron Spectrosc. Related Phenom. **64/65**, 715-723 (1993)
45. Over, H., Schwegmann, S., Ertl, G., Cvetko, D., De Renzi, V., Floreano, L., Gotter, R., Morgante, A., Peloi, M., Tommasini, F., Zennaro, S., *Temperature behavior of the ($\sqrt{3}\times\sqrt{3}$)R30°-1CO and the (2×2)-3CO overlayers on Rh(111): a combined HAS and LEED investigation*, Surf. Sci. **376**, 177-184 (1997)
46. Beutler, A., Lundgren, E., Nyholm, R., Andersen, J.N., Setlik, B. Heskett, D., *On the adsorption sites for CO on the Rh(111) single crystal surface*, Surf. Sci. **371**, 381-389 (1997)
47. Beutler, A., Lundgren, E., Nyholm, R., Andersen, J.N., Setlik, B. Heskett, D., *Coverage- and temperature-dependent site occupancy of carbon monoxide on Rh(111) studied by high-resolution core-level photoemission*, Surf. Sci. **396**, 117-136 (1998)
48. Van Hove, M.A., Koestner, R.J., Somorjai, G.A., *Low-energy electron-diffraction intensity analysis of a surface structure with three CO molecules in the unit cell, Rh(111)-(2×2)-3CO: compact adsorption in simultaneous bridge and nonsymmetric near-top sites*, Phys. Rev. Lett. **50**, 903-906 (1983)
49. Schweitzer, M. G.: *Laserspektroskopische Untersuchungen zur Adsorption und Reaktion von CO auf katalytisch aktiven Oberflächen*, PhD thesis, University of Heidelberg, (2003)
50. Dubois, L. H., Somorjai, G. A.: *The chemisorption of CO and CO_2 on Rh(111) studied by high resolution electron energy loss spectroscopy*, Surf. Sci. **91**, 514-532 (1980)
51. Yates, J.T. Jr, Williams, E. D., and Weinberg, W. H.: *Does chemisorbed carbon monoxide dissociate on Rhodium*, Surf. Sci. **91**, 562-570 (1980)
52. see e.g. Ertl, G., Freund, H.-J.: *Catalysis and surface science*, Phys. Today **52**, 32-38 (1999) and references therein.
53. Dacapo: A total energy program based on density functional theory, http://dcwww.camp.dtu.dk/campos//Dacapo/
54. Hammer, B., Hansen, L. B., Nørskov, J. K.: *Improved adsorption energetics within density-functional theory using revised Perdew-Burke-Ernzerhof functionals*, Phys. Rev. B **59**, 7413 (1999)
55. Perdew, J. P., Chevary, J. A., Vosko, S. H., Jackson, K. A., Pederson, M. R., Singh, D. J., Fiolhais, C.: *Atoms, molecules, solids, and surfaces: Applications of the generalized gradient approximation for exchange and correlation*, Phys. Rev. B **46**, 6671-6687 (1992)
56. Wei, D. H., Skelton, D. C., Kevan, S. D.: *Desorption and molecular interactions on surfaces: C/Rh(110), CO/Rh(100) and CO/Rh(111)*, Surf. Sci. **381**, 49-64 (1997)
57. Polanyi, J. C., Wong, W. H.: *Location of energy barriers. I. Effect on the dynamics of reactions A + BC*, J. Chem. Phys. **51**, 1439-50 (1969)

58. Polanyi, J. C., Mok, M. H.: *Location of energy barriers. II. Correlation with barrier height*, J. Chem. Phys. **51**, 1451-1469 (1969)
59. Polanyi, J. C., Mok, M. H.: *Location of energy barriers. III. Effect on the dynamics of reactions $AB + CD \rightarrow AC + BD$*, J. Chem. Phys. **53**, 4588-4604 (1970)
60. Laidler, K. J.: *Chemical Kinetics*, 3rd ed. Harper & Row, New York (1987)
61. Van der Laan, G. P., Beenackers, A. A. C. M.: *Kinetics and selectivity of the Fischer Tropsch synthesis: A literature review*, Catal. Rev.-Sci. Eng., **41**, 255-318 (1999).

Stochastic Modeling and Deterministic Limit of Catalytic Surface Processes[*]

J. Starke[1,2], C. Reichert[2], M. Eiswirth[3], and K. Oelschläger[1]

[1] Institut für Angewandte Mathematik, Universität Heidelberg.
[2] Interdisziplinäres Zentrum für Wissenschaftliches Rechnen (IWR), Universität Heidelberg.
[3] Fritz-Haber-Institut der Max Planck Gesellschaft, Berlin.

Summary. Three levels of modeling, microscopic, mesoscopic and macroscopic are discussed for the CO oxidation on low-index platinum single crystal surfaces. The introduced models on the microscopic and mesoscopic level are stochastic while the model on the macroscopic level is deterministic. It can be derived rigorously for low-pressure conditions from the microscopic model, which is characterized as a moderately interacting many-particle system, in the limit as the particle number tends to infinity. Also the mesoscopic model is given by a many-particle system. However, the particles move on a lattice, such that in contrast to the microscopic model the spatial resolution is reduced. The derivation of deterministic limit equations is in correspondence with the successful description of experiments under low-pressure conditions by deterministic reaction-diffusion equations while for intermediate pressures phenomena of stochastic origin can be observed in experiments. The models include a new approach to the platinum phase transition, which allows for a unification of existing models for Pt(100) and Pt(110). The rich nonlinear dynamical behavior of the macroscopic reaction kinetics is investigated and shows good agreement with low pressure experiments. Furthermore, for intermediate pressures, noise-induced pattern formation, which has not been captured by earlier models, can be reproduced in stochastic simulations with the mesoscopic model.

1 Introduction

Stochastic modeling with many-particle systems and derivation of the corresponding deterministic limit is important in many areas of science. The present work concentrates on the CO oxidation on Pt single crystal surfaces but the techniques presented here can certainly be used in other areas as well.

---

[*]This work has been supported by the German Research Foundation (DFG) through SFB 359 (Project B3) at the University of Heidelberg.

1.1 CO Oxidation on Pt Single Crystal Surfaces

Pattern formation under non-equilibrium conditions has been studied using a number of catalytic surface reactions on a large variety of different catalysts [9, 10, 29]. In order to distinguish genuine self-organized patterns from influences of catalyst inhomogeneities, it is necessary to work with uniform surfaces; best suited are oriented single crystal surfaces. The system studied most extensively under these conditions is the CO oxidation on Pt single crystals [27, 8, 29, 40]. In ultra-high vacuum experiments, the elementary processes were elucidated. The reaction proceeds via the classical Langmuir-Hinshelwood mechanism,

$$CO + \star \rightleftharpoons CO_{ad} \qquad (1)$$
$$O_2 + 2\star \rightarrow 2O_{ad}$$
$$CO_{ad} + O_{ad} \rightarrow CO_2 + 2\star,$$

where $\star$ is a vacant adsorption site on the Pt surface. It is important to note that there is asymmetric inhibition of adsorption, i.e. pre-adsorbed CO blocks oxygen adsorption but not vice versa. In addition, the adsorbate-induced phase transition

$$1 \times 2 \rightleftharpoons 1 \times 1 \quad \text{for Pt(110), and} \qquad (2)$$
$$\text{hex} \rightleftharpoons 1 \times 1 \quad \text{for Pt(100)},$$

has to be taken into account because the surface structure influences the reactivity. The considered experimental situation assumes constant partial pressures of CO and O_2 in the well-mixed gaseous phase. The produced CO_2 disappears immediately from the Pt surface. It is therefore sufficient to model the adsorbed CO molecules and oxygen atoms on the Pt surface in addition to the platinum phase.

On a molecular level, the relevant elementary processes are of stochastic nature. Thus, fluctuations were shown to strongly influence the behavior in experiments with field emitter tips and corresponding Monte-Carlo simulations [46, 45]. In contrast, pattern formation on extended single crystal surfaces at low pressure ($\lesssim 10^{-4}$ mbar) did not reveal any effects which would suggest a stochastic origin, and could indeed be successfully modeled with (deterministic) reaction-diffusion equations [8]. The reason is that at low pressures there occur due to the diffusion of adsorbed CO on the Pt surface about 10^6 site changes per adsorption event, i.e. the surface is well mixed on a length scale of about 1μm and fluctuations on the molecular level are averaged out. With increasing pressure smaller and smaller patches can be regarded as well mixed, the size of a critical nucleus decreases [2]. Models based on reaction-diffusion equations are expected to fail and stochastic effects can become relevant. An experimental observation at an intermediate oxygen pressure ($p_{O_2} = 10^{-2}$ mbar) is reproduced in Fig. 1. The CO pressure p_{CO} had been stepwise increased to a point shortly before the whole surface would switch to

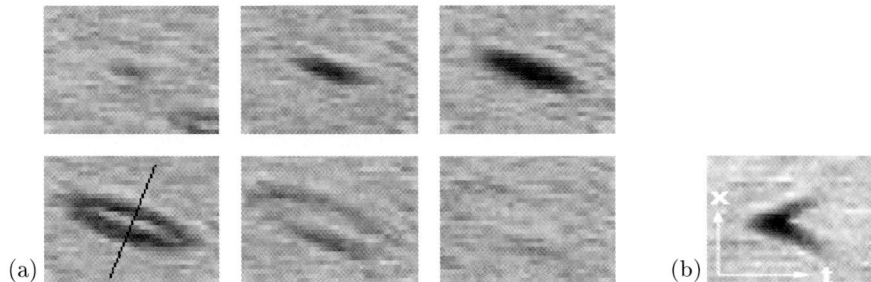

Fig. 1. (a) Snapshots of a Pt(110) single crystal surface showing the so-called raindrop patterns using EMSI (Ellipso Microscopy for Surface Imaging) [40]. The time between the snapshots is 160 ms. The length scale is 100 μm $\times$ 70 μm. The partial pressures are $p_{CO} = 7 \times 10^{-3}$ mbar and $p_{O_2} = 2.2 \times 10^{-2}$ mbar. (b) Space-time diagram of the raindrop [40] showing 1.6 s $\times$ 100 μm. (Reprinted with permission of H. H. Rotermund.)

the CO-covered state. CO nuclei were observed to originate at various places, forming a ring-shaped pattern, but were subsequently destroyed (propagation failure). These phenomena are called raindrop patterns due to the similarity to damped out waves on a water surface in starting rain. They seem to appear randomly distributed all over the catalyst surface [41].

Consequently, presented in terms of three levels of modeling, a unified stochastic model was developed which reproduces the mean-field limit, i.e. the deterministic behavior at low pressures, but also describes the stochastic effects observed in experiments at intermediate pressures. In addition, for these intermediate pressures, the reaction is no longer isothermal because the elevated turnover releases more heat, which is also included in the presented modeling.

1.2 Stochastic Modeling and Deterministic Limit

Various types of many-particle models have been proposed to describe and analyze phenomena on small scales in science including fluctuations and other stochastic effects. To analyze the behavior of those microscopically defined systems on macroscopic scales, it is useful to study the many-particle models in the limit as the particle number tends to infinity. A class of models with strong local interaction essentially between immediately neighboring particles leads to the so-called hydrodynamic limit [42, 5, 31]. On the other hand, models where any particle has a substantial albeit weak influence on essentially all other particles, i.e. models with global interactions lead to the so-called McKean-Vlasov limit [36]. To describe the experimental conditions of the CO oxidation on Pt considered in the present paper the limit behavior of many-particle models exhibiting a local interaction with moderate interaction range which is much larger than the typical distance between neighboring particles but much smaller than the total system size has to be studied. This results in a situation in between the hydrodynamic and the McKean-Vlasov limit.

In the following we present and investigate two types of these stochastic models with local interactions between many mesoscopically neighboring particles, namely a stochastic lattice model, also known as box model or Gillespie model [16, 17] and a moderately interacting many-particle model in continuous space. In contrast to microscopic Monte-Carlo or stochastic cellular automaton models the considered stochastic lattice model allows for fast simulations of large particle numbers for realistic parameters while the moderately interacting many-particle model is well suited for a rigorous derivation of the corresponding macroscopic limit equation.

Examples of stochastic lattice models using cells as spatial discretization can be found in [21] and [16]. Some analysis of these models has been based on the Master equation while a mathematically rigorous derivation of the deterministic limit was performed for chemical models assuming either specific reaction kinetics with at most quadratic terms or polynomial kinetics for systems with only one species [32, 33]. A rigorous derivation of the Smoluchowski equation for similar lattice models can be found in [20].

A rigorous derivation of the deterministic limit for particular many-particle models with moderate interaction, which model the evolution of biological populations, can be found in [37]. Applications of this approach have been employed to model the spatio-temporal pattern formation of myxobacteria [44] and to derive the Smoluchowski equation in the context of an astrophysical application [18].

The present paper is organized as follows: A mesoscopic lattice model for isothermal CO oxidation and its deterministic limit for the spatially homogeneous case are presented in Section 2. In Section 3 a limit theorem for a microscopic model, namely a moderately interacting many-particle system, is presented, which gives a rigorous derivation of the macroscopic reaction-diffusion equations as deterministic limit equations. The stochastic and thermo-kinetic effects which come into play at intermediate pressures are modeled in Section 4.

2 Stochastic Lattice Model for Isothermal CO Oxidation

The presented stochastic lattice model describes CO oxidation on Pt on a mesoscopic level with stochastic birth-death processes for the number of adsorbed CO molecules, oxygen atoms and adsorption sites in (1×1) surface structure.

2.1 Model Formulation and Birth-Death Processes

Instead of distinguishing every single adsorption site like in microscopic Monte-Carlo simulations or in lattice gas approaches, the Pt surface is divided into a two-dimensional array or lattice of cells of mesoscopic size each containing N adsorption sites. Each cell is assumed to be well-mixed. The spatial resolution can thus be varied by changing N which automatically rescales the characteristic length. The state of the system with cells (i,j) at time $t \geq 0$ is described by

$$\mathbf{X}_N(t) = \left(\mathbf{X}_N^{(i,j)}(t)\right)_{i,j} \text{ with } \mathbf{X}_N^{(i,j)}(t) = \begin{pmatrix} N_{\mathrm{CO}}^{(i,j)}(t) \\ N_{\mathrm{O}}^{(i,j)}(t) \\ N_{1\times 1}^{(i,j)}(t) \end{pmatrix} \in S_N^{(i,j)} = \{0,\ldots,N\}^3, \tag{3}$$

where $N_{\mathrm{CO}}^{(i,j)}$, $N_{\mathrm{O}}^{(i,j)}$, and $N_{1\times 1}^{(i,j)}$ denote for each cell (i,j) the numbers of adsorbed CO molecules, adsorbed oxygen atoms, and adsorption sites in a non-reconstructed (1×1) surface structure. In the following the specific labels (i,j) of each cell are only displayed for the considered interactions between cells. In order to characterize a stochastic process corresponding to the reaction mechanisms, transition probabilities to other states $P\big[\mathbf{X}_N(t+h) = \mathbf{X}_N(t) + \delta\mathbf{X}_N \mid \mathbf{X}_N(t)\big]$ for small h have to be specified. We introduce the concentrations $u_N := \frac{N_{\mathrm{CO}}}{N}$, $v_N := \frac{N_{\mathrm{O}}}{N}$, and $w_N := \frac{N_{1\times 1}}{N}$; in the limit $N \to \infty$ they are denoted by u, v, and w.

The terms appearing below in the transition probabilities concerning N_{CO} and N_{O} essentially correspond to those used for the (deterministic) reconstruction model for Pt(110) in [34]. The parameters and rate constants are defined in Table 1. The parameters which can be varied in experiments are the partial pressures of CO and oxygen in the gas phase p_{CO}, p_{O_2} and the surface temperature T. For further discussions see [34, 12, 28].

Adsorption

- Birth of a CO molecule on the surface.
  $N_{\mathrm{CO}} \to N_{\mathrm{CO}} + 1$.

  $$P\big[\mathbf{X}_N(t+h) = \mathbf{X}_N(t) + (1,0,0) \mid \mathbf{X}_N(t)\big] = \\ Np_{\mathrm{CO}}\kappa_{\mathrm{CO}} s_{\mathrm{CO}} \left(1 - u_N^\xi(t)\right) h + o(h) \tag{4}$$

- Birth of two adsorbed oxygen atoms.
  $N_{\mathrm{O}} \to N_{\mathrm{O}} + 2$.

  $$P\big[\mathbf{X}_N(t+h) = \mathbf{X}_N(t) + (0,2,0) \mid \mathbf{X}_N(t)\big] = \\ Np_{\mathrm{O}_2} \tfrac{1}{2}\kappa_{\mathrm{O}} \left(s_{\mathrm{O}}^{\mathrm{rec}}(1 - w_N(t)) + s_{\mathrm{O}}^{1\times 1} w_N(t)\right) \\ \times \left((1 - u_N(t))(1 - v_N(t))\right)^2 h + o(h) \tag{5}$$

By assuming the adsorption rate of CO to be proportional to $1-u^\xi$ independent of the oxygen coverage as in [34], one implicitly drops the conservation constraint imposed on N_O, N_{CO}, and the number of vacant sites by the pure Langmuir-Hinshelwood mechanism. On the other hand, it has been observed experimentally that the presence of oxygen on the surface does not noticeably influence the adsorption of CO [29, 13]. Consequently, we assume that CO molecules can also be adsorbed at O-covered sites, whereas the dissociative adsorption of an oxygen molecule can only take place at two neighboring free sites. The probability that a particular site is occupied neither by CO nor by O is then $(1-u)(1-v)$ instead of $1-u-v$ like it is used in [34]. Since the first term is larger only by uv, the difference is negligible if u or v is small (i.e. at low pressure). At intermediate pressure the new term is in better agreement with experiments [39].

The exponent ξ is introduced formally to model a precursor effect in the adsorption of CO as in [34], but it was always set to 1 in the present computations. Note that the sticking coefficient of oxygen on the 1×1 phase is higher than on the reconstructed phase which can lead to oscillatory, doubly metastable, and excitable behavior of the reaction kinetics.

Desorption

- Death of a CO molecule through desorption.
  $N_{CO} \to N_{CO} - 1$.

$$P[\mathbf{X}_N(t+h) = \mathbf{X}_N(t) + (-1,0,0) \mid \mathbf{X}_N(t)] = \\ N\left(k_{des}^{rec}(1-w_N(t)) + k_{des}^{1\times1}w_N(t)\right)u_N(t)h + o(h) \quad (6)$$

The desorption of CO molecules from the reconstructed and the 1×1 phase has been distinguished here because the difference in the binding energies cannot be neglected in the case of Pt(100).

Reaction

- Death of a CO molecule and an oxygen atom through reaction.
  $N_{CO} \to N_{CO} - 1$, $N_O \to N_O - 1$.

$$P[\mathbf{X}_N(t+h) = \mathbf{X}_N(t) + (-1,-1,0) \mid \mathbf{X}_N(t)] = \\ Nk_{re}u_N(t)v_N(t)h + o(h) \quad (7)$$

For a well-mixed patch the reaction probability is proportional to the product of the concentrations (mass action kinetics).

Phase transition

- Death of an adsorption site in a 1×1 patch and birth of a site in a reconstructed patch.
$N_{1\times 1} \to N_{1\times 1} - 1$.

$$P\big[\mathbf{X}_N(t+h) = \mathbf{X}_N(t) + (0,0,-1) \mid \mathbf{X}_N(t)\big] = \\ Nk_{\mathrm{rec}} f_{\mathrm{rec}}\left(u_N(t), w_N(t)\right) w_N(t) h + o(h) \qquad (8)$$

- Death of a site in a reconstructed patch and birth of a site in a 1×1 patch.
$N_{1\times 1} \to N_{1\times 1} + 1$.

$$P\big[\mathbf{X}_N(t+h) = \mathbf{X}_N(t) + (0,0,1) \mid \mathbf{X}_N(t)\big] = \\ Nk_{1\times 1} f_{1\times 1}\left(u_N(t), v_N(t)\right) \left(1 - w_N(t)\right) h + o(h), \qquad (9)$$

where

$$f_{\mathrm{rec}}(u_N, w_N) = (1-\epsilon)\left(1-u_N\right)^\lambda + \epsilon\left(1-w_N\right)^\lambda, \qquad (10)$$

and

$$f_{1\times 1}(u_N, w_N) = (1-\epsilon) u_N^\lambda + \epsilon w_N^\lambda, \quad \lambda > 1, \; \epsilon \in [0,1]. \qquad (11)$$

This ansatz is motivated by the fact that the phase transition proceeds via nucleation and growth. The probability for nucleation is determined solely by the CO coverage, but the growth of a phase can to some extent be autocatalytic which leads to a dependence of the rate of growth on the concentration of the phase itself. Therefore we choose e.g. for the growth rate of the 1×1 phase on a reconstructed surface a weighted sum of u and w to some power λ, respectively (cf. equations (10) and (11)). A highly nonlinear dependence of this rate on u has been obtained experimentally by Hopkinson et al. [26]. They measured the growth rate of the 1×1 phase on a hex-R reconstructed Pt(100) surface to depend on the CO coverage on the hex-R phase to a power of about 4.5. It is plausible that the exponent for w should be of the same order of magnitude. The reverse transition is modeled in an analogous way. In order not to introduce too many parameters, the same ϵ and λ are used, however, the rate constants are allowed to be different.

The effect of the weight ϵ is shown in Fig. 2. For $\epsilon = 0.1$, which describes well the situation of Pt(110), the equilibrium portion of 1×1 phase can be expressed as a function of u_{eq}, as expected (cf. Fig. 2a). At higher values of ϵ both phases are stable within a certain range of u_{eq} (cf. Fig. 2b), which is the case for Pt(100) where the larger weight ϵ captures the larger influence of the neighboring phase for the Pt phase transition. Pt(111) does not undergo a structural phase transition under the considered conditions. It can therefore be described by the same model after neglecting this process.

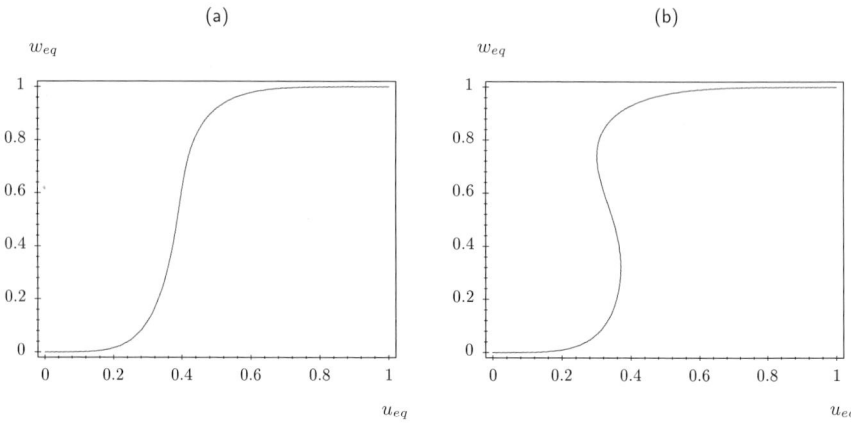

Fig. 2. Equilibria of the kinetics for $v \equiv 0$, $\frac{k_{\text{rec}}}{k_{1\times 1}} = 0.2$, $\lambda = 4$, $k_{\text{des}}^{\text{rec}} = k_{\text{des}}^{1\times 1}$, at different values of ϵ. (a) $\epsilon = 0.1$. (b) $\epsilon = 0.3$. Introducing different binding energies for CO on 1×1- and reconstructed phase and thus a coupling between u and w, had almost no effect on the curves.

Diffusion

- Diffusive jump processes of a CO molecule in cell (i,j) to neighboring cells.

$$\begin{pmatrix} N_{\text{CO}}^{(i,j)} \\ N_{\text{CO}}^{(i+\Delta i, j+\Delta j)} \end{pmatrix} \to \begin{pmatrix} N_{\text{CO}}^{(i,j)} - 1 \\ N_{\text{CO}}^{(i+\Delta i, j+\Delta j)} + 1 \end{pmatrix} \text{ with } \begin{pmatrix} \Delta i \\ \Delta j \end{pmatrix} \in \left\{ \begin{pmatrix} \pm 1 \\ 0 \end{pmatrix}, \begin{pmatrix} 0 \\ \pm 1 \end{pmatrix} \right\}.$$

$$P\left[\begin{pmatrix} \mathbf{X}_N^{(i,j)} \\ \mathbf{X}_N^{(i+\Delta i, j+\Delta j)} \end{pmatrix}(t+h) = \begin{pmatrix} \mathbf{X}_N^{(i,j)} \\ \mathbf{X}_N^{(i+\Delta i, j+\Delta j)} \end{pmatrix}(t) + \begin{pmatrix} -1 \\ 0 \\ 0 \\ 1 \\ 0 \\ 0 \end{pmatrix} \right.$$

$$\left. \bigg| \begin{pmatrix} \mathbf{X}_N^{(i,j)} \\ \mathbf{X}_N^{(i+\Delta i, j+\Delta j)} \end{pmatrix}(t) \right] = N \frac{1}{4} D_{\text{CO},N} u_N^{(i,j)}(t) + o(h) \qquad (12)$$

Suitable boundary conditions have to be added. The diffusion of adsorbed oxygen is negligible.

Transition probabilities to other states behave like $o(h)$. (It is also assumed that the probability for a transition is zero if it would lead out of the state space.) Given an initial distribution for $\left(\mathbf{X}_N^{(i,j)}(0)\right)$, there is a Markov jump process in continuous time with state space S_N containing the states $S_N^{(i,j)}$ of all cells whose transition probabilities satisfy the above equations. Its paths can be chosen right-continuous with left limits. Because the state space S_N and transition intensities are bounded for fixed N, explosions of particle numbers cannot occur, and the process is well-defined for $t \in [0, \infty)$ [3]. The time

Table 1. Kinetic parameters for Pt(110) and Pt(100).

		Pt(110)	Pt(100)
CO adsorption			
impingement rate	κ_{CO}	3.135×10^5 mbar^{-1} s^{-1}	2.205×10^5 mbar^{-1} s^{-1}
sticking coefficient	s_{CO}	1	1
saturation coverage	u_s	1	1
precursor exponent	ξ	1	1
O$_2$ adsorption			
impingement rate	κ_O	5.858×10^5 mbar^{-1} s^{-1}	3.75×10^5 mbar^{-1} s^{-1}
sticking coefficient	s_O^{rec}	0.3	0.001
	$s_O^{1\times 1}$	0.6	0.3
saturation coverage	v_s	1	1
Arrhenius rates	$\nu \exp(-\frac{E}{k_B T})$		
CO desorption	k_{des}^{rec}	$\nu_{des}^{rec} = 5 \times 10^{13}$ s^{-1}	$\nu_{des}^{rec} = 3.323 \times 10^{12}$ s^{-1}
		$E_{des}^{rec} = 32.3$ kcal/mol	$E_{des}^{rec} = 27.5$ kcal/mol
CO desorption	$k_{des}^{1\times 1}$	$\nu_{des}^{1\times 1} = 5 \times 10^{13}$ s^{-1}	$\nu_{des}^{1\times 1} = 8.640 \times 10^{14}$ s^{-1}
		$E_{des}^{1\times 1} = 32.3$ kcal/mol	$E_{des}^{1\times 1} = 35.0$ kcal/mol
reaction	k_{re}	$\nu_{re} = 5 \times 10^5$ s^{-1}	$\nu_{re} = 1.185 \times 10^9$ s^{-1}
		$E_{re} = 8.1$ kcal/mol	$E_{re} = 14$ kcal/mol
phase transition	$k_{1\times 1}$	$\nu_{1\times 1} = 10^3$ s^{-1}	$\nu_{1\times 1} = 2.417 \times 10^{11}$ s^{-1}
		$E_{1\times 1} = 6.9$ kcal/mol	$E_{1\times 1} = 25$ kcal/mol
	k_{rec}	$\nu_{rec} = 0.2 \times 10^3$ s^{-1}	$\nu_{rec} = 4.833 \times 10^{10}$ s^{-1}
		$E_{rec} = 6.9$ kcal/mol	$E_{rec} = 25$ kcal/mol
phase transition			
exponent	λ	4	4
weight	ϵ	0.1	0.3

evolution of $P[\mathbf{X}_N(t) = \mathbf{Y} \mid \mathbf{X}_N(0) = \mathbf{X}_0]$, $\mathbf{Y}, \mathbf{X}_0 \in S_N$, is governed by a master equation [23, 22, 15, 25].

2.2 Deterministic Limit

To investigate the dynamical properties of the model it is convenient to use the deterministic limit of the introduced many-particle model. The deterministic limit is given for the homogeneous case in the following.

We denote the stochastic processes describing the dynamics of the densities by

$$\mathbf{x}_N(t) = \frac{1}{N} \mathbf{X}_N(t) = \begin{pmatrix} u_N(t) \\ v_N(t) \\ w_N(t) \end{pmatrix}. \tag{13}$$

Note that all transition intensities (cf. equations (4–9)) are proportional to a product of the system size N and a function of the concentrations only. Because of this scaling property it can be proved rigorously that, if $\lim_{N\to\infty} \mathbf{x}_N(0) = \mathbf{x}_0 \in [0,1]^3$ holds with probability one. The paths of the stochastic processes $\mathbf{x}_N(t)$ approximate the solution $\mathbf{x}(t)$ of the initial value problem given by the system of ODEs

$$\dot{u} = p_{CO}\kappa_{CO}s_{CO}\left(1 - u^\xi\right) - \left(k_{des}^{rec}(1-w) + k_{des}^{1\times 1}w\right)u - k_{re}uv$$
$$\dot{v} = p_{O_2}\kappa_O\left(s_O^{rec}(1-w) + s_O^{1\times 1}w\right)\left((1-u)(1-v)\right)^2 - k_{re}uv$$
$$\dot{w} = k_{1\times 1}f_{1\times 1}(u,w)(1-w) - k_{rec}f_{rec}(u,w)w, \tag{14}$$

and the initial condition $\mathbf{x}(0) = \mathbf{x}_0$ at large particle numbers in the following sense:

For arbitrary $t \geq 0$: $\lim_{N\to\infty} \sup_{0 \leq s \leq t} |\mathbf{x}_N(s) - \mathbf{x}(s)| = 0$ with probability one.

In addition, a central limit theorem can be shown which states that fluctuations of the concentrations vanish like $O(\frac{1}{\sqrt{N}})$. The proofs follow directly from theorems by Kurtz [35, 14]. A non-rigorous approach to these results is known in the physical literature as van Kampen's system size expansion [15].

Bifurcation analysis

Subsequently we present some aspects of the bifurcation structure of system (14) with parameters appropriate for Pt(110) and Pt(100). For a more complete treatment see [39]. The computations were performed using algorithms contained in the AUTO 97 package by Doedel et al. [6]. Abbreviations of the bifurcations found are listed in Table 2, for details see [24, 47, 19].

For Pt(110), bifurcation diagrams in p_{CO} and p_{O_2} have been computed for several fixed sample temperatures. The kinetic parameters used are listed in Table 1.

At higher temperatures (500 K–560 K) the bifurcation diagram is mainly organized by a cusp and two Takens-Bogdanov points. In total there are 12 parameter regions with different dynamical behavior (cf. Fig. 3), but only regions 1–5 are physically relevant because the remaining regions are too small to be detected in typical experiments. In areas 1, 2, and 3 there is only one attractor, a stable node and an asymptotically stable periodic orbit, respectively. The maximal width of the oscillatory region 2 at 540 K amounts to ≈ 10% of the value of p_{CO} at the supercritical Hopf bifurcation. It decreases towards higher as well as lower temperatures. In 4 and 5 two attractors coexist; in 5 there are two stable nodes, while in 4 the system reaches either a stable node or a small asymptotically stable periodic orbit. The width of region 4 is $\lesssim 1\%$ of p_{CO} at 540 K, so it could possibly be detected experimentally. The areas with nontrivial dynamics, such as bistability or oscillations, move towards higher partial pressures as temperature is increased.

Phase portraits of the dynamics in the different parameter regions are sketched in Fig. 3. In the pictures containing three fixed points the lower

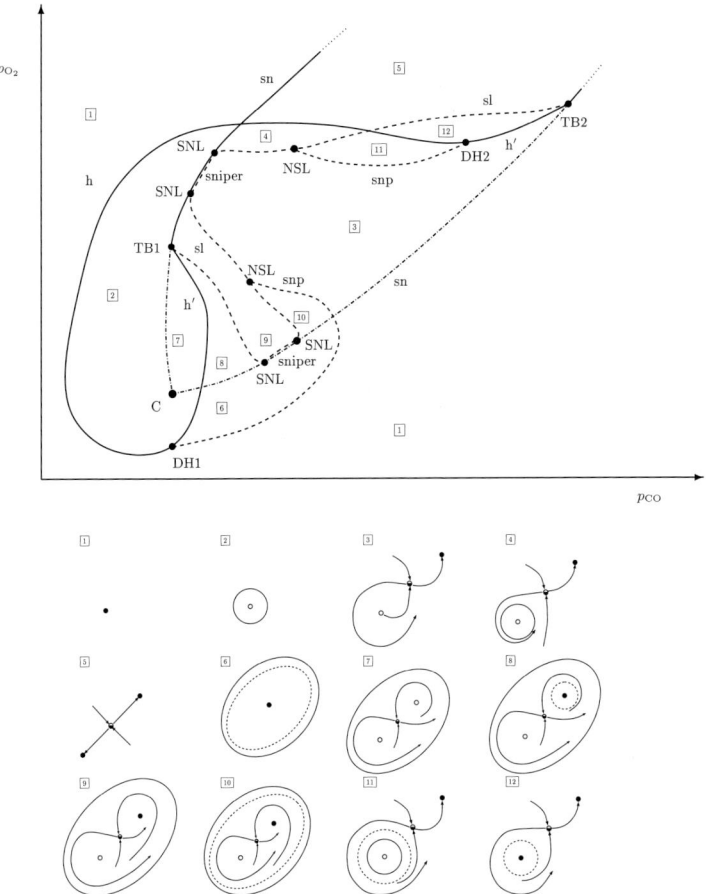

Fig. 3. Sketches of the complete bifurcation diagram at 500 K–560 K (top) and phase portraits in different parameter regions (bottom) for Pt(110). Hopf bifurcations, and saddle-node bifurcations involving a stable node and a saddle with a one-dimensional unstable manifold are drawn with solid lines, saddle-node bifurcations involving a saddle with a one-dimensional and another with a two-dimensional unstable manifold are drawn with dash-dotted lines. The dashed curves indicate global bifurcations, cf. Table 2. In the phase portraits stable nodes are represented by filled circles, saddle points with a two-dimensional unstable manifold by empty circles, and saddle points with a one-dimensional unstable manifold by half-filled circles. Asymptotically stable periodic orbits are indicated by solid lines, unstable ones by dashed lines.

one can always be identified with a reactive, only partially non-reconstructed surface with a relatively high oxygen coverage (reactive state), the upper one with a mainly CO-covered 1×1 surface (poisoned state).

Table 2. Commonly used denotations of bifurcations and their abbreviations. By a degenerate Hopf bifurcation we mean the one described in [19].

Bifurcation	Co-dimension	Abbreviation
Hopf bifurcation (supercritical)	1	h
Hopf bifurcation (subcritical)	1	h'
saddle-node	1	sn
saddle loop homoclinic bifurcation	1	sl
saddle-node/infinite period saddle-node on a loop	1	sniper
saddle node of periodic orbits	1	snp
cusp	2	C
Takens-Bogdanov bifurcation	2	TB
degenerate Hopf bifurcation	2	DH
saddle-node loop	2	SNL
neutral saddle loop trace 0 saddle loop	2	NSL
Takens-Bogdanov-cusp	3	TBC

From a physical point of view the model presented here yields almost the same results as the one proposed in ref. [34]. The most important distinction is that here the two curves of saddle-node bifurcations do not merge for the investigated parameter region in a second cusp when p_{CO} and p_{O_2} are increased, rather bistability persists even at atmospheric pressures, in agreement with experiment. In fact, this is due to the change of the adsorption kinetics of oxygen, as was checked by repeating the computation of the saddle-node curves with the term that was used in former models [34]. As in [34], the dynamics seems to be essentially two-dimensional; for instance a period doubling transition to chaos which was observed experimentally [11], could not be found.

In the case of Pt(100), the difference in the binding energies of CO on 1×1- and hex phase must be taken into account. Moreover, some kinetic parameters have to be changed, most importantly the sticking coefficient of oxygen on the hex phase is much smaller than on the 1×2 phase of Pt(110). The weight ϵ is increased to $\epsilon = 0.3$, in order to model hysteresis in the phase transition; the influence of adsorbed oxygen on the restructuring processes has been neglected. The kinetic parameters used have been adapted from an older model [28]; they are listed in Table 1. For details of the bifurcation structure see [39].

The oscillatory region is much broader than in the case of Pt(110). It compares favorably with experiments (cf. Figs. 4a and 4b). In contrast to earlier models [28, 1] no arbitrary parameters have to be introduced through so-called defect terms to get the required behavior. In contrast to Pt(110), Pt(100) exhibits two distinct bistable regions. The one at low pressures is due to the

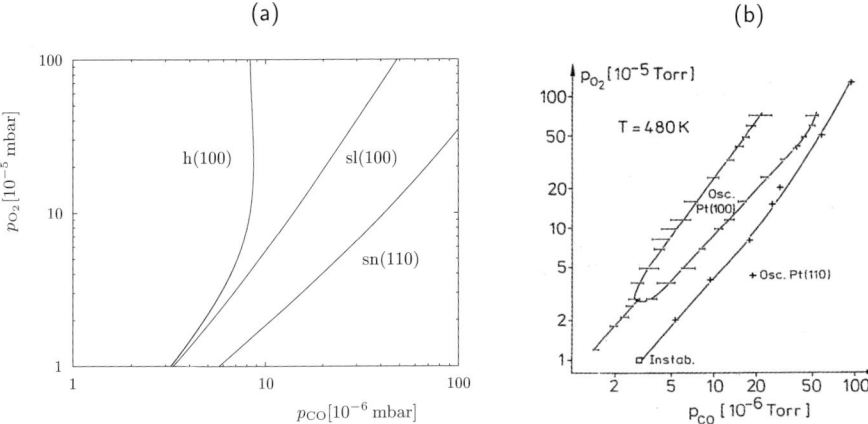

Fig. 4. Occurrence of oscillations at 480 K. (a) The curves of Hopf- and sl bifurcations are plotted for Pt(100) using a logarithmic scale. The line on the right is the curve of saddle-node bifurcations corresponding to the loss of stability of the reactive state for Pt(110). (b) Corresponding experimental diagram [12].

hysteresis in the phase transition and extends to zero oxygen pressure, while the one at high pressures is controlled by the reaction dynamics and again does not close. Overall, the bifurcation fine structure turned out somewhat simpler than with Pt(110).

3 Limit Theorem for Many-Particle Systems in Continuous Space

In this section, which summarizes [38], the relationship between two different approaches to the mathematical modeling of general chemical systems with intermediate particle interactions, i.e. with microscopically large and macroscopically small range, is investigated. In particular, the rigorous derivation of macroscopic limit equations (reaction-diffusion equations) from microscopic moderately interacting many-particle models is discussed.

3.1 Microscopic and Macroscopic Modeling of Chemical Reactions

For a large number of complex chemical systems, like the CO oxidation on Pt, the elementary reaction steps are understood, i.e., the microscopic dynamics of the chemical components is known. In contrast to this, many models describing macroscopic phenomena like spatio-temporal pattern formation should be improved, or, at least, their foundation on the microscopic description has to be supplemented.

In order to model particular chemical reactions on a microscopic level we consider a class of interacting many-particle systems. Especially, we focus on the use of moderately interacting many-particle systems, where for large particle numbers the mean distance between neighboring particles is much smaller than the range of the interaction, which in turn is much smaller than the spatial system size. This is fulfilled for many important chemical reactions, where the dynamics is reaction- rather than diffusion-controlled. These systems can be regarded as being locally well-mixed. To distinguish that model from the stochastic lattice model described in Section 2 we use the term "microscopic model" for the approach in this section because the positions of individual particles are considered here in continuous space and not on a discrete lattice which corresponds to a reduced spatial resolution. Nevertheless, by using densely populated cells or a moderately scaled interaction, both models are associated with a local averaging in connection with the respective interaction.

Similarly as in [37] we study for any $N \in \mathbb{N}$ a system of $\approx N$ particles in $\mathbb{R}^d$. To model different reacting species the set of all particles is divided into a finite collection of K subsets, where K is independent of N. The dynamics of the individual particles is determined by three contributions, namely

- immigration, i.e., the entering of new particles into the system e.g. by adsorption of a gaseous species on a surface,
- reaction, i.e., the transformation of pairs of interacting particles into particles of another species, and
- diffusion.

As far as the interaction among the particles is concerned, we suppose similarly as in [37] that immigration- and reaction rates depend on regularized versions of the empirical processes (16) of the various species. For fixed $N \in \mathbb{N}$ these regularizations are defined as convolutions of the respective empirical processes with a kernel ψ_N. Additionally, in our model the reaction rates involve some interaction potential φ_N describing the influence of the distance between the reacting particles, i.e., pair-interaction is included. With δ_0 denoting Dirac's δ-distribution we assume that

$$\lim_{N \to \infty} \psi_N = \lim_{N \to \infty} \varphi_N = \delta_0 \quad \text{in } \mathcal{S}'(\mathbb{R}^d) \text{ sufficiently slow.} \tag{15}$$

The rate of convergence in (15) has to be slow enough, in particular, in order that the various contributions to the interaction scale in a moderate way and hence moderately interacting many-particle systems are obtained, cf. [37]. In contrast to [37], in the model discussed here the diffusion may vanish for certain species. Moreover, no drift is acting on the particles. Last not least, the analysis in [37] is simplified considerably by the absence of explicit pair-interaction.

We are aiming at the study of the many-particle systems for large particle numbers, i.e., we investigate the limit $N \to \infty$. Especially, the convergence of the systems of empirical processes to the solution of a system of reaction-diffusion equations will be proved. This limit system represents a macroscopic model of the respective chemical reaction.

As an important application the particular example of the heterogeneous catalytic CO oxidation on Pt is considered for low pressures. The already mentioned 10^6 site changes of an adsorbed CO molecule per adsorption event due to diffusive hopping results in a locally well-mixed system. The asymmetric inhibition as well as the dynamical adsorbate dependent reconstruction of the Pt structure can fully be considered also in this microscopic moderately interacting many-particle model. The description of this interesting surface reaction with its complicated spatio-temporal dynamical behavior (cf. Section 2.2 for a numerical bifurcation analysis of the kinetics) is a suitable test for the proposed abstract chemical model. The mathematical modeling in this example including specific rates for the physical processes is presented in Section 3.5. Additionally, the corresponding system of partial differential equations for the macroscopic model is given there.

Sections 3.2 and 3.3 contain mathematical details for the general microscopic model and a limit theorem describing its relation to the corresponding macroscopic model. In Section 3.4 some remarks on the proof of the limit theorem can be found.

3.2 Mathematical Details of the Microscopic Model for General Reactions

For $N \in \mathbb{N}$ and $t \geq 0$ we use disjoint subsets $M_N^r(t)$, $r = 1, ..., K$, of $\mathbb{N}$ to describe the particles in the various subpopulations marked with r.

Let $X_N^k(t) \in \mathbb{R}^d$ be the position of particle k in the N-th system at time t. For fixed $N \in \mathbb{N}$ the distribution of the particles in space is characterized by empirical processes

$$\mathbb{X}_N^r(t) = \frac{1}{N} \sum_{k \in M_N^r(t)} \delta_{X_N^k(t)}, \quad t \geq 0, \ r = 1, \ldots, K, \qquad (16)$$

which obviously take values in the space of nonnegative measures on $\mathbb{R}^d$.

The empirical processes will be used for two purposes:

- The reaction- and immigration rates describing the interaction between the particles depend on these processes.
- They are studied in the limit as $N \to \infty$ to determine the asymptotic behavior of the many-particle systems.

To quantify the interaction the empirical processes $\mathbb{X}_N^r$ are not employed directly. Instead, they are replaced by regularized versions ρ_N^r, which are obtained by convolving $\mathbb{X}_N^r$ with some suitable kernel ψ_N, i.e.,

$$\rho_N^r(.,t) = \mathbb{X}_N^r(t) * \psi_N, \quad t \geq 0, \ r = 1, \ldots, K, \ N \in \mathbb{N}. \qquad (17)$$

We suppose that the family ψ_N, $N \in \mathbb{N}$, is obtained by the scaling

$$\psi_N(x) = \theta_N \psi_1(\theta_N^{1/d} x), \quad x \in \mathbb{R}^d, \ N \in \mathbb{N}, \tag{18}$$

where the function ψ_1 satisfying

$$\psi_1 \in C_b(\mathbb{R}^d; [0, \infty)), \quad \int_{\mathbb{R}^d} dx \, \psi_1(x) = 1, \quad \psi_1(-y) = \psi_1(y), \ y \in \mathbb{R}^d, \tag{19}$$

is fixed. For the scaling coefficients θ_N, $N \in \mathbb{N}$, in (18) we suppose

$$\theta_1 = 1, \quad \theta_N \in [1, \infty), \ N \in \mathbb{N}, \quad \lim_{N \to \infty} \theta_N = \infty \quad \text{sufficiently slow.} \tag{20}$$

It will be convenient to assemble ρ_N^r, $r = 1, \ldots, K$, in the function $\rho_N : \mathbb{R}^d \times [0, \infty) \to [0, \infty)^K$ defined by

$$\rho_N(x, t) = (\rho_N^1(x, t), \ldots, \rho_N^K(x, t)), \quad x \in \mathbb{R}^d, \ t \geq 0. \tag{21}$$

As a further abbreviation we employ

$$\rho_N(x, y, t) = \frac{1}{2}(\rho_N(x, t) + \rho_N(y, t)), \quad x, y \in \mathbb{R}^d, \ t \geq 0, \tag{22}$$

which will be utilized in the reaction rates.

Remark 1. Obviously, the transition from an empirical process X_N^r to its regularization ρ_N^r may be considered as a spatial blurring of the point masses describing the positions of the individual particles. If this blurring is strong enough, i.e., if $\theta_N \to \infty$ sufficiently slow, cf. (20), it may be expected that asymptotically as $N \to \infty$ local fluctuations of mass concentrations vanish in ρ_N^r, and therefore $\rho = (\rho^1, \ldots, \rho^K) = \lim_{N \to \infty} \rho_N$ should exist under rather general circumstances.

Now, for fixed $N \in \mathbb{N}$ the different contributions to the dynamics of the N-th many-particle system are described in detail.

To put our considerations involving random variables and stochastic processes on a firm basis we can work with a suitable filtered probability space $(\Omega, \mathcal{F}, (\mathcal{F}_t)_{t \geq 0}, \mathbf{P})$, where $(\mathcal{F}_t)_{t \geq 0}$ is some filtration satisfying the usual conditions, cf. [30]. We assume that any stochastic process appearing subsequently in this section is adapted to $(\mathcal{F}_t)_{t \geq 0}$.

Diffusion

We suppose that during their lifetimes the particles perform independent Brownian motions, where the diffusion constant is $\sigma_r \geq 0$, if the respective particle belongs to species r. Since the particles of certain species may be immobile, $\sigma_r = 0$ is allowed. To model the resulting paths of the particles independent standard $\mathbb{R}^d$-valued Brownian motions B_N^k, $k \in \mathbb{N}$, may be employed.

Immigration

It is assumed that particles may enter the system with density-dependent rates. More precisely, we suppose the existence of sufficiently regular functions

$$\beta^r \in C_b^1\big(\mathbb{R}^d \times [0,\infty)^K; [0,\infty)\big), \quad r = 1, \ldots, K,$$

such that

$$\mathbf{P}\big[\text{a particle of species } r \text{ is immigrating in } U \text{ during } (t, t+\delta] \big| \mathcal{F}_t\big] \quad (23)$$

$$= \delta N \int_U dx\ \beta^r(x, \rho_N(x,t)) + o(\delta), \quad \text{as } \delta \searrow 0,$$

$$t \geq 0, \ r = 1, \ldots, K,\ U \subseteq \mathbb{R}^d \text{ measurable}.$$

Reaction

In nature various different types of reactions may occur. For notational simplicity we consider in our general model only the cases, where different particles of species r and r' can react to turn either into one particle of species p, or two particles of species p and p', or three particles of species p, p' and p''. Whereas r and r' are allowed to coincide, we assume that the species p, p' and p'' are different. As a consequence, for that type of reactions the species of the reaction products can be completely described by a set $B \in \Theta_3^K$, where Θ_3^K is the set of all subsets of $\{1, \ldots, K\}$ with $1 \leq \mathrm{card}(B) \leq 3$.

To model in the N-th system the influence of the spatial distance between two reacting particles on the reaction rate we use some kernel φ_N. As in (18) we assume that φ_N is obtained by a scaling

$$\varphi_N(x) = \tau_N \varphi_1(\tau_N^{1/d} x), \quad x \in \mathbb{R}^d,\ N \in \mathbb{N}, \quad (24)$$

from some fixed function φ_1. This function and the sequence τ_N, $N \in \mathbb{N}$, are supposed to satisfy similar conditions as in (19) and (20).

To take into account catalytic effects we introduce functions

$$\beta_{r,r'}^B \in C_b^2\big([0,\infty)^K; [0,\infty)\big), \quad r, r' = 1, \ldots, K,\ B \in \Theta_3^K, \quad (25)$$

where the set B arising as upper index in $\beta_{r,r'}^B$ describes the species of the particles appearing after the reaction.

As in (23) we then assume

$$\mathbf{P}\left[\begin{array}{l}\text{particles } k \neq l \text{ of species } r \text{ and } r' \text{ located in } x \text{ and } y \text{ react} \\ \text{during } (t, t+\delta] \text{ to yield particles of species given by } B\end{array}\bigg| \mathcal{F}_t\right] \quad (26)$$

$$= \frac{\delta}{N} \varphi_N(x-y) \beta_{r,r'}^B\big(\rho_N(x,y,t)\big) + o(\delta), \quad \text{as } \delta \searrow 0,$$

$$x, y \in \mathbb{R}^d,\ t \geq 0,\ r, r' = 1, \ldots, K,\ B \in \Theta_3^K,$$

where ρ_N is introduced in (17), (21) and (22). Moreover, we suppose that all particles resulting from a reaction between particles in $x, y \in \mathbb{R}^d$ emerge at the same point $(x+y)/2$.

Remark 2. In particular cases it can happen that certain reaction products do not contribute to further reactions, i.e., actually, they leave the system under consideration. This situation would correspond to $B = \emptyset$.

To drop the restriction to at most three particles resulting from any reaction or the assumption that the species of these particles are different we should assume that B is an arbitrary subset or an unordered tuple of elements of $\{1, \ldots, K\}$.

Also splitting reactions, where one particle is divided into several fragments belonging to different species, are possible. For such reactions we may introduce additional reaction rates β_r^B, $r = 1, \ldots, K$, B a subset or an unordered tuple of elements of $\{1, \ldots, K\}$, in (25).

Apart from some trivial modifications our mathematical considerations also remain valid for the generalizations mentioned in this remark.

3.3 A Limit Theorem

The relationship between the microscopic model, i.e., the many-particle system described in Section 3.2, and its macroscopic equivalent, i.e., a nonlinear system of partial differential equations, is specified by the subsequent result.

Theorem 1. *Suppose that suitable regularity conditions are satisfied. Then,*

$$\lim_{N \to \infty} \sup_{t \leq T^*} |\langle \mathbb{X}_N^r(t), f \rangle - \langle \rho^r(., t), f \rangle| = 0, \quad \text{in probability}, \tag{27}$$

$$r = 1, \ldots, K, \quad f \in C_b^1(\mathbb{R}^d) \cap L^2(\mathbb{R}^d),$$

with $\langle \mu, f \rangle = \int_{\mathbb{R}^d} f(x) \mu(dx)$ *for* $f \in C_b(\mathbb{R})$ *and a measure* μ. *The limit densities collected in the* $[0, \infty)^K$-*valued function* $\rho = (\rho^1, \ldots, \rho^K)$ *solve*

$$\frac{\partial}{\partial t} \rho^r(x, t) = \beta^r(x, \rho(x, t)) - \sum_{r'=1}^{K} \sum_{B \in \Theta_3^K} \rho^r(x, t) \rho^{r'}(x, t) \beta_{r,r'}^B(\rho(x, t)) \tag{28}$$

$$+ \frac{1}{2} \sum_{r', r''=1}^{K} \sum_{\{B \in \Theta_3^K : r \in B\}} \rho^{r'}(x, t) \rho^{r''}(x, t) \beta_{r', r''}^B(\rho(x, t)) + \frac{\sigma_r^2}{2} \Delta \rho^r(x, t),$$

$$x \in \mathbb{R}^d, \ t \in [0, T^*], \ r = 1, \ldots, K.$$

The regularity conditions needed for this theorem are stated in detail in [38]. First, we need various conditions on the scaling coefficients θ_N, τ_N, $N \in \mathbb{N}$, which appear in (18) and (24), and further coefficients a_N, $N \in \mathbb{N}$ (cf. Section 3.4). Next, the interaction kernels, ψ_1, φ_1, cf. (18) and (24), and the kernel W_1, cf. (30) and (31) in Section 3.4, have to be regular enough. Finally, the solution $(\rho^1, \ldots, \rho^K)$ of the limit equation (28) should be sufficiently smooth in the time interval $[0, T^*]$.

We note that a_N, $N \in \mathbb{N}$, and W_1 are needed for technical reasons, i.e., they have no meaning in the modeling but are employed to deduce Theorem 1 mathematically rigorously.

3.4 Some Remarks on the Proof of the Limit Theorem

The proof is based on systems of stochastic differential equations

$$\langle \mathbb{X}_N^r(t), f(.,t) \rangle = \frac{1}{N} \sum_{k \in M_N^r(t)} f(X_N^k(t), t) \tag{29}$$

$$= \frac{1}{N} \sum_{k \in M_N^r(0)} f(X_N^k(0), 0)$$

$$+ \frac{1}{N} \int_0^t \int_{\mathbb{R}^d} \mathbb{J}_N^r(dx, ds) f(x, s)$$

$$- \frac{1}{N} \int_0^t \sum_{r'=1}^K \sum_{B \in \Theta_3^K} \sum_{\substack{k \in M_N^r(s) \\ l \in M_N^{r'}(s)}} \boldsymbol{\rho}_{N;r',r';k,l}^B(ds) f(X_N^k(s), s)$$

$$+ \frac{1}{2N} \int_0^t \sum_{r',r''=1}^K \sum_{\{B \in \Theta_3^K : r \in B\}} \sum_{\substack{l \in M_N^{r'}(s) \\ l' \in M_N^{r''}(s)}} \boldsymbol{\rho}_{N;r',r'';l,l'}^B(ds) f\left(\frac{X_N^l(s) + X_N^{l'}(s)}{2}, s\right)$$

$$+ \frac{\sigma_r^2}{2N} \int_0^t ds \sum_{k \in M_N^r(s)} \Delta f(X_N^k(s), s)$$

$$+ \frac{\sigma_r}{N} \int_0^t \sum_{k \in M_N^r(s)} \nabla f(X_N^k(s), s) \cdot B_N^k(ds)$$

$$+ \frac{1}{N} \int_0^t ds \sum_{k \in M_N^r(s)} \left.\frac{\partial}{\partial u} f(X_N^k(s), u)\right|_{u=s},$$

$$t \geq 0, \; r = 1, \ldots, K, \; f \in C_b^{2,1}(\mathbb{R}^d \times [0, \infty)), \; N \in \mathbb{N},$$

for the empirical processes to describe the dynamics of the many-particle systems.

In (29) we denote for $r = 1, \ldots, K$ and $N \in \mathbb{N}$ by $\mathbb{J}_N^r$ some random field in $\mathbb{R}^d \times [0, \infty)$, which describes the instants in space-time, where in the N-particle system particles of species r are immigrating. Moreover, for any $N \in \mathbb{N}$, $r, r' = 1, \ldots, K$, $k \in M_N^r(.)$, $l \in M_N^{r'}(.)$ and $B \in \Theta_3^K$ the process $\boldsymbol{\rho}_{N;r,r';k,l}^B$ marks that instant, where the particles k and l of the r-th and the r'-th species, respectively, interact to produce particles described by the set B.

As a first application of (29) some formal arguments can be used to guess the limit dynamics (28). In this context, we suppose that for any $t \geq 0$ the empirical processes $\mathbb{X}_N^1(t), \ldots, \mathbb{X}_N^K(t)$ indeed converge to limits having smooth densities $\rho^1(\cdot, t), \ldots, \rho^K(\cdot, t)$. Next, we represent the stochastic integrals on the right side of (29) as sums of a martingale and a process of bounded variation of first order. We then observe that the martingales vanish as $N \to \infty$ and take into account (15). As a formal result, (29) turns into a weak version of (28).

Now, the empirical processes $\mathbb{X}_N^1, \ldots, \mathbb{X}_N^K$ and their supposed respective limits $\rho^1, \ldots, \rho^K$ can be compared directly. For that purpose we introduce with

$$\mathbb{Q}_N(t) = \sum_{r=1}^{K} \|(\mathbb{X}_N^r(t) - \rho^r(.,t)) * W_N\|_2^2, \quad t \geq 0, \ N \in \mathbb{N}, \quad (30)$$

a family of stochastic processes, which describe for fixed N some L^2-distance between the N-particle system and the limit ρ. In (30) we use kernels

$$W_N(x) = a_N W_1(a_N^{1/d} x), \quad x \in \mathbb{R}^d, \ N \in \mathbb{N}, \quad (31)$$

where W_1 is a sufficiently regular probability density and $a_N \nearrow \infty$ as $N \to \infty$.

To investigate the asymptotics as $N \to \infty$ we employ (28) and (29) to deduce for any $\mathbb{Q}_N$, $N \in \mathbb{N}$, a stochastic differential equation describing its time evolution. After extensive, but fairly straightforward estimates of the various terms of this stochastic differential equation essentially Gronwalls Lemma can be used to verify that $\mathbb{Q}_N$ vanishes as $N \to \infty$. In particular,

$$\lim_{N \to \infty} \mathbf{E}\left[1 \wedge r_N \sup_{t \leq T^*} \mathbb{Q}_N(t)\right] = 0 \quad (32)$$

for some suitable sequence r_N, $N \in \mathbb{N}$, with $\lim_{N \to \infty} r_N = \infty$ is obtained. Finally, (27) is a simple consequence of (32).

For the details of the proof of Theorem 1 we refer to [38].

3.5 Application to Heterogeneous Catalysis: CO Oxidation on Pt

As indicated in Section 3.1 the general considerations on the modeling of chemical reactions with many-particle models in continuous space is applied here to the example of the CO oxidation on low-index Pt-surfaces in a low-pressure atmosphere, where the partial pressures of O_2 and CO are kept constant.

In the present context it is particularly important that for modeling purposes the use of moderately interacting many-particle systems is justified, as under the considered conditions a locally well-mixed system is given.

The description in Section 3.2 is specified to a moderately interacting many-particle system in continuous space $\mathbb{R}^2$ with three species, i.e., $d = 2$ and $K = 3$. For the N-particle system we may employ $\mathbb{X}_N^1$, $\mathbb{X}_N^2$ and $\mathbb{X}_N^3$ for the empirical processes associated to the CO-molecules, O-atoms and Pt-1×1 adsorption sites, respectively. As far as the Langmuir-Hinshelwood mechanism with asymmetric inhibition and the adsorbate-induced Pt phase transition given in (1) and (2) are concerned, some particular physical processes are modeled in terms of immigration, namely the adsorption of CO and O_2 and the transition from Pt-rec to Pt-1×1. The corresponding immigration rates are β^1, $\beta^{2,2}$ and β^3. Since the empirical processes related to oxygen describe

O-atoms, the adsorption of O_2-molecules is related to another trivial extension of the model introduced in Section 3.2, namely the simultaneous immigration of two particles at the same place. For this reason the notation $\beta^{2,2}$ instead of β^2 is employed. The remaining reactions in (1) and (2), i.e., the desorption of CO, the creation of CO_2 and the transition from Pt-1×1 to Pt-rec, are described as reactions according to Section 3.2. For those cases, after taking into account in particular Remark 2, the reaction rates $\beta_1^\emptyset$, $\beta_{1,2}^\emptyset$ and $\beta_3^\emptyset$ have to be specified.

Remark 3. The modelling of the phase transition Pt-rec $\to$ Pt-1 × 1 in terms of immigration is a trick, which is used to simplify the description. If also the Pt-rec phase is modelled as a species, this component of the dynamics could be described in terms of a reaction, too.

To quantify the above-mentioned rates we use the considerations in Section 2 which is based on [39]. In particular, after some slight modifications we choose

$$\beta^1(x,\rho^1,\rho^2,\rho^3) = p_{CO}\kappa_{CO}s_{CO}H_1(1-(\rho^1)^\xi)S(x), \quad (33a)$$

$$\beta^{2,2}(x,\rho^1,\rho^2,\rho^3) = p_{O_2}\kappa_O(s_O^{1\times 1}\rho^3 + s_O^{rec}H_1(1-\rho^3)) \quad (33b)$$
$$(H_1(1-\rho^1)H_1(1-\rho^2))^2 S(x)/2,$$

$$\beta^3(x,\rho^1,\rho^2,\rho^3) = k_{1\times 1}f_{1\times 1}(\rho^1,\rho^3)H_1(1-\rho^3)S(x), \quad (33c)$$

$$\beta_1^\emptyset(\rho^1,\rho^2,\rho^3) = k_{des}^{1\times 1}\rho^3 + k_{des}^{rec}H_1(1-\rho^3), \quad (33d)$$

$$\beta_{1,2}^\emptyset(\rho^1,\rho^2,\rho^3) = k_{re}, \quad (33e)$$

$$\beta_3^\emptyset(\rho^1,\rho^2,\rho^3) = k_{rec}f_{rec}(\rho^1,\rho^3), \quad x \in \mathbb{R}^2,\ \rho^1,\rho^2,\rho^3 \in [0,\infty). \quad (33f)$$

The kinetic parameters κ_{CO}, s_{CO}, ξ, κ_O, $s_O^{1\times 1}$, s_O^{rec}, $k_{1\times 1}$, $k_{des}^{1\times 1}$, k_{des}^{rec}, k_{re} and k_{rec}, which are strictly positive, are given in Table 1. p_{CO} and p_{O_2} denote the partial pressures of CO and O_2, respectively, in the gaseous phase above the Pt-surface. The function $f_{1\times 1}$ defined in (11) is used with the densities ρ^1,ρ^2,ρ^3 which results in

$$f_{1\times 1}(\rho^1,\rho^3) = (1-\epsilon)(\rho^1)^\lambda + \epsilon(\rho^3)^\lambda, \quad \rho^1,\rho^3 \in [0,\infty), \quad (34a)$$

whereas for f_{rec} we have (cf. equation (10))

$$f_{rec}(\rho^1,\rho^3) = (1-\epsilon)(H_1(1-\rho^1))^\lambda + \epsilon(H_1(1-\rho^3))^\lambda, \quad \rho^1,\rho^3 \in [0,\infty). \quad (34b)$$

In (34) the constants $\epsilon \in [0,1]$ and $\lambda > 1$ have to be chosen suitably, as it is described in Section 2. The function $S(x)$ with spatial variable x in (33a) - (33c) is used to model the size of the Pt-surface, which is finite. We suppose

$$S \in C_b^1(\mathbb{R}^d;[0,1])\ \text{with}\ S(x) = \begin{cases} 1, & \text{if } |x| \leq R, \\ 0, & \text{if } |x| \geq R', \end{cases}$$

for suitable constants $R' > R > 0$. In order to avoid negative rates in (33) and (34) the function $[0,1] \ni u \to 1-u$ from Section 2 is replaced by $[0,\infty) \ni u \to H_1(1-u)$, where $H_1(v) = \max\{v,0\}$, $v \in \mathbb{R}$.

In [39], only the spatially homogeneous situation was considered. In particular, the diffusive hopping of adsorbed CO-molecules mentioned in Section 3.1 is neglected. Here, we model that phenomenon in terms of independent Brownian motions acting on the CO-molecules with diffusion constant $\sigma_1 = \sqrt{2D}$ for some $D > 0$. Both adsorbed O-atoms and Pt-1×1 adsorption sites are considered as immobile, i.e., $\sigma_2 = \sigma_3 = 0$ is supposed.

To complete our mathematical model of this particular chemical reaction we still have to choose suitable kernels ψ_1 and φ_1 and associated scaling coefficients θ_N, $N \in \mathbb{N}$, and τ_N, $N \in \mathbb{N}$. To describe the considered chemical situation, it suffices to take into account the regularity conditions mentioned in Theorem 1 in Section 3.3.

By the particular form of the rates in (33) and (34) we obtain as limit dynamics the system

$$\frac{\partial}{\partial t}\rho^1 = p_{CO}\kappa_{CO} s_{CO}(1-(\rho^1)^\xi)S \qquad (35a)$$
$$- (k_{des}^{1\times 1}\rho^3 + k_{des}^{rec}(1-\rho^3))\rho^1 - k_{re}\rho^1\rho^2 + D\Delta\rho^1,$$

$$\frac{\partial}{\partial t}\rho^2 = p_{O_2}\kappa_O(s_O^{1\times 1}\rho^3 + s_O^{rec}(1-\rho^3))((1-\rho^1)(1-\rho^2))^2 S - k_{re}\rho^1\rho^2, \qquad (35b)$$

$$\frac{\partial}{\partial t}\rho^3 = k_{1\times 1}f_{1\times 1}(\rho^1,\rho^3)(1-\rho^3)S - k_{rec}f_{rec}(\rho^1,\rho^3)\rho^3. \qquad (35c)$$

We note that the function H_1 does not appear in (35) as a consequence of a maximum principle, which implies

$$0 \leq \rho^1(x,t), \rho^2(x,t), \rho^3(x,t) \leq 1, \quad x \in \mathbb{R}^2,\ t \geq 0,$$

if this condition is imposed for the initial time $t=0$.

Obviously, the system (35) of partial differential equations is an immediate extension of the system (14) of ordinary differential equations. The additional term $D\Delta\rho^1$ on the righthand-side of (35a) allows for the description of spatial inhomogeneities. The obtained reaction-diffusion equation (35) can now be used as microscopically justified macroscopic model for subsequent analysis and numerical simulation. While this model is in excellent agreement with low pressure experiments over a wide parameter range, it will fail for intermediate pressures, where stochastic effects become important macroscopically. This will be discussed in the next section.

4 Stochastic and Thermokinetic Effects in CO Oxidation

For increasing pressure, there are fewer and fewer diffusion sites changes of adsorbed species per adsorption event. Therefore the characteristic length over which the system can be regarded as well mixed decreases until the corresponding number of sites becomes so small that fluctuations have to be taken into account and the mean-field limits, as in Sections 2 and 3, are not applicable any more.

4.1 Heat Balance and Deterministic Thermo-Kinetic Model

At intermediate pressures, the system is no longer isothermal due to heat release by the reaction. The temperature T can be modeled using the heat balance equation [4, 43] for a volume V of the catalyst with heat capacity C:

$$CV\dot{T} = Q_{\text{ad}}^{\text{CO}} + Q_{\text{des}}^{\text{CO}} + Q_{\text{ad}}^{\text{O}} + Q_{\text{re}} + Q_{\text{rad}} + Q_{\text{cond}}. \tag{36}$$

$Q_{\text{ad}}^{\text{CO}}, Q_{\text{ad}}^{\text{O}}, Q_{\text{des}}^{(i,j)\,\text{CO}}$, and Q_{re} are the rates of production of heat energy through adsorption, desorption and reaction processes, Q_{rad} and Q_{cond} are the balances for the gain and loss of heat through radiation, conduction and the feedback system regulating the temperature during the experiments by electrical heating.

Linearizing the equation around the temperature field $T(x_1, x_2) \equiv T_0$ with space variables x_1, x_2 in the absence of any reaction and rescaling the temperature to $\theta = \frac{T-T_0}{T_0}$ leads to the equation [43]

$$\begin{aligned}\dot{\theta} = \gamma \Big\{ &H_{\text{ad}}^{\text{CO}} \kappa_{\text{CO}} p_{\text{CO}} s_{\text{CO}} (1 - u^{\xi}) \\ &- (H_{\text{des,rec}}^{\text{CO}} k_{\text{des}}^{\text{rec}}(\theta)(1-w) + H_{\text{des},1\times1}^{\text{CO}} k_{\text{des}}^{1\times1}(\theta) w) u \\ &+ \frac{1}{2} H_{\text{ad}}^{\text{O}_2} \kappa_{\text{O}} p_{\text{O}_2} (s_{\text{O}}^{\text{rec}}(1-w) + s_{\text{O}}^{1\times1} w) \big((1-u)(1-v)\big)^2 \\ &+ H_{\text{re}} k_{\text{re}}(\theta) uv \Big\} - \mu\theta + D_\theta \Delta\theta. \end{aligned} \tag{37}$$

Here γ^{-1} represents a heat capacity and the parameter μ describes the strength of the effective temperature feedback:

$$\gamma(T_0) = \frac{\rho_s}{C\rho_b l_T T_0}, \tag{38}$$

$$\mu(T_0) = \frac{4\sigma e_s T_0^3 + \lambda_h/l_T}{C\rho_b l_T}. \tag{39}$$

The parameters entering these equations are summarized for Pt(110) in Table 3.

The full deterministic model is obtained by the combination of equation (37) for the rescaled temperature with the reaction-diffusion equations

$$\dot{u} = \kappa_{\text{CO}} p_{\text{CO}} s_{\text{CO}} (1 - u^{\xi}) - \big(k_{\text{des}}^{\text{rec}}(\theta)(1-w) + k_{\text{des}}^{1\times1}(\theta) w\big) u - k_{\text{re}}(\theta) uv$$
$$+ D_{\text{CO}}(\theta) \Delta u \tag{40}$$
$$\dot{v} = \kappa_{\text{O}} p_{\text{O}_2} \big(s_{\text{O}}^{\text{rec}}(1-w) + s_{\text{O}}^{1\times1} w\big) \big((1-u)(1-v)\big)^2 - k_{\text{re}}(\theta) uv \tag{41}$$
$$\dot{w} = k_{1\times1}(\theta) f_{1\times1}(u,w)(1-w) - k_{\text{rec}}(\theta) f_{\text{rec}}(u,w) w. \tag{42}$$

Table 3. Parameters required for heat balance equation for Pt(110). We assume $H_{\text{des,rec}}^{CO} = H_{\text{des},1\times1}^{CO} = H_{\text{des}}^{CO}$.

H_{ad}^{CO}	heat of adsorption of CO	135 kJ mol^{-1}
H_{des}^{CO}	heat loss by desorption of CO	$\approx H_{\text{ad}}^{CO}$
H_{re}	heat of reaction	20 - 30 kJ mol^{-1}
$\frac{1}{2}H_{\text{ad}}^{O_2}$	heat of adsorption of an oxygen atom	115 kJ mol^{-1}
ρ_s	density of surface atoms	8.84×10^{18} atoms/m^2
σ	Stefan-Boltzmann constant	5.6705×10^{-8} Jm^{-2}K^{-1}s^{-1}
e_s	integral emissivity of platinum	$0.05 - 0.1$
λ_h	heat conduction coefficient	72 W m^{-1} K^{-1}
l_T	characteristic thickness of surface layer	5×10^{-5} m
C	specific heat capacity of platinum	130 J kg^{-1}K^{-1}
ρ_b	bulk mass density of platinum	21.09×10^3 kgm^{-3}
D_θ	diffusion constant of temperature	$\lambda_h/(C\rho_b)$

and suitable boundary conditions. This macroscopic model (40) – (42) resulted in Section 3.5 in the interior of the bounded Pt surface (i.e. for $S(x) = 1$) as limit (35) from the moderately interacting many-particle model using temperature dependent Arrhenius rates $k(\theta) = \nu \exp(-\frac{E}{k_B T_0}) \exp(\frac{E}{k_B T_0} \frac{\theta}{1+\theta}) = \nu \exp(-\frac{E}{k_B T})$.

For large μ and not too high reaction rates the system would remain isothermal so that thermal effects can be analyzed using μ as bifurcation parameter. Thermo-kinetic effects are a consequence of the asymmetric inhibition of adsorption (i.e. pre-adsorbed CO blocks oxygen adsorption, but not vice versa) and the strong temperature dependence of CO desorption. An O-covered surface exhibits a high reaction rate and therefore becomes hot whereas a high CO coverage keeps the catalyst relatively cool. Since in turn lower T favors CO coverage through reduced desorption, the effect is autocatalytic.

A partial bifurcation analysis of the model is reproduced in Fig. 5. For large μ the O-covered surface is excitable close to the Hopf bifurcation and stable pulses exist. With decreasing μ this bifurcation shifts to higher p_{CO}, so that the O-covered branch becomes sub-excitable. Pulses can still be triggered, but they shrink and eventually vanish (Fig. 6). Physically this can be readily explained by the temperature effects of the changes in reaction rate. The rate drops sharply on the (predominantly) CO-covered areas, because here oxygen adsorption is blocked. However, behind the pulses the reconstruction has been lifted and the reactivity is increased to values even higher than on the original O-covered 1×2 surface because of the enhanced sticking coefficient of oxygen on the 1×1 surface. Therefore, the temperature locally rises to values higher than at the beginning. Since the temperature spreads rapidly via heat

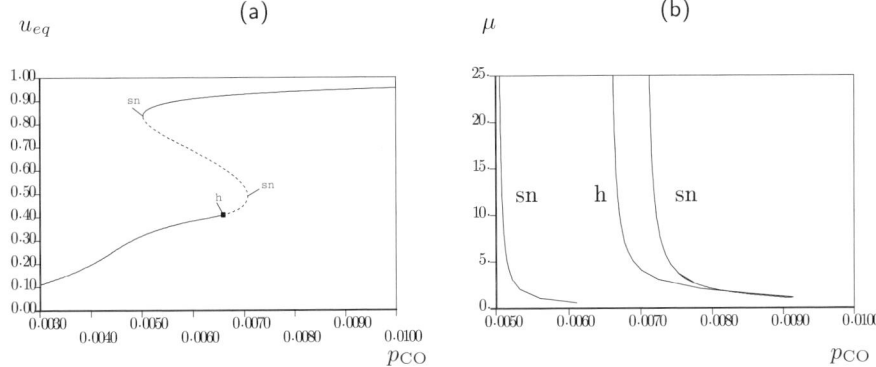

Fig. 5. (a) Continuation of CO coverage in p_{CO} for the thermo-kinetic model. There is a Hopf bifurcation point (h) between two saddle-node bifurcation points (sn). (b) Continuation of Hopf and sn bifurcations in p_{CO} and μ.

conduction, the CO pulses are overrun from the inside (a hotter surface cannot maintain a high CO coverage due to enhanced desorption).

4.2 Stochastic Thermo-Kinetic Model and Raindrop Patterns

The stochastic model described in Section 2 was combined with thermo-kinetic effects using equation (37). Since heat conduction is much faster than diffusion, it was treated deterministically. Moreover, since the patch considered in the stochastic simulation is relatively small, the temperature diffusion was assumed infinitely fast. For small pressures, the model behaved practically isothermally and followed the deterministic path very closely. In particular, no spontaneous nucleation in the bistable or excitable region was observed. On the other hand, for intermediate pressure (with parameters $N = 10^3$, hopping rate corresponding to $D_{CO} = 1.4 \times 10^{-14} \frac{m^2}{s}$) significant fluctuations become visible and critical nuclei did form spontaneously. A computer simulation for $p_{O_2} = 10^{-2}$ mbar of nucleation, pulse formation and propagation failure is reproduced in Fig. 7. the rate of such events turned out to be extremely sensitive with respect to p_{CO}. The nucleation was obviously random, the subsequent pulse annihilation occurred not too far from the deterministic path (cf. Fig. 6), although relatively small random variations in the maximum local CO coverage led to noticeable variations in the time it took to return to the initial state.

The homogeneous nucleation rate on an ideal surface can be estimated using large-deviations analysis [7], assuming that a site in a patch of critical size is CO-covered with probability $\bar{u}$. For an equilibrium CO coverage $\bar{u}$, a critical coverage u^* and a critical nucleus containing n_{cr} sites, the system spends a time fraction of $\left[\left(\frac{\bar{u}}{u^*}\right)^{u^*}\left(\frac{1-\bar{u}}{1-u^*}\right)^{1-u^*}\right]^{n_{cr}}$ in a state with CO coverage $u \geq u^*$. $\bar{u}$ and u^* can be taken from the null-clines of the deterministic model, n_{cr}

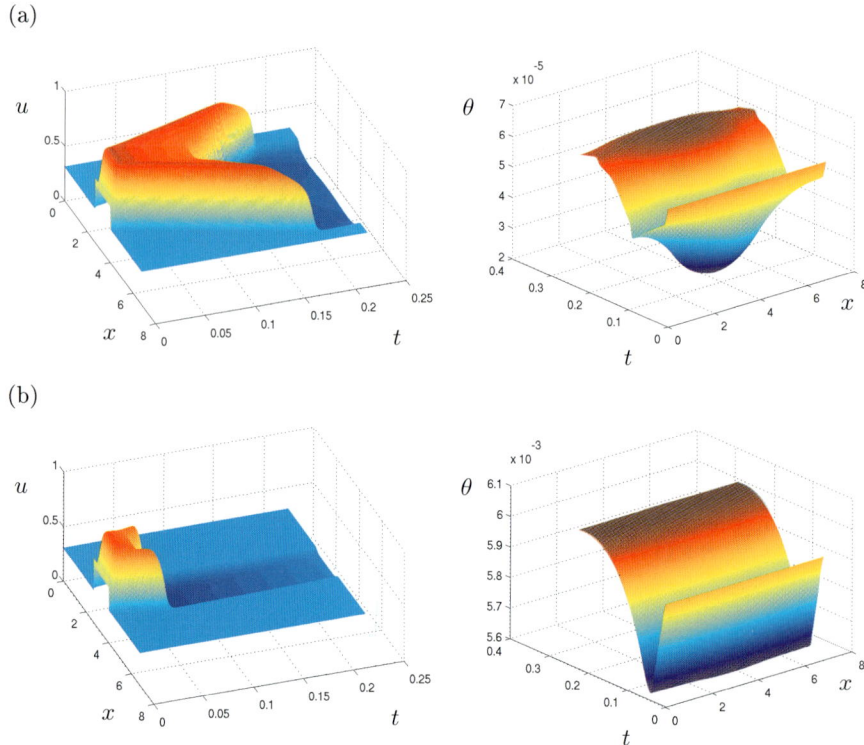

Fig. 6. Simulation of pulse propagation using thermo-kinetic model equations (40) – (42) and (37) in one dimension with rescaled temperature and no-flux boundary conditions (cf. Fig. 1b). The space variable x is given in μm, the time t is given in s. A CO nucleus was put in the initial conditions on an otherwise predominantly O-covered surface. (a) For $\mu = 10^3\,\text{s}^{-1}$ pulses form, propagate but finally die due to rising temperature. (b) For $\mu = 10\,\text{s}^{-1}$ pulses can still be formed, but die quickly. Parameter are $p_{CO} = 5.0 \times 10^{-3}$ mbar, $p_{O_2} = 1.55 \times 10^{-2}$ mbar, $D_{CO} = 10^{-12}\,\text{m}^2/\text{s}$, $T_0 = 520\,\text{K}$, the pre-exponential factors $\nu_{\text{des}}^{\text{rec}}$, $\nu_{\text{des}}^{1\times1}$, ν_{re}, $\nu_{1\times1}$ and ν_{rec} have been multiplied by 10^2 compared to those given in Table 1. Other parameters as in Tables 1 and 3. In the simulation shown heat conductance was chosen unrealistically small in order to visualize where heat production takes place. The effect of nucleation and propagation failure, however, persists even for a realistic heat diffusion (i.e. very large thermal conductivity).

decreases linearly with the pressure and is between 10^3 and 10^4 at 10^{-2} mbar [2].

The expected number of events can be obtained by multiplying with the density of adsorption sites and dividing by the characteristic time, for which we chose the time required for the impingement of one monolayer CO. The resulting function is obviously very sensitive to n_{cr} but also depends crucially on the excitation threshold $u^* - \bar{u}$. Reasonable values [40] of 1 to 100 mm^{-2}s^{-1}

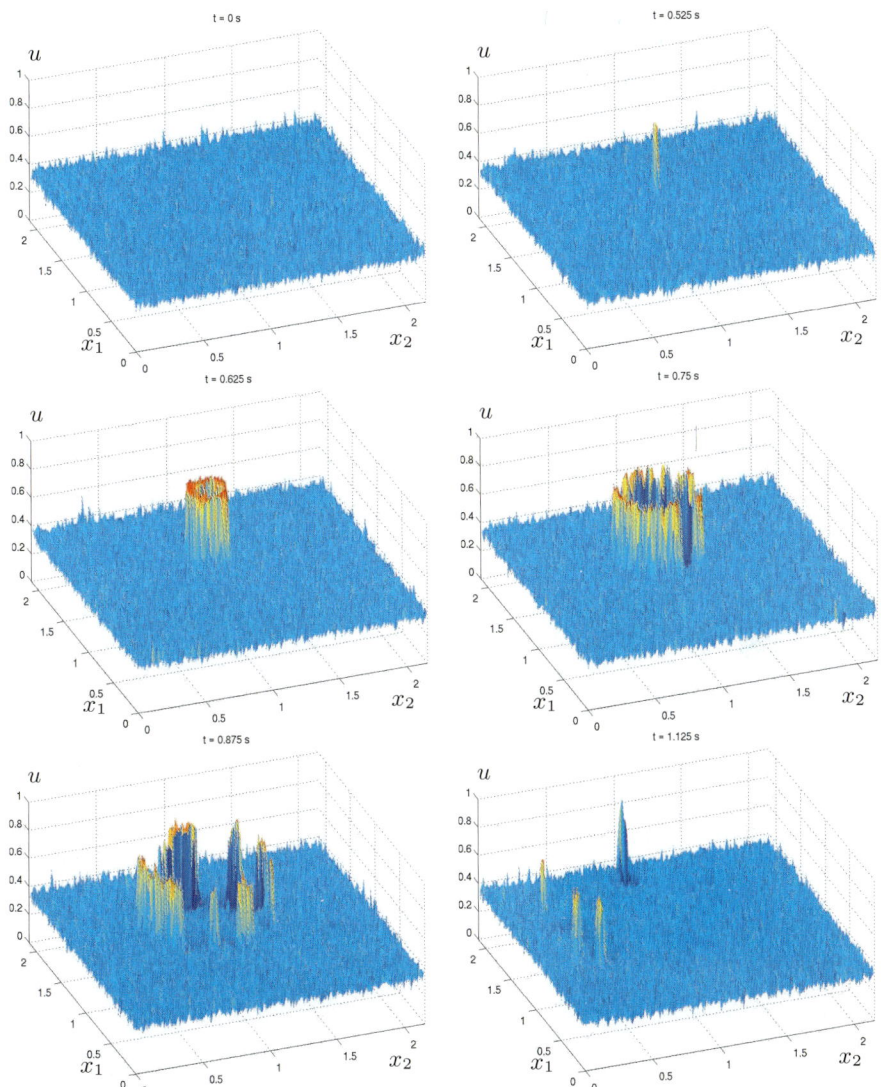

Fig. 7. Two-dimensional stochastic simulation (x_1, x_2 in μm) of raindrop patterns on a predominantly O-covered Pt(110) surface using the stochastic lattice model with periodic boundary conditions and thermo-kinetic effects included (cf. Fig. 1a). CO nucleation occurred due to coverage fluctuations. We used 200×200 cells, $N = 10^3$ adsorption sites for each cell and $p_{CO} = 5.22 \times 10^{-3}$ mbar, $p_{O_2} = 1.55 \times 10^{-2}$ mbar, $D_{CO} = 1.4 \times 10^{-14} \frac{m^2}{s}$, $T_0 = 520$ K, $\mu = 10^3$ s^{-1}. Other parameters as in Fig. 6.

were obtained with $n_{cr} = 3000 - 4000$, $u^* = 0.41$ and $0.33 < \bar{u} < 0.35$. Increasing n_{cr} by one order of magnitude (which would correspond to a pressure decrease by one order) resulted in values indistinguishable from zero.

Obviously, surface inhomogeneities always play a role on real catalysts. Nevertheless, the experimental observations and their close correspondence to a realistic model clearly suggest that at least a significant fraction of the observed nuclei form uniformly distributed over the surface. The presented effect therefore constitutes the first example of mesoscopic pattern formation (10 – 100 μm) in a surface reaction that is initiated by microscopic fluctuations and cannot be fully captured in a deterministic description.

5 Conclusion

Spatio-temporal pattern formation in the CO oxidation on Pt has been studied experimentally over a wide range of parameters. The observed phenomena mostly appear deterministic, except for very small catalyst areas or at sufficiently high pressure where also temperature variations become observable. In order to develop a consistent description of the large variety of experimental patterns, a unified model of CO oxidation on all low-index plane Pt surfaces was developed. It consists of three levels of modeling, microscopic, mesoscopic, and macroscopic and includes thermal effects. The macroscopic model can be rigorously derived from the microscopic one; it applies when the surface can be regarded as locally well mixed, i.e., for low pressures in the gas phase. Stochastic effects at intermediate pressures, such as random nucleation, can be reproduced in simulations with the mesoscopic model.

References

1. R.F.S. Andrade, G. Dewel, and P. Borckmans. Modelling of the kinetic oscillations in the CO oxidation on Pt(100). *J. Chem. Phys.*, 91:2675–2682, 1989.
2. M. Bär, C. Zülicke, M. Eiswirth, and G. Ertl. Theoretical modeling of spatiotemporal self-organization in a surface catalyzed reaction exhibiting bistable kinetics. *Journal of Chemical Physics*, 96:8595 – 8604, 1992.
3. L. Breiman. *Probability*. Addison-Wesley, 1968.
4. J. Cisternas, P. Holmes, I. G. Kevrekidis, and X. Li. CO oxidation on thin Pt crystals: Temperature slaving and the derivation of lumped models. *Journal of Chemical Physics*, 118(7):3312 – 3328, 2003.
5. A. De Masi and E. Presutti. *Mathematical methods for hydrodynamic limits*. Lecture Notes in Mathematics, 1501. Springer-Verlag, 1991.
6. E. Doedel, A.R. Champneys, T.F. Fairgrieve, Y.A. Kusnetsov, B. Sandstede, and X. Wang. AUTO 97: Continuation and Bifurcation Software for Ordinary Differential Equations. http://ftp.cs.concordia.ca/pub/doedel/auto.
7. R. Durrett. *Probability: Theory and Examples*. Duxbury Press, Belmont, 1996.
8. M. Eiswirth, M. Bär, and H.H. Rotermund. Spatiotemporal selforganization on isothermal catalysts. *Physica D*, 84:40 – 57, 1995.
9. M. Eiswirth and G. Ertl. Kinetic oscillations in the catalytic oxidation of CO on Pt(110). *Surf. Sci.*, 90:177, 1986.

10. M. Eiswirth and G. Ertl. Pattern formation on Catalytic Surfaces. In R. Kapral and K. Showalter, editors, *Chemical Waves and Patterns*, page 447. Kluwer, Dordrecht, 1995.
11. M. Eiswirth, K. Krischer, and G. Ertl. Transition to chaos in an oscillating surface reaction. *Surfacce Science*, 202(3):565–591, 1988.
12. Eiswirth, M. Phänomene der Selbstorganisation bei der Oxidation von CO an Pt(110). Dissertation, Ludwig-Maximilians-Universität München, 1987.
13. G. Ertl. Reaktionen an Festkörper-Oberflächen. *Berichte der Bunsengesellschaft*, 98:1413 – 1420, 1994.
14. Ethier, S.N., Kurtz, T.G. *Markov Processes: Characterization and Convergence.* Addison-Wesley, 1986.
15. C.W. Gardiner. *Handbook of Stochastic Methods.* Springer-Verlag, 1985.
16. D. Gillespie. Exact stochastic simulation of coupled chemical reactions. *The Journal of Physical Chemistry*, 81(25):2340 – 2361, 1977.
17. D. T. Gillespie. Approximate accelerated stochastic simulation of chemically reacting systems. *The Journal of Physical Chemistry*, 81(25):2340 – 2361, 1977.
18. S. Großkinsky, C. Klingenberg, and K. Oelschläger. A rigorous derivation of Smoluchowski's equation in the moderate limit. *Stochastic Anal. Appl.*, 22(1):113 – 141, 2004.
19. J. Guckenheimer. Multiple Bifurcation Problems for Chemical Reactors. *Physica D*, 20(1):1–20, 1986.
20. Flavius Guias. Convergence properties of a stochastic model for coagulation-fragmentation processes with diffusion. *Stochastic Anal. Appl.*, 19(2):245–278, 2001.
21. H. Haken. Statistical physics of bifurcation, spatial structures, and fluctuations of chemical reactions. *Zeitschrift für Physik B*, 20:413 – 420, 1975.
22. H. Haken. *Advanced Synergetics.* Springer Series in Synergetics. Springer-Verlag, Heidelberg, Berlin, New York, 1983.
23. H. Haken. *Synergetics, An Introduction.* Springer Series in Synergetics. Springer-Verlag, Heidelberg, Berlin, New York, 1983.
24. J. Hale and H. Koçak. *Dynamics and Bifurcations.* Springer-Verlag, Heidelberg, Berlin, New York, 1991.
25. J. Honerkamp. *Stochastic Dynamical Systems.* VCH, 1994.
26. A. Hopkinson, J.M. Bradley, X.-C. Guo, and D.A. King. Nonlinear Island Growth Dynamics in Adsorbate-Induced Restructuring of Quasihexagonal Reconstructed Pt(100) by CO. *Phys. Rev. Letters*, 71(10):1597–1600, 1993.
27. R. Imbihl. Temporal and spatial patterns in catalytic reactions on single crystal surfaces. *Heterogeneous Chemistry Reviews*, 1:125, 1994.
28. R. Imbihl, M.P. Cox, G. Ertl, H. Müller, and W. Brenig. Kinetic Oscillations in the Catalytic CO Oxidation on Pt(100): Theory. *J. Chem. Phys.*, 83(4):1578–1587, 1985.
29. R. Imbihl and G. Ertl. Osicallatory kinetics in heterogeneous catalysis. *Chemical Reviews*, 95(3):697 – 733, 1995.
30. I. Karatzas and S. E. Shreve. *Brownian Motion and Stochastic Calculus.* Springer-Verlag, Heidelberg, Berlin, New York, 1991.
31. Claude Kipnis and Claudio Landim. *Scaling limits of interacting particle systems.* Springer-Verlag, Heidelberg, Berlin, New York, 1999.
32. P. Kotelenez. Law of large numbers and central limit theorem for linear chemical reactions with diffusion. *The Annals of Probability*, 14(1):173 – 193, 1986.

33. P. Kotelenez. High density limit theorems for nonlinear chemical reactions with diffusion. *Probability Theory and Related Fields*, 78:11 – 37, 1988.
34. K. Krischer, M. Eiswirth, and G. Ertl. Oscillatory CO oxidation on Pt(110): Modeling of temporal self-organization. *Journal of Chemical Physics*, 96(12):9161 – 9172, 1992.
35. Kurtz, T.G. *Approximation of Population Processes*. Society for Industrial and Applied Mathematics, 1981.
36. S. Méléard. Asymptotic behaviour of some interacting particle systems – McKean-Vlasov and Boltzmann models. In D. Talay and L. Tubaro, editors, *Probabilistic Models for Nonlinear Partial Differential Equations*, Lecture Notes in Mathematics 1627, pages 42 – 95. Springer-Verlag, 1995.
37. K. Oelschläger. On the derivation of reaction-diffusion equations as limit dynamics of systems of moderately interacting stochastic processes. *Probability Theory and Related Fields*, 82:565 – 586, 1989.
38. K. Oelschläger and J. Starke. Many-particle models and reaction diffusion equations for chemical systems. Manuskript, 2004.
39. C. Reichert, J. Starke, and M. Eiswirth. Stochastic model of CO-oxidation on platinum surfaces and deterministic limit. *Journal of Chemical Physics*, 115(10):4829 – 4838, 2001.
40. H. H. Rotermund. Imaging of dynamic processes on surfaces by light. *Surface Science Reports*, 29:265 – 364, 1997.
41. H. H. Rotermund. Imaging pattern formation in surface reactions from ultrahigh vacuum up to atmospheric pressures. *Surface Science*, 386:10 – 23, 1997.
42. H. Spohn. *Large Scale Dynamics of Interacting Particles*. Springer-Verlag, 1991.
43. J. Starke, C. Reichert, M. Eiswirth, and H. H. Rotermund. Fluctuation-induced pattern formation in a surface reaction. in preparation, 2005.
44. Angela Stevens. The derivation of chemotaxis equations as limit dynamics of moderately interacting stochastic many-particle systems. *SIAM J. Appl. Math.*, 61(1):183–212, 2000.
45. Yu. Suchorski, J. Beben, R. Imbihl, E.W. James, D.-J. Liu, and J.W. Evans. Fluctuations and Critical Phenomena in Catalytic CO-Oxidation on Pt Facets. *Physical Review B*, 63(16):165417, 2001.
46. Yu. Suchorski, J. Beben, E.W. James, J.W. Evans, and R. Imbihl. Fluctuation-induced transitions in a bistable surface reaction: Catalytic CO oxidation on a Pt field emitter tip. *Phys. Rev. Letters*, 82(9):1907–1910, 1999.
47. S. Wiggins. *Introduction to Applied Nonlinear Dynamical Systems and Chaos*. Springer-Verlag, Berlin, Heidelberg, New York, 1990.

Part V

Turbulent Flow and Combustion

Preamble

This chapter contains three articles considering aspects of modelling and simulation of turbulent reactive flows.

The first article *Multigrid methods for large-eddy simulation* discusses the Large-Eddy Simulation (LES) method which can be used to overcome the multi-scale complexity within turbulent flow simulations by an appropriate subgrid model, avoiding the necessity to resolve all turbulent length scales. Beside this filtering in space, a filtering in time allows for larger time steps as well and gives rise to implicit methods, where an algebraic system of equations has to be solved. Multigrid as a numerical multi-scale approach matches LES quite well with that respect and will be applied. It is essential to control the numerical error introduced by the discretisation and the numerical solver in order to minimize the influence on the turbulent solution and, hence, being able to identify the model error of the subgrid model. The solutions obtained are compared with benchmark solutions found in literature.

The second article *Modeling and simulation of turbulent non-reacting and reacting spray flows* describes the modeling and simulation of turbulent spray flames where the emphasis is on the modeling of the interaction of the vaporization and the turbulent flow field. In particular, the mixing process as well as the interaction of the turbulent flame with the vaporization are studied. The chemical reactions are incorporated using a novel spray flamelet model which accounts for the interaction of the flame with the spray. These processes are least understood when it comes to the simulation of dilute spray flames which play an important role in combustion applications.

The third article *Transport and diffusion in boundary layers of turbulent channel flow* investigates transport and diffusion of scalars in a turbulent channel flow by use of Direct Numerical Simulations (DNS). Experiments with the same geometry have been set up to examine details of the near-wall region. Transport and diffusion are observed with flow-tagging methods. In one setup the deformation of a line of tracer-molecules, placed perpendicular to the flow direction, is investigated. Furthermore the seeding and detecting of tracer-molecules into the boundary layer has been realized in two different configurations, in order to investigate species transport between the bulk flow and the boundary layer.

Multigrid Methods for Large-Eddy Simulation*

A. Gordner, S. Nägele, and G. Wittum

Technische Simulation, Interdisziplinäres Zentrum für Wissenschaftliches Rechnen (IWR), Universität Heidelberg.

Summary. The Large-Eddy simulation (LES) method can be used in order to break the multi-scale complexity within turbulent flow simulations, since not all turbulent length scales have to be resolved, but will be given by an appropriate subgrid model. Beside this filtering in space, a filtering in time allows for larger time steps as well and gives rise to implicit methods, where an algebraic system of equations has to be solved. Multigrid as a numerical multi-scale approach matches LES quite well with that respect and will be applied. It is essential to control the numerical error introduced by the discretisation and numerical solver in order to minimize the influence on the turbulent solution and hence, being able to identify the model error of the subgrid model. Two different stabilization methods, that are used within the collocated Finite Volume dicretisation for unstructured grids, are investigated with respect to their mass conservation error. The obtained solutions will be compared with benchmark solutions found in literature. The used subgrid model takes advantage of mesh dependent parameters. A practical solution within the multigrid procedure is to derive the model parameter on the finest grid level and inject it successively to the coarser grid levels. By this strategy good convergence rates result.

1 Introduction

Flows of incompressible fluids are modelled by the Navier–Stokes equations

$$\frac{\partial \mathbf{u}}{\partial t} - \nabla(\nu(\nabla \mathbf{u} + (\nabla \mathbf{u})^T)) + \nabla \cdot (\mathbf{u}\mathbf{u}^T) + \nabla p = 0 \qquad (1)$$

$$\mathrm{div}(\mathbf{u}) = 0 \qquad (2)$$

with viscosity ν, velocity $\mathbf{u} = (u_1, ..., u_d)^T$, d the spatial dimension and pressure p. For Newton fluids, the viscosity ν is constant. However, for small

---

*This work has been supported by the German Research Foundation (DFG) through SFB 359 (Project A5) at the University of Heidelberg.

viscosities and large velocities the flow develops unordered small scale fluctuations of velocity and pressure. In this case the flow is called turbulent.

Turbulence has a slowdown and a mixing effect for the flow. It occurs everywhere in nature as well as in technology. Since the first description of turbulence as a phenomenon by Reynolds in 1893, turbulence and its generation is still not fully understood.

The multi-scale character of turbulence makes simulation of turbulent flows a difficult business. To account for the full nonlinear multi-scale effect of turbulence, all relevant scales in time and space have to be resolved accurately (see e.g. Hinze [6]). This is not possible for flows in complex technical applications. Thus, depending on the scale of interest and the complexity of the problem, different modelling approaches exist.

Direct Numerical Simulation (DNS) of turbulent flows resolves all scales involved. DNS on a technologically interesting scale is not possible and will not be possible in the near future. Statistical averaging models, so-called Reynolds-averaged Navier–Stokes-Equations (RANS), are derived and closed by some empiric equations for additional unknown quantities, like the turbulent kinetic energy and dissipation. Although these models only give rough approximations of the flow, RANS models offer a cheap and simple way to obtain time averaged coarse scale behaviour of turbulent flows.

A third way to model turbulence is the Large-Eddy Simulation (LES). The large scales which can be represented by the computational grid are resolved directly, while structures smaller than the grid size have to be approximated with subgrid-scale models, see e.g. for an overview Wilcox [19]. LES computations require finer grids than RANS simulations do. However, LES allows for much better accuracy for critical turbulence quantities and thus LES is more and more used for simulations on technical scales.

For DNS most often high-order explicit schemes are used for time discretisation. Although the turbulent solutions are not smooth due to high small-scale fluctuations, high-order or even spectral methods are used for the discretisation in space. Since it is assumed and required that all scales are resolved in DNS, often structured equidistant Cartesian grids are used. Explicit schemes in time avoid the necessity of constructing sophisticated solvers, but they require a huge number of time-steps.

With the RANS approach, standard methods for the Navier–Stokes equations are applied, like finite volume discretisations in space, implicit schemes in time, and also multigrid methods.

The LES approach is relatively new in comparison to the DNS and RANS methods and can be classified between the two others. Since the idea in LES is to apply filtering in space, mainly approaches from DNS are used so far, i.e. explicit discretisation in time and higher order methods on structured grids etc. However, it may be reasonable to apply a filter in time too, and LES may be viewed as an improvement of RANS models, since they are only modelling few scales in contrast to modelling all scales. The opportunity of using implicit methods and larger time steps requires the efficient solution of an algebraic system of equations, which gives rise to the investigation of solvers for LES. Multigrid as numerical multi-scale approach matches LES quite well in that respect. However, the coarse-grid operator needs a discussion, since the subgrid-model depends on the computational grid.

We focus on unstructured or not necessarily structured grids. The most convenient and flexible way for this is to use a collocated arrangement of the variables. By this, one can use the same ansatz functions for all unknowns, but the resulting system is unstable and an unphysical checkerboard pressure distribution can result. Schneider and Raw [16] introduced a very interesting stabilisation which we will briefly describe in one of the following sections. For compressible flows enhancements have been introduced by Karimian [8]. This modification has also some advantages for the incompressible case which will be addressed. In turbulent flow calculations the error in mass conservation is one major issue. In Nägele [13] the properties of the discretisation methods in use are thoroughly discussed for the laminar flow regime.

Here, the behaviour of the various stabilisation and discretisation possibilities and their consequences for the quality of the turbulent solution is investigated. LES subgrid-models depend on the filter size. Special treatment is required, if the mesh size is used as filter size within a multigrid procedure.

2 LES: Framework, Filtering, Subgrid-Scale Models

In a Large-Eddy Simulation large scales are resolved directly, while the unresolved small scales are modelled. Thus, one needs a specific length scale, with which the unknowns are decomposed in large and small scales. Therefore, each unknown is split in a local average $\overline{f}_i$, representing the large scales, and a subgrid-scale component f'_i where $f_i = \overline{f}_i + f'_i$ stands for an unknown component. The local averages are generated by the application of a filter. This filter is a convolution integral of the form:

$$\overline{f}(x,t) = \int_\Omega G_\Delta(x,y) f(y,t) \, dV$$

We use a volume-average box-filter with

$$G_\Delta(x,y) = \begin{cases} \frac{1}{|\Omega_\Delta(x)|} & y \in \Omega_\Delta(x) \\ 0 & else \end{cases} \quad \text{and filter width } \Delta := \sqrt[d]{|\Omega_\Delta(x)|}$$

where Ω_Δ denotes the support of the filter function. For the subgrid-scale components f'_i a model has to be defined.

To transform the governing equations system into one depending only on local averages $\overline{f}_i$, the filter operator is applied to the incompressible Navier–Stokes equations. Under the assumption that integration and differentiation commute, equation (2) can be written in component-wise form:

$$\frac{\partial \overline{u}_i}{\partial t} - \frac{\partial}{\partial x_j}\left(\nu\left(\frac{\partial \overline{u}_i}{\partial x_j} + \frac{\partial \overline{u}_j}{\partial x_i}\right)\right) + \frac{\partial}{\partial x_j}(\overline{u_i u_j}) + \frac{\partial \overline{p}}{\partial x_i} = 0 \quad (3)$$

$$\frac{\partial \overline{u}_j}{\partial x_j} = 0 \quad (4)$$

The Einstein summation convention is used unless stated otherwise.

The unclosed convection term $\frac{\partial}{\partial x_j}(\overline{u_i u_j})$ is then replaced introducing the subgrid-scale stress tensor $\tau_{ij} := \overline{u_i u_j} - \overline{u}_i \overline{u}_j$. The momentum equation then becomes:

$$\frac{\partial \overline{u}_i}{\partial t} - \frac{\partial}{\partial x_j}\left(\nu\left(\frac{\partial \overline{u}_i}{\partial x_j} + \frac{\partial \overline{u}_j}{\partial x_i}\right)\right) + \frac{\partial}{\partial x_j}(\overline{u}_i \overline{u}_j) + \frac{\partial \overline{p}}{\partial x_i} + \frac{\partial}{\partial x_j}\tau_{ij} = 0 \quad (5)$$

It remains to specify a model for τ_{ij}. The first and oldest model has been introduced by Smagorinsky [17]. Several other models exist, i.e. developed by Germano [4] and slightly modified by Lilly [10] and a model developed by Zang et.al. [20]. Some models are based on a pure eddy viscosity assumption [4],[10], [17], where it is assumed that the anisotropic part of the subgrid-scale stress tensor is proportional to the shear stress tensor. In the model developed by Zang et.al. [20], a slightly different approach is introduced. Their model consists of a mixture between an eddy viscosity and a scale similarity model. The model terms of these models for the anisotropic part of the subgrid-scale stress tensor read:

$$\text{eddy viscosity model: } \tau_{ij} - \frac{1}{3}\delta_{ij}\tau_{kk} = -2C\Delta^2|\overline{S}|\overline{S}_{ij} \quad (6)$$

$$\text{mixed model: } \tau_{ij} - \frac{1}{3}\delta_{ij}\tau_{kk} = -2C\Delta^2|\overline{S}|\overline{S}_{ij} + \mathcal{L}_{ij}^m - \frac{1}{3}\delta_{ij}\mathcal{L}_{kk}^m \quad (7)$$

with $\overline{S}_{ij} = \frac{1}{2}\left(\frac{\partial \overline{u}_i}{\partial x_j} + \frac{\partial \overline{u}_j}{\partial x_i}\right)$, $|\overline{S}| = \sqrt{2\overline{S}_{ij}\overline{S}_{ij}}$ and $\mathcal{L}_{ij}^m = \overline{\overline{u}_i \overline{u}_j} - \overline{\overline{u}}_i \overline{\overline{u}}_j$.

In [4], [10] and [20], for the determination of the model parameter C a dynamic approach is applied. For this, equation (3) is filtered with a second filter $G_{\hat{\Delta}}$ with $\hat{\Delta} > \Delta$ (compare Figure 1) resulting in:

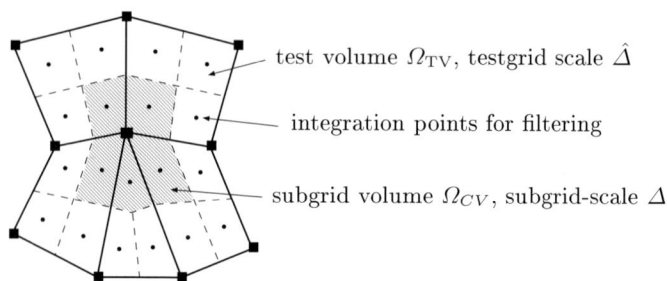

Fig. 1. Comparison between testgrid and subgrid-filter scales.

$$\frac{\partial \hat{\bar{u}}_i}{\partial t} - \frac{\partial}{\partial x_j}\left(\nu\left(\frac{\partial \hat{\bar{u}}_i}{\partial x_j} + \frac{\partial \hat{\bar{u}}_j}{\partial x_i}\right)\right) + \frac{\partial}{\partial x_j}(\hat{\bar{u}}_i\hat{\bar{u}}_j) + \frac{\partial \hat{\bar{p}}}{\partial x_i} + \frac{\partial}{\partial x_j}T_{ij} = 0 \quad (8)$$

Analogous to equation (5), the unclosed term is replaced by introducing the testgrid-scale stress tensor $T_{ij} := \widehat{\bar{u}_i \bar{u}_j} - \hat{\bar{u}}_i \hat{\bar{u}}_j$. The same model as for τ_{ij} is applied for T_{ij} depending on twice filtered variables but with the same parameter C. To compare these two representations, equation (5) is filtered with $G_{\hat{\Delta}}$ and subtracted from equation (8), resulting into:

$$L_{ij} := \widehat{\bar{u}_i \bar{u}_j} - \hat{\bar{u}}_i \hat{\bar{u}}_j = T_{ij} - \widehat{\tau_{ij}}. \quad (9)$$

The Leonard term L_{ij} represents the resolved turbulent stress and relates the subgrid-scale and testgrid-scale stress tensors with each other.

The insertion of an eddy viscosity model (6) for T_{ij} and τ_{ij} leads to:

$$T_{ij} - \frac{1}{3}\delta_{ij}T_{kk} = -2C\hat{\Delta}^2|\hat{\bar{S}}|\hat{\bar{S}}_{ij}$$

$$\tau_{ij} - \frac{1}{3}\delta_{ij}\tau_{kk} = -2C\Delta^2|\bar{S}|\bar{S}_{ij}$$

and with (9) reveals the relation:

$$L_{ij} - \frac{1}{3}\delta_{ij}L_{kk} = -2C\hat{\Delta}^2|\hat{\bar{S}}|\hat{\bar{S}}_{ij} + 2C\Delta^2\widehat{|\bar{S}|\bar{S}_{ij}} =: 2CM_{ij}.$$

Since this is a tensorial equation for a single parameter, a least squares approach is used to determine C.

$$\text{minimise:} \quad Q = (L_{ij} - \frac{1}{3}\delta_{ij}L_{kk} - 2CM_{ij})^2$$

$$\Rightarrow \quad C = \frac{L_{ij}M_{ij}}{M_{kl}M_{kl}}.$$

For the mixed model (7) the result is very similar except that a slightly more difficult expression has to be minimised. First the models for the stress tensors are:

$$T_{ij} - \frac{1}{3}\delta_{ij}T_{kk} = -2C\hat{\Delta}^2|\hat{\bar{S}}|\hat{\bar{S}}_{ij} + \mathcal{L}_{ij}^M - \frac{1}{3}\delta_{ij}\mathcal{L}_{kk}^M$$

$$\tau_{ij} - \frac{1}{3}\delta_{ij}\tau_{kk} = -2C\Delta^2|\bar{S}|\bar{S}_{ij} + \mathcal{L}_{ij}^m - \frac{1}{3}\delta_{ij}\mathcal{L}_{kk}^m$$

with $\mathcal{L}_{ij}^M = \widehat{\bar{u}_i\bar{u}_j} - \hat{\bar{u}}_i\hat{\bar{u}}_j$. To describe the expression for the model parameter C a few auxiliary tensors are defined:

$$H_{ij} := \widehat{\overline{\bar{u}_i\bar{u}_j}} - \hat{\bar{\bar{u}}}_i\hat{\bar{\bar{u}}}_j$$

$$I_{kk} := L_{kk} + \widehat{\mathcal{L}_{kk}^m} - \mathcal{L}_{kk}^M \qquad \text{isotropic part}$$

$$M_{ij} := -\hat{\Delta}^2|\hat{\bar{S}}|\hat{\bar{S}}_{ij} + \Delta^2\widehat{|\bar{S}|\bar{S}_{ij}}$$

In this case the insertion of the terms in expression (9) results in:

$$\text{minimise:} \quad Q = (L_{ij} - H_{ij} - \frac{1}{3}\delta_{ij}I_{kk} - 2CM_{ij})^2$$

$$\Rightarrow C = \frac{(L_{ij} - H_{ij})M_{ij}}{M_{kl}M_{kl}}.$$

In general, for this mixed model the parameter C is smaller than for the pure eddy viscosity model since the tensors L_{ij} and H_{ij} are approximately of the same size which can be easily seen by their definitions. To prevent strong oscillations of the model parameter C, it is averaged over the test volume Ω_{TV} belonging to filter $G_{\hat{\Delta}}$.

After determination of the model parameter C the insertion of the model term in equation (5) results in:

$$-\nu\left(\frac{\partial \bar{u}_i}{\partial x_j} + \frac{\partial \bar{u}_j}{\partial x_i}\right) + \delta_{ij}\bar{p} + \tau_{ij} = \begin{cases} -(\nu + \nu_t)\left(\frac{\partial \bar{u}_i}{\partial x_j} + \frac{\partial \bar{u}_j}{\partial x_i}\right) + \delta_{ij}p^* & \text{eddy viscosity} \\ \underbrace{-(\nu + \nu_t)}_{:=\nu_{\text{eff}}}\left(\frac{\partial \bar{u}_i}{\partial x_j} + \frac{\partial \bar{u}_j}{\partial x_i}\right) + \mathcal{L}_{ij}^m + \delta_{ij}p^* & \text{mixed} \end{cases}$$

with $p^* := \bar{p} + \frac{1}{3}\tau_{kk}^{eddy}$ and τ_{kk}^{eddy} the trace of the eddy viscosity part of the model.

The Navier–Stokes equations are thus modified substituting the viscosity ν by the effective viscosity ν_{eff} and a modified pressure p^*.

3 Discretisation

The equations are discretised by a collocated finite volume method, where all unknowns are located at the nodal points. The control volumes are defined via dual boxes of the underlying finite element grid. A simple sketch of the control volume in 2d can be seen in Figure 2. The construction, however, is general and applies to 3d as well. After applying Gauss' theorem and splitting the integration over the whole control volume surface into a sum of integrations over subsurfaces, the resulting system in discretised form reads:

$$\frac{\delta}{\delta t}|CV|U_i + \sum_{ip=1}^{\#ip(CV)}\left(u_i u_j n_j + p n_i - \nu\left(\frac{\partial u_i}{\partial x_j} + \frac{\partial u_j}{\partial x_i}\right)n_j\right)\bigg|_{ip} = 0 \quad (10)$$

$$\sum_{ip=1}^{\#ip(CV)}(u_j n_j)\bigg|_{ip} = 0. \quad (11)$$

$\#ip(CV)$ denotes the number of subsurfaces of the control volume surface and is therefore equal to the number of integration points of the control volume

CV. $|CV|$ denotes the area of the control volume in 2d and accordingly the volume in 3d. In Figure 2, an example for a control volume is shown with 10 integration points or subsurfaces respectively. The outer normal of each subsurface $\mathbf{n} = (n_1, ..., n_d)^T$ is scaled by the subsurface area to get a shorter notation.

For the time discretisation $\frac{\delta}{\delta t}$ a diagonally implicit Runge-Kutta Method of second order is applied, which is described in the paper of Alexander [1]. If everything but the time derivative is summarised in the operator $K(\mathbf{u}, p, t)$:

$$K(\mathbf{u}, p, t) = \begin{pmatrix} \sum_{ip=1}^{\#ip(CV)} \left(u_i u_j n_j + p n_i - \nu \left(\frac{\partial u_i}{\partial x_j} + \frac{\partial u_j}{\partial x_i} \right) n_j \right)\Big|_{ip} \\ \sum_{ip=1}^{\#ip(CV)} (u_j n_j)\Big|_{ip} \end{pmatrix}$$

and the time derivative are given by the operator $T(\mathbf{u}, p, t) = \begin{pmatrix} |CV| \\ 0 \end{pmatrix}$, the time stepping scheme can be formulated in two steps:

$$T(\mathbf{u}, p, t + \alpha \Delta t) - T(\mathbf{u}, p, t) + \alpha \Delta t K(\mathbf{u}, p, t + \alpha \Delta t) = 0$$
$$T(\mathbf{u}, p, t + \Delta t) - T(\mathbf{u}, p, t + \alpha \Delta t) + (1 - 2\alpha) \Delta t K(\mathbf{u}, p, t + \alpha \Delta t) + \alpha \Delta t K(\mathbf{u}, p, t + \Delta t) = 0,$$

where $\alpha = 1 - \frac{1}{2}\sqrt{2}$. This scheme is of order 2 and stable.

A quasi-Newton linearisation of the convection term yields

$$\sum_{j=1}^{d} u_i u_j n_j \approx \sum_{j=1}^{d} u_i \tilde{u}_j n_j \qquad (12)$$

where $\tilde{u}_j$ stands for the latest approximation of the integration point velocity u_j. The assembly of the discretised system can be done element-wise since mainly subsurface integrals have to be computed.

All unknowns are located in the nodes, thus the discretisation would be unstable if the integration point quantities are interpolated via the ansatz functions only, because the LBB-condition is not fulfilled in this case, see

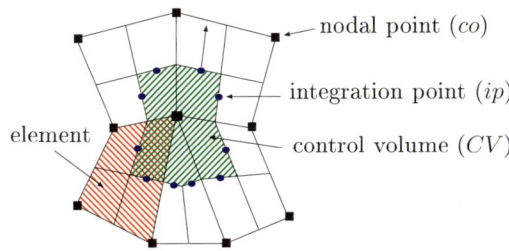

Fig. 2. Sketch of a 2d control volume, consisting of 5 subcontrol volumes (SCV).

Brezzi [3]. To stabilise the system, a special interpolation for the integration point velocities is constructed. This interpolation is based on the idea to determine the relation between velocity and pressure by the use of the momentum equation. It was originally developed by Schneider and Raw [16] and further modified by Schneider and Karimian [8]. Both methods will be briefly described in the following section. For more details we refer to Nägele [13].

4 Stabilisation and Convection Treatment

The stabilisation developed by Schneider and Raw [16] is called FIELDS (<u>Fi</u>nite <u>E</u>lement <u>D</u>ifferential <u>S</u>cheme) and will be explained in subsection 4.1. The stabilisation introduced by Karimian [8] is designated for distinction by FLOW. The differences of FLOW in comparison to FIELDS will be addressed in Subsection 4.2.

4.1 FIELDS

The idea of FIELDS is to derive the velocities u_i in each element and in each integration point using the momentum equation. The momentum equation is approximated by a very simple finite difference approach, where the diffusion part is assumed to be a Laplacian. The convection term is linearised and afterwards discretised by an upwind method:

$$\sum_{j=1}^{d} u_j \frac{\partial u_i}{\partial x_j} \approx \sum_{j=1}^{d} \tilde{u}_j \frac{\partial u_i}{\partial x_j} = \|\tilde{\mathbf{u}}\| \frac{\partial u_i}{\partial s} \quad \text{where} \quad s = \frac{1}{\|\tilde{\mathbf{u}}\|} \tilde{\mathbf{u}}. \tag{13}$$

Thus, the simplified form of the momentum equation is

$$\frac{\partial u_i}{\partial t} - \nu \Delta u_i + \|\tilde{\mathbf{u}}\| \frac{\partial u_i}{\partial s} + \frac{\partial p}{\partial x_i} = 0 \tag{14}$$

To explain the detailed form of the stabilisation, the position of all integration points as well as the local flow direction at integration point ip_4 are shown in Figure 3. Starting from equation (14) the finite difference approximation

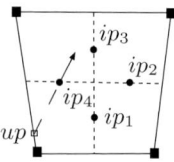

Fig. 3. Sketch of element with integration points.

for one integration point reads:

$$\frac{u_i - u_i^0}{\Delta t} - \nu \frac{\sum_{k=1}^{n_N} N_k(ip) U_i^k - u_i}{L_d^2} + \|\tilde{\mathbf{u}}\| \frac{u_i - u_i^{up}}{L_c} + \sum_{k=1}^{n_N} \frac{\partial N_k}{\partial x_i} P_k = 0 \bigg|_{ip_l} \quad (15)$$

where u_i^0 denotes the integration point velocity at the previous time point $t - \Delta t$, L_d^2 is the diffusion length scale, U_i^k corresponds to the value of the i-th velocity component in node k, u_i^{up} denotes the upwind velocity, L_c is the distance between the integration point ip and the corresponding upwind position up, and $\tilde{\mathbf{u}}$ refers to the last approximation of the integration point velocity $\mathbf{u}$. N_k are the nodal ansatz functions with $k = 1, ..., n_N$ and n_N the number of nodes of the element.

This leads to a system of equations depending on integration point velocities, nodal velocities and nodal pressures which can be solved directly in each element. The resulting approximation for the integration point velocities

$$u_i = \left(\frac{1}{\Delta t} + \frac{\nu}{L_d^2} + \frac{\|\tilde{\mathbf{u}}\|}{L_c}\right)^{-1} \left(\frac{1}{\Delta t} u_i^0 + \frac{\nu}{L_d^2} \sum_{k=1}^{n_N} N_k U_i^k + \frac{\|\tilde{\mathbf{u}}\|}{L_c} u_i^{up} - \sum_{k=1}^{n_N} \frac{\partial N_k}{\partial x_i} P^k\right)$$

are put into the continuity equation (11) to stabilise the system of equations. By doing so, a pressure dependence in form of a Laplacian, scaled with a constant times the mesh size squared, is introduced in the continuity equation, as known from other stabilised schemes.

4.2 FLOW

Karimian introduced a modification, which also takes the continuity equation into account to determine the special interpolation for the stabilisation. His modification demands that the momentum equation error $\dot{\epsilon}$ and the error of the continuity equation $\ddot{\epsilon}$ are balanced in all integration points in the following way:

$$\underbrace{\frac{\partial u_i}{\partial t} - \nu \Delta u_i + \|\tilde{\mathbf{u}}\| \frac{\partial u_i}{\partial s} + \frac{\partial p}{\partial x_i}}_{:= \dot{\epsilon}} = \underbrace{u_i \operatorname{div}(\mathbf{u})}_{:= \ddot{\epsilon}}\bigg|_{ip_l}.$$

The mass error $\ddot{\epsilon}$ times u_i can be reformulated to:

$$u_i \ddot{\epsilon} = \sum_{j=1}^{d} u_j \frac{\partial u_i}{\partial x_j} - \sum_{j=1, j \neq i}^{d} \left(u_j \frac{\partial u_i}{\partial x_j} - u_i \frac{\partial u_j}{\partial x_j}\right).$$

This can be approximated and discretised by:

$$u_i \ddot{\epsilon} \approx \|\tilde{\mathbf{u}}\| \frac{u_i^{dn} - u_i^{up}}{L_{dn}} - \sum_{j=1, j \neq i}^{d} \sum_{k=1}^{n_N} \left(\tilde{u}_j \frac{\partial N_k}{\partial x_j} U_i^k - \tilde{u}_i \frac{\partial N_k}{\partial x_j} U_j^k\right),$$

where L_{dn} denotes the length between the upwind and downwind point. This leads to the FLOW approximation for the integration point velocities:

$$\frac{u_i - u_i^o}{\Delta t} - \nu \frac{\sum_k N_k U_i^k - u_i}{L_d^2} + \|\tilde{\mathbf{u}}\| \frac{u_i - u_i^{up}}{L_c} + \sum_{k=1}^{n_N} \frac{\partial N_k}{\partial x_i} P^k$$
$$- \|\tilde{\mathbf{u}}\| \frac{u_i^{dn} - u_i^{up}}{L_{dn}} + \sum_{j=1, j \neq i}^{d} \sum_{k=1}^{n_N} \left(\tilde{u}_j \frac{\partial N_k}{\partial x_j} U_i^k - \tilde{u}_i \frac{\partial N_k}{\partial x_j} U_j^k \right) = 0 \quad (16)$$

In contrast to the approximation of FIELDS in equation (15), the resulting integration point velocities of the FLOW-ansatz depend on all velocity components in the nodes which leads to a tighter coupling of the components and to a better nonlinear behaviour and nonlinear convergence rates, see examples in Section 6.1. Furthermore, the error in the continuity equation tends to be lower than for FIELDS.

4.3 Convection Treatment

Various upwinding methods can be applied to discretise the convection term (13) and (12). They differ in the choice of the upwind point up. The simplest and most diffusive choice is the full upwind scheme illustrated in Figure 4(c). The upwind point for this example is $\mathbf{u}^{up} = \mathbf{U}_3$ which is clearly a very bad approximation and far away from the streamline. Within the skewed upwind, the closest upwind node of the element serves as approximation. This leads to the choice $\mathbf{u}^{up} = \mathbf{U}_0$ as upwind point, which is an improvement but still very diffusive. A better approximation can be obtained by the Linear Profile Skewed method (LPS) displayed in Figure 4(a). This scheme determines the upwind point by intersection of the streamline and the element boundary. The resulting approximation for the upwind point in the example would be:

$$\mathbf{u}^{up} = \frac{b}{a+b} \mathbf{U}_0 + \frac{a}{a+b} \mathbf{U}_1. \quad (17)$$

PHYSICAL ADVECTION CORRECTION (PAC)

PAC-upwinding can be written in the form

$$u_i(ip) \approx \breve{u}_i^{up} + \delta u_i \quad (18)$$

which is a generalised skewed upwinding technique introduced by Raithby [15], depending on a scaled upwind velocity plus a correction term. In the context of this paper the special interpolation already described above in section 4.1 can be used leading to the following approximations for the upwind point and the correction term:

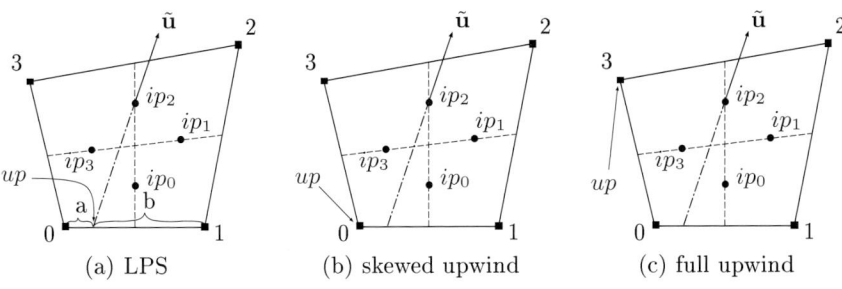

Fig. 4. Upwind methods.

$$\tilde{u}_i^{up} = \left(\frac{1}{\Delta t} + \frac{\nu}{L_d^2} + \frac{\|\tilde{\mathbf{u}}\|}{L_c}\right)^{-1} \frac{\|\tilde{\mathbf{u}}\|}{L_c} u_i^{up}$$

$$\delta u_i = \left(\frac{1}{\Delta t} + \frac{\nu}{L_d^2} + \frac{\|\tilde{\mathbf{u}}\|}{L_c}\right)^{-1} \left(\frac{u_i^o}{\Delta t} + \frac{\nu}{L_d^2} \sum_{k=1}^{n_{co}} N_k U_i^k - \sum_{k=1}^{n_{co}} \frac{\partial N_k}{\partial x_i} P^k\right).$$

This method shows a very good behaviour and increases the order of the upwind technique inside the interpolation clearly as will be shown by means of the Driven Cavity problem.

Additionally, depending on the local Peclet number Pe ($Pe = \frac{u_n L}{\nu}$), normal velocity u_n, local length scale L) a blending between central differences and one of the two upwind techniques described above is always used to increase the approximation order of the convection operator.

4.4 Diffusion Length Scale L_d^2

The diffusion operator mainly determines the stabilisation constant introduced in the continuity equation. The effects of that term are thoroughly discussed in [13]. Thus, only the different expressions for the diffusion length scale are presented:

$$L_{d_{Raw}}^2 = \frac{|SCV|^2}{2\|\mathbf{n}\|^2} + \frac{3}{8}\|\mathbf{n}\|^2$$

$$L_{d_{(2)}}^2 = \left(\frac{2\|\mathbf{n}\|^2}{|SCV|^2} + \frac{8}{3\|\mathbf{n}\|^2}\right)^{-1}$$

$$L_{d_{(2)corr}}^2 = \left(\frac{2\|\mathbf{n}_{\min}\|^2}{|SCV|^2} + \frac{8}{3\|\mathbf{n}_{avg}\|^2}\right)^{-1}$$

In these expressions the normal vector $\mathbf{n}$, belonging to that integration point, is scaled with the corresponding area of the face. $|SCV|$ denotes the subcontrol volume. In the stabilised (or corrected) form of (2) which will be further referred to by method $(2)_{corr}$, also $\|\mathbf{n}_{\min}\|^2 = \min_{n \in \text{element}} \|\mathbf{n}\|^2$ the minimal norm of all normals of that element and $\|\mathbf{n}_{avg}\|^2 = \frac{1}{n_N}\sum_{k=1}^{n_N}\|\mathbf{n}_k\|^2$ an average measure of the normals is introduced. In 3d similar expressions arise which are presented in Nägele [12], including investigations of the different behaviour of these expressions.

5 Solver Settings and Realisation

The solving strategy will be described very briefly since mainly standard components even in the multigrid regime can be used which will be demonstrated numerically.

A Quasi-Newton iteration is used as nonlinear solver. In each nonlinear step the resulting fully coupled linear equation system is solved by a standard geometric multigrid method. The full weighting operator serves as restriction, while for prolongation bilinear interpolation is used. The multigrid employs point-block ILU_β as smoother, based on $(d+1) \times (d+1)$-blocks at each grid point whereas the ordering of the unknowns in these blocks is u_1, $u_2(, u_3)$, p. This smoother worked very well since the coupling of the different equations in these blocks is very strong and is taken into account by the ILU-method. A V-cycle with different settings for β, pre- and post-smoothing steps is used.

Within the multigrid procedure on each grid level a turbulent model for the coarse grid operator has to be formed. Hence, the turbulent viscosity ν_t has to be determined on the coarser grids to set up the coarse grid operators. Based on the model parameter C on the finest grid level, the turbulent viscosity is computed. Only on the finest grid level the assumptions for the turbulence models can be assumed to be true, because on coarser levels it would result in undetermined effects.

Therefore, a simple strategy is used to transport the model part in the equations to the coarser grid levels. The injection operator is used, since then it is assured that on each point in space the same turbulent viscosity, based on the fine grid physics, is used within the multigrid process. Other grid transfer operators than the injection would average the turbulent viscosity leading to undetermined physical effects.

Beside the turbulent viscosity ν_t, also the scale similarity terms $\mathcal{L}_{ij}^m$ are used within the turbulence models. However, this term contributes to the right hand side and has an influence on the defect but not on the matrix itself. The transfer of the defect is handled within the solving procedure by the transfer operators for the defect, therefore no special grid transfer treatment for $\mathcal{L}_{ij}^m$ is necessary.

6 Numerical Results

The first example is a laminar flow which demonstrates the influence of the convection term treatment. Since the accuracy of turbulence modelling severely depends on the discretisation of this term, the first example shows

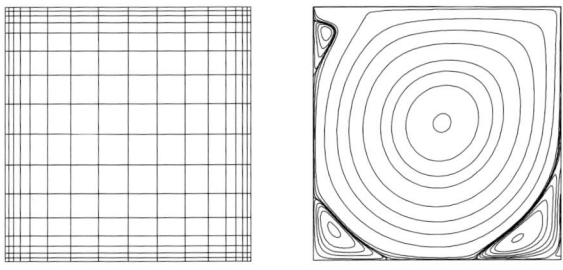

Fig. 5. Coarse grid and streamlines of the Driven Cavity problem.

the different introduced numerical error. Especially the influence of the upwind techniques can be shown for this problem. Then turbulent examples will be given where the effects of the discretisations for the resulting solutions and solving behaviour in the turbulent case are discussed.

6.1 Driven Cavity

This test case is dedicated to the investigation of the convection term treatment in the momentum equation. In this term a lot of numerical diffusion can be introduced which destroys the approximation quality severely.

The problem is defined in the unit square with no-slip boundary conditions everywhere except the top lid. There the velocity is prescribed to be $\mathbf{u} = (1, 0)^T$. A relative high Reynolds number of Re=3200 based on the lid velocity and the side length of the domain is chosen. LPS upwind schemes (see Section 4.3) in combination with physical advection correction or standard upwinding are investigated. The results are compared to reference values from Ghia et al., cited in [5], the values of u_1 along $x = \frac{1}{2}$ and u_2 along $y = \frac{1}{2}$. The coarsest grid and the resulting streamlines can be seen in Figure 5. The coarse grid is uniformly refined 5 times and the results on each level are illustrated in Figures 6-7. First, the PAC-scheme from equation (18) in combination with the LPS-scheme (see equation (17)) leads to a very good approximation already on level 3. Further refinement is not necessary for this scheme, since the results match the reference data very well. If standard upwinding in combination with LPS is used, the results are much worse and a much higher resolution is necessary to get approximately the same accuracy.

This demonstrates that advection correction increases the approximation order explicitly, whereas a simple upwind technique leads to a rather high numerical diffusivity as Figure 7 demonstrates. All of the above results have been obtained using the FLOW stabilisation. The FIELDS method did not lead to an equation system which could be solved properly, at least not if PAC-upwinding was used. Therefore, the FIELDS-results are omitted for this example.

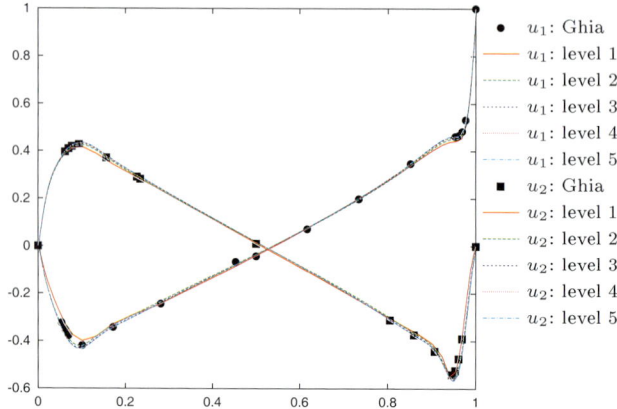

Fig. 6. Comparison with reference data for PAC-upwinding with LPS.

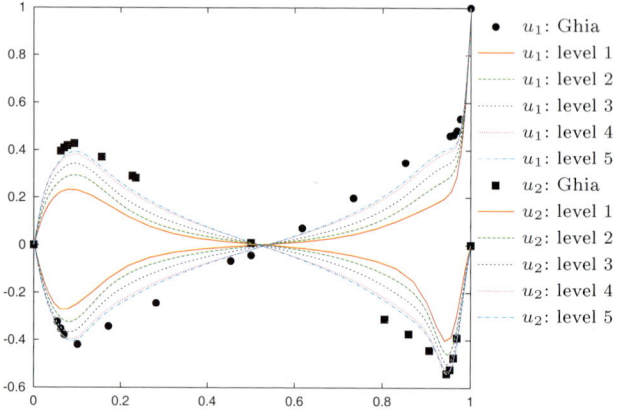

Fig. 7. Comparison with reference data for upwinding with LPS.

6.2 Mixing Layer Problem in Two Dimensions

The mixing layer problem is a shear flow problem where the driving velocities are in opposite direction. It is a popular test problem for studying turbulence models and their differences. This problem was investigated in Boersma et al [2], John [7], Lesieur et al. [9] and also in Nägele/Wittum [14].

The shear flow is defined in $\Omega = (0,1) \times (-0.5, 0.5)$ with periodic boundary conditions in x-direction and symmetry boundary conditions at $y = -0.5$ and $y = 0.5$. The initial solution is defined as

$$\mathbf{u} = \begin{pmatrix} U_\infty \tanh(\frac{2y}{\delta}) \\ 0 \end{pmatrix}.$$

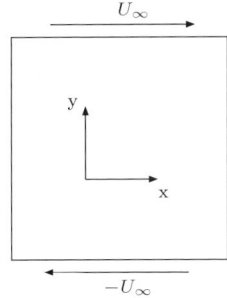

Fig. 8. Problem definition of the 2d Mixing Layer Problem.

depending on the free-stream velocity U_∞ and the vorticity thickness δ:

$$\delta := \frac{2U_\infty}{\max\limits_{y}|\langle\omega_z\rangle(y,t)|},$$

where ω_z is the z-component of $\omega = \nabla \otimes u$ and $\langle.\rangle$ is an average in periodic direction, see Figure 8. To enforce the mixing process a divergence-free initial disturbance $\mathbf{u}'$ is added to the starting solution:

$$\mathbf{u}' = 0.001 \begin{pmatrix} \frac{\partial \Psi}{\partial y} \\ -\frac{\partial \Psi}{\partial x} \end{pmatrix} \quad \text{with} \quad \Psi = e^{-(\frac{2y}{\delta})^2}(\cos(16\pi x) + \cos(40\pi x)).$$

Thus, the initial vorticity thickness of the undisturbed initial solution results in

$$\delta_{t=0} = \frac{2U_\infty}{\max\limits_{y}|\langle\omega_z\rangle(y,t)|} = \frac{2U_\infty}{\max\limits_{y}(U_\infty \frac{2}{\delta}(1-\tanh^2(\frac{2y}{\delta})))} = \frac{2U_\infty}{U_\infty \frac{2}{\delta}} = \delta.$$

As reference values $U_\infty = 1$, $\delta = \frac{1}{28}$ and $Re = \frac{U_\infty \delta}{\nu} = 10000$ are chosen, so that the simulations were run with a viscosity of $\nu = \frac{1}{280000}$ corresponding to the values used by John in [7].

As turbulence model the dynamic model of Germano/Lilly, equation (6), as well as the mixed model from Zang et al., see equation (7), is used to compare the discretisation methods also in view of their solving behaviour. A full Newton approximation scheme is used to take all nonlinear effects into account. For upwinding the PAC procedure, illustrated on page 384, is applied. The nonlinear reduction rate is set to 10^{-5}. The equidistant coarse grid contains 25 nodes with a mesh size $h = \frac{1}{4}$, while the grid on level 5 has a 2^5 times finer mesh size $h = \frac{1}{128}$.

Within the implicit time discretisation, the constant time step size is adjusted to $\Delta t = \frac{\delta}{U_\infty}$, resulting into CFL-numbers in the range $4 < \text{CFL} < 8$. The linear algebraic system of equations is solved with a linear multigrid cycle.

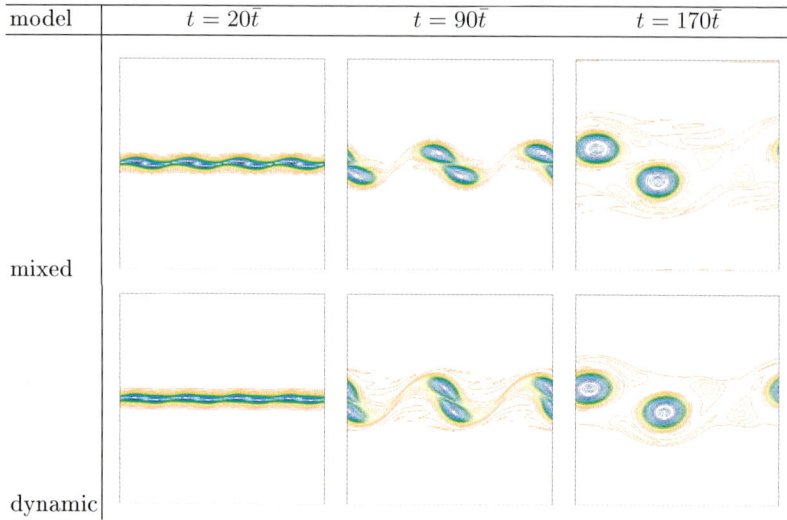

Fig. 9. Vorticity ω_z for the FLOW-stabilisation.

The linear V-cycle multigrid procedure (V(2,2)) uses an ILU_0 smoother with $\lambda = 0.7$ damping and 2 pre- and postsmoothing steps. A linear reduction rate of 10^{-1} is required. In Figure 9, the vorticity ω_z for the described turbulence models in (6), (7) based on the FLOW stabilisation are plotted for different time points.

Both results are similar, i.e. the time history of the development of the vortices, and their recombination to larger vortices occur to almost identical time points. Also the dissipation rate of the total kinetic energy is almost the same. This is not the case for the FIELDS stabilisation method. Especially for the mixed model, where the scale similarity term acts as a source term in the mass equation. In contrast to the FLOW method, where the additional term $u_i \text{div}(\mathbf{u})$ is introduced, the FIELDS stabilisation leads to higher total kinetic energy. Comparisons with the results, obtained by [7], show almost the same dissipation rate for the FLOW method. Hence the FLOW stabilisation leads to better results for turbulent flows than FIELDS, where the total kinetic energy is disturbed by the influence of the stabilisation.

This effects also the nonlinear convergence rate, which is almost twice as fast for the FLOW method in comparison to the FIELDS scheme, see Table 1, whereas the linear convergence rates of the multigrid cycles are almost identical with a small advantage for the FLOW scheme.

6.3 Mixing Layer Problem in Three Dimensions

The mixing layer problem in three dimensions has been investigated by Vreman et al. [18] in order to compare subgrid-models, by Moser and Rogers

Table 1. Linear and non-linear convergence rates, averaged over 320 timesteps.

model	stabilisation		lin. conv.		non-lin. conv.	
			$\kappa^t(u)$	$\kappa^t(p)$	$\psi^t(u)$	$\psi^t(p)$
mixed	FIELDS	Raw	0.018862	0.055175	0.512351	0.521532
		(2)	0.019113	0.054372	0.511071	0.513979
	FLOW	Raw	0.016566	0.045848	0.283442	0.269224
		(2)	0.016471	0.045079	0.280470	0.266567
dynamic	FIELDS	Raw	0.019643	0.098656	0.496130	0.428898
		(2)	0.019792	0.100009	0.499238	0.426275
	FLOW	Raw	0.017458	0.053130	0.277070	0.235629
		(2)	0.016929	0.052587	0.277128	0.236590

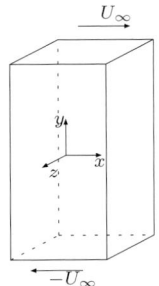

Fig. 10. Mixing Layer 3d: problem definition.

[11] to determine turbulent structures and by John [7] to compare different strategies of turbulence modelling. The 3d mixing layer set up is similar to the 2d case. The computational domain $\Omega = (0,2) \times (0,2) \times (0,4)$ has periodic boundaries in x- and z- direction and symmetric boundaries at $y = 0$ and $y = 4$, see Figure 10.

The initial solution is defined as

$$\mathbf{u} = \begin{pmatrix} U_\infty \tanh(\frac{2y-2}{\delta}) \\ 0 \\ 0 \end{pmatrix}$$

with the vorticity thickness $\delta = \delta_0 = \frac{1}{14}$ and the free stream velocity $U_\infty = 1$. The time step is restricted to $\Delta t \in [\frac{1}{16}\bar{t}, \frac{1}{2}\bar{t}]$ with $\bar{t} = \frac{\delta_0}{U_\infty}$, resulting into CFL-numbers between 1 and 2.4. The setup of the linear and nonlinear solver is identical to the 2d case, described in 6.2. The coarse grid on level 0 consists of $5 \times 9 \times 5$ nodes, with equidistant mesh size in x- and z direction. The mesh size in y-direction is continously increased by a factor of 1.5 starting from the middle towards the upper and lower border. After 4 regular grid refinements, one obtains on level 4 the smallest mesh size of 0.01538, and 0.05192 as the largest mesh size. The resulting problem with 2 113 536 unknowns is solved parallel on 32 processors based on an RCB load distribution.

In order to generate turbulence, first a 2d perturbation $\mathbf{u}'_{2d}$ is used

$$\mathbf{u}'_{2d} = 0.0005 \begin{pmatrix} \frac{\partial \Psi_{2d}}{\partial y} \\ -\frac{\partial \Psi_{2d}}{\partial x} \\ 0 \end{pmatrix} \quad \text{with} \quad \Psi_{2d} = e^{-\left(\frac{2y-2}{\delta}\right)^2} \sum_{k=0}^{3} \cos(2^k \pi x)$$

followed by a 3d perturbation $\mathbf{u}'_{3d}$ of the form

$$\mathbf{u}'_{3d} = 0.0002 \begin{pmatrix} 2\frac{\partial^2 \Psi_{3d}}{\partial y \partial z} \\ -\frac{\partial^2 \Psi_{3d}}{\partial x \partial z} \\ -\frac{\partial^2 \Psi_{3d}}{\partial x \partial y} \end{pmatrix} \quad \text{with} \quad \Psi_{3d} = e^{-\left(\frac{2y-2}{\delta}\right)^2} \sum_{k=0}^{3} \cos(2^k \pi x) \cos(2^k \pi z).$$

Both velocity perturbations are divergence free. Under this conditions, the solution is very similar to the 2d case. The vorticity ω_z is plotted in Figure 11.

The total kinetic energy is plotted in Figure 12 for the two stabilisation methods FIELDS and FLOW and the mixed and dynamic subgrid-model, see equation (6), (7). The simulations with the FIELDS method did not converge for $\bar{t} > 20$ respectively $\bar{t} > 40$ for the dynamic model, which can be hardly recognised in Figure 12 and 14. In accordance to the 2d case, the FLOW stabilisation works better. However, the dissipation rate of the total kinetic energy in the mixed model is larger than that of the dynamic model, which corresponds to the results obtained by Vreman et al. [18]. The formation of the vortices ($\bar{t} \approx 20$) and the reformation of the vortices to larger vortices ($\bar{t} \approx 40$), can be clearly recognized in the momentum thickness μ illustrated in Figure 13. The momentum thickness μ is typically used in 3d instead of the vorticity

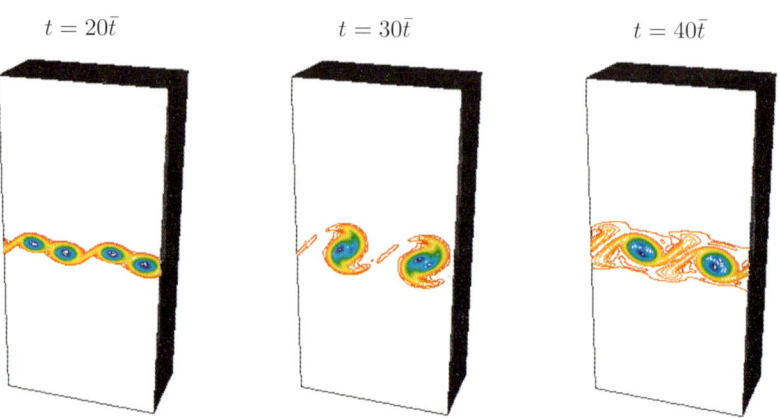

Fig. 11. ω_z in the $z = 1$ plane with the mixed model and the FLOW stabilisation.

Table 2. Over $\bar{t} = 110$ averaged convergence rates.

model	stabilisation	lin. conv.		non-lin. conv.	
		$\kappa^t(u)$	$\kappa^t(p)$	$\psi^t(u)$	$\psi^t(p)$
mixed	FIELDS[2]	0.037229	0.050160	0.243842	0.263019
	FLOW	0.034460	0.057098	0.165111	0.189170
dynamic	FIELDS[3]	0.032495	0.080685	0.155020	0.131084
	FLOW	0.030020	0.075093	0.150837	0.132699

thickness as in 2d. The definition of μ is i.e. given in [18]. Both turbulence models give a similar time history for the momentum thickness μ in 3d, which qualitatively corresponds to the evolution of the vorticity thickness of the 2d-problem. In Figure 14, the subgrid-scale dissipation rate $\bar{\epsilon}_{SGS}$ ($\bar{\epsilon}_{SGS} = \int_\Omega \tau_{ij}\bar{S}_{ij}\,dV$) is plotted, which indicates that the energy flow from the resolved scales to the subgrid-scales. Backscatter, which is the energy flow from the subgrid-scales to the resolved scales might occur, when the small vortices recombine to larger vortices, which can be identified as positive values for $\bar{\epsilon}_{SGS}$ in Figure 14. The mixed model is able to model "backscatter" phenomena better than the dynamic model, which is more dissipative.

In Table 2 the averaged linear and nonlinear convergence rates are given. In principle, they show the same behaviour as in the 2d case, where the nonlinear convergence rate of the FLOW stabilisation was almost twice as good.

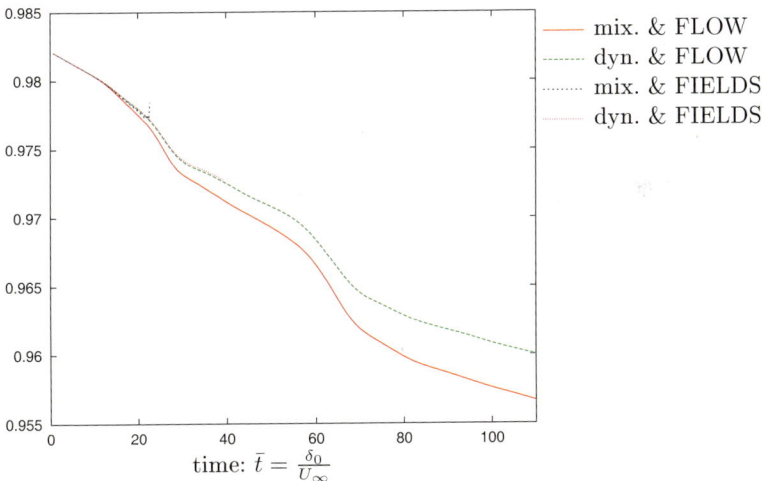

Fig. 12. Total kinetic energy of mixed and dynamic models with different stabilisation methods FIELDS and FLOW.

---

[2] values are averaged over $t \approx 22\bar{t}$.
[3] values are averaged over $t \approx 39\bar{t}$.

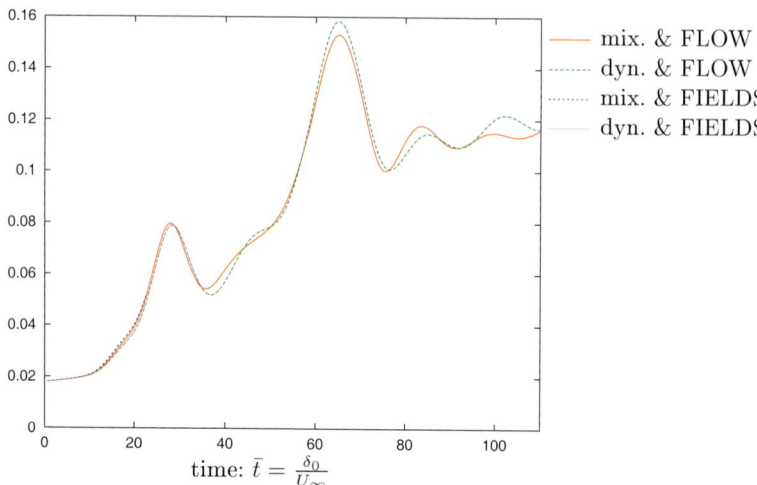

Fig. 13. Momentum thickness μ for the mixed and dynamic model in combination with FIELDS and FLOW stabilisation.

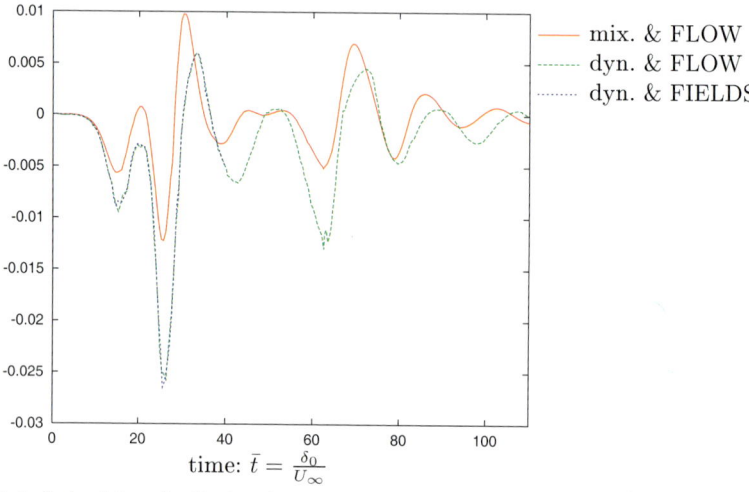

Fig. 14. Subgrid-scale dissipation rate $\bar{\epsilon}_{SGS}$ for the mixed and dynamic model.

7 Conclusion

Within numerical simulations of turbulent flows it is important that the discretisation and the numerical solving procedure does not influence the turbulence modelling. A mass conservative discretisation is one aspect. Two stabilisation techniques for a collocated finite volume method were presented. Both are based on similar ideas, but lead to different couplings between the components. The first one called FIELDS couples the velocities with the pressure and vice versa. Whereas the second one called FLOW also includes a coupling of the velocities to one another, which leads to a reduced error of mass conservation in comparison to FIELDS and furthermore, to a much better nonlinear convergence. The missing velocity coupling in FIELDS in connection with the mixed model leads to a divergence error, disturbing the turbulent solution, which again make the FLOW discretisation more favourable for turbulent simulations. In 3d turbulence applications, the FLOW stabilisation show better stability, while the FIELDS method diverged within the linear iteration cycle using the mixed model.

A multigrid procedure can be used to solve the linear system of equations that occurred for implicit time discretisations, allowing for CFL-numbers greater than 1. However, the mesh size dependent modelling parameter have to get special treatment within the multigrid cycle. A practical solution is to derive the model parameter on the finest grid level and inject it to the coarser grid level. By this strategy good convergence rates result.

References

1. R. Alexander. Diagonally implicit Runge-Kutta methods for stiff o.d.e.'s. *SIAM Journal on Numerical Analysis*, 14:1006–1021, 1977.
2. B.J. Boersma, M.N. Kooper, F.T.M. Nieuwstadt, and P. Wesseling. Local grid refinement in large-eddy simulations. *Journal of Engineering Mathematics*, 32:161–175, 1997.
3. F. Brezzi. On the existence uniqueness and approximation of saddle-point problems arising from Lagrangian multipliers. *RAIRO Anal Numer*, 8:129–151, 1974.
4. M. Germano. Turbulence: the filtering approach. *Journal of Fluid Mechanics*, 238:325–336, 1992.
5. U. Ghia, K.N. Ghia, and C.T. Shin. High-Re solutions for incompressible flow using the Navier-Stokes equations and a multigrid method. *Journal of Computational Physics*, 48:387–411, 1982.
6. J.O. Hinze. *Turbulence*. McGraw Hill, 1975.
7. Volker John. Large eddy simulation of turbulent incompressible flows. analytical and numerical results for a class of LES models. Habilitationsschrift, 2002.
8. S.M.H. Karimian and G. Schneider. Pressure-based control-volume finite element method for flow at all speeds. *AIAA Journal*, 33:1611–1618, 1995.
9. Marcel Lesieur, Chantal Staquet, Pascal Le Roy, and Pierre Comte. The mixing layer and its coherence examined from the point of view of two-dimensional turbulence. *Journal of Fluid Mechanics*, 192:511–534, 1988.
10. D.K. Lilly. A proposed modification of the Germano subgrid-scale closure method. *Physics of Fluids A*, 4:633–634, 1992.
11. Robert D. Moser and Michael M. Rogers. The three-dimensional evolution of a plane mixing layer: pairing and transition to turbulence. *Journal of Fluid Mechanics*, 247:275–320, 1993.
12. Sandra Nägele. *Mehrgitterverfahren für die inkompressiblen Navier-Stokes Gleichungen im laminaren und turbulenten Regime unter Berücksichtigung verschiedener Stabilisierungsmethoden*. PhD thesis, Universität Heidelberg, 2003.

13. Sandra Nägele and Gabriel Wittum. On the influence of different stabilisation methods for the incompressible Navier-Stokes equations. *to be submitted*.
14. Sandra Nägele and Gabriel Wittum. Large-eddy simulation and multigrid methods. *Electronic Transactions on Numerical Analysis*, 15:152–164, 2003.
15. G.D. Raithby. Skew upstream differencing schemes for problems involving fluid flow. *Computer Methods in Applied Mechanics and Engineering*, 9:153–164, 1976.
16. G.E. Schneider and M.J. Raw. Control volume finite-element method for heat transfer and fluid flow using colocated variables - 1. computational procedure. *Numerical Heat Transfer*, 11:363–390, 1987.
17. J. Smagorinsky. General circulation experiments with the primitive equations i. the basic experiment. *Monthly Weather Review*, 9:99–164, 1963.
18. Bert Vreman, Bernard Geurts, and Hans Kuerten. Large-eddy simulation of the turbulent mixing layer. *Journal of Fluid Mechanics*, 339:357–390, 1997.
19. D.C. Wilcox. *Turbulence Modeling for Computational Fluid Dynamics*. DCW Industries, 1993.
20. Y. Zang, R.L. Street, and J.R. Koseff. A dynamic mixed subgrid-scale model and its application to turbulent recirculating flows. *Physics of Fluids A*, 12:3186–3195, 1993.

Modeling and Simulation of Turbulent Non-Reacting and Reacting Spray Flows*

H.-W. Ge, D. Urzica, M. Vogelgesang, and E. Gutheil

Interdisziplinäres Zentrum für Wissenschaftliches Rechnen (IWR), Universität Heidelberg

Summary. The scope is the modeling and simulation of turbulent spray flames where the emphasis is the modeling of the interaction of the vaporization and the turbulent flow field. In particular, the mixing process as well as the interaction of the turbulent flame with the vaporization are studied. These processes are least understood when it comes to the simulation of dilute spray flames.

1 Introduction

Since the single droplet vaporization process is fundamental for the spray models, the first section deals with the vaporization of single droplets where emphasis is given to comparison of simulations with results from new experimental methods. The models aim to validate the experimental study.

The interaction of the chemical reactions with the flow field is modeled using a new flamelet approach for turbulent spray diffusion flames. The model includes sets or libraries of laminar spray flames in the counterflow configuration which are presented in the second section. The third part addresses the improvement of turbulent spray flow modeling where the interaction of the flow field with the vaporization is studied. The widely used k–ϵ turbulence model is known to fail in regimes of recirculation and counter gradient diffusion. Here, a Reynolds stress model is developed and presented to relax these shortcomings. Moreover, the simulation of an ethanol/air spray jet flame has been studied which is experimentally investigated in Ref. [1].

2 Modeling of Droplet Vaporization

The understanding of the process of single droplet vaporizing is crucial and forms the base for spray combustion which has many important applications in

---

*This work has been supported by the German Research Foundation (DFG) through SFB 359 (Project B2) at the University of Heidelberg.

physical sciences and engineering. Examples are the combustion processes of direct-injection engines, gas turbine combustors and liquid rocket propulsion systems. The objective of this part is a numerical study of the vaporization of a single spherically fuel and water droplets. A new experimental method where tracer species are added to the liquid and then are screened with laser-optical methods are used to verify the numerical results.

First, a cold ethanol droplet chain is injected into the hot exhaust of a premixed methane/air flame [2], and the vaporiztion process is studied. A second experiment is performed where stagnant cold ethanol droplets in cold air are exposed to a laser beam that heats up the liquid and leads to droplet vaporization [3]. The mathematical model needs to be modified to account for this setup.

2.1 Experimental setup

The left part of Fig. 1 shows the experimental setup of the McKenna burner where a premixed methane/air flame is generated. An cold ethanol droplet chain is injected from the left side into the hot exhaust and burnt after heating and vaporization.

The second experiment [3] is shown in the right part of Fig. 1. Here, a rhodamine 6G–doped water droplet of variable starting volume (0.5-1.3 μl) is suspended on two glass fibers in a focal plane of a high-resolution CCD camera. The suspending fibers distort the droplet's spherical shape. Thereofore, the volume information from the shadowgraph image using back-lit images of droplets of known volumes that are distorted by the suspending fibers are evaluated. The area and the x-axis diameter of the shadowgraph droplet image are determined, using a segmentation image-processing method.

The droplets are heated with a CO_2 laser beam focused by a $f = 1000$ mm spherical ZnSe lens. The CO_2 laser is a pulsed laser with an intensity of

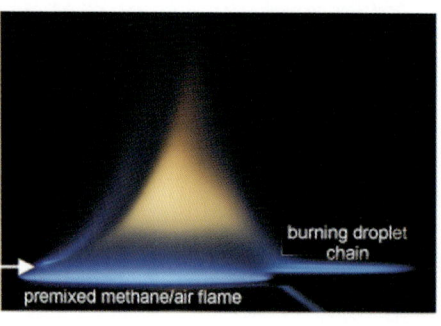
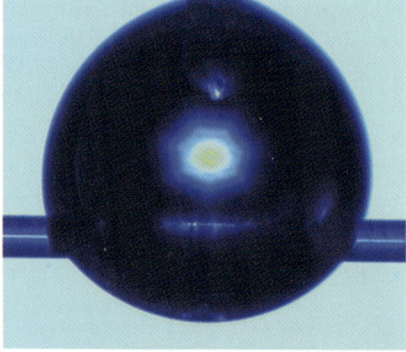

Fig. 1. Burning a droplet in the hot exhaust of a premixed methane/air flame (left) [2] and a suspended droplet on two glass fibers (right) [3].

600 W/cm². Different heating rates are realized by varying the laser-pulse width between 2 ms and 15 ms.

2.2 Mathematical model

The mathematical model used to perform the numerical simulations is Abramzon–Sirignano model [4] for convective droplet vaporization using a uniform temperature model. The droplets in the experiment are separated so that they do not influence each other. The following liquid-phase equations are solved [4]:

- Vaporization Rate:

$$\dot{m} = 2\pi \rho D R \widetilde{Sh} \ln(1 + B_M) \tag{1}$$

- Droplet Heating:

$$M_l C_{pl} \frac{dT_l}{dt} = \dot{m} \left[\frac{C_p(T - T_l)}{B_H} - H_v \right] + W_{abs} \tag{2}$$

- Droplet Motion:

$$\frac{d\mathbf{v}}{dt} = \frac{3}{8} \frac{\rho}{\rho_l} \frac{1}{R} (\mathbf{u} - \mathbf{v}) \mid \mathbf{u} - \mathbf{v} \mid C_D + \mathbf{g}. \tag{3}$$

Here $\dot{m}$ denotes the mass vaporization rate, C_p is the heat capacity at constant pressure, ρ is the density, R the droplet radius, D mass diffusivity, H_v the latent heat of vaporization, and T the temperature. The subscript l denotes liquid phase properties and variables without subscript refer to the gas phase; $\mathbf{u}$ and $\mathbf{v}$ are gas and liquid velocity, resecptively. The modified Sherwood number, $\widetilde{Sh}$, the mass transfer number, B_M, and the energy transfer number, B_H, are given in [4].

The last term in the energy equations is needed for the simulation of the second experiment, and it describes the absorbed energy flux, W_{abs}, of the laser beam. If I_0 (W/m²) is the intensity of the incident light of the laser beam and the spherical water droplet radius is R, then the absorbed energy flux is given by [5]

$$W_{abs} = Q_{abs} \, I_0 \, \pi R^2 \tag{4}$$

where Q_{abs} is the efficiency factor, which can be computed via Mie scattering theory from

$$Q_{abs} = Q_{ext} - Q_{sca} \tag{5}$$

with

$$Q_{ext,sca} = \frac{C_{ext,sca}}{\pi R^2} \tag{6}$$

and

$$C_{sca} = \frac{2\pi}{k^2} \sum_{n=1}^{\infty} (2n+1) \left(|a_n|^2 + |b_n|^2 \right) \quad (7)$$

$$C_{ext} = \frac{2\pi}{k^2} \sum_{n=1}^{\infty} (2n+1) \text{Re}\{a_n + b_n\}. \quad (8)$$

Here, Q_{abs} and Q_{sca} denote the efficiency factors for extinction and scattering, respectively, C_{ext} and C_{sca} are the extinction cross section and scattering cross section, respectively, n is the real refractive index, a_n and b_n denote the scattering coefficients from Mie theory, and Re denotes the real value. For the present conditions, Q_{abs} is computed to be close to unity ($Q_{abs} = 0.98$). Almost all of the light that enters in a sufficiently large sphere is absorbed, none of the non-reflected light will be transmitted [6].

2.3 Vaporization of droplets in a hot exhaust of a flame

For the numerical simulation of vaporizing droplets in a hot gas stream, several modifications of standard models that treat the vaporization of cold droplets in hot air are necessary. The hot exhaust of the methane/air flame requires the use of different gas-phase properties[7, 8]. In the present simulation, the initial radius of the ethanol droplet ranges from 34.4 μm to 91.8 μm, and the initial droplet velocities are 12.4 m/s and 17.8 m/s, respectively.

In Fig. 2, the results of the simulation are compared with the experimental data of the evaporation of a cold ethanol droplet in a hot (1100 K) exhaust of a methane/air premixed flame, for two different initial droplet radii. The dashed curves symbolizes simulation results and symbols show the experimental data which are in good agreement in particular for the bigger droplet.

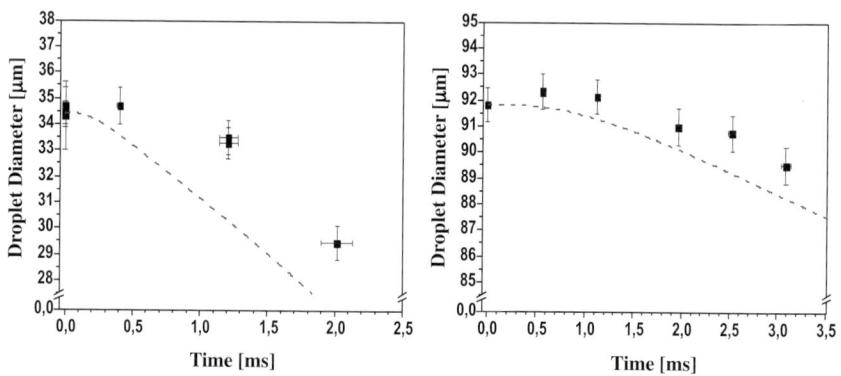

Fig. 2. Evaporation of a cold ethanol droplet in a hot (1100 K) exhaust methane.

2.4 Vaporization of water droplets by pulsed laser heating

The evaporation of a single droplet by irradiation with a pulsed laser is extensively studied in the literature. In [10], a set of coupled equations for the droplet temperature, radius and beam intensity in the diffusive regime, assuming low-mass-flux conditions, was set up. Another study of vaporization of laser irradiated droplets was done in [9, 10, 11]. Experimental studies of vaporization of droplets by irradiation with a pulsed laser can be found in [11, 12, 14]. In the case considered here, the vaporization of a large droplet by irradiation with a long-time laser pulse of low energy, the vaporization process is governed by the processes of heat conduction and diffusion of the vapor [10]. In this case one can assume isobaric vaporization.

In this section, a H_2O droplet is heated by a pulsed CO_2 laser [3] and the vaporization process is studied. The initial radius of water droplet is about 1.3 mm, and it varies somewhat from experiment to experiment. The conditions in the experiment are standard conditions. Four different pulse duration times are used: 2 ms, 5 ms, 8 ms, and 15 ms.

Figure 3 shows the comparison between simulation results and experimental results for different pulse widths. The simulation results (lines) are in very good agreement with experimental results (symbols). The small discrepancies between the simulation results and experiments are mainly due to distortions of the droplet by suspending fibers, experimental errors or heat conduction by fibers, so that the droplet is not perfect spherical. The droplet life time stongly depends on the laser pulse duration. Figure 4 shows the droplet temperature versus time obtained from the simulations, for different laser pulse durations of 2, 5, 8, and 15 ms. It can be seen that for longer laser pulses, the droplet temperature is higher and closer to the boiling temperature of

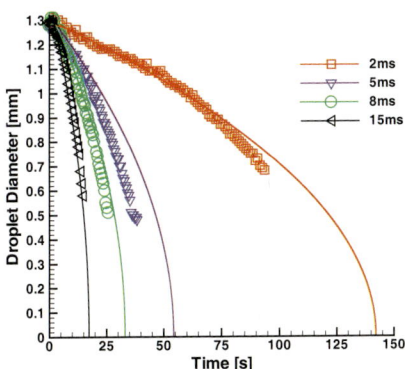 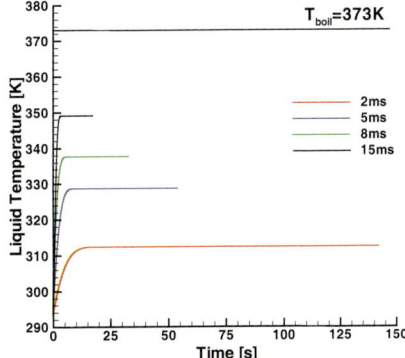

Fig. 3. Vaporization of spherical water droplet in cold air, heated by a pulsed laser beam. Pulse duration times: 15 ms (△), 8 ms (○), 5 ms (▽), and 2 ms (□).

Fig. 4. Computed droplet temperature versus time for different laser pulse duration times [13].

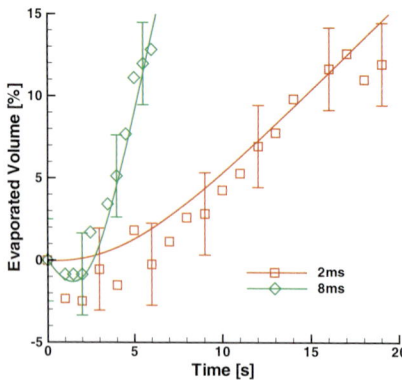
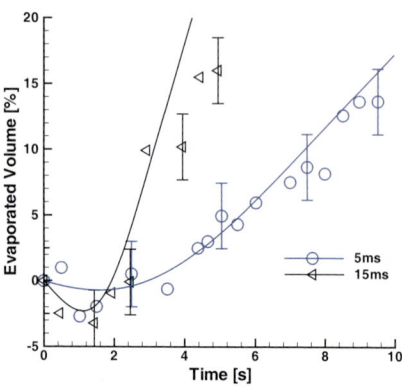

Fig. 5. Simulated and experimental droplet evaporated volume versus time for different pulse duration times [13].

Fig. 6. Simulated and experimental droplet evaporated volume versus time for different pulse duration times [13].

the liquid. Also, the total heating time is longer for shorter laser pulses (i.e. smaller duty cycles). Unfortunately, the liquid temperature inside the droplet could not be measured, so that a validation of the calculated liquid-phase temperatures is not possible.

In Figs. 5 and 6, the simulated and experimental results of the droplet evaporated volume versus time for different pulse duration times are presented. The percentage of evaporated volume is computed from

$$V_{evap} = 100 \left(1 - \left(\frac{R}{R_0}\right)^3\right), \qquad (9)$$

where R denotes the instantaneous radius of the droplet during the evaporation process, and R_0 is the initial droplet radius. Figs. 5 and 6 show an enlarged view of the evaporated droplet volume versus time at the beginning of the evaporation process. It can be seen that the numerical simulations and experimental results of evaporated volume are in good agreement. The experimental error is about 2-3%, and some error bars are introduced into the figures. It is observed that at the beginning of the evaporated process, there is an expansion of the droplet.

2.5 Conclusions

In summary, it can be concluded that the basic process of droplet vaporization is understood by both the experimental method and the model and the latter one will be used in the following sections.

3 Modeling of Laminar Counterflowing Spray Flames

Counterflow spray flames have been investigated in the last years both by means of experiment [15, 16, 17], numerical simulations [18, 19, 20, 21] as well as asymptotical methods [22]. The counterflow configuration is convenient for the investigation of both reacting and non-reacting (spray) flows since the flow field is well defined and boundary conditions can easily be modified. For numerical studies, a similarity transformation [18] is suitable that transfers the two-dimensional governing gas phase equations into one-dimensional form which accesses the use of detailed gas-phase processes such as chemical reactions and detailed transport modeling [19, 20].

Continillo and Sirignano [18] postulate that there may be multiple solutions of structures of counterflowing spray flames. It is known that for gas flames, there is a solution with and without a flame, the second one presents the cold solution. For spray flames, the same situation exists, and typically the solution with a flame is presented. For spray flames in the counterflow configuration, there may be two reaction zones where one resides on the spray side of the configuration and the second one on the gas side. At elevated strain, the gas-sided flame extinguishes due to the low residence time of reactants associated with high strain. Here, structures of methanol/air spray flames with multiple solutions of low strained spray flames are presented that have not been reported in the literature so far.

3.1 Mathematical model

The mathematical model is identical to the formulation in previous papers [19, 20] and the equations to describe the problem are summarized briefly.

The two-dimensional gas-phase equations are non-dimensionalized [18, 19, 20, 23] and transformed into one-dimensional equations using a similarity transformation that accounts for density changes [19, 20]:

$$\eta = \int_0^y \rho \, dy \quad \text{and} \quad f = \int_0^\eta \frac{u}{x} \, d\eta. \tag{10}$$

The resulting gas-phase equations are given in terms of the gas-phase velocity, v, the stream function, f', the dimensionless gas temperature, θ, and the diffusion velocity, V_{ky}, for the k-th chemical species in axial direction, y:

$$v = -1/\rho \left([\alpha + 1]f + f_v \right) \quad \text{with} \quad f_v = -\int_0^\eta 1/\rho \, S_v \, d\eta \tag{11}$$

$$\frac{d}{d\eta} \left(\mu \rho \frac{df'}{d\eta} \right) + ([\alpha + 1]f + f_v)f'' = (f')^2 - \frac{1}{\rho} - \frac{S_m}{\rho x} \tag{12}$$

$$\frac{d}{d\eta}\left(\lambda\rho\frac{d\theta}{d\eta}\right)+c_P([\alpha+1]f+f_v)\frac{d\theta}{d\eta}=\rho\sum_{k=1}^{K}V_{ky}c_{P_k}\frac{d\theta}{d\eta}+\frac{1}{\rho}\sum_{k=1}^{K}h_k\dot{w}_k-\frac{1}{\rho}S_e \quad (13)$$

$$-\frac{d}{d\eta}(\rho V_{ky})+([\alpha+1]f+f_v)\frac{dY_k}{d\eta}=-\frac{1}{\rho}\dot{w}_k-(\delta_{ik}-Y_k)\frac{1}{\rho}S_v \quad (14)$$

$$V_{ky}=-\frac{\rho D_k}{\bar{W}}\frac{d(\bar{W}Y_k)}{d\eta}-\frac{\rho D_{T_k}}{\theta}\frac{d\theta}{d\eta} \quad (15)$$

The boundary conditions for Eqs. (12–14) are

$$\eta=-\infty: \quad f=f_{-\infty}; \quad f'=1; \quad Y_k=Y_{k-\infty}; \quad \theta=1;$$
$$\eta=+\infty: \quad f'=\sqrt{\rho_{-\infty}/\rho_{\infty}}; \quad Y_k=Y_{k\infty}; \quad \theta=T_{\infty}/T_{-\infty}.$$

The liquid phase equations are transformed using the following relations [18, 19, 20, 23]:

$$\xi=r/r(t); \quad \xi_s=r(t)/r_0; \quad \tau=\frac{1}{t_l^*}\int_0^t\frac{dz}{\xi_s} \quad (16)$$

where $r(t)$ and r_0 are the time-dependent droplet radius and the initial droplet radius, respectively. The droplet vaporization, heating, and motion are described by the following set of equations:

Vaporization:

$$\frac{d\xi_s}{d\tau}=-1/9\,a\rho_f D_f\tilde{Sh}(1+B_M) \quad (17)$$

Droplet heating:

$$\frac{\partial\theta_l}{\partial\tau}-\frac{\xi}{\xi_s}\frac{d\xi_s}{d\tau}\frac{\partial\theta_l}{\partial\xi}=\frac{1}{\xi_s\xi^2}\frac{\partial}{\partial\xi}\left(\xi^2\frac{\partial\theta_l}{\partial\xi}\right) \quad (18)$$

Droplet velocity equation in x-direction:

$$\frac{\partial^2 s}{\partial\tau^2}-\frac{1}{\xi_s}\left(\frac{d\xi_s}{d\tau}-a\mu\right)\frac{ds}{d\tau}=ab\mu\frac{df}{d\eta}s+b^2\xi_s^2 g_x \quad (19)$$

with $s(0)=x_{l0}/u_{l0}=s_0$ and $s'(0)=1$.

Droplet velocity equation in η-direction:

$$\frac{\partial^2\eta_l}{\partial\tau^2}+\rho\frac{d\rho^{-1}}{d\tau}\frac{d\eta_l}{d\tau}-\frac{1}{\xi_s}\left(\frac{d\xi_s}{d\tau}-a\mu\right)\frac{d\eta_l}{d\tau}=ab\mu(-([\alpha+1]f+f_v))+\rho b^2\xi_s^2 g_\eta.$$

In the above equations, a and b are constants that result from non-dimensionalization [19, 20]. The boundary conditions for the liquid phase equations (17-20) are

$$\xi_s(0) = 1; \quad \frac{\partial \theta_l}{\partial \xi}\Big|_{\xi=0} = 0; \quad \frac{\partial \theta_l}{\partial \xi}\Big|_{\xi=1} = \frac{\dot{q}}{3\xi_s}; \quad \eta_l(0) = \eta_{l0}; \quad \eta'_0(0) = \eta'_{l0} \quad (20)$$

$$\dot{q} = \dot{m} \left[\frac{c_{Pf}}{c_{Pl}} \left(\frac{T^\star}{T_l^\star} \theta - \theta_{ls} \right) / B_T - L \right]. \quad (21)$$

The source terms, S_v, S_m, and S_e, for the mass, momentum, and energy equations, respectively, needed in Eqs. (11-14) are given in [20].

Finally, the conservation of droplets number assuming no production or destruction of droplets is given by the following equation [18, 19, 20]:

$$n = n_0 s_0 \eta'_{l0} \rho / (s \eta'_l \rho_0); \quad s = x_l / u_{l0}. \quad (22)$$

The detailed chemical reaction mechanism for the methanol/air system comprises 23 chemical species and 168 elementary reactions [24, 19].

The given set of equations is strongly coupled, and it is solved using a hybrid solution method applying the Thomas algorithm for the gas-phase equations. The number of iterations for the gas phase is typically 100, then the liquid-phase equations are solved. The solution is continued until the residuals are below 10^{-4}.

The procedure so far is not new and has been applied in the previous work. The multiple solutions, however, are found using the following approach: The numerical procedure to obtain the double flame starts at low strain from scratch whereas the single-flame solution is obtained using a start profile from a high-strain result where only one chemical reaction zone exists. Starting from low strain, the double flame structure is obtained. A continuous increase of strain leads to extinction of the gas-sided flame as will be discussed in the next section. The reduction of strain starting from the high-strain results then leads to a flame structure with one chemical reaction zone only. This approach leads to the following results.

3.2 Results and discussion

A methanol spray in air is studied where air is injected on both sides of the configuration and the LHS stream is laden with the mono-disperse spray. The initial gas and droplet temperatures are 300 K, and the initial droplet radius, r_0, is 25 μm. Figures 7 and 8 show two flame structures with identical initial conditions. In Fig. 7, two reaction zones exist whereas in Fig. 8, a structure with a single reaction zone is displayed. The latter case corresponds to the cold solution of the gas flame structure and only the spray-sided flame persists.

Figures 7 and 8 reveal that the reaction zones on the spray side of the flame essentially are the same: This applies to the gas temperature profile as well as to the concentrations and spray characteristics. The differences occur on the gas side of the configuration where the gas-side reaction zone is absent or present, respectively. In Fig. 7, the droplet vaporization is enhanced due to the high gas phase temperature in the second reaction zone strongly affecting droplet vaporization. The vaporization again enhances the chemical reactions, and the outer flame structures considerably differ. Thus, different outer flame structures are obtained for identical boundary and initial conditions.

The possible existence of multiple solutions was discussed by Continillo and Sirignano [18] who first applied the similarity transformation to the governing gas-phase equations. They demonstrated the validity of the approach, the existence of solutions and postulated multiple solutions.

As gas strain rate is increased to 500/s on the left hand side of the configuration, the gas-sided chemical reaction zone extinguishes due to decreased residence time of the species, and the resulting structure is displayed in Fig. 9. The chemical reaction zone is narrowed compared to Fig. 8 by 25% and about a factor of two compared to Fig. 7. The flame temperature decreases. This is typical for flame structures at elevated strain rate [18, 19, 20].

The finding of the solution shown in Fig. 8 is attributable to the fact that the present model is not suitable to model flame ignition even though the chemical reaction mechanism is suitable to predict correct ignition times [24, 25]. If the gas-phase equations were formulated in time-dependent form, then the reduction of strain rate starting from the flame structure shown in Fig. 9 would include the ignition of the gas-sided reaction zone leading to the structure shown in Fig. 7. The advantage of the present formulation is that it enables the finding of the flame structure displayed in Fig. 8. There is no doubt that the latter flame structure is physical since the existence of cold solutions of both spray and gas flames in the counterflow configuration is obvious: They present the inert mixing of two jets that may be gas jets or droplet laden jets.

The gas-sided flame of the spray flame shown in Figure 7 can be compared to a pure gas flame with LHS boundary conditions obtained from the spray flame at the stagnation plane, $z = 0$ mm. Figures 10 and 11 show a

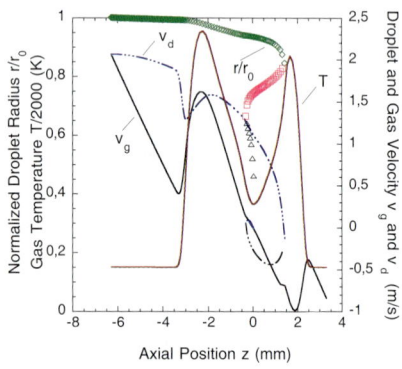
Fig. 7. Structure of a methanol/air spray flame with two reaction zones, $a = 300/s$ [23].

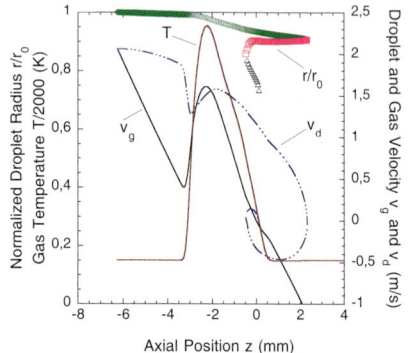
Fig. 8. Structure of a methanol/air spray flame with a single reaction zone, $a = 300/s$ [23].

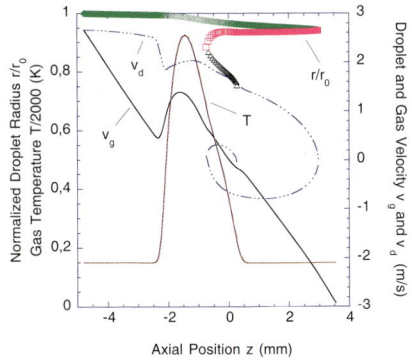

Fig. 9. Structure of a methanol/air spray flame, $a = 500/s$ [23].

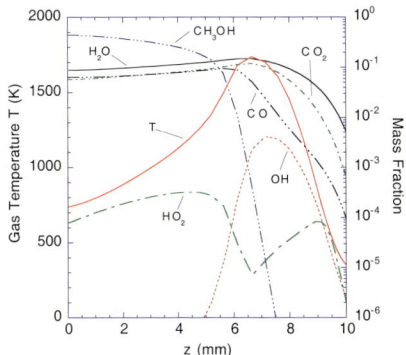

Fig. 10. Structure of a methanol/air spray flame: Gas side of the spray flame, a=300/s [23].

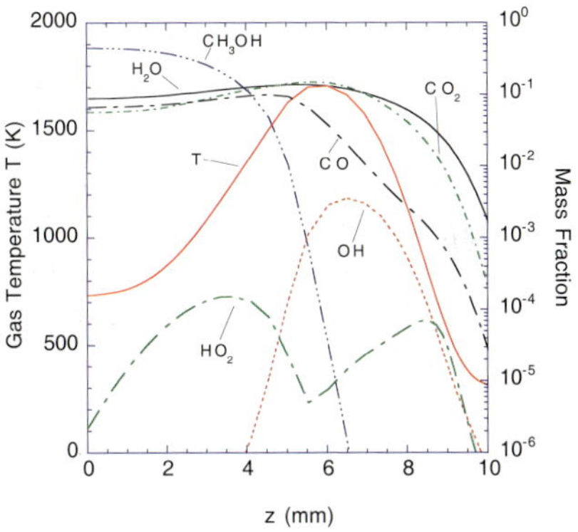

Fig. 11. Structure of a methanol/air gas flame, a=300/s [23].

comparison of the chemical species profiles of the gas-sided reaction zone of the spray flame and the pure gas flame computation. The figure reveals that the flame structures are essentially the same except for minor differences in the HO_2 profile at the left boundary as well as a somewhat steeper increase of gas temperature on the LHS in Fig. 10. This is attributable to the fact that the zero-gradient boundary condition for the gas temperature is fulfilled (enforced) in Fig. 11 due to the physical boundary in this situation whereas the structure shown in Fig. 10 is a cut on the LHS at the stagnation point of the flame shown in Fig. 7. These differences, however, are minor and do not affect the principal finding.

Figures 10 and 11 show profiles of some of the chemical species. The methanol vapor decreases as it is approaching the flame zone where the OH radical identifies the position of the flame front. The characteristics of gas

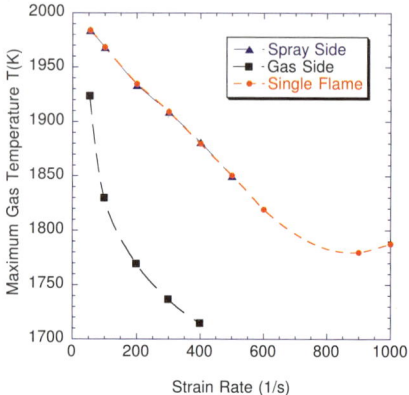

Fig. 12. Maximum gas temperature versus strain rate for the reactions zones in both the gas- and the spray flames [23].

flames are found such as CO formation before CO_2 formation in the laminar flame front.

The multiple flame structures for the present conditions persist up to a strain rate of 400/s; at $a = 500$/s, the gas-side flame is extinguished as shown in Fig. 9. An overview of obtained flame structures is displayed in Fig. 12 where the maximum flame temperature is plotted versus strain rate for both the gas and the spray flames. It can be seen that the spray-sided reaction zone (triangles) and the single spray flame solution (circles) coincide within the range of existence between strain rates of 55/s through 400/s. Squares show the maximum gas temperature of the gas-side reaction zone of the double-flame structure which typically lies below the spray-sided reaction zone. As strain rate is increased above a certain value, the droplets oscillate between the opposed jets as discussed in Refs. [19, 20], and the flame temperature may increase again as shown in Fig. 12 (circles). The current computation has not been carried to extinction.

The present findings of multiple flame structures are also found for planar flame structures and for different liquid fuels such as ethanol, and the range of strain rate in which they occur is similar as for the conditions shown in this study. The question may arise if the set of structures found here is complete or if it is possible to find even more different structures for the same inlet and boundary conditions. The third possibility of a potential structure where the spray-side flame is extinguished and the gas-side flame persists may not be physical at low strain since this situation does not allow for enough fuel vapor to sustain a flame, evaporation is reduced if no flame exists in this region. At elevated strain, the gas-side flame does not exist leading to the conclusion that there may not be other numerical solutions.

In the context of the flamelet model for turbulent spray flames, the laminar spray flame structures may be used for the generation of a laminar flamelet library [26]. The present finding raises the question of how a spray flamelet may be chosen in the situation of multiple structures.

The procedure followed by Hollmann et al. [26] includes a splitting of the flame structure with two reaction zones into a spray part and a gas part. If there is spray in the computational cell of the turbulent flow field, the spray part of the laminar counterflow flamelet is chosen, and the droplet size of the laminar and turbulent flow field need to be matched. In case of a pure gas cell, the gas-sided part of the reaction zone is taken. The present result shows that this procedure may be simplified by replacing the gas-phase side of the spray flame by a flamelet resulting from a pure gas flame since the flamelets shown in Fig. 8 do not differ considerably. The result of the present study simplifies the use of the flamelet model in spray flames since gas flames only depend on strain rate and inlet composition and temperature whereas the spray flames need to account for differences in initial droplet size and velocity in addition to the gas-flame inlet conditions. Therefore, this study is not only interesting in terms of a basic finding, but it also strongly simplifies the formulation and use of a laminar spray flame library in turbulent spray combustion.

3.3 Conclusions

The present section presents multiple solutions of structures of counterflowing spray flames for the first time. These structures exist in the low-strain regime where there may be two reaction zones: One reaction zone close to the spray injector is promoted by the evaporating spray and the gas flame again sustains evaporation whereas the second reaction zone is a gas flame located on the gas side of the counterflow configuration. In the low-strain situation, this flame structure with a double flame is found and a second one with one spray-sided reaction zone only. At high strain, the gas-sided flame extinguishes due to the reduced residence time of the reactants, and a unique solution is obtained.

The study also discusses the completeness of the found structures as well as the question of how the present finding affects models for turbulent spray combustion such as the flamelet model. In particular, a novel method of including laminar spray flame structures into the flamelet model for turbulent spray flames is suggested.

4 Simulation of Turbulent Spray Flames

The section concerns the flamelet modeling of turbulent spray flames where emphasis is given to the implementation of the laminar spray flamelets discussed in the previous section. Moreover, a Reynolds stress model is derived to improve the standard k–ϵ turbulence model for spray flows that suffers from failure in regimes where counter-gradient diffusion and recirculation zones exist. Moreover, the simulation of an ethanol/air spray jet flame has been studied which is experimentally investigated in Ref. [1].

4.1 Flamelet Modeling of Turbulent Spray Flames

The flamelet structures that have been presented in the previous section may be used to simulate turbulent spray flames [26]. The inclusion of gas flamelets into models for turbulent gas flames is a well established method, c.f. [27] where both steady and unsteady systems have been considered. The present study employs the flamelet model for turbulent spray flames where the consideration of gas-phase flamelets is compared to the model using the laminar spray flamelets computed in the previous section.

The procedure of the implementation of laminar spray flamelets is displayed in Fig. 13. The flamelet structure shwon in the previous section yields two reaction zones the left of which is a spray flame and the right part resides on the gas-side of the configuration. Therefore, the spray-sided flamelet is used for situations where spray is present in the control volume of the turbulent spray flame under consideration whereas the gas flamelet is used elsewhere. The results from the previous section motivate to replace the right wing (gas-sided flame structure) marked with a green box through a pure gas flamelet.

Figure 14 shows results of the simulation of a methanol/air spray diffusion flame that has been studied experimentally by McDonell and Samuelsen [28]. The left part of the temperature contour plot shows the result with gas-phase flamelets and the right-hand side displays the computation with the spray flamelet. It can be seen the result using the spray flamelet does not show flame

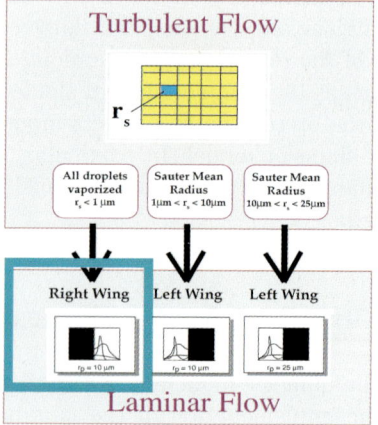

Fig. 13. Evaluation of spray flamelets for use in the turbulent spray model.

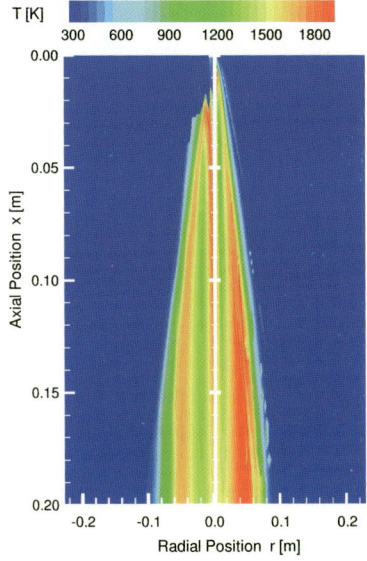

Fig. 14. Structure of a methanol/air flame: Gas side of the spray flame, a=300/s.

lift-off which is also not seen in the experiment. Another major difference is the corrugation at the spray flamelet that stems from flame penetration of spray particles. This behavior cannot be described by use of the gas flamelets. Therefore, the use of laminar spray flamelets is strongly encouraged.

4.2 Reynolds Stress Model for Turbulent Spray Flames

The Reynolds stress model has been successfully developed for turbulent recirculating gas flows [29], and it is extended here for use in spray flows.

The major difference in gas-phase versus liquid/gas systems is the vaporization process that needs to be considered if atomization as well as droplet breakup and coalescence are neglected. Thus, a dilute spray is considered here. Moreover, the turbulent fluctuations caused by the flow field in reacive flows require the consideration of Favre–averaged equations [30]. The formulation here is given in three-dimensional coordinates, for generality of the equations.

Eq. 23 shows the transport equation for the Reynolds stress term [31]

$$\frac{\partial}{\partial t}\left(\overline{\rho}\widetilde{u_i''u_j''}\right) + \frac{\partial}{\partial x_k}\left(\overline{\rho}\widetilde{u_k}\widetilde{u_i''u_j''}\right) = \frac{\partial}{\partial x_k}\left(C_S\overline{\rho}\frac{\widetilde{k}}{\widetilde{\varepsilon}}\widetilde{u_k''u_l''}\frac{\partial \widetilde{u_i''u_j''}}{\partial x_l}\right) + P_{ij} - \frac{2}{3}\delta_{ij}\overline{\rho}\widetilde{\varepsilon}$$
$$- C_1\overline{\rho}\widetilde{\varepsilon}\left(\frac{\widetilde{u_i''u_j''}}{\widetilde{k}} - \frac{2}{3}\delta_{ij}\right) - C_2\left(P_{ij} - \frac{1}{3}P_{ll}\right)$$
$$+ C_3\left(\overline{\rho}g_i\overline{u_j''} + \overline{\rho}g_j\overline{u_i''} - \frac{2}{3}\delta_{ij}\overline{\rho}g_l\overline{u_l''}\right) \quad (23)$$
$$- \frac{1}{4.3\overline{\rho}}\frac{\widetilde{k}}{\widetilde{\varepsilon}}\left(\frac{\partial \overline{p}}{\partial x_i}\widetilde{u_j''u_l''} + \frac{\partial \overline{p}}{\partial x_j}\widetilde{u_i''u_l''}\right)\frac{\partial \overline{p}}{\partial x_l}$$
$$- \overline{\left(u_i''u_j'' + \widetilde{u}_i u_j'' + \widetilde{u}_j u_i''\right)L_v} + \overline{u_j''L_{m,i}} + \overline{u_i''L_{m,j}}$$

where the terms in the last line describe the interaction of the turbulent, two-dimensional flow field with the vaporization source terms in the mass conservation equations, L_v, and the momentum equation, L_m. All other terms are standard terms that also appear in the gas-phase equations. These terms are closed following the procedure discussed in Ref. [26]. In the above equation, $P_{ij} = -\overline{\rho}\widetilde{u_j''u_k''}\frac{\partial \widetilde{u}_i}{\partial x_k} - \overline{\rho}\widetilde{u_i''u_k''}\frac{\partial \widetilde{u}_j}{\partial x_k}$ describes the production of Reynolds-stresses through shear stress. The constants C_S, C_1, C_2, and C_3 are 0.22 [32], 3.0 [33], 0.33 [33], and 0.5 [34], respectively. δ_{ij} denotes the Dirac–Delta function, and g_i is the gravitational force in i–direction.

The Reynolds-stress equation is solved together with the conservation equations for the momentum, the energy, and mass fractions of chemical species. Special care needs to be taken in discretizing the source terms of Eq. 23 in order to assure convergence [31].

Results and discussion

The simulations concern a methanol/air spray flame that has been experimentally investigated by McDonell and Samuelsen [28] referred to in the previous section. Figure 15 shows a comparison of calculated and measured radial profiles of the root mean square (RMS) of the velocity correlations. The two-dimensional configuration enables comparison of $\sqrt{\widetilde{u''^2}}$ and $\sqrt{\widetilde{v''^2}}$ whereas the third component $\sqrt{\widetilde{w''^2}}$ is only available for the simulation. Moreover, the Favre averaged values of the axial velocity component, $\widetilde{u}$, is displayed.

It is seen that the agreement of experimental and computational axial velocity component on the axis is poor. This is attributable to uncertainties in the experimental initial conditions because of some swirl that is visible in the experimental data. Since there are no experimental data of the radial velocity component in the inlet regime of the nozzle, the simulation assumed zero values here leading to a faster spread associated with lower values in the axial component.

A comparison of the similarity profiles of the axial gas velocity shown in Fig. 17 shows excellent agreement so that the discrepancies shown in Figs. 15

and 16 are attributable to the discussed issue. The standard deviations of the velocity components, however, are well predicted by the simulation except for some values in the initial region of the jet.

In the experiment studied by McDonell and Samuelsen [28] has only a very small recirculation zone and a hard test of the Reynolds stress model is not possible. Therefore, an artificial strongly swirling flow filed was generated to test the model. Figure 18 shows that the model is capable to simulate a strongly swirling flow, a situation where the k–ϵ model fails. The next step will be the simulation of a flow field with swirl where sufficient experimental data are available for a reliable simulation.

4.3 Simulation of an Ethanol/Air Spray Flow

Ethanol is a prospective fuel since it can be obtained from biomass. Moreover, it is a cleaner fuel compared to higher hydrocarbons because of its lower tendency to produce soot.

This section concerns the simulation of a turbulent ethanol/air spray flow that has been investigated both experimentally in the group of Profs. Wolfrum and Schulz [1]. The focus here is the application of a new laser-optical method described in Section 2 together with the mathematical model for turbulent spray flows described in the previous section.

Figure 19 shows the experimental setup. A perforated plate with a honeycomb grid between which glas balls are positioned is used to generate a uniform air flow in which the ethanol spray is injected. The experimental positions are displayed in Fig. 20. The figure also shows the simulated droplet velocity field with the spray angle. The positions marked with circles display the experimental positions where the first one is used to generate initial profiles for the liquid phase characteristics. The remaining positions at 5 mm and

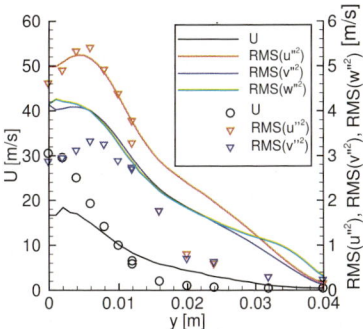

Fig. 15. Comparison of simulated (lines) and measured [28] (symbols) Reynolds stress terms at $x = 25$ mm.

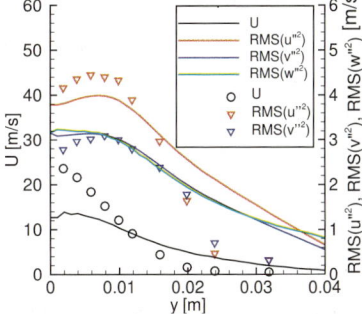

Fig. 16. Comparison of simulated (lines) and measured [28] (symbols) Reynolds stress terms at $x = 75$ mm.

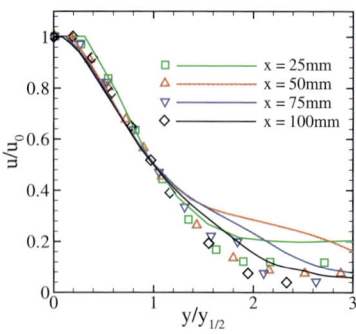

Fig. 17. Similarity profiles of the computed (lines) and experimental [28] (symbols) axial gas velocity at various axial positions.

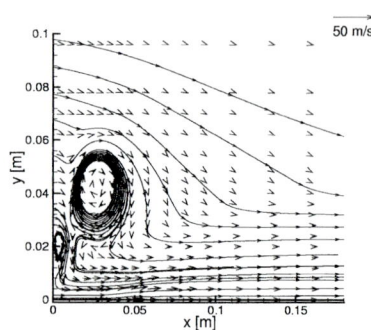

Fig. 18. Structure of a methanol/air spray flame: Gas side of the spray flame, a=300/s.

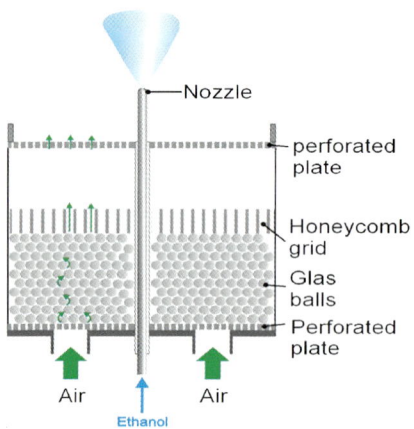

Fig. 19. Experimental setup of a turbulent ethanol/air spray flow [1].

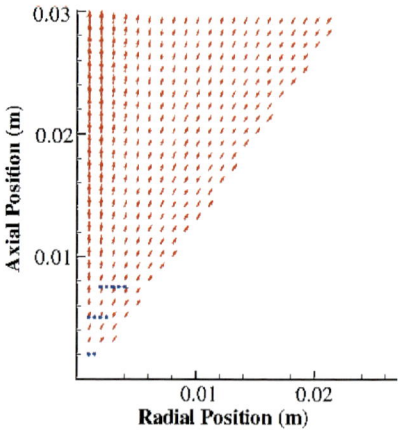

Fig. 20. Vector plot of ensemble averaged droplet velocity. Filled circles: experimental positions.

7.5 mm from the nozzle exit marked with 0 mm may be used for comparison with the experimental data.

The experimental data comprise spray characteristics such as the droplet velocities and the Sauter mean radius of the spray which are shown in Figs. 21 and 22. The lines in Figs. 19 and 22 show simulated and the symbols measured values. Considering the fact that the initial gas velocity distribution has not been measured, the agreement is good. The simulated values of the droplet velocities are systematically higher than the measured values by up to 30%. This causes a higher droplet evaporation rate leading to up to 10%

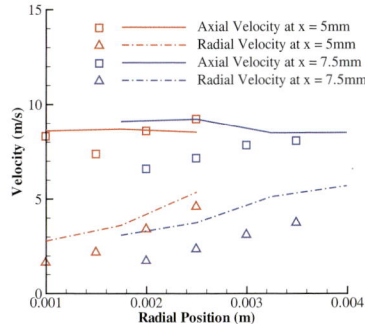

Fig. 21. Radial profiles of ensemble-averaged droplet velocities at $x = 5$ mm and 7.5 mm. Symbols: experiment, lines: simulation.

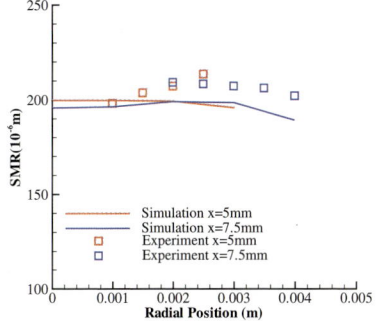

Fig. 22. Radial profiles of Sauter mean radius of droplets at section $x = 5$ mm and 7.5 mm. Symbols: experiment, lines: simulation.

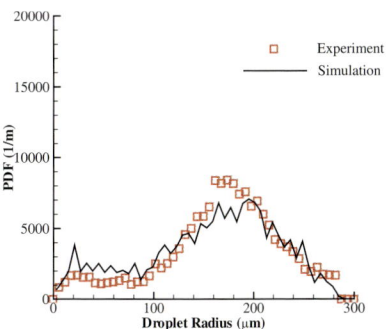

Fig. 23. Droplet size distribution at $x = 5$ mm, $r = 1$ mm. Solid lines: Simulation. Symbols: Experiment.

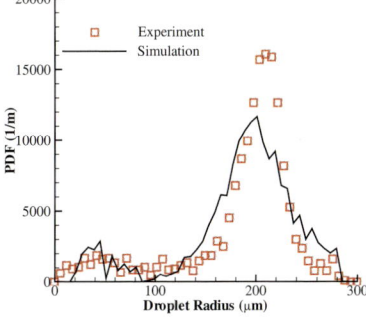

Fig. 24. Droplet size distribution at $x = 7.5$ mm, $r = 3.5$ mm. Solid lines: Simulation. Symbols: Experiment.

smaller droplet radii compared to the experimental values. The discrepancies between experiment and simulation in the droplet velocities can be attributed to the unknown initial gas flow velocities, the coarse initial droplet size in the experiment and to experimental uncertainties.

Figure 23 and 24 show the comparison of droplet size distribution from numerical simulation and experiment. The size of the droplet radii ranges from 0 to 300 m. In the present computation, the whole range is split into 50 equal-sized intervals. The number fraction of each interval is calculated and then the PDFs of droplet radii are evaluated. The numerical results are in good agreement with experimental data and show the shift of the peak value as vaporization proceeds.

4.4 Conclusions

This work shows that turbulent spray flows can be well described by means of numerical simulations. Detailed models are suitable to predict the evolution of both droplet size and velocity distributions in both non-reactive ethanol and methanol sprays if initial conditions after the spray breakup are known from measurements. In the ethanol/air spray, the probability density functions for droplet size distribution are in good agreement with experiment.

If recirculation zones are present, the Reynolds stress model enables the simulation of these regimes even in presence of a turbulent spray. The simpler k-ϵ model is not suitable to predict these regimes.

For turbulent spray flames, a flamelet model has been successfully derived, and the implementation of laminar spray flamelets strongly improve the prediction of regimes where both chemical reactions and droplet vaporization is present. In particular, gas flamelets are not suitable to predict spray particles to penetrate the turbulent spray flame front.

References

1. Ge, H.-W., Vogelgesang, M., Gutheil, E., Düwel, I., Kronemayer, H., Schulz, C.: Experimental and numerical investigation of turbulent spray flows, *Spray-05*, Antalya, Turkey, June 2005.
2. Schorr, J.: *Entwicklung und Anwendung von Fluoreszenztracer-Verfahren für die lasergestützte, abbildende Spraydiagnostik*, Ph.D. Thesis, Heidelberg (2004).
3. Düwel, I., Schorr, J., Wolfrum, J., Schulz, C.: Laser induced fluorescence of tracers dissolved in evaporating droplets. *Appl. Phys. B*, 78, 127–131 (2004).
4. Sirignano, W. A. : *Fluid dynamics and transport of droplets and sprays.* Cambrigde University Press, New York (1999).
5. van de Hulst, H. C.: *Light Scattering by small particles.* Wiley, New York (1957).
6. Bohren, C., Huffman, D.: *Absorption and scattering of light by small particles.* Wiley, New York (1998).
7. Reid, R. C., Prausnitz, J.M., Poling, B.E.: *The properties of gases and liquids.* McGraw-Hill, New York (1987).
8. Lisal, M., Smith, W., Nezbeda, I.: Accurate vapour-liquid equilibrium calculations for complex systems using the reaction Gibbs ensemble Monte Carlo simulation method. *Fluid Phase Equilib.*, 181(1-2), 127–146 (2001).
9. Armstrong R. L., Zardecki A.: Diffusive and convective vaporization of irradiated droplets. *Appl. Phys.*, 62, 4571–4578 (1987).
10. Armstrong, R. L., O'Rourke, P.J., Zardecki A.: Vaporization of irradiated droplets. *Phys. Fluids*, 29, 3573–3577 (1986).
11. Kucherov, A. N.: Sublimation and vaporization of an ice aerosol particle in the form of thin cylinder by laser radiation. *Int. J. Heat Mass Transf.*, 43, 2793–2806 (2000).
12. Kafalas, P., Ferdinand, A. P. Jr.: Fog droplet vaporization and fragmentation by a 10.6-micrometer laser pulse. *Appl. Opt.*, 12, 29–33 (1973).

13. Urzica, D., Düwel, I., Schulz, C., Gutheil, E.: Laser-induced evaporation of a single droplet - An experimental and computational investigation. *Proc. 20th Ann. Conf. ILASS-Europe*, 241–246 (2005).
14. Kafalas, P., Herrmann, J.: Dynamics and energetics of the explosive vaporization of fog droplets by a 10.6-micrometer laser pulse. *Appl. Opt.*, 12, 772–776 (1973).
15. Chen, Z. H., Liu, T. H., Sohrab, S. H.: Combustion of liquid fuel spray in stagnation-point flow. *Combust. Sci. Tech.*, 60, 63–77 (1988).
16. Li, S. C.: Spray stagnation flames. *Prog. Energy Combust. Sci.*, 23, 303–347 (1997).
17. Santoro, V. S., Gomez, A.: Extinction and reignition of counterflow spray diffusion flames Interacting with laminar vortices. *Proc. Combust. Inst.*, 29, 585–592 (2002).
18. Continillo, G., Sirignano, W. A.: Counterflow spray combustion modeling. *Combust. Flame*, 81, 325–340 (1990).
19. Gutheil, E., Sirignano, W. A.: Counterflow spray combustion modeling including detailed transport and detailed chemistry. *Combust. Flame*, 113/2, 92 (1998).
20. Gutheil, E.: Structure and extinction of laminar ethanol/air spray flames. *Combust. Theory Modeling*, 5, 131–145 (2001).
21. Chen, N.-H., Rogg, B., Bray, K. N. C.: Modelling laminar two-phase counterflow flames with detailed chemistry and transport. *Proc. Combust. Inst.*, 24, 1513–152 (1992).
22. Greenberg, J. B., Sarig, N.: An analysis of multiple flames in counterflow spray combustion. *Combust. Flame*, 104, 431–459 (1996).
23. Gutheil, E.: Multiple solutions for structures of laminar counterflow spray flames. *Prog. Comput. Fluid Dyn.*, 7(5), 414–419 (2005).
24. Chevalier, C.: *Entwicklung eines detaillierten Reaktionsmechanismus zur Modellierung der Verbrennungsprozesse von Kohlenwasserstoffen bei Hoch- und Niedertemperaturbedingungen.* Ph.D. Thesis, Universität Stuttgart, (1993).
25. Gutheil, E.: Numerical investigation of the ignition of dilute fuel sprays including detailed chemistry. *Combust. Flame*, 93, 239–254 (1993).
26. Hollmann, C., Gutheil, E.: Flamelet-modeling of turbulent spray diffusion flames based on a laminar spray flame library. *Combust. Sci. Tech.*, 135, 1-6, 175–192 (1998).
27. Peters, N.: *Turbulent combustion.* Cambridge University Press, Cambridge, UK (2000).
28. McDonell, V. G., Samuelsen, G. S.: An experimental data base for the computational fluid dynamics of reacting and nonreacting methanol sprays. *J. Fluids Engin.* 117, 145–153 (1995).
29. Landenfeld, L., Kremer, A., Hassel, E., Janicka, J.: Comparison of Reynolds stress closures for strongly swirling combusting jets. *11th Symp. on Turbulent Shear Flows*, Grenoble (1997)
30. Libby, P. A., Williams, F. A. (Eds.): *Turbulent reactive flows.* Academic Press, London (1994).
31. Vogelgesang, M.: *Entwicklung eines Reynolds-Spannungsmodells für turbulente Sprayflammen,* Ph.D. Thesis, Heidelberg University (2005).
32. Daly, H. J., Harlow, H.: Transport equations in turbulence. *Phys. Fluids*, 13, 2634–2649 (1070).
33. Gibson, M. M., Younis, B. A.: Calculation of swirling jets with a Reynolds stress closure. *Phys. Fluids*, 29(1), 38 (1986).
34. Launder, B. E.: On the Effects of a gravitational field on the turbulent transport of heat and momentum. *J. Fluid Mech.*, 67(3), 569 (1975).

Transport and Diffusion in Boundary Layers of Turbulent Channel Flow[*]

S. Sanwald[1], J. v. Saldern[2], U. Riedel[1], C. Schulz[2], J. Warnatz[1], and J. Wolfrum[2]

[1] Interdisziplinäres Zentrum für Wissenschaftliches Rechnen (IWR), Universität Heidelberg.
[2] Physikalisch-Chemisches Institut (PCI), Universität Heidelberg.

Summary. To investigate the transport and the diffusion of scalars in a turbulent channel flow, direct numerical simulations (DNS) are performed and experiments with the same geometry have been set up to examine details of the near-wall region. Transport and diffusion are observed with flow-tagging methods. In one setup the deformation of a line of tracer-molecules, placed perpendicular to the flow direction, is investigated. Furthermore the seeding and detecting of tracer-molecules into the boundary layer has been realized in two different configurations, in order to investigate species transport between the bulk flow and the boundary layer.

1 Introduction

Turbulent reactive flows are of high interest for many industrial processes. Due to the significant increase of the mixing rates of the species in turbulent flows, the reaction time is much shorter than in laminar flows. To obtain a better understanding of turbulent flows, the transition from a fully developed turbulent flow to a laminar flow in near-wall regions needs further investigation of heat and mass transfer in the flow and through the boundary layer into the main flow. With catalytic walls, which promote chemical reactions at the surface, the situation gets even more complicated.

To investigate the heat and mass transfer in the near-wall flow, both, a mutually agreed experiment and simulation have been set up under the same conditions. Thus for the comparison of the experimental results with the simulation of the experiment, all relevant physical and chemical scales of the near-wall flow have to be resolved in the simulation. Due to the lack of appropriate models for the computation of boundary layers with heat and mass transfer for large eddy simulations (LES, large eddies are resolved and small

---

[*]This work has been supported by the German Research Foundation (DFG) through SFB 359 (Project B2) at the University of Heidelberg. The calculations presented in this paper have been performed on HELICS [19] at the IWR.

scales are modeled), direct numerical simulation (DNS) is used to explore the channel flow. The final aim is to find a possibility to model smaller scales and use LES for the calculations of larger domains.

The experimental investigation of the near-wall flow and the transport phenomena between the wall layer and the turbulent flow addresses three different fields.

1.1 Flow Diagnostics with Two-Time-Step Imaging

Flow-field imaging is most frequently carried out via particle-imaging velocimetry (PIV) where particles, i.e. droplets or solid particles are added to the flow. Double exposure with a time delay adjusted to the flow velocity then allows to investigate the local flow vectors [7]. Close to surfaces, this technique, however, fails due to the interaction of the particles with the surface. Furthermore, particles do not perfectly follow the smallest turbulence elements. Therefore, we investigate the near-wall flow with molecular tracers that are photolytically generated within the flow. They have virtually the same transport phenomena as the fluid of interest. Different flow-tagging techniques have been proposed over the last decade using photodissociation of vibrationally hot H_2O [6], or vibrational excitation of O_2 [8, 9]. Nitric oxide (NO) was formed from photodissociation of tert-butyl nitrate [10] or from photolytically generated oxygen atoms with subsequent reaction with nitrogen [13]. We use photodissociation of nitrogen dioxide with subsequent laser-induced fluorescence imaging (LIF) of NO. We expand this technique to allow the investigation of the temporal development of the turbulent flows [11, 12]. The first possibility is by either photolytically tagging in two subsequent time steps with a single detection of both marked volumes. Second is with one tagging and subsequent detection of each individual flow situation after two variable delays.

1.2 Temperature Measurements Close to the Wall

The temperature can be imaged via a tracer-based density measurement by homogeneously adding acetone to the air flow and observing laser-induced fluorescence [17, 16]. This technique has been applied in various experiments and has been adopted to investigations in near-wall layers [14].

1.3 Investigation of Species Transport

The species transport between the boundary layer and the turbulent flow requires a species that is generated close to the wall. Alternatively, the species can be doped selectively either into the boundary layer or into the bulk flow. We use three different approaches to study the species transport between the wall and the turbulent flow [15]. These are (i) sublimation of naphthalene from a coated, heated surface, (ii) doping the boundary layer with NO by leaking NO into the laminar boundary flow through a small hole and (iii) adding NO to the main flow by a pulsed nozzle. In all cases the spacial distribution of the respective species is imaged via LIF in a two-dimensional plane with high spatial resolution. In this paper we report on the flow-tagging measurements and the NO-based species transport measurements.

2 Channel Setup

In order to gain more insight into the turbulent flow, we focus on a simple profile in a straight channel. It is important for the setup of the channel to keep in mind the needs for the simulation and for the experiments. The dimensions should be kept small to allow a direct numerical simulation and big enough to access the turbulent flow in the channel by measurement techniques.

To avoid the influence of a secondary flow the dimensions of the cross-section (60×5 mm^2) have an aspect ratio of 12 [1, 5]. The geometry and the main setup of the channel are shown in Fig. 1. The calculated domain is marked by the read area.

The Reynolds number (Re) (equation 1) is based on the channel half-width (δ) and the mean velocity ($\bar{U}_0$) over the time at the centerline of the channel.

$$Re = \frac{\bar{U}_0 \delta}{\nu} \quad (1)$$

We chose moderate Reynolds numbers of $Re = 5000$ and $Re = 7000$ for the simulation and for the experiments. Due to the high resolution needed to sufficiently represent the boundary layers, the number of grid points would increase dramatically with higher Reynolds numbers. For the temperature, the simulations are initialized with the room temperature (293 K) of the experiment. The resulting value for the viscosity was $1.51 \cdot 10^{-5}$ m^2/s.

In the *simulations* the time-dependent solution of the compressible Navier-Stokes equation is calculated. The set of coupled partial differential equations has to be solved. Conservation of total mass, momentum, energy and in a

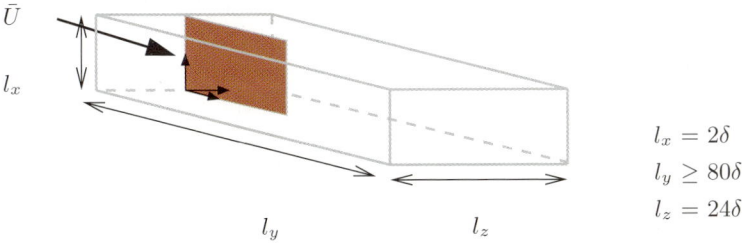

Fig. 1. Geometry of the channel ($\delta = 2.5$ mm)

multicomponent case the conservation of the chemical species [4] has to be guaranteed. The spatial discretization is performed using a finite-difference scheme with sixth-order central-derivatives. The time-integration is calculated by higher order explicit Runge-Kutta algorithms. Computational requirements for a DNS are very high and the use of parallel computers is required. Hence, the message passing interface (MPI) is used for the communication between the nodes.

All simulations are performed with 400 integration points in the x-direction and 600 integration points in the y-direction. The computational domain is marked in red (see Fig. 1). To keep the calculated domain small, periodic boundary conditions in the flow direction were used. The velocity was initialized with the mean velocity profile perturbed with random fluctuations. The span-wise direction is bounded by infinite plates. At the walls the velocities vanish and the temperature is kept at a constant level, equal or higher than the flow media.

The experiments were carried out in an optically accessible flow channel. The flow is observed through windows at the short side of the channel. One wall at the long side is heatable to 300 K above the gas temperature and can be replaced by a catalytically active (Pt-coated ceramic) wall. The main flow can be adjusted to a flow-rate of 80 m^3/h resulting in mean flow velocities of $v = 74$ m/s associated with a Reynolds number of $Re_b = 24500$ at room temperature. The following results are all generated at a Reynolds number of $Re = 5000$ and $Re = 7000$. The channel flow is therefore fully turbulent, no further turbulence-generating elements were required.

For the flow-tagging experiments the flow was seeded with 1185 ppm NO_2. In order to avoid handling and doping of NO_2, the NO_2 is generated from NO + air in a 55 l reaction vessel prior to mixing the seeding gas to the main flow. The residence time of approximately 10 min in the reaction vessel leads to complete equilibrium ($> 99\% NO_2$). Mass flow controllers (Bronkorst and Tylan) control the flow rate [18].

For the flow-tagging measurements the beam of a frequency-tripled Nd:YAG laser (Thomson BMI) is focused to a line with (10 mJ/pulse, focused with a $f = 400$ mm lens) and introduced into the flow channel perpendicular to the flow and the investigated wall. In experiments with double generation of the "line" a double-pulse Nd:YAG laser is used (Thomson BMI). The transport of the photolytically generated spatial "line" of NO is then observed at various delays (up to 300 μs) with LIF imaging. NO is excited within the A-X(0,0) band at 226 nm with a light sheet 5 mm high and 0.1 mm thick. For detection of single events after the generation of the line a excimer-pumped frequency-doubled dye laser (Lambda Physik Scanmate II with coumarin 2 as dye) is used. For the detection of the "line" at two subsequent time steps the frequency-tripled laser light of the double pulse Nd-YAG laser is used to pump a midband OPO (GWU). It is tuned to generate a signal wavelength at 450 nm which is then extra-cavity frequency doubled in a BBO crystal. The resulting UV-beam has a pulse energy of approximately 1 mJ per laser

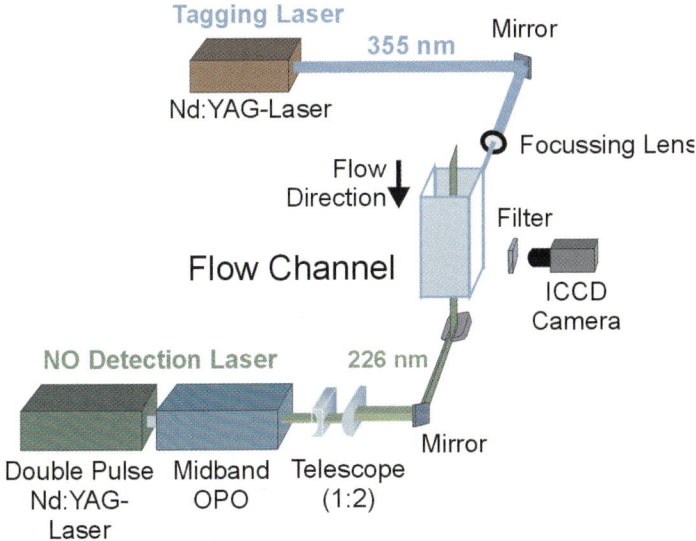

Fig. 2. Optical setup of the laser-based investigations in the flow channel

pulse and a spectral width of ~ 4 cm^{-1}. It therefore excites a large portion of the NO A-X(0,0) bandhead. The overall excitation is weaker by a factor of four compared to the (0.1cm^{-1} half-width) dye-laser excitation tuned to the strongest NO transition. The optical setup is shown in Fig. 2. In order to maximize the NO-LIF signal, we operate the flow channel with pure nitrogen (technical grade). This increases the signal due to the lack of quenching by O_2 by a factor of 75. NO-LIF signals are discriminated against scattered light by a 2 mm UG5 filter (Schott) and focused (Halle UV lens, $f = 100$ mm, $f_\# = 2$) onto the chip of an image-intensified interline camera (LaVision flowmaster with relay-intensifier). The image intensifier ensures short detection gate times for both frames, the maximum repetition rate of 200 ns of the double frame camera is no restriction for our experiments. For the investigation of species transport small quantities of pure NO (Messer Griesheim) are doped into the flow channel either through a pulsed nozzle (Bosch) or through a hole (1 mm diameter) with low leak rate. The observation of the resulting distribution of NO is carried out in two subsequent time steps using the same laser/camera setup described above.

3 Results

In the experiments as well as in the simulations the setup leads to a Reynolds-number (Re) of 5000 and 7000, based on the channel half-width (δ) and the mean velocity ($\bar{U}_0$) at the center of the channel. With the friction velocity (u_τ) (equation 2), which can be calculated by the wall shear stress (τ_w), it is

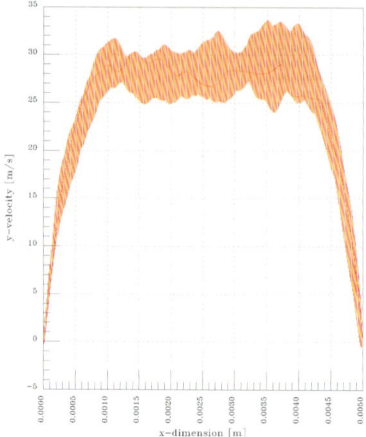

Fig. 3. Variation of the y-velocities for each x-position (the actual position is slightly highlighted)

Fig. 4. Mean displacement of the "NO-line" after 100 μs (experiment). The dimensions of the frame are 5.0×5.8 mm and the flow propagates from bottom to the top

possible to generate the Reynolds-number (Re_τ) for turbulent flows [2] with equation 3.

$$u_\tau = \sqrt{\frac{\tau_w}{\rho}} = \sqrt{\nu \left(\frac{\partial \bar{U}_0}{\partial x}\right)_{x=0}} \qquad (2)$$

$$Re_\tau = \frac{u_\tau \delta}{\nu} \qquad (3)$$

In the experiments the wall shear stress is estimated from the mean displacement of the generated NO-concentration in Fig. 4 (error: 0.1 mm). Between

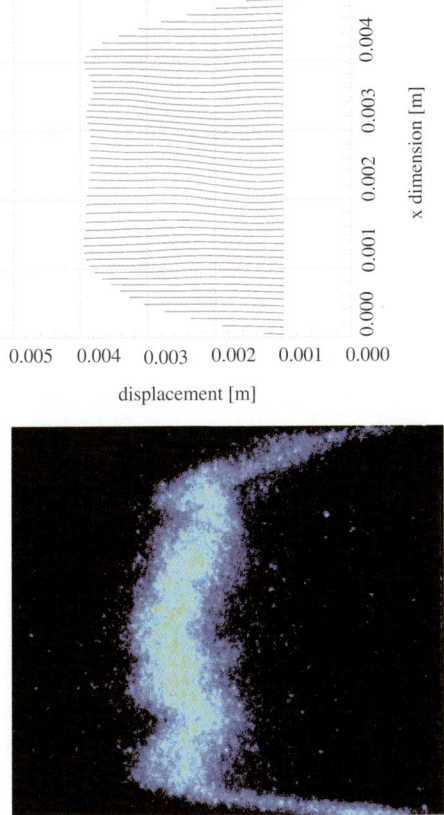

Fig. 5. Displacement of the marked points after 100 μs. Upper frame: Simulation based tracing of marked particles. Lower frame: Experiment (instantaneous image)

the generation and detection of the NO-line was a delay of 100 μs. The displacement at two different distances from the wall (0.01 mm an 0.11 mm) divided by the time between generation and detection of the generated NO-concentration results in the flow velocity at the corresponding distance from the wall. These two different velocities at two different distances from the wall allow the estimation of the velocity gradient dependent on the distance from the wall (see equation 2).

	τ_w	u_τ	Re_τ
experiment	1.027	0.924	153
simulation	1.129	0.968	160

In the table above the values of the calculated parameters are shown. The results of the simulation are in accordance with the experimental results. Fig. 5 shows the instantaneous displacement of the "NO-line". It is set in relation to the displacements of the simulation. The upper part of Fig. 5 shows the streakline trace of a time segment over 100 μs. The trace has been produced by following a fluid volume over the time with the results from the simulation. Between the written plot states the trace has been interpolated linearly.

In the experiments the displacement is conceived by detecting the "NO-line" after a certain time with LIF. Here, the delay also was 100 μs. The comparison of the lower and upper frame shows a very good agreement between the simulation and the experiment.

The observation of the photolytically generated structure ("NO-line") at two time-steps after generation is shown in Fig. 6. The delay between the generation and the first detection was 50 – 250 μs and the delay to the second detection was 50 – 100 μs. The original position of the "NO-line" is shown in the lower right frame. The total flow had a mass flow of 338 l/min through the 300 mm^2 cross-section channel. The flow propagates from the right to the left. The position of the channel walls are marked by the two horizontal lines.

Fig. 7 shows the distorted "NO-line" from the experiment and the simulation. Here, the Reynolds-number of the flow is 7000. On the left-hand side the results from the simulations at different times, after the "NO-line" has been initialized, are presented. The first picture is directly plotted after the initialization ($\Delta t = 0$ μs) of the "NO-line". The concentration is evident from the color-bar located at the right side. The pictures below are at $\Delta t = 50$ μs and $\Delta t = 130$ μs.

On the right-hand side of Fig. 7 the results from one series of the experiments is compared with the simulation. The first picture also indicates the undeformed "NO-line". The pictures below display the deformed lines at $\Delta t = 50$ μs and $\Delta t = 125$ μs.

Also with the time evolution the flow field of the simulations indicate good conformance with the experimental results.

The investigation of the species transport in the boundary layers was performed in two different experimental setups.

At first, the species transport from the NO-seeded boundary layer into the turbulent main flow was observed. The results are presented in Fig. 8. The image pairs represent two typical situations, imaged in two time-steps ($\Delta t = 100$ μs). While in the left images (a) no large eddies lead to large-scale mixing, in the right images (b) turbulent mixing over long distances (~ 2 mm) is active. Both situations occur intermittently in the flow channel. The arrow marks the position where NO is seeded into the boundary layer.

In Fig. 9 the transport out of the boundary layer is presented as NO concentration profiles obtained at different positions downstream from the seeding point. The positions are marked in the upper left frame in Fig. 8. The apparent decrease of signal close to the wall is an artifact of the measurement and due to laser attenuation.

The second experiment was set up to observe the species transport from the seeded main flow into the boundary layer. NO is added to the main flow through a pulsed nozzle leading to steep NO concentration gradients at the

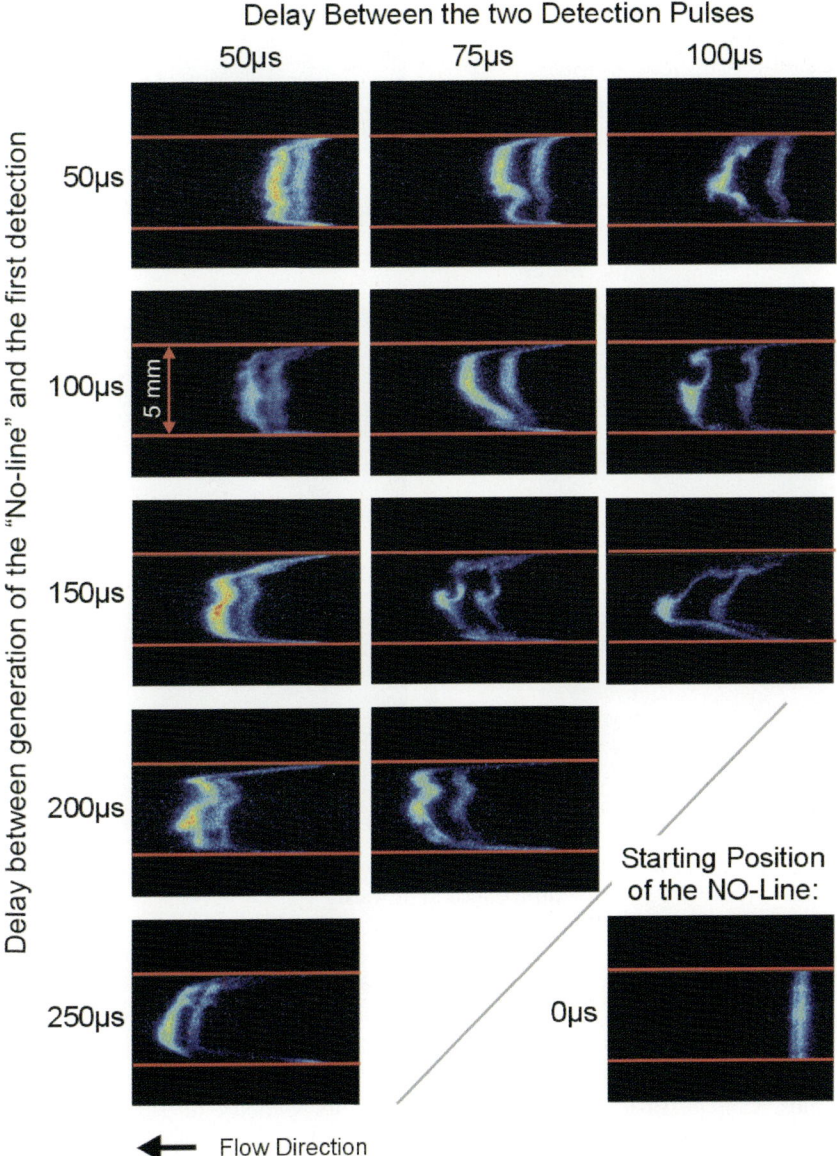

Fig. 6. Experimentally measured "NO-line" series with different time delays ($Re = 5000$)

leading edge of the NO-cloud. The NO-cloud in the main flow "overtakes" the NO in the boundary layer due to the difference in average velocity. Therefore, after a certain distance, the NO-doped main flow is in contact with an undoped boundary layer.

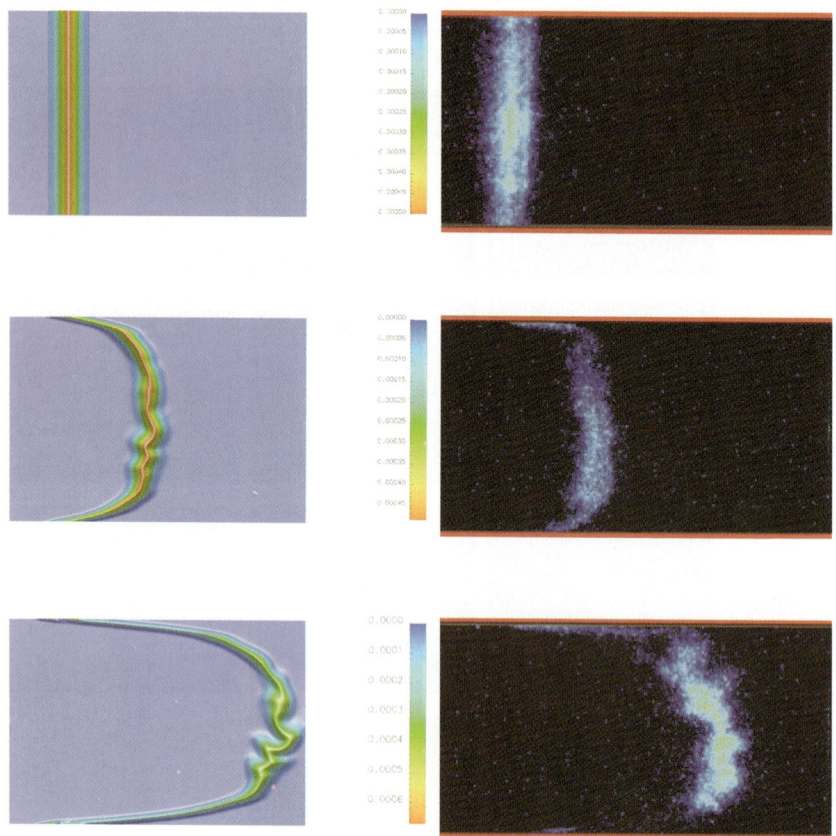

Fig. 7. Comparison of the simulations and the experimental deformations of the "NO-line" ($Re = 7000$), flow direction is rightwards

The species transport from the seeded main flow into the boundary layer is shown in Fig. 10. The NO concentration profiles are shown for two time-steps (solid lines (red) taken at $\Delta t = 100$ μs after the dotted lines) at different distances to the wall. The lines in the first picture show the location of the profiles. Concentration gradients and transport velocity decreases with decreasing distance to the wall. As in Fig. 8 the boundary layer in Fig. 11 is disturbed by a vortex and the discharge of the temperature into the main flow is observable.

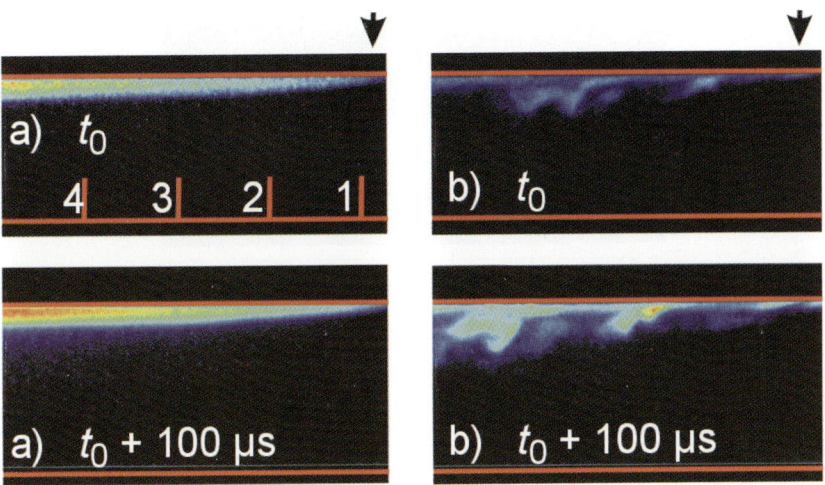

Fig. 8. Experimental results of the NO-seeded boundary layers. Flow direction is from the right to the left ($Re = 5000$)

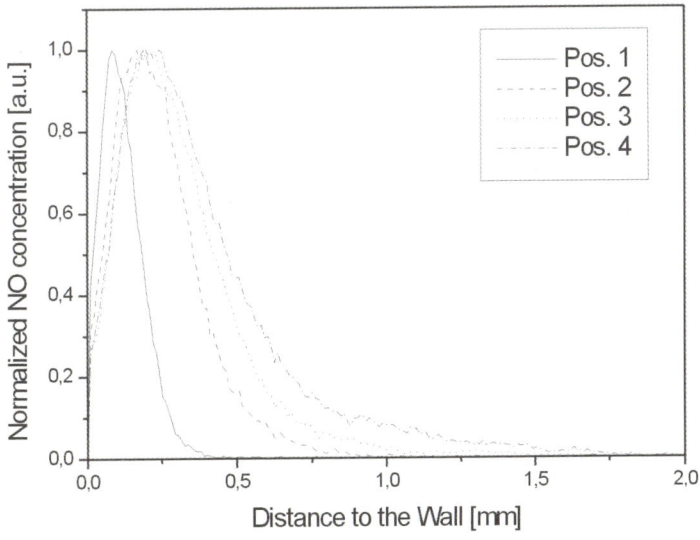

Fig. 9. NO concentration profiles of the upper left frame a) in Fig. 8

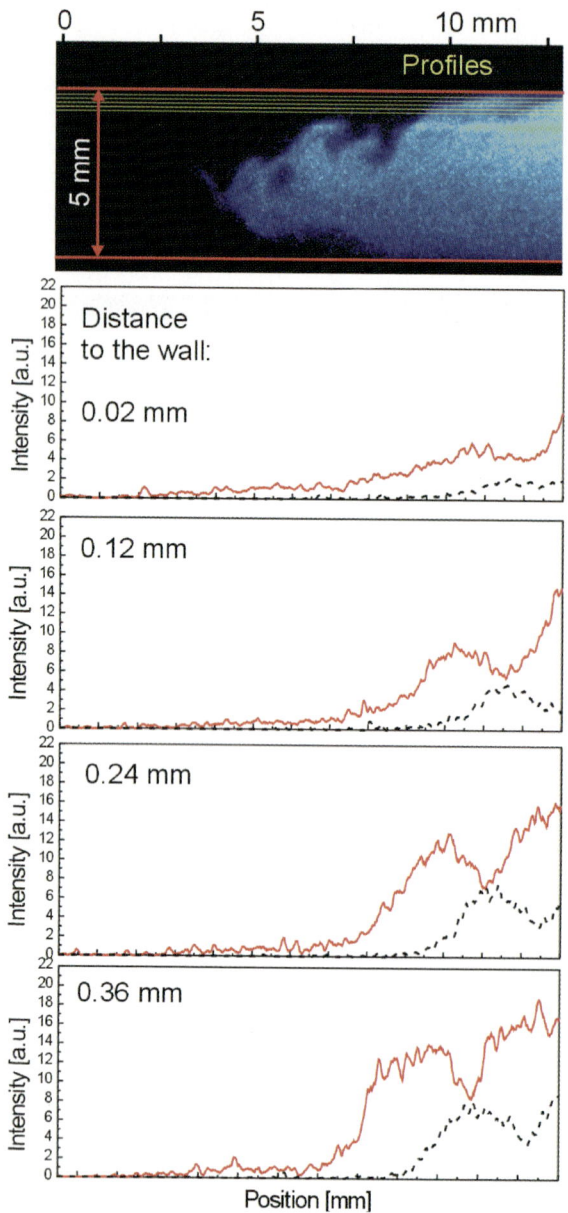

Fig. 10. Measured intensity profiles of the transport into the boundary layer

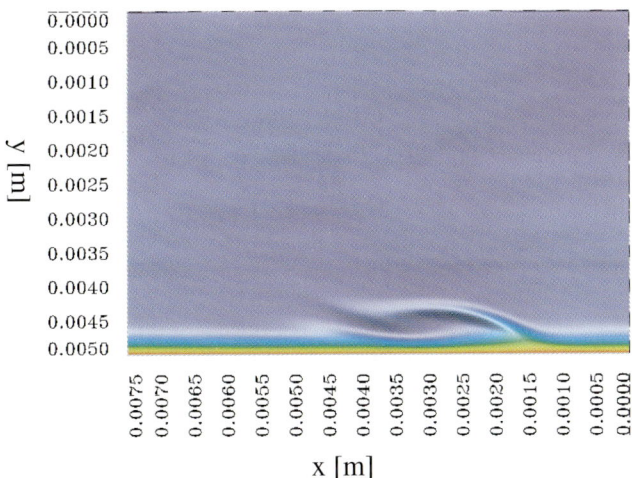

Fig. 11. Simulation of the lower wall heated to $\Delta T = 20$ K above the mean flow temperature of $T = 293$ K. The flow propagates from the right to the left

4 Conclusions

In this paper, the heat and mass transfer in the transition region from a fully developed turbulent flow to a laminar flow in near-wall regions is investigated experimentally as well as numerically. In the experiments photolytically generated molecular tracers are used and in the numerical model of the flow time-dependent direct numerical simulation is employed.

Velocity gradients and wall shear stress function – obtained from the displacement of the NO concentration in the experiment and obtained from the velocity field resolving all turbulent scales – agree quite well. Detailed information on the interaction of the NO line profile and the turbulent flow field is presented.

Finally, the species transport in the boundary layer is analyzed. In a first experimental setup the transport out of an NO-seeded boundary layer into the turbulent main flow is measured. Two characteristic situations are identified occurring intermittently: Time periods of large scale turbulent mixing and time periods where no large eddies interact with the laminar sub-layer. In a second setup the main flow is seeded with NO and the transport into the boundary layer is investigated. The characteristic contact length of the NO molecules (streamwise distance where NO is found at a very small distance from the wall) is determined.

References

1. H. Schlichting: Grenzschicht-Theorie. 8. Auflage, G. Braun, Karlsruhe, 1982.
2. S.B. Pope: Turbulent Flows. Cambridge University Press, Cambridge, 2000.

3. J. Kim & P. Moin: Transport of Passive Scalars in a Turbulent Channel Flow. Turbulent Shear Flows 6, edited by J-C Andre et al., SpringerVerlag Berlin, Heidelberg, 85 – 96, 1989.
4. T. Poinsot, D. Veynante: Theoretical and numerical combustion. Edwards Ed., 2001.
5. J. Laufer: Investigation of Turbulent Flow in a Two-Dimensional Channel. NACA Technical Report, 1053, 1951.
6. R.W. Pitz, J.A. Wehrmeyer, L.A. Ribarov, D.A. Oguss, F. Batliwala, P.A. DeBarber, S. Deutsch and P.E. Dimotakis: Unseeded molecular flow tagging in cold and hot flows using ozone and hydroxyl tagging velocimetry. Meas. Sci. Technol. 11:1259-1271, 2000.
7. J. v. Saldern and C. Schulz: Investigation of small-scale flow structures by simultaneous flow-tagging and PIV experiments. Eurotherm Seminar 71, Visualization, imaging and data analysis in convective heat and mass transfer, 169-173, Reims, 2002.
8. R. Miles, C. Cohen, J. Conners, P. Howard, S. Huang, E. Markovitz and G. Russell: Velocity measurements by vibrational tagging and fluorescent probing of oxygen. Opt. Lett. 12:861, 1987.
9. R.B. Miles, W. Lempert and B. Zhang: Turbulence structure measurement by RELIEF flow tagging. Fluid. Dyn. Res. 8:9-17, 1991.
10. S. Krüger and G. Grünefeld: Gas-phase velocity field measurements in dense sprays by laser-based flow tagging, Appl. Phys. B 70:463-466, 2000.
11. C. Orlemann, C. Schulz, and J. Wolfrum: NO-flow tagging by photodissociation of NO2. A new approach for measuring small-scale flow structures. Chem. Phys. Lett. 307:15-20, (1999).
12. S. Doose, C. Orlemann, C. Schulz, J. Wolfrum, P. Geißler, and B. Jähne: NO-flow tagging by photodissociation of NO2: applications of a new technique to visualize turbulent flow structures. Proceedings of the Joint Meeting of the British, German and French sections of the Combustion Institute, 427-429, Nancy, 1999.
13. T. Elenbaas, N.M. Sijtsema, R.A.L. Tolboom, N.J. Dam, W. van de Water and J.J. ter Meulen: Characterization of turbulence by air photolysis and recombination tracking (APART). AIAA 2002-0694, 2002.
14. S. Doose: Laserspektroskopische Untersuchungen zur hochaufgelösten Visualisierung von wandnahen Geschwindigkeits- und Temperaturfeldern in einer turbulenten Strömung. Diplomarbeit, Ruprecht-Karls-Universität Heidelberg, 1999.
15. J. v. Saldern: Laserspektroskopische Untersuchungen laminarer Grenzschichten turbulenter Strömungen mit Hilfe molekularer Marker. Dissertation, Ruprecht-Karls-Universität Heidelberg, 2003.
16. M. Thurber, F. Grisch and R. Hanson: Temperature imaging with single- and dual-wavelength acetone planar laser-induced fluorescence. Optics Letters 22:251-253, 1997.
17. F. Großmann, P.B. Monkhouse, M. Ridder, V. Sick and J. Wolfrum: Temperature and pressure dependences of the laser-induced fluorescence of gas-phase acetone and 3-pentanone. Appl. Phys. B 62:249-253, 1996.
18. S. Doose, C. Orlemann, and C. Schulz: Investigation of small-scale flow structures using NO-flow tagging by photodissociation of NO2. CLEO, Pacific Rim, Seoul, Korea, 1999.
19. HELICS, IWR - Universität Heidelberg (HBFG funds, hww cooperation), helics.uni-hd.de

Part VI

Diffusion and Transport in Accretion Discs

Preamble

Processes which heavily rely on fluid motions, chemical reactions, diffusion, and transport of matter and radiation are common in nature and technology. The complex interplay of these individual processes has mainly been studied for laboratory conditions, but there is growing evidence that important phenomena in space also are ruled by the joint action of these basic processes. The two articles of this chapter are devoted to studying two types of such problems, both of which have become of central interest to astronomical and space research during the last decade.

The first article *"Evolution of protoplanetary disks including detailed chemistry and mineralogy"* deals with the evolution of protoplanetary accretion disks being the birthplaces of new planetary systems that are now observed to exist around a major fraction of solar-type stars. Protoplanetary accretion disks are flat rotating gaseous disks with a considerable content of mineral dust around newly born stars. They are the sites where planets form by agglomeration of the solids to planetary sized bodies and by capture of gaseous envelopes by the more massive protoplanets. The formation of planets starts after a period of excessive chemical and mineralogical processing of the gaseous and solid material and of intense mixing by large scale flows and turbulent mixing. If one is interested in understanding how planets form in space, in particular, what is the origin of the planetary system we live in, and what determines the ultimate composition of the planetary bodies and their atmospheres and oceans, one has to study in detail all the chemical and mixing processes and their interplay which are operating in the dusty gas disk which revolves around a young star during the first few million years of its existence.

The second article *"Numerical methods for multidimensional radiative transfer"* considers the problem of multidimensional radiation transport in hypervelocity flows, which is of paramount importance for the interpretation of observations in the context of cosmological observations. Photons are of particular importance for astrophysics mainly for two reasons. Since they have energy and momentum they may contribute significantly to the balance equation for these quantities. Therefore, they have to be considered carefully in radiation-hydrodynamical calculations as mentioned above. Secondly, photons provide most of the information we have on celestial objects: the directions from where they come indicate the positions in the sky, from their number one can infer the energy production in the object of origin, the temporal dependence of the observed photon numbers allows insights into the stability of the emitting object, and finally the spectral distribution of the photons makes the determination of many parameters of the originating plasma (e. g., its temperature, pressure, chemical composition) possible. For the modeling of celestial radiation fields it is usually sufficient to consider the time averaged square of the electric field, the specific intensities. Quantities as the energy density of the radiation field or the corresponding energy flux can then be derived by simple integrations. The specific intensities are governed by the radiative transfer equation, a linearized Boltzmann equation, which connects the macroscopic properties of radiation with microscopic properties that determine the thermodynamical state of the matter. The equation as such is linear, but it is usually of high dimension (up to three space, two angle, one frequency and one time dimensions have to be considered), it involves unusual boundary conditions and, in important cases, an intricate coupling of the variables. These facts make the solution extremely difficult and expensive. Analytical solutions are available only for very special cases so that one has to rely mainly on numerical solutions.

Evolution of Protoplanetary Disks Including Detailed Chemistry and Mineralogy[*]

H.-P. Gail and W. M. Tscharnuter

Institut für Theoretische Astrophysik, Universität Heidelberg

Summary. The earliest phase of evolution of planetary system is determined by the intimate coupling of a number of physical and chemical processes: turbulent hydrodynamic flows under the action of gravity, radiative transfer in a dusty gas, extensive mixing of material by large scale flows and turbulent duffusion, vaporization and condensation of ices and minerals, chemical reactions in the gas phase and at the surface of dust grains, and agglomeration of dust into planets. The project attempted to develop the first numerical model calculations which account for at least the most important ones of these processes in a consistent way. The chemistry and mineralogy in the planetary formation region of protoplanetary disks was investigated and the basic processes which have to be included in a numerical model were determined. Stationary and time dependent models for protoplanetary disks including the basic chemistry and mineralogy were constructed in different approximations in one and two space dimensions. These allow to determine the composition of the disk material which determines the raw material for planet formation. Preliminary comparison with observations of primordial material in comets and meteorites from the formation time of the solar system shows that the models already reproduce successfully some basic observational features.

1 Introduction

Protoplanetary accretion disks are the sites of formation of planetary systems around newly born stars. If one is interested in understanding how planets form and what determines the ultimate composition of the planetary bodies and their atmospheres and oceans, one has to study in detail the processes operating in the flat rotating dusty gas disk which revolves around a young star during the first few million years of its existence.

This problem has been attracting the interest of the astronomical and space science community for several decades. It suddenly received a dramatically expanding attention after the first extrasolar planetary systems around

---

[*]This work has been supported by the German Research Foundation (DFG) through SFB 359 (Project C1) at the University of Heidelberg.

nearby stars had been detected. The interest has concentrated up to now mainly on the question of how nature manages to collect roughly 10^{41} tiny dust grains with a size of about 0.1 μm into an about 10^4 km sized planetary body (or core of a gaseous planet). The basic features of this process, at least for the formation of the planetary system we live in, now seem to be clarified. The nature of the dust material generally was not considered in such investigations, since this is of only secondary importance for the agglomeration process itself, and the very first phases of the process, which determine the composition of the planets, were ignored.

On the other hand, geochemists have always been interested in meteorites because since the early days of chronology of the solar system it has been clear that meteorites represent surviving material from the formation time of the solar system. A tremendous body of information on details of the processes operating at the time of formation of the planetary bodies and on the composition of bodies in the solar system and on compositional variations between bodies has accumulated during the last decades. This information presently has not yet helped much, however, to arrive at the envisaged goal of reconstructing the formation history of the solar system.

The power of present day computers now allows to tackle the problem from a different side. The basic physical processes involved in the formation of planetary systems generally are processes which are also operating in common terrestrial processes. Most of them are well studied because of their importance for technical processes or for processes operating in nature. During the formation phase of the solar system they only work compared to terrestrial conditions on vastly different length and time scales which implies high demands on the computing facilities, but this should be no problem in view of the ever increasing computer power. This makes it worth trying to develop a complete numerical simulation of the early evolutionary phases of protoplanetary disks where the crucial processes are operating which determine the properties of the emerging planetary system. By comparing results of computer simulations with data obtained from the analysis of meteorites and planets in our solar system and with observations of extrasolar planetary systems it will be possible to develop a complete picture of how planetary systems form.

In the present project the first steps have been undertaken to develop such a computer simulation of the planetary formation process. The project concentrated on the coupling of models for the structure and evolution of protoplanetary disks with the important transport and diffusion processes and combined them with detailed modeling of the chemistry and mineralogy of the disk material in order to determine the composition of the material from which planetary systems form.

2 Formation of Planetary Systems

Planetary systems today are thought to be by-products of the star formation process. Stars form in dense molecular clouds in a galaxy which contain about 10^4 to 10^6 solar masses. At the beginning of the star formation process a small region in such a cloud, containing, for instance, around one solar mass, for some reason becomes unstable against gravitational contraction and starts to collapse toward the mass center. Because the matter in the parent molecular cloud is in a state of vivid turbulence the matter in the gravitationally unstable region usually carries a net angular momentum. For this reason the unstable parcel of matter cannot directly collapse into a star but only into a flat rotating disk-like configuration. Viscous frictional processes resulting either from convectively driven turbulence or magnetic instabilities result in a redistribution of matter and angular momentum in the disk: angular momentum is transported outward by frictional processes while matter is transported inward toward the center. Within one to two times 10^5 years a star forms at the center, which contains nearly all of the mass, and a very extended and thin disk which initially contains about 10 % of the total mass. The onset of a strong stellar wind at this stage then disperses the remaining molecular cloud material and further mass-infall ceases.

The viscous frictional processes continue to operate in the remaining material of the disk and matter continues to migrate toward the star while angular momentum is transported outward and transferred to the ever expanding outer parts of the disk. Since there is no further supply of matter to the disk by infalling material from the molecular cloud the disk slowly empties into the star with a gradually decreasing mass-accretion rate.

With decreasing mass and temperature in the disk at some instant the fine dust material in the disk starts to agglomerate into bigger units. The first phase of Brownian coagulation from about 0.1 μm-sized interstellar dust grains to about 100 μm-sized dust flocks lasts for quite a long time, but once turbulence driven coagulation overturns, the further growth to km-sized bodies, the planetesimals, is very rapid and lasts for only a few 10^3 years. This sudden precipitation of the solid component from the gas-dust mixture in the disk marks the onset of the planetary formation process. The process occurs at different times in different regions of the disk; in the region where in our solar system the terrestrial planets did form this happens probably about 5×10^5 to 10^6 years after the end of mass-infall.

Since the planetesimals form very rapidly, their composition is a snapshot of the composition of the disk material at the time and location of their formation. In the warm inner parts of the accretion disk the planetesimals contain the mineral inventory of the disk material; in the cold outer parts they additionally contain considerable amounts of frozen gases, mainly H_2O ice. The nuclei of comets which enter from time to time the inner solar system from outer reservoirs are surviving planetesimals from the planetary formation

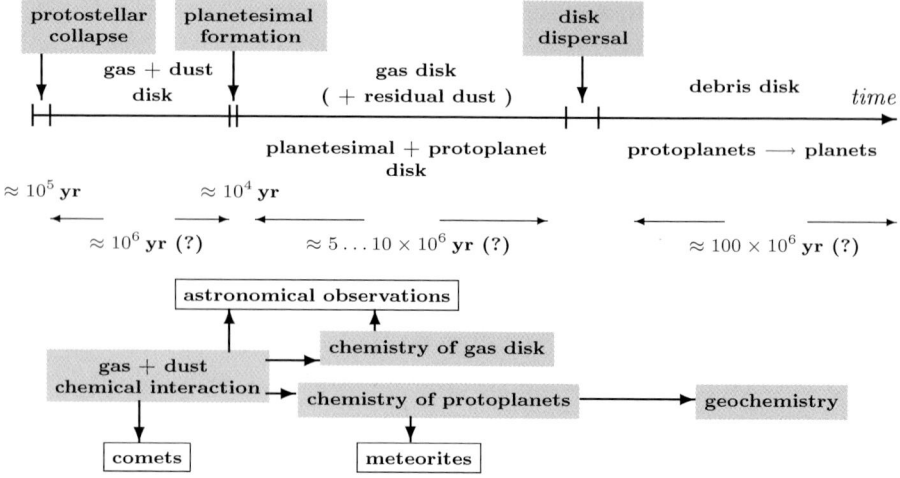

Fig. 1. Processes and timescales relevant for the planetary formation process.

phase. They offer the possibility to study directly the 4.6×10^9 years old material from the very first phase of formation of our solar system.

Up to a size of a few km the dynamics of the planetesimals is determined by the gravitational field of the star and by a frictional coupling to the gas component of the disk. For this, planetesimals are born with circular orbits with very low inclinations with respect to the equatorial plane. Above a size of a few km the planetesimals contain sufficient mass, such that their mutual gravitational attraction becomes important for their dynamics. The further evolution of the planetesimal swarm then is determined by the gravitational interaction in the planetesimal gas, which leads to significant eccentricities and inclinations of their orbits. This results in crossings of orbits and planetesimal-planetesimal collisions, which lead to further growth of the bodies. Within a period of about annother million years, a swarm of a few hundred protoplanets develops, which collects nearly all of the available planetesimal material. In the inner region of the disk the protoplanets are moon to mercury sized while in the outer region where water ice exists the protoplanets have several earth masses.

The massive protoplanets in the cold regions then start to collect massive gaseous envelopes and develop to Jupiter-type planets. The much smaller protoplanets in the inner disk region grow on timescales of the order of 30 to 100 million years into a few earth-like planets.

This picture (here somewhat simplified) of the formation of our solar system emerged during about the last forty years of intense research, starting with the classical work of Safronov [42]. The details may be found in the proceedings of the Protostar & Planets conference series (Black & Matthews [3], Levy & Lunine [34], Manning et al. [35]) and in Kerridge & Matthews [48], and in Canup & Righter [9]). Figure 1 schematically indicates the important steps in the planetary system formation process.

The detection of a now considerable number of extrasolar planetary systems confirms the picture that planetary formation is an ubiquitous phenomenon associated with star formation. These systems, however, are different from our own solar system in so far as they mostly contain Jupiter-type planets close to their central star. Such systems are the only ones which can be detected by the presently available detection methods, but they do not fit easily into the above picture of the formation of our solar system, which predicts Jupiter-type planets to form in the icy outer zones of the accretion disk. Possibly there exist more than one route to planetary formation. We do not discuss this point further since our main interest is our own solar system.

In the following sections we concentrate on the very first phase of the formation of the planetary system, the evolution of the accretion disk from the end of mass infall to the formation of the planetesimal swarm. This phase is of particular interest because the composition of the bodies in the solar system is determined during *this* phase. Since the planetesimals precipitate rapidly from the dusty gas in the accretion disk and thereafter form closed systems which exchange most likely only small amounts of material with the residual matter in the disk, the composition of the planetesimals is fixed after their formation. During later growth to protoplanets and planets, only limited mixing of the more massive bodies between different zones of the planetary system is possible. In other words: the planets and small bodies in our planetary system essentially reflect the composition of the disk material at the time of formation of their precursor planetesimals.

Clearly, in the more massive bodies we do not observe the pristine material from the planetesimals, because these bodies warmed up by radioactive heating and the bigger ones become molten in their center. However, their elemental and isotopic composition remains unchanged, and from the outer layers of bodies from the asteroid belt we have in the form of carbonaceous meteorites a relatively little metamorphosed material from the formation time of the solar system which can be studied in the laboratory.

Additionally we have the possibility to study very pristine material in cometary nuclei. For these it is not simply to determine the location of formation of individual bodies, but this shortcoming is more than outweighed by the fact that their material survived completely unchanged.

The study of the chemical and mineralogical composition of the bodies of the planetary system, in particular of comets and meteorites, opens the possibility to test the validity of models for the formation of the planetary system by comparing model predictions with observed compositions, if it is possible to accurately model the chemical and mineralogical evolution of the disk material prior to planetesimal formation. The development of such models was the aim of the present project of the 'Sonderforschungsbereich'.

3 Dust Input from Interstellar Space

The material from the parent molecular cloud, from which a star and its planetary system forms, is a mixture of gaseous species and tiny dust grains with a typical mass fraction of solids of about 2%. The composition of the gas phase can in principle be determined by microwave observations of molecular rotational lines. The composition of the solid component can only be determined indirectly. By direct observations one can only determine the absorption properties of the dust from the far ultraviolet to the far infrared spectral region. This does not allow to determine uniquely the composition of the mixture of solids forming the interstellar dust. This is because solids either show pure continuum absorption, for instance, metals or soot, which cannot be associated with any specific material, or they show marked infrared absorption bands as, for instance, silicate minerals with their strong $\approx 10\,\mu$m and $\approx 18\,\mu$m absorption features which are, however, more characteristic for local groupings of atoms in the crystal lattice and their bond properties rather than for specific materials.

One then tries to pin down the composition by taking advantage of the knowledge of the cosmic abundance of elements. Since abundant dust species can only be formed by the most abundant elements which form very stable (refractory) solids, this leaves one with a rather limited number of solid compounds which are the possible candidate members of the interstellar dust mixture. From abundance determinations of elements in the gaseous component of the interstellar medium one knows that the elements forming refractory compounds are strongly depleted from the gas phase and hence are almost completely condensed into dust (e.g., Savage & Sembach [43]). Therefore one tries to find a mixture from the candidate dust compounds which (i) reproduces the observed absorption properties of the interstellar dust and (ii) locks up all of the least abundant elements required for their formation. This procedure clearly is not free of ambiguities, but the only way in which one presently can proceed.

Figure 2 shows the abundance of the most abundant elements in the cosmic element mixture. This is essentially identical with the solar element abundance. It is derived from the analysis of a big number of stellar spectra (and from some other sources) and is the universal elemental composition of the stars at the time of their birth. There is some systematic variation of the abundances of elements heavier than H and He to the light elements H and He with the time of formation of the stars since the heavy elements are slowly synthesized in stars and returned into the interstellar medium at the end of their life. So very old stars have an only very small abundance of heavy elements, but during the last 10 billion years the increase of heavy element abundances has significantly slowed down. The heavy element inventory of stars which like the sun are born during th last few billion years is the same within a factor of about a few (e.g., Goswami & Prantzos [22]). There are also only minor variations of the relative abundances between the heavy elements.

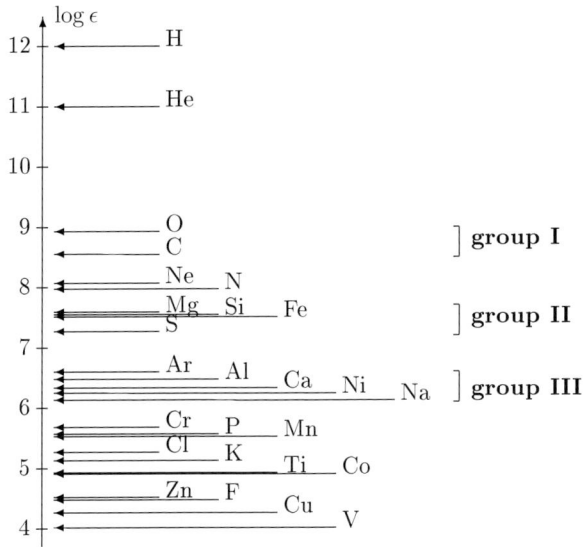

Fig. 2. Cosmic standard element abundances ϵ according to Anders and Grevesse [1] and Grevesse and Noels [23] in the convention of stellar atmosphere theory that $\log \epsilon = 12$ for hydrogen.

Planetary systems therefore all are born with the same mixture of heavy elements. Only if a dying star pollutes the molecular cloud material with its products of element synthesis and if a new star (and its associated planetary system) is born from such material before this is too much diluted by mixing processes in the cloud, there may be some modifications of the elemental composition. For instance, it is generally assumed that the high abundance of the long lived radioactive ^{26}Al in the early solar system is due to a small contribution of material from a dying star to the solar system material.

>From Fig. 2 it is obvious, that only refractory compounds of the rock forming elements Si, Fe, Mg, Al, Ca, Na, Ni and their compounds with O, S (and possibly N), and of carbonaceous compounds may contribute to the interstellar dust mixture. In the cosmic element mixture O is more abundant than C. The carbon then in principle would be completely bound in the extremely stable CO molecule, however part of the interstellar material comes from dying carbon stars where C is more abundant than O and these stars form carbon dust. This carbon dust survives in the interstellar medium despite its oxygen richness because oxidation reactions are much too slow in this cool environment.

Pollack et al. [41] carefully discussed the question which types of dust, formed from the most abundant dust forming elements (C, Mg, Si, and Fe), can be expected to exist in the parent molecular cloud of freshly formed stars, and in the cool outer parts of the resulting protostellar and protoplanetary accretion disks. They arrived at the conclusion that there exists a multicomponent mixture of several kinds of dust species, which is dominated most likely by the few species listed in Table 1. This mixture best satisfies the restrictions of (i) completely locking up the most abundant rock forming el-

Table 1. Composition of the dust mixture in the outer regions of a protoplanetary accretion disk according to Pollack et al. [41], and fraction of the most abundant elements condensed into the different dust species. x denotes the mole fraction of the pure magnesium silicate end members of the solid solution series forming the magnesium-iron-silicates olivine and pyroxene, respectively.

Dust	composition	Mg	Fe	Si	S	C	x
olivine	$Mg_{2x}Fe_{2(1-x)}SiO_4$	0.83	0.42	0.63			0.7
pyroxene	$Mg_xFe_{1-x}SiO_3$	0.17	0.09	0.27			0.7
quartz	SiO_2			0.10			
iron	Fe		0.10				
troilite	FeS		0.39		0.75		
kerogen	HCNO					0.55	

ements and (ii) reproducing the observed absorption properties of the dust. The mixture consists of:

- Olivine with composition $Mg_{2x}Fe_{2(1-x)}SiO_4$. This is a solid solution of forsterite Mg_2SiO_4 and fayalite Fe_2SiO_4 with mixing fraction $0 \leq x \leq 1$.
- Pyroxene with composition $Mg_xFe_{1-x}SiO_3$. This is a solid solution of enstatite $MgSiO_3$ and ferrosilite $FeSiO_3$ with mixing fraction $0 \leq x \leq 1$.
- Quartz with composition SiO_2.
- Metallic iron.
- Troilite with composition FeS.
- Carbon dust, which most likely contains significant fractions of H, O, and N. It is not clear if this material is identical with, or similar to, the kerogen observed in carbonaceous chondrites, also a H, O, and N bearing carbonaceous material.

The solid iron and the troilite probably are solid solutions with Ni or NiS, respectively. Compounds of less abundant elements like the very refractory Ca-Al-compounds most likely are also present but have significantly lower abundances and need presently not be considered.

The silicate dust components have an amorphous lattice structure since the two pronounced absorption bands of interstellar dust at $\approx 10\,\mu m$ and $\approx 18\,\mu m$ due to bending and stretching vibrations in the SiO_4 tetrahedron show a structureless bell shaped profile, which is characteristic for amorphous materials. For crystalline materials the bands would show a characteristic structure, which is absent from the interstellar dust absorption spectrum (Kemper et al. [30]). The interstellar silicates therefore are not the true minerals olivine and pyroxene but a material with composition like that of olivine and pyroxene. The lattice structure of quartz probably also is amorphous, though from observations nothing is known about its structure.

This kind of material of interstellar origin is found in the cold outer parts of a protoplanetary accretion disk.

4 Structure and Evolution of Protoplanetary Disks

With regard to our 1- and (1+1)-dimensional models of protoplanetary accretion disks, which follow essentially from averaging the physical quantities (density, ρ, pressure, P, kinematic viscosity, ν, ...) over the vertical (z-) direction and from the assumption that the disk be hydrostatic in the z-direction, respectively, it is most convenient to formulate the basic equations of disk evolution in cylindrical coordinates (r, φ, z). Since we have restricted ourselves to axial symmetry, the physical quantities do not depend on the azimuthal direction, φ, in our models. Up to now this approach has also been good practice in the literature, in particular, concerning the material transport and mixing mainly in the radial (r-) direction. However, a consistent description of these processes in both the radial and vertical (z-) direction is only possible within the framework of at least spatially 2-dimensional models. For a first step in this direction see Section 8.

4.1 Basic Equations

As motivated above, we write our basic model equations expressing the conservation laws of physics in cylindrical coordinates. With axial symmetry the two independent spatial coordinates of the position vector are the radius r and the vertical height z. Denoting by v_r and v_z the velocity components in the radial and vertical direction, the continuity equation which expresses bulk mass conservation then reads

$$\frac{\partial \rho}{\partial t} + \frac{1}{r}\frac{\partial}{\partial r}(r\rho v_r) + \frac{\partial}{\partial z}(\rho v_z) = 0 \ . \tag{1}$$

As already emphasized several times before, protoplanetary "fluids" are necessarily multi-component flows consisting of many gaseous species and embedded microscopically small solid dust particles. In different parts of the disk the concentrations of the various species will differ due to chemical reactions and other processes like combustion of soot particles or condensation and evaporation of the dust particles. Both advective transport and diffusive mixing therefore play a fundamental role in the evolution of protostellar disks. Hence, in addition to the continuity equation (1), for each species, an advection-diffusion equation with source term of the type

$$\frac{\partial}{\partial t}(\rho c_i) + \frac{1}{r}\frac{\partial}{\partial r}(r\rho v_r c_i) + \frac{\partial}{\partial z}(\rho v_z c_i) = \\ \frac{1}{r}\frac{\partial}{\partial r}\left(r\rho D \frac{\partial c_i}{\partial r}\right) + \frac{\partial}{\partial z}\left(\rho D \frac{\partial c_i}{\partial z}\right) + R_i \ , \tag{2}$$

holds. c_i, R_i are the concentration ($\sum_i c_i = 1$) and the source or sink terms ($\sum_i R_i = 0$) of the i-th species, respectively, and D is the diffusion coefficient. More details are discussed in Sect. 6.

As long as relative flow velocities between the various species – in particular, the dust grains – are negligibly small, conservation of momentum is expressed by three equations of motion in the radial and vertical direction (linear momentum) and in the azimuthal direction (angular momentum). If Φ is the gravitational potential and $\eta := \rho \nu$ the dynamical viscosity, and if v_φ denotes the angular velocity, the Navier-Stokes equations are given in explicit form as

$$\frac{\partial}{\partial t}(\rho v_r) + \frac{1}{r}\frac{\partial}{\partial r}(r\rho v_r\, v_r) + \frac{\partial}{\partial z}(\rho v_z\, v_r) =$$
$$\rho\left(\frac{v_\varphi^2}{r} - \frac{\partial \Phi}{\partial r}\right) - \frac{\partial P}{\partial r} + \frac{4\pi}{c}\int_0^\infty \rho\kappa_\nu H_{r,\nu}\, d\nu$$
$$+ \frac{4}{3}\frac{\partial}{\partial r}\left[\eta r \frac{\partial}{\partial r}\left(\frac{v_r}{r}\right)\right] + \frac{2}{3}\frac{\partial}{\partial r}\left[\eta\left(\frac{v_r}{r} - \frac{\partial v_z}{\partial z}\right)\right] + 2\eta\frac{\partial}{\partial r}\left(\frac{v_r}{r}\right)$$
$$+ \frac{\partial}{\partial z}\left[\eta\left(\frac{\partial v_r}{\partial z} + \frac{\partial v_z}{\partial r}\right)\right] \tag{3}$$

for the radial component of the momentum; the equation for its vertical component reads

$$\frac{\partial}{\partial t}(\rho v_z) + \frac{1}{r}\frac{\partial}{\partial r}(r\rho v_r\, v_z) + \frac{\partial}{\partial z}(\rho v_z\, v_z) =$$
$$-\rho\frac{\partial \Phi}{\partial z} - \frac{\partial P}{\partial z} + \frac{4\pi}{c}\int_0^\infty \rho\kappa_\nu H_{z,\nu}\, d\nu$$
$$+ \frac{2}{3}\frac{\partial}{\partial z}\left[\eta\left(2\frac{\partial v_z}{\partial z} - \frac{1}{r}\frac{\partial}{\partial r}(rv_r)\right)\right] + \frac{1}{r}\frac{\partial}{\partial r}\left[r\eta\left(\frac{\partial v_r}{\partial z} + \frac{\partial v_z}{\partial r}\right)\right]. \tag{4}$$

Though of minor importance for protostellar accretion disks radiation pressure is included. It is given via the integral of the product extinction coefficient times the radiative flux over all frequencies $\nu \geq 0$, i.e., the radiation pressure force $\vec{f}_{\rm rad} \propto \int_0^\infty \kappa_\nu \cdot \vec{H}_\nu\, d\nu$, with $\vec{H}_\nu = (H_{r,\nu}, 0, H_{z,\nu})$ having non-trivial components only in the r- and z-direction. More details concerning the treatment of the radiation field are given below.

The third Navier-Stokes equation relates to the conservation of angular momentum:

$$\frac{\partial}{\partial t}(\rho r v_\varphi) + \frac{1}{r}\frac{\partial}{\partial r}(r\rho v_r\, r v_\varphi) + \frac{\partial}{\partial z}(\rho v_z\, r v_\varphi) =$$
$$\frac{1}{r}\frac{\partial}{\partial r}\left[r^3 \eta \frac{\partial}{\partial r}\left(\frac{v_\varphi}{r}\right)\right] + \frac{\partial}{\partial z}\left[\eta \frac{\partial}{\partial z}(rv_\varphi)\right]. \tag{5}$$

If necessary, self-gravitation of the disk is accounted for by solving Poisson's equation

$$\frac{1}{r}\frac{\partial}{\partial r}\left(r\frac{\partial \Phi_{\text{disk}}}{\partial r}\right) + \frac{\partial^2 \Phi_{\text{disk}}}{\partial z^2} = 4\pi G\rho, \tag{6}$$

where G is the gravitational constant and Φ_{disk} denotes the gravitational potential of the disk. The total gravitational potential, Φ, appearing in the Navier-Stokes equations (3) and (4) is the sum of the central protosun's and the disk's potential, that is, $\Phi = \Phi_{\text{sun}} + \Phi_{\text{disk}}$.

Two energy balance equations, one for the matter and the other one for the radiation, together with the equations of state, complete the set of the basic equations. Let ε, ε_{rad} be the specific internal energy of the disk material and the radiation energy density, respectively, $\vec{F}_{\text{rad}} := (F_r, F_\varphi \equiv 0, F_z)$ the (total) radiation flux, κ_ν the extinction coefficient, and $a_\nu, 0 \leq a_\nu \leq 1$, the albedo, the last two quantities being dependent on the frequency of the radiation field, $I_\nu(t, \vec{x}, \vec{\Omega})$, at the time, t, the spatial position, $\vec{x}$, and in the direction $\vec{\Omega}$. Then, in the simplest case of isotropic scattering, the source function of radiation, S_ν, is the sum of the albedo-weighted thermal emission and pure scattering:

$$S_\nu = (1 - a_\nu) B_\nu(T) + a_\nu J_\nu, \tag{7}$$

where

$$B_\nu(T) := \frac{2h}{c^2} \frac{\nu^3}{\exp\left(\frac{h\nu}{k_B T}\right) - 1}$$

is the Planck function at the temperature, T, and

$$J_\nu := \frac{1}{4\pi} \oint_{4\pi} I_\nu \left|d\vec{\Omega}\right|$$

the average of the specific intensity over the unit sphere, i.e., the "zeroth" moment of the radiation field I_ν. Furthermore, the relation

$$\varepsilon_{\text{rad}} = \frac{4\pi}{c} \int_0^\infty J_\nu \, d\nu$$

holds. The radiation flux vector, $\vec{F}_{\text{rad}}$, then is, in this terminology, equal to 4π times the frequency-integrated "first" moment, $\vec{H} = \int_0^\infty \vec{H}_\nu d\nu$ with

$$\vec{H}_\nu = \frac{1}{4\pi} \oint_{4\pi} I_\nu \vec{\Omega} \left|d\vec{\Omega}\right|,$$

and the "second" moment of the radiation field, $\overleftrightarrow{K} = \int_0^\infty \overleftrightarrow{K}_\nu \, d\nu$, with

$$\overleftrightarrow{K}_\nu = \frac{1}{4\pi} \oint_{4\pi} I_\nu \vec{\Omega} \otimes \vec{\Omega} \left|d\vec{\Omega}\right|,$$

relates to the radiation pressure in the form $\overleftrightarrow{P}_{\text{rad}} = (4\pi/c) \overleftrightarrow{K}$. The quantities h, c, and k_B, have the usual meaning of Planck's constant, the speed of light, and Boltzmann's constant, respectively. Now the frequency-integrated balance equations for the radiative energy and flux, being first order accurate with respect to the (small) ratio $|\vec{v}|/c$, read in compact form (see, e.g., [10])

$$\frac{\partial \varepsilon_{\rm rad}}{\partial t} + \vec{\nabla} \cdot (\vec{v}\, \varepsilon_{\rm rad}) + \vec{\nabla} \cdot \vec{F}_{\rm rad} + \overleftrightarrow{P}_{\rm rad} : \vec{\nabla}\vec{v} = -4\pi \int_0^\infty \rho \kappa_\nu \left(J_\nu - S_\nu\right) {\rm d}\nu \quad (8)$$

and

$$\frac{1}{c} \frac{\partial \vec{F}_{\rm rad}}{\partial t} + c\vec{\nabla} \cdot \overleftrightarrow{P}_{\rm rad} = -4\pi \int_0^\infty \rho \kappa_\nu \vec{H}_\nu \,{\rm d}\nu \,, \quad (9)$$

respectively. Here the abbreviation "$\vec{\nabla}\cdot$" means the divergence operator applied to a vector or tensor, yielding a scalar or vector quantity, "$\vec{\nabla}$" is short for the gradient operator yielding a vector or tensor quantity when applied to a scalar or vector. The colon operator, ":", indicates the twofold contraction of the tensor product of two 2-tensors to a scalar, e.g., the contraction of the radiation pressure tensor, $\overleftrightarrow{P}_{\rm rad}$, with the gradient of the velocity field, $\vec{\nabla}\vec{v}$ (i.e., the covariant derivative of $\vec{v}$), in Eq. (8) above.

Finally, the first law of thermodynamics governing the balance of the internal energy is expressed as

$$\frac{\partial}{\partial t}(\rho \varepsilon) + \frac{1}{r}\frac{\partial}{\partial r}(r\rho v_r \varepsilon) + \frac{\partial}{\partial z}(\rho v_z \varepsilon) + P\left[\frac{1}{r}\frac{\partial}{\partial r}(rv_r) + \frac{\partial v_z}{\partial z}\right] = \\ 4\pi \int_0^\infty \rho \kappa_\nu (J_\nu - S_\nu)\,{\rm d}\nu + Q_{\rm vis} + Q_{\rm chem} \quad (10)$$

where the source term on the right hand side consists of three contributions, the radiation term, the energy dissipation term due to (turbulent) viscosity, and the "thermochemical" term arising from the chemical reactions. The two (algebraic) equations of state

$$P = P(\rho, T), \qquad \varepsilon = \varepsilon(\rho, T) \quad (11)$$

relate the gas pressure, P, and the internal energy, ε, with the mass density, ρ, and the temperature, T, respectively. Aside from model-dependent initial and boundary conditions, the system of equations is now complete.

A few remarks concerning the various approaches of how radiative transfer is treated in the various model calculations should be made:

1. The time derivatives of the radiative flux, $\vec{F}_{\rm rad}$, on the left hand side of Eq. (9) are always small enough to be entirely neglected for practical purposes. This is reasonable, since the overall evolution of protoplanetary disks is expected to proceed rather smoothly and quasi-stationary.
2. The whole modeling efforts then focus on the determination of the radiation pressure, $\overleftrightarrow{P}_{\rm rad}$. This can be accomplished to varying degrees of approximation.

(a) *The Eddington-factor method*: Go back to an appropriate version of the basic radiative transfer equation (which is discussed by Meinköhn [37] in this book) for the specific intensity, I_ν. From the solution compute the 0th, 1st, and 2nd moment of I_ν, that is, J_ν, $\overrightarrow{H}_\nu$, and $\overleftrightarrow{K}_\nu$, respectively. Define the (symmetric) Eddington-factor tensor, $\overleftrightarrow{f}$, by the ratio $\overleftrightarrow{f} := \int_0^\infty \overleftrightarrow{K}_\nu \, d\nu / \int_0^\infty J_\nu d\nu$ and close the moment equations (8) and (9) by setting $\overleftrightarrow{P}_{\text{rad}} = \overleftrightarrow{f} \, \varepsilon_{\text{rad}}$.

(b) *The Eddington approximation* holds in optically thick regions of the disk. The radiation field there can be assumed to be fairly isotropic, leading to an isotropic Eddington factor characterized by one single number $f = 1/3$ and, hence, to an isotropic radiation pressure $\overleftrightarrow{P}_{\text{rad}} = (\varepsilon_{\text{rad}}/3) \overleftrightarrow{e}$, with $\overleftrightarrow{e}$ denoting the unity tensor.

3. In both cases (a) and (b) the integral on right hand side of Eq. (9) is replaced by introducing flux-averaged opacities, $\kappa_{\overrightarrow{H}} \overrightarrow{H} := \int_0^\infty \kappa_\nu \overrightarrow{H}_\nu d\nu$ (component-wise). If local thermodynamical equilibrium can be assumed, the stronger relation, $J_\nu = B_\nu(T)$ holds and the Rosseland mean of the extinction coefficient, κ_ν, is the correct flux-averaged opacity.

4. In a similar fashion the Planck mean, κ_p, of the "true" absorption coefficient is obtained: $\kappa_P := \int_0^\infty \kappa_\nu \cdot (1 - a_\nu) B_\nu(T) \, d\nu / \int_0^\infty B_\nu(T) \, d\nu$. Finally, the mean value of the absorption coefficient with respect to the radiation energy density, κ_J, is defined as $\kappa_J := \int_0^\infty \kappa_\nu \cdot (1 - a_\nu) J_\nu \, d\nu / \int_0^\infty J_\nu \, d\nu$. Both mean values, κ_P and κ_J are needed to solve the radiation energy equation (8).

5. The simplest, most popular and thus most frequently used, though rather rough, approximation is to restrict oneself to the so-called grey case. Then, by definition, both the extinction coefficient and the albedo are independent of the frequency. The Rosseland mean and Planck mean are usually taken as the respective substitutes.

Existing disk models are based on several further approximations, in particular, if emphasis is laid on the investigation of the complex chemical and mineralogical processes in protoplanetary disks. One salient feature of accretion disks is the fact that they are geometrically thin ("pancake"-like) objects. Hence, the simplest way to construct disk models is to use quantities that are, together with the structure equations, properly averaged over the vertical (z-) direction. The simplest non-trivial disk models are the spatially 1-dimensional ones. Only the radial coordinate, r, is retained, the z-dependence is completely eliminated and thus all relations referring to the vertical direction are omitted. Since the viscous timescale is, in general, orders of magnitude larger than the dynamical ("Keplerian") revolution time, the Keplerian rotation law can be assumed to be established at every instant of time. The slow radial drift is completely governed by the efficiency of turbulent friction in redistributing angular momentum. So, in this 1-dimensional approach only the detailed conservation of mass and angular momentum has to be observed. Nevertheless, assured a thermally steady state, it is even possible to define a radial distribution of an "effective" temperature as an equivalent measure of the radiative flux by demanding this flux which is emitted from the disk's surface to be equal to the locally dissipated energy within the disk. This temperature definition in the 1-dimensional approximation relies entirely on an energy balance argument, no detailed radiative transfer needs to be solved. Incidentally, here and in the following, too, it is tacitly assumed that energy transport takes place exclusively in the vertical direction where the gradients are much steeper than in the radial direction.

Within the framework of the so-called (1+1)-approximation the disk is assumed to be vertically in hydrostatic equilibrium. The basic evolutionary equations are, as in the pure 1-dimensional case, the z-averaged Eqs. (1) and (5) for mass and angular momentum conservation. The method of determining the hydrostatic vertical stratification of the density, temperature, and pressure at each radial distance is very much the same as that used for stellar structure calculations. Of course, to be able to proceed in this manner, gradients in the radial direction are neglected. The connection between the basic 1-dimensional (r-) time evolution modeling and the z-hydrostatic assumption is the obvious demand that the z-integrated density, i. e., the column density, is always equal to the surface density as a result of the 1-dimensional evolutionary equations.

First steps toward constructing fully 2-dimensional models are discussed in Sect. 8. Of course, making most realistic 3-dimensional disk models available will be the final goal of all our efforts. For the time being, however, conducting such an ambitious venture would demand an amount of computer power that exceeds by far the one available today.

5 Dust Metamorphosis

Viscous torques in the accretion disk induce a slow inward migration of the disk material by which the material is transported from the cold outer disk into increasingly warmer zones. Additionally, turbulent mixing in the convectively unstable parts of the disk intermingles material from different disk regions. At sufficiently high temperature, solid diffusion and annealing processes are activated in the dust grains which tend (i) to exsolve impurities from the grain lattice and (ii) to form a regular crystalline lattice structure (Duschl, Gail, Tscharnuter [12]; Gail [17]; Hallenbeck, Nuth & Daukantas [24]; Fabian et al. [14]). The grain material develops from the dirty and amorphous composition, which is responsible for the observed extinction properties of interstellar grains, to a more clean and crystalline lattice structure. The basic chemical composition of individual grains as given in Table 1, however, is not changed by these processes. The assemblage of minerals forming the dust mixture essentially remains preserved at this stage, since there operate only

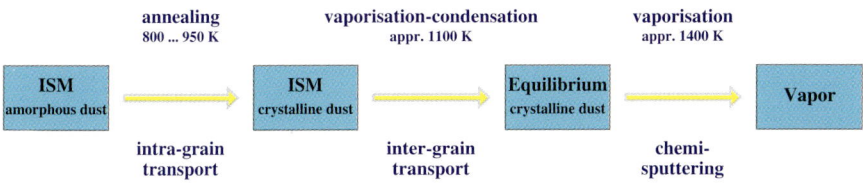

Fig. 3. Schematic representation of the metamorphic processes suffered by dust particles in the accretion disk.

intra-grain transport processes by solid diffusion, but no *inter-grain* transport processes via the gas phase.

This change in the lattice structure of the dust grains occurs in the region where the timescale for solid diffusion and annealing roughly equals the timescale for radial inward migration of the disk material. The model calculations in Gail [18] and Wehrstedt & Gail [52, 53] have shown this to occur at about 800 K for glassy silicates. If the structure of the silicate dust is more similar to that of the smokes prepared in the annealing experiments of Hallenbeck et al. [24] one determines from the data in Hallenbeck, Nuth & Nelson [25] a somewhat higher annealing temperature of about 950 K (cf. Gail [18]). At temperatures above the annealing temperature the grains have a crystalline lattice structure, which at the same time increases their stability against evaporation.

At a temperature significantly higher than that required for annealing, transport processes between the surfaces of different grains via the gas phase start to operate (evaporation and re-condensation). These *inter-grain* transport processes via the gas phase tend to remove the thermodynamically less stable condensed components from the interstellar medium (ISM) mineral mixture in favor of the thermodynamically most stable materials. Since the ISM dust mixture of the outer disk contains dust species which have been formed in chemically such diverse environments as circumstellar shells of AGB-, RSG-, WC-stars, and supernovae, or by destruction and re-condensation behind shocks in the ISM, part of the mineral components of the mixture are thermodynamically unstable for the element mixture of the protoplanetary disk material. The element mixture in the protoplanetary disk results from mixing material from a big number of different sources with widely different element compositions. Though the dust grains from the different sources were completely stable in the special environment in which they originally had been formed, part of them are unstable in the resulting average element mixture of the protoplanetary accretion disk. Grains consisting of such unstable materials slowly vanish by evaporation or chemi-sputtering while their material is transported via the gas phase and finally is consumed by the growth of those mineral grains which are the most stable ones for the average element

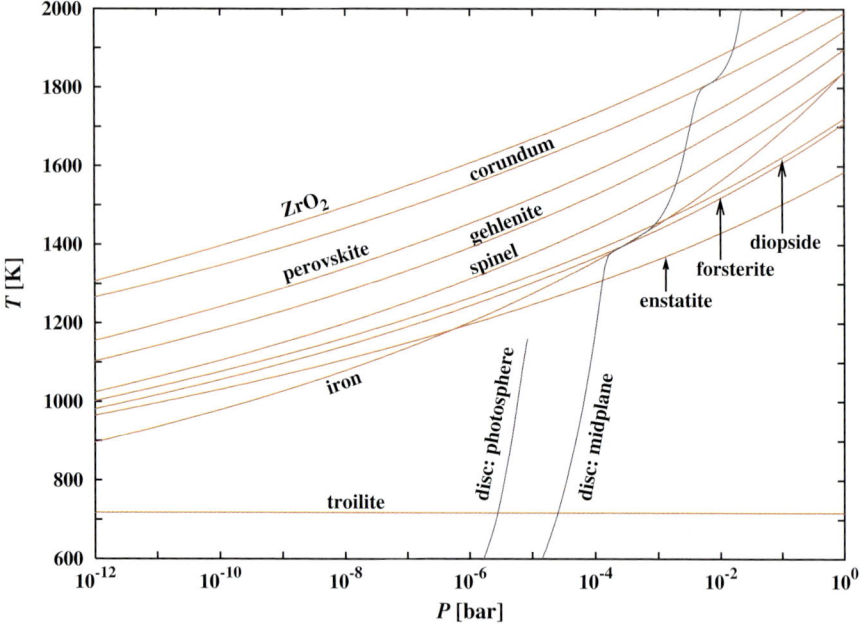

Fig. 4. Stability limits in the P-T plane of the minerals formed by the most abundant refractory elements Si, Mg, Fe, Al, Ca and of Ti and Zr in a Solar System element mixture. The blue lines are P-T trajectories corresponding to the photosphere at $\tau = \frac{2}{3}$ (left) and the midplane of a stationary protoplanetary accretion disk (right) of a solar like protostar with an accretion rate of $10^{-7}\,M_\odot\,\mathrm{yr}^{-1}$.

mixture in the protoplanetary disk and at the local temperature and pressure conditions.

These processes, i. e., evaporation, transport via the gas phase, re-condensation, drive the non-equilibrium mineral assemblage of the dust originating from the parent molecular cloud toward an equilibrium composition that corresponds to the average element mixture in the protoplanetary disk. This change in the dust composition occurs in the zone where the timescale for this kind of chemical metamorphism is comparable with the timescale of radial inward migration of the disk material. Inside this zone, the composition of the mineral assemblage corresponds to a chemical equilibrium state.

Figure 3 gives a schematical sketch of the dust metamorphosis in a protostellar disk and the processes involved.

5.1 Dust Mixture in the Inner Disk

The mineral mixture existing in chemical equilibrium for standard cosmic element abundances and the stability limits against evaporation/decomposition of the most abundant minerals in a chemical equilibrium mixture are shown in Fig. 4. In the warm disk zone the most abundant rock forming elements (Mg, Si, Fe) form the following mineral species:

- Forsterite with composition Mg_2SiO_4.
- Enstatite with composition $MgSiO_3$.
- Solid iron (which forms a solid solution with all the available Ni).
- Troilite with composition FeS (which forms a solid solution with NiS).

The mole fraction of the Fe-bearing end members of the silicate solid solutions olivine and pyroxene is negligibly small in chemical equilibrium at temperatures well above 500 K.

There are two significant differences in the chemical composition between the equilibrium mineral mixture in the inner disk and the ISM dust mixture in the outer disk proposed by Pollack et al. [41]. First, since the average element mixture of the disk is oxygen rich there exists no carbon dust in the equilibrium mixture. The carbon dust is destroyed during the evolution toward the equilibrium mixture. Second, the silicate dust components olivine and pyroxene in the ISM are iron rich, while in chemical equilibrium the silicates are nearly iron free and can be assumed to be essentially pure forsterite and enstatite. Additionally, quartz is absent in chemical equilibrium for the standard cosmic element mixture.

The lattice structures of the silicates in both cases also are completely different. The ISM dust has an amorphous lattice structure and probably a considerable fraction of impurities is build into the lattice. In chemical equilibrium, however, the substances are crystalline and the concentration of impurities solved in the solids is small.

5.2 The Three Main Dust Mixtures

The transport, diffusion, annealing, evaporation, and re-condensation processes in a protoplanetary accretion disk result in the existence of *three completely different dust mixtures in the inner, middle and outer part of the accretion disk:*

1. Amorphous dust with a strong non-equilibrium composition in the cold outer parts of the disk.
2. A mixture of crystalline dust with a strong non-equilibrium composition and of equilibrium dust from the inner disk zone. This mixture is found in a certain zone of the inner parts of the disk.
3. Crystalline dust with chemical equilibrium composition in the innermost parts of the disk down to the evaporation limits of the solids.

Carbon dust is only part of the first two mixtures. The equilibrium mixture is free of carbon dust.

In the first mixture the isotopic mixture of the elements and the abundances of impurities in the individual grains may be very different since the

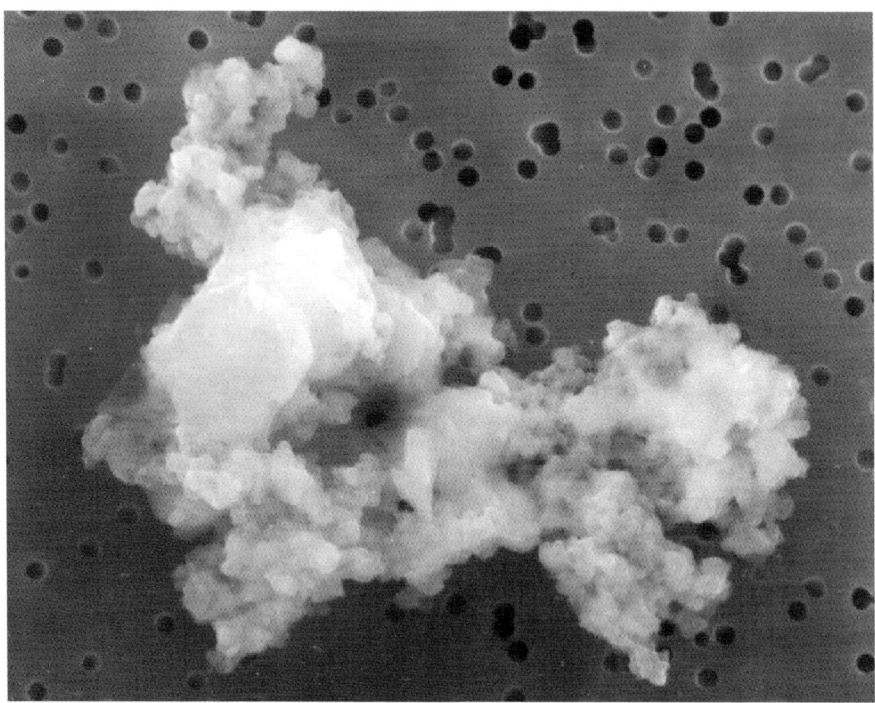

Fig. 5. An interplanetary dust particle (IDP). The typical size of such particles is about $10\,\mu$m, the sub-units are of $\approx 100\,n$m size. Such particles are collected by specially prepared high flying aircrafts in the stratosphere. The IDPs are almost certainly dust agglomerates from the protoplanetary accretion disk from which the solar system emerged, which are preserved in cometary nuclei. They are liberated during the flyby of comets at the sun. They contain a complex mixture of minerals and refractory carbonaceous material from the warm inner zone of the accretion disk, and of interstellar dust. (From NASA's stardust web-page).

grains preserve the vastly different isotopic and elemental compositions of their formation sites. In meteorites they correspond to the presolar grains.

The grains of the second mixture are already processed. They probably lost all or nearly all of their noble gas and other highly volatile element content by outgassing at high temperatures. Also the isotopic anomalies or their impurity content may be partially erased by internal diffusion and exchange processes with the gas phase. In meteorites, this dust component will hardly be recognized as grains of interstellar origin.

The third component is home-made dust of the accretion disk. The isotopic composition of individual grains corresponds to the average isotopic mixture of the disk material after complete mixing.

The three different mixtures do not exist separately in clearly distinct zones of the disk. Radial mixing processes transport material from inner disk regions into outer disk regions such that one finds some fraction of annealed ISM dust and dust with chemical equilibrium composition also in the outer disk. The true composition of the dust mixture, thus, is rather complex, even if only the most abundant dust species are considered.

Dust particles from the formation time of the solar system are conserved in cometary nuclei and can be studied in the laboratory even today. A picture of such a grain is shown in Fig. 5.

The opacities of these dust mixtures are strongly different, the opacity of the equilibrium mixture being lower than that of the ISM mixture by a factor of more than ten. The reason is that the carbon in the ISM mixture accounts for nearly one half of the total opacity, that the opacity of crystalline material usually is much lower than that of amorphous material, and that iron poor silicates have a lower opacity than iron rich silicates. The strong change of the opacity by the evolution of the dust mixture from the ISM mixture to the equilibrium mixture clearly has significant consequences for the disk structure.

A realistic modeling of the structure and evolution of protoplanetary accretion disks and of the composition of the mineral mixture in the disk needs to include in the model computations the gradual change in dust composition and lattice structure

- from the ISM mixture of grains of interstellar origin, encountered in the cool outer parts of the disk,
- through the intermediate crystalline non-equilibrium mixture
- into the chemical equilibrium mixture existing in the warm inner parts of the disk.

6 Transport and Mixing Processes in Accretion Disks

6.1 Observational Indications for Mixing Processes

There are a number of observational indications that the early solar nebular material was rather well mixed over length scales extending over tens of AU. One observation, which was the first hint for the existence of large scale mixing, is the homogeneity of the isotopic abundance ratios of most elements as it is determined for the Earth, for material returned from the Moon, and for meteorites from Mars and several regions of the Asteroid belt. There are some exceptions which have not yet fully been explained, e.g., the small but systematic varying $^{16}O/^{17}O$ and $^{16}O/^{18}O$ isotopic excesses across the inner part of the Solar Nebula, which seemingly cannot be explained as normal isotopic effects of chemical reactions (e.g., Thiemes [48]). These raised some doubts on the efficiency of mixing processes in the solar nebula, but possibly they, in fact, may be explained by isotopic effects of a particular class of chemical reactions (Thiemes [49]). Since infrared spectral observations, both from ground based instruments and from outside the earth's atmosphere with the

ISO satellite, became possible, observations of dust from cometary nuclei and of accretion disks around newly formed stars provided new and strong hints for the existence of large scale mixing processes in accretion disks.

For cometary dust, analysis of the profile of the 10 μm silicate dust feature of comets (Hanner et al. [26]) and of the far infrared ISO spectrum of comet Hale-Bopp (Wooden et al. [55, 56]) showed the silicate dust to be an amorphous silicate material with an admixture of up to 30 % of crystalline silicates. Comets are left-over bodies from the time of planetary formation which must have accumulated in a very cold region of the Solar Nebula, since they contain frozen gases with very low evaporation temperatures. During the past 4.6×10^9 yr they have never become hotter than about 30 K, since otherwise their significant Ar content would have been outgassed. The crystalline silicates in comets therefore also must have been present in the formation zone of the comets. On the other hand, the molecular cloud material which builds up the outer regions of the accretion disk does not contain observable quantities of crystalline material. It seems also completely unlikely that there exists a process which converts amorphous material into crystalline one at temperatures well below 100 K. The only plausible explanation for the presence of copious amounts of crystalline material in cometary nuclei is that this material was formed under high temperature conditions in the inner parts of the disk and subsequently mixed outward, though shocks have also been proposed as the origin of the crystalline material (Harker & Desh [27]).

Analysis of infrared spectra from accretion disks around protostars also revealed the existence of crystalline material within them (e. g., Bouwman [5]). Nuth [40] pointed out that one seems to observe an evolutionary sequence in the sense that older disks contain more crystalline material than very young disks. This is just what one expects if crystalline material is formed in the hot central parts of the disk and then is slowly transported outward. The present day instruments allow only to observe the outer regions of accretion disks, while the innermost parts are invisible due to insufficient instrumental resolution. As a result, the crystalline material becomes visible only if it arrives in the cold outer parts of the disk.

On the nature of the mechanism responsible for the large scale mixing can presently only be speculated. In principle two types of processes are possible:

- Mixing by turbulent flows, as they are likely to exist in protoplanetary accretion disks.
- Large scale circulations flows, as they are common in rotating systems.

Both types of mixing processes were studied in this project of the SFB.

6.2 Turbulent Mixing and Transport in Accretion Disks

Turbulent mixing processes of dust in protoplanetary accretion disks have been studied by Morfill & Völk [38], Gail [18, 21], Wehrstedt & Gail [52, 53], Bockelée-Morvan et al. [4] within the approximation of one-zone Keplerian

α-disk models, which are believed to describe with reasonable accuracy the vertically averaged structure of disks in late evolutionary phases where the planetary formation process is assumed to commence. The main purpose was to study the transport of dust from the hot inner zones of the disk where crystalline material is formed to the cold outer parts of the disk where the nuclei of comets are formed.

In late evolutionary phases accretion disks are flat rotating structures. Turbulent flows in such disks mix the dusty gas between different disk regions in the radial and vertical directions, while the Keplerian shear flow serves for a rapid mixing in the azimuthal direction. The vertical extension of a disk at some distance r from the star is small compared to r. For this reason turbulent mixing time scales for vertical mixing are much shorter than for radial mixing. It is then assumed that the disk matter is well mixed in the vertical and azimuthal direction such that there are no concentration gradients of species in these two directions. Only the much slower radial mixing is studied.

The basic equation for the particle density n_i of some tracer of kind i in cylindrical coordinates and assuming rotational symmetry is (cf. also equation 2)

$$\frac{\partial n_i}{\partial t} + \frac{1}{r}\frac{\partial}{\partial r}r\left(n_i v_{i,r} - nD_{i,r}\frac{\partial c_i}{\partial r}\right) + \frac{\partial}{\partial z}\left(n_i v_{i,z} - nD_{i,z}\frac{\partial c_i}{\partial z}\right) = R_i, \quad (12)$$

where $v_{i,r}$, $v_{i,z}$ and $D_{i,r}$, $D_{i,z}$ are velocities and turbulent diffusion coefficients of tracer i in r and z direction, respectively, and c_i is the concentration of tracer i with respect to the carrier gas of density n. R_i are formation or destruction rates of tracer i. In order to obtain the tracer transport equation in the radial direction in the one zone approximation one has to integrate over the vertical z-direction. Assuming no infall of matter from outside one obtains after some manipulations

$$\frac{\partial \bar{c}_i}{\partial t} + \bar{v}_{i,i}\frac{\partial \bar{c}_i}{\partial r} = \frac{1}{hn_0 r}\frac{\partial}{\partial r}hn_0\bar{D}_{i,r}\frac{\partial \bar{c}_i}{\partial r} + \frac{\bar{R}_i}{\bar{n}}. \quad (13)$$

The bar denotes averages with respect to the vertical direction. n_0 is the density of the carrier gas in the midplane of the disk and h is the equivalent height of the disk, i.e., the vertical thickness of a disk with constant density n_0 in the vertical direction which has the same vertically integrated surface density as that calculated with the true density distribution

$$hn_0 = \int_0^\infty dz\, n(z). \quad (14)$$

For turbulent mixing the diffusion coefficient D is related to the turbulent viscosity ν by the relation

$$D_{i,r} = \frac{\nu}{Sc} \quad (15)$$

with the Schmidt number Sc (e.g., McComb [36]) with typically $Sc \approx 0.7$. For ν the viscosity from the α-disk model is used which also describes the viscous

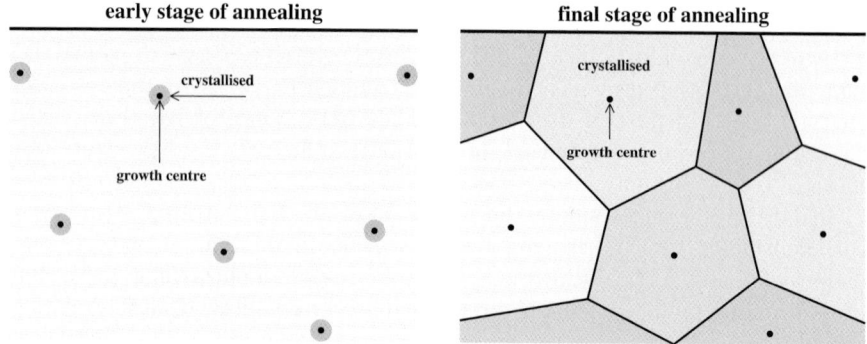

Fig. 6. Schematic representation of the annealing process.

accretion of the disk. For the velocity $v_{i,r}$ it is assumed that this equals the velocity with which the disk matter slowly moves toward the star (accretion).

The application of Eq. (15) with a fixed number Sc and the identification of $v_{i,r}$ with the accretion flow velocity limits the calculation of tracer transport to species which are dynamically closely coupled to the gas, e. g., gas particles or small dust particles with sizes less than about 1 cm. Bigger particles are decoupled from the gas and their velocities and diffusion coefficients have to be determined, for instance, as described in Schräpler & Henning [44].

The equations (13) for the transport, diffusion, and reaction of the species of interest have to be solved simultaneously with the set of equations for the structure and evolution of the accretion disk. Different processes relevant for protoplanetary disks have been treated in this way: annealing of amorphous dust, combustion of carbon dust grains, and evaporation and condensation of dust.

6.3 Annealing of Amorphous Dust

The dust formed under astrophysical conditions often has an amorphous lattice structure, where the basic building blocks of the minerals, e. g., the SiO_4-tetrahedrons in the silicates, show a disordered arrangement in the solid. In fact, most of the cosmic silicate dust does not show any evidence for crystallinity (e. g., Kemper et al. [30]). In some cases, however, if the material is heated for sufficiently long times to a sufficiently high temperature, internal rearrangement processes are activated by which the material evolves into a regular crystalline lattice structure (annealing). Starting from some crystallization centers, crystallized regions grow into the amorphous surrounding until the crystallized regions coalesce. This process is schematically indicated in Fig. 6. Such annealing of the cosmic dust is thought to occur if the disk material slowly moves from the cold outer regions into the hot inner parts of the accretion disk.

Annealing has been studied in the laboratory for the most important dust components formed in an oxygen rich environment by a number of groups, e.g., Hallenbeck, Nuth & Daukantas [24], Brucato et al. [7], Fabian et al. [14]. These investigations are reviewed in Gail [20] and Colangeli et al. [11]. For some of the relevant mineral structures the annealing properties are now known from the laboratory investigations and this can be used to model the dust annealing in accretion disks.

The annealing of the amorphous silicate grains in protoplanetary disks and its implications for the disk structure is discussed in Duschl et al. [12], Gail [17, 18] and Wehrstedt & Gail [52, 53, 54], and Bockelée-Morvan et al. [4]. In all the model calculations the annealing process is modeled by solving an equation for the growth of a crystallized region, described by its volume $V_{\rm cr}$, starting from some growth center. The equation for the growth in the most simple case is

$$\frac{dV_{\rm cr}}{dt} = 6V_0^{\frac{1}{3}} V_{\rm cr}^{\frac{2}{3}} \nu_{\rm vib} e^{-E_a/kT}. \tag{16}$$

$\nu_{\rm vib}$ is a characteristic vibration frequency in the lattice and E_a is the activation energy for crystallization growth. V_0 is defined as the volume of one SiO_4 tetrahedron in the lattice. Details are described in Gail [17]. A similar equation has been used by Sogawa & Kozasa [46].

It is shown in Gail [18] how this simple model can be used to determine the distribution function of the degrees of crystallization for an ensemble of dust grains. One defines a grid $0 \leq x_i \leq 1$ $(i = 1, \ldots, I)$ of totally I discrete degrees of crystallization for the dust grains. The gain and loss term in the advection-diffusion-reaction equation (13) for the concentration c_i of grains with a degree of crystallization between x_i and x_{i+1} is shown in Gail [18] to be given by

$$R_i = \left(\frac{c_{i-1}}{x_i - x_{i-1}} - \frac{c_i}{x_{i+1} - x_i} \right) \frac{dV_{\rm cr}^{\frac{1}{3}}}{dt} \tag{17}$$

where $dV_{\rm cr}/dt$ is defined by (16). A solution of the system of differential equations (13) with the appropriate boundary conditions (all grains amorphous in the cold disk, all grains crystalline in the warm inner part of the disk) then yields the probability distribution c_i for the different degrees of crystallization from which the average degree of crystallization $x_{\rm cr}$ can be determined.

Calculations have been performed for stationary disk models by Gail [18], for time dependent models by Wehrstedt and Gail [52], and for simple semi-analytic disk models and assuming only two states (either crystalline or amorphous) by Bockelée-Morvan et al. [4]. Some results are shown in Fig. 7 for a time dependent model. The model calculations that annealing occurs at temperatures of about 800 K for olivine and pyroxene dust grains, and at about 900 K for quartz grains. The low average annealing temperature found in the calculations is important in so far, as dust destruction processes by evaporation occur at a much higher temperature. Both processes, crystallization and evaporation, occur in different zones of a protoplanetary disk.

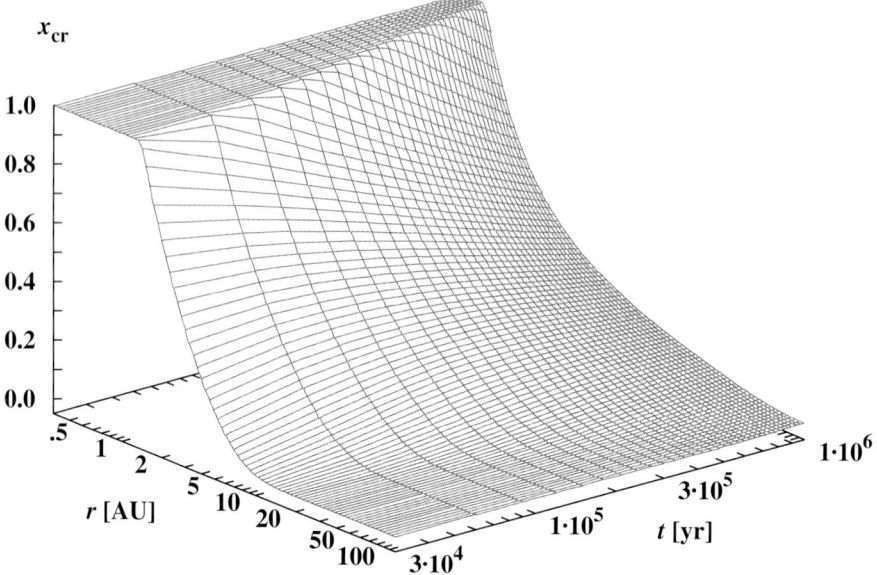

Fig. 7. Radial and temporal variation of the degree of crystallization $x_{\rm cry}$ of olivine grains due to annealing and radial mixing of crystallized grains for a time dependent disk model (Wehrstedt & Gail [52]). The model calculation shows that crystalline dust is mixed into fairly distant outer regions of the disk during its evolution. At the same time the inner zone of completely crystallized dust shrinks as the disk is depleted and generally becomes cooler.

An important result of the model calculations is that crystalline dust formed at about 800 K is transported by diffusion into cool outer parts of the disk, even into the region between Jupiter and Saturn (5 ... 10 AU) where it may be incorporated into the planetesimals formed in the cold outer disk. Probably this mechanism is responsible for the considerable fraction of crystalline silicates observed to exist in cometary nuclei (e. g., Wooden et al. [55, 56, 57]).

6.4 Combustion of Carbon Dust

The interstellar dust mixture contains a significant fraction of carbon dust material (see Table 1). This material is unstable in the oxygen rich element mixture of the protoplanetary disk; in chemical equilibrium the carbon would be present as CO molecule. Chemical reactions between the carbonaceous dust material and the gas phase then tend to convert the solid carbon into gaseous CO. The basic processes for the conversion of the carbon dust into gas phase species have been studied in Duschl et al. [12], Bauer et al. [2], Finocchi et al. [15, 16]. The basic reaction step initializing the oxidation process is the attack of a OH radical with a six-ring at the periphery of a large PAH:

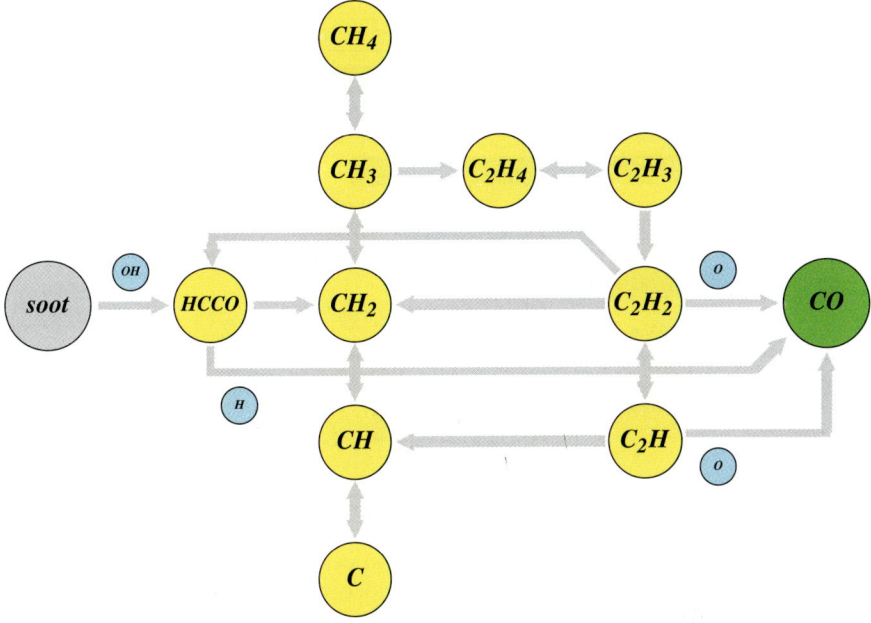

Fig. 8. Gas phase reaction for the conversion of carbon dust into CO.

$$\text{(structure)} + \text{OH} \longrightarrow \text{(structure)} + \text{HCCO} \tag{18}$$

The ketenyl radical HCCO then triggers a sequence of gas phase reactions which ultimately ends up with CO. The essential reaction steps are indicated in Fig. 8. The first step is the reaction

$$\text{HCCO} + \text{H} \longrightarrow \text{CO} + \text{CH}_2$$

which results in the prompt formation of one CO molecule. The second C atom then requires a rather complex reaction sequence for the formation of the second CO molecule. The carbon combustion occurs in the accretion disk at temperatures slightly above 1 000 K. This is at much lower temperature than the combustion temperature of carbonaceous material under terrestrial conditions. This low temperature results from the long timescale available for the process in an accretion disk (about 10^3 yrs).

As a by-product of these gas phase reactions a considerable concentration of CH_4 and C_2H_2 molecules builds up in the exhaust gases. The products of the carbon combustion are mixed by transport and diffusion processes from the

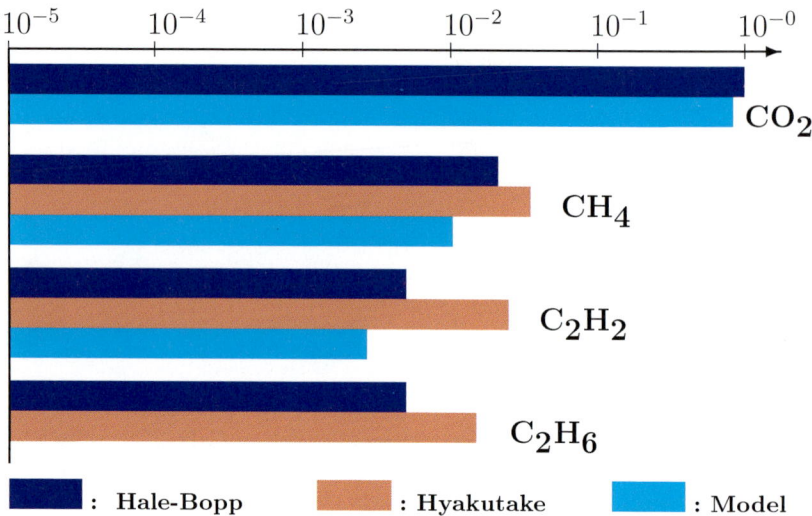

Fig. 9. Comparison of calculated (Gail [19]) and in comets Hale-Bopp and Hyakutake observed molecular abundances of the products of carbon combustion. Concentrations are relative to CO.

warm region of the disk into the cold outer regions. This process was modeled in Gail [19] and it was found that by this process a significant concentration of CH_4 and C_2H_2 is build up in the region of the accretion disk where the nuclei of comets are formed. These gases will freeze out together with water vapor and may be found in comets. Figure 9 compares the result of the model calculation with observed abundances of some molecular species in two recent comets. The predicted and the observed abundances relative to CO are in reasonable agreement for CO_2, CH_4, and C_2H_2 which indicates that at least part of the high concentrations of hydrocarbons found in cometary nuclei may result from mixing the exhaust gases of carbon dust combustion into the cold outer parts of the disk. The discrepancy between a high observed abundance of C_2H_6 and a calculated very low abundance in the exhaust gases shows, that additional chemical processes determine part of the abundances of the hydrocarbons. A more detailed modeling of the chemistry considering mixing processes and the ion-molecule and dust-surface chemistry in the cold parts of the disk is required.

6.5 Large-Scale Flows

In models of the chemistry and of transport processes in the disk only diffusive transport has been considered in the past. In real disks there may exist, however, large scale circulation currents, which may serve for a much more efficient radial transport of matter in the disk than diffusional transport. Urpin [50]

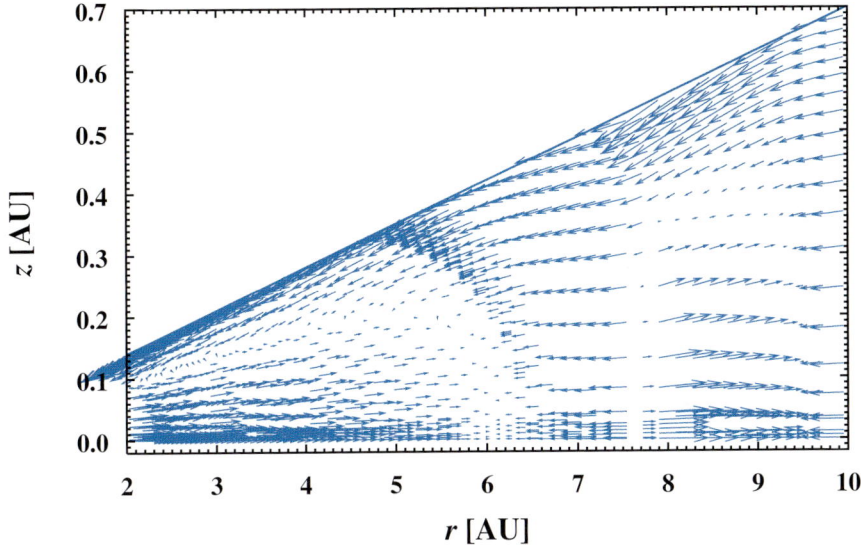

Fig. 10. 2-dimensional flow field in the disk at $t = 10^5$ yr between 2 and 10 AU. The solid line marks the upper boundary of the polar grid ($\theta_1 = 4.01°$). For details see text.

has shown that such flows exist and that mass-inflow in the disk by accretion is not simply an inward directed flow but that in fact only the net flow is inward directed. Actually the flow is outward directed in a thick layer containing the midplane and is inward directed in higher layers. This has later also be found, for instance, in numerical 2D hydrodynamic calculations (e.g., Kley & Lin [32]). In Keller & Gail [29] and Wehrstedt & Gail [54] the effect of such large-scale circulation flows on mixing material between different regions of the accretion disk is studied.

The analytic solutions of Urpin [50] and Takeuchi & Lin [47] are generalized (Keller & Gail [29] and Wehrstedt & Gail [54]) and the resulting semi-analytic model for the flow structure is implemented in a time dependent (1+1)-dimensional disk model. Figure 10 shows the flow structure of such large scale flows which are outward directed close to the midplane and inward directed in higher layers. The details of the flow structure are significantly influenced by the complicated temperature structure of the disk.

Transport of tracers has been studied for this type of flow field combined with diffusional transport. The equations (12) for transport, diffusion and reaction are solved for the problems of silicate dust annealing and carbon combustion. Some results are shown in Fig. 11. The calculations show that after about 10^6 yrs one has a significant fraction of annealed crystalline dust transported outward into the region between Jupiter and Saturn where the Oort cloud comets are believed to have been formed 4.6×10^6 years ago. The

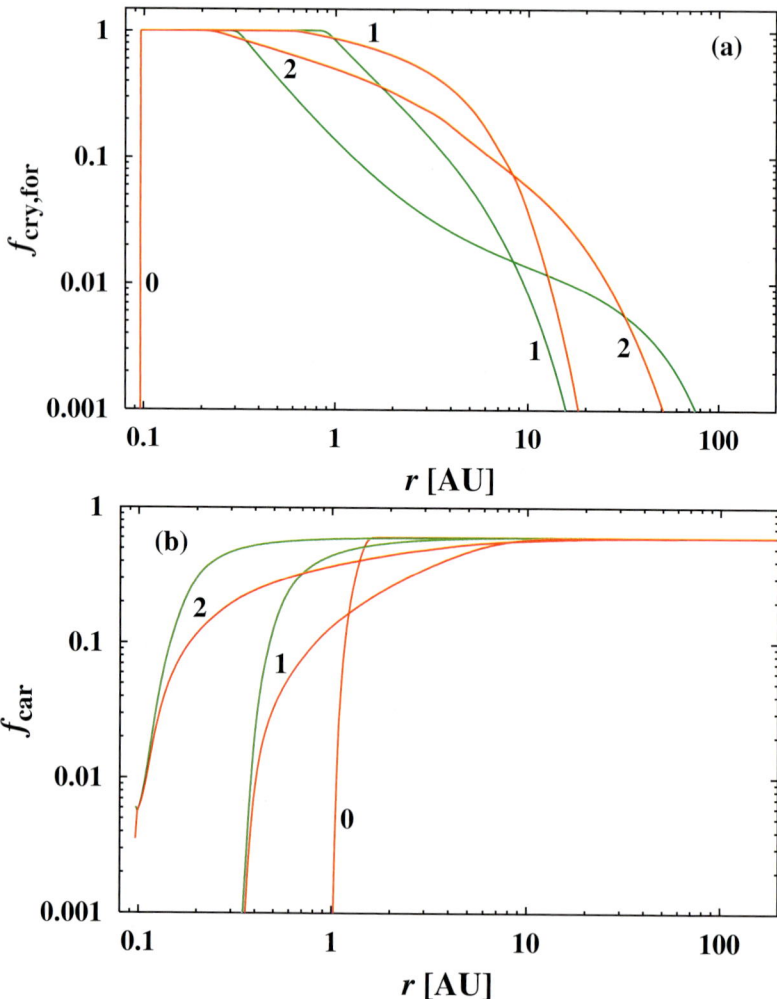

Fig. 11. Tracer transport in the (1+1)-dimensional model with the meridional flow field (red lines) and for comparison with the one-zone velocity field (green lines) at times $t = 0$ (0), 10^5 (1) and 10^6 yr (2). Models of Wehrstedt & Gail [54]. **(a)** Degree of crystallization of forsterite $f_{\rm cry,for}$. **(b)** Fraction of C condensed into solid carbon $f_{\rm car}$.

concentration of the crystalline dust is much higher if one considers the large scale circulations as if only diffusional transport alone is considered. This is in better agreement with observations which find a rather high degree of crystallinity of silicate dust in comets (e.g., Wooden et al. [55, 56, 57]).

7 Chemical and Mineralogical Evolution of the Dust

7.1 Vaporization of the Interstellar Dust Component

It is important to note that the pristine silicate dust components olivine, orthopyroxene and quartz are not stable in an environment with standard cosmic element abundances:

1. Quartz does not exist in chemical equilibrium in such an environment because the vapor pressure of SiO molecules is much higher for quartz than for olivine and pyroxene. In an equilibrium state the silicon is condensed into forsterite and enstatite and not into quartz.
2. The interstellar olivine and orthopyroxene are assumed to have a rather high Fe/(Fe+Mg) ratio (cf. Table 1), while in chemical equilibrium the iron content is low. Since the upper stability limit of the iron-magnesium silicates considerably decreases with increasing iron content, the iron rich interstellar silicates are unstable with respect to evaporation and re-condensation of their material into the iron poor silicates forsterite and enstatite and into solid iron.

Therefore, if accretion transports the dust material of interstellar origin into warm inner disk regions where the dust starts to evaporate, the interstellar dust mixture is converted into a mixture corresponding to chemical equilibrium which is different from the interstellar dust mixture.

Due to annealing the silicates of interstellar origin have a crystalline lattice structure at the instant when they start to evaporate. This allows to use thermodynamic data of crystalline silicates for calculating their evaporation, which are well defined contrary to the thermodynamic data of amorphous materials. The reactions responsible for the evaporation of silicates in the presence of a strong excess of hydrogen are:

$$Mg_{2x}Fe_{2(1-x)}SiO_4(s) + 3H_2 \longrightarrow 2xMg + 2(1-x)Fe + SiO + 3H_2O$$
$$Mg_xFe_{(1-x)}SiO_3(s) + 2H_2 \longrightarrow xMg + (1-x)Fe + SiO + 2H_2O$$
$$SiO_2(s) + H_2 \longrightarrow SiO + H_2O$$

The corresponding reactions for the condensation/evaporation of the minerals existing in chemical equilibrium are

$$2Mg + SiO + 3H_2O \longleftrightarrow Mg_2SiO_4(s) + 3H_2$$
$$Mg + SiO + 2H_2O \longleftrightarrow MgSiO_3(s) + 2H_2$$
$$Fe(s) \longleftrightarrow Fe$$

Since the characteristic timescales for a change of pressure and temperature conditions in a parcel of material moving inward due to accretion onto the star is of the order of $10^3 \ldots 10^4$ yrs, all evaporation and condensation processes occur very slowly. Under these conditions the equilibrium condensates forsterite, enstatite, and solid iron will be in a quasistationary state with the gas phase abundances of Mg, Fe, SiO, and H_2O, while the non-equilibrium components olivine, pyroxene, and quartz will not. For them the partial pressure of, for

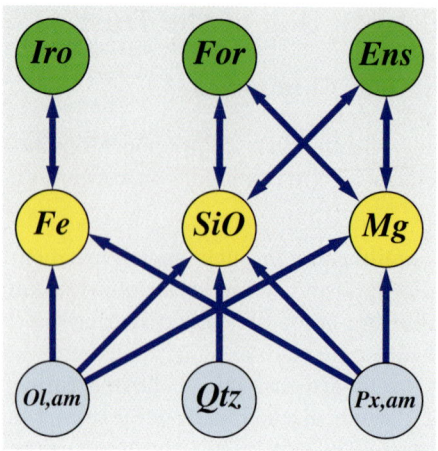

Fig. 12. Exchange of material between the different dust components (denoted by obvious abbreviations) and the vapor species (Fe, Mg, SiO). The exchange with the H_2O reservoir is not shown.

instance, SiO due to vaporization will be higher than the vapor pressure of the equilibrium condensates. Provided that evaporation and growth is not limited by diffusion in the gas phase all material evaporated off from the grains of interstellar origin will precipitate on the iron poor silicates and on solid iron. The exchange of material between interstellar dust and the disk-made dust is depicted in Fig. 12. In this way the interstellar dust mixture is converted into the chemical equilibrium mixture.

In order to include these processes into model calculations for the structure and evolution of protoplanetary disks one introduces for each dust species a set grain radii a_i $(i = 1, \ldots, I)$ where a_1 and a_I are the smallest and biggest dust grains considered, respectively. The concentration of dust grains with respect to the carrier gas with radii between a_i and a_{i+1} is c_i. One then calculates the evolution of the size distribution c_i by transport, diffusion and evaporation or condensation processes. The equations which have to be solved for each i and for all of the different dust species are Eqs. 13 for one-zone models or Eqs. (12) for 2-dimensional models. The corresponding rate terms are given in Gail [21].

7.2 Conversion of Silicates into the Equilibrium Mixture

Figure 13 shows the fraction of the Si bound into olivine, orthopyroxene, and quartz, as calculated from the solutions of the advection-diffusion-reaction equations for these dust species, as function of the midplane temperature in a stationary disk model. The last ISM grains evaporate at about 1 100 K. At this temperature they are already completely crystallized (cf. Fig. 11) and their vapor pressure can and must be calculated for crystalline materials. Further, by interdiffusion all spatial compositional inhomogeneities with respect to the cations Fe^{2+} and Mg^{2+} within silicate grains are erased for $T \gtrsim 700$ K for olivine and $T \gtrsim 900$ K for pyroxene (cf. Gail [20]). If the grains arrive at the

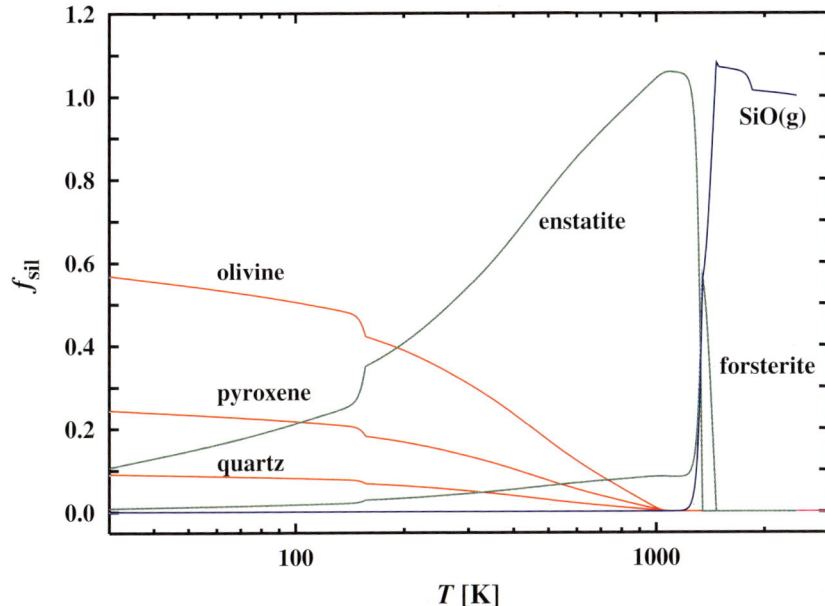

Fig. 13. Radial variation of the fraction of Si condensed into the individual Si-bearing condensates: ISM silicates (red lines), disk made silicates (green lines), and the fraction of Si bound in SiO molecules (blue line). The abundances are normalized to the interstellar Si abundance

evaporation zone the silicate grains can be assumed to show a homogeneous chemical composition within a grain and a crystalline structure.

For the growth and evaporation of the equilibrium silicates there holds that the characteristic timescales for a change of temperature and pressure experienced by a dust grain in the disk are so long that their growth and evaporation occurs under near equilibrium conditions. The deviations of the gas phase partial pressures of the vapor compounds of the silicates, i. e., SiO, Mg, and Fe[2], then are very small and essentially there exists chemical equilibrium between the solids and the gas phase. The stable silicates in this case are iron-poor olivine, i. e., essentially pure forsterite, and iron-poor orthopyroxene, i. e., essentially pure enstatite.

The relative abundance of forsterite and enstatite for that fraction of the element abundances of Si, Mg, O which is not bound in ISM-dust, can be calculated for the chemical equilibrium state from chemical thermodynamics. The resulting fraction of Si bound in forsterite and enstatite dust also is shown in Fig. 13. As one sees the region of evaporation of the equilibrium dust components forsterite and enstatite is well separated from the region

---

[2] H_2O in any case is so abundant that its abundance variations with varying degree of condensation of the silicates can be neglected.

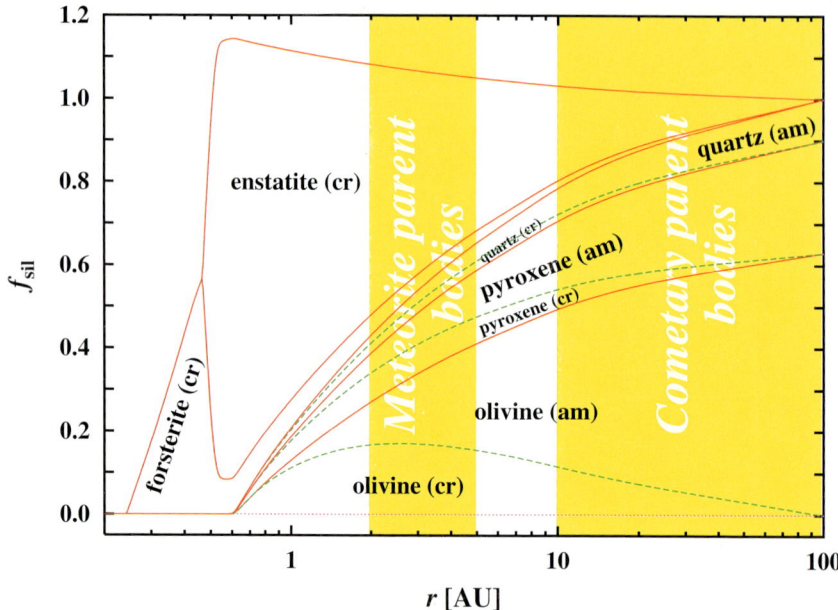

Fig. 14. Cumulative representation of the distribution of the silicon condensed into dust between the different silicate dust components (interstellar and equilibrium ones) within the protoplanetary disk, and the fraction of amorphous (am) and crystalline (cr) interstellar silicates. The equilibrium dust components forsterite and enstatite are always crystalline. The total Si abundance is normalized to the interstellar Si abundance. An enhancement of the Si abundance in the region of evaporation of the interstellar silicates is due to diffusive mixing effects. The shaded areas roughly indicates the regions where the parent bodies of meteorites are located and the formation zone of comets. Stationary disk model with an accretion rate $\dot{M} = 1 \times 10^{-7}$ $M_\odot$ yr^{-1}.

where the interstellar silicates evaporate. These processes operate in different regions of the disk.

Close to the upper stability limit of forsterite, the forsterite is the dominating silicate mineral because of its high stability. At a somewhat lower temperature enstatite dominates because in the solar system element mixture the Mg abundance only slightly exceeds the Si abundance. In this case complete condensation of both elements Mg and Si is only possible if enstatite dominates in the mixture, since the formation of forsterite consumes two Mg atoms per Si atom while the formation of enstatite consumes only one.

Figure 14 shows in a cumulative representation the radial variation of the distribution of the Si condensed between the different silicate dust species, calculated for a stationary accretion disk model. This is the mixture of silicate dust species which is expected to exist in a protoplanetary accretion disk

Table 2. Composition of the silicate mixture at 3 and 20 AU in a stationary protoplanetary disk model with $\dot{M} = 1 \times 10^{-7}\,\mathrm{M_\odot\,yr^{-1}}$. The numbers are fractions of the silicon contained in the different dust species. 'cr' and 'am' denote crystalline or amorphous dust, respectively.

r	olivine		forsterite	pyroxene		enstatite	quartz	
AU	cr	am	cr	cr	am	cr	cr	am
3	0.159	0.158	0.037	0.067	0.070	0.458	0.023	0.027
20	0.073	0.469	0.010	0.031	0.202	0.129	0.011	0.076

from which the planetesimals are formed. Note the local enrichment of Si by diffusive mixing effects in the region where the interstellar silicates evaporate.

7.3 Expected Composition of Planetesimals

Table 2 shows the composition of the silicate mixture at two radii, 3 AU and 20 AU, which are roughly representative for the regions where the parent bodies of meteorites and the cometary nuclei, respectively, have been formed in our solar system. The results presently are obtained on the basis of a rather crude modeling of the accretion disk (one-zone approximation, stationary, α-disk) and cannot be considered as completely realistic, but they probably outline the trends which are to be expected for the composition of the silicate mineral mixture:

- A high fraction of equilibrated silicate dust from the warm inner region and a rather high degree of crystallization in the zone where the parent bodies of meteorites are formed.
- Mostly interstellar dust with an admixture of up to about 20 % annealed interstellar dust and equilibrated dust from the warm inner disk region in the zone where cometary nuclei are formed.

These trends are in accord with what is observed for dust from comets (e. g., Wooden et al. [55, 56]). For meteoritic matrix material the predicted fraction of enstatite in the mixture seems to be somewhat high, but the general trends fit with the observed composition of the matrix material of primitive meteorites (e. g., Scott et al. [45], Brearley et al. [6], Buseck & Hua [8]). The somewhat high pyroxene content in this model results from the assumption of complete chemical equilibrium between forsterite and enstatite, which possibly cannot be attained since a conversion of forsterite to enstatite is slow (e. g., Gail [20]) because of a high activation energy barrier. This process is presently not yet included in the model calculations.

The result for the composition of the condensed material for a time dependent (1+1)-dimensional model calculation (Wehrstedt & Gail [54]) is shown in Fig. 15. The silicate components are calculated in less detail in this model. Also in this model one obtains a significant fraction of crystalline silicates in

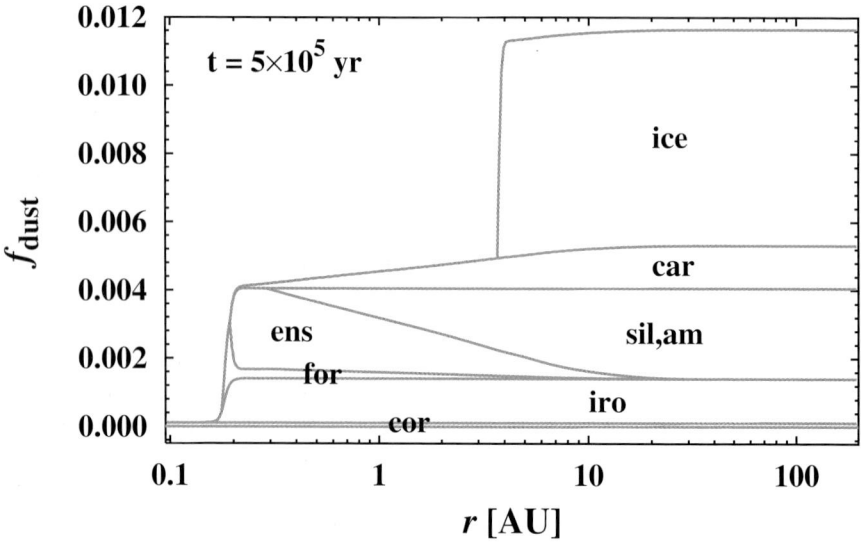

Fig. 15. Snapshot of a time-dependent (1+1)-dimensional model calculation of the evolution of a protoplanetary disk for the composition of planetesimals at $t = 5 \times 10^5$ yrs. Shown are the mass fractions of the dust components relative to the gas component in a cumulative fashion (Wehrstedt & Gail [54]). The dust species are denoted by obvious abbreviations.

the zone between 5 and 10 AU where the Oort cloud comets are assumed to have been formed.

The model calculations of stationary and of time dependent one-zone and (1+1)-dimensional models show that radial mixing processes play an important role for the evolution of the composition of the condensed material in the protoplanetary disk. The results obtained up to now are not in perfect agreement with what one observes for the composition of the material in cometary nuclei and in meteoritic matrix material but reproduces the general observed trends. The shortcomings of the models can be attributed to two causes: Firstly, some important chemical processes are not yet considered in the models because of a lack of data required to calculate the reaction kinetics of the processes, and secondly a realistic modeling of the transport and diffusion processes in the accretion disk requires time dependent 2-dimensional (better 3-dimensional) models of the hydrodynamics, coupled with radiative transfer and chemistry, which is currently under development (see section 8).

However, recent observations with the new interferometer at the ESO 8-m telescopes which now observationally resolve the innermost parts of nearby accretion disks around young stellar objects show much better agreement with the predictions of the model calculations for the mineral composition in the 1–2 AU region (van Boeckel et al. [51]). Therefore, it seems to be quite obvious that the most severe shortcomings of the model calculations carried out so far refer most likely to the modeling of the hydrodynamics and the mixing processes.

8 Two-Dimensional Axisymmetric Disk Models

As has been shown in the preceding Sections, 1- and (1+1)-dimensional models of protoplanetary accretion disks are very useful for exploring the overall evolutionary features. In these approximations, the basic effects of turbulence within the disk, comprising both an efficient redistribution of angular momentum and diffusive *radial* mixing of the disk material, can be well studied. However, there remain specific shortcomings which can be removed only if the assumption of the disk to be hydrostatic in the vertical direction is abandoned. For instance, within the framework of the (1+1)-dimensional disk models it is, by definition, impossible to determine the real structure of the disk's velocity field, e. g., its dependence on the vertical coordinate, z, nor can mixing processes in the z-direction be described adequately: only a z-averaged radial net "drift" of material – in the outward or inward direction, depending only on the radial distance from the center – is available. In particular, the investigation of detailed *advective* mass transport and mixing is clearly beyond the scope of the (1+1)-dimensional models.

Thus, in order to be able to describe the various physical and chemical processes going on in protoplanetary disks in a more realistic way, honest 2-dimensional models with axial symmetry must be taken into consideration. This means, together with solving the extensive chemical network including not only pure gas-phase reactions, but also combustion, chemi-sputtering as well as sublimation and condensation processes of microscopic dust particles, to set about an extremely computer time consuming exercise. It does not seem appropriate to aim at even more involved, fully 3-dimensional, disk models in the first step. Rather, what we are going to present in the following are results of a first pilot study for disk models with axial symmetry.

8.1 Numerical Strategy

Global hydrodynamical models of protoplanetary disks extending from about 0.1 astronomical units (AU) up to, say, 100 AU would strictly demand the use of implicit numerical schemes, owing to the fact that accretion disks exhibit largely dispersed characteristic scales. This leads to prohibitively small Courant timesteps, $\delta t_{\mathrm{Courant}}$, which sets a severe constraint for the efficiency of explicit schemes. However, implicit methods are in general very costly and have been successfully applied so far only to the purely hydrodynamical aspect of accretion disks (Keller [28]).

As a compromise, mixed explicit-implicit numerical schemes seem to be best suited for simulating selected parts of the disk at the expense of being forced to invent well-educated, but inevitably somewhat artificial, boundary conditions. Nevertheless, there is good hope that the general numerical strategy just briefly outlined will translate into a robust computer code that could serve as the working horse for many years to come.

Equations and discretisation

We use a finite-volume scheme and discretise the Navier-Stokes-Equations (1)–(5) for compressible fluids on a staggered (r, z)-grid of a cylindrical coordinate system, (r, φ, z), denoting again the radial, azimuthal, and vertical coordinate, respectively. Scalar quantities are defined in the interior, the individual components of vectorial quantities are tied to the respective boundary surfaces of the numerical cells. Turbulent viscosity is parameterized according to the β-disk prescription by [13]. This is an important modification of the classic α-viscosity. The new prescription warrants physical consistency if the disk's self-gravitation in the vertical direction were to exceed the vertical component of the gravitational acceleration which the central protosun exerts on the disk material. However, self-gravitation, i.e., a suitable Poisson solver, is currently not yet implemented in the existing code. Since we restrict ourselves to axially symmetric geometry, shock fronts in the azimuthal direction (spiral- or bar-like structures) are excluded. The artificial tensor viscosity term is tailored in a way as to smooth out shock fronts in the (r, z)-plane without changing the local distribution of angular momentum. A chemical network for the C, H, O chemistry is implemented and the various species undergo diffusive mixing according to Eq. (2). Depending on whether the dust grains are ice-coated or not grey opacities are calculated from respective fit formula, the albedo is assumed to be zero. To model the (grey) radiation transport (Eq. (9)) we employ the Eddington approximation with flux delimiter (the absolute value of the 1st moment $|\vec{H}|$ must not exceed the 0th moment J!). Finally, we take the two energy balance equations (8) and (10) into account.

Two-step solution procedure

The equations are numerical solved by an two-step operator-splitting procedure according to Norman & Winkler [39]. In the first *source* step yielding intermediate updates of the physical variables (the velocity components, internal energy, concentrations, ...) after the time interval $\delta t < \delta t_{\text{Courant}}$, only the source terms in the equations are taken into account. The second, *transport* step the remaining advection terms are evaluated to perform the final updating. A nominal second-order van-Leer scheme is applied to carry out the transport step.

As already mentioned, for the sake of simplicity, energy transport by radiation is modeled according to the Eddington approximation. The energy balance equation for the radiation field then reduces to a spatially 2-dimensional diffusion equation with a source term that couples to the energy balance equation for the disk material. In optically thin regions where the density (or the opacity) is low and the Eddington approximation becomes worse, the respective Courant timestep

$$\delta t_{\text{Courant}} \simeq \frac{\min(\Delta r, \Delta z)}{c} \Delta \tau \qquad (19)$$

gets very low. Here the effective optical depth $\Delta \tau = \min(\Delta r, \Delta z)\kappa\rho$ and $\min(\Delta r, \Delta z)/c$ is the minimum light travel time through the numerical cell of size $\Delta r \times \Delta z$. As a consequence, the source step for the energy equation – with the mass density, ρ kept constant – must be solved implicitly. The combination of a Newton-Raphson scheme combined with the GMRES algorithm to solve the linearized equation iteratively turned out to be a very efficient strategy to determine the temperature, pressure, and radiation energy density in a consistent fashion.

Chemistry

For this first, explorative, study we adopted a minimum set of 92 gas-phase reactions between the 12 species, H, H_2, O, O_2, OH, H_2O, HO_2, H_2O_2, CO, CO_2, HCO, and H_2CO. The combustion of soot/graphite particles which is the starting point for the synthesis of hydrocarbons has not yet been included. However, to test the robustness of the code with respect to the radiative energy transport term, the sublimation and condensation of water ice and silicates has been taken into account. Here we assumed that equilibrium between the solid and the gas phase is always instantaneously established. This particular model assumption implied a crucial test for the robustness of the code. In the inner, hot zones of the disk where the silicates evaporate and, to a lesser extent, within the ice-forming regions, the opacity can drop abruptly several orders of magnitude, in this way giving rise to dramatic changes of the radiation field. In this highly non-stationary phase the timesteps have become small, but never caused a lethal stop of the calculation.

The chemical network is solved by a simple backward Euler scheme for each numerical cell during the source step. At this interface the code can easily be parallelized. This is independent of the question of whether or not a more accurate and/or efficient solver ought to be implemented eventually.

8.2 Results

Initial conditions

In order to test the robustness of our explicit-implicit 2-dimensional radiation hydrodynamical code we chose an initial configuration which forces the "disk" to evolve into a pronounced dynamical regime in the first place. Only after this non-stationary "switch on" phase our disk model would asymptotically

Table 3. The initial data.

quantity	symbol	value	dimension
central mass	$M_\star$	1	$M_\odot$
disk mass	M_{disk}	0.01	$M_\odot$
radial extension	r	$0.8 \leq r \leq 5.8$	AU
vertical extension	z	$-1.5 \leq z \leq 1.5$	AU
radial velocity	v_r	0	km s^{-1}
vertical velocity	v_z	0	km s^{-1}
azimuthal velocity	v_φ	$\sim 33 \ldots 12$ (Keplerian rotation)	km s^{-1}
density	ρ	1.91×10^{-11}	g cm^{-3}
gas temperature	T_{kin}	20	K
radiation temperature	T_{rad}	20	K
mean molecular weight	μ	2.377	–
gas pressure	P	1.336×10^{-2}	dyn cm^{-2}

species		abundances (concentrations) per weight	
molecular hydrogen	c_{H_2}	0.708933	
helium	c_{He}	0.274615	
molecular nitrogen	c_{N_2}	0.000920	
carbon monoxide	c_{CO}	0.006997	
water	$c_{\text{H}_2\text{O}}$	0.003020	
"dust"	c_{dust}	0.005515	
	$\sum c \equiv 1$		

approach a quasi-stationary state that is *assumed* by the (1+1)-dimensional description. This situation is achieved by choosing an initially *homogeneous* distribution of the disk material. Table 3 contains a compilation of the initial data we used in the simulation. We discretise the equations on a (48×64) Eulerian tensor-product (r, z)-grid. In the z-direction the gridpoints z_j, $j = 0, 1, 2, \ldots, 64$ with $z_0 = 0\,\text{AU}$ are chosen to be equally spaced, i.e., $\Delta z = \text{const} = 1.5/64 \approx 0.0234\,\text{AU}$ (because of the assumed equatorial symmetry we need to cover only the "upper half" of the disk); the distribution of the radial gridpoints, r_i, $i = 0, 1, 2, \ldots, 48$ with $r_0 = 0.8\,\text{AU}$, is tailored according to a geometrical law, $r_i = r_{i-1} \cdot q$ with $q = (5.8/0.8)^{1/48} \approx 1.042134$, i.e., the radial mesh sizes, $\Delta r_i := r_i - r_{i-1}$, $i = 1, 2, \ldots, 48$, expand monotonically from $\Delta r_1 \approx 0.0337\,\text{AU}$ to $\Delta r_{48} \approx 0.2345\,\text{AU}$ (cf. Table 3 for the actual numbers inserted).

The cosmical abundances of the chemical elements, H, He, C, O, N, Mg, Si, S, Fe, according to Population I is adopted. The heavier elements, Mg, Si, S, Fe are tied up in solid grains (mainly in forsterite, Mg_2SiO_4, and troilite, FeS), and 70 % of the carbon is present in the form of condensed soot/graphite particles, with diameters typically smaller than $0.1\,\mu$m. The remaining part of the carbon is assumed to be entirely in gaseous CO. The surplus of oxygen is tied up in water, H_2O. Thus, initially the gaseous component contains only the species H_2, He, N_2, CO, H_2O; helium and nitrogen do not participate in the chemical reaction network. All other species listed in Subsect. 8.1 above are synthesized later in the evolution.

Boundary conditions

Besides axial symmetry, we also assume symmetry of the disk with respect to the "equatorial" plane, $z = 0$. There the z-derivatives of the scalars like density, temperature, pressure, etc., and the vertical velocity component, v_z, must vanish. Both at the inner and outer boundary we fix the radial velocity component, v_r, by taking the respective stationary drift velocity as a result of the redistribution of angular momentum according to the β-viscosity, $\eta/\rho = \beta r v_\varphi$, with β (set equal to 10^{-3} in our simulation) being the inverse of the "critical" Reynolds number (cf. [13]). At the inner boundary material leaks out at a total rate, $\dot{M}_{\mathrm{ib}}$, while at the outer boundary the total incoming mass flux, $\dot{M}_{\mathrm{ob}} = 10^{-6}\,M_\odot\,\mathrm{yr}^{-1}$, is kept constant throughout the simulation. These particular boundary conditions allow for a stationary disk model where the equality $\dot{M}_{\mathrm{ib}} = \dot{M}_{\mathrm{ob}} = 10^{-6}$ holds. Mass infall from "above" the disk is neglected. Concerning the thermal boundary conditions, the incoming radiation is assumed to have an equivalent temperature of 20 K, and there is no radiative energy transport across both the inner and the outer radial boundary. The "vacuum" density above the disk proper is given arbitrarily by a lower limit of about $10^{-15}\,\mathrm{g\,cm}^{-3}$ in our simulation. The velocity field is artificially reset to zero after each timestep if the density shows a tendency to decrease further below this lower limit. Briefly, this – physically motivated – strategy is a simplistic way to successfully cope with a "disguised" form of an otherwise always nasty free boundary value problem.

Simulation

Since we start out with a homogeneous and isothermal disk-like configuration, vertical pressure gradients have not yet built up to compensate the vertical gravitational pull the central protosun exerts on the disk material. As a consequence, the initial configuration commences to collapse essentially in the z-direction, while the radial direction is less affected. This is due to the fact that the initial disk is a Keplerian one (cf. Table 3) and is therefore radially stabilized by centrifugal forces.

Figure 16 displays the contour lines of several important physical quantities according to their distribution in the (r, z)-plane. It represents a snapshot of the non-stationary collapse-like "switch-on" phase of our model disk after one revolution period as measured at a radial distance of 1 AU. As expected, the highest compression occurs in the innermost regions of the disk near the

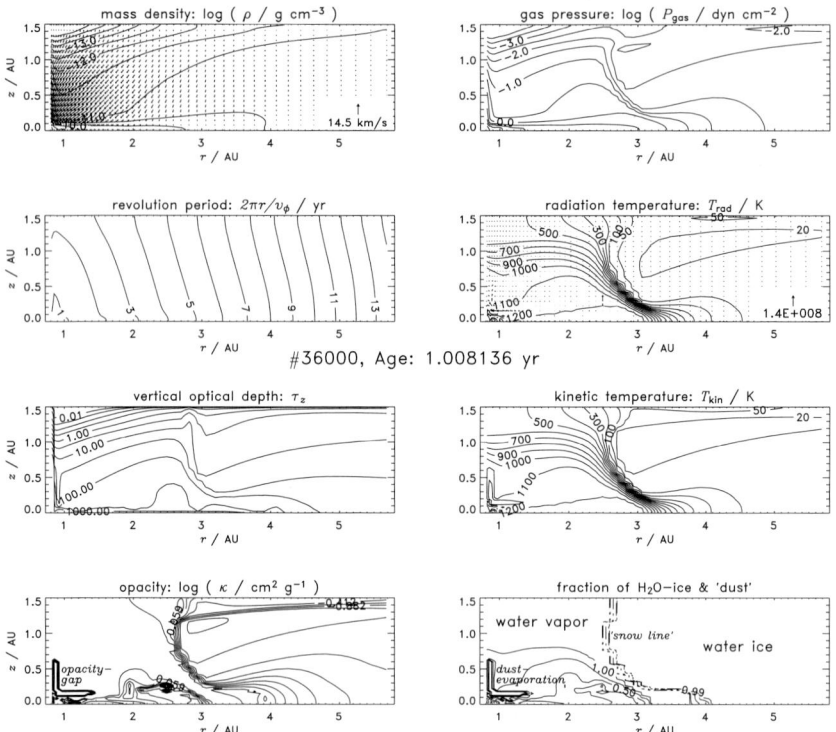

Fig. 16. Contour lines of the different physical parameters after 1 yr, i. e., after one revolution of the disk around the central protosun at a radial distance of 1 AU. The arrow fields drawn in the left uppermost panel showing the equi-contours of the mass density, and in the second panel from above on the right hand side, showing the distribution of the radiation temperature, represent the velocity field (in units of km s^{-1}) and the radiative flux (in erg cm^{-2} s^{-1}), respectively. The scales are indicated by the single arrow in the lower right corner of the respective panels. The number of timesteps to cover this first year amounts to 36000.

equatorial plane. This happens also at the inner boundary, which is an artifact of our adopted "rigid-wall" condition: the mass flow across the inner border of the disk ceases with vanishing viscosity. Though highly artificial, mastering such a complex dynamical situation is, nevertheless, again a firm test for the robustness of the code. In the very early phases (within about the first 0.2 yr) Mach numbers of above 40 were present, declining to about 13 after 1 yr and to 4.5 after 12 yr which is the revolution time of the outermost disk layers ($r \gtrsim 5$ AU). In other words, the flow exhibits strong shock fronts almost from the beginning. However, they become weaker and eventually disappear when after several dynamical time scales, i. e., several tens of years, a viscosity-

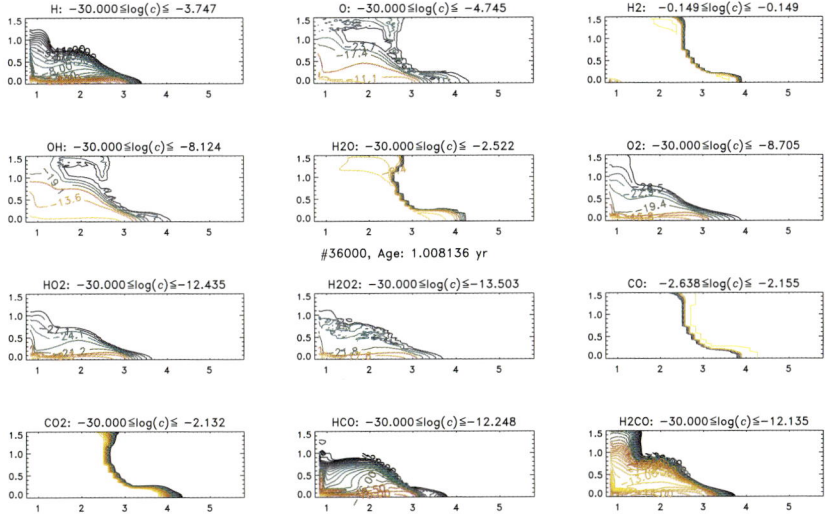

Fig. 17. Distribution of the concentration, c, that is, the relative abundance (in mass) of the gaseous species after 1 yr, i.e., after one revolution of the disk around the central protosun at a radial distance of 1 AU. The headline contains the chemical formula of the species whose concentration contour lines are shown and, in addition, the information about the range in between the concentrations vary over the (r, z)-plane. The lower limit of $\log(c) = -30$ means zero abundance. Colors indicate different values of concentration, coded yellow-red-green-blue from the highest to the smallest level in descending order. In the panel showing the distribution of water vapor (H_2O, second panel from above in the middle column) the snow line (cf. Fig. 16) is again visible as a distinct feature.

driven accretion flow remains which is asymptotically approaching a quasi-stationary state.

The temperatures in the maximum compression zones rise well above the silicate dust sublimation temperature, thus creating a deep opacity gap and, accordingly, a noticeable difference between the kinetic and the radiation temperature. The hot gas cannot cool fast enough because the dust, as the principal absorber, does not exist anymore, while the radiation energy quickly diffuses away (cf. the panels of Fig. 16 showing the distribution of the opacity, radiation temperature, and kinetic temperature, respectively). A heat wave originating from the hot interior and a pressure wave propagate radially toward the outer regions of the disk. As a consequence, the "snow line", that is, the border line between "cold" and "warm" zones of the disk where water is frozen out on the silicate grains or is sublimated in the form of water vapor, respectively, is continually shifted to larger radii.

Along with the rather violent dynamical and thermal evolution of our model disk the various gas-phase reactions in the chemical network comprising

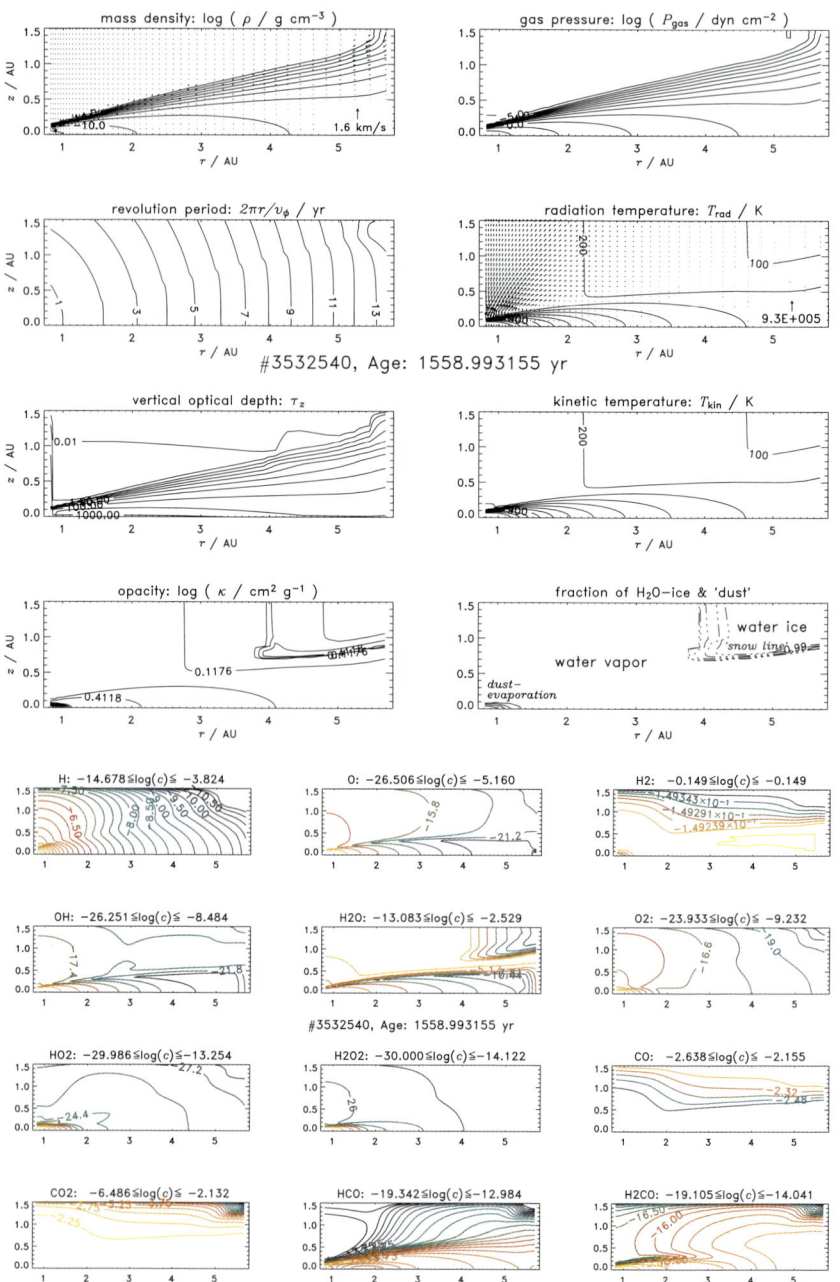

Fig. 18. The same drawings of the physical quantities and the distribution of the chemical species as in Figs. 16 and 17, but at the end of the simulation. In total 3 532 540 timesteps were necessary to cover the entire evolutionary period of 1559 yr.

92 reactions are activated in the "warm" zones. This is illustrated by Fig. 17. It shows the distribution of all *reactive* gaseous species that we have taken into consideration, again at an age of 1 yr. The salient feature is the obviously rapid production of CO_2. Since only the (reactive) species H_2, H_2O, and CO are present initially, there must exist an extremely efficient reaction path within the chemical network, which ends at CO_2. As a matter of fact, one obvious possibility refers to the forth ("→") and back ("←") reactions

$$CO + H_2O \rightleftarrows CO_2 + H_2 \qquad (20)$$

with their respective rate constants, $k_\rightarrow$ and $k_\leftarrow$, which differ by several orders of magnitude independent of the temperature. The numbers actually used in our calculations are (in units of $cm^3 s^{-1}$)

$$k_\rightarrow = 8.90 \times 10^{-11} \sqrt{T} \exp(-2600/T),$$
$$k_\leftarrow = 2.94 \times 10^{-14} \sqrt{T} \exp(-7550/T).$$

Applied to our initial mixture of the disk material this yields, relative to the number of CO molecules, a formation time for CO_2 of about 10^5 s already at low temperatures around 170 K and a density $\rho = 10^{-10}$ g cm^{-3}, scaling with the inverse of the density, ρ^{-1}, and decreasing exponentially with the temperature down to 1 s at 1000 K. Thus, in every region where the temperature is higher than the sublimation temperature of water ice CO is almost instantaneously – compared with the evolution time of 1 yr $\cong 3 \times 10^7$ s – transformed into CO_2. According to the reactions (20) a transient equilibrium between H_2O (vapor), CO, and CO_2 adjusts while the concentration of H_2 as the principal constituent of the gas mixture stays practically constant.

Figure 18 shows the distribution of the physical quantities and chemical species after 1559 yr, at the end of the simulation. The disk is gradually evolving into a stationary state, hydrostatic equilibrium in the vertical direction has already been established. The chemistry is still dominated by the equilibrium state of the fast reactions (20) dictating that the disk material becomes the more devoid of water the lower the temperature is, in fact, the outer parts of the disk where the temperature is in the range 300 K $< T <$ 200 K are already very dry. At still lower temperatures either the timescales to reach the equilibrium are too long or the water vapor condensates to form ice-coated grains. The "snow line" no longer touches the equatorial plane, but has retracted to a vertical distance of somewhat less than twice the half-thickness of the disk, $h \simeq 0.1r$, thus forming thin "ice clouds" in the disk's upper atmosphere. Notice also the vertical contour lines of the water-vapor concentration, indicating rising "humidity" of the disk material again near the outer boundary. This feature reflects nothing but our boundary condition that still unprocessed material of initial composition (see Table 3) flows in at the constant total rate, $\dot{M}_{ob} = 10^{-6} M_\odot$ yr^{-1}. The spatial extent of the transition zone between the boundary and the disk proper may serve as a measure of the time a mass element takes to arrive at the aforementioned chemical equilibrium state.

However, there is a big caveat about the applicability of our findings to real nature. As a matter of fact, the reactions (20) governing the relative abundances of CO, CO_2, and particularly H_2O in the disk are very uncertain. It is by no means clear whether these reactions take place at all, and if so, whether the assumed rate coefficients are correct. Nevertheless, it would be interesting to know if there is observational evidence for "dry" regions in contemporary protostellar nebulae or if there is fossil evidence of a dry environment in our own former Solar Nebula, which then ought to be recorded in meteoritic material that formed there some 4.6 billion years ago.

9 Concluding Remarks

In the last couple of years the work done in this project has centered on the physics and chemistry in protoplanetary accretion disks and on their coupling with flow fields in theses disks and the associated transport processes. In a sequence of papers the most important processes determining the mineral composition in such disks have been identified. Methods to include these processes in model calculations for the structure and evolution of protoplanetary accretion disks were developed. These models are in many respects rather crude, first and foremost, because the fundamental problem of modeling the hydrodynamics of such disks is not yet completely solved and first reliable results are only just becoming available. Nevertheless, the model predictions seem to be already in reasonable agreement with the still rather sparse observational material with which it can be compared. The prospect to obtain now observations of protoplanetary accretion disks which resolve the central parts of the disk to scales below 1 AU, by means of interferometry using the new 8-m class telescopes, will make detailed investigations of the dust properties and mineral composition in the terrestrial planet formation zone possible. The first results confirm the theoretical models developed in the present project (van Boeckel et al. [51]). This should be a stimulus to further develop the models for the complex interplay between hydrodynamics, radiative transfer, gas phase chemistry, chemistry of the mineral components, and transport and mixing process in protoplanetary disks.

References

1. Anders, E., Grevesse, N.: Abundances of the elements: Meteoritic and solar. Geochimica et Cosmochimica Acta, **53**, 197–214 (1989)
2. Bauer, I., Finocchi, F., Duschl, W. J., Gail, H.-P., Schlöder, J. P.: Simulation of chemical reactions and dust destruction in protoplanetary accretion disks. Astronomy & Astrophysics, **317**, 273–289 (1997)
3. Black, D. C., Matthews M. S.: Protostars and Planets II. (University of Arizona Press, Tucson 1985)

4. Bockelée-Morvan, D., Gautier, D., Hersant, F., Huré, J.-M., Robert, F.: Turbulent radial mixing in the solar nebula as the source of crystalline silicates in comets. Astronomy & Astrophysics, **384**, 1107–1118 (2002)
5. Bouwman, J., de Koter, A., van den Ancker, M. E., Waters, L. B. F. M.: The composition of the circumstellar dust around the Herbig Ae stars AB Aur and HD 163296. Astronomy & Astrophysics, **360**, 213–226 (2000)
6. Brearley, A. J., Scott, E. R. D., Keil, K., Clayton, R. N., Mayeda, T. K., Boynton, W. V., Hill, D. H.: Chemical, isotopic and mineralogical evidence for the origin of matrix in ordinary chondrites. Geochimica et Cosmochimica Acta, **53**, 2081–2093 (1989)
7. Brucato, J. R., Colangeli, L., Mennella, V., Palumbo, P., Bussoletti, E.: Mid-infrared spectral evolution of thermally annealed amorphous pyroxene. Astronomy & Astrophysics, **348**, 1012–1019 (1999)
8. Buseck, P. R., Hua, X.: Matrices of carbonaceous chondrite meteorites. Ann. Rev. Earth Planet. Sci., **21**, 255–305 (1993)
9. Canup, R. M., Righter, K.: Origin of the Earth and Moon. (University of Arizona Press, Tucson 2000)
10. Castor, J. I.: Problems in astrophysical radiation hydrodynamics. In: Astrophysical Radiation Hydrodynamics, NATO ASI Series C: Mathematical and Physical Sciences Vol. **188**, ed. by K.-H. Winkler, M. L. Norman. (Reidel, Dordrecht 1986) pp. 1–43
11. Colangeli, L., Henning, T., Brucato, J. R., Clément, D., Fabian, D. and 15 co-authors: The role of laboratory experiments in the characterisation of silicon-based cosmic material. Astronomy & Astrophysics Rev., **11**, 97–152 (2003)
12. Duschl, W. J., Gail, H.-P., Tscharnuter, W. M.: Destruction processes for dust in protoplanetary accretion discs. Astronomy & Astrophysics, **312**, 624-642 (1996)
13. Duschl, W. J., Strittmatter, P. A., Biermann, P. L.: A note on hydrodynamic viscosity and selfgravitation in accretion disks. Astronomy & Astrophysics, **357**, 1123–1132 (2000)
14. Fabian, D., Jäger, C., Henning, Th., Dorschner, J., Mutschke, H.: Steps toward interstellar silicate mineralogy. V. Thermal Evolution of Amorphous Magnesium Silicates and Silica. Astronomy & Astrophysics, **364**, 282–292 (2000)
15. Finocchi, F., Gail, H.-P., Duschl, W. J.: Chemical reactions in protoplanetary disks II. Carbon dust oxidation. Astronomy & Astrophysics, **325**, 1264–1279 (1997)
16. Finocchi, F., Gail, H.-P.: Chemical reactions in protoplanetary accretion disks III. The role of ionisation processes. Astronomy & Astrophysics, **327**, 825–844 (1997)
17. Gail, H.-P.: Chemical reactions in protoplanetary accretion disks IV. Multi component dust mixture. Astronomy & Astrophysics, **332**, 1099–1122 (1998)
18. Gail, H.-P.: Radial mixing in protoplanetary accretion disks I. Stationary disc models with annealing and carbon combustion. Astronomy & Astrophysics, **378**, 192–213 (2001)
19. Gail H.-P.: Radial mixing in protoplanetary accretion disks III. Carbon dust oxidation and abundance of hydrocarbons in comets. Astronomy & Astrophysics, **390**, 253–265 (2002)
20. Gail, H.-P.: Formation and Evolution of Minerals in Accretion Disks and Stellar Outflows. In: Astromineralogy. Lecture Notes in Physics Vol. **609**, ed. by Th. Henning. (Springer, Heidelberg 2003) pp. 55–120

21. Gail, H.-P.: Radial mixing in protoplanetary accretion discs IV. Metamorphosis of the silicate dust complex. Astronomy & Astrophysics, **413**, 571–591 (2004)
22. Goswami, A., Prantzos, N.: Abundance evolution of intermediate mass elements (C to Zn) in the Milky Way halo and disk. Astronomy & Astrophysics, **359**, 191–212 (2000)
23. Grevesse, N., & Noels, A.: Cosmic abundances of the elements. In: Origin and Evolution of the Elements, ed. by N. Prantzos, E. Vangioni-Flam, M. Cassé. (Cambridge University Press, Cambridge 1993) pp. 15–25
24. Hallenbeck, S. L., Nuth, J. A., Daukantas, P. L.: Mid-Infrared Spectral Evolution of Amorphous Magnesium Silicate Smokes Annealed in Vacuum: Comparison to Cometary Spectra. Icarus, **131**, 198–209 (1996)
25. Hallenbeck, S. L., Nuth, J. A., Nelson, R. N.: Evolving Optical Properties of Annealing Silicate Grains: From Amorphous Condensate to Crystalline Mineral. Astrophysical J., **535**, 247–255 (2000)
26. Hanner, M. S., Hackwell, J. A., Russel, R. W., Lynch, D. K.: Silicate emission feature in the spectrum of comet Mueller 1993a. Icarus **112**, 490–495 (1994)
27. Harker, D. E., Desch, S. J.: Annealing of Silicate Dust by Nebular Shocks at 10 AU. *Astrophysical J. Lett.*, **565**, L109–L112 (2002)
28. Keller, Ch.: Zeitliche Entwicklung von Akkretionsscheiben mit chemischen Reaktionen. PhD Thesis, University of Heidelberg (2003)
29. Keller, Ch., Gail, H.-P.: Radial mixing in protoplanetary accretion disks. VI. Mixing by large-scale radial flows. Astronomy & Astrophysics, **415**, 1177–1185 (2004)
30. Kemper, F., Vriend, W. J., Tielens, A. G. G. M.: The Absence of Crystalline Silicates in the Diffuse Interstellar Medium. Astrophysical J., **609**, 826–837 (2004)
31. Kerridge, J. F., Matthews, M. S.: Meteorites and the Solar System. (University of Arizona Press, Tucson 1988)
32. Kley, W., Lin, D. N. C.: Two-dimensional viscous accretion disk models. I On meridional circulations in radiative regions. Astrophysical J., **397**, 600–612 (1992)
33. Lenzuni, P., Gail, H.-P., Henning, Th.: Dust evaporation in protostellar cores. Astrophysical J., **447**, 848–862 (1995)
34. Levy, E. H., Lunine, J. I.: Protostars and Planets III. (University of Arizona Press, Tucson 1993)
35. Mannings, V., Boss, A. P., Russel, S. S.: Protostars and Planets IV. (University of Arizona Press, Tucson 2000)
36. McComb, W. D.: The Physics of Fluid Turbulence. (Oxford Science Publ., Oxford 1990)
37. Meinköhn, E., Kanschat, G., Rannacher, Wehrse, R.: Numerical methods for multidimensional radiative transfer. In: Reactive Flow, Diffusion and Transport, ed. by R. Rannacher et. al. (Springer, Berlin-Heidelberg 2005).
38. Morfill, G. E., Völk, H. J.: Transport of dust and vapor and chemical fractionation in the early protosolar cloud. Astrophysical J., **287**, 371–395 (1984)
39. Norman, M. L., Winkler, K.-H. A.: 2-D Eulerian hydrodynamics with fluid interfaces, self-gravity and rotation. In: Astrophysical Radiation Hydrodynamics, NATO ASI Series C: Mathematical and Physical Sciences Vol. **188**, ed. by K.-H. Winkler, M. L. Norman. (Reidel, Dordrecht 1986) pp. 187–221
40. Nuth, J. A.: Constraint on nebular dynamics based on observations of annealed magnesium silicate grains in comets and disks around Herbig Ae and Be stars. Lunar & Planetary Sci. Conf., **30**, 1726 (1999)

41. Pollack, J. B., Hollenbach, D., Beckwith, S., Simonelli, D. P., Roush, T., Fong W.: Composition and radiative properties of grains in molecular clouds and accretion disks. Astrophysical J., **421**, 615–639 (1994)
42. Safronov, V. S.: Evolution of the Protoplanetary Cloud and Formation of the Earth and Planets. (Nauka Press, Moskow 1969)
43. Savage, B. D., Sembach, K. R.: Interstellar Abundances from Absorption-Line Observations with the Hubble Space Telescope. Annual Review Astron. & Astrophys. **34**, 279–330 (1996)
44. Schräpler, R., Henning, Th.: Dust diffusion, sedimentation, and gravitational instabilities in protoplanetary disks. Astrophysical J., **614**, 960–978 (2004)
45. Scott, E. R. D., Barber, D. J., Alexander, C. M., Hutchinson, R., Peck J. A.: Primitive material surviving in chondrites: matrix. In: Meteorites and the Early Solar System, ed. by J. F. Kerridge, M. S. Matthews. (University of Arizona Press, Tucson 1988) pp. 718–745
46. Sogawa, H., Kozasa T.: On the Origin of Crystalline Silicate in Circumstellar Envelopes of Oxygen-rich Asymptotic Giant Branch Stars. Astrophysical J., **516**, L33–L36 (1999)
47. Takeuchi, T., Lin, D. N. C.: Radial Flow of Dust Particles in Accretion Disks. Astrophysical J., **581**, 1344–1355 (2002)
48. Thiemes, M. H.: Heterogeneity in the nebula: Evidence from stable isotopes. In: [31], pp. 899–923
49. Thiemes, M. H.: Mass-Independent Isotopic Effects in Chondrites: The role of Chemical Processes. In: Chondrules and the Protoplanetary Disk, ed. by R. H. Hewins, R. H. Jones, E. R. D. Scott. (Cambridge University Press, Cambridge 1996) pp. 107–118
50. Urpin, V. A.: Hydrodynamic Flows in Accretion Disks. Soviet Astron., **28**, 50 (1984)
51. van Boeckel, R., Min, M., Leinert, Ch., Waters, L. B. F. M., Richichi, A. et al.: The building blocks of planets within the 'terrestrial' region of protoplanetary disks. Nature, **432**, 479–482 (2004)
52. Wehrstedt, M., Gail, H.-P.: Radial mixing in protoplanetary accretion disks II. Time dependent disk models with annealing and carbon combustion. Astronomy & Astrophysics, **385**, 181–204 (2002)
53. Wehrstedt, M., Gail, H.-P.: Radial mixing in protoplanetary accretion disks. V. Models with different metallicities. Astronomy & Astrophysics, **410**, 917–935 (2003)
54. Wehrstedt, M., Gail, H.-P.: Radial mixing in protoplanetary accretion disks VII. 2-dimensional transport of tracers. (in preparation) (2005)
55. Wooden, D. H., Harker, D. E., Woodward, C. E., Butner, H. M., Koike, C., Witteborn, F. C., McMurtry, C. W.: Silicate Mineralogy of the Dust in the Inner Coma of Comet C/1995 01 (Hale-Bopp) Pre- and Postperihelion. Astrophysical J., **517**, 1034–1058 (1999)
56. Wooden, D. H., Butner, H. M., Harker, D. E., Woodward, C. E.: Mg-Rich Silicate Crystals in Comet Hale-Bopp: ISM Relics or Solar Nebula Condensates?. Icarus, **143**, 126–137 (2000)
57. Wooden, D. H., Woodward, C. E., Harker, D. E.: Discovery of Crystalline Silicates in Comet C/2001 Q4 (NEAT). Astrophysical J., **612**, L77–L80 (2004)

Numerical Methods for Multidimensional Radiative Transfer[*]

E. Meinköhn[1,2], G. Kanschat[1], R. Rannacher[1], and R. Wehrse[2]

[1] Institut für Angewandte Mathematik, Universität Heidelberg
[2] Institut für Theoretische Astrophysik, Universität Heidelberg

Summary. This paper presents a finite element method for solving the resonance line transfer problem in moving media. The algorithm is capable of dealing with three spatial dimensions, using hierarchically structured grids which are locally refined by means of duality-based a posteriori error estimates. Application of the method to coherent isotropic scattering and complete redistribution gives a result of matrix structure which is discussed in the paper. The solution is obtained by way of an iterative procedure, which solves a succession of quasi-monochromatic radiative transfer problems. It is therefore immediately evident that any simulation of the extended frequency-dependent model requires a solution strategy for the elementary monochromatic transfer problem, which is fast as well as accurate. The present implementation is applicable to arbitrary model configurations with optical depths up to 10^3–10^4. Additionally, a combination of a discontinuous finite element method with a superior preconditioning method is presented, which is designed to overcome the extremely poor convergence properties of the linear solver for optically thick and highly scattering media. The contents of this article is as follows:

- Introduction
- Overview: numerical methods
- Monochromatic 3D radiative transfer
- Polychromatic 3D line transfer
- Test calculations
- Applications
- Multi-model preconditioning
- Conclusion

1 Introduction

Energy transfer via radiation plays a key role in various scientific applications, such as combustion physics, thermonuclear fusion and astrophysics. The equation describing the transport of photons or neutrons through a medium and

---

[*]This work has been supported by the German Research Foundation (DFG) through SFB 359 (Project C2) at the University of Heidelberg.

thereby accounting for absorption and emission effects is generally known as radiative transfer equation or radiation transport equation. It takes the form of a partial integro-differential equation that turns out to be equivalent to a linear Boltzmann equation if certain simplifications can be made. These simplifications arise, for instance, from an a-priori knowledge of the temperature field or from the assumption of local thermodynamic equilibrium. But as soon as these simplifications are dropped, non-linearities occur by way of the extinction coefficients or opacities, which depend on the temperature or on the number density of the energy-level population.

In astrophysics, particularly, it is obvious that next to the sensitivity improvements of the telescope equipment in the 1980s the current telescope development is characterized by an extraordinary improvement in spatial resolution: of course, the positioning of telescopes on satellites in space is one factor for this improvement, but the main fact is the use of so-called new generation telescopes with their computer controlled active and adaptive optical components. Especially, in combining these new instruments to very powerful modern interferometers provide spatially excellent resolved observations, revealing the geometrical complexity of many celestial objects. Thus, numerical codes for spherical-symmetrical or plane-parallel geometry, commonly used for astrophysical applications for decades, are not sufficient anymore. Modeling these multidimensional transfer problems makes the use of codes exploiting at least the 2D – in many cases the full 3D – structure indispensable.

In this paper, a summary of a collaboration between the numerics division of the Institute of Applied Mathematics and the Institute for Theoretical Astrophysics, both part of the University of Heidelberg, is presented. This interdisciplinary project started more than a decade ago, resulting in a continuous finite element method for solving multidimensional resonance line transfer problems in moving media. The appearance of an emission line profile is predominantly determined by complex angle-frequency couplings and by the macroscopic velocity field. Thus, taking scattering and Doppler effects into consideration, the corresponding three-dimensional radiative transfer equation is a partial integro-differential equation for the invariant radiation intensity $\mathcal{I} = \mathcal{I}(\mathbf{x}, \mathbf{n}, \nu, t)$ (cf. [49]), depending on the spatial variable $\mathbf{x}$, photon propagation direction $\mathbf{n}$, frequency ν and time variable t:

$$\frac{1}{c}\frac{\partial}{\partial t}\mathcal{I} + \mathbf{n} \cdot \nabla_x \mathcal{I} - \frac{\nu}{1+\mathbf{n}\cdot\frac{\mathbf{v}}{c}} \mathbf{n} \cdot \nabla_x \left(\mathbf{n} \cdot \frac{\mathbf{v}}{c}\right) \frac{\partial \mathcal{I}}{\partial \nu} = \\ -(\kappa + \sigma)\mathcal{I} + \sigma \int\int_{\mathbb{R}+\ S^2} R\mathcal{I} d\hat{\omega} d\hat{\nu} + \kappa B. \qquad (1)$$

The parameters entering this equation are:

$$
\begin{aligned}
S^2 &: \text{unit sphere within } \mathbb{R}^3 \\
\omega &: \text{solid angle} \\
c &: \text{speed of light} \\
\mathbf{v} &: \text{velocity field} \\
\kappa = \kappa(\mathbf{x}, \nu) &: \text{absorption coefficient} \\
\sigma = \sigma(\mathbf{x}, \nu) &: \text{scattering coefficient} \\
R = R(\hat{\nu}, \hat{\mathbf{n}}; \nu, \mathbf{n}) &: \text{redistribution function} \\
B = B(\nu, T(\mathbf{x})) &: \text{Planck function} \\
T(\mathbf{x}) &: \text{temperature field}
\end{aligned}
$$

If we assume that the velocity field causes no change of sign of the Doppler term, i.e. the term including the frequency derivative, Eq. (1) is generally solved with the constraint that the initial values of both the time and the frequency variable are fixed, and that the intensity of the incident radiation originating from outside the spatial domain is a given function at the boundary. More complex velocity fields give rise to a frequency boundary value problem rather than an initial value problem.

The algorithm discussed here works in three spatial dimensions on hierarchically structured grids which are locally refined by means of duality-based a posteriori error estimates ("DRW method"; see [6] and the articles [7, 8, 10, 11, 15] in this volume). The solution is obtained using an iterative procedure, where quasi-monochromatic radiative transfer problems are solved successively. Thus, a fast and accurate solution strategy for the monochromatic transfer problem is crucial to simulate the extended frequency-dependent model efficiently (for more details see [36, 55, 46, 47]). A detailed overview of recent numerical methods for multidimensional radiative transfer problems is given in Sect. 2. Section 3 provides a detailed description and solution strategy of the monochromatic 3D radiative transfer problem. In Sect. 4 an algorithm for the solution of the frequency-dependent line transfer problem is presented. The code accounts for arbitrary velocity fields up to approximately 10% of the speed of light and complete redistribution. The latter is critical, e.g., for the correct modeling of Lyα line profiles in optically thick media. A sample of test calculations for line transfer problems is presented in Sect. 5 illustrating the underlying physics. In Sect. 6 results for more complex multidimensional configurations are presented. Finally, in Sect. 7 a preconditioner for monochromatic radiative transfer problems is derived. It is based on a smoothing operator for specific intensities and on the diffusion approximation of the mean intensity. The scheme presented combines these two preconditioners in a way related to two-level multigrid algorithms. It is demonstrated that the method allows for fast solution of the discrete linear problems in optically thick media where scattering dominates as well as when transition layers occur.

2 Overview: Numerical Methods

The development of advanced discretization methods for the transport equation is of fundamental importance, since the numerical effort of modeling increasingly complex multidimensional problems with increasing accuracy is still extremely challenging:

- The radiation intensity is in general (neglecting polarization) a function of seven variables: three spatial variables $\mathbf{x} = (x, y, z)$, two angular variables describing the photon propagation direction $\mathbf{n}$, one frequency variable ν, and one time variable t. Even if only a moderate discretization of 10^2 grid points for each independent variable is taken into consideration, result in a huge discrete problem with 10^{14} unknowns. This amount of data demands not only extremely powerful computers, but also efficient numerical methods to reduce the memory and CPU requirements. Especially for the latter point, an appropriate discretization method is of fundamental importance.
- The intensity is usually a rapidly changing function of the spatial, angular and frequency variables yielding jumps of the intensity or its derivatives within small parts of the corresponding computational domain. These jumps usually cause a considerable loss in accuracy for many discretization methods.
- Depending on the coefficient ranges, the linear Boltzmann equation behaves like totally different equation types: in material free areas it behaves like a hyperbolic wave equation; in scattering dominant, optically thick media it behaves like an elliptic (steady-state) or parabolic (time-dependent) diffusion equation; and in regions with highly forward-peaked phase function, it can behave like a parabolic equation. It is extremely difficult to find a discretization method efficiently dealing with these different types of behavior.

Numerical algorithms solving multidimensional radiative transfer problems are described in numerous publications since the late 1980s. These algorithms may be roughly classified in three categories: Monte-Carlo methods, Discrete-Ordinate methods and Angle-Moment methods. The stochastic approach of *Monte-Carlo* codes is extremely flexible, since the concept of following photon packages is in principal applicable to arbitrary multidimensional configurations, as, e.g., to the ultraviolet continuum transfer [58], the optical and infrared continuum transport [68], the molecular line transfer [30, 35, 53], and the Compton scattering [54, 26]. If the optical depth of the configuration is not too large, Monte-Carlo methods converge reasonably fast. In the case of large optical depths, this method is usually extraordinarily time consuming. The latter case is, however, favorable for the deterministic *Angle-Moment* method, which usually assumes the so-called diffusion approximation [69, 57, 50]. Unfortunately, this technique is not applicable within optically thin regions, where the diffusion approximation is not valid anymore and the radiation field may significantly deviate from isotropy. Another deterministic approach to solve multidimensional radiative transfer problems, the *Discrete-Ordinate* method, is characterized by the additional discretization of the photon propagation direction. In order to avoid numerical artefacts, the subdivision of the ordinate domain has to be preferably uniform. The spatial discretization, i.e. the discretization of the transport operator of the radiative transfer equation, is often performed by using approaches based on finite differences [61] or by using the method of characteristics [20].

These methods usually work on structured grids, yielding algebraic systems that are solved very fast for homogeneous media and smooth data, but fail in the case of complex geometries or steep gradients of the solution and coefficients. The resulting system is by far too large to be solved even on supercomputers, since very high resolution is needed to produce accurate solutions for the challenging configurations described above. For many radiative transfer problems the steep gradients of the coefficients or the solution is confined to small regions within the computational domain. Only these small parts need the high resolution and the rest of the spatial domain must be resolved moderately, since the transport of photons is smooth. Solution adapted, locally refined grids offer the possibility to achieve highly resolved small areas of rapidly changing solutions, whereas in regions of smooth transport the grid remains coarse. Unfortunately, the generation of adaptively refined grids on the basis of a finite difference or characteristics method is extremely difficult. Alternatively, the finite element method (FEM) is exceedingly suitable to deal with adaptively refined grids and, therefore, approximates complex geometries and steep gradients of the coefficients and of the solution very well. A FEM upwind discretization on pre-refined grids is presented in [62]. In [36] the FEM discretization is stabilized via the streamline diffusion method yielding locally refined and, therefore, problem adapted grids by means of duality-based a posteriori error estimates ("DRW method"; for details see [6] and the articles [7, 8, 10, 11, 15] in this volume). Despite its evident advantages, the FEM is rarely used up to now within the astronomical community, since the code development is more complex than for the other commonly used discretization methods described above.

The discretization of multidimensional radiative transfer problems yield extremely large linear systems of algebraic equations, making direct solution strategies impossible. Usually, an iterative scheme is applied to solve these large systems of equations (e.g., see [29, 64]). The standard iterative method used in astrophysics to solve these systems is the so-called Λ- or Source Iteration method [48]. Considering the whole discrete system, it is a Richardson method with nearly block-Jacobi preconditioning. Using a full Jacobi preconditioner is a first step to better convergence rates [62]. Since the transport operator is inverted explicitly, these methods converge extremely fast for transport dominated problems. Exploiting the triangular matrix structure of an upwind discretization, the inversion is indeed very cheap (one matrix-vector-multiplication). In the case of highly scattering, optically thick media this method – like other stationary iterations – breaks down, since the condi-

Table 1. A radiative transfer model hierarchy. For details see text below.

Geometry	Scattering	Motion	Thermodyn.	Time
1D plane-parallel	none	none	LTE	stationary
1D spher.-symm.	coherent, isotropic	slow (v/c < 10%)	NLTE, 2-level	non-stationary level pop.
2D euclidian	coherent, non-isotropic	fast (v/c > 10%)	NLTE, multi-level	non-stationary radiation
3D euclidian	complete redistribution, isotropic		general distribution	
3D non-euclidian	partial redistribution, isotropic			
	partial redistribution, non-isotropic			

tion number of the iteration matrix becomes very large. Since the convergence rate of preconditioned Richardson iteration methods only depend on the condition number, these are not suited for the scattering dominated case. But here [36] comes in handy, showing that the eigenvalue distribution of the transport matrix is clustered, with one of the eigenvalues vanishing whilst the others are either close to unity or at least bounded away from zero. Krylov space methods like GMRES or bi-CGSTAB (for details see [56] and references therein) are particular suitable to deal with this eigenvalue distribution [36]. Being superior to stationary iteration methods, Krylov type solvers still show poor convergence properties for scattering dominant transfer problems, necessitating the use of convergence acceleration methods.

During the last several decades, significant progress has been achieved in the development of acceleration methods, i.e. preconditioning methods, for discrete-ordinate methods that sped up the iterative convergence of these problems. A comprehensive review article was recently published (see [2]), which covers practically all of the main methods that have been discussed above. When comparing the various acceleration methods, the Diffusion Synthetic Acceleration (DSA) turns out to be a very efficient preconditioning method guaranteeing rapid convergence for all values of optical depths and scattering coefficients. DSA exploits the diffusion approximation, a well-known approximation of the Boltzmann equation for highly scattering, optically thick media, so as to establish a solution of the full radiative transfer problem in a very efficient way. Preconditioning leads to acceleration in the discretized source iteration, which corresponds to a Richardson iteration, and the algorithm makes use of DSA efficiently as well as robustly. Discrete transport problems, for which a convergence in source iteration would ordinarily require hundreds, thousands, or even millions of iteration steps, are now solved with

DSA in less than a few tens iterations. Unfortunately, this method shows a loss in the effectiveness for multidimensional Cartesian grids in the presence of material discontinuities. In [65] it is shown that a Krylov iterative method, preconditioned with DSA, is an effective remedy that can be used to efficiently compute solutions for this class of problems. Results from numerical experiments show that replacing source iteration with a preconditioned Krylov method can efficiently solve transfer problems that are virtually intractable with accelerated source iteration (see [65]).

Next to Monte-Carlo and extremely slow Angle-Moment methods (see [59, 21]), only the finite difference method described in [61] existed prior to the Heidelberg project of developing a robust FEM code modeling multidimensional radiation fields in heterogeneous and highly scattering media. Table 1 displays a radiative transfer model hierarchy. The marked boxes correspond to the models the codes developed at Heidelberg University are able to deal with. The abbreviations LTE and NLTE account for local and non-local thermodynamic equilibrium, respectively.

3 Monochromatic 3D Radiative Transfer

3.1 The Radiative Transfer Problem

Our aim is the calculation of the radiation field in diffuse matter in space. Assuming the matter is surrounded by a vacuum, we will only consider a convex domain containing the area of interest. Radiation leaving this domain will not enter it again. Inside this 3D domain $\Omega \subset \mathbb{R}^3$, the specific intensity I satisfies the monochromatic radiative transfer equation

$$\mathbf{n} \cdot \nabla_x I(\mathbf{x}, \mathbf{n}) + \kappa(\mathbf{x}) I(\mathbf{x}, \mathbf{n}) \\ + \sigma(\mathbf{x}) \left(I(\mathbf{x}, \mathbf{n}) - \frac{1}{4\pi} \int_{S^2} P(\mathbf{n}', \mathbf{n}) I(\mathbf{x}, \mathbf{n}') \, d\omega' \right) = f(\mathbf{x}), \qquad (2)$$

where $\mathbf{x} \in \Omega$ is location in space and $\mathbf{n}$ is the unit vector pointing in the direction of the solid angle $d\omega$ of the unit sphere S^2. The optical properties of the matter are given by the absorption coefficient $\kappa(\mathbf{x})$ and the scattering coefficient $\sigma(\mathbf{x})$. The angular phase function P occurring in the scattering integral is normalized, such that $\frac{1}{4\pi} \int P(\mathbf{n}', \mathbf{n}) \, d\omega' = 1$. The source term

$$f(\mathbf{x}) = \kappa(\mathbf{x}) B(T(\mathbf{x})) + \epsilon(\mathbf{x}) \qquad (3)$$

consists of thermal emission depending on a temperature distribution $T(\mathbf{x})$ and an additional emissivity $\epsilon(\mathbf{x})$ of a point source or an extended object. B is the Planck-Function. To be able to solve Eq. (2), boundary conditions of the form

$$I(\mathbf{x}, \mathbf{n}) = g(\mathbf{x}, \mathbf{n}) \qquad (4)$$

must be imposed on the "inflow boundary"

$$\Gamma_- = \{(\mathbf{x}, \mathbf{n}) \in \Gamma \,|\, \mathbf{n}_\Gamma \cdot \mathbf{n} < 0\}, \tag{5}$$

where $\mathbf{n}_\Gamma$ is the normal unit vector to the boundary surface Γ. The sign of the product $\mathbf{n}_\Gamma \cdot \mathbf{n}$ describes the "flow direction" of the photons across the boundary. If there are no light sources outside the modeled domain, the function g will be zero everywhere.

The left hand side of Eq. (2) will be abbreviated as an operator $\mathcal{A}$ applied to the intensity function I, yielding the very compact operator form of the radiative transfer equation

$$\mathcal{A}I(\mathbf{x}, \mathbf{n}) = f(\mathbf{x}). \tag{6}$$

As already mentioned in the introduction the resulting linear system of algebraic equations is usually too large and this amount of data cannot be handled even on most advanced supercomputers. The application of efficient error estimation and grid adaption techniques is necessary to obtain reliable quantitative results. Finite element methods, in particular so-called Galerkin methods, are most suitable for these techniques (for details see [6] and the articles [7, 8, 10, 11, 15] in this volume).

3.2 Finite Element Discretization

Equation (2) is analyzed in [18] and the natural space for finding solutions is

$$W = \{I \in L^2(\Omega \times S^2) \,|\, \mathbf{n} \cdot \nabla_x I \in L^2(\Omega \times S^2)\}, \tag{7}$$

where L^2 is the Lebesgue space of the square integrable functions. If we consider homogeneous vacuum boundary condition, i.e. $g(\mathbf{x}, \mathbf{n}) = 0$, the solution space is

$$W_0 = \{I \in W \,|\, I = 0 \text{ on } \Gamma_-\}. \tag{8}$$

In order to apply a finite element method, we have to use a weak formulation of Eq. (2). Therefore, we multiply both sides of Eq. (2) by a trial function $\varphi(\mathbf{x}, \mathbf{n})$ and integrate over the whole domain $\Omega \times S^2$. Thus, the weak formulation reads: Find $I \in W_0$, such that

$$\int_\Omega \int_{S^2} \mathbf{n} \cdot \nabla_x I(\mathbf{x}, \mathbf{n}) \varphi(\mathbf{x}, \mathbf{n}) \, d\omega \, d^3x$$
$$+ \int_\Omega \int_{S^2} (\kappa(\mathbf{x}) + \sigma(\mathbf{x})) I(\mathbf{x}, \mathbf{n}) \varphi(\mathbf{x}, \mathbf{n}) \, d\omega \, d^3x$$
$$- \int_\Omega \int_{S^2} \int_{S^2} \sigma(\mathbf{x}) P(\mathbf{n}', \mathbf{n}) I(\mathbf{x}, \mathbf{n}') \varphi(\mathbf{x}, \mathbf{n}) \, d\omega' \, d\omega \, d^3x$$
$$= \int_\Omega \int_{S^2} f(\mathbf{x}) \varphi(\mathbf{x}, \mathbf{n}) \, d\omega \, d^3x \qquad \forall \varphi \in W_0. \tag{9}$$

By extending the definition of the L^2-scalar product we introduce the abbreviation

$$(I, \varphi) = (I, \varphi)_{\Omega \times S^2} = \int_\Omega \int_{S^2} I\varphi \, d\omega \, d^3x. \tag{10}$$

Thus, the weak formulation of the operator form in Eq. (6) is

$$(AI, \varphi) = (f, \varphi) \quad \forall \, \varphi \in W_0. \tag{11}$$

If there is no scattering, i.e. $\sigma(\mathbf{x}) = 0$ on Ω, the problem decouples to a system of stationary advection equations on Ω. These equations are hyperbolic. If the solutions are not smooth, standard finite element techniques applied to this type of equations are known to produce spurious oscillations. We can achieve stability by applying the streamline diffusion modification (see [31] and [33]):

$$(AI, \varphi + \delta \mathbf{n} \cdot \nabla_x \varphi) = (f, \varphi + \delta \mathbf{n} \cdot \nabla_x \varphi) \quad \forall \, \varphi \in W_0. \tag{12}$$

The cell-wise constant parameter function δ depends on the local mesh width and the coefficients κ and σ. Note that the solution of Eq. (11) also solves Eq. (12). No additional consistency error is induced since the stabilization term is simply added to Eq. (11). In the following, the streamline diffusion discretization term will be omitted but implicitly assumed.

Applying standard Galerkin finite elements to solve Eq. (11), we choose a finite dimensional subspace W_h of W consisting of functions that are piecewise polynomials with respect to a subdivision or *triangulation* $\mathbb{T}_h$ of $\Omega \times S^2$. The mesh size h is the piecewise constant function defined on each triangulation cell K by the diameter of the cell. The discrete analogon of Eq. (11) is finding $I_h \in W_h$, such that

$$(AI_h, \varphi_h) = (f, \varphi_h) \quad \forall \, \varphi_h \in W_h. \tag{13}$$

The construction of the subspace W_h needs some further consideration (see [36]). The discretized domain is a tensor product of two sets of completely different length scales: While Ω represents a domain in geometric space, S^2 is the unit sphere in the Euclidean space $\mathbb{R}^3$. Therefore, we use a tensor product splitting of the five-dimensional domain $\Omega \times S^2$, such that a grid cell of the five-dimensional grid will be the tensor product of a two-dimensional cell K_ω and a three-dimensional cell K_x. Accordingly, the mesh sizes with respect to $\mathbf{x}$ and ω will be different.

On S^2 we use fixed discretization based on a refined icosahedron. Quadrature points are the cell centers projected on S^2. This projection method guarantees equally distributed quadrature points, each associated with a solid angle $d\omega$ of the same size. In comparison to other numerical quadrature methods, additional symmetry conditions are redundant and discretization artifacts at the poles are avoided. Furthermore, we use piecewise constant trial functions. That way, the seven-dimensional integration of the scattering term in the weak formulation of Eq. (9) can be calculated very efficiently. For example, the integration of the intensity over the whole unit sphere is simply replaced by a sum divided by the number of discrete ordinates M

$$\frac{1}{4\pi} \int_{S^2} I(\mathbf{n})\, d\omega \rightarrow \frac{1}{M} \sum_{j=1}^{M} I_j \,. \tag{14}$$

Nevertheless, our scheme is second order accurate in the evaluated ordinate points due to super-convergence (see [36]).

For the space domain Ω we use locally refined hexahedral meshes. The mesh size of the spatial cell K_x is denoted by h. Since the boundaries are arbitrary for our astrophysical applications, we can choose a unit cube for Ω and place the simulation object within this cube. We only have ensure that the boundary of Ω is far away from the simulation object where the photons may freely escape towards the observer. Thus, we do not have to worry about boundary approximations. We use continuous piecewise trilinear trial functions in space.

3.3 Error Estimation and Adaptivity

The calculation of complex radiation fields in astrophysics often requires high resolution in parts of the computational domain, for example in regions with strong opacity gradients. Then, reliable error bounds are necessary to rule out numerical errors. Due to the high dimension of the radiative transfer problem, a well suited method for error estimation and grid adaptation is necessary to achieve results of sufficient accuracy even with the storage capacities of parallel computers.

In addition, computational goals in astrophysics are often more specific. The result of a simulation is to be compared with observations to develop a physical model for the celestial object. For instance, in the case of a distant unresolved object, the radiative energy emanating the domain Ω in *one* particular direction, i.e. the direction towards the observer, is of interest. Generally, a measured quantity like the radiative energy can be expressed as a linear functional $\mathcal{M}(I)$. By linearity, the error of the measured quantity is $\mathcal{M}(I) - \mathcal{M}(I_h) = \mathcal{M}(e)$, where $e = I - I_h$. In the following, we will show that it is possible to obtain an *a posteriori* estimate for $\mathcal{M}(e)$, even if the exact solution I is unknown.

Suppose, $z(\mathbf{x}, \mathbf{n})$ is the solution of the dual problem

$$\mathcal{M}(\varphi) = (\varphi, \mathcal{A}^* z) \qquad \forall \varphi \in W_0, \tag{15}$$

where the dual radiative transfer operator is defined by

$$\mathcal{A}^* z(\mathbf{x}, \mathbf{n}) = -\mathbf{n} \cdot \nabla_x z(\mathbf{x}, \mathbf{n}) + \big(\kappa(\mathbf{x}) + \sigma(\mathbf{x})\big) z(\mathbf{x}, \mathbf{n}) \\ - \sigma(\mathbf{x}) \int_{S^2} P(\mathbf{n}', \mathbf{n}) z(\mathbf{x}, \mathbf{n}')\, d\omega'. \tag{16}$$

The boundary conditions for the dual problem are complementary to those in the primal problem, i.e. $I = 0$ on the "outflow boundary" $\Gamma_+ = \{(\mathbf{x}, \mathbf{n}) \in \Gamma \mid \mathbf{n}_\Gamma \cdot \mathbf{n} > 0\}$. Then, we get the formal error representation

$$\begin{aligned}\mathcal{M}(e) = (e, \mathcal{A}^* z) &= (\mathcal{A}e, z) \\ &= (\mathcal{A}e, z - z_i) \\ &= \sum_{K \in \mathbb{T}_h} (f - \mathcal{A}I_h, z - z_i)_K\end{aligned} \qquad (17)$$

for arbitrary $z_i \in W_h$. In the second line of Eq. (17) a characteristic feature of finite element methods is used, the so-called Galerkin orthogonality

$$(\mathcal{A}I - \mathcal{A}I_h, \varphi_h) = 0 \qquad \forall \varphi_h \in W_h. \qquad (18)$$

Since the dual solution z is unknown, it is a usual approach to apply Hölder's inequality and standard approximation estimates of finite element spaces [3] to obtain the duality-based a posteriori error estimate ("DRW method"; see [6] and the articles [7, 8, 10, 11, 15] in this volume)

$$\mathcal{M}(e) \leq \eta = \sum_{K \in \mathbb{T}_h} \eta_K, \qquad \eta_K = C_K h_K^2 \|\varrho\|_K \|\nabla^2 z\|_K, \qquad (19)$$

where the constant C_K is determined by local approximation properties of W_h. The residual function ϱ of I_h is defined by $\varrho = f - \mathcal{A}I_h$. Since the dual solution z is not available analytically, it is usually replaced by the finite element solution z_h to the dual problem in Eq. (15). This involves a second solution step of the same structure as the primal problem. It is clear, that by this replacement the error estimate in Eq. (19) is not strictly true anymore. Experience shows that the additional error is small and the estimate may be used if multiplied with a modest security factor larger than one (for details see [6]).

A first approach in the development of strategies for grid refinement based on a posteriori error estimates is the control of the global energy or L^2-error involving only local residuals of the computed solution (see [25]). Formally, using the functional $\mathcal{M}(\varphi) = \|e\|^{-1} (e, \varphi)$ in Eq. (15), we obtain $\mathcal{M}(e) = \|e\|^{-1} (e, e) = \|e\|$ for the left hand side of (17). An estimate of the right hand side of Eq. (17) yields

$$\|e\| \leq \tilde{\eta} = \sqrt{\sum_{K \in \mathbb{T}_h} \eta_{L^2}^2}, \qquad \eta_{L^2} = C_K C_s h_K^2 \|\varrho\|_K. \qquad (20)$$

In this estimate, the information about the global error sensitivities contained in the dual solution is condensed into just one "stability constant"

---
[3] for details see [6] and the standard Bramble-Hilbert theory in [13]

$C_s = \|\nabla^2 z\|$. This is appropriate in estimating the global L^2-error in case of constant coefficients. For more general situations with strongly varying coefficients, it may be advisable rather to follow the "DRW approach" in which the information from the dual solution is kept within the a posteriori error estimator defined in Eq. (19). Nevertheless, the L^2-indicator η_{L^2} in Eq. (20) provides a reasonable grid refinement criterion to study the qualitative behavior of the solution I everywhere in Ω (see [66] for an illustration of this approach).

The grid refinement process on the basis of an a posteriori error estimate is organized in the following way: Suppose that some error tolerance TOL is given. The aim is to find the most economical grid $\mathbb{T}_h$ on which

$$|\mathcal{M}(e)| \leq \hat{\eta}(I_h) = \sqrt{\sum_{K \in \mathbb{T}_h} \eta_K^2(I_h)} \leq TOL, \qquad (21)$$

with *local refinement indicators* η_K taken from Eq. (19). A qualitative investigation of grids obtained via different adaptive refinement criteria, e.g. Eq. (19) and Eq. (20), is published in [55]. Having computed the solution on a coarse grid, the so-called *fixed fraction* grid refinement strategy (see [36], [6]) is applied: The cells are ordered according to the size of η_K and a fixed portion ν (say 30%) of the cells with largest η_K is refined. This guarantees, that in each refinement cycle a sufficient large number of cells is refined. Then, a solution is computed on the new grid and the process is continued until the prescribed tolerance is achieved. This algorithm is especially valuable, if a computation "as accurate as possible" is desired, that is, the tolerance is not reached, but computer memory is exhausted. Then, the parameter ν has to be determined by the remaining memory resources.

3.4 Resulting Matrix Structure

Given a discretization with N vertices in Ω and M ordinates in S^2, the discrete system has the form

$$\mathbf{A} \cdot \mathbf{u} = \mathbf{f}, \qquad (22)$$

with the vector $\mathbf{u}$ containing the discrete specific intensities and the vector $\mathbf{f}$ the values of the source term. Both vectors are of length $(N \cdot M)$. $\mathbf{A}$ is a $(N \cdot M) \times (N \cdot M)$ matrix. Applying the tensor product structure proposed above, we may write

$$\mathbf{A} = \mathbf{T} + \mathbf{K} + \mathbf{S} \qquad (23)$$

with the block structure

$$\mathbf{T} = \begin{pmatrix} \mathbf{T}_1 & & 0 \\ & \ddots & \\ 0 & & \mathbf{T}_M \end{pmatrix}, \quad \mathbf{K} = \begin{pmatrix} \mathbf{K}_1 & & 0 \\ & \ddots & \\ 0 & & \mathbf{K}_M \end{pmatrix},$$

$$\mathbf{S} = \begin{pmatrix} \omega_{11}\mathbf{S}_1 & \cdots & \omega_{1M}\mathbf{S}_1 \\ \vdots & \ddots & \vdots \\ \omega_{M1}\mathbf{S}_M & \cdots & \omega_{MM}\mathbf{S}_M \end{pmatrix},$$

where $\omega_{il} = P(\mathbf{n}_i, \mathbf{n}_l)/M$. If we account for the finite element streamline diffusion discretization as described in Eq. (12), the entries of the $N \times N$ blocks are defined by

$$\mathbf{T}_i^{jk} = (\varphi_j + \delta \mathbf{n}_i \cdot \nabla_x \varphi_j, \mathbf{n}_i \cdot \nabla_x \varphi_k)$$
$$\mathbf{K}_i^{jk} = (\varphi_j + \delta \mathbf{n}_i \cdot \nabla_x \varphi_j, (\kappa + \sigma)\varphi_k)$$
$$\mathbf{S}_i^{jk} = (\varphi_j + \delta \mathbf{n}_i \cdot \nabla_x \varphi_j, \sigma \varphi_k),$$

where $j = 1, ..., N$ and $k = 1, ..., N$.

3.5 Iterative Methods

The linear system of equations resulting from the discretization described above is large (10^7 unknowns at least), sparse, and strongly coupled due to the integral operator. Usually, an iterative scheme is applied to solve such large systems of equations (e.g., see [29, 64]). The standard algorithm used in astrophysics in this case is the so-called Λ- or Source Iteration method [48] which can be viewed as a form of the Richardson method with nearly block-Jacobi-preconditioning. Use of a full Jacobi preconditioner is a first step towards better convergence rates [62]. Since the transport operator is inverted explicitly, these methods converge very fast for transport dominated problems. If, further, the triangular matrix structure of upwind-discretizations is exploited, the inversion is indeed very cheap (one matrix-vector-multiplication).

Unfortunately, the present interest is in opaque and highly scattering media where this method – like other stationary iterations – breaks down, because the condition number of the iteration matrix becomes very large. Since the convergence rate of preconditioned Richardson iteration methods exclusively depends on the condition number, stationary iterations are not suited for the scattering dominated case. It can be shown, though, that the eigenvalue distribution of the transport matrix is clustered, with one eigenvalue located at the origin, while the others are close to unity or at least bounded away from zero (for details see [36]). Krylov space methods like GMRES or bi-CGSTAB described in [56] are particular suited to deal with such a type of eigenvalue distribution. While Krylov-type solvers are superior to stationary iteration methods, they still have poor convergence properties for scattering-dominated and opaque transfer problems, so that convergence acceleration methods need to be brought in. During the last several decades, significant progress has been achieved in the development of acceleration methods, i.e. preconditioning methods, which were designed to speed up the iterative convergence of discrete-ordinates approaches. A comprehensive review recently appeared in the literature (see [2]), covering the main methods that have been proposed. Of all the acceleration methods, the Diffusion Synthetic Acceleration (DSA) is one of the most efficient preconditioning methods, guaranteeing rapid convergence for all optical thicknesses and scattering values. In Sect. 7 a DSA-type preconditioner is presented that acts like a smoothing operator for specific intensities and which corresponds to a multilevel method for the diffusion approximation of the mean intensity. The word "multilevel" or "multigrid" often connotes a sequence of increasingly coarse spatial meshes. However, for transport problems, space and angle are variables, and the method we consider involves "collapsing" the transport problem onto "coarser grids" independent of the photon direction. It is demonstrated that the method allows to arrive at a fast solution of discrete linear problems in those cases where scattering is dominant as well as when transition regions occur.

3.6 Parallelization

Transport dominated problems differ in one specific point from elliptic problems: There is a distinct direction of information flow. This has to be considered in the development of parallelization strategies. While a domain decomposition for Poisson's equation should minimize the length of interior edges, this does not yield an efficient method for transport equations.

In [37], it was concluded that parallelization strategies for transport equations should not divide the domain across the transport direction. The solution of the radiative transfer equation consists of a number of transport inversions for different photon propagation directions. The construction of an efficient domain decomposition method would require a direction dependent splitting of the domain Ω. Since this causes immense implementation problems, we decided to use ordinate parallelization.

This strategy distributes the ordinate space S^2 of the radiative transfer equation. Since we use discontinuous shape functions for the ordinate variable and there is no local coupling due to the integral operator, this results in a true non-overlapping parallelization. Clearly, it has disadvantages, too. As the scattering integral is a global operator, ordinate parallelization involves global communication. In add, the resolution in space is restricted by the per node memory and not by the total memory of the machine. In [55] memory and CPU time requirements are investigated for a test model configuration distributing the ordinate space S^2 over 1 to 16 processors. The results show that the code scales optimal (i.e. double the number of processors, halves the required CPU time and memory) for parallel applications.

4 Polychromatic 3D Line Transfer

In the following an algorithm modeling polychromatic, i.e. frequency-dependent, 3D radiation fields in moving media is presented. This extension of the monochromatic model described in the previous section is required, since the radiation field is of major importance for the energy transfer in many differentially moving celestial objects like novae, supernovae, collapsing molecular clouds, collapsing and expanding or rotating gas clouds (halos), and accretion discs. Additionally, resonance line profiles are strongly affected by radiative transfer processes, providing deep insight into the density structure and the nature of the macroscopic velocity field of the emanating object.

The present work aims at the modeling of 3D radiation fields in gas clouds from the early universe, in particular as to the influence of varying distributions of density and velocity. In observations of high-redshift gas clouds (halos), the Lyα transition from the first excited energy level to the ground state of the hydrogen atom is usually found to be the only prominent emission line in the entire spectrum. It is a well-known assumption that high-redshifted hydrogen clouds are the precursors of present-day galaxies. Thus, the investigation of the Lyα line is of paramount importance for the theory of galaxy formation and evolution. The observed Lyα line – or rather, to be precise, its profile – reveals both the complexity of the spatial distribution and of the kinematics of the interstellar gas, and also the nature of the photon source.

The transfer of resonance line photons is profoundly determined by scattering in space and frequency. Analytical (see [51]) as well as early numerical methods [1, 32] were restricted to one-dimensional, static media. Only recently, codes based on the Monte Carlo method were developed which are capable to investigate the more general case of a multi-dimensional medium (see [3, 4, 70]).

4.1 Line Transfer in Moving Media

The frequency-dependent radiation field in moving media is obtained by solving the non-relativistic radiative transfer equation in a co-moving frame (for details see [67]). For a three-dimensional domain Ω the operator form of the equation is

$$(\mathcal{T} + \mathcal{D} + \mathcal{S} + \chi(\mathbf{x}, \nu))\mathcal{I}(\mathbf{x}, \mathbf{n}, \nu) = f(\mathbf{x}, \nu). \quad (24)$$

$\mathcal{T}$ is the transfer operator, $\mathcal{D}$ the "Doppler" operator, and $\mathcal{S}$ the scattering operator, which are defined as follows

$$\mathcal{T}\mathcal{I}(\mathbf{x}, \mathbf{n}, \nu) = \mathbf{n} \cdot \nabla_x \mathcal{I}(\mathbf{x}, \mathbf{n}, \nu),$$

$$\mathcal{D}\mathcal{I}(\mathbf{x}, \mathbf{n}, \nu) = w(\mathbf{x}, \mathbf{n})\nu \frac{\partial}{\partial \nu} \mathcal{I}(\mathbf{x}, \mathbf{n}, \nu),$$

$$\mathcal{S}\mathcal{I}(\mathbf{x}, \mathbf{n}, \nu) = -\frac{\sigma(\mathbf{x})}{4\pi} \int_0^\infty \int_{S^2} R(\hat{\mathbf{n}}, \hat{\nu}; \mathbf{n}, \nu) \mathcal{I}(\mathbf{x}, \hat{\mathbf{n}}, \hat{\nu}) \, d\hat{\omega} \, d\hat{\nu}.$$

The relativistic invariant specific intensity $\mathcal{I}$ is a function of six variables, three of which give the spatial location $\mathbf{x}$, while two variables give the direction $\mathbf{n}$ (pointing in the direction of the solid angle $d\omega$ of the unit sphere S^2), and one variable gives the frequency ν.

The extinction coefficient $\chi(\mathbf{x},\nu) = \kappa(\mathbf{x},\nu) + \sigma(\mathbf{x},\nu)$ is the sum of the absorption coefficient $\kappa(\mathbf{x},\nu) = \kappa(\mathbf{x})\varphi(\nu)$ and the scattering coefficient $\sigma(\mathbf{x},\nu) = \sigma(\mathbf{x})\varphi(\nu)$. The frequency-dependence is given by a normalized profile function $\varphi \in L^1(R^+)$. The core of the Lyα line is dominated by Doppler broadening. The effects of radiation and resonance damping may be important in the wings of the line at low column densities. Under the assumption that these mechanisms are all uncorrelated, one can account for their combined effects by a convolution procedure. The folding of the Doppler profile with the Lorentz profiles from radiation and resonance damping gives a Voigt profile, which is the general description of Lyα line profiles (see [48]). For all applications presented in this paper $v_\mathrm{turb} \gg v_\mathrm{therm}$ is adopted. This results in a very broad Doppler core dominating the Lorentzian wings of the Voigt profile. Thus, a Doppler profile

$$\varphi(\nu) = \frac{1}{\sqrt{\pi}\,\Delta\nu_D} \exp\left[-\left(\frac{\nu-\nu_0}{\Delta\nu_D}\right)^2\right] \qquad (25)$$

is a reasonable description of a Lyα line profile, where ν_0 is the frequency of the line center and c is the speed of light. The Doppler width $\Delta\nu_D$ and the Doppler velocity v_D are determined by a thermal velocity v_therm and a small scale turbulent velocity v_turb,

$$\Delta\nu_D = \frac{\nu_0}{c} v_D = \frac{\nu_0}{c}\sqrt{v_\mathrm{therm}^2 + v_\mathrm{turb}^2}. \qquad (26)$$

For the source term

$$f(\mathbf{x},\nu) = \kappa(\mathbf{x},\nu) B(T(\mathbf{x}),\nu) + \epsilon(\mathbf{x},\nu), \qquad (27)$$

we can consider thermal radiation and non-thermal radiation. In the case of thermal radiation, f is calculated from a temperature distribution $T(\mathbf{x})$, where $B(T,\nu)$ is the Planck function.

The "Doppler" operator $\mathcal{D}$ causes the Doppler shift of the photons. A derivation for non-relativistic moving media [4] can be found in [67]. The basis of the derivation is a "simplified" Lorentz transformation, which, in fact, is a Galilei transformation in combination with a linear Doppler formula. We restrict ourselves to sufficiently small velocity fields, and furthermore neglect aberration and advection effects, i.e. the photon propagation direction $\mathbf{n}$ is kept unchanged during the transformation. The function

$$w(\mathbf{x},\mathbf{n}) = -\mathbf{n}\cdot\nabla_x\left(\mathbf{n}\cdot\frac{\mathbf{v}(\mathbf{x})}{c}\right) \qquad (28)$$

---

[4]This includes velocities fields up to approximately 10% of the speed of light

is the gradient of the velocity field $\mathbf{v}(\mathbf{x})$ in direction $\mathbf{n}$. Note that the sign of w may change depending on the complexity of the velocity field $\mathbf{v}$.

Line transfer problems often include scattering processes where both the direction and the frequency of a photon may be changed. These changes are described by a general redistribution function $R(\hat{\mathbf{n}}, \hat{\nu}; \mathbf{n}, \nu)$ which gives the joint probability that a photon will be scattered from direction $\hat{\mathbf{n}}$ in solid angle $d\hat{\omega}$ and frequency range $(\hat{\nu}, \hat{\nu} + d\hat{\nu})$ into solid angle $d\omega$ in direction $\mathbf{n}$ and frequency range $(\nu, \nu + d\nu)$. If we assume that the radiation field is nearly isotropic, the scattering process may be approximated by

$$\mathcal{SI} = -\frac{\sigma(\mathbf{x})}{4\pi} \int_0^\infty R(\hat{\nu}, \nu) \int_{S^2} \mathcal{I}(\mathbf{x}, \hat{\mathbf{n}}, \hat{\nu}) \, d\hat{\omega} \, d\hat{\nu}, \qquad (29)$$

where $R(\hat{\nu}, \nu)$ is the angle-averaged redistribution function

$$R(\hat{\nu}, \nu) = \frac{1}{(4\pi)^2} \int_{S^2} \int_{S^2} R(\mathbf{x}, \hat{\mathbf{n}}, \hat{\nu}; \mathbf{n}, \nu) \, d\hat{\omega} \, d\omega. \qquad (30)$$

The function defined by Eq. (30) is normalized such that

$$\int_0^\infty \int_0^\infty R(\hat{\nu}, \nu) \, d\hat{\nu} \, d\nu = 1. \qquad (31)$$

For the calculations in Sect. 4.3, we consider two limiting cases: strict coherence and complete redistribution in the comoving frame. In the former case, we have

$$R(\hat{\nu}, \nu) = \varphi(\hat{\nu})\delta(\nu - \hat{\nu}), \qquad (32)$$

and in the latter

$$R(\hat{\nu}, \nu) = \varphi(\hat{\nu})\varphi(\nu). \qquad (33)$$

Thus, for coherent isotropic scattering, the scattering term simplifies to

$$\mathcal{S}^{\text{coh}} \mathcal{I}(\mathbf{x}, \mathbf{n}, \nu) = -\frac{\sigma(\mathbf{x}, \nu)}{4\pi} \int_{S^2} \mathcal{I}(\mathbf{x}, \hat{\mathbf{n}}, \nu) \, d\hat{\omega} \qquad (34)$$

In the case of complete redistribution, the photons are scattered isotropically in angle, but are randomly redistributed over the line profile. Then, the scattering term reads

$$\mathcal{S}^{\text{crd}} \mathcal{I}(\mathbf{x}, \mathbf{n}, \nu) = -\frac{\sigma(\mathbf{x}, \nu)}{4\pi} \int_0^\infty \varphi(\hat{\nu}) \int_{S^2} \mathcal{I}(\mathbf{x}, \hat{\mathbf{n}}, \hat{\nu}) \, d\hat{\omega} \, d\hat{\nu}. \qquad (35)$$

4.2 Boundary Conditions

Formally, the frequency derivative in the Doppler term of (24) can be viewed as being similar to the time derivative for non-stationary radiative transfer or heat transfer problems, so that a parabolic system of initial value problems is obtained. Unfortunately, the Doppler term also contains a function $w(\mathbf{x}, \mathbf{n})$, which is essentially the gradient of the macroscopic velocity field, multiplied by the photon propagation direction $\mathbf{n}$ (cf. (28)). Depending on the complexity of the velocity field, this function $w(\mathbf{x}, \mathbf{n})$ may change its sign at different points $\mathbf{x}$ and for different directions $\mathbf{n}$ when setting up the initial value either on the lower or upper frequency boundary. In this paper we take this into account by defining an additional frequency boundary value problem, which involves a "frequency inflow boundary" $\Sigma^- = \Omega \times S^2 \times \partial \Lambda$ depending on the sign of $w(\mathbf{x}, \mathbf{n})$. In order to solve (24) correctly, boundary conditions of the form

$$\mathcal{I}(\mathbf{x}, \mathbf{n}, \nu) = p(\mathbf{x}, \mathbf{n}, \nu) \quad \text{on } \Gamma^- \times \Lambda \quad \text{and} \tag{36}$$

$$\mathcal{I}(\mathbf{x}, \mathbf{n}, \nu) = q(\mathbf{x}, \mathbf{n}, \nu) \quad \text{on } \Sigma^-. \tag{37}$$

must be imposed. The "spatial inflow boundary" is

$$\Gamma^- \times \Lambda = \Big\{ (\mathbf{x}, \mathbf{n}, \nu) \in \Gamma \ \Big| \ \mathbf{n}_\Gamma \cdot \mathbf{n} < 0 \Big\}, \tag{38}$$

where $\mathbf{n}_\Gamma$ is the normal unit vector to the boundary surface Γ of the spatial domain Ω. The sign of the product $\mathbf{n}_\Gamma \cdot \mathbf{n}$ describes the "flow direction" of the photons across the boundary of the spatial domain. For our modeling of the spectral Lyα line, we assume that there is no continuum emission "outside of the line" and that there are no light sources "outside the modeled domain". In this case the two boundary conditions for the solution of the transfer equation (24) in moving media are

$$\mathcal{I}(\mathbf{x}, \mathbf{n}, \nu) = 0 \quad \text{on } \Sigma^-, \tag{39}$$

$$\mathcal{I}(\mathbf{x}, \mathbf{n}, \nu) = 0 \quad \text{on } \Gamma^- \times \Lambda. \tag{40}$$

4.3 Finite Element Discretization

The simultaneous Galerkin discretization in space-frequency is, from a theoretical point of view, an extremely appealing approach. Considering the memory requirement, the solution of the complete algebraic system resulting from this discretization is by far too "expensive". With a view to reducing the memory demand, a discretization first in space and then in frequency is performed. This approach is called method of lines and finally results in a system of ordinary differential equations. Contrary to the continuous Galerkin discretization in space, the discretization in frequency is a discontinuous Galerkin (DG) method. It is important to note that the above discretization in space and in frequency can be interpreted as a simultaneous Galerkin discretization in the space-frequency domain. The frequency DG method is performed by using piecewise polynomials of degree zero, which corresponds to an implicit Euler method for N equidistantly distributed frequency points

$\nu_i \in \{\nu_1, \nu_2, ..., \nu_N\} \subset \Lambda$ (see [9]). DG methods are not only used for discretizing ordinary differential equations, especially initial value problems, but also for the time discretization and mesh control of partial differential equations as described in [23, 24] and [22].

Discretization for coherent scattering

In order to solve the radiative transfer equation (24), we first carry out the frequency discretization including coherent scattering. With the abbreviation

$$\mathcal{A}^{\text{coh}} = \mathcal{T} + \mathcal{S}^{\text{coh}} + \chi(\mathbf{x}, \nu) \tag{41}$$

Eq. (24) can be written as

$$\mathcal{A}^{\text{coh}} \mathcal{I}(\mathbf{x}, \mathbf{n}, \nu) + w(\mathbf{x}, \mathbf{n}) \nu \frac{\partial}{\partial \nu} \mathcal{I}(\mathbf{x}, \mathbf{n}, \nu) = f(\mathbf{x}, \nu). \tag{42}$$

Since the function $w(\mathbf{x}, \mathbf{n})$ may change its sign, this simple difference scheme for the Doppler term reads

$$w(\mathbf{x}, \mathbf{n}) \nu \frac{\partial \mathcal{I}}{\partial \nu} \longrightarrow w \nu_i \frac{\mathcal{I}_i - \mathcal{I}_{i-1}}{\Delta \nu} \qquad (w_i > 0), \tag{43}$$

and

$$w(\mathbf{x}, \mathbf{n}) \nu \frac{\partial \mathcal{I}}{\partial \nu} \longrightarrow w \nu_i \frac{\mathcal{I}_{i+1} - \mathcal{I}_i}{\Delta \nu} \qquad (w_i < 0), \tag{44}$$

where $\Delta \nu$ is the constant frequency step size. All quantities referring to the discrete frequency point ν_i are denoted by an index "i". Employing the Euler method, we get a semi-discrete representation of Eq. (42)

$$\left(\mathcal{A}_i^{\text{coh}} + \frac{|w|\nu_i}{\Delta \nu}\right) \mathcal{I}_i = f_i + \frac{|w|\nu_i}{\Delta \nu} \begin{cases} \mathcal{I}_{i-1} & (w > 0) \\ \mathcal{I}_{i+1} & (w < 0) \end{cases}. \tag{45}$$

The additional term on the left side of Eq. (45) can be interpreted as an artificial opacity, which is advantageous for the solution of the resulting linear system of equations. The additional term on the right side of Eq. (45) is included as an artificial source term. Equation (45) can be written in a compact operator form

$$\tilde{\mathcal{A}}_i^{\text{coh}} \mathcal{I}_i = \tilde{f}_i. \tag{46}$$

Given a discretization with L degrees of freedom in Ω, M directions on the unit sphere S^2 and N frequency points, the overall discrete system has the matrix form

$$\mathbf{A}^{\text{coh}} \mathbf{u} = \mathbf{f}, \tag{47}$$

with the vector $\mathbf{u}$ containing the discrete intensities and the vector $\mathbf{f}$ the values of the source term for all frequency points ν_i. Both vectors are of length $(L \cdot M \cdot N)$ and $\mathbf{A}^{\text{coh}}$ is a $(L \cdot M \cdot N) \times (L \cdot M \cdot N)$ matrix. The fact that the function $w(\mathbf{x}, \mathbf{n})$ may change its sign, results in a block-tridiagonal structure of $\mathbf{A}^{\text{coh}}$ and we get

$$\begin{pmatrix} \tilde{\mathbf{A}}_1^{\text{coh}} & \mathbf{R}_1 & 0 & \cdots & 0 \\ \mathbf{B}_2 & \tilde{\mathbf{A}}_2^{\text{coh}} & \mathbf{R}_2 & \ddots & \vdots \\ 0 & \ddots & \ddots & \ddots & 0 \\ \vdots & \ddots & \ddots & \ddots & \mathbf{R}_{N-1} \\ 0 & \cdots & 0 & \mathbf{B}_N & \tilde{\mathbf{A}}_N^{\text{coh}} \end{pmatrix} \begin{pmatrix} \mathbf{u}_1 \\ \mathbf{u}_2 \\ \vdots \\ \vdots \\ \mathbf{u}_N \end{pmatrix} = \begin{pmatrix} \mathbf{f}_1 \\ \mathbf{f}_2 \\ \vdots \\ \vdots \\ \mathbf{f}_N \end{pmatrix}. \tag{48}$$

According to the sign of $w(\mathbf{x}, \mathbf{n})$ the block matrices $\mathbf{R}_i$ and $\mathbf{B}_i$ hold entries of $w(\mathbf{x}, \mathbf{n})\nu_i/\Delta\nu$ causing a redshift and blueshift of the photons in the medium, respectively. Requiring a reasonable resolution, the resulting linear system of equations of the total system is too large to be solved directly. Hence, we are treating N "monochromatic" radiative transfer problems

$$\tilde{\mathbf{A}}_i^{\text{coh}} \mathbf{u}_i = \tilde{\mathbf{f}}_i, \tag{49}$$

with a slightly modified right hand side

$$\tilde{\mathbf{f}}_i = \mathbf{f}_i + \mathbf{R}_i \mathbf{u}_{i+1} + \mathbf{B}_i \mathbf{u}_{i-1}. \tag{50}$$

Note that using simple velocity fields (e.g. collapse or expansion of a gas cloud) the sign of $w(\mathbf{x}, \mathbf{n})$ is fixed and the matrix $\mathbf{A}^{\text{coh}}$ has a block-bidiagonal structure. This is favorable for our solution strategy, since we only need to solve Eq. (49) once for each frequency point.

Discretization for complete redistribution

In order to include complete redistribution, we use a simple quadrature method for the frequency integral in the scattering operator $\mathcal{S}^{\text{crd}}$ in Eq. (35). The Doppler term is discretized using an implicit Euler scheme as described in the previous section. Starting from N equidistantly distributed frequency points $\nu_i \in \{\nu_1, \nu_2, ..., \nu_N\} \subset \Lambda$ and N weights $q_1, q_2, ..., q_N$, we define a quadrature method

$$Q(\nu_i) := \sum_{j=1}^{N} q_j \xi(\nu_j) \tag{51}$$

for integrals $\int_\Lambda \xi(\nu') d\nu'$. In the case of complete redistribution the kernel is

$$\xi(\nu_j) = \frac{\varphi(\nu_j)}{4\pi} \int_{S^2} \mathcal{I}(\mathbf{x}, \hat{\mathbf{n}}, \nu_j) d\hat{\omega}, \tag{52}$$

and the semi-discretized scattering integral in Eq. (35) for a particular frequency point $\ni_i$ reads

$$\frac{\sigma_i}{4\pi}\varphi_i q_i \int_{S^2} \mathcal{I}_i d\hat{\omega} + \frac{\sigma_i}{4\pi}\sum_{j\neq i}^{N}\varphi_j q_j \int_{S^2} \mathcal{I}(\mathbf{x},\hat{\mathbf{n}},\nu_j)d\hat{\omega}. \tag{53}$$

With this scattering operator the semi-discrete formulation of the transfer problem for each frequency point $\ni_i$ is

$$\left(\mathcal{A}_i^{\mathrm{crd}} + \frac{|w|\nu_i}{\Delta\nu}\right)\mathcal{I}_i = \tilde{f}_i + \frac{\sigma_i}{4\pi}\sum_{j\neq i}^{N}\varphi_j q_j \int_{S^2}\mathcal{I}(\mathbf{x},\hat{\mathbf{n}},\nu_j)d\hat{\omega}, \tag{54}$$

where

$$\mathcal{A}_i^{\mathrm{crd}} = \mathcal{T} + \chi_i + \varphi_i q_i \mathcal{S}^{\mathrm{coh}}. \tag{55}$$

The additional terms on the right hand side of Eq. (54) must be interpreted as artificial source terms. Eq. (54) can also be written in a compact operator form

$$\tilde{\mathcal{A}}_i^{\mathrm{crd}}\mathcal{I}_i = \hat{f}_i. \tag{56}$$

The total discrete system has the matrix form (cf. Eq. (47))

$$\mathbf{A}^{\mathrm{crd}}\mathbf{u} = \mathbf{f}. \tag{57}$$

Unfortunately, the global frequency coupling via the scattering integral in Eq. (35) results in a full block matrix and we get

$$\mathbf{A}^{\mathrm{crd}} = \begin{pmatrix} \tilde{\mathbf{A}}_1^{\mathrm{crd}} & \mathbf{R}_1+\mathbf{Q}_2 & \mathbf{Q}_3 & \cdots & \mathbf{Q}_N \\ \mathbf{B}_2+\mathbf{Q}_1 & \tilde{\mathbf{A}}_2^{\mathrm{crd}} & \mathbf{R}_2+\mathbf{Q}_3 & \ddots & \vdots \\ \mathbf{Q}_1 & \ddots & \ddots & \ddots & \vdots \\ \vdots & \ddots & & \ddots & \vdots \\ \mathbf{Q}_1 & \cdots & \cdots & \cdots & \tilde{\mathbf{A}}_N^{\mathrm{crd}} \end{pmatrix}. \tag{58}$$

$\mathbf{B}_i$ and $\mathbf{R}_i$ again contain the contribution of the Doppler factor $w(\mathbf{x},\mathbf{n})\nu_i/\Delta\nu$, whereas $\mathbf{Q}_j$ holds the terms from the quadrature scheme. As we already explained in the case of coherent scattering, we do not directly solve the total system, but successively solve N "monochromatic" radiative transfer problems

$$\tilde{\mathbf{A}}_i^{\mathrm{crd}}\mathbf{u}_i = \hat{\mathbf{f}}_i, \tag{59}$$

successively, with a modified right hand side

$$\hat{\mathbf{f}}_i = \mathbf{f}_i + \mathbf{R}_i\mathbf{u}_{i+1} + \mathbf{B}_i\mathbf{u}_{i-1} + \sum_{j\neq i}\mathbf{Q}_j\mathbf{u}_j. \tag{60}$$

4.4 Full Solution Algorithm

Equation (46) and (56) are of the same form as the monochromatic radiative transfer equation, cf. Eq. (6), for which we propose a solution method based on a finite element technique in Sect. 3. The finite element method employs hierarchically structured grids which are locally refined by means of duality-based a posteriori error estimates. Now, we apply this method to Eq. (46) or Eq. (56). The solution is obtained using an iterative procedure, where quasi-monochromatic radiative transfer problems are solved successively. In brief, the full solution algorithm reads:

1. Start with $\mathcal{I} = 0$ for all frequencies.
2. Solve Eq. (47) or Eq. (57) for $i = 1, .., N$.
3. Repeat step 2 until convergence is reached.
4. Refine grid and repeat steps 2 and 3.

We start with a relatively coarse grid, where only the most important structures are pre-refined, and assure that the frequency interval $[\nu_1, \nu_N]$ is wide enough to cover the total line profile. Then, we solve Eq. (46) or Eq. (56) for each frequency several times depending on the choice of the redistribution function and the velocity field. During this fixed point iteration, we monitor the changes of the resulting line profile in a particular direction $\mathbf{n}_{\text{out}}$. When the line profile remains unchanged, we turn to step 4 and refine the spatial grid. Again, we apply the fixed fraction grid refinement strategy: The cells are ordered according to the size of the local refinement indicator $\eta_K = \max(\eta_K(\nu_i))|_{\nu_i}$ and a fixed portion of the cells with largest η_K is refined. $\eta_K(\nu_i)$ is an indicator for the error contribution of the solution in cell K at frequency ν_i.

5 Test Calculations

The simple 3D test calculations presented in this section use a spherically symmetric density distribution of the form

$$\chi(\mathbf{x}) = \begin{cases} \chi_0/(1+\alpha r_c^2) & \text{for } r \leq r_c, \\ \chi_0/(1+\alpha r^2) & \text{for } r_c < r \leq r_h, \\ \chi_0/(1+\alpha r_h^2)/10^3 & \text{for } r > r_h, \end{cases} \quad (61)$$

where r_c is a constant core radius in the center of the halo. The halo radius is r_h and the square of the radius is $r^2 = x^2 + y^2 + z^2$. The constant opacity χ_0 is determined by the line center optical depth

$$\tau = \int_{r_c}^{r_h} \chi(\mathbf{x}) \varphi(\nu_0) \, \mathbf{n}_{\text{thick}} \, d\mathbf{x} \quad (62)$$

between r_c and r_h along the most optical thick photon direction $\mathbf{n}_{\text{thick}}$. In total, the spatial distribution of χ is determined by three parameters: the radii r_c and r_h, the dimensionless parameter α, and the optical depth τ. For r_c, r_h and α we use the values given in Table 2.

Since we are predominantly interested in the transfer of radiation in resonance lines like Lyα, we assume $\sigma(\mathbf{x}) = \chi(\mathbf{x})$ and $\kappa(\mathbf{x}) = 0$ for all calculations presented here. This restricts us to the use of a given source function. In particular, we consider a single spatially confined source region located at the origin of the unit cube at the position $\mathbf{x}_i = 0$ with radius r_s:

$$f(\mathbf{x}, \nu) = \begin{cases} \varphi(\nu) & \text{for } |\mathbf{x} - \mathbf{x}_i| \leq r_s \\ 0 & \text{for } |\mathbf{x} - \mathbf{x}_i| > r_s \end{cases}. \tag{63}$$

The function $\varphi(\nu)$ is the Doppler profile defined in Eq. (25).

In general, the finite element code is able to consider arbitrary velocity fields. For velocity fields defined on a discrete grid, e.g. resulting from hydrodynamical simulations, the velocity gradient in direction $\mathbf{n}$ must be obtained numerically. Here, we use two simple velocity fields and calculate the function w analytically.

The first velocity field describes a spherically symmetric collapsing ($v_0 < 0$) or expanding ($v_0 > 0$) medium and is of the form

$$\mathbf{v}_{\text{io}} = v_0 \left(\frac{r_0}{r}\right)^l \frac{\mathbf{x}}{r}, \tag{64}$$

where $r = |\mathbf{x}|$ and v_0 the scalar velocity at radius r_0. The corresponding w function is

$$w(\mathbf{x}, \mathbf{n}) = v_0 \left(\frac{r_0}{r}\right)^l \left(\frac{1}{r} - (l+1)\frac{|\mathbf{nx}|}{r^3}\right). \tag{65}$$

For the second velocity field, we assume rotation around the z-axis

$$\mathbf{v}_{\text{rot}} = v_0 \left(\frac{R_0}{R}\right)^l R^{-1} \begin{pmatrix} y \\ -x \\ 0 \end{pmatrix}, \tag{66}$$

where $R^2 = x^2 + y^2$ is the distance from the rotational axis and v_0 the scalar velocity at distance R_0. For $\mathbf{n} = (n_x, n_y, n_z)$, the w function reads

$$w = v_0 \left(\frac{R_0}{R}\right)^l (l+1) \left(\frac{xy(n_y^2 - n_x^2) + n_x n_y(x^2 - y^2)}{R^3}\right). \tag{67}$$

Fig. 1 shows the results of the finite element code for different optical depths, velocity fields and redistribution functions. We used 41 frequencies equally spaced in the interval $(\nu - \nu_0)/\Delta\nu_D = [-4, 6]$ and 80 directions. Starting with a grid of 4^3 cells and a pre-refined source region, we needed 3–5 spatial refinement steps.

For the simplest case of a static model configuration with coherent isotropic scattering even analytical solutions exist (see [55]). Fig. 1a displays the line

Table 2. Parameters used for all calculations. Distances are given in units of the computational unit cube.

r_h	r_c	α	r_s	v_D	v_0	r_0	R_0
1.0	0.2	10^3	0.2	$10^{-3}c$	$-10^{-3}c$	0.2	1.0

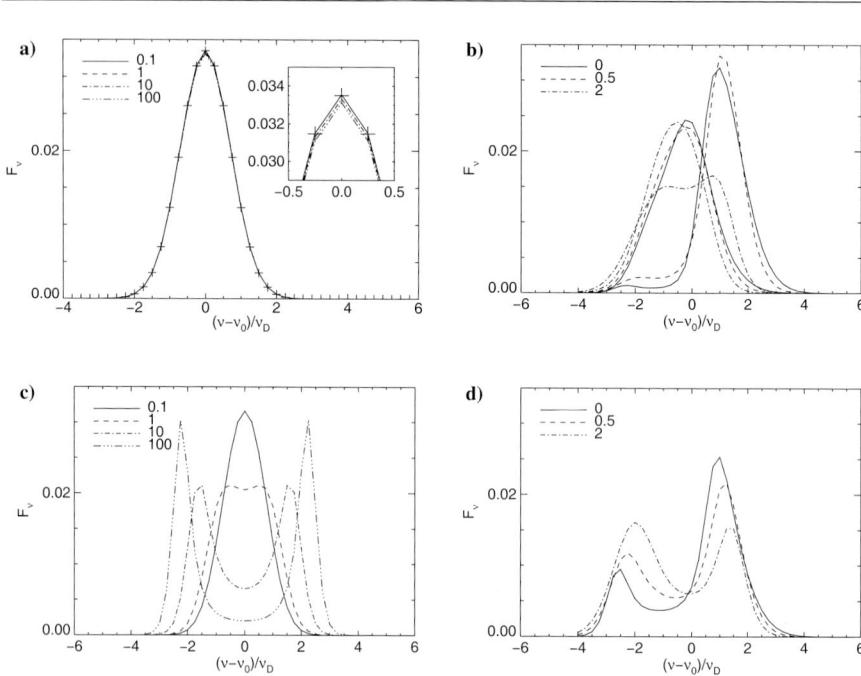

Fig. 1. Lyα line profiles calculated with the finite element code for a spherically symmetric model configuration: a) a static halo with coherent scattering, b) an collapsing halo with coherent scattering, c) a static halo with complete redistribution, and d) an collapsing halo with complete redistribution. For the static cases a) and c) the line styles refer to calculations with different optical depth τ as indicated. The small window in a) enlarges the peak of the line. The crosses mark the results of the analytical solution. For the moving halos we show in b) the results for $\tau = 1$ (thin lines) and $\tau = 10$ (thick lines) and in d) only for $\tau = 10$. Here, the line styles refer to the exponent l used for the velocity fields.

profiles for different τ. As expected, the Doppler profile is preserved and the flux F_ν is independent on τ. The deviation of the numerical results from the analytical solution indicated with crosses is very small. The line profiles for $\tau = 0.1$ and $\tau = 1$ are identical even in the little window which shows the peak of the line in more detail. For $\tau = 100$, the total flux is still conserved better than 99%. This result demonstrates that the frequency-dependent version of our finite element code works correctly.

Next, we consider a collapsing halo with coherent scattering and show the effects of frequency coupling due to the Doppler term. The line profiles in Fig. 1b are plotted for different exponents l of the velocity field $\mathbf{v}_{io}$ defined in Eq. (64). The line profiles are redshifted for $\tau = 1$ (thin lines). Most of the photons directly travel through the part of the halo moving away from the observer. Since the Doppler term is proportional to the gradient of the velocity field, the redshift is larger for a greater exponent l. For $\tau = 10$ (thick lines), the line profiles are blueshifted. Before photons escape from the optically thick part of the halo in front of the source, they are back-scattered and blueshifted in the approaching part of the halo behind the source. The blueshift is less pronounced for the accelerated inflow with $l = 2$, because the strong gradient of the velocity field leads to a slight redshift in the very inner parts of the halo. In this region, the total optical depth is still small. Further out, where the total optical depth increases, the line profile becomes blueshifted.

Complete redistribution leads to an even stronger coupling than the Doppler effect (see Sect. 3). The line profiles obtained for a static model with complete redistribution are displayed in Fig. 1c for different τ. With increasing optical depth the photons more and more escape through the line wings. For $\tau \geq 1$, we get a double-peaked line profile with an absorption trough in the line center. The greater τ the larger the distance between the peaks and the depth of the absorption trough. Since our frequency resolution is too poor for the narrow wings, the flux conservation is only 96% for $\tau = 100$.

Fig. 1d shows the results for an collapsing halo with complete redistribution for $\tau = 10$ and different exponents l. For $l = 0$ and $l = 0.5$ the collapsing motion of the halo enhances the blue wing of the double-peaked line profile. Equally, an expanding halo would enhance the red peak. But for $l = 2$, the red peak is slightly enhanced for an collapsing halo due to the strong velocity gradient, as explained above. This example allows an insight into the mechanisms of resonance line formation and shows the necessity of a profound multidimensional treatment.

6 Applications

All calculations discussed in this section were performed with complete redistribution. We used 49 equidistant frequencies covering the interval $(\nu - \nu_0)/\Delta\nu_D = [-6, 6]$, 80 directions and started from a grid with 4^3 cells and pre-refined source regions, which results in several 10^3 cells for the initial mesh. 3–7 refinement steps were performed leading to approximately 10^5 cells for the finest grid.

6.1 Elliptical Halos

As a first step towards a full three-dimensional problem without any symmetries, we investigated an axially symmetric, disk-like model configuration with

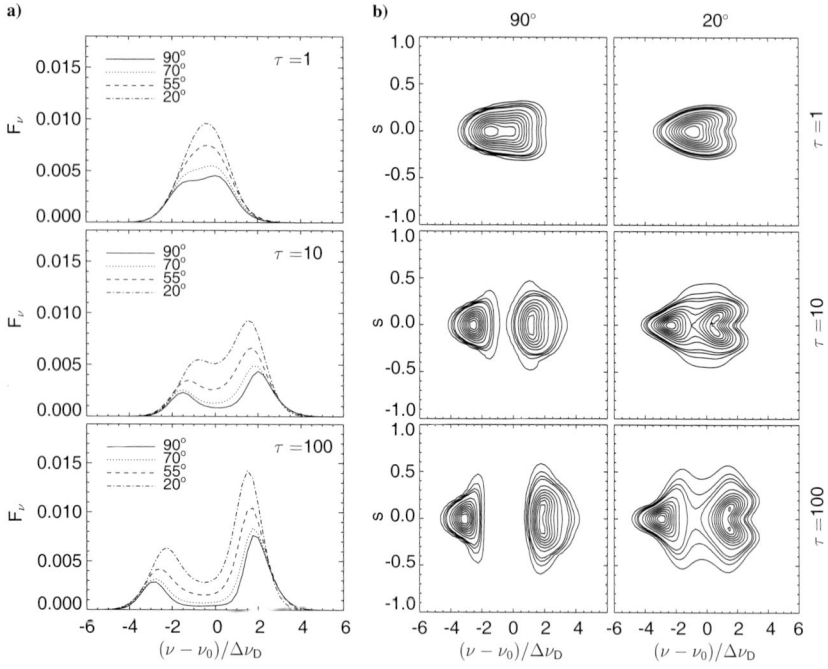

Fig. 2. Results obtained for a disk-like model configuration with global inflow motion ($l = 0.5$): a) Line profiles for different viewing angles and optical depths $\tau = \tau(\mathbf{n}_{\text{thick}})$. b) Two-dimensional spectra for an edge-on view (90°) and a nearly face-on view (20°) and different optical depths. The parameter s gives the spatial position within a slit covering the "visible" part of the disk-like halo. The contours are given for 2.5%, 5%, 7.5%, 10%, 20%, ..., 90% of the maximum value.

$r^2 = (x/3)^2 + (y/3)^2 + z^2$ in Eq. (61), and with a single source located at $\mathbf{x} = 0$ for several $\tau = \tau(\mathbf{n}_{\text{thick}})$ and for different velocity fields. The most optically thick direction $\mathbf{n}_{\text{thick}}$ is the direction within the equatorial plane of the disk. The direction perpendicular to the equatorial plane is the z-axis which we also call rotational axis even for cases without rotation.

First, we consider an inflow velocity field with $l = 0.5$ suitable for a gravitational collapse. Fig. 2a displays the calculated line profiles for different τ and viewing angles. As expected, the blue peak of the line is enhanced. The higher τ the stronger the blue peak. Two-dimensional spectra from high-resolution spectroscopy provide frequency-dependent data for only one spatial direction. This is achieved using so-called slit masks for telescope observations. In order to compare our results with these observations we calculated two-dimensional spectra in Fig. 2b using the data within a slit covering the "visible" part of the disk-like halo, i.e. the observable spatial distribution of the frequency integrated intensity. For low optical depth, the shape of the contour lines is a triangle. Photons changing frequency in order to escape via the blue wing

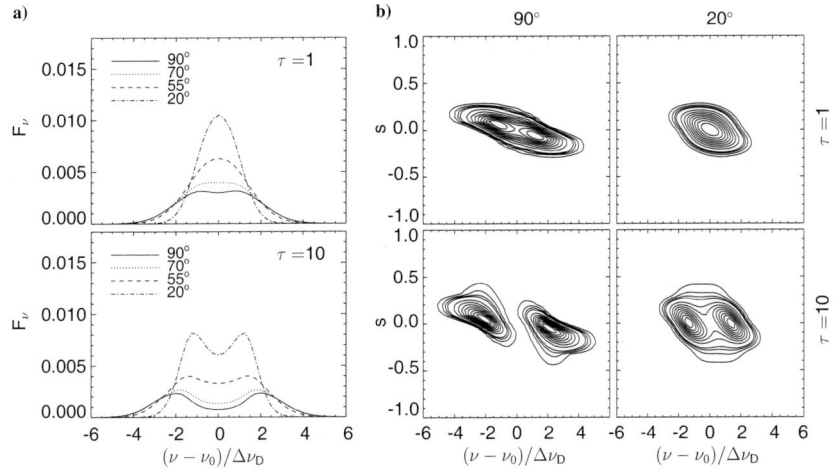

Fig. 3. Results obtained for a disk-like model configuration with Keplerian rotation ($l = 0.5$) around the z-axis: a) Line profiles for different viewing angles and optical depths $\tau = \tau(\mathbf{n}_{\text{thick}})$. b) Two-dimensional spectra for an edge-on view (90°) and a nearly face-on view (20°) and different optical depths. The parameter s gives the spatial position within a slit covering the "visible" part of the disk-like halo. The contours are given for 2.5%, 5%, 7.5%, 10%, 20%, ..., 90% of the maximum value.

are also scattered in space. The consequence is the greater spatial extent of the blue wing. Apart from a growing gap between the two peaks with higher optical depth, the general triangular shape in conserved.

Next, we investigated the disk-like model configuration with Keplerian rotation ($l = 0.5$), where the z-axis is the axis of rotation. The results are plotted in Fig. 3 for $\tau = 1$ and $\tau = 10$. The calculated line profiles appear to be symmetric with respect to the line center and show the same behavior with increasing optical depths as in the static case. However, the extension of the line wings towards higher and lower frequencies is strongly increasing with increasing viewing angle because of the growing effect of the velocity field. Rotation is clearly visible in the two-dimensional spectra (Fig. 3b). The shear of the contour lines is the characteristical pattern indicating rotational motion. For an edge-on view at $\tau = 10$, Keplerian rotation produces two banana-shaped emission regions.

Finally, Fig. 6.1 depicts the spatial intensity distribution for particular frequency points within the interval $(\nu - \nu_0)/\Delta\nu_D = [-2, +2]$ (rows) and for different viewing angles (columns). The left figure shows the results obtained for a disk-like model configuration with global inflow motion ($l = 0.5$). The right figure shows the results obtained for a disk-like model configuration with Keplerian rotation ($l = 0.5$) around the z-axis. The intensity increases from blue to white.

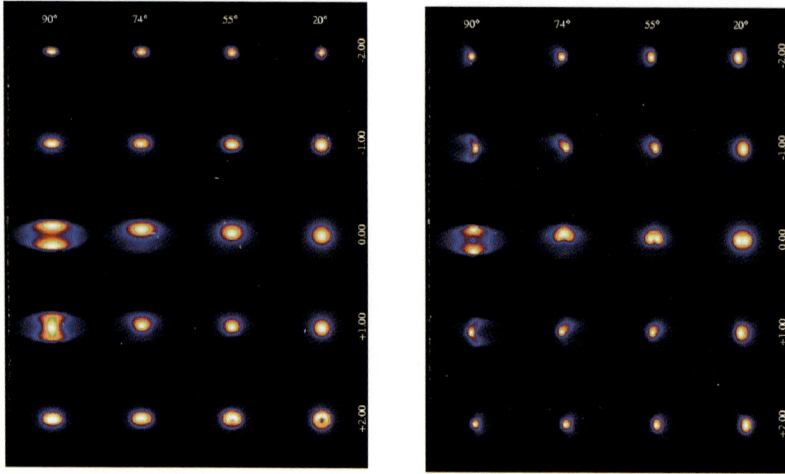

Fig. 4. Spatial intensity distribution for particular frequency points within the interval $(\nu - \nu_0)/\Delta\nu_D = [-2, +2]$ (rows) and for different viewing angles (columns). *Left:* Results obtained for a disk-like model configuration with global inflow motion ($l = 0.5$). *Right:* Results obtained for a disk-like model configuration with Keplerian rotation ($l = 0.5$) around the z-axis. The intensity increases from blue to white.

In spite of the relatively low optical depth of the simple model configurations, our results reflect the form of line profiles and the patterns in two-dimensional spectra of many high redshift galaxies. For example, the two-dimensional spectra of the Lyα blobs published in [60] (see Fig. 8) are comparable to the results obtained for collapsing (Fig. 2b) and rotating (Fig. 3b) halos. In the sample presented in [63] are many high redshift radio galaxies with single-peaked and double-peaked Lyα profiles. The corresponding two-dimensional spectra show asymmetrical emission regions which are more or less extended in space. The statistical study of emission lines from high redshift radio galaxies in [19] indicates that the triangular shape of the Lyα emission is a characteristical pattern in the two-dimensional spectra of high redshift radio galaxies. Since the emission of the blue peak of the line profile is predominately less pronounced, most of the associated halos should be in the state of expansion.

6.2 Multiple Sources

High redshift radio galaxies are found in the center of proto clusters. In such an environment, it could be possible that the Lyα emission of several galaxies is scattered in a common halo. To investigate this scenario, we started with a spherically symmetric distribution for the extinction coefficient and with

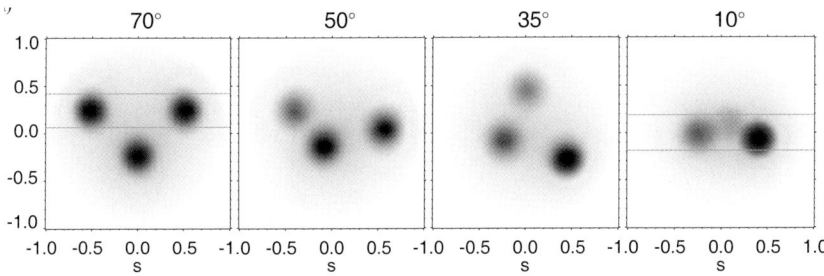

Fig. 5. Spatial distribution of the frequency integrated intensity for a static spherically symmetric halo configuration with $\tau = 10$ and three source regions for different viewing angles. The slit position is indicated for two different viewing angles (for details see text).

three source regions located at $x_1 = [0.5, 0.25, 0]$, $x_2 = [-0.5, 0.25, 0]$ and $x_3 = [0, -0.25, 0]$ forming a triangle in the x-y-plane. In the following figures, the results for $\tau = 10$ are presented.

Figure 5 shows Lyα images for four selected viewing directions. The specified angle is the angle between the viewing direction and the x-y-plane. Note that the orientation of the source positions within the plane is different for each image. Viewing the configuration almost perpendicular to the x-y-plane (70°), renders all three source regions visible, because the source regions are situated in the more optically thin, outer parts of the halo. Remember that most of the scattering matter is in and around the center of the system. For other angles, some of the source regions are located behind the center and therefore less visible on the images.

The corresponding line profiles and two-dimensional spectra are displayed in Fig. 6 for the static case as well as for a halo with global inflow and Keplerian rotation. Two-dimensional spectra from high-resolution spectroscopy provide frequency-dependent data for only one spatial direction. This is achieved using so-called slit masks for telescope observations. In order to compare our results with these observations we calculated two-dimensional spectra using the data within a slit covering the most "visible" parts of the halo, i.e. as many observable source regions as possible. Width and position of the slits are depicted in Fig. 5. We get double-peaked line profiles for almost all cases, except for the rotating halo, where the line profiles are very broad for viewing angles lower than 70°. Additional features, dips or shoulders, are visible in the red or blue wing. They arise because the three sources have significantly different velocities components with respect to the observer.

The slit for a viewing angle of 70° contains two sources. They show up as four emission regions in the two-dimensional spectra (Fig. 6b). The pattern is very symmetric, even for the moving halos. For a viewing angle of 10°, the slit covers all source regions. Nevertheless, the two-dimensional spectra are dominated by two pairs of emission regions resulting from the sources

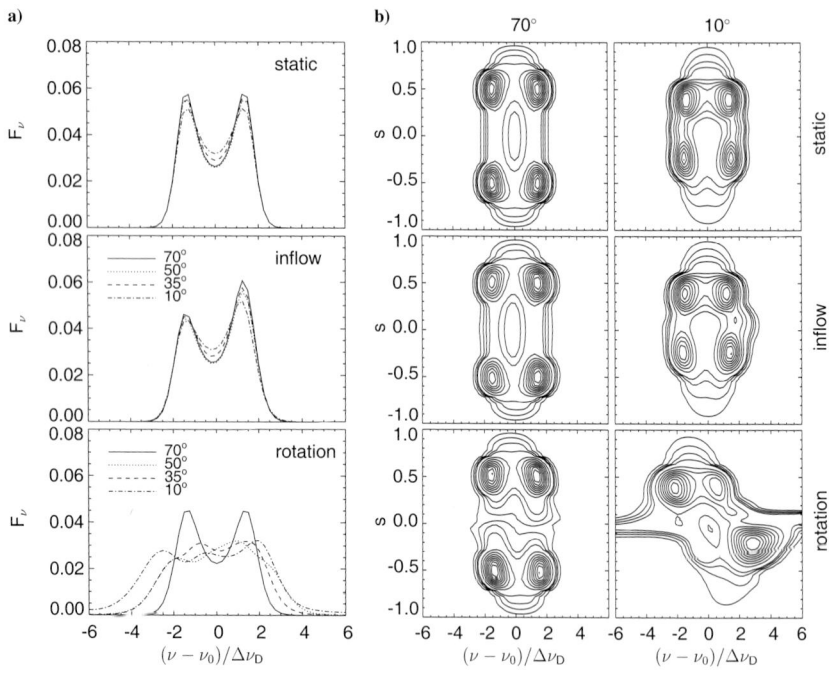

Fig. 6. Results obtained for a spherically symmetric halo configuration with $\tau = 10$ and three source regions for the static case and two different velocity fields (inflow and rotation): a) Line profiles for different viewing angles. b) Two-dimensional spectra for a nearly face-on view (70°) and a nearly edge-on view (10°). The position and width of the slit is indicated in Fig. 5. The contours are given for 2.5%, 5%, 7.5%, 10%, 20%, ..., 90% of the maximum value.

located closer to the observer. The third source region only shows up as a faint emission in the blue part in the case of global inflow. In the case of rotation, emission regions from a third source are present. But the overall pattern is very irregular and prevents a clear identification of the emission regions.

Finally, Fig. 6.2 depicts the spatial intensity distribution for particular frequency points within the interval $(\nu - \nu_0)/\Delta\nu_D = [-2, +2]$ (rows) and for different viewing angles (columns). The left figure shows the results obtained for a spherically symmetric halo configuration with $\tau = 10$ and three source regions for global inflow motion. The right figure shows the results obtained for a spherically symmetric halo configuration with $\tau = 10$ and three source regions for Keplerian rotation around the z-axis. The intensity increases from blue to white.

This example demonstrates the complexity of three dimensional problems. In a clumpy medium, the determination of the number and position of Lyα sources would be difficult by means of frequency integrated images alone.

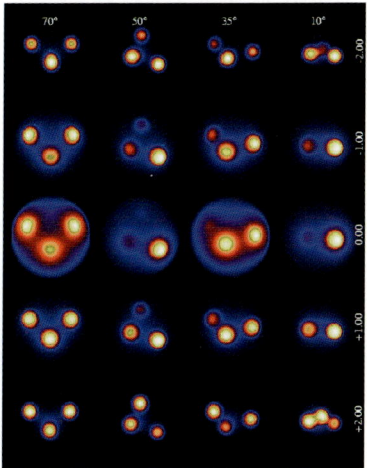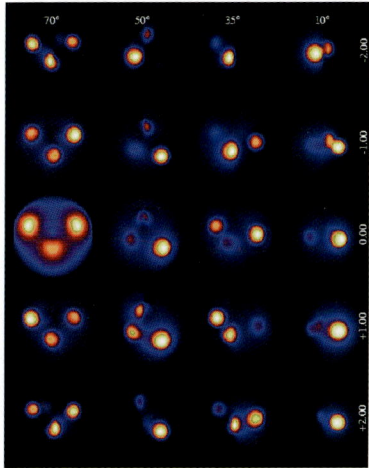

Fig. 7. Spatial intensity distribution for particular frequency points within the interval $(\nu - \nu_0)/\Delta\nu_D = [-2, +2]$ (rows) and for different viewing angles (columns). *Left:* Results obtained for a spherically symmetric halo configuration with $\tau = 10$ and three source regions for global inflow motion. *Right:* Results obtained for a spherically symmetric halo configuration with $\tau = 10$ and three source regions for Keplerian rotation around the z-axis. The intensity increases from blue to white.

Two-dimensional spectra may help, but could prove to be too complicated. More promising are images obtained from different parts of the line profile or information from other emission lines of OIII or Hα in a manner proposed in [39].

7 Multi-Model Preconditioning

It is often necessary to solve radiation transfer problems in composite domains with mixed transport regimes, so that in some parts of the domain there is diffusion control, whereas radiative transfer dominates elsewhere. The character of the transfer changes strongly, giving a change in the type of equation, if diffusion control changes to radiation control. In case of such a mixed regime, therefore, the application of standard solvers and of standard preconditioning methods for the discrete problem is usually marred by an extremely slow rate of convergence. In this section, we propose a solution method combining the radiative transfer model with its diffusion approximation, which is designed to yield good convergence rates for both types of the equation. For various reasons, the diffusion approximation alone is not sufficient, and the more general radiation transfer equation must be solved. First, we do not assume a lower bound for the scattering cross section and, therefore, the approximation may become inaccurate in parts of the domain. Furthermore, the question arises as to what boundary conditions for the diffusion problem are physically acceptable. These problems can be fully resolved only by solving the radiative transfer equation. Since the spectrum of the discrete operator becomes degenerate in the limit when scattering dominates, the solution of the corresponding discrete linear system is particularly challenging (see [36]).

When the parameters of the equation are constant, efficient solution techniques can be found easily. If scattering is small, a simple Gauss-Seidel process yields good convergence rates (see [34]). In the limit of high scattering, the solution is isotropic in the interior of the domain yielding an elliptic second order equation, which can be solved efficiently using multilevel preconditioning. The scheme suggested in this section combines these two preconditioners in a way related to two-level multigrid algorithms. It exploits the good convergence properties of both methods in different regimes. Similar schemes have been proposed under the names of quasi diffusion and diffusion-synthetic acceleration method (for details see [44, 2] and the literature cited therein).

7.1 Discretization and Diffusion Approximation

Let us first consider monochromatic radiative transfer problems and rewrite the corresponding partial integro-differential equation (2) for a multidimensional spatial domain $\Omega \subset \mathrm{R}^d$ $(d = 2, 3)$

$$\mathbf{n} \cdot \nabla_x I(\mathbf{x}, \mathbf{n}) = -\chi(\mathbf{x}) \Big(I(\mathbf{x}, \mathbf{n}) - \frac{1-\epsilon}{2(d-1)\pi} \int_{S^{d-1}} I(\mathbf{x}, \mathbf{n}') \, d\omega' - \epsilon f(\mathbf{x}) \Big), \quad (68)$$

with the thermalization factor $\epsilon = \kappa/\chi$ and the albedo $1 - \epsilon = \sigma/\chi$. For the sake of simplicity, isotropic scattering (i.e., phase function $P(\mathbf{n}', \mathbf{n}) \equiv 1$), a constant source function $f(\mathbf{x}) = 1$ and – if not stated otherwise – zero boundary conditions are adopted. The test calculations which follow are exclusively performed for 2D configurations (i.e., $d = 2$). For the discretization of the integral operator in (68), a collocation method is used, again as described in Sect. 3.2, using a set of equally spaced points on the unit circle. Then, the transport equation (68) gives for an arbitrary fixed direction $\mathbf{n}_i$

$$\mathbf{n}_i \cdot \nabla_x I(\mathbf{x}, \mathbf{n}_i) + \chi(\mathbf{x}) I(\mathbf{x}, \mathbf{n}_i) = \chi(\mathbf{x}) \Big((1-\epsilon) \sum_{k=1}^{k=m} \omega_{ik} I(\mathbf{x}, \mathbf{n}_k) + \epsilon f(\mathbf{x}) \Big), \quad (69)$$

where the ω_{ik} are suitable quadrature weights with $\sum_k \omega_{ik} = 1$. Equation (69) is discretized by the discontinuous Galerkin finite element method (DGFEM) proposed for neutron transport without scattering in [45] and for radiative transfer problems in [38]. Let $\mathbb{T}_h$ be a mesh of quadrilaterals K covering the domain Ω exactly. On $\mathbb{T}_h$ the finite element space V_h used for the specific intensity is defined by

$$V_h = \{v \in L^2(\mathbb{T}_h) \mid v|_K \in \mathcal{P},\ \forall K \in \mathbb{T}_h\}. \tag{70}$$

$\mathcal{P}$ is either the space of polynomials of degree k or the space of isoparametric tensor product polynomials of degree k on K. We assume $k \geq 1$, yielding a discretization of second order at least.

At large optical depths ($\chi \gg 1$) radiative transfer in scattering dominated media ($\epsilon \to 0$) becomes simple because all scale lengths become larger than the photon mean free path length ($\sim 1/\chi(\mathbf{x})$). Simple approximations to this transport problem, based either on ad hoc physical assumptions such as Fick's law (cf. [40, 18]) or on a polynomial approximation, which assumes that the flux has an expansion into the first two Legendre Polynomials (cf. [16]), result in a diffusion equation

$$-\nabla_x \cdot \left(\frac{1}{d\chi} \nabla_x J(\mathbf{x})\right) + \kappa J(\mathbf{x}) = \kappa f(\mathbf{x}) \tag{71}$$

for the mean intensity $J(\mathbf{x}) = 1/(2(d-1)\pi) \int_{s^2} I(\mathbf{x}, \mathbf{n}) d\omega$. Usually, fixed boundary conditions of the type $J(\mathbf{x}) = g(\mathbf{x})$ are posed on the spatial boundary surface Γ. To derive the basic asymptotic (diffusion) limit for scattering dominant, optically thick transport problems, the point $\mathbf{x}$ under consideration must be far from the boundary (optical depth $\tau \gg 1$) and far from the zones of the domain Ω where the coefficients κ and σ vary strongly. In the 1970s and 1980s, the mathematical relationship between transport and diffusion theory has been clarified in a large body of work showing that transport theory transitions into diffusion theory in a certain asymptotic limit. A sampling of this work is given in [41, 27, 52, 42, 43]. As for the radiative transfer equation (69) the DGFEM discretization of the diffusion equation (71) is achieved by the (symmetric) interior penalty method due to [5] with polynomials of degree $k \geq 1$, yielding a discretization of second-order at least. The linear system of equations resulting from the DGFEM discretization of Eq. (68) has the block structure

$$\mathbf{A}_{\text{diff}}\, \mathbf{J} = \mathbf{f}. \tag{72}$$

$\mathbf{J}$ is the vector of discrete mean intensities and $\mathbf{f}$ is any given data (thermal radiation and boundary data). The $(n \times n)$-matrix $\mathbf{A}_{\text{diff}}$ corresponds to the DGFEM discretization of the left hand side of Eq. (71).

7.2 Preconditioning

The linear system of equations resulting from the DGFEM discretization of Eq. (68) has the block structure

$$\left(\begin{pmatrix} A_1 & & \\ & \ddots & \\ & & A_M \end{pmatrix} - \begin{pmatrix} \omega_{11} X & \cdots & \omega_{1M} X \\ \vdots & & \vdots \\ \omega_{M1} X & \cdots & \omega_{MM} X \end{pmatrix}\right) \begin{pmatrix} u_1 \\ \vdots \\ u_M \end{pmatrix} = \begin{pmatrix} f_1 \\ \vdots \\ f_M \end{pmatrix}. \tag{73}$$

Here, u_i is the vector of discrete intensities and f_i is any given data (thermal radiation and boundary data) for any fixed direction $\mathbf{n}_i$ ($i = 1, \ldots, M$). Each $(n \times n)$-matrix A_i corresponds to the DGFEM discretization of the left hand side of Eq. (69) for a single radiation direction $\mathbf{n}_i$ and X discretizes the multiplication with $(1 - \epsilon)\chi$. The operator form of Eq. (73) reads

$$\mathbf{A}_{\text{rte}}\, \mathbf{u} = \mathbf{f}, \tag{74}$$

where $\mathbf{A}_{\text{rte}} = \mathbf{T} - \mathbf{S}$ is the pure transport operator $\mathbf{T}$ minus the scattering operator $\mathbf{S}$.

Given a system of linear equations (74) and an iterate $\mathbf{u}^i$, the *iteration error* is defined as

$$\mathbf{e}^i := \mathbf{u} - \mathbf{u}^i. \tag{75}$$

An equation for the error is given by the defect equation

$$\mathbf{A}_{\text{rte}}\, \mathbf{e}^i = \mathbf{A}_{\text{rte}}\left(\mathbf{u} - \mathbf{u}^i\right) = \mathbf{f} - \mathbf{A}_{\text{rte}}\, \mathbf{u}^i = \mathbf{d}^i. \tag{76}$$

An approximation $\mathbf{g}(\mathbf{x}, \mathbf{n})$ for $\mathbf{e}^i$ can be computed by solving

$$\mathbf{M}\mathbf{g} = \mathbf{d}^i \quad \longleftrightarrow \quad \mathbf{g} = \mathbf{M}^{-1}\left(\mathbf{f} - \mathbf{A}_{\text{rte}}\, \mathbf{u}^i\right). \tag{77}$$

$\mathbf{M}$ should be an approximation of $\mathbf{A}_{\text{rte}}$ that is easy to invert, such as:

$\mathbf{M}$ identity matrix (Richardson method)
$\mathbf{M}$ diagonal of $\mathbf{A}_{\text{rte}}$ (Jacobi method)
$\mathbf{M}$ lower triangle of $\mathbf{A}_{\text{rte}}$ (Gauss-Seidel method).

Now a new iterate is obtained by

$$\mathbf{u}^{i+1} = \mathbf{u}^i + \mathbf{g}. \tag{78}$$

Renumbering the degrees of freedom according to the photon transport direction $\mathbf{n}_i$, the matrices A_i in Eq. (73) can be inverted easily by a front solving algorithm (see [34]). Therefore, a simple, well-known preconditioner for the system (73) is the matrix

$$\mathbf{\Lambda} = \begin{pmatrix} A_1^{-1} & & \\ & \ddots & \\ & & A_M^{-1} \end{pmatrix}. \tag{79}$$

It is this preconditioner, that is used in the well-known Λ- or Source Iteration method. Thus, with $\mathbf{M} \equiv \mathbf{\Lambda}$ the new iterate in Eq. (78) reads

$$\mathbf{u}^{i+1} = \mathbf{\Lambda}\left(\mathbf{f} + \begin{pmatrix} \omega_{11}X & \cdots & \omega_{1M}X \\ \vdots & & \vdots \\ \omega_{M1}X & \cdots & \omega_{MM}X \end{pmatrix} \mathbf{u}^i\right). \tag{80}$$

Table 3. GMRES steps required for the reduction of residual by a factor of 10^{-5} for constant coefficients using the preconditioner Λ.

χ	0.10	1	10	10^2	10^3	10^4
ϵ	0.91	0.1	10^{-2}	10^{-4}	10^{-6}	10^{-8}
cells						
256	2	5	15	88	423	1424
1024	2	5	16	108	1427	4929
4096	2	5	16	122	2871	13759
16384	2	5	17	131	4718	—

This preconditioner performs well only if the extinction χ is small or the albedo $1 - \epsilon$ is bounded away from 1. It can be enhanced by replacing the blocks A_i in $\mathbf{\Lambda}$ by $A_i - \omega_{ii} X$ (cf. [36, 62]). Still, this version suffers from the same limitations as the Λ-preconditioner in the case of dominant scattering. In fact, acceleration methods for the Λ- or Source Iteration method that are based on matrix algebra, such as Jacobi and Gauss-Seidel preconditioning, have not proved to be as useful as methods based on direct manipulation of the transport equation itself. This latter class of methods is described in great detail in [44, 2].

We are now going to formulate a preconditioning scheme for the more general radiation transport problem in Eq. (74), which uses the solution of the diffusion problem in Eq. (72) to improve the extremely poor convergence for configurations which are optically thick and dominated by scattering. Transport-dominated problems are also solved by the preconditioning front-solving algorithm described above, which performs efficiently. The scheme suggested here combines these two preconditioners in a way related to two-level multigrid algorithms (see [28]). It exploits the good convergence properties of both methods in different regimes. The general framework of a two-level multigrid algorithm places much of the earlier work on transport acceleration methods described in [2] in a common, unified framework (for details see [44]). For such algorithms, one has a "fine mesh" on which the original problem defined in Eq. (74) is to be solved, and a "coarse mesh" on which "corrections" to the transport iterative solution are obtained (cf. Eq. (72)). The word "multigrid" often connotes a sequence of increasingly coarse spatial meshes. However, for transport problems, space and angle are variables, and the method we consider involves "collapsing" the transport problem onto "coarser grids" independent of the photon direction. Since only two "grids" are used, the framework of a two-level multigrid algorithm is adequate, for which we propose the acronym MMP, which stands for multi-model preconditioner.

In comparison to standard Λ- or Krylov-type iteration methods with acceleration by Jacobi or Gauss-Seidel preconditioning, our MMP algorithm is designed to give less of an iteration error $\mathbf{e}^i$ as defined by Eq. (75), while requiring a significantly smaller number of iteration steps. Like any other two-level multigrid method, our MMP algorithm combines the smoothing operation in

Eq. (80) with a "coarse grid" correction, which is obtained from solving the diffusion problem defined in Eq. (72). When this is done, the vector $\mathbf{d}^i(\mathbf{x}, \mathbf{n})$ can be composed, which is the defect iterate of the discretized radiation transport problem (cf. Eq. (74)–(77)). The MMP method then employs the following succession of steps:

1. "fine grid" pre-smoothing by transport preconditioner $\mathbf{\Lambda}$

$$\mathbf{g}_1(\mathbf{x}, \mathbf{n}) = \mathbf{\Lambda}\, \mathbf{d}^i(\mathbf{x}, \mathbf{n}), \tag{81}$$

2. projection of "fine grid" residual to "coarse grid" (restriction)

$$\bar{\mathbf{g}}_2(\mathbf{x}) = \frac{1}{2\pi(d-1)} \int_{S^{d-1}} \left(\mathbf{d}^i(\mathbf{x}, \mathbf{n}) - \mathbf{A}_{\text{rte}}\, \mathbf{g}_1(\mathbf{x}, \mathbf{n}) \right) d\omega \tag{82}$$

3. "coarse grid" solution of diffusion problem (cf. Eq. (72))

$$\mathbf{A}_{\text{diff}}\, \bar{\mathbf{g}}_3(\mathbf{x}) = \bar{\mathbf{g}}_2(\mathbf{x}) \tag{83}$$

4. "coarse grid" correction (prolongation)

$$\mathbf{g}_4(\mathbf{x}, \mathbf{n}_j) = \mathbf{g}_1(\mathbf{x}, \mathbf{n}_j) + \bar{\mathbf{g}}_3(\mathbf{x}) \quad \forall \text{ directions } j = 1, \ldots, M \tag{84}$$

5. "fine grid" post-smoothing by transport preconditioner $\mathbf{\Lambda}$

$$\mathbf{g}_5(\mathbf{x}, \mathbf{n}) = \mathbf{g}_4(\mathbf{x}, \mathbf{n}) + \mathbf{\Lambda} \left(\mathbf{d}^i(\mathbf{x}, \mathbf{n}) - \mathbf{A}_{\text{rte}}\, \mathbf{g}_4(\mathbf{x}, \mathbf{n}) \right). \tag{85}$$

The accuracy of the high-frequency eigenvectors improves due to error reduction by what is known as the smoothing property of the iteration (for details see [28]). These high-frequency eigenvectors are determined by the local part of the transfer operator $\mathbf{A}_{\text{rte}}$, i.e. they arise from pure transport and absorption. The idea of the MMP scheme is to construct an iteration that is complementary to the smoother, so that a reduction in the error for the low-frequency eigenvectors is obtained. These errors are determined by the non-local part of the transfer operator, i.e., they arise from the scattering contribution to $\mathbf{A}_{\text{rte}}$. A reduction of the low-frequency errors can be obtained from using a "coarser grid", so that less computational effort is needed.

7.3 Numerical Results

We test the performance of the multi-model preconditioner with a model problem on the unit square $\Omega = [-1, 1]^2$ that uses constant coefficients χ and ϵ on the whole domain.

All MMP examples are computed using the GMRES method with right preconditioning and, therefore, convergence is measured by the norm of the original residual, not the preconditioned one. Table 3 and 4 display the number of iteration steps required to reduce the Euclidean norm of the initial

Table 4. GMRES steps required for the reduction of residual by a factor of 10^{-5} for constant coefficients using the exact MMP preconditioner.

χ	0.10	1	10	10^2	10^3	10^4
ϵ	0.91	0.1	10^{-2}	10^{-4}	10^{-6}	10^{-8}
cells						
256	2	4	7	12	24	26
1024	2	4	7	10	21	24
4096	2	4	7	9	16	23
16384	2	4	7	8	14	24

Table 5. Computation time [sec] required to reduce the residual by a factor of 10^{-5}.

χ	0.10	1	10	10^2	10^3	10^4
ϵ	0.91	0.1	10^{-2}	10^{-4}	10^{-6}	10^{-8}
cells	\multicolumn{6}{c}{only Λ preconditioner}					
4096	0.5	1.3	4.8	103	3319	16457
16384	2.6	7.4	23.5	—	—	—
	\multicolumn{6}{c}{exact MMP preconditioner}					
4096	4.5	7.2	13	15	24.5	36.7
16384	19.4	33	55.3	62.6	104	181.5

residual by a factor of 10^{-5}. Since one smoothing step is sufficient for the MMP method, only the post-smoothing in Eq. (85) is applied. The results displayed in Table 4 show clearly that the number of iteration steps depends only moderately on the extinction coefficients and the thermalization parameters, which was the primal goal. In particular, these numbers are considerably smaller than the ones reported for the Λ-iteration as preconditioner in Table 3. There, the GMRES method was restarted every 150 steps; we chose such a large basis size to ensure a fair comparison.

Since Eq. (71) gives a good approximation to the radiative transfer problem in regions with dominant scattering, this is not true in the transport dominated case. Moreover, the model problem we consider contains parts where $\chi(\mathbf{x}) = 0$. In order to solve Eq. (71), we define a cut-off $\widehat{\chi}$ and replace $\chi(\mathbf{x})$ by $\max\{\widehat{\chi}, \chi(\mathbf{x})\}$. The results in this section show that this is a feasible approach. Having this modification in mind, we use χ throughout the paper in order not to confuse the notation.

Since the two preconditioners, Λ and exact MMP require different amounts of work in each iteration step, the number of steps required is not sufficient for an assessment of their efficiency. Therefore, we compare computation times for the two schemes in Table 5. The longer computation times for the Λ-preconditioner on the finer grid are missing, because the large basis chosen required a bigger machine with different runtime characteristics. Still, the tables show that the exact MMP preconditioner is not much slower when not needed (a factor of about 5 per step), while it clearly outperforms the Λ-method for diffusive problems.

8 Conclusion

A finite element code for solving the resonance line transfer problem in moving media is presented. Simple velocity fields and complete redistribution are considered. The code is applicable to any three-dimensional model configuration with optical depths up to 10^3–10^4.

The application to the hydrogen Lyα line originating from slightly optically thick model configurations ($\tau \leq 10^2$) are shown and the resulting line profiles are discussed in detail. The systematic approach from very simple to more complex models gave the following results:

- An optical depth of $\tau \geq 1$ leads to the characteristic double peaked line profile with a central absorption trough as expected from analytical studies (e.g., see [51]). This form of the profile is the result of scattering in space and frequency. Photons escape via the line wings where the optical depth is much lower.
- Global velocity fields destroy the symmetry of the line profile. Generally, the blue peak of the profile is enhanced for models simulating a collapse motion and the red peak for expanding media. But there are certain velocity fields (e.g. with steep gradients) and spatial distributions of the extinction coefficient, where the formation of a prominent peak is suppressed.
- Double-peaked line-profiles show up as two emission regions in the two-dimensional spectra. Global collapse or expansion leads to an overall triangular shape of the emission. Rotation produces a shear pattern resulting in banana-shaped emission regions for optical depths ≥ 10.
- For non-symmetrical model configurations, the optical depth varies with the line of sight. Thus, the flux, the depth of the absorption trough and the pattern in the two-dimensional spectra strongly depend on the viewing direction.

The applications clearly demonstrate the capacity of the finite element code and show that the three-dimensional structure, the kinematics and the total optical depth of the model configurations are very important. The latter point is crucial, since the convergence of the algorithm is extremely poor for optical depths $\geq 10^3 - 10^4$. These large depths refer to photons at the line center. Photons in the line wings or those emerging close to the boundary of the halo configuration hardly interact with the cold hydrogen gas and will escape almost freely towards the observer. These objects are especially challenging from a numerical point of view. They are simulated by the multi-model preconditioning (MMP) scheme presented in this paper, which combines two distinct preconditioners that are designed to accelerate the solution of transport dominated and highly scattering, optically thick radiative transfer problems. The development of the MMP method was stimulated by the theory of two-level multigrid algorithms which exploit the good convergence properties of both transport and scattering acceleration methods in different spatial regimes. The monochromatic model problems clearly demonstrate the considerable improvement in run-time efficiency of the MMP scheme.

References

1. Adams, T.F.: *The escape of resonance-line radiation from extremely opaque media.* Astrophys. J., **174**, 439 (1972)
2. Adams, M.L., Larsen, E.W.: *Fast iterative methods for discrete-ordinates particle transport calculations.* Prog. Nucl. Energy, **40**, No. 1, 3–159 (2002)
3. Ahn, S.-H., Lee, H.-W., Lee, H. M.: *Lyα line formation in starburst galaxies. I. Moderately thick, dustless and static H I media.* Astrophys. J., **554**, 604–614 (2001)
4. Ahn, S.-H., Lee, H.-W., Lee, H. M.: *Lyα line formation in starburst galaxies. II. Extremely thick, dustless and static H I media.* Astrophys. J., **567**, 922–930 (2002)
5. Arnold, D.N.: *An interior penalty finite element method with discontinuous elements.* SIAM J. Numer. Anal., **19**, 742–760 (1982)
6. Becker, R., Rannacher, R.: *An optimal control approach to a posteriori error estimation in finite element methods.* In: Iserles, A. (ed.) Acta Numerica 2001, **37**, p. 1–102, Cambridge University Press (2001)
7. Becker, R., Braack, M., Richter, T.: *Parallel multigrid on locally refined meshes.* In: Rannacher, R. (ed.) Reactive Flows, Diffusion and Transport. Springer, Berlin (2005)
8. Becker, R., Kapp, H., Meidner, D., Rannacher, R., Vexler, B.: *Adaptive finite element methods for PDE-constrained optimal control problems.* In: Rannacher, R. (ed.) Reactive Flows, Diffusion and Transport. Springer, Berlin (2005)
9. Böttcher K.: *Adaptive Schrittweitenkontrolle beim Unstetigen Galerkin-Verfahren für Gewöhnliche Differentialgleichungen.* Diploma Thesis, University of Heidelberg (1996)
10. Braack, M., Richter, T.: *Mesh and model adaptivity for flow problems.* In: Rannacher, R. (ed.) Reactive Flows, Diffusion and Transport. Springer, Berlin (2005)
11. Braack, M., Richter, T.: *Solving multidimensional reactive flow problems with adaptive finite elements.* In: Rannacher, R. (ed.) Reactive Flows, Diffusion and Transport. Springer, Berlin (2005)
12. Braess, D.: Finite Elemente. Springer, Berlin (1997)
13. Bramble, J.H.,Hilbert, S.R.: Bounds for a class of linear functionals with applications to Hermite interpolation. Numer. Math., **16**, 362–369 (1971)
14. Brenner, S.C., Scott, L.R.: The Mathematical Theory of Finite Element Methods. Springer, New York (1996)
15. Carraro, T., Heuvelin, V., Rannacher, R., Waguet, C.: *Determination of kinetic parameters in laminar flow reactors. I. Numerical aspects.* In: Rannacher, R. (ed.) Reactive Flows, Diffusion and Transport. Springer, Berlin (2005)
16. Case, K.M., Zweiffel, P.F.: *Linear Transport Theory.* Addison-Wesley Publishing Company, Reading, Massachusetts (1967)
17. Ciarlet, P.G.: The Finite Element Method for Elliptic Problems. North Holland, New York (1978)
18. Dautray R., Lions J.-L.: *Mathematical Analysis and Numerical Methods for Science and Technology.* Vol. 6, Springer, Berlin Heidelberg New York (2000)

19. De Breuck, C., Röttgering, H., Miley, G., van Breugel, W., Best, P.: *A statistical study of emission lines from high redshift radio galaxies.* Astron. Astrophys., **362**, 519–543 (2000)
20. Dullemond, C.P., Turolla, R.: *An efficient algorithm for two-dimensional radiative transfer in axisymmetric circumstellar envelopes and disks.* Astron. Astrophys., **360**, 1187–1202 (2000)
21. Efstathiou, A., Rowan-Robinson, M.: *Radiative transfer in axisymmetric dust clouds.* Mon. Not. Roy. Astron. Soc., **245**, 275–288 (1990)
22. Eriksson K., Johnson C., Thomée C.: *Time discretization of parabolic problems by the discontinuous Galerkin method.* RAIRO, Modélisation Math. Anal. Numér., **19**, 611–643 (1985)
23. Eriksson K., Johnson C.: *Error estimates and automatic time step control for nonlinear parabolic problems. I.* SIAM J. Numer. Anal., **24**, 12–23 (1987)
24. Eriksson K., Johnson C.: *Adaptive finite element methods for parabolic problems. I: A linear model problem.* SIAM J. Numer. Anal., **28**, 43–77 (1991)
25. Führer C., Kanschat G.: *A posteriori error control in radiative transfer.* Computing, **58**, 317–334 (1997)
26. Haardt, F., Maraschi, L.: *A two-phase model for the X-ray emission from Seyfert galaxies.* Astrophys. J., **380**, L51–L54 (1991)
27. Habetler, G.J., Matkowsky, B.J.: *Uniform asymptotic expansions in transport theory with small mean free paths, and the diffusion approximation.* J. Math. Phys., **16**, 846–854 (1975)
28. Hackbusch, W.: *Multi-Grid Methods and Applications.* Springer, Heidelberg (1985)
29. Hackbusch, W.: *Iterative Lösung Grosser Schwachbesetzter Gleichungssysteme.* Teubner, Stuttgart (1993)
30. Hogerheijde, M.: *The Molecular Environment of Low-Mass Protostars.* Ph.D. Thesis, Rijks Univesiteit Leiden (1998)
31. Hughes, T.J.R., Brooks, A.N.: Streamline upwind/Petrov-Galerkin formulations for convection dominated flows with particular emphasis on the incompressible Navier-Stokes equations. Comput. Methods Appl. Mech. Eng., **32**, 199–259 (1982)
32. Hummer, D.G., Kunasz, P.B.: *Energy loss by resonance line photons in an absorbing medium.* Astrophys. J., **236**, 609–618 (1980)
33. Johnson, C.: Numerical Solution of Partial Differential Equations by the Finite Element Method. Cambridge University Press, Cambridge (1987)
34. Johnson, C., Pitkäranta, J.: *An analysis of the discontinuous Galerkin method for a scalar hyperbolic equation.* Math. Comput., **46**, 1–26 (1986)
35. Juvela, M.: *Non-LTE radiative transfer in clumpy molecular clouds.* Astron. Astrophys., **322**, 943–961 (1997)
36. Kanschat, G.: *Parallel and Adaptive Galerkin Methods for Radiative Transfer Problems.* Ph.D. Thesis, University of Heidelberg (1996) http://www.iwr.uni-heidelberg.de/sfb359/Preprints1996.html
37. Kanschat, G.: *Solution of multi-dimensional radiative transfer problems on parallel computers.* In: Björstad P. and Luskin M. (eds.) Parallel Solution of PDE, IMA Vol. in Math. and its Appl., **120**, Springer, Berlin Heidelberg New York, 85–96 (2000)
38. Kanschat, G., Meinköhn, E.: *Multi-model preconditioning for radiative transfer problems.* Preprint SFB 359, University of Heidelberg (2004) http://www.sfb359.uni-heidelberg.de/Preprints2004.html

39. Kurk, J.D., Pentericci, L., Röttgering, H.J.A., Miley, G.K.: *Observations of radio galaxy MRC 1138-262: Merging galaxies embedded in a giant Lyα halo*. In: Henney, W.J., Steffen, W., Raga, A.C., Binette, L. (eds) Emission Lines from Jet Flows. RMxAA (Serie de Conferencias), Vol. **13**, 191–195 (2002)
40. Lamarsh, J.R.: *Introduction to Nuclear Reactor Theory*. Addison-Wesley Publishing Company, Reading, Massachusett (1965)
41. Larsen, E.W., Keller, J.B.: *Asymptotic solution of neutron transport problems for small scale mean free paths*. J. Math. Phys., **15**, 75–81 (1974)
42. Larsen, E.W., Pomraning, G.C., Badham, V.C.: *Asymptotic analysis of radiative transfer problems*. J. Quant. Spectrosc. Radiat. Transfer, **29**, 285–310 (1983)
43. Larsen, E.W., Morel, J.E., Miller, W.F.jun.: *Asymptotic solutions of numerical transport problems in optically thick, diffusive regimes*. J. Comput. Phys., **69**, 283–324 (1987)
44. Larsen, E.W.: *Transport acceleration methods as two-level multigrid algorithms*. Oper. Theory, Adv. Appl., **51**, 34–47 (1991)
45. LeSaint, P., Raviart, P.-A.: *On a finite element method for solving the neutron transport equation*. In: de Boor (ed.) Mathematical Aspects of Finite Elements in Partial Differential Equations. Academic Press, New York, 89–123 (1974)
46. Meinköhn, E., Richling, S.: *Radiative transfer with finite elements. II. Lyα line transfer in moving media*. Astron. Astrophys., **392**, 827–839 (2002)
47. Meinköhn, E.: *Modeling Three-Dimensional Radiation Fields in the Early Universe*. Ph.D. Thesis, University of Heidelberg (2002) http://www.iwr.uni-heidelberg.de/sfb359/Preprints2002.html
48. Mihalas, D.: *Stellar Atmospheres*. W.H. Freeman and Company, Second Edition, San Francisco (1978)
49. Mihalas, D., Weibel-Mihalas, B.: *Foundation of Radiation Hydrodynamics*. Oxford University Press, New York (1984)
50. Murray, S., Castor, J., Klein, R., McKee, C.: *Accretion disk coronae in high-luminosity systems*. Astrophys. J., **435**, 631–646 (1994)
51. Neufeld, D.A.: *The transfer of resonance-line radiation in static astrophysical media*. Astrophys. J., **350**, 216–241 (1990)
52. Papanicolaou, G.C.: *Asymptotic analysis of transport processes*. Bull. Am. Math. Soc., **81**, 330–392 (1975)
53. Park, Y.-S., Hong, S.S.: *Three-dimensional Non-LTE Radiative Transfer of CS in Clumpy Dense Cores*. Astrophys. J., **494**, 605 (1998)
54. Pozdniakov, L.A., Sobol, I.M., Suniaev, R.A.: *The profile evolution of X-ray spectral lines due to Comptonization - Monte Carlo computations*. Astron. Astrophys., **75**, 214–222 (1979)
55. Richling, S., Meinköhn, E., Kryzhevoi, N., Kanschat, G.: *Radiative transfer with finite elements. I. Basic method and tests*. Astron. Astrophys., **380**, 776–788 (2001)
56. Sleijpen, G.L.G., Fokkema, D.R.: *Bicgstab(L) for linear equations involving unsymmetric matrices with complex spectrum*. Electronic Transactions on Numerical Analysis, Vol. **1**, 11–32 (1993)
57. Sonnhalter, C., Preibisch, T., Yorke, H.: *Frequency dependent radiation transfer in protostellar disks*. Astron. Astrophys., **299**, 545 (1995)
58. Spaans, M.: *Monte Carlo models of the physical and chemical properties of inhomogeneous interstellar clouds*. Astron. Astrophys., **307**, 271 (1996)

59. Spagna Jr., G.F., Leung, C.M.: *Numerical solution of the radiation transport equation in disk geometry.* J. Quant. Spectrosc. Radiat. Transfer, **37**, 565–580 (1987)
60. Steidel, C.C., Adelberger, K.L., Sharply, A.E., Pettini, M., Dickinson, M., Giavalisco, M.: *Lyα Imaging of a Proto-Cluster Region at $<z>=3.09$.* Astrophys. J., **532**, 170–182 (2000)
61. Stenholm, L.G., Störzer, H., Wehrse, R.: *An efficient method for the solution of 3-D radiative transfer problems.* J. Quant. Spectrosc. Radiat. Transfer, **45**, 47–56 (1991)
62. Turek, S.: *An efficient solution technique for the radiative transfer equation.* Imp. Comput. Sci. Eng., **5**, No. 3, 201–214 (1993)
63. van Ojik, R., Röttgering, H.J.A., Miley, G.K., Hunstead, R.W.: *The nature of the extreme kinematics in the extended gas of high redshift radio galaxies.* Astron. Astrophys., **317**, 358–384 (1997)
64. Varga, R.S.: *Matrix Iterative Analysis.* Springer, Berlin Heidelberg (2000)
65. Warsa, J.S., Wareing, T.A., Morel, J.E.: *Krylov Iterative Methods and the Degraded Effectiveness of Diffusion Synthetic Acceleration for Multidimensional S_n Calculations in Problems with Material Discontinuities.* Nucl. Sci. Eng., **147**, 218–248 (2004)
66. Wehrse R., Meinköhn E., Kanschat G.: *A review of Heidelberg radiative transfer equation solutions.* In: Stee P. (ed.) Radiative Transfer and Hydrodynamics in Astrophysics. EAS Publication Series, Vol. 5, 13–30 (2002)
67. Wehrse R., Baschek B., von Waldenfels, W.: *The diffusion of radiation in moving media: I. Basic assumptions and formulae.* Astron. Astrophys., **359**, 780–787 (2000)
68. Wolf, S., Henning, T., Stecklum, B.: *Multidimensional self-consistent radiative transfer simulations based on the Monte-Carlo method.* Astron. Astrophys., **349**, 839–850 (1999)
69. Yorke, H., Bodenheimer, P., Laughlin, G.: *The formation of protostellar disks. I - 1 M(solar).* Astrophys. J., **411**, 274–284 (1993)
70. Zheng, Z., Miralda-Escudé, J.: *Monte Carlo simulation of Lyα scattering and application to damped Lyα systems.* Astrophys. J., **578**, 33–42 (2002)

Part VII

Flows in Porous Media

Preamble

This chapter contains five articles which report on new models, simulation methods and experimental techniques for environmental flows. Four of them are concerned with modeling and simulation of flow in porous media while one article investigates the wind-induced flow in a small natural lake.

Homogenization is a mathematical tool that enables a rigorous derivation of macroscopic models from microscopic processes and geometry. One of the achievements of homogenization theory is the mathematical understanding of fluid flow in porous media. The first article *"Multiscale analysis of processes in complex media"* starts from model equations at microscopic level and uses multiscale techniques for determining macroscopic descriptions for reactive flow in porous media, flow at interfaces between free flow and filtration flow, flow along rough boundaries, and flow through a network of channels.

The fully saturated case treated phenomenologically by Darcy in 1856 can now be derived rigorously from the Stokes equation. However, the case of two immiscible fluids in a porous medium is much less understood. Macroscopic models currently used in engineering practice are nonlinear extensions of the fully saturated case. More elaborate models have been developed but none of them is derived rigorously. Here the second article *"Microscopic interfaces in porous media"* offers new insights into the upscaling of two-phase flow in porous media.

The third article *"High-accuracy approximation of effective coefficients"* is concerned with the accurate solution of so-called cell problems arising in homogenization theory and which allow the computation of coefficients for the macroscopic model. It is shown theoretically and numerically that higher-order finite element methods can be used to compute the coefficients with high accuracy.

The fourth article *"Numerical simulation and experimental studies of unsaturated water flow in heterogeneous systems"* then uses higher-order finite element methods to solve numerically the (engineering-type) macroscopic models for two-phase flow in porous media. New experimental methods developed in the project allow a quantitative comparison of experiment and simulation in the case of heterogeneous media.

Finally, the fifth article *"Lake dynamics: observation and high-resolution numerical simulation"* investigates mixing and density stratification in a small lake in the Rhinegraben. High resolution temperature and velocity data have been acquired over a period of two years, and a variety of hydrodynamic phenomena are demonstrated. Numerical simulation of the flow in the lake is carried out with high resolution on massively parallel computers in order to further investigate some of these phenomena.

Multiscale Analysis of Processes in Complex Media[*]

W. Jäger[1] and M. Neuss-Radu[1]

Institut für Angewandte Mathematik willi.jaeger@iwr.uni-heidelberg.de

Summary. In this report we present results concerning the study of processes in complex media. Starting from model equations at microscopic level and using the techniques of the multi-scale analysis, we determine macroscopic descriptions for reactive flow in porous media, flow at interfaces between free flow and filtration flow, flow along rough boundaries and flow through a network of channels.

1 Introduction

One of the challenges of mathematics and modeling is to describe multiple scale systems at each scale and to describe the transfer between the different scale levels quantitatively. Physical and chemical processes in porous media, in soil, in biological tissues or technological textiles, in networks of vessels and in composite materials generally involve several scales. Starting from micro-

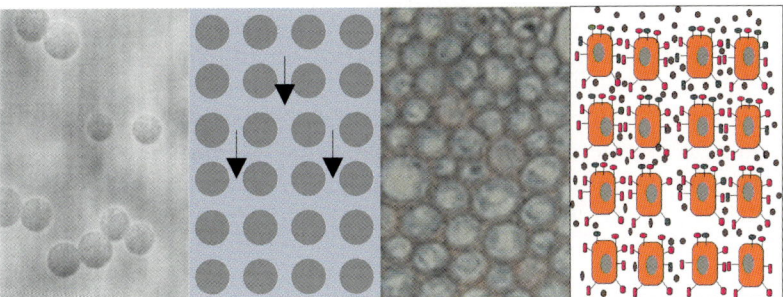

Fig. 1. Complex media: a chromatographic column, a periodic approximation for a chromatographic column, a section through a plant root, and a sketch of an array of cells.

---

[*]This work has been supported by the German Research Foundation (DFG) through SFB 359 (Project D1) at the University of Heidelberg.

scopic model equations for flow, diffusion, transport and reactions in media with periodic and, in special cases, with random structures in geometry and physical properties, effective macroscopic equations could be determined and the approximation errors could be estimated. It is important that the system parameters can be calculated directly by solving microscopic cell problems in the periodic case, and problems for realizations of the process in the stochastic case. The following problems will be covered in this report:

- Derivation of macroscopic equations for reactive flow in porous media. Modeling the chromatographic columns
- Derivation of the effective transmission laws for Navier–Stokes flow at interfaces
- Effective laws for rough boundaries
- Rough surfaces and drag reduction
- Flow through filters and membranes
- Flow through networks of channels

Formulation of the mathematical problem

We assume that the geometrical structures, the material properties as well as the processes in these structures are well defined and quantitatively determined on the microscopic level. Especially, the dependence on the scale parameters are given. Hereby, the scales $\varepsilon, \delta, \ldots$ may describe e.g. the characteristic grain or pore size, the thickness of a channel, the frequency or amplitude of oscillations of the boundary. Our approach consists in the study of the asymptotic limit when the scale parameter ε tends to zero. In order to simplify the analysis, one assumes very often that there is underlying periodic structure, described with a periodic lattice. However, in many cases the results can be transferred to stochastic structures of special type.

The main problem in the analysis is to control the solutions of the systems for fixed ε and their dependence on the scale parameter. In order to study the limit $\varepsilon \to 0$, one may at first try a formal expansion with respect to the scale parameter. This leads to an equation for a macroscopic approximation. The coefficients in this equation can be calculated by solving local problems depending on the microscopic variable and averaging. However, the necessary error estimate can get more tedious. Using concepts of multiscale convergence

Fig. 2. The periodic structure and the standard cell.

it is easier to show the convergence to limit functions depending on both, the macroscopic and the microscopic variable and to estimate the errors using e.g. energy estimates. We refer to the literature as far as concepts of asymptotic expansion and multi-scale convergence are concerned e.g. [26], [1].

Before we are going to give a survey on the main results obtained in this project, we want to stress an experience we made: In general, by homogenization we will reduce the complexity with respect to the geometry for instance, however we increase the size of the system, since in each macroscopic point, a coupled microscopic problem has to be solved. Even in the case, where the arising cell problems can be solved and a macroscopic system with effective coefficients can be derived by averaging over the microscopic variable, it may be better for computing the solution to keep the larger system. This system has however a special structure, which can be exploited in the numerical algorithms.

Let us now finish the introduction by an important example for a macroscopic law which can be derived by asymptotic methods.

Darcy's law for flow through porous media

We consider a column filled with periodically distributed grains (solid phase) whose boundary we denote by Γ^ε and fluid (fluid phase) occupying the domain Ω^ε. Let us start to consider just a flow through this column. For simplicity we assume on the microscopic scale a stationary Stokes flow

$$-\Delta u^\varepsilon + \nabla p^\varepsilon = 0 \text{ in } \Omega^\varepsilon$$
$$\text{div } u^\varepsilon = 0 \text{ in } \Omega^\varepsilon$$
$$u^\varepsilon = 0 \text{ on } \Gamma^\varepsilon$$

Using the homogenization method, the following result can be shown, see e.g. [26]

$$\frac{u^\varepsilon}{\varepsilon^2} \to u_0, \quad \text{weakly for } \varepsilon \to 0,$$

where the filtration velocity u_0 satisfies Darcy's law

$$u_0(x) = K(-\nabla p + f)(x).$$

The permeability tensor K can be computed by solving the following cell problem

$$-\Delta_y w^j + \nabla_y \Pi^j = e_j \text{ in } X$$
$$\text{div}_y w^j = 0 \text{ in } X$$
$$w^j = 0 \text{ on } \Gamma$$

and by averaging over the microscopic variable y

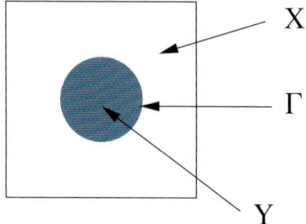

Fig. 3. The standard cell in a porous medium.

$$K_i^j = \frac{1}{|X|} \int_X w_i^j(y) dy.$$

The same result can be obtained for the stationary Navier–Stokes equations. For non-stationary flow the cell problem will be also non-stationary and the macroscopic equation is going to include an integration over time of the flux.

2 Derivation of Macroscopic Equations for Reactive Flow in Porous Media. Modeling the Chromatographic Colums

2.1 The Chromatographic Column

In the following we consider a column with porous grains (chromatographic column), see Fig.1, assuming that the flow of the fluid is zero in the solid phase. Chemical substances are transported and diffuse in the column. In general we will have a system of equations coupling the dynamics of the substances and the flow. At first let us consider the case where there are no chemical reactions in the phases. The substances will penetrate into the grain (adsorption) and leave the solid phase again (desorption) depending on the concentration differences and the occupation of the sites in the solid phase.

Formulation of the microscopic model

In the fluid phase Ω^ε we assume the Stokes equations for the fluid:

$$\begin{aligned}
-\Delta u_\varepsilon + \nabla p_\varepsilon &= \gamma && \text{in } \Omega^\varepsilon \\
\operatorname{div} u_\varepsilon &= 0 && \text{in } \Omega^\varepsilon \\
u^\varepsilon &= 0 && \text{on } \Gamma^\varepsilon
\end{aligned}$$

The transport of the substances in the fluid phase is given by the following equation for the concentrations v_ε, where we drop the index indicating the specific substance:

$$\partial_t v_\varepsilon + u_\varepsilon \cdot \nabla v_\varepsilon = D_f \triangle v_\varepsilon \quad \text{in } \Omega^\varepsilon$$

We denote the concentrations in the solid phase by w_ε, and we assume for simplicity a diffusion equation in the solid phase:

$$\partial_t w_\varepsilon = \varepsilon^2 D_s \triangle w_\varepsilon.$$

Here the scaling of the diffusion by the square of the scaling factor is chosen in order to keep in the scale limit $\varepsilon \to 0$ the observed physical effects. The transmission between solid and fluid phase on the interface is modelled by the following conditions. First, the natural condition for the continuity of the normal fluxes on the interface between the solid and the fluid phase

$$D_f \nabla v_\varepsilon \cdot n = \varepsilon^2 D_s \nabla w_\varepsilon \cdot n$$

and second, the following jump condition for the concentrations which is still causing problems for the mathematical analysis:

$$h(v_\varepsilon) = w_\varepsilon.$$

Here, h is a function which has to satisfy specific conditions, guaranteeing not only existence but also proper estimates of the solution. These estimates, which must control the dependence of the solution on the scale parameter ε, are necessary for the limiting process $\varepsilon \to 0$. The structure of h can be derived by a statistical physics analysis. Rather often the assumption

$$h(v) = Kv$$

is made, where K is the so called calibration constant. In this case, by rescaling the concentration in the solid phase, the jump in the concentrations can be removed in a simple way. This simpler case was solved by Vogt in a thesis (1982). In case of a nonlinear h the situation is getting more complicated, especially if one has to consider a system of substances, which may interact through the transmission condition and additional chemical reaction in both phases. Here already the existence and the estimates for the microscopic problem pose problems, since the standard methods cannot be applied. Before we discuss this topic, let us formulate the resulting macroscopic problem. Here we include chemical reactions in the solid and the fluid phase and assume that the reaction terms verify conditions sufficient to show that the solutions are bounded in L^∞, uniformly with respect to ε. Since formulating sufficient conditions on the reactive terms is an independent problem, which is not solved to an extent needed in applications, we do not state an assumption covering a special class of problems.

Formulation of the macroscopic model

Since the flow is independent from the transported substances, we have Darcy's law for the filtration flow

$$u(x) = K(-\nabla p + \gamma)(x). \tag{1}$$

The following equation models the evolution of substances on the local level in a grain

$$\partial_t w(t, x, y) = D_s \triangle_y w(t, x, y) + g(x, y, w(t, x, y)), \tag{2}$$

with fixed initial data and $w(t, x, y) = h(v(t, x))$ on the boundary Γ, see Fig.3 for the notation. Here we included also a reactive term in the solid phase. The concentrations v of the substances on the macroscopic level satisfy

$$|Y|\partial_t v(t, x) + S(t, x) = \sum d_{jk} \partial_j \partial_k v(t, x) - u \nabla v(t, x) + f(x, v(t, x)) \tag{3}$$

$$S(t, x) = \int_\Gamma D_s \nabla_y w(t, x, y) \cdot n \, d\Gamma(y). \tag{4}$$

The effective diffusion coefficients d_{jk} can be computed solving a boundary problem for the Laplace operator just in the fluid part X of the cell, f takes into account possible reactions, and the sink-source term S is nothing but the contribution of the microstructure to the macroscopic level. Using integration by parts and assuming that there are no reactions in the solid phase, and that the initial values of w are zero, the sink-source term S can be reformulated as follows

$$\begin{aligned} S(t, x) &= \int_\Gamma c \nabla_y w(t, x, y) \cdot n \, d\Gamma(y) \\ &= \int_Y c \triangle w(t, x, y) dy \\ &= \int_Y \partial_t w(t, x, y) dy \\ &= \int_0^t \rho(t - s) \partial_s h(v(s, x)) ds \end{aligned}$$

Here ρ is the spatial average of the corresponding Greens function to the initial-boundary value problem for the diffusion in the grain. This representation shows that there is a memory term included representing the exchange between the scales.

The case of one substance and more general jump function h (isotherm) was studied and solved in [14]. The model was used for investigations of technological chromatographic columns in cooperation with petrol industry. There the inverse problem of determining the "isotherm" h from real data was treated. The determined nonlinearities gave good agreement with the real measurements. However, for many applications several substances are involved and a system arises which up to now could not be treated well enough.

The microscopic problem – a problem by itself

The arising problem is of general interest and has the following form

$$\partial_t w_j - \sum_{k,l \leq n} \partial_k (A_j^{kl} \partial_l w_j) = f_j, \text{ in } \Omega_j$$

$j = 1, 2$, where $\Omega = \Omega_1 \cup \Omega_2$,

and w satisfies the no flux or Dirichlet condition on the exterior boundary Γ_e and the transmission condition

$$A_1^{kl} \partial_l w_1 \cdot n_k = A_2^{kl} \partial_l w_2 \cdot n_k$$

and the jump condition

$$w_2 = h(w_1)$$

on the interior boundary Γ_i. In this formulation we skipped the dependence on the scale.

To solve this transmission problem, we considered the transformed problem: Set

$$v = \begin{cases} w_1 & \text{in } \Omega_1 \\ h^{-1}(w_2) & \text{in } \Omega_2 \end{cases}$$

and obtain the following differential equation

$$\partial_t \bar{h}(v) - \sum_{k,l \leq n} \partial_k (\bar{A}^{kl} \partial_l \bar{h}(v)) = \bar{f} \tag{5}$$

which has to be interpreted in the weak sense. Whereas this equation can be solved in the scalar case for a strictly monotone increasing, differentiable h, the case of systems is still under investigation. The scalar case was treated by Jäger and Simon for more general equations in [15]. Also for the scalar case and the stationary case, the existence and regularity of the solutions was analyzed by Jäger and Kutev in [4]. For systems the problem arises to prove the a-priori estimates, necessary to show existence with help of the Rothe technique and in case of homogenization, to perform the limit $\varepsilon \to 0$. The situation is similar to the system treated by Alt and Luckhaus, [2]

$$\partial_t b(v) = \sum_j \partial_j a_j(\cdot, v, \nabla v) + f(\cdot, v, \nabla v)$$

There the assumption which has to be made is that b is a gradient. However the same assumption for h is not satisfied in applications. A statistical physics analysis shows that h has the following structure:

$$h_l(v) = v_l \partial_{v_l} \log \Phi(v), \quad l = 1, \ldots, k$$

where
$$\Phi_q(v) = \sum_{|\alpha| \leq q} m_\alpha v^\alpha$$

Examples

(1) $q = 1$, k species

$$\Phi_1(v) = m_0 + \sum_{0 < l \leq k} m_l v_l \qquad h_j(v) = \frac{m_j}{m_0 + \sum_{0 < l \leq k} m_l v_l} v_j.$$

(2) one species (scalar case $k = 1$)

$$\Phi_q(v) = \sum_{0 \leq l \leq k} m_l v^l \qquad h(v) = \frac{\sum_{0 < l \leq q} m_l l v^l}{\sum_{0 \leq l \leq q} m_l v^l}$$

Remark 1. In new variables $\zeta_j = \log v_j$ the mapping h is gradient of $\log \Phi$. The function is logarithmic convex, uniformly only if v is bounded. This observation is crucial for the analysis of the system, which is not yet finished since the bounds for v still have to be proven.

2.2 Generalizations for Modeling of "Cell" Systems

Systems with surface reactions

Several generalizations in several directions were treated. These mainly concern the modelling of catalytic reactions on the surfaces of grains and processes in cell tissue. The first contribution to surface reactions was given under restrictive assumptions by Hornung and Jäger [5], e.g. the reaction-diffusion on the surface was assumed to be linear. Neuss-Radu developed a generalization of two-scale convergence including processes on surfaces [20]. This concept could be generalized to other lower dimensional structures, also to sets with fractal structure (see e.g. Kolumban). The result for diffusion and reactions on the surfaces can be roughly described as follows: The microscopic systems contains model equations for the processes on the surface of the grains, e.g.

$$\partial_t w_\varepsilon = \varepsilon^2 D_s \triangle_{\Gamma^\varepsilon} w_\varepsilon + g(\cdot, w_\varepsilon) + f(\cdot, v_\varepsilon, w_\varepsilon) \tag{6}$$

where $\triangle_{\Gamma^\varepsilon}$ is the Laplace-Beltrami operator on the grain surface Γ^ε. The equations for the fluid phase are kept. The transmission between the fluid and the solid phase is modelled by

$$D_f \nabla v_\varepsilon \cdot n = \varepsilon f(\cdot, v_\varepsilon, w_\varepsilon) \tag{7}$$

Here, two critical points have to be treated as usual: The a-priori estimates for the micro problem have to be derived and the homogenization limit has to be performed using the concept of multiscale convergence. Here, the nonlinearities are causing serious difficulties. First, the estimates for the microscopic systems have to be given, basic for proving existence and uniqueness of solutions for the ε−problems, and for the compactness of the solution set in a topology adjusted to the multiscale convergence. Then, the identification of the limits of nonlinear terms has to be performed. As far as the a-priori estimates are concerned, we decided to separate the problem of proving a-priori bounds and performing the homogenization limit. We assume the a-priori estimates necessary for the convergence. In general, there is a need for further research aiming for appropriate conditions on the structure and the magnitudes in reaction networks allowing the control of the solutions to the diffusion-reaction system. We obtain the following macroscopic model equation for the case of surface reactions. The local problem

$$\partial_t w(t,x,y) = D_s \triangle_{\Gamma_y} w(t,x,y) + g(t,y,w(t,x,y)) + f(t,y,v(t,x),w(t,x,y)),$$

is coupled through the source f to the macroscopic equation for the concentrations, defined on the whole domain

$$|Y|\partial_t v(t,x) + S(t,x) = \sum d_{jk}\partial_j\partial_k v(t,x) - u\nabla v(t,x).$$

u is the filtration velocity. The sink-source-term S, describing the flux between the levels, has the form as expected:

$$S(t,x) = \alpha \int_\Gamma f(t,y,v(t,x),w(t,x,y))d\Gamma(y).$$

Applications

Using homogenization arguments a system of diffusion-reaction equations coupled with reaction equations on surfaces was derived modelling the morphogenesis of organisms. Hydra was chosen as a test organism. This system describes the diffusion of chemical substances in the intercellular space of the tissue interacting with the receptors on the cell surface and enzymes produced by the cells. The pattern in formation of head and foot observed or produced by experiments is explained by receptor distribution in space and time. Simulations proved that these models show results, which are at least in qualitative agreement with experiments [27], [17]. Also the following important feature could be achieved: patterns can be achieved by stochastic perturbation of homogenous initial data. In a recent investigation, the homogenization limit could be analytically justified in the case that there are only reactions and no diffusion on the cell surface [18].

Systems with semi-permeable interfaces

Hornung, Jäger, Mikelic treated in [6] the reactive transport through an array of cells with semi-permeable membranes. Here the transmission conditions between the fluid and solid phase are of the following type, covering conditions arising in real applications:

$$(av_\varepsilon(t,x) - w_\varepsilon(t,x), -\varepsilon Dv_\varepsilon \cdot \nabla w_\varepsilon(t,x))_j \in M_j \text{ where } (s,q) \in M_j \text{ iff}$$

1. case: $s = 0$ (Dirichlet)
2. case: $q = 0$ (Neumann)
3. case: $q = +b_j s^+$ (inflow)
4. case: $q = -b_j s^-$ (outflow)
5. case: $s \leq 0$ and $q \geq 0$ and $sq = 0$ (inflow)
6. case: $s \geq 0$ and $q \leq 0$ and $sq = 0$ (outflow)

Again, the main difficulties arise in passing to the homogenization limit in the nonlinear terms. However, an essential progress in the modeling of biological tissue could be achieved in this paper. Due to the increased information on the processes in membranes, the cytoplasm and the nuclei, multiscale analysis will be applied more and more to modeling of biological cells and tissues.

3 Derivation of the Effective Transmission Laws for Navier–Stokes Flow at Interfaces

Despite the huge number of homogenization papers for viscous incompressible flow in porous media, there existed only very few contributions addressing the behaviour of the flow at interfaces or boundaries. Analytic difficulties may arise at interfaces in media with different scales in different regions. A typical situation was analyzed in several papers in this team: fluid flow close to the interface between a porous media and a domain of free flow. Whereas the free flow may be described by a Navier–Stokes system, the flow through porous media may satisfy Darcy's law. With the exception, that the region of free flow is enclosed in the porous media, we expect different orders in velocities. As in the analysis of behavior of solutions close to the boundary of a medium with scale dependent structure, one has to study layers at the boundary depending on the scale. In the case of a periodic structure geometrically characterized by a periodic lattice, difficulties with the construction of boundary layers arise even for plain boundaries, if this boundary is in a location, which is in an irrational relationship to the underlying lattice. A detailed analysis of this phenomenon was given by Neuss-Radu [21]. She could show, that in the irrational situation, the boundary layers have no exponential decay at infinity. This negative statement is related to the fact, that the corresponding boundary layer problem involves halfspaces, where the spectrum for the Laplace operator cannot be bounded away from 0. In the case of layered media, boundary layers also for curved domains could be constructed in [22]. These facts should be taken into account for the following part of the report. Considering periodic structures, we only choose plain interfaces in a rational position with respect to the lattice.

In the following we consider the Navier–Stokes equation

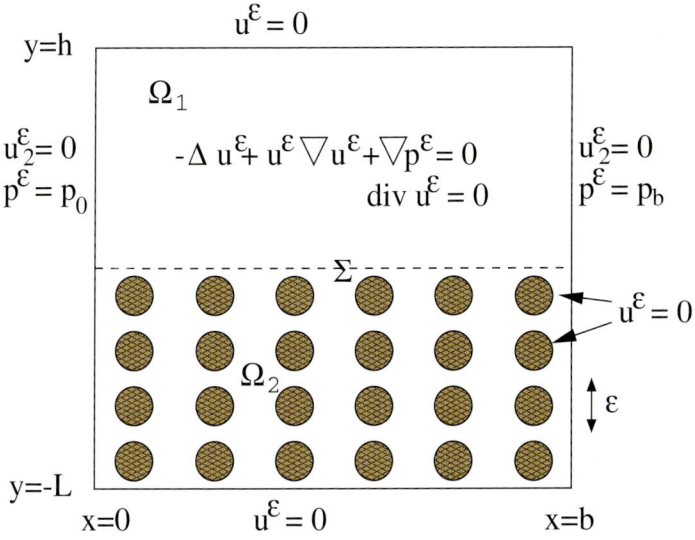

Fig. 4. The partially porous domain.

$$\partial_t u^\varepsilon + (u^\varepsilon \nabla) u^\varepsilon - \mu \Delta u^\varepsilon + \nabla p^\varepsilon = f, \quad \text{div } u^\varepsilon = 0$$

(or the Stokes equation, stationary and instationary) in partial porous domains. The notations and the boundary conditions can be taken from Fig.4. In the following we only can sketch ideas of the analysis, the most important part of which is given in [7]. Let us consider for simplicity the stationary equation, in order to avoid additional difficulties with initial layers, and the Stokes equation in 2 dimensions. The results are not essentially changed if we consider the Navier–Stokes equation in a situation, where the existence and uniqueness of the solutions are guaranteed. We expect as zero order approximation for the free flow the solution of the following system,

$$-\mu \Delta u^0 + \nabla \pi^0 = 0 \text{ in } \Omega_1$$
$$\text{div } u^0 = 0 \text{ in } \Omega_1$$
$$u^0 = 0 \text{ on } \Sigma$$

which surely has to be corrected due to the filtration flow given by

$$u_D = -\varepsilon^2 K \nabla p^0 \text{ in } \Omega_2$$
$$\text{div } u_D = 0 \text{ in } \Omega_2$$
$$p^0 = ? \text{ on } \Sigma$$

u_D only can be determined if the boundary values on the boundary of the domain are given and the pressure at Σ is known. The main problem consists in determining the unknown pressure and to construct terms to control the

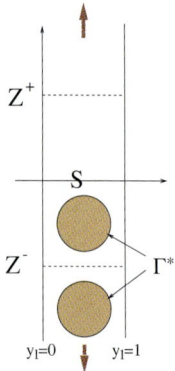

Fig. 5. The boundary layer cell $Z^{bl} = Z^+ \cup S \cup Z^-$.

error. In order to achieve this aim, we have to solve a boundary layer system of the following or similar type, formulated on a boundary layer cell, see Fig.5.

$$-\Delta_y \beta^{bl} + \nabla_y \omega^{bl} = 0 \quad \text{in } Z^+ \cup Z^-$$
$$\text{div}_y \beta^{bl} = 0 \quad \text{in } Z^+ \cup Z^-$$
$$[\beta^{bl}]_S = 0 \quad \text{on } S$$
$$[(\nabla_y \beta^{bl} - \omega^{bl} I)e_2]_S = e_1 \quad \text{on } S$$
$$\beta^{bl} = 0 \quad \text{on hole boundaries}$$
$$\text{periodic boundary condition in } y_1$$

The solution β^{bl}, ω^{bl} to this layer problem exists (β^{bl} is unique, ω^{bl} is unique mod. constants). They have exponential behavior at y_2 tending to ∞

$$\exp(\gamma_0 |y_2|)\nabla_y \beta^{bl} \in L^2(Z^{bl})$$
$$\exp(\gamma_0 |y_2|)\beta^{bl}, \exp(\gamma_0 |y_2|)\omega^{bl} \in L^2(Z^-)$$

and C_1^{bl} and C_ω^{bl} are such that

$$|\beta^{bl}(y) - (C_1^{bl}, 0)| \leq \text{const.} \exp(-\gamma_0 y_2)$$
$$|\omega^{bl}(y) - C_\omega^{bl}| \leq \text{const.} \exp(-\gamma_0 y_2)$$

for $y_* < y_2$. Using this layer solutions, Jäger and Mikelic could prove in [7] that the following relation holds answering the question mark set above:

$$p^0(x_1, 0-) = p^0(x_1, 0+) + \mu C_\omega^{bl} \partial_2 u_1^0(x_1, 0+).$$

This relation differs from all statements in the literature before, where continuity of the pressure was stated using physical reasoning, not taking into account that there is a singular perturbation in the problem and that in the limit jumps are possible.

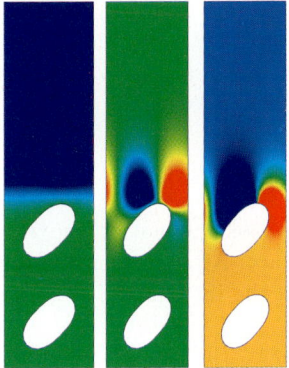

Fig. 6. Numeric results: horizontal and tangential components of the velocity β^{bl} and the pressure ω^{bl}.

The question if the decisive asymptotic constant C_ω^{bl} does not vanish could not be answered by a pure analytic proof. In the symmetric situation it was possible to prove $C_\omega^{bl} = 0$. Using numerical methods based on good error estimates, it could be proven in [12] that in general we have to expect a jump of the pressure. The simulation results in Fig.6 show that the pressure tends to different values at $+\infty$ and $-\infty$. The estimates for the numerical errors, especially the reduction to a bounded region, given in [12], and used for determining the pressure jump, can be applied also in other situations.

The natural question arises if the effect of a porous medium to the free flow could be modeled by a boundary condition on the interface. Such a law was formulated and experimentally checked by Beavers and Joseph. This law states that the normal derivative of the tangential component of the velocity at the interface is proportional essentially to the tangential velocity. In [9] this law could be derived with help of boundary layer techniques and the

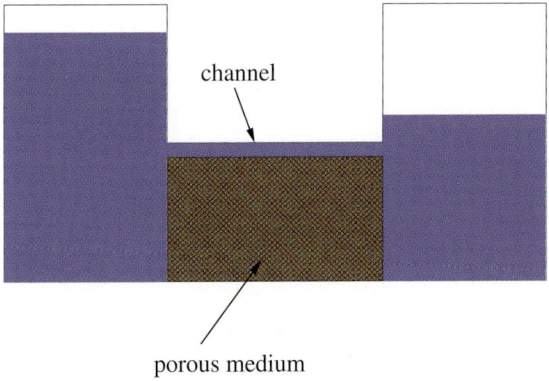

Fig. 7. The experiment by Beavers and Joseph.

energy method in estimating the ε approximation errors. Here we are giving a heuristic argument for the boundary condition. The solution u^ε can be viewed as a function of the macroscopic variable x and a microscopic variable $\frac{x}{\varepsilon}$. A formal expansion motivates the ansatz

$$u^\varepsilon = u^0 - \varepsilon \beta^{bl}\left(\frac{x}{\varepsilon}\right) \partial_2 u_1^0 + \varepsilon C_1^{bl}(\partial_2 u_1^0 e_1 + d^1) + O(\varepsilon^2)$$

Here d^1 is the solution of Stokes equation in Ω^1 with boundary condition $d^1 = -\partial_2 u_1^0 e_1$ on the interface. After formal differentiation and reordering of the expansion we obtain on the interface Σ

$$\partial_2 u_1^\varepsilon = \partial_2 u_1^0 \left(1 - \partial_{y_2}\beta^{bl}\left(\frac{x}{\varepsilon}\right)\right) + O(\varepsilon)$$
$$u_1^\varepsilon = -\varepsilon \beta_1^{bl}\left(\frac{x}{\varepsilon}\right) \partial_2 u_1^0 + O(\varepsilon^2)$$

Replace $\partial_2 u_1^0$ in the second equation (on the interface) by $\partial_2 u_1^\varepsilon$ and obtain formally an interface law which still contains an oscillatory factor. This factor converges, if x_2 is not zero, to the asymptotic velocity C_1^{bl}. Thus we can formulate the Beavers-Joseph law:

$$\varepsilon C_1^{bl} \partial_{x_2} u_1^{eff} + u_1^{eff} = 0 \text{ on } \Sigma.$$

The following error estimate is proven in [9]

$$\|\varepsilon C_1^{bl} \partial_{x_2} u_1^\varepsilon + u_1^\varepsilon\|_\Sigma \leq C \varepsilon^{\frac{3}{2}} \log \varepsilon$$

Here $\|\cdot\|_\Sigma$ is a proper norm on the surface Σ. This boundary condition is also known as Navier condition. The effective equations can be formulated as follows in the case of the stationary Navier–Stokes,

$$-\mu \Delta u^{eff} + (u^{eff}\nabla)u^{eff} + \nabla p^{eff} = 0 \text{ in } \Omega_1$$
$$\operatorname{div} u^{eff} = 0 \text{ in } \Omega_1$$
$$u_1^{eff} + \varepsilon C_1^{bl} \partial_{x_2} u_1^{eff} = 0 \text{ on } \Sigma$$

Additional boundary conditions have to be posed, e.g. in case of a channel, the pressure at the end may be described. In case of a flat channel ($0 \leq x_1 \leq b, 0 \leq x_2 \leq h$), one obtains as perturbation of the Hagen-Poiseulle flow in an explicit form:

$$u^{eff} = \left(\frac{p_b - p_0}{2b\mu}\left(x_2 - \frac{\varepsilon C_1^{bl} h}{h - \varepsilon C_1^{bl}}\right)(x_2 - h), 0\right).$$

4 Effective Laws for Rough Boundaries

Rough boundaries can be treated analytically very similar to the case of an interface to a porous medium. Here roughness is modeled by rapidly oscillatory

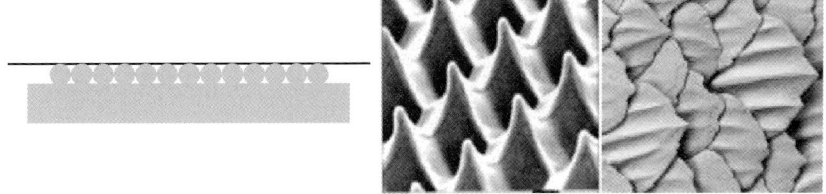

Fig. 8. Rough boundaries: sandpaper, laser generated skin, and shark skin.

surfaces. Whereas for the Laplace equation rapidly oscillatory boundaries and oscillatory conditions were considered in the literature, there was a gap for systems modeling flow. The main aim of the investigations in this project was to replace the rough boundary by a smooth one and change the boundary condition. The new boundary conditions are taking into account the effects of the roughness similarly as the Navier condition is modelling the influence of a porous medium on the free flow.

Roughness is influencing the friction forces at the boundary. Nature and technology are using these effects to reduce the friction. Multiscale analysis is providing the effective model for the roughness, which is necessary for simulations.

For simplicity we formulate here the problem in 2d treated in [10], whereas the full 3d case is considered in [11]. Since we are applying perturbation methods, we have to consider situations, where we know that the original problem can be solved uniquely, that means we have to assume size conditions if we consider the stationary Navier–Stokes system. That meant in the beginning, that we had to assume at first laminar flow with small Reynolds number. Following the concept for the analysis of boundary layers we proceed as above considering now as boundary layer cell an half infinite cylinder, see Fig.9.

$$-\Delta_y \beta^{bl} + \nabla_y \omega^{bl} = 0 \text{ in } Z \setminus S$$
$$div_y \beta^{bl} = 0 \text{ in } Z \setminus S$$
$$[\beta^{bl}]_S = 0, [(\nabla_y \beta^{bl} - \omega^{bl} I)e_2]_S = e_1 \text{ on } S$$
$$\beta^{bl} = -y_2 e_1 \text{ on } \Gamma$$

The problem is in some aspects simpler than the transition problem for a porous medium. Again β^{bl} converges to $(C_1^{bl}, 0)$ for $y_n \to \infty$. The limit C_1^{bl} is the decisive coefficient in the Navier condition to be posed on the smooth surface:

$$\varepsilon C_1^{bl} \partial_{x_2} u_1^{\mathit{eff}} + u_1^{\mathit{eff}} = 0 \text{ on } \Sigma$$

The difference of u^{ε} and the approximation u^{eff} and also the dependence of the results on the selection of the position of Σ can be estimated. We remark that u^{eff} is depending on ε. The effective solutions can be used to approximate

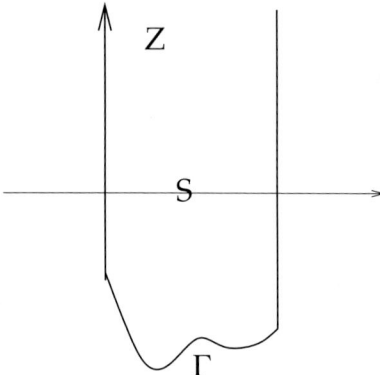

Fig. 9. The boundary layer cell in the case of domains with rough boundaries.

and compute drag forces. Here a roughened flat channel was considered. We get for the tangential drag (skin friction)

$$F_t^\varepsilon = \int_0^b \mu \partial_{x_2} u_1^\varepsilon(x_1, 0) dx_1$$

and the corresponding approximation F_t^{eff} the estimate:

$$|F_t^\varepsilon - F_t^{eff}| \leq C \varepsilon^{\frac{3}{2}}$$

Since for the flat channel the effective flow can be explicitly computed, the drag F_t^{eff} is explicitly given by

$$F_t^{eff} = -\frac{1}{2} \frac{(p_b - p_0)h^2}{h - \varepsilon C_1^{bl}}.$$

It can easily be seen that C_1^{bl} is negative, therefore, F_t^{eff} decreases with ε. This observation stirs the project to analyse the dependence of C_1^{bl} on the geometric stucture of the roughness.

Since the assumptions in [10] were to restrictive, Jäger and Mikelic considered situations of higher Reynolds numbers, assuming that the theory of Schlichting is accepted. This theory states that even in turbulent flow there exists a laminar layer at the boundary of thickness $\delta = \frac{\mu}{v_t}$, where μ is the kinematic viscosity and v_t the tangential velocity at the inner boundary of the layer. In [11] the turbulent Couette flow over rough boundary was studied in 3d. Here the decisive parameter for the expansion is the relative height of the riblets with respect to the thickness δ of the viscous layer system. If this scale ε is small, one can show the same results as for the laminar case. If it is of order 1, nonlinear boundary laws are expected, see also Fig.10.

The following results could be proven for small ε:

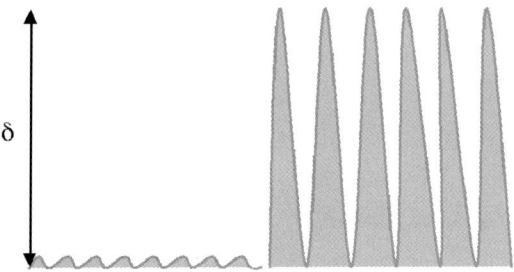

Fig. 10. Orders of magnitude for the height of the riblets.

- The effective boundary condition is a Navier-slip condition with a matrix as coefficient in front of the shear stress, which can be computed by solving a boundary cell problem,
- The effective Couette–Navier flow is an $O(\varepsilon^2)$-approximation in the mass flow, an $O(\varepsilon^2) - L1-$approximation in the velocity, and an $O(\varepsilon^2)$-approximation in the tangential drag.

5 Rough Surfaces and Drag-Reduction

The results obtained in Section 4 could be applied to drag reduction, by using methods of shape optimization. Friedman solved in [3] the problem to find the optimal shape of microstructures on the surface of bodies in a Couette flow. Using the formula derived by asymptotic analysis and pre-analysis of the objective functional and its dependence on the design parameters, the computational costs could be reduced essentially. In 3 dimensions the Navier-condition obtained in [11] have the form

$$\varepsilon \sum_{i=1}^{2} M_{ij} \frac{\partial u_i^{eff}}{\partial x_3} + u_j^{eff} = 0 \quad \text{for } j = 1, 2.$$

In the case of a channel with a rough boundary one obtains two components of the drag

$$(F_t^{eff})_1 = \frac{\mu}{L^3} \frac{1}{1 - \frac{\varepsilon}{L_3} C_\parallel^{bl}} U_1$$

$$(F_t^{eff})_2 = \frac{\mu}{L^3} \frac{1}{1 - \frac{\varepsilon}{L_3} C_\perp^{bl}} U_2$$

where the prescribed inflow is $U = (U_1, U_2, 0)$.

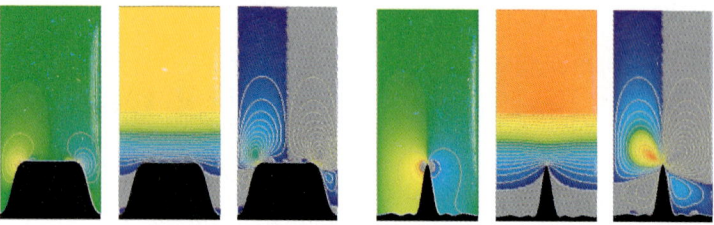

Fig. 11. Numeric results: Starting shape and optimized shape of riblets

Friedmann treated mainly the 2d case and was able to transfer the results to special geometries in 3d. In 2d it is possible with help of the representation formula to determine the variations of the objective functional. It has the form

$$\delta I = \int_\Gamma \alpha(s) |\partial_3 \beta^{bl}(s)|^2 ds + o\left(\|\alpha\|_{C^2}\right)$$

where the variation of the rough surface is given by

$$\Gamma_\lambda = \{s + \lambda \alpha(s) n(s) | s \in \Gamma\}$$

with n denoting the normal field.

6 Flow Through Filters and Membranes

In [8] Jäger and Mikelic considered an incompressible, non stationary fluid flow, governed by a given pressure drop, in a domain that contains a filter of finite thickness. The filter consists of a big number of tiny, axially symmetric tubes with non-constant sections. Global existence for the ε-problem and the effective behaviour of the velocity and the pressure field is determined. The effective velocity v in the filter part is constant in the axial direction, and the effective pressure obeys the so-called fourth-power law.

Estimates for the approximation errors are derived. The flamelet structures observed in a gas burner with such a filter illustrate the oscillation of the flow in the layer close to the filter, which mathematically have to be expected.

This is another example of practical importance, where the techniques of asymptotic analysis can help to reduce the computational complexity of a problem. In modelling catalytic processes it will be necessary to add chemical reactions on the surface and in the interior of the channels. Furthermore, temperature transport in the solid phase is coupling the channels in the filter. To increase surface area the solid walls in a catalyst are made of porous material carrying the catalytic active substances. Scale analysis based on real data shows that the Peclet number is large and the transport term has stronger effect on the homogenization limit, as shown in an ongoing investigation.

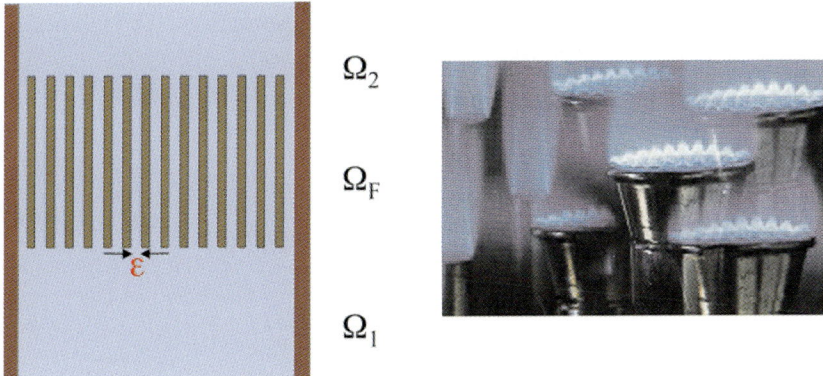

Fig. 12. The mathematical representation of the filter domain and a gas burner

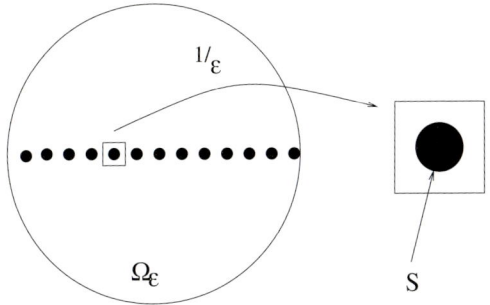

Fig. 13. A domain with a sieve.

In the case discussed before the filter had finite thickness. For thin filters, where both the thickness of the channels and that of the filter are very small, the question arises, which effective transmission conditions have to be posed in the limit on the interface. Jäger, Oleinik and Shaposhnikova analysed in [13] sieve structures, see Fig.13 in the case of the Laplace operator.

$$\triangle u^\varepsilon = f \quad \text{in } \Omega_\varepsilon$$
$$\nabla u^\varepsilon \cdot n + b u^\varepsilon = 0 \quad \text{on } S_\varepsilon$$
$$u^\varepsilon = 0 \quad \text{on } \Gamma_\varepsilon$$

Here Γ_ε denotes the outer boundary of the domain Ω_ε and S_ε denotes the boundary of the obstacles along a hypersurface (modeling the sieve) inside the domain. The diameter of the obstacles is a_ε. For the homogenization limit $\varepsilon \to 0$ the following effective models could be derived

- Case 1: $a_\varepsilon = o(\varepsilon)$

$$\triangle u^0 = f \quad \text{in } \Omega$$
$$u^0 = 0 \quad \text{on } \partial \Omega$$

This means, the sieve is not felt at all in the full domain Ω. The following estimate for the approximation error holds

$$\|u^\varepsilon - u^0\|_{H^1(\Omega)} \leq c \frac{a_\varepsilon^{\frac{n}{2}}}{\varepsilon} \quad \text{for } n > 3$$

$$\|u^\varepsilon - u^0\|_{H^1(\Omega)} \leq c \frac{a_\varepsilon}{\varepsilon}(\log \frac{a_\varepsilon}{\varepsilon})^{\frac{1}{2}} \quad \text{for } n = 2$$

- Case 2: $a_\varepsilon = C_0 \varepsilon$

$$\triangle u^0 = f \quad \text{in } \Omega$$
$$[u^0]_\Sigma = 0, \; [\partial_\nu u^0]_\Sigma = \mu_0 u^0 \quad \text{on } \Sigma$$

where $\mu_0 = -2C_0^{(n-1)} b |S|$ The approximation error can be estimated by

$$\|u^\varepsilon - u^0\|_{H^1(\Omega)} \leq c\sqrt{\varepsilon}$$

Recently Jäger and Neuss-Radu in [23], [24], [25] analyzed diffusion reaction systems and ion transport in domains separated by membranes and derived transmission conditions on the interface. In order to pass to the homogenization limit, new methods of asymptotic analysis had to be developed and applied in a situation where two phenoma are involved: rapid oscillations and reduction of dimension. Introducing locally microscopic coordinates and properly scaled norms including macroscopic and microscopic variables, it is possible to use the apriori estimates of the model systems to pass to the homogenization limit. Two situations were considered:

- Transport of ions, modelled by the Nernst-Planck equations, through a membrane with channels carrying charges which control the permeability of the membrane.
- Reactions and diffusion in a domain separated by a thin membrane with oscillatory physical and chemical properties.

In an ongoing work, Jäger, Neuss-Radu and Mikelic derived a system of equations modelling reactive flow through a system of biological cells and its interaction with the mechanics of the cell system and analyzed the homogenisation limit for a tissue in 3d.

7 Flow Through Networks of Channels

In many situations we have to deal with network structures and processes on these networks. So far, there is no general method available to pass from a detailed fine scale model to a coarse scale model using for instance phase field concepts to describe these processes on the macroscopic level. There is an urgent need for new mathematical tools for such situations, e.g. in physiology (blood vessels, vessels in plant leaves, nets of neurons). As seen in the scan of a plant leaf it seems to be necessary to describe the processes on the first levels of branching in geometric detail whereas the vessel systems arising after few branches may be described by model "averaging" over the finer channel system. The channel system, which has to be resolved in detail is complex by itself and, therefore, the complexity has to be reduced as far as possible. Here reduction to lower dimension is one possibility suggesting to replace the vessels in 3d by curves, especially by straight lines, connecting the bifurcation points. In this approach one has to assume that the vessels are almost straight and the ratio of their thickness and their length is small, which is a rather restricted assumption for vessels in physiology.

The main result to this subject was achieved by Lenzinger [16], where the flow of a viscous Newtonian fluid is investigated in bifurcations of thin three-dimensional pipes with a diameter-to-length ratio of order $O(\varepsilon)$. The model is based on the steady-state Navier–Stokes equations with pressure conditions on the in- and outflow boundaries. Existence and local uniqueness is proven under the assumption of small data represented by a Reynolds number Re_ε of order $O(\varepsilon)$. An asymptotic expansion in powers of ε and Re_ε for the solution of this Navier–Stokes problem is constructed and a method of computing the pressure drop and the flux based on Poiseuille flow is presented. The influence of the bifurcation geometry on the fluid flow is analyzed by introducing local Stokes problems in the junction. It is proven that the solutions of these Stokes problems in the junction of diameter $O(M)$ approximate the solutions of the corresponding Leray problems in the infinite bifurcation up to an error decaying exponentially in M. The construction of the approximation for the Navier–Stokes solution is presented and its properties are discussed. The approximation is based on the idea of a continuous matching of the Poiseuille velocity to the solution of the junction problem on each pipe-junction interface. The error estimates derived for the approximation in powers of ε and Re_ε improve results obtained by Marusic-Paloka. Furthermore, it could be shown that Kirchhof's law of the balancing fluxes has to be corrected in $O(\varepsilon)$ in order to obtain an adequate error estimate for the gradient of velocities.

So far, only the first steps are made to achieve a better modelling of flow, transport and reactions in such networks. Additional difficulties arise due to the fact that model equations valid on larger scale have to be replaced on finer scales. This fact is known in modelling e.g. the complex "fluid" blood, where on finer scale the diameter of a vessel is essential for the effective viscosity. Since this subject is very important for many applications, it deserves more attention in the future research.

8 Final Remark

New technologies are providing more and more information on processes at the micro and even nano level, which has to be integrated in modelling on meso or macro scale. Mathematical modelling and simulation in general, multiscale analysis and numerics are confronted with new challenges: models on these scales have to be linked and the transfer between the scales has to be made possible in an efficient way. The multiscale methods for discrete and stochastic model systems will become increasingly important. However, also the already established methods will have to be developed, so that more complex systems integrating reactive flow, diffusion and transport, mechanics and other physical interactions and processes can be handled. This team could help to develop expertise necessary for the next steps in the research.

References

1. Allaire, G.: Homogenization and two-scale convergence. SIAM J. Math. Anal **23**, 1482-1518 (1992)
2. Alt, H.-W., Luckhaus, S.: Quasilinear elliptic-parabolic differential equations. Math. Z., **183**, 311-341 (1983)
3. Friedmann, E.: Riblets in the Viscous Sublayer. Optimal Shape Design of Microstructures. Heidelberg, Thesis (2005)
4. Jäger, W., Kutev, N.: Discontinuous Solutions of the Nonlinear Transmission Problem for Quasilinear Elliptic Equations. Preprint **22**, Universität Heidelberg.
5. Hornung, U., Jäger, W.: Diffusion, Convection, Adsorption, and Reaction of Chemicals in Porous Media. Journal of Differential Equations **92**, 199-225 (1991)
6. Hornung, U., Jäger, W., Mikelic,A.: Reactive Transport Trough an Array of Cells with Semipermeable Membranes. M2AN, **28**, 59–94 (1994)
7. Jäger, W., Mikelic, A.: On the Boundary Conditions at the Contact Interface Between a Porous Medium and a Free Fluid. Annali della Scuola Normale Superiore di Pisa, **23**, 403–465 (1996)
8. Jäger, W., Mikelic, A.: On the Effective Equations of a Viscous Incompressible Fluid Flow Through a Filter of Finite Thickness. Comm. Pure. Appl. Math., **51**, 1073–1121 (1998)
9. Jäger, W., Mikelic, A.: On the Interface Boundary Condition by Beavers, Joseph and Saffman. SIAM J. Appl. Math., **60**, 1111-1127 (2000)
10. Jäger, W., Mikelic, A.: On the Rouhgness-Induced Effective Boundary Conditions for an Incompressible Viscous Flow. Journal of Differential Equations, **170**, 96–122 (2001)
11. Jäger, W., Mikelic, A.: Couette Flows over a Rough Boundary and Drag Reduction. Comm. Math. Physics, **232**, 429–455 (2003)
12. Jäger, W., Mikelic, A., Neuß, N.: Analysis of the Laminar Viscous Flow over a Porous Bed. SIAM J. Sci. Comp. **22**, 2006-2028 (2001)
13. Jäger, W., Oleinik, O. A., Shaposhnikova, T. A.: Homogenization of solutions of the Poisson Equation in a Domain Perforated Along a Hypersurface, with Mixes Boundary Conditions on the Boundary of the Cavities. Trans. Moskow. Math. Soc., **59**, 135-157 (1998)
14. Jäger, W., Postel, M., Sepulveda, M., Valentin, P.: Numerische Simulation und Identifizierung reaktiver Flüssigkeiten in einer Chromatographiesäule. Mathematik- Schlüsseltechnologie für die Zukunft, Springer Verlag, 219–230 (1996)

15. Jäger, W., Simon, L.: On Transmission Problems for Nonlinear Parabolic Differential Equations. Annales Univ.Sci. Budapest, **44**, 143-158 (2003)
16. Lenzinger, M.: Viscous Fluid Flow in Bifurcating Channels and Pipes. Heidelberg, Thesis (2005)
17. Marciniak-Czochra, A.:Receptor-based models with hysteresis for pattern formation in hydra, Math. Biosci., **199**, 97-119 (2006)
18. Marciniak-Czochra, A., Ptashnyk, M.:Derivation of a macroscopic receptor-based model using homogenization techniques. Submitted
19. Neuß, N., Neuss-Radu, M., Mikelic, A.: Effective laws for the Poisson equation on domains with curved oscillating boundaries. Applicable Analysis, **85**, 479–502 (2006)
20. Neuss-Radu, M.: Some extensions of two-scale convergence. C. R. Acad. Sci. Paris Ser. I Math., **322**, 899–904 (1996)
21. Neuss-Radu, M.: A result on the decay of the boundary layers in the homogenization theory. Asymptotic Analysis,**23**, 313–328 (2000)
22. Neuss-Radu, M.: The boundary behavior of a composite material. Mathematical Modelling and Numerical Analysis, **35**, 406–435 (2001)
23. Neuss-Radu, M., Jäger, W.: Homogenization of thin porous layers and applications to ion transport through channels of biological membranes. Oberwolfach Reports, **49**, 2809–2812 (2005)
24. Neuss-Radu, M., Jäger, W.: Effective transmission conditions for reaction-diffusion processes in domains separated by an interface. Submitted.
25. Neuss-Radu, M., Jäger, W.: A Multi-Scale Approach to Ion Transport Through Channels. In preparation.
26. Sanchez-Palencia, E.: Non-Homogenous Media and Vibration Theory. Springer Lecture Notes in Physics 127, Springer-Verlag Berlin Heidelberg New-York (1980)
27. Sherratt, J. A., Maini, P. K., Jäger, W., Müller, W. A.: A Receptor Based Model for Pattern Formation in Hydra. Forma **10**, 77–95 (1995)

Microscopic Interfaces in Porous Media*

W. Jäger and B. Schweizer[†]

Institut für Angewandte Mathematik, Universität Heidelberg

Summary. Homogenization techniques have been applied successfully to pore scale models of porous media, at least in the description of single-phase fluids. In the case that two phases are present (e.g. water and oil) one expects upscaled equations in the form of the Leverett two-phase flow system. Unfortunately, due to the appearance of many interfaces inside the porous medium and a lack of control of the geometric properties of the domains, no rigorous derivation of the Leverett system can be expected in the near future. In this contribution we report on rigorous derivations for geometries that allow a uniform description of the interfaces.

1 Introduction

One of the main achievements of homogenization theory was the mathematical understanding of fluid flow in porous media. Initiated by studies of Sanchez-Palencia [8], tools were developed that allowed to start from microscopic models for flows in porous media, and to derive averaged (homogenized) equations that do not resolve the fine structure any more, but nevertheless describe averaged quantities well. The mathematical homogenization theory was in particular successful in deriving from Stokes- or Navier–Stokes equations in a microscopic (periodic or stochastic) geometry the Darcy-law in the macroscopic geometry.

Nevertheless, until today, one of major subjects regarding flows in porous media can not be treated mathematically: Two-phase flows. Here, two fluids (e.g. water and air or water and oil) fill the pores of the medium. Together with the flow of the two components, also the interfaces will in general move through the medium. Within this process, the front may change its topology and air bubbles may be created in the water-region (compare Figure 1). Due to the unpredictability of the geometry, it seems to be out of reach to homogenize

---

*This work has been supported by the German Research Foundation (DFG) through SFB 359 (Project D1) at the University of Heidelberg.

[†]Actual affiliation: Mathematisches Institut, Universität Basel

Fig. 1. A front inside a porous medium

a microscopic model for two fluids in the pore-space with a free boundary separating them.

It is important to note that, dispensing with mathematical rigor, one can guess equations that describe the homogenized situation, where both fluids are distributed with varying density over the whole macroscopic domain. These equations go back to Leverett [7] and are also known as 'generalized Darcy-law' or 'two-phase flow equations'. Collecting various physical parameters in three fundamental coefficient functions $k_A(s), k_B(s)$ (permeabilities), and $p_c(s)$ (capillary pressure), the equations read

$$v_A = -k_A(s)\nabla p_A, \qquad v_B = -k_B(s)\nabla p_B, \qquad (1)$$
$$\partial_t s + \nabla \cdot v_A = 0, \qquad \partial_t(1-s) + \nabla \cdot v_B = 0, \qquad (2)$$
$$p_B - p_A = p_c(s). \qquad (3)$$

Here v_A and v_B are the velocity fields of the two fluids, p_A and p_B are their pressures, s is the saturation, i.e. the local volume fraction of fluid A. With equations (1) one assumes that a Darcy law remains valid for both fluids, equations (2) express the conservation of mass. Equation (3) is motivated from the study of a capillary in which the two fluids are in contact over a fluid-fluid interface. For a small diameter of the tube, the interface is close to a spherical cap, the curvature of which is determined by the diameter of the tube and the physical constant of the contact angle. Surface tension creates a pressure difference across the free boundary. The pressure jump is proportional to the mean curvature of the cap and its absolute value decreases with the tube diameter. The dependence of $p_c = p_c(s)$ on the saturation expresses the following mechanism. Assuming that water is the wetting fluid 'A', a decreasing water saturation s means that water is pushed out of the pores. It will first leave the large pores in which fluid 'B' enters easily. With this process also the free interfaces are shifted from larger to smaller pores and thus the capillary pressure is increased.

In this contribution we want to distinguish three scales. The pores constitute a microscopic scale, the domain of the entire porous medium the macroscopic scale. The Leverett equations describe the macroscopic scale in which neither the pores nor the position of interfaces is resolved. Additionally to these two scales, one may consider a mesoscopic scale, where pores are not resolved, but interfaces are. In sections 2 and 3 we study the homogenization from the microscopic scale to the mesoscopic scale, in section 4 we consider the step to the macroscopic scale.

Within the SFB 359 a rigorous mathematical analysis was achieved in two situations describing a two-phase flow in a porous medium. The first regards the stationary Stokes equations in a special geometry, where the material of the porous medium consists of an array of parallel cylinders. The second regards the instationary case in a filter geometry. In both cases the geometry keeps the topology of the interface fixed and allows a uniform description of the interface.

2 Homogenization of a Free Boundary Problem – the Stationary case

The geometry

We first desribe the geometry and fix notations. The independent variables of the macroscopic domain are $(x_1, x_2, x_3) = (\tilde{x}, x_3)$. The geometry is determined by the cross-section of the cylinders $Q \subset\subset (0,1)^2$ with smooth boundary. We set

$$\Pi = (0,1)^2 \subset \mathbb{R}^2, \qquad \Pi_\varepsilon = \Pi \setminus \varepsilon(\mathbb{Z}^2 + Q),$$
$$\Omega = \Pi \times (0, \infty) \subset \mathbb{R}^3, \qquad \Omega_\varepsilon = \Pi_\varepsilon \times (0, \infty).$$

We consider a free boundary problem in Ω_ε. For given functions $h^\varepsilon : \Pi_\varepsilon \to \mathbb{R}$ and $h_0 : \Pi \to \mathbb{R}$, we study the domains $G_\varepsilon := \Omega_\varepsilon \cap \{x_3 < h^\varepsilon(\tilde{x})\}$ and $G_0 := \Omega \cap \{x_3 < h_0(\tilde{x})\}$. The unknown functions are the physical variables height, velocity, and pressure,

$$h^\varepsilon : \Pi_\varepsilon \to \mathbb{R}, \qquad v^\varepsilon : G_\varepsilon \to \mathbb{R}^3, \qquad p^\varepsilon : G_\varepsilon \to \mathbb{R}.$$

We now formulate the equations for $(v^\varepsilon, p^\varepsilon, h^\varepsilon)$. To simplify the notation we consider periodic boundary conditions on the macroscopic lateral boundaries and assume $\varepsilon = \frac{1}{N}$ for some natural number N. Following the normalization of Allaire we can assume that the viscosity is $\nu = \varepsilon^2$, the surface-tension parameter is chosen to be $\beta = 1$. To state the boundary conditions we write $\partial_u G_\varepsilon := \text{graph}(h^\varepsilon)$ for the upper boundary of G_ε, and $\partial_l G_\varepsilon := (\partial G_\varepsilon \cap \partial \Omega_\varepsilon) \setminus \partial \Omega$ for the microscopic lateral boundaries of G_ε. Normal vectors are denoted by n. The system constitutes a free boundary problem: the domain G_ε is defined by h^ε, which is one of the unknown functions.

 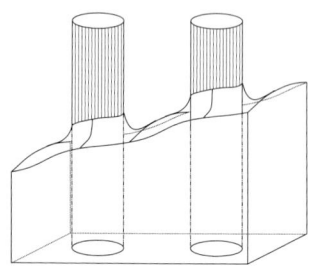

Fig. 2. Domain of h^ε and the local geometry

$$-\varepsilon^2 \Delta v^\varepsilon + \nabla p^\varepsilon = f \quad \text{in } G_\varepsilon,$$
$$\nabla \cdot v^\varepsilon = 0 \quad \text{in } G_\varepsilon,$$
$$v^\varepsilon \cdot n = 0 \quad \text{on } \partial G_\varepsilon,$$
$$\varepsilon^m \tau \cdot \partial_n v^\varepsilon + \gamma \tau \cdot v^\varepsilon = 0 \quad \text{on } \partial_l G_\varepsilon,$$
$$T(v^\varepsilon, p^\varepsilon) \cdot n = \mathcal{H}(h^\varepsilon) \cdot n \quad \text{on } \partial_u G_\varepsilon,$$
$$v^\varepsilon = 0 \quad \text{on } \{x_3 = 0\},$$
$$\sin(\Theta) = \alpha \quad \text{on } \partial \Pi_\varepsilon.$$

The fourth equation must hold for all tangential vectors τ; it expresses a general slip condition and we always assume $\gamma \geq 0$ and $m \geq 0$. In the fifth equation we used the stress tensor $T(v^\varepsilon, p^\varepsilon)_{ij} := -p^\varepsilon \delta_{ij} + \varepsilon^2 (Dv^\varepsilon)_{ij}$, where δ_{ij} is the Kronecker symbol and $Dv^\varepsilon := \nabla v^\varepsilon + (\nabla v^\varepsilon)^T$ is the symmetric gradient. The scalar

$$\mathcal{H}(h) = \nabla \cdot \left(\frac{\nabla h}{\sqrt{1 + |\nabla h|^2}} \right)$$

is the mean curvature of the surface $\Gamma = graph(h)$. Finally, $\Theta = \Theta(\nabla h^\varepsilon, n)$ is the angle between cylinder wall and free boundary, and $\alpha \in (-1, 1)$ is a given constant that prescribes the contact angle at the triple lines.

The homogenized system

It is shown in [10] that for small forces f and $m = 1$, solutions $(v^\varepsilon, p^\varepsilon, h^\varepsilon)$ to the above system exist. After a normalization of the pressure we find bounded sequences v^ε, $p^\varepsilon - \hat{p}(\varepsilon)$, and h^ε, the latter bounded in the Lipschitz norm. In particular, after extending the solutions to the entire domain, we find limit functions $h_0 : \Pi \to \mathbb{R}$ and $P : G_0 \to \mathbb{R}$, and a subsequence $\varepsilon \to 0$ with $(p^\varepsilon - \hat{p}(\varepsilon), h^\varepsilon) \to (P, h_0)$. The functions P and h_0 are characterized by the following theorem.

Theorem 1. *For $m = 1$ and small f, weak limits $(P, h_0) = \lim_{\varepsilon \to 0}(p^\varepsilon - \hat{p}(\varepsilon), h^\varepsilon)$ satisfy the homogenized free boundary problem*

$$V := M \cdot (f - \nabla P) \qquad \text{in } G_0,$$
$$\nabla \cdot V = 0 \qquad \text{in } G_0,$$
$$(V \cdot n)(\tilde{x}, h_0(\tilde{x})) = 0 \qquad \forall \tilde{x} \in \Pi,$$
$$-P(\tilde{x}, h_0(\tilde{x})) = \nabla \cdot E(\nabla h_0(\tilde{x})) \qquad \forall \tilde{x} \in \Pi,$$
$$V_3(\tilde{x}, 0) = 0 \qquad \forall \tilde{x} \in \Pi,$$

together with periodicity conditions on the macroscopic lateral boundaries. The matrix M is determined by a Stokes problem with slip condition in a three-dimensional cell.

The non-linear map E is determined via a cell-problem on $F := (0,1)^2 \setminus Q$ for $h_1(x,.)$. With $T[\xi] := \xi/\sqrt{1+|\xi|^2}$ we must solve for given $\nabla h_0(\tilde{x})$

$$\nabla_y \cdot T\left[\nabla h_0(\tilde{x}) + \nabla_y h_1(y)\right] = \alpha \frac{|\partial Q|}{|F|} \qquad \forall y \in F, \tag{4}$$

$$\tilde{n} \cdot T\left[\nabla h_0(\tilde{x}) + \nabla_y h_1(y)\right] = \alpha \qquad \forall y \in \partial Q, \tag{5}$$

with periodicity conditions, where $\tilde{n}$ is the normal to ∂Q. The map E is then given by

$$E(\nabla h_0(\tilde{x})) := \frac{1}{|F|} \int_F T\left[\nabla h_0(\tilde{x}) + \nabla_y h_1(y)\right] \, dy. \tag{6}$$

For details regarding function spaces and the result in the case $m < 1$ we refer to [9]. For the case of a no-slip condition we refer to [11]. With the above theorem we correct a typing error: Proposition 2.13 of [9] is correctly stated only for $m < 1$, where a two-dimensional cell problem appears for the velocity field. Proposition 2.13 of [9] does not include the case $m = 1$ correctly: in this case the standard three-dimensional cell problem for the velocity field determines the positive matrix $M \in \mathbb{R}^{2 \times 2}$.

Main steps in the proof

1. Estimates for the solution sequence. One first has to notice that, for $\alpha \neq 0$, the above system does not have a bounded sequence of pressure functions. Assuming that gradients of the velocity are bounded, we can neglect the strain-part of the stress-tensor, and a pressure average over the free boundary can be calculated as

$$\frac{1}{|F|} \int_{\Pi_\varepsilon} p^\varepsilon(\tilde{x}, h^\varepsilon(\tilde{x})) \, d\tilde{x} \sim -\frac{1}{|F|} \int_{\Pi_\varepsilon} n \cdot T(v^\varepsilon, p^\varepsilon) \cdot n \, d\tilde{x}$$

$$= -\frac{1}{|F|} \int_{\Pi_\varepsilon} \mathcal{H}(h^\varepsilon(\tilde{x})) \, d\tilde{x} = -\frac{1}{|F|} \int_{\Pi_\varepsilon} \nabla \cdot T[\nabla h^\varepsilon(\tilde{x})] \, d\tilde{x}$$

$$= -\frac{1}{|F|} \int_{\partial \Pi_\varepsilon} \tilde{n} \cdot T[\nabla h^\varepsilon(\tilde{x}))] \, dS(\tilde{x})$$

$$= -\frac{1}{|F|} \alpha |\partial \Pi_\varepsilon| = -\frac{|\partial Q|}{|F|} \alpha \frac{1}{\varepsilon} =: \hat{p}(\varepsilon).$$

We exploited here, that the normal component of $T[\nabla h^\varepsilon]$ is the sine of the contact angle Θ, $\alpha = \tilde{n} \cdot T[\nabla h^\varepsilon]$.

The first step in the proof of the homogenization result is to find bounds for the solution sequence. For $m = 1$, a uniform bound for

$$\|\varepsilon \nabla v^\varepsilon\|_{L^2} + \|p^\varepsilon - \hat{p}(\varepsilon)\|_{L^2} + \|h^\varepsilon\|_{\text{Lip}}$$

is shown in [10]. In particular, we have a good control on how the domains G_ε approximate G_0, and we find two-scale weak limits for $\varepsilon \nabla v^\varepsilon$, $\nabla h^\varepsilon \to \nabla_x h_0(x) + \nabla_y h_1(x, y)$, and $p^\varepsilon - \hat{p}(\varepsilon) \to P$. Note how the normalization with $\hat{p}(\varepsilon)$ appears as the right hand side in (4) and makes system (4)–(5) solvable.

2. Derivation of the force-balance on the free boundary. Our aim here is to explain why it is natural to expect the above stated homogenized system. In the rigorous derivation one starts from the weak formulation of the free boundary system and takes limits of the integrals using the method of two-scale convergence.

For the map $T : \mathbb{R}^2 \to \mathbb{R}^2$, $\xi \mapsto \xi/\sqrt{1 + |\xi|^2}$ as above, we consider the functions

$$\eta^\varepsilon := T[\nabla h^\varepsilon] \xrightarrow{2-scale} \eta_0(x, y)$$

in the sense of two-scale convergence. We now start from the free boundary equation

$$-p^\varepsilon \sim n \cdot T(v^\varepsilon, p^\varepsilon) \cdot n = \mathcal{H}(h^\varepsilon) = \nabla \cdot T[\nabla h^\varepsilon] = \nabla \cdot \eta^\varepsilon.$$

The order of ε^{-1} together with (7) below yields (4). To order ε^0 we find $-P|_{\partial_u G_0} = \nabla \cdot (1/|F|) \int_F \eta_0(., y) \, dy$. This must be justified with a careful study of the fact that, and in what sense, the functions $p^\varepsilon(., h^\varepsilon(.)) - \hat{p}(\varepsilon)$ converge indeed to the trace of the limit function $P : G_0 \to \mathbb{R}$. We recall here that we have only L^2-bounds for the pressure functions.

3. Identification of the two-scale limit function η_0. The homogenized system with the nonlinear function E of definition (6) is derived, once we can show

$$\eta_0(x, y) = T[\nabla_x h_0(x) + \nabla_y h_1(x, y)]. \tag{7}$$

This is what we formally expect from the equality $\eta^\varepsilon = T[\nabla h^\varepsilon]$. In order to show the limit expression, we observe that T is monotone as the gradient of the convex function $\mathbb{R}^2 \ni \xi \mapsto \sqrt{1 + |\xi|^2}$. Using the monotonicity trick as in [1], and exploiting the bulk equations once more, we can verify (7).

Interpretation of the results in terms of the capillary pressure

One may interpret the above result as follows. Stationary fronts in the special geometry of cylindrical obstacles with cross-section $Q \subset (0, 1)^2$ and size ε, and with a contact angle determined by the number $\alpha \in (-1, 1)$, lead to the capillary pressure

$$p_c(s) = p_c = -\frac{|\partial Q|}{1 - |Q|}\alpha\frac{1}{\varepsilon}.$$

The dependence on s vanishes: an increase in fluid volume results in a motion of the front in vertical direction. Due to the invariance of the geometry in vertical direction, this process does not effect the typical pore-size and does therefore not alter the capillary pressure.

The above results justify the concept of a capillary pressure in a special geometry. The justification succeeds, since the geometry prescribes a 'typical pore-size' independently of the position of the front.

3 Homogenization for a Variable Pore-Size – Moving Fronts

We now ask:

> What is the corresponding result, if the 'typical pore-size' changes with the position of the front?

The reasoning which led to the capillary pressure relation (3) assumed that in the single pore the free interface spans a fixed, given diameter. If the front is situated in a pore as indicated in Figure 3, it is not clear which diameter determines the capillary pressure.

We further notice that a system of stationary equations can not provide us a law for the capillary pressure: In the filter geometry indicated in the right part of Figure 3, we can find a family of stationary solutions, all without fluid motion and with constant pressure, but with different pressure constants. In two dimensions, for given pressure constant, it suffices to chose in each tube an interface which has the desired constant curvature and the desired contact angle. We refer to [4] for an analysis of the stationary situation in the case $\alpha = 0$ in more complex geometries.

We see that it is necessary to analyze the instationary situation. In order to extract the relevant processes, we study in [14] and [15] the following model system.

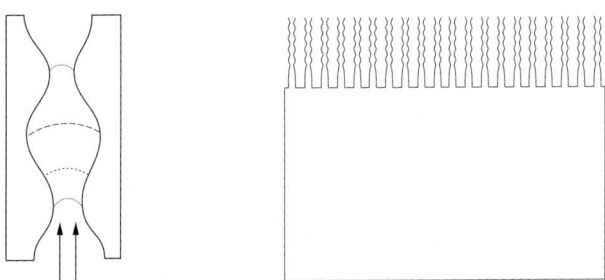

Fig. 3. With the motion of the front, the typical diameter changes.

The model equations.

The fluid occupies the two-dimensional domain $\Omega := (-1,1) \times (-1,0)$ and is described by a Darcy law. All quantities are assumed to be periodic across the macroscopic lateral boundaries. The interface is located along the upper boundary $\Gamma := (-1,1) \times \{0\}$, which models the filter of Figure 3. For $\gamma \in (0, 1/2)$ we set

$$\Gamma_1^\varepsilon := \varepsilon \cdot (\mathbb{Z} + (-\gamma, \gamma)) \cap \Gamma, \qquad \Gamma_2^\varepsilon := \varepsilon \cdot (\mathbb{Z} + (\gamma, 1-\gamma)) \cap \Gamma.$$

Γ_2^ε models the part of the upper boundary, where the fluid is in contact with material. We will impose a no-flux condition along this part of the boundary. Along Γ_1^ε we have the fluid-gas interphase. The position of the front in the single tube is modeled with a height function $h^\varepsilon : \Gamma_1^\varepsilon \to \mathbb{R}$. Interpreting h^ε as the average height we allow for h^ε only piecewise constant functions and we use the L^2-projection Q_ε onto the space of piecewise constant functions; Q_ε consists in replacing a function by its piecewise averages.

With the Darcy law $v^\varepsilon = -\varepsilon \nabla p^\varepsilon$ we find the model equations

$$\Delta_x p^\varepsilon = 0 \qquad \text{in } \Omega \times (0,T), \qquad (8)$$

$$\partial_t h^\varepsilon(x_1, t) = -\varepsilon Q_\varepsilon \partial_2 p^\varepsilon(x_1, 0, t) \qquad \forall (x_1, 0) \in \Gamma_1^\varepsilon, \forall t, \qquad (9)$$

$$p^\varepsilon(x_1, 0, t) = \mathcal{P}_0\left(\frac{x_1}{\varepsilon}, \frac{h^\varepsilon(x_1, t)}{\varepsilon}\right) \qquad \forall (x_1, 0) \in \Gamma_1^\varepsilon, \forall t, \qquad (10)$$

$$\partial_2 p^\varepsilon(x_1, 0, t) = 0 \qquad \forall (x_1, 0) \in \Gamma_2^\varepsilon, \forall t. \qquad (11)$$

The second equation expresses that the height in the single tube increases with the total inflow into the tube and exploits that the vertical derivative $-\varepsilon \partial_2 p^\varepsilon$ is the vertical velocity of the fluid. Equation (10) expresses that the pressure in the single tube depends on the diameter and thus on the position. The dependence on $h^\varepsilon/\varepsilon$ expresses that the diameter is oscillating on the pore-scale ε. The above equations must be complemented with the ε-independent boundary conditions on bottom (inflow condition) and lateral boundaries (periodicity). As initial values one has to prescribe the height function. We set $h^\varepsilon = 0$ on Γ_2^ε for all times.

It remains to specify the material law $\mathcal{P}_0$ which prescribes the the pressure in dependence of the height in each cell. A simple choice is a sawtooth function, $\mathcal{P}_0(x_1, s) = \mathcal{P}_0(k, s)$ for all $x_1 \in (k - \gamma, k + \gamma)$, $k \in \mathbb{Z}$, with

$$\mathcal{P}_0(k, s) = a_0(k) \cdot s \mod p_{max}(k). \qquad (12)$$

While the free boundary advances (i.e. h^ε increases), the pressure varies; the maximal pressure is $p_{max}(k)$. Equations (8)–(12) have a unique solution $(p^\varepsilon, h^\varepsilon)$, see [12].

The homogenized system

Let us assume that the maximal pressures $p_{max}(k) = p_{max}$ are independent of k. We then expect that every weak limit p^0 of the pressure functions p^ε satisfies the following equations

$$\Delta_x p^0 = 0 \quad \text{in } \Omega, \tag{13}$$
$$p^0 \leq p_{max} \quad \text{on } \Gamma, \tag{14}$$

together with periodicity conditions and inflow conditions on the macroscopic boundaries. In order to find an evolution equation for p^0 we differentiate equation (10) with respect to time. In a point $x_1 \in (k-\gamma, k+\gamma)$ we calculate

$$\partial_t p^\varepsilon(x_1, 0, t) \leq \partial_s \mathcal{P}_0(k, h^\varepsilon) \cdot \frac{1}{\varepsilon} \partial_t h^\varepsilon = -\partial_s \mathcal{P}_0(k, h^\varepsilon) \cdot Q_\varepsilon \partial_2 p^\varepsilon(x_1, 0, t).$$

The inequality holds as equality if the pressure has no jumps. If jumps occur, the strict inequality holds. In the case of a nonlinear law instead of (12), the right hand side contains two oscillating terms. We therefore introduce a function $\Theta_k : \mathbb{R} \to \mathbb{R}$ with

$$\Theta_k'(\rho) = \frac{2\gamma}{\partial_s \mathcal{P}_0(k, s_k(\rho))}, \quad \text{where } s_k(\rho) \text{ is defined by } \mathcal{P}_0(k, s_k(\rho)) = \rho.$$

With this function we may write

$$\frac{1}{2\gamma} \partial_t \Theta_k(p^\varepsilon(x_1, 0, t)) \leq -Q_\varepsilon \partial_2 p^\varepsilon(x_1, 0, t).$$

One can show that the values of p^ε are not oscillatory. This allows to take weak limits on both sides of the last equation. The factor 2γ takes into account that the equation holds only for $x_1 \in \Gamma_1^\varepsilon$. We introduce a function $\Theta : \mathbb{R} \to \mathbb{R}$, defined with mean values by

$$\Theta'(\rho) = \left\langle \frac{2\gamma}{\partial_s \mathcal{P}_0(k, s_k(\rho))} \right\rangle, \tag{15}$$

the average is taken over $k \in \mathbb{Z}$. In the limit $\varepsilon \to 0$ we arrive at the relations

$$\partial_t(\Theta \circ p^0) \leq -\partial_2 p^0, \tag{16}$$
$$\partial_t(\Theta \circ p^0) = -\partial_2 p^0 \quad \text{on } \{(x_1, 0) \in \Gamma | p^0(x_1, 0) < p_{max}\}. \tag{17}$$

Both relations are understood in the sense of distributions; in the derivation one has to exploit that p^ε does not have large oscillations. For the derivation of (17) one has to verify that, if the averaged pressure is below p_{max}, then in no cell a jump of the pressure can occur.

Relations (13)–(17) together with a continuity condition form indeed a complete system of evolution equations. It was shown in [14] and [15] that, in many situations, this is indeed the upscaled system characterizing every weak limit p^0 of solutions p^ε to the ε-problems. But the above **is not the upscaled system in general!** Indeed, even in the simplest case that a_0 and p_{max} are independent of k, we can find a counter-example: let p^ε be a solution which is ε-periodic in x_1. If the front advances, the pressure drops in all cells at the same time. In particular, also the averaged pressure p^0 has jumps and is not the unique solution of the upscaled system above, since it does not satisfy the continuity condition.

Nevertheless, it turns out that the periodic behavior is not typical. If we consider a stochastic perturbation of the coefficients, all the cells are slightly different. The following effect occurs: If the pressure drop occurs in a cell $\varepsilon(k-\gamma, k+\gamma)$, this creates a back-flow in the neighboring cells. This inhibits further pressure jumps. This effect of mutual blocking leads to the following

Theorem 2. *For stochastic laws $\mathcal{P}_0(k,.)$ and corresponding solutions p^ε, the upscaled equations (13)–(17) hold in the sense of distributions almost surely.*

This Theorem is shown in [15] for a specific choice of stochastic laws $\mathcal{P}_0$. The function Θ of (15) is then defined with the expected value on the right hand side.

4 From Meso- to Macroscopic Models: Homogenizing Hysteresis

The upscaled system of the last section may be interpreted in the scale of the Leverett-equations as follows. If fluid enters the filter, then, as long as equality (17) holds, the average pressure p^0 increases, since Θ is strictly monotone. But, due to (13), p^0 can not exceed the maximal pressure p_{max}. We conclude that advancing fronts (or $\partial_t s > 0$) eventually lead to a pressure difference $p_c^+ = p_{max}$. Since the typical diameter spanned by the front changes with the saturation s, we may write $p_c^+ = p_c^+(s)$. If instead the fronts move in the opposite direction ($\partial_t s < 0$), then the evolution lowers the pressure until a minimal pressure $p_c^- = p_c^-(s) = p_{min}$ is achieved. The third situation is that the fronts do not move ($\partial_t s = 0$). In this case, the pressure jump across the front is free to take any value in the interval $[p_c^-(s), p_c^+(s)]$ (compare our reasoning regarding the stationary case). Assuming a fixed difference $2\rho = p_c^+(s) - p_c^-(s)$ we find the hysteresis model

$$p \in p_c^\circ(s) + \rho \, \text{sign}(\partial_t s) \qquad (18)$$

with the multi-valued sign function, $\text{sign}(0) = [-1, 1]$. This relation for the capillary pressure is coupled with the conservation of mass, $\partial_t s = \Delta p$.

The above hysteresis model was studied in [3] for a linear function $s \mapsto p_c^\circ(s)$. Varying material parameters were homogenized in [2] in the deterministic, and in [13] in the stochastic case. It is interesting to note that, in this homogenization of hysteresis equations, the averaged equations simplify in the stochastic case.

References

1. G. Allaire. Homogenization and two-scale convergence. *SIAM J. Math. Anal.*, 23, No.6:1482–1518, 1992.
2. A. Beliaev. Porous medium flows with capillary hysteresis and homogenization. In *Homogenization and Applications to Material Sciences*, pages 23–32, Timisoara, 2001.
3. A. Beliaev. Unsaturated porous flows with play-type capillary hysteresis. *Russian J. Math. Phys.*, 8(1):1–13, 2001.
4. G. Buttazzo and B. Schweizer. On the Γ-limit of the one-dimensional Hausdorff-measure for a sequence of distance functions. *J. Convex Analysis*, 12:239–253, 2005.
5. S. Conti and B. Schweizer. A sharp-interface limit for the geometrically linear two-well problem in the gradient theory of phase transitions in two dimensions. *Arch. Rat. Mech. Anal.*, 179:413–452, 2006.
   in two dimensions. *Preprint 87 des MPI Leipzig*, 2003.
6. S. Conti and B. Schweizer. Rigidity and Gamma convergence for solid-solid phase transitions with $SO(2)$ invariance. *Comm. Pure Appl. Math.* (to appear)
7. M.C. Leverett. Steady flow of gas-oil-water mixtures through unconsolidated sands. *Trans. AIME*, 132(149), 1938.
8. E. Sanchez Palencia. *Non Homogenieous Media and Vibration Theory*. Number 17 in Lecture Notes in Physics. Springer, 1980.
9. B. Schweizer. Homogenization of a fluid problem with a free boundary. *Comm. Pure Appl. Math.*, 53:1118–1152, 2000.
10. B. Schweizer. Uniform estimates in two periodic homogenization problems. *Comm. Pure Appl. Math.*, 53:1153–1179, 2000.
11. B. Schweizer. Homogenization of a free boundary problem: The no-slip condition. In N. Antonić et al., editor, *Dubrovnik conference: Multiscale problems in science and technology*. Springer, 2002.
12. B. Schweizer. *Laws for the Capillary Pressure via the Homogenization of Fronts in Porous Media*. Habilitationsschrift an der Ruprecht-Karls-Universität Heidelberg, 2002.
13. B. Schweizer. Averaging of flows with capillary hysteresis in stochastic porous media. *submitted*, 2004.
14. B. Schweizer. Laws for the capillary pressure in a deterministic model for fronts in porous media. *SIAM J. Math. Anal.*, 36, no. 5, 1489–1521, 2005.
15. B. Schweizer. A stochastic model for fronts in porous media. *Ann. Mat. Pura Appl. (4)*, 184, no. 3, 375–393, 2005.

High-Accuracy Approximation of Effective Coefficients*

N. Neuss

Interdisziplinäres Zentrum für wissenschaftliches Rechnen (IWR), Universität Heidelberg

Summary. In this contribution we study the calculation of effective coefficients for media with periodic heterogeneities. We use finite element methods of high order which allows to obtain high accuracy with relatively low computational effort. This is shown both theoretically and practically.

1 Introduction

When simulating heterogeneous media, it is often impossible to simulate the microscopic processes directly. Instead, it is necessary to fall back on so-called homogenised models, which describe the behaviour of a reduced representation of the solution, usually obtained by some kind of averaging procedure. Depending on the smallness of the ratio between the microscopic and macroscopic scale, the error of the homogenised model tends to zero, and often it is a completely satisfactory way to formulate the problem under consideration. If sufficient information about the microstructure is available, the effective properties of the medium can be calculated by solving suitable problems on representative cells and averaging.

In this article, we restrict ourselves to the special case of periodic heterogeneities occurring in two model systems of partial differential equations, namely linear elasticity and Stokes flow. After introducing the model problems, the homogenisation process is described including convergence results and error estimates. Finally, numerical results are presented in Section 5.

---

*This work has been supported by the German Research Foundation (DFG) through SFB 359 (Project D1) at the University of Heidelberg.

2 Model Problems

This section introduces the two model problems considered in this article. The first problem is linear elasticity with a heterogeneous coefficient, the second is Stokes flow through a porous domain.

Let us describe the first problem. Let $\mathbf{A} : \mathbb{R}^n \to (\mathbb{R}^{n \times n})^{n \times n}$ be a 4-tensor with 1-periodic components $\mathbf{A}_{ij}^{kl} \in L^\infty(\mathbb{R}^n, \mathbb{R})$. The tensor $\mathbf{A}$ can be chosen to satisfy for all $i, j, k, h \in \{1, \ldots, n\}$ the symmetry conditions

$$\mathbf{A}_{ij}^{hk}(y) = \mathbf{A}_{ji}^{kh}(y) = \mathbf{A}_{hj}^{ik}(y), \quad y \in \mathbb{R}^n, \tag{1}$$

and appropriate ellipticity and continuity are the existence of constants $0 < \lambda_1 \leq \lambda_2$ with

$$\lambda_1 \sum_{i,j=1}^n |\eta_{ik}|^2 \leq \mathbf{A}_{ij}^{hk}(y) \eta_{ih} \eta_{jk} \leq \lambda_2 \sum_{i,j=1}^n |\eta_{ik}|^2 \tag{2}$$

for all $y \in \mathbb{R}^n$ and all symmetric matrices $\eta \in \mathbb{R}^{n \times n}$.

First we introduce some notation for function spaces: for $\Omega \subset \mathbb{R}^n$ being a bounded Lipschitz domain, we denote with $L^2(\Omega)$ the set of all square-integrable functions, and set

$$L_0^2(\Omega) = \{u \in L^2(\Omega) : \int_\Omega u(x)\,dx = 0\}, \tag{3}$$

$$H^1(\Omega) = \{u \in L^2(\Omega) : \nabla u \in L^2(\Omega, \mathbb{R}^n)\}, \tag{4}$$

$$H_0^1(\Omega) = \{u \in H^1(\Omega) : u = 0 \text{ on } \partial\Omega\}. \tag{5}$$

Spaces for vector-valued functions are denoted with $L^2(\Omega, \mathbb{R}^n) \cong (L^2(\Omega))^n$, $H^1(\Omega, \mathbb{R}^n) \cong (H^1(\Omega))^n$, etc.

Now the elasticity problem is the following: given $\mathbf{f} \in L^2(\Omega, \mathbb{R}^n)$, we search for $\mathbf{u}^\varepsilon \in H_0^1(\Omega, \mathbb{R}^n)$ such that

$$\mathbf{a}^\varepsilon(\mathbf{u}^\varepsilon, \mathbf{v}) := \int_\Omega \nabla \mathbf{v}(x) : (\mathbf{A}(\frac{x}{\varepsilon}) \nabla \mathbf{u}^\varepsilon(x))\,dx = \int_\Omega \mathbf{f}(x) \cdot \mathbf{v}(x)\,dx \tag{6}$$

for all $\mathbf{v} \in H_0^1(\Omega, \mathbb{R}^n)$. Here, $a \cdot b$ denotes the Euclidean scalar product between vectors $a, b \in \mathbb{R}^n$, and $M : N = \text{trace}(M^t N)$ generalises this scalar product to arbitrary rectangular matrices $M, N \in \mathbb{R}^{m \times n}$. Equation (6) is the variational form of the equation

$$\begin{aligned} -\text{div}\left(\mathbf{A}(\frac{x}{\varepsilon}) \nabla \mathbf{u}^\varepsilon(x)\right) &= \mathbf{f}(x), & x \in \Omega \\ \mathbf{u}^\varepsilon(x) &= 0, & x \in \partial\Omega. \end{aligned} \tag{7}$$

Korn's inequality shows that the Lemma of Lax-Milgram is applicable and yields existence and uniqueness of $\mathbf{u}^\varepsilon \in H_0^1(\Omega, \mathbb{R}^n)$.

The second model problem is (generalised) Stokes flow in a medium with periodically distributed holes. Let $\square = (0,1)^n$. For a smoothly bounded compact subset $\emptyset \neq Z \subset\subset \square$, we set $Y = \square - Z$ and

$$Z^\varepsilon = \bigcup_{\mathbf{k} \in \mathbb{Z}^n, \varepsilon(\mathbf{k}+\square) \subset \Omega} \varepsilon(\mathbf{k}+Z), \quad \Gamma^\varepsilon = \partial Z^\varepsilon, \quad \Omega^\varepsilon = \Omega \setminus Z^\varepsilon. \tag{8}$$

Given then $\mathbf{f}^\varepsilon \in L^2(\Omega^\varepsilon, \mathbb{R}^n)$ and $g^\varepsilon \in L_0^2(\Omega^\varepsilon)$, we search for a pair $(\mathbf{u}^\varepsilon, p^\varepsilon) \in H_0^1(\Omega^\varepsilon, \mathbb{R}^n) \times L_0^2(\Omega^\varepsilon)$ such that

$$\int_{\Omega^\varepsilon} \nabla \mathbf{u}^\varepsilon : \nabla \mathbf{v}\, dx - \int_{\Omega^\varepsilon} p^\varepsilon \operatorname{div} \mathbf{v}\, dx = \int_{\Omega^\varepsilon} \mathbf{f}^\varepsilon \cdot \mathbf{v}\, dx, \quad \mathbf{v} \in H_0^1(\Omega^\varepsilon, \mathbb{R}^n),$$
$$\int_{\Omega^\varepsilon} \operatorname{div} \mathbf{u}^\varepsilon\, q\, dx = \varepsilon^2 \int_{\Omega^\varepsilon} g^\varepsilon q\, dx, \quad q \in L_0^2(\Omega), \tag{9}$$

which is the weak form of

$$-\Delta \mathbf{u}^\varepsilon(x) + \nabla p^\varepsilon(x) = \mathbf{f}^\varepsilon(x), \quad x \in \Omega^\varepsilon,$$
$$\operatorname{div} \mathbf{u}^\varepsilon(x) = \varepsilon^2 g^\varepsilon(x), \quad x \in \Omega^\varepsilon, \tag{10}$$
$$\mathbf{u}^\varepsilon(x) = 0, \quad x \in \partial \Omega^\varepsilon.$$

Theorem 1. *On Ω^ε, the Poincaré estimate*

$$\|\mathbf{u}^\varepsilon\|_{L^2(\Omega^\varepsilon)} \lesssim \varepsilon \|\nabla \mathbf{u}^\varepsilon\|_{L^2(\Omega^\varepsilon)} \tag{11}$$

is valid, and the Babuvska-Brezzi stability constant satisfies

$$\beta_{\Omega^\varepsilon} = \inf_{0 \neq q \in L^2(\Omega^\varepsilon)} \sup_{\mathbf{v} \in H_0^1(\Omega^\varepsilon, \mathbb{R}^n)} \frac{(\operatorname{div} \mathbf{v}, q)_{L^2(\Omega^\varepsilon)}}{\|q\|_{L^2(\Omega^\varepsilon)} \|\mathbf{v}\|_{H_0^1(\Omega^\varepsilon, \mathbb{R}^n)}} \gtrsim \varepsilon. \tag{12}$$

An immediate consequence is the a-priori estimate

$$\frac{1}{\varepsilon^2}\|\mathbf{u}^\varepsilon\|_{L^2(\Omega^\varepsilon)} + \frac{1}{\varepsilon}\|\nabla \mathbf{u}^\varepsilon\|_{L^2(\Omega^\varepsilon)} + \|p^\varepsilon\|_{L^2(\Omega)} \lesssim \|\mathbf{f}^\varepsilon\|_{L^2(\Omega^\varepsilon, \mathbb{R}^n)} + \|g^\varepsilon\|_{L^2(\Omega^\varepsilon)} \tag{13}$$

for solutions of (9) resp. (10).

Proof. Estimate (11) can easily be proved in each cell due to the Dirichlet condition on Γ^ε. Now we give a sketch of a proof for (12). For a given $q \in L_0^2(\Omega^\varepsilon)$, one has to construct a velocity field $\mathbf{v} \in H_0^1(\Omega^\varepsilon)$ such that $(\operatorname{div} \mathbf{v}, q)_{L^2(\Omega^\varepsilon)}$ is reasonably large. This is done as follows. First, one defines $\lambda \in H^1(\Omega)$ as solution to

$$\Delta \lambda(x) = q(x), \quad x \in \Omega^\varepsilon, \quad \frac{\partial \lambda}{\partial n} = 0, \quad x \in \partial \Omega^\varepsilon.$$

Then $\mathbf{v}_1 := \nabla \lambda \in H^1(\Omega^\varepsilon, \mathbb{R}^n)$ for which due to the smoothness of ∂Z the estimate

$$\|\mathbf{v}_1\|_{L^2(\Omega^\varepsilon)} + \varepsilon \|\nabla \mathbf{v}_1\|_{L^2(\Omega^\varepsilon)} \lesssim \|q\|_{L^2(\Omega^\varepsilon)}$$

is valid. A second correction eliminates the wrong boundary values of $\mathbf{v}_1$ along Γ^ε: let $\mathbf{v}_2$ solve for each cell $Y_\mathbf{k}^\varepsilon$ with interior boundary $\Gamma_\mathbf{k}^\varepsilon$ a local Stokes problem with div $\mathbf{v}_2 = 0$ and boundary values $\mathbf{v}_2 = -\mathbf{v}_1$ on $\Gamma_\mathbf{k}^\varepsilon$ and $\mathbf{v}_2 = 0$ on $\partial Y_\mathbf{k}^\varepsilon - \Gamma_\mathbf{k}^\varepsilon$. A scaling argument shows that

$$\|\nabla \mathbf{v}_2\|_{L^2(\Omega^\varepsilon, \mathbb{R}^n)} \lesssim \frac{1}{\varepsilon} \|\mathbf{v}_1\|_{L^2(\Omega, \mathbb{R}^n)} + \|\nabla \mathbf{v}_1\|_{L^2(\Omega, \mathbb{R}^n)} \lesssim \frac{1}{\varepsilon} \|q\|_{L^2(\Omega)}.$$

A third correction $\mathbf{v}_3$ is a boundary layer along $\partial \Omega$ which does not meet Γ^ε and corrects the boundary values of $\mathbf{v}_1$ along $\partial \Omega$. Since dist $(\partial \Omega, \Gamma^\varepsilon) \gtrsim \varepsilon$, $\mathbf{v}_3$ can be shown to satisfy the same estimate as $\mathbf{v}_2$. Therefore, $\mathbf{v} = \mathbf{v}_1 + \mathbf{v}_2 + \mathbf{v}_3$ satisfies div $\mathbf{v} = q$ and

$$\frac{(\operatorname{div} \mathbf{v}, q)_{L^2(\Omega^\varepsilon)}}{\|q\|_{L^2(\Omega^\varepsilon)} \|\mathbf{v}\|_{H_0^1(\Omega^\varepsilon, \mathbb{R}^n)}} \gtrsim \frac{\|q\|_{L^2(\Omega^\varepsilon)}^2}{\frac{1}{\varepsilon} \|q\|_{L^2(\Omega^\varepsilon)}^2} = \varepsilon$$

which proves (12). Finally, estimate (13) is an easy consequence of (11) and (12).

3 Homogenisation

The idea of homogenisation theory is that each of the problems presented in Section 2 allows, for small ε, the approximation by the solution of a certain homogenised problem without fine-scale features.

For problem (6),(7), the homogenised problem is to find $\mathbf{u}^0 \in H_0^1(\Omega, \mathbb{R}^n)$, such that

$$a^0(\mathbf{u}^0, \mathbf{v}) := \int_\Omega \nabla \mathbf{v} : \mathbf{A}^0 \nabla \mathbf{u}^0 \, dx = \int_\Omega \mathbf{v} \cdot \mathbf{f} \, dx, \quad \mathbf{v} \in H_0^1(\Omega, \mathbb{R}^n). \quad (14)$$

where the matrix-valued homogenised tensor $\mathbf{A}^0 \in (\mathbb{R}^{n \times n})^{n \times n}$ is computed as

$$\mathbf{A}_{ij}^0 = \int_\square \left(\mathbf{A}_{ij}(y) + \sum_{l=1}^n \mathbf{A}_{il}(y) \frac{\partial \mathbf{N}_j}{\partial y_l}(y) \right) dy \quad (15)$$

with matrix-valued functions $\mathbf{N}_k \in H_{\text{per},0}^1(\square, \mathbb{R}^{n \times n})$, $k = 1, \ldots, n$ satisfying

$$a(\mathbf{N}, \mathbf{v}) := \int_\square \nabla \mathbf{v}(y) \mathbf{A}(y) \nabla \mathbf{N}_k(y) \, dy = -\int_\square \nabla \mathbf{v}(y) \mathbf{A}(y) e_k \otimes \mathbf{I}^{n \times n} \, dy \quad (16)$$

for all $\mathbf{v} \in H_{\text{per},0}^1(\square, \mathbb{R}^n)$. It can be shown that $\mathbf{A}^0$ inherits ellipticity and continuity from $\mathbf{A}$, such that the Lemma of Lax-Milgram is again applicable and yields the existence of a unique solution $\mathbf{u}^0$ of (14).

Theorem 2. Let $\mathbf{u}^\varepsilon \in H_0^1(\Omega, \mathbb{R}^n)$ be the solution of (6), $\mathbf{N}_k \in H_{\mathrm{per},0}^1(\Box, \mathbb{R}^{n \times n})$ be the solutions of (16), and $\mathbf{u}^0 \in H_0^1(\Omega, \mathbb{R}^n)$ be the solution of (14). Under the assumption $u^0 \in H^2(\Omega)$ the first-order corrector

$$\mathbf{u}^{1,\varepsilon}(x) := \mathbf{u}^0(x) + \varepsilon \sum_{k=1}^n \mathbf{N}_k\left(\frac{x}{\varepsilon}\right) \frac{\partial \mathbf{u}^0}{\partial x_k}(x) \in H^1(\Omega) \tag{17}$$

together with the boundary layer correction $\theta^\varepsilon \in u^{1,\varepsilon} + H_0^1(\Omega)$ on $\partial \Omega$ and

$$\int_\Omega \nabla \mathbf{v}^\varepsilon(x) : \mathbf{A}^\varepsilon(x) \nabla \theta^\varepsilon(x) dx = 0, \quad \mathbf{v}^\varepsilon \in H_0^1(\Omega) \tag{18}$$

satisfies the error estimate

$$\left\| \nabla(\mathbf{u}^\varepsilon - \mathbf{u}^{1,\varepsilon} - \theta^\varepsilon) \right\|_{L^2(\Omega)} \leq C\varepsilon \left\| D^2 \mathbf{u}^0 \right\|_{L^2(\Omega)}. \tag{19}$$

Furthermore, $\|\theta^\varepsilon\|_{H^1(\Omega)} \lesssim \varepsilon^{\frac{1}{2}}$ and $\|\theta^\varepsilon\|_{L^2(\Omega)} \lesssim \varepsilon$.

Proof. Although a similar estimate can be found in [10], it requires higher regularity of $\mathbf{u}^0$. Therefore, we give an alternative proof which follows [5],[9] for the diffusion case.

Since, for all $\mathbf{v} \in H_0^1(\Omega, \mathbb{R}^n)$,

$$\int_\Omega \nabla \mathbf{v} : \mathbf{A}^\varepsilon \nabla \mathbf{u}^\varepsilon \, dx = \int_\Omega \mathbf{v} \cdot \mathbf{f} \, dx = \int_\Omega \nabla \mathbf{v} : \mathbf{A}^0 \nabla \mathbf{u}^0 \, dx,$$

we have (note the renaming of the index k to j in a part of the expression in step 2)

$$\int_\Omega \nabla \mathbf{v} : \mathbf{A}^\varepsilon \nabla \left(\mathbf{u}^\varepsilon - \mathbf{u}^0 - \varepsilon \sum_{k=1}^n \mathbf{N}_k\left(\frac{x}{\varepsilon}\right) \frac{\partial \mathbf{u}^0}{\partial x_k} \right) dx =$$

$$\int_\Omega \nabla \frac{\partial \mathbf{v}}{\partial x_i} \left(\mathbf{A}_{ij}^0 \frac{\partial \mathbf{u}^0}{\partial x_j} - \mathbf{A}_{ij}^\varepsilon \frac{\partial \mathbf{u}^0}{\partial x_j} - \mathbf{A}_{ij}^\varepsilon \varepsilon \frac{\partial}{\partial x_j} \left(\sum_{k=1}^n \mathbf{N}_k\left(\frac{x}{\varepsilon}\right) \frac{\partial \mathbf{u}^0}{\partial x_k} \right) \right) dx =$$

$$\int_\Omega \frac{\partial \mathbf{v}}{\partial x_i} \left(\mathbf{A}_{ij}^0 - \mathbf{A}_{ij}^\varepsilon - \mathbf{A}_{ij}^\varepsilon \nabla_y \mathbf{N}_j \right) \frac{\partial \mathbf{u}^0}{\partial x_j} dx + O(\varepsilon \left\| D^2 \mathbf{u}^0 \right\|_{L^2(\Omega)} \|\nabla \mathbf{v}\|_{L^2(\Omega)}).$$

Now, because of the cell problems (16) each component of the matrix-valued tensor $\mathbf{T}_{ij} := \mathbf{A}_{ij}^0 - \mathbf{A}_{ij} - \mathbf{A}_{ik} \frac{\partial \mathbf{N}_j}{\partial y_k}$ is divergence free with $\langle \mathbf{T}_{ij} \rangle_\Box = 0$. By applying the subsequent Lemma 1 component-wise, there is a matrix-valued 3-tensor $\mathbf{S}_{ik}^j$ fulfilling $\mathbf{S}_{ik}^j = -\mathbf{S}_{ki}^j$ and $\sum_{k=1}^n \frac{\partial \mathbf{S}_{ik}^j}{\partial y_k} = \mathbf{T}_{ij}$. Since $\frac{\partial}{\partial y_k} = \varepsilon \frac{\partial}{\partial x_k}$, we have

$$\int_\Omega \sum_{i,j=1}^n \frac{\partial \mathbf{v}}{\partial x_i} \mathbf{T}_{ij} \frac{\partial \mathbf{u}^0}{\partial x_j} dx = \varepsilon \int_\Omega \sum_{i,j,k=1}^n \frac{\partial \mathbf{v}}{\partial x_i} \frac{\partial \mathbf{S}_{ik}^j}{\partial x_k} \frac{\partial \mathbf{u}^0}{\partial x_j} dx$$

$$= \varepsilon \int_\Omega \sum_{i,j,k=1}^n \frac{\partial^2 \mathbf{v}}{\partial x_i \partial x_k} \mathbf{S}_{ik}^j \frac{\partial \mathbf{u}^0}{\partial x_j} dx + O(\varepsilon \left\| D^2 \mathbf{u}^0 \right\|_{L^2(\Omega)} \|\nabla \mathbf{v}\|_{L^2(\Omega)}).$$

Due to the antisymmetry $\mathbf{S}^j_{ik} = -\mathbf{S}^j_{ki}$ the first term in the result vanishes, and the theorem is proved.

Lemma 1. *For $1 < q < \infty$, assume that the vector field $\mathbf{t} \in L^q(\square, \mathbb{R}^n)$ satisfies div $\mathbf{t} = 0$ in distributional sense together with $\int_\square \mathbf{t}\, dy = 0$. Then there is a skew-symmetric matrix $(S_{ik})^n_{i,k=1} \in (W^{1,q}(\square))^{n \times n}$ with*

$$\sum_{k=1}^n \frac{\partial S_{ik}}{\partial y_k} = t_i, \qquad i = 1, \ldots, n. \tag{20}$$

Proof. Define $S_{ik} = \frac{\partial \sigma_i}{\partial y_k} - \frac{\partial \sigma_k}{\partial y_i}$, where σ_i is the periodic solution of $\Delta \sigma_i = t_i$ on $\square$ which exists because $\int_\square t_i\, dy = 0$. It is easy to check that this S_{ik} satisfies all desired properties.

Finally, the homogenised problem for Stokes equation (9),(10) is the following Darcy problem. Given $\mathbf{f} \in L^2(\Omega, \mathbb{R}^n)$, and $g \in L^2_0(\Omega)$, we search for $p \in H^1(\Omega)$ and $u \in L^2(\Omega, \mathbb{R}^n)$ with div $u \in L^2(\Omega)$ such that

$$\mathbf{u}(x) = \mathbf{K}\,(\mathbf{f}(x) - \nabla p(x)) \quad \text{in } \Omega,$$
$$\text{div } \mathbf{u}(x) = g(x) \text{ in } \Omega, \tag{21}$$
$$\mathbf{u} \cdot \mathbf{n} = 0 \text{ on } \partial\Omega.$$

Here, the permeability tensor $\mathbf{K}$ is computed as

$$\mathbf{K}_{ij} = \int_Y (\nabla \mathbf{w}_i(y))^t \nabla \mathbf{w}_j(y)\, dy = \int_Y w^j_i(y)\, dy, \tag{22}$$

where, for $i = 1, \ldots, n$, the $(\mathbf{w}_i, \pi_i) \in H^1_{\text{per}}(Y, \mathbb{R}^n) \times L^2_0(Y)$ are weak solutions of the cell problems

$$-\Delta_y \mathbf{w}_i(y) + \nabla \pi_i(y) = \mathbf{e}_i \quad \text{in } Y,$$
$$\text{div } \mathbf{w}_i(y) = 0 \quad \text{in } Y, \tag{23}$$
$$\mathbf{w}_i(y) = 0 \quad \text{on } \partial Y.$$

The following convergence result then links problem (9) with problem (21).

Theorem 3. *We assume that g^ε and $\mathbf{f}^\varepsilon$ are extended to Ω such that they converge strongly in L^2 to functions $g \in L^2_0(\Omega)$ and $\mathbf{f} \in L^2(\Omega, \mathbb{R}^n)$ for $\varepsilon \to 0$. Then the solutions $\mathbf{u}^\varepsilon$ of (9) (extended to Ω by 0) satisfy*

$$\frac{\mathbf{u}^\varepsilon}{\varepsilon^2} \rightharpoonup \mathbf{u} \quad (\varepsilon \to 0) \qquad \text{weakly in } L^2(\Omega) \tag{24}$$

where $\mathbf{u}$ denotes the solution to (21).

Proof. The proof is a straightforward adaption of [11] where the divergence-free case $g \equiv 0$ was treated. $\square$

Remark 1. In the divergence-free case, it is proved in [1] that $\frac{\mathbf{u}^\varepsilon}{\varepsilon^2}$ converges strongly in L^2 towards $\sum_{i=1}^n (f_i - \frac{\partial p}{\partial x_i})\mathbf{w}_i$. Furthermore, the error estimate $\left\|\frac{\mathbf{u}^\varepsilon}{\varepsilon^2} - \sum_{i=1}^n (f_i - \frac{\partial p}{\partial x_i})\mathbf{w}_i\right\|_{L^2(\Omega^\varepsilon)} \lesssim \varepsilon^{\frac{1}{6}}$ is proved in [6].

4 Numerical Approximation of Effective Coefficients

In this section, we describe how to compute the effective coefficients (15) and (22) by solving the corresponding cell problems (16) and (23) numerically. Our method of choice are conforming finite element spaces of order p which allows for a good approximation of smooth problems, and a comparatively easy proof of error estimates.

For $h > 0$, let $T_h^\square$ denote a mesh of the unit cell $\square$ which fits across the identified boundary $\partial \square$. It has a set of cells $\mathcal{K}(T_h^\square)$, where each cell $K \in \mathcal{K}(T_h^\square)$ is an image of reference cells $\hat{K}_K$ under a smooth map $\Phi_K : \hat{K}_K \to K$. We consider only conforming meshes without hanging nodes. The reference cells $\hat{K}_K$ are allowed to be arbitrary products of unit-simplices of different dimensions, which allows both simplex and cube meshes as a special cases. We ensure a reasonable quality of $T_h^\square$ by assuming that there is some $d_T \in \mathbb{N}, d_T \geq 2$ and some constant $C_T > 0$ such that for all cells $K \in T_h^\square$ with associated map $\Phi_K : \hat{K} \to K$ to its reference cell $\hat{K}$, we have

$$\|D^\alpha \Phi_K\|_{L^\infty(\hat{K})} \leq C_T h^{|\alpha|}, \quad \forall \alpha : 0 \leq |\alpha| \leq d_T \tag{25}$$

$$\|\Phi_K^{-1}\|_{C^{0,1}(K)} \leq C_T h^{-1}. \tag{26}$$

For $1 \leq p \leq d_T - 1$, we now define conforming finite element spaces of order $p \in \mathbb{N}$ on T_h. Writing a reference cell $\hat{K} = s_1 \times \cdots \times s_l$ as a product of simplices of dimensions $d_1, \ldots, d_l$, we define

$$S^p(\hat{K}) = \{p = p_1 \otimes \ldots \otimes p_l : p_i \in P^p(s_i)\} \tag{27}$$

as the space of polynomials which are tensor products of polynomials on the factor simplices s_i with degree p. This is used to define the following finite element spaces

$$S^p(T_h^\square) = \{\varphi \in H^1(\square) : \forall K \in \mathcal{K}(T_h^\square) : \varphi \circ \Phi_K \in S^p(\hat{K}_K)\}, \tag{28}$$

$$S^p_{\text{per}}(T_h^\square) = S^p(T_h^\square) \cap H^1_{\text{per}}(\square), \quad S^p_{\text{per},0} = S^p_{\text{per}} \cap L^2_0(\square). \tag{29}$$

Analogous to the notation of function spaces, vector-valued finite element spaces are denoted as $S^p(T_h^\square, \mathbb{R}^n) \cong (S^p(T_h^\square))^n$, etc.

For computing the effective elastic tensor (15) we then have the following approximation result.

Theorem 4. *Let* $\mathbf{N}_h \in S^p_{\text{per},0}(T_h^\square, \mathbb{R}^{n^3})$ *denote the finite element approximation of the tensor* $\mathbf{N}$ *from problem (16). Then, for all* $1 \leq k \leq p$, *we have the error estimate*

$$\|\nabla(\mathbf{N} - \mathbf{N}_h)\|_{L^2(\square)} \leq C(n, p, C_T) h^k \|\mathbf{N}\|_{H^{k+1}(\square)}. \tag{30}$$

Computing the homogenised coefficients using $\mathbf{N}_h$ instead of $\mathbf{N}$ in formula (15) then yields an approximation $\mathbf{A}_h^0$ which satisfies

$$\|\mathbf{A}_h^0 - \mathbf{A}^0\|_\infty \lesssim h^{2k}. \tag{31}$$

Proof. (30) is a consequence of standard finite element approximation theory. Inequality (31) follows because

$$\mathbf{A}_h^0 - \mathbf{A}^0 = \int_\Box \mathbf{A}(y)\nabla(\mathbf{N}_h - \mathbf{N}) = \int_\Box \nabla \mathbf{N}\mathbf{A}(y)\nabla(\mathbf{N} - \mathbf{N}_h)$$

and due to Galerkin orthogonality

$$\int_\Box \nabla \mathbf{N}\mathbf{A}(y)\nabla(\mathbf{N} - \mathbf{N}_h) = \int_\Box \nabla(\mathbf{N} - \mathbf{N}_h)\mathbf{A}(y)\nabla(\mathbf{N} - \mathbf{N}_h) \lesssim h^{2k}.$$

$\Box$

For discretising the Stokes cell problems (23) we use again finite elements on a mesh T_h^Y of the cell Y. Our method of choice are Taylor-Hood elements of order p, where

$$\mathbf{V}_h = \{\mathbf{v} \in S_{\text{per}}^{p+1}(T_h^Y, \mathbb{R}^n) : \mathbf{v} = 0 \text{ on } \Gamma\} \tag{32}$$

and

$$\Pi_h = L_0^2(Y) \cap S_{\text{per}}^p(T_h^Y) \tag{33}$$

are ansatz spaces for velocity, resp. pressure. For $p \geq 1$ and under mild assumptions on T_h^Y, this pair of spaces is known to be stable uniformly in h, see [3]. An immediate consequence is the following error estimate:

Theorem 5. *Let $\mathbf{w}_{i,h} \in \mathbf{V}_h, \pi_{i,h} \in \Pi_h$ be the finite element solution of (9). If $\mathbf{w}_i \in H^{k+1}(Y)$ and $\pi_i \in H^k(Y)$ for some $1 \leq k \leq p$, then*

$$\|\nabla(\mathbf{w}_i - \mathbf{w}_{i,h})\|_{L^2(Y)} + h\,\|\nabla(\pi_i - \pi_{i,h})\|_{L^2(Y)} \lesssim h^k(\|\mathbf{w}_i\|_{H^{k+1}(Y)} + \|\pi_i\|_{H^k(Y)}). \tag{34}$$

If problem (23) is H^2-regular, we have also

$$\|\mathbf{w}_i - \mathbf{w}_{i,h}\|_{L^2(Y)} + h\,\|\pi_i - \pi_{i,h}\|_{L^2(Y)} \lesssim h^{k+1}(\|\mathbf{w}_i\|_{H^{k+1}(Y)} + \|\pi_i\|_{H^k(Y)}). \tag{35}$$

The components of the permeability $\mathbf{K}$ are obtained as the right-hand sides of (23) evaluated on the solutions $\mathbf{w}_i$, such that we obtain in a similar way to the proof of Theorem 4 the following estimate.

Corollary 1. *Using the finite element solutions $\mathbf{w}_{i,h}$ in formula (22) leads to an approximation $\mathbf{K}_h$ of $\mathbf{K}$. Under the assumption of H^2-regularity of the cell problems (23), and if for some $1 \leq k \leq p$ we have $\mathbf{w}_i \in H^{k+1}(\Omega, \mathbb{R}^n)$, $\pi_i \in H^k(\Omega)$ for $i = 1, \ldots, n$, then*

$$\|\mathbf{K} - \mathbf{K}_h\|_\infty \lesssim h^{2k}. \tag{36}$$

Remark 2. Of course, high regularity of the cell solutions $\mathbf{N}_i$ (resp. $\mathbf{w}_i$ and π_i) can only be expected, if the elasticity tensor $\mathbf{A}$ (resp. the pore boundary ∂Z) are sufficiently smooth. If this is not the case, mesh adaption becomes necessary, and a more sophisticated error analysis has to be performed, cf. [12], [2].

5 Numerical Results

In this section, we first demonstrate the high-accuracy calculation of effective coefficients corresponding to Section 4. Then we demonstrate the use of effective equations for constructing optimal preconditioners. All calculations in this section were done with the object-oriented interactive finite element toolbox FEMLISP, see [7], [4]. More information can be found in [8].

First, we want to calculate the effective coefficient for two-dimensional linear elasticity where the isotropic elasticity tensor

$$A_{ij}^{kl}(x) = \lambda(x)\delta_{ik}\delta_{jl} + \mu(x)(\delta_{ij}\delta_{kl} + \delta_{kj}\delta_{il}) \tag{37}$$

is determined by the Lamé parameters $\lambda, \mu > 0$ and describes a circular inlay, see the left-hand side of Fig. 1. The right-hand side of Fig. 1 shows the coarsest mesh which resolves the inlay exactly.[2] This coarse mesh is then uniformly refined to yield improved approximations. Using a polynomial degree $p = 5$ for approximating the tensor $\mathbf{N}$ as described in Section 4, we obtain the following results for approximating $\mathbf{A}^0$:

N_{cell}	N_{dof}	N_{entries}	CPU	A_{11}^{11}	A_{12}^{12}	A_{12}^{21}
9	1800	43060	5.4	4.1458940638	1.3176717343	1.2966840277
36	7200	176224	23.7	4.1412496929	1.3139564023	1.2979726371
144	28800	705600	74.9	4.1412384319	1.3139473004	1.2979716831
576	115200	2822400	293.8	4.1412383854	1.3139472825	1.2979716903

Here N_{cell} denotes the number of cells, N_{dof} denotes the number of the degrees of freedom, N_{entries} denotes the number of matrix entries, and CPU gives the number of seconds the calculation needed on a PC (2.4 GHz Pentium 4 processor, 1 Gigabyte main memory). The last three columns give three components of the effective tensor $\mathbf{A}_h^0$ which uniquely determine the whole tensor due to symmetry properties. We observe a very fast convergence which is in accordance with the theoretical estimate of Theorem 4. Note that

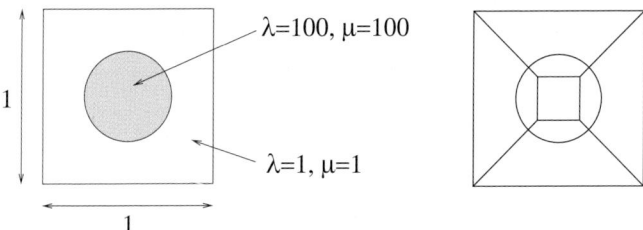

Fig. 1. Lamé constants and coarsest mesh.

---

[2] An isoparametric approximation of the interface would have led to a suboptimal order of convergence because the integration of $\mathbf{A}$ over cells occurring in (15) would not have been calculated accurately enough.

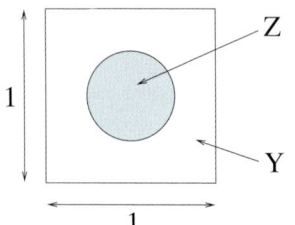
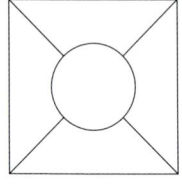

Fig. 2. Porous cell and coarsest mesh.

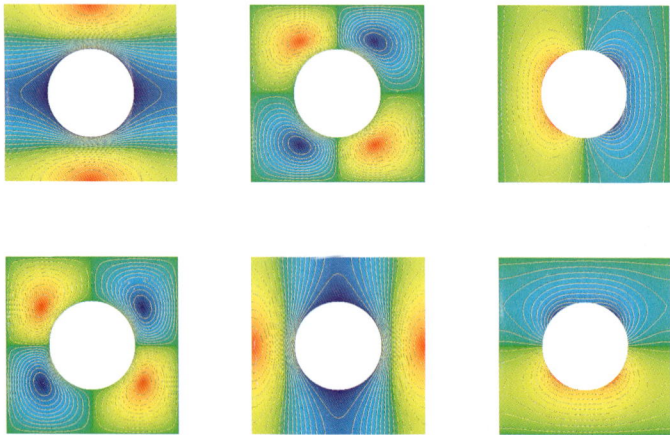

Fig. 3. Solutions $\mathbf{w_i}, \pi_i$ of (23).

this elasticity tensor does not describe an isotropic elastic medium in spite of the Cartesian symmetry of the cell.

Next, we consider the calculation of a permeability tensor as described in Section 4. We consider the two-dimensional geometry shown in Fig. 2 We use the pair $\mathbf{V}_h \times \Pi_h$ from (32), (33) for $p=4$ and obtain the following results:

N_{cell}	N_{dof}	N_{entries}	CPU	K
4	578	33823	1.2	1.9943087655d-02
16	2218	142179	12.2	1.9901508352d-02
64	8666	567991	41.0	1.9901435210d-02
256	34234	2267511	158.5	1.9901435350d-02

Also in this case we observe a rapid convergence which is in accordance with Theorem 5. Note that because of the symmetric geometry, the permeability is actually a scalar multiple of the identity. Fig. 3 shows the solution $(\mathbf{w}_i, \pi_i)$ for $i=1$ (first row) and $i=2$ (second row).

References

1. G. Allaire. Homogenization of the Navier–Stokes equations i. *Arch. Rat. Mech. Anal.*, 113:209–259, 1991.
2. W. Bangerth and R. Rannacher. *Adaptive Finite Element Methods for Differential Equations*. Birkhäuser Verlag, 2003.
3. F. Brezzi and R. S. Falk. Stability of higher order Taylor-Hood methods. *SIAM J. Numer. Anal.*, 28:581–590, 1991.
4. Femlisp. Homepage. http://www.femlisp.org.
5. V. V. Jikov, S. M. Kozlov, and O. A. Oleinik. *Homogenization of Differential Operators and Integral Functionals*. Springer–Verlag, Berlin, 1994.
6. E. Marusic-Paloka and A. Mikelić. An error estimate for correctors in the homogenization of the stokes and navier–stokes equations in a porous medium. *Boll. Unione Mat. Ital.*, 10-A:661–671, 1996.
7. N. Neuss. Femlisp – a multi-purpose tool for solving partial differential equations. Technical report, IWR, Universität Heidelberg, 2003.
8. N. Neuss. *Schnelle numerische Approximation von effektiven Parametern mit einer interaktiven Finite-Elemente Umgebung*. Habilitationsschrift. Universität Heidelberg, Heidelberg, 2003.
9. M. Neuss-Radu. The boundary behavior of a composite material. *M2AN*, 35:407–435, 2001.
10. O. A. Oleinik, A. S. Shamaev, and G. A. Yosifian. *Mathematical Problems in Elasticity and Homogenization*. North–Holland, Amsterdam, 1992.
11. L. Tartar. Incompressible flow in a porous medium - convergence of the homgenization process. In E. Sanchez-Palencia, editor, *Non-Homogenous Media and Vibration Theory*, volume 127 of *Lecture Notes in Physics*, pages 368–377. Springer–Verlag, Berlin, 1980.
12. R. Verfürth. *A Review of A Posteriori Error Estimation and Adaptive Mesh-Refinement Techniques*. Wiley Teubner, Chichester, Stuttgart, 1996. ISBN 0-471-96795-5.

Numerical Simulation and Experimental Studies of Unsaturated Water Flow in Heterogeneous Systems*

P. Bastian[1], O. Ippisch[1], F. Rezanezhad[2], H. J. Vogel[2], and K. Roth[2]

[1] Interdisziplinäres Zentrum für Wissenschaftliches Rechnen (IWR), Universität Heidelberg
[2] Institute für Umweltphysik, Universität Heidelberg

Summary. In this paper we develop higher-order Discontinuous Galerkin finite element methods for the simulation of two-phase flow in heterogeneous porous media and experimental methods to determine structure and saturation in two-dimensional Hele-Shaw cells with high resolution in space and time. Together with additional measurements of governing parameters this allows us to compare experiments with numerical simulations. Results are shown to be in good qualitative agreement.

1 Introduction

Natural porous media are heterogeneous on all scales, a fact that makes modelling and simulation of the flow of immiscible fluids in such a porous medium very difficult. The main focus of this paper is the development of accurate and efficient numerical methods for the solution of the two-phase flow and Richards equations. Then we apply these methods to the simulation of experiments conducted to answer the question: Given the structure of a porous medium (in the form of a gray-scale image as shown in Figure 6) can we simulate the movement of water and air during drainage of an initially fully water saturated medium? The answer to this question requires an experimental method that allows us to determine the structure of the porous medium and the water saturation with sufficiently high resolution in space and time. This is achieved by a quasi two-dimensional setup in a Hele-Shaw cell enabling the use of the light transmission method [23] as a proxi for water content.

In the numerical part of this paper we present the application of higher order discontinuous Galerkin (DG) methods to the equations of two-phase flow in porous media. Due to their flexibility, DG methods have been popular among the finite element community and they have been applied to a wide

---

*This work has been supported by the German Research Foundation (DFG) through SFB 359 (Project D3) at the University of Heidelberg.

range of computational fluid problems. Since the first DG method introduced in [19] the methods have been developed for hyperbolic problems known as the Runge-Kutta DG method [11, 10, 7, 8, 13] and for elliptic problems in [25, 18, 12, 20, 21]. A unified analysis for many DG methods for elliptic problems has been given recently in [2]. A general overview is available in [9]. Advantages of DG methods are their higher order convergence property, local conservation of mass and flexibility with respect to meshing and hp-adaptive refinement. Their uniform applicability to hyperbolic, elliptic and parabolic problems as well as their robustness with respect to strongly discontinuous coefficients renders them very attractive for porous medium flow and transport calculations.

The experimental investigation of flow and transport in heterogeneous porous media is typically limited to the measurement of fluxes across the boundaries of a sample as a response to certain boundary conditions. The internal dynamics of state variables is only accessible at a small number of locations using instruments like e.g. pressure sensors. An attractive approach to get a high spatial and temporal resolution of phase saturations is a Hele–Shaw cell. It is made of two parallel glass plates a small distance apart with the porous material placed inbetween. The thickness of the cell is small enough to transmit visible light while the lengths of the other two dimensions are only limited by mechanical stability. In such a pseudo two-dimensional porous medium the intensity of transmitted light is directly related to phase saturations – in our experiments water and air – and hence, the dynamics of the water content can be measured at a high spatial and temporal resolution using a digital camera. To directly compare our numerical simulations to corresponding experiments, we used a Hele-Shaw cell filled with heterogeneous sand. The structure of the sand, which is required for the numerical simulations, could be deduced from light transmission through completely dry material. As typical experiments we investigate a slow and a fast drainage of the heterogeneous material.

This paper is organized as follows: Section 2 introduces the governing partial differential equations modeling two-phase flow in porous media. Section 3 presents the Discontinuous Galerkin scheme together with two validating numerical examples showing the effectiveness of the method. Section 4 gives details on the experimental materials and methods and Section 5 compares the experimental results with numerical simulations of the experiment. Conclusions are given in Section 6.

2 Governing Equations

Let Ω be a polygonal domain in $\mathbb{R}^n, n = 2, 3$. The conservation of mass for each phase is given by:

$$\varphi \frac{\partial(\rho_i s_i)}{\partial t} + \nabla \cdot (\rho_i \boldsymbol{u}_i) = \rho_i q_i, \quad i = n, w, \tag{1}$$

$$\boldsymbol{u}_i = -\lambda_i \boldsymbol{K}(\nabla p_i - \rho_i \boldsymbol{g}), \quad i = n, w. \tag{2}$$

Here φ is the porosity of the porous medium, ρ_i is the density of each phase, s_i is the phase saturation, $\boldsymbol{u}_i$ is the phase velocity, $\lambda_i = k_{ri}(s_i)/\mu_i$ is the phase mobility, μ_i is the viscosity of the i-th fluid, $k_{ri}(s_i)$ is the nonlinear relative permeability function for each phase, $\boldsymbol{K}$ is the absolute permeability tensor and $\boldsymbol{g}$ is the gravity vector. The following closure relations are added:

$$s_w + s_n = 1, \tag{3}$$

$$p_n - p_w = p_c(s_w). \tag{4}$$

The last equation gives the macroscopic capillary pressure as a nonlinear function of saturation. If we assume incompressiblity of the fluids, we can rewrite the above system in the pressure-saturation formulations which is presented below.

2.1 Pressure Equation

A formulation based on wetting-phase pressure is given by

$$\nabla \cdot \boldsymbol{u} = q, \tag{5}$$

$$\boldsymbol{u} = -\lambda \boldsymbol{K} \left(\nabla p_w - \boldsymbol{G} \right) - \lambda_n \boldsymbol{K} \nabla p_c \tag{6}$$

where $\boldsymbol{u} = \boldsymbol{u}_w + \boldsymbol{u}_n$ is the total velocity, λ is the total mobility ($\lambda = \lambda_w + \lambda_n$) and $\boldsymbol{G} = \frac{1}{\lambda}(\lambda_n \rho_n + \lambda_w \rho_w)\boldsymbol{g}$.

A formulation based on global pressure is given by

$$\nabla \cdot \boldsymbol{u} = q, \tag{7}$$

$$\boldsymbol{u} = -\lambda \boldsymbol{K} \left(\nabla p - \boldsymbol{G} \right) \tag{8}$$

where the global pressure is defined as

$$p(\boldsymbol{x}, t) = p_n(\boldsymbol{x}, t) - \int_{1-S_{nr}}^{S_w(\boldsymbol{x},t)} f_w(\xi) p_c'(\xi)\, d\xi + p_c(1 - S_{nr})$$

with S_{nr} the residual saturation of the non-wetting phase and $f_w = \lambda_w/\lambda$ the fractional flow function. Mathematically, the global pressure is well defined for all values of $S_w \in [S_{wr}, 1 - S_{nr}]$.

Let the boundary of the domain be decomposed into disjoint open sets Γ_{pD} and Γ_{pN} corresponding to Dirichlet and Neumann boundary conditions for the pressure. Then the flow equation is subject to the following boundary conditions:

$$p_w = p_{wd} \quad \text{or} \quad p = p_d \quad \text{on } \Gamma_{pD}, \tag{9}$$

$$\boldsymbol{u} \cdot \boldsymbol{n} = U \quad \text{on } \Gamma_{pN} \tag{10}$$

2.2 Saturation Equation

Assuming incompressibility, the saturation equation for phase $i = w, n$ is given by

$$\varphi \frac{\partial s_i}{\partial t} + \nabla \cdot \boldsymbol{u}_i = q_i \tag{11}$$

and the phase velocities can be computed from the total velocity and capillary pressure via

$$\boldsymbol{u}_w = f_w \left(\boldsymbol{u} - \lambda_n (\rho_n - \rho_w) \boldsymbol{K} \boldsymbol{g} \right) + \frac{\lambda_w \lambda_n}{\lambda} \boldsymbol{K} \nabla p_c \tag{12}$$

$$\boldsymbol{u}_n = f_n \left(\boldsymbol{u} - \lambda_w (\rho_w - \rho_n) \boldsymbol{K} \boldsymbol{g} \right) - \frac{\lambda_w \lambda_n}{\lambda} \boldsymbol{K} \nabla p_c \tag{13}$$

Any of the two saturation equations can be combined with any of the two flow equations since the saturation equation is formulated independently of the pressure being used. This then results in two nonlinear coupled equations. The flow equation is elliptic and the saturation equation is degenerate parabolic in general or hyperbolic of first order in the case of zero capillary pressure.

To simplify the writing, throughout the paper, we fix the primary variable for the saturation to be the wetting phase saturation s_w. The proposed methods can be rewritten for the non-wetting phase saturation in the same way. The saturation equation is equipped with the boundary conditions

$$s_w = s_{wd} \quad \text{on} \quad \Gamma_{sD}, \tag{14}$$

$$\boldsymbol{u}_w \cdot \boldsymbol{n} = U_w \quad \text{on} \quad \Gamma_{sF}, \tag{15}$$

where we assumed that the boundary of the domain has been partitioned into the segments Γ_{sD} and Γ_{sF}. Our numerical models allow for any combination of the boundary conditions for the two equations.

In the incompressible case only an initial value s_{w0} for saturation is required. In case of a vertical or three-dimensional reservoir where gravity is included the initial saturation is usually obtained by assuming hydrostatic equilibrium for both phases.

2.3 Richards Equation

Richards equation is is obtained from the two-phase flow system under the assumption of constant non-wetting phase pressure $p_n = p_a = const.$ This assumption is a good approximation if the non-wetting phase is air which is always in contact with the surface (no entrapped air). In that case wetting-phase saturation can be computed from wetting-phase pressure via $s_w = p_c^{-1}(p_a - p_w)$ and the system reduces to Richards equation in its so-called mixed form:

$$\varphi \frac{\partial p_c^{-1}(p_a - p_w)}{\partial t} + \nabla \cdot \left\{ -\lambda_w (p_c^{-1}(p_a - p_w)) \boldsymbol{K} (\nabla p_w - \rho_w \boldsymbol{g}) \right\} = q_w \tag{16}$$

Dirchlet and flux boundary conditions as well as an initial condition have to be specified. The numerical scheme described below for the saturation equation immediately applies for Richards equation.

3 Discontinuous Galerkin Scheme

The domain Ω is subdivided into a non-degenerate quasi-uniform partition $\mathcal{E}_h = \{E\}_E$, where E is a triangle or a quadrilateral in 2D, or a tetrahedron, hexahedron or prism in 3D. Let h denote the maximum diameter of the elements in $\mathcal{E}_h$. Let Γ_h be the union of the open sets that coincide with interior edges (faces in 3D) of elements of $\mathcal{E}_h$. Let e denote a segment of Γ_h shared by two elements E^k and E^l of $\mathcal{E}_h$; we associate with e, once and for all, a unit normal vector $\boldsymbol{n}_e$ directed from E^k to E^l ($k > l$) and we define formally the jump and average of a function ψ on e by:

$$[\psi] = (\psi|_{E^k})|_e - (\psi|_{E^l})|_e, \quad \{\psi\} = \frac{1}{2}(\psi|_{E^k})|_e + \frac{1}{2}(\psi|_{E^l})|_e.$$

If e is adjacent to $\partial\Omega$, then the jump and the average of ψ on e coincide with the trace of ψ on e and the normal vector $\boldsymbol{n}_e$ coincides with the outward normal $\boldsymbol{n}$.

We will approximate the pressure and saturation by discontinuous polynomials of total degrees r_p^E and r_s^E respectively on each mesh element E. Let $r_p = \max\{r_p^E : E \in \mathcal{E}_h\}$ and $r_s = \max\{r_s^E : E \in \mathcal{E}_h\}$. We denote by $\mathcal{D}_{r_p}$ and $\mathcal{D}_{r_s}$ the discontinuous finite element spaces.

3.1 Spatial Discretization for Phase Pressure Equation

Rewriting the phase pressure equation (5), we have

$$-\nabla \cdot (\lambda \boldsymbol{K} \nabla p_w) = q - \nabla \cdot \boldsymbol{\chi}, \quad \boldsymbol{\chi} = \lambda \boldsymbol{K} \boldsymbol{G} - \lambda_n \boldsymbol{K} \nabla p_c.$$

Then, the scheme due to [18] for the pressure equation is: Find $p_w \in \mathcal{D}_{r_p}$ such that for all $v \in \mathcal{D}_{r_p}$

$$a(\lambda; p_w, v) + g(\lambda; p_w, v) = l_1(\lambda, \boldsymbol{\chi}; v). \tag{17}$$

with the bilinear form

$$a(\mu; p, v) = \sum_{E \in \mathcal{E}_h} \int_E \mu \boldsymbol{K} \nabla p \cdot \nabla v$$
$$- \sum_{e \in \Gamma_h} \int_e \{\mu \boldsymbol{K} \nabla p \cdot \boldsymbol{n}_e\}[v] + \sum_{e \in \Gamma_h} \int_e \{\mu \boldsymbol{K} \nabla v \cdot \boldsymbol{n}_e\}[p]. \tag{18}$$

Moreover we have the boundary term

$$g(\mu; p, v) = \sum_{e \in \Gamma_{pD}} \int_e (\lambda \boldsymbol{K} \nabla v \cdot \boldsymbol{n}_e) p - (\lambda \boldsymbol{K} \nabla p \cdot \boldsymbol{n}_e) v \tag{19}$$

and the right hand side

$$l_1(\lambda, \chi; v) = \sum_{E \in \mathcal{E}_h} \int_E \chi \cdot \nabla v - \sum_{e \in \Gamma_h \cup \Gamma_{pD}} \int_e \{\chi \cdot \boldsymbol{n}_e\}[v] \qquad (20)$$
$$+ \sum_{e \in \Gamma_{pD}} \int_e \lambda \boldsymbol{K} \nabla v \cdot \boldsymbol{n}_e p_{wd} + \int_\Omega qv - \int_{\Gamma_{pN}} Uv.$$

3.2 Spatial Discretization for Global Pressure Equation

The scheme for the global pressure equation is: Find $p \in \mathcal{D}_{r_p}$ such that for all $v \in \mathcal{D}_{r_p}$
$$a(\lambda; p, v) + g(\lambda; p, v) = l_1(\lambda, \lambda \boldsymbol{K} \boldsymbol{G}; v). \qquad (21)$$

3.3 Conservative Projection

The total velocity $\boldsymbol{u}$ computed in the DG scheme is discontinuous at element boundaries and does not even have continuous normal component $\boldsymbol{u} \cdot \boldsymbol{n}_e$. The average flux $\{\boldsymbol{u} \cdot \boldsymbol{n}_e\}$ is the conserved flux but is inconsistent with the fluxes evaluated from left and right. A velocity field with continuous normal component, i. e. $\boldsymbol{u} \in H(\text{div}; \Omega)$ is, however, required by most transport simulations such as the schemes described below. In [5] we describe a simple projection scheme $\Pi^* : (\mathcal{D}_{r_p})^n \to H(\text{div}; \Omega)$ using the BDM finite element spaces from [6] and prove that this projected velocity is as accurate as the DG velocity. This projected velocity $\Pi^* \boldsymbol{u}$ will be used in the transport simulation where required.

3.4 Spatial Discretization for Saturation Equation

We choose the wetting phase saturation to be the primary variable. The formulation for the non-wetting phase can be obtained the same way. Consider the continuous-in-time scheme: Find $s_w \in \mathcal{D}_{r_s}$ such that for all $z \in \mathcal{D}_{r_s}$
$$\frac{\partial}{\partial t}(\varphi s_w, z)_\Omega + b(\frac{\lambda_w \lambda_n}{\lambda}; s_w, z)$$
$$+ c(\boldsymbol{u}; s_w, z) + J(\sigma, \beta; s_w, z) = r(\boldsymbol{u}, \sigma, \beta; z), \qquad (22)$$

where we have used the L_2 inner product
$$(y, z)_\Omega = \int_\Omega yz, \qquad (23)$$

and the nonlinear form related to degenerate diffusion
$$b(\mu; s_w, z) = - \sum_{E \in \mathcal{E}_h} \int_E (\mu \boldsymbol{K} |p'_c| \nabla s_w) \cdot \nabla z - \sum_{e \in \Gamma_{sD}} \int_e \mu \boldsymbol{K} \nabla z \cdot \boldsymbol{n}_e p_c(s_{wd})$$
$$+ \sum_{e \in \Gamma_h \cup \Gamma_{sD}} \int_e \{\mu \boldsymbol{K} \nabla z \cdot \boldsymbol{n}_e\} [p_c] - \{\mu \boldsymbol{K} |p'_c| \nabla s_w \cdot \boldsymbol{n}_e\} [z]. \qquad (24)$$

Here we used
$$\nabla p_c = p'_c \nabla s_w = -|p'_c| \nabla s_w \qquad (25)$$
which is always true *for the discrete saturation on an element*. The nonlinear form related to convection is
$$c(\boldsymbol{u}; s_w, z) = \sum_{E \in \mathcal{E}_h} \int_E f_w \boldsymbol{w} \cdot \nabla z + \sum_{e \in \Gamma_h \cup \tilde{\Gamma}} \int_e f_w^{up} \{\boldsymbol{w} \cdot \boldsymbol{n}_e\}[z] \qquad (26)$$
where $\boldsymbol{w}(\boldsymbol{u}, s_w) = \boldsymbol{u} - \lambda_n(\rho_n - \rho_w)\boldsymbol{Kg}$. The upwinding of the fractional flow f_w^{up} on an edge $e = \partial E_1 \cap \partial E_2$, with $\boldsymbol{n}_e$ outward to E_1 is done in the standard way:
$$f_w^{up} = \begin{cases} f_w|_{E_1} & \text{if } \{\boldsymbol{w} \cdot \boldsymbol{n}_e\} \geq 0, \\ f_w|_{E_2} & \text{else} . \end{cases}$$
On boundary edges upwind evaluation is just the identity. The boundary has been decomposed into $\Gamma_{in} = \{\boldsymbol{x} \in \partial \Omega : \boldsymbol{u}_w(\boldsymbol{x}) \cdot \boldsymbol{n} < 0\}$ and $\Gamma_{out} = \partial \Omega \setminus \Gamma_{in}$ and we set $\tilde{\Gamma} = \Gamma_{sD} \cap \Gamma_{out}$.
$$J(\sigma, \beta; s_w, z) = \sum_{e \in \Gamma_h} \frac{\sigma}{|e|^\beta} \int_e [s_w][z] + \sum_{e \in \Gamma_{sD}} \frac{\sigma}{|e|^\beta} \int_e s_w z \qquad (27)$$
is the interior penalty term with two user-defined parameters σ and β. Finally, the right hand side is given by
$$r(\boldsymbol{u}, \sigma, \beta; z) = \int_\Omega q_w z - \sum_{e \in \Gamma_{sF}} \int_e U_w z$$
$$+ \sum_{e \in \Gamma_{sD}} \frac{\sigma}{|e|^\beta} \int_e s_{wd} z - \sum_{e \in \Gamma_{sD} \cap \Gamma_{in}} \int_e f_w(s_{wd}) \boldsymbol{w}(s_{wd}) \cdot \boldsymbol{n}_e z \qquad (28)$$

3.5 Temporal Discretization and Global Solution Algorithm

Let $0 = t^0 < t^1 < \cdots < t^N = T$ be a subdivision of the time interval $[0, T]$. Denote the time step by $\Delta t^i = t^{i+1} - t^i$. The pressure at time t^i is denoted by p^i or p_w^i and the saturation by s_w^i. We now define a solution algorithm based on a decoupling of the pressure and saturation equations.

The algorithm uses one of the pressure formulations and combines it with an appropriate time discretization scheme for the saturation equation. Choosing a basis of the finite element space we can write the saturation in terms of coefficients: $s_w = \mathcal{P}(\boldsymbol{S}_w)$, where $\mathcal{P} : \mathbb{R}^n \to \mathcal{D}_{r_s}$ is the usual finite element isomorphism. Inserting this representation into (22) we arrive at a system of nonlinear ordinary differential equations (ODE)
$$\frac{d}{dt} \boldsymbol{S}_w(t) = \boldsymbol{L}(\boldsymbol{u}; t, \boldsymbol{S}_w(t)) \qquad (29)$$
for the coefficients $\boldsymbol{S}_w$. Note that the inversion of the mass matrix is very cheap since it is block-diagonal. The complete algorithm is then as follows:

1. Initialize the reservoir with $s_w^0 = s_{w0}$. Set $i = 0$. Choose a parameter $R \in \mathbb{N}$.
2. If $i \bmod R = 0$ solve the pressure equation for p^i else set $p^i = p^{i-1}$. This requires the solution of a linear system of equations every R^{th} time step which is done by a multigrid method described in [4]. Note that in the case of Richards equation no pressure equation needs to be solved.
3. Select Δt^i and compute new saturation $s_w^{i+1} = \mathcal{P}(S_w^{i+1})$ by applying an s-stage Runge-Kutta method to (29):

   a) $S_w^{(0)} = S_w^i$

   b) $S_w^{(j)} = \Lambda\Pi\left(\sum_{k=0}^{j}\left[a_{jk}S_w^{(k)} + b_{jk}\Delta t^i \boldsymbol{L}(\Pi^*\boldsymbol{u}^i; t^i + d_k\Delta t^i, S_w^{(k)})\right]\right)$, for all $i = 1(1)s$

   c) $S_w^{n+1} = S_w^{(s)}$

   The Runge-Kutta method can either be fully explicit or diagonally implicit. We have implemented the explicit TVD Runge-Kutta methods of order 2 and 3 from [22], the second and third order strongly S-stable implicit methods of [1] and the fourth order L-stable implict method of [15]. In the implicit case the nonlinear algebraic equations are solved by Newton's method using a multigrid solver for the linear systems. Note that in this case only the projected total velocity $\Pi^*\boldsymbol{u}^i$ is taken from the previous time step in all stages of the Runge-Kutta method.

   Here $\Lambda\Pi$ denotes the application of the slope limiting procedure described in [13], which is a modification of the minmod procedure. The explicit methods of [22] assure that the cell averages are monotone. The implicit Runge-Kutta methods (nor any other method we know of) do not possess this property. Thus the slope limiting in the case of implicit time discretization does not yield a monotone method (and may be omitted).
4. Set $i = i + 1$. If $t^i < T$ go to step 2 else quit.

3.6 Comparison of Numerical Schemes

Fivespot

This example demonstrates the case of dominant nonlinear convection in the saturation equation. Setup is shown in Fig. 1 and parameters are:

Domain: $\Omega = (0, 300[m])^2$, set $\Gamma_{in} = \{\boldsymbol{x} \in \partial\Omega : \|\boldsymbol{x} - (0,0)^T\| \leq h\}$, $\Gamma_{out} = \{\boldsymbol{x} \in \partial\Omega : \|\boldsymbol{x} - (300,300)^T\| \leq h\}$, $\Gamma = \partial\Omega \setminus \Gamma_{in} \setminus \Gamma_{out}$

Fluid properties: $\mu_w = 0.001[Pa\ s]$, $\mu_n = 0.01[Pa\ s]$

Rock: $\varphi = 0.2$, permeability see below

Parametrizations: Brooks-Corey model with $s_{rw} = 0.2$, $s_{rn} = 0.15$, $\theta = 2$ and $p_e = 10^4[Pa]$

Boundary cond.: $\boldsymbol{u}\cdot\boldsymbol{n} = \boldsymbol{u}_w\cdot\boldsymbol{n} = \frac{1}{h}\cdot 1.05\cdot 10^{-4}[m/s]$ on Γ_{in}, $p = 10^5[Pa]$, $s_w = s_{rw}$ on Γ_{out}, $\boldsymbol{u}\cdot\boldsymbol{n} = \boldsymbol{u}_w\cdot\boldsymbol{n} = 0$ on Γ

Initial condition: $s_{w0} = s_{rw}$

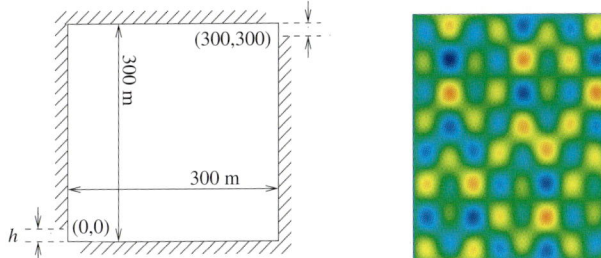

Fig. 1. Left: Domain for the Fivespot problem with h denoting mesh size. Right: Permeability field for the heterogeneous but smooth case.

Table 1. Run times of the schemes on a Pentium IV 2.2 GHz machine.

grid	RKDG(2,1), triangles	RKDG(2,1) quad	FCFV quad
40 · 40	20m	9m	10m
80 · 80	2h 37m	1h 18m	1h 10m
160 · 160	21h 30m	11h 24m	12h 31m

For the permeability field we consider two different isotropic cases with $K = k \cdot I$ and:

I Homogeneous: $k = 10^{-11}[m^2]$.
II Continuous heterogeneous: $k = 10^{-11+\delta}[m^2]$ where

$$\delta = \sin(10\pi\bar{x})\sin(8\pi\bar{y}) + \frac{7}{10}\sin(3\pi\bar{x})\cos(6\pi\bar{y}) + \frac{3}{10}\sin(\frac{\pi\bar{x}}{2})\sin(\pi\bar{y})$$

and $\bar{x} = x/300$, $\bar{y} = y/300$. This permeability field is shown in Figure 1 (right).

Case I

Figure 2 shows results for the homogeneous computation at 300 days. A comparison of DG using explicit Runge-Kutta scheme ($r_p = 2, r_s = 1$) with a fully implicit and fully coupled finite volume method as described in [3] is given.

Clearly, the front is much sharper in the DG scheme which makes it easier to determine break-through time. Although grid convergence is difficult to obtain due to the singularity of the solution in the corners we can clearly see that the 40^2 DG solution is as good as the 80^2 FV solution. For more accurate solutions DG saves between one and two levels of mesh refinement. Computation times are given in Table 1. For the same accuracy the DG code is about 10 times faster.

Case II

A convergence study for the continuous heterogenous permeability field is shown in Figure 3. The DG solution on the 40^2 quadrilateral mesh is already

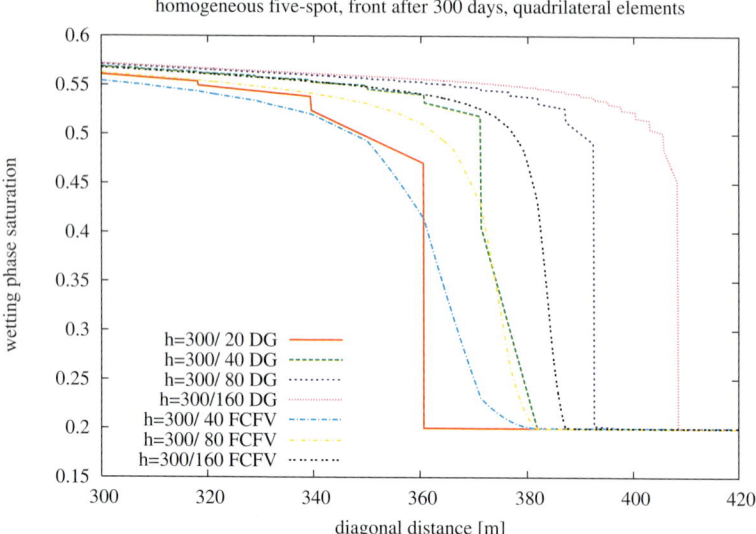

Fig. 2. Comparison of front along diagonal cut for homogeneous five-spot problem showing DG and fully coupled vertex centered finite volume scheme on quadrilateral elements.

comparable to the FV solution on the 160^2 mesh which results in similar or better savings as in the homogeneous case.

DNAPL infiltration problem

This test problem illustrates the performance of the method in the case where capillary diffusion is dominating. The setup and a plot of the saturation at final time is shown in Fig. 4. The parameters are given in the following table:

Domain: $\Omega = (0, 0.9[m]) \times (0, 0.65[m])$, $\Omega' = (0.3, 0.45[m]) \times (0325, 0.4875[m])$, $\Gamma_{in} = \{(x,y) \in \partial\Omega : 0.3 < x < 0.6, y = 0.65\}$, $\Gamma_P = \{(x,y) \in \partial\Omega : x = 0 \lor x = 0.9\}$, $\Gamma_S = \{(x,y) \in \partial\Omega : y = 0 \land 0 < x < 0.9\}$, $\Gamma_0 = \partial\Omega \backslash \Gamma_{in} \backslash \Gamma_P$

Fluid properties: $\mu_w = 0.001[Pa\ s]$, $\mu_n = 0.0009[Pa\ s]$, $\rho_w = 1000[kg/m^3]$, $\rho_n = 1460[kg/m^3]$

Rock: $\varphi = 0.4$, $\boldsymbol{K}(\boldsymbol{x}) = k(\boldsymbol{x}) \cdot \boldsymbol{I}$ where $k(\boldsymbol{x}) = 6.64 \cdot 10^{-13}[m^2]$ if $\boldsymbol{x} \in \Omega'$ and $k = 6.64 \cdot 10^{-11}[m^2]$ else

Parametrizations: Brooks-Corey model with $s_{rw} = 0.1$, $s_{rn} = 0.0$, $\theta = 2.7$ and $p_e = 755[Pa]$

Bnd. conditions: $\Gamma_{in} : \boldsymbol{u} \cdot \boldsymbol{n} = \boldsymbol{u}_n \cdot \boldsymbol{n} = -0.075/\rho_n[m/s]$, $\Gamma_P : p(x,y) = \frac{0.65-y}{0.65} \cdot 6376.5[Pa]$ and $\boldsymbol{u}_n \cdot \boldsymbol{n}$, $\Gamma_S : \boldsymbol{u} \cdot \boldsymbol{n} = 0$ and $s_n = 0$, $\Gamma_0 : \boldsymbol{u} \cdot \boldsymbol{n} = \boldsymbol{u}_n \cdot \boldsymbol{n} = 0$

Initial condition: $s_{n0} = 0$

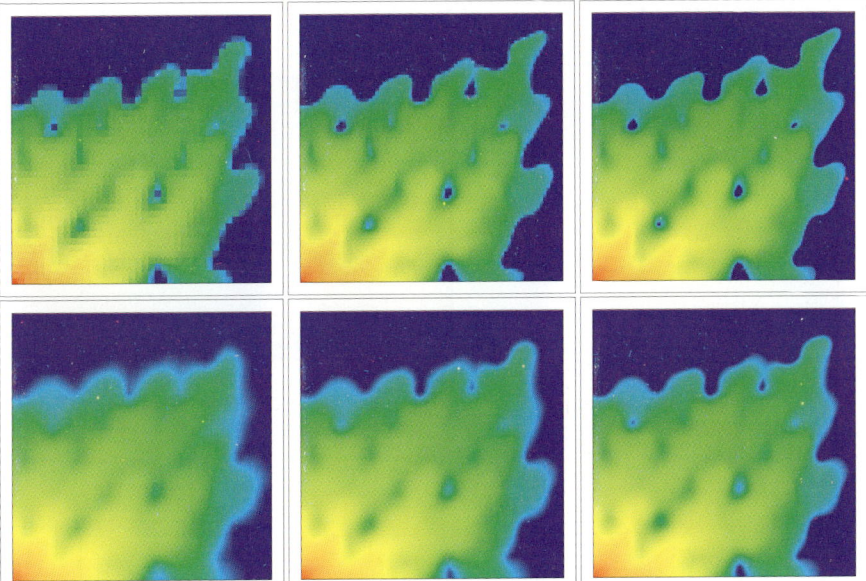

Fig. 3. Results for the continuous heterogeneous case. DG on quadrilaterals in the top row and FV on quadrilaterals in the bottom row. Meshes are 40^2, 80^2 and 160^2 (from left to right).

Note that in the subdomain $\Omega' \subseteq \Omega$ only the absolute permeability is reduced by a factor of 100. The constitutive relations are the same in the whole domain.

Fig. 5 shows the saturation along the line $x = 0.5$ for different schemes. The vertex centered finite volume method is now second order accurate since central differencing and Crank-Nicolson time-integration can be used in the diffusion-dominated case. The second order FV scheme exhibits no oscillations. With the DG space discretization we now use the implicit Runge-Kutta method *without* applying any limiter. The solution shows small undershoots of about 1% at the foot of the front in this case. Besides that, the results of the DG scheme are very promising when compared with the finite volume scheme: The FV solution on a 384×256 mesh is already obtained on a 48×32 mesh with the DG scheme using second order polynomials in space and a second order Runge-Kutta scheme in time. In wall clock time the DG scheme has been 75 times faster for the same accuracy.

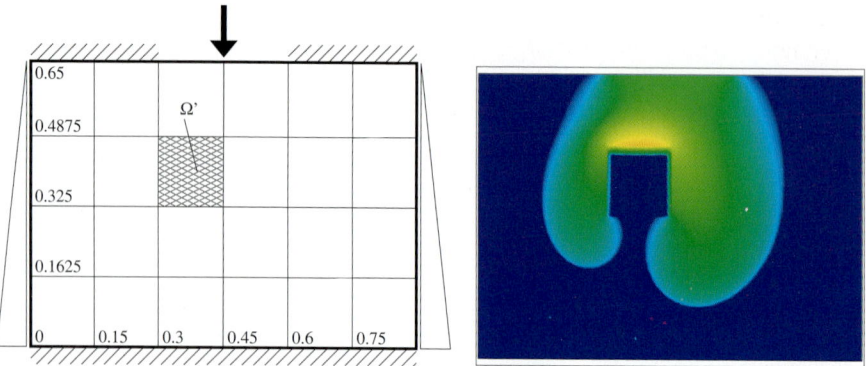

Fig. 4. Problem setup (left) and saturation plot after 2000 s (right) in DNAPL infiltration problem.

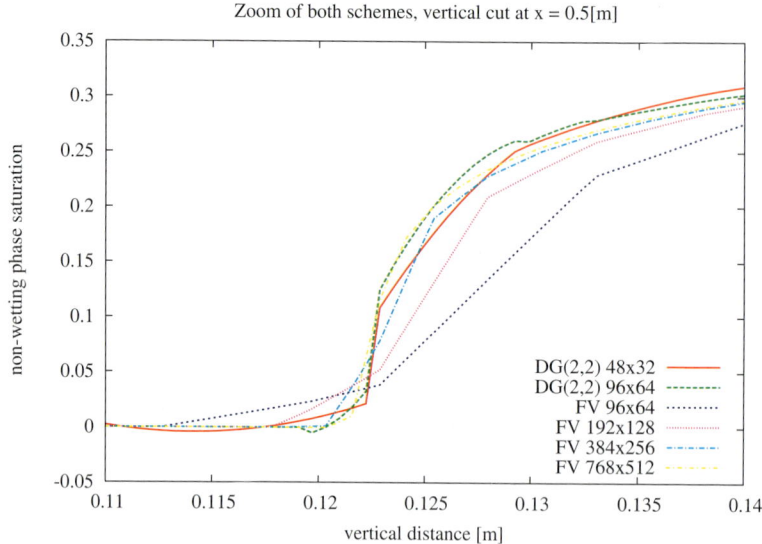

Fig. 5. Solution along a vertical cut for the DNAPL infiltration problem.

4 Experiment

4.1 Hele-Shaw Cell

A porous medium of different sands was filled in a Hele-Shaw cell with a height of 160 cm and a width of 60 cm. The thickness of the cell, i.e. the separation of the glass plates, was 0.3 cm. The distance between the parallel glass plates was fixed by a metal frame and additionally by a number of 14 screws evenly distributed inside the cell. The latter was necessary to avoid any bending of the glass plates due to the hydrostatic pressure at water saturation.

The dry sand was filled in 3 different layers: i)a homogeneous coarse textured sand at the bottom (0-30 cm, grain diameter 0.63-1.25 mm), ii) a heterogeneous layer in the middle (30-120cm, grain diameter 0.25-1.25 mm), and iii) another homogeneous coarse textured sand on top (120-160 cm). The lower horizon allowed for a homogeneous drainage of the heterogeneous domain and the upper horizon ascertained its mechanical stability. The layer in the middle was filled in a way that the grains of different size were allowed to separate during the filling procedure to produce a heterogeneous structure of different textures and herewith different hydraulic properties.

At the bottom, the cell was connected to a water reservoir, through which the water table inside the cell could be adjusted. Water flow across the lower boundary could be measured by placing the water reservoir on a computer controlled balance. The upper end of the cell was open to air, i.e. the non-wetting phase. The vertical boundaries were closed for water flow.

4.2 Light Transmission Method (LTM)

To measure the heterogeneity of the material and the dynamics of water content we used the light transmission method according to [23]. The cell was placed in front of a light box with four fluorescent lamps as a stable and low-temperature light source. A homogeneous illumination of the entire surface of the Hele-Shaw cell was achieved using a diffusion foil between the lamps and the cell.

The transmitted light was recorded by a digital camera with a spatial resolution of 1376×2272 pixel. Inside the heterogeneous layer this gives 0.4 mm^2 per pixel which is in the size range of the sand grains. The temporal resolution was one minute including an exposure time of 13 s (aperture F8) and image storage.

First, an image of the dry material was taken to distinguish the regions of different grain size or texture. This image is shown in the lower right corner of Figure 6. The intensity of transmitted light increases with grain size. The mean hydraulic properties $p_c(s_w)$ and $\lambda_w(s_w)$ of the sands were measured in separate column experiments.

The images were recorded in RGB-format and the transmitted intensity was calculated as $I = \frac{1}{3}(R+G+B)$. Previous experiments have shown that in air-water systems this intensity is related to water content [16],[14] and hence, it can be used as a proxi variable which can be measured at a high spatial and temporal resolution. Based on the images of the intensities measured for the completely dry material I_{dry} and the water saturated material I_{wet} we estimated the water saturation from the measured intensities through

$$s_w = \frac{I - I_{\text{dry}}}{I_{\text{wet}} - I_{\text{dry}}} . \tag{30}$$

This is a first approximation assuming linearity. However, we emphasize that the intensity-saturation relation is almost certainly nonlinear, especially in the range of small saturations. An accurate determination of saturation from light intensities requires a careful calibration depending on the material. This has not been done yet for heterogeneous materials.

4.3 Boundary Conditions

The experiments were performed inside a section of the heterogeneous part of the Hele-Shaw cell (75-120cm depth) starting from complete water saturtion. For convenience z is defined to be 0 cm at the lower boundary of this area and z is 45 cm at the upper boundary. We did two simple experiments at a constant temperature of 20°C:

- A quick drainage where the water table was decreased from $z = 45$ to 0 in one step. The dynamic of water saturation was recorded during a period of 20 minutes (one image per minute).
- A slow drainage where the water table was gradually decreased from $z = 45$ to 0 within a period of 180 minutes. Images were taken at a rate of one image per minute.

In Fig. 6 a sketch and lab photo of the used experimental Setup are shown.

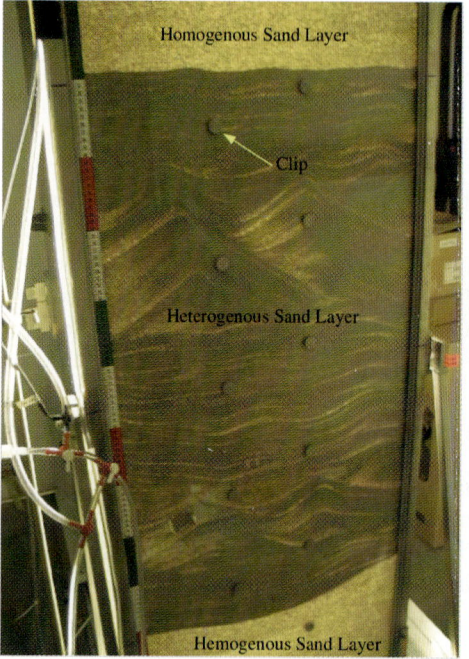

Fig. 6. Lab photo of the Hele-Shaw cell and the proxy image describing the structure of the porous medium.

5 Numerical Simulation of the Experiment

Fast and slow drainage from the Hele-Shaw cell described in the previous section is modelled by Richards equation which is then solved using the Discontinuous Galerkin method and also with a standard cell-centered finite volume scheme. The parameters of the soil water characteristic were transferred to the heterogeneous structure using the assumption of Miller similarity. The concept of Miller similarity ([17]) implies that at two different locations the geometry of the pore space is exactly the same and only the size of the elements which compose the pore space is different. If we can associate a scaling parameter with a typical scale of the pore space (e.g. average or maximal pore diameter), the hydraulic parameters measured at a certain scale χ^* can be used at a different scale χ, if capillary pressure and permeability are scaled by the relation:

$$S_l(\psi_l) = S_l^*\left(\psi_l \cdot \frac{\chi}{\chi^*}\right) \qquad (31)$$

$$K = K^* \cdot \left[\frac{\chi}{\chi^*}\right]^2 \qquad (32)$$

By definition, porosity is constant in Miller similar media. We assumed that the pixel values of an image taken from the dry sand could be used as proxy for the scaling parameter. We calculated the scaling parameter from the pixel value $p(x,y)$ of a material map by the relation $\chi/\chi^* = exp((p(x,y)-128)/128 \cdot \sigma)$, where σ is the standard deviation of the distribution of scaling parameters. The capillary pressure and relative permeability functions used were of van Genuchten type ([24]) with parameters at the reference scale χ^* measured by seperate column experiments.

Figure 7 shows a comparison of the saturation obtained by numerical simulations with the experimentally determined saturation at a given time. The left column of pictures shows the slow drainage while the right column corresponds to the fast drainage. The top row shows the result obtained with the Discontinuous Galerkin scheme presented in Sections 3.4 and 3.5 using polynomial degree 2 and the second-order Runge-Kutta scheme in time. The interior penalty parameter was set to 10^{-8}. The grid size was 64×128 and subsampling to a grid of size 512×1024 was used to create the plots. The middle row shows the results obtained with the cell-centered finite volume scheme using harmonic averaging of permeabilities, full upwinding and backward Euler in time on a grid of size 512×1024 (no subsampling). Total compution times for the two different numerical methods was about the same. The experimentally determined saturation (using the linear intensity-saturation relation Eq. (30)) is shown in the bottom row.

Qualitative agreement is found comparing simulation and experiment in the sense that a) at the given final times the depth of full saturation is about the same and b) the same structures have been drained. The agreement is

Fig. 7. Comparison of simulation and experiment for slow drainage after 180 minutes in left column and fast drainage after 24 minutes in right column. Top row shows simulation with DG method on 64×128 mesh (cropped to interesting region), middle row shows simulation with FV method on 512 × 1024 mesh and bottom row shows experimental results with resolution 1376 × 1136.

better for the case of slow drainage than for fast drainage. In the experimental results for the fast drainage boundary effects are visible that are not present in the simulation. However, a quantitative comparison of saturation values is currently not possible, especially for low saturation values, due to the missing calibration of the intensity-saturation relation.

Comparing the two different methods we can state that due to the highly heterogeneous structure the low-order finite volume scheme is the method of choice. Still, the results of the high-order scheme on a coarser mesh are quite satisfactory which is due to the fact that the pressure (the primary variable, not shown) is much smoother than the saturation. Larger differences between the two numerical schemes can only be seen where fine-textured sand is surrounded by coarse-textured sand and patches of high saturation remain.

6 Conclusions

In this paper we presented a discontinuous Galerkin scheme for Richards and two-phase flow equations in porous media. For model problems we could show numerically that the new scheme is superior to low-order finite volume schemes. The examples included convection and diffusion dominated cases.

In the experimental part fast and slow drainage of a Hele-Shaw cell containing an artificial heterogeneous porous medium build up from different sands was visualized using the light transmission method. This allowed us to measure the heterogeneity and the dynamics of water movement with high spatial and temporal resolution.

Using the approach of Miller similarity the DG scheme and a FV scheme were used to simulate the drainage experiments. The two numerical schemes were comparable in cost and accuracy despite the fact that the low-order scheme is better suited for the highly heterogeneous problem. Qualitatively the experiment and the numerical simulation were in agreement but a quantitative comparison was not possible due to the missing calibration of the intensity-saturation relation.

We conclude that the DG method is very well suited to simulate flow and transport in highly heterogeneous porous media. After a suitable calibration of the light transmission method a detailed quantative comparison with experimental results will be possible.

References

1. Alexander, R.: 1977, 'Diagonally implicit Runge-Kutta methods for stiff O.D.E.'s'. *SIAM Journal Numer. Anal.* **14**, 1006–1021.
2. D. N. Arnold, F. Brezzi, B. Cockburn, and L. D. Marini, *Unified analysis of discontinuous Galerkin methods for elliptic problems*, SIAM J. Numer. Anal. **39** (2002), no. 5, 1749–1779.

3. Bastian, P. and R. Helmig: 1999, 'Efficient fully-coupled solution techniques for two-phase flow in porous media. Parallel multigrid solution and large scale computations'. *Adv. Water Res.* **23**, 199–216.
4. Bastian, P. and V. Reichenberger: 2000, 'Multigrid for higher order discontinuous Galerkin finite elements applied to groundwater flow'. Technical Report 2000-37, SFB 359.
5. Bastian, P. and B. Rivière: 2003, 'Superconvergence and H(div)-projection for discontinuous Galerkin methods'. *Int. J. Numer. Meth. Fluids.* **42**, 1043–1057.
6. Brezzi, F., J. Douglas, and L. Marini: 1985, 'Two families of mixed finite elements for second order elliptic problem'. *Numerische Mathematik* **47**, 217–235.
7. B. Cockburn, S. Hou, and C.-W. Shu, *TVB Runge-Kutta local projection discontinuous Galerkin finite element method for conservation laws III: One-dimensional systems*, J. Comput. Phys. **84** (1989), 90.
8. _____, *TVB Runge-Kutta local projection discontinuous Galerkin finite element method for conservation laws IV: The multidimensional case*, Math. Comput. **54** (1990), 545.
9. B. Cockburn, S. Y. Lin, and C.-W. Shu (eds.), *Discontinuous Galerkin methods. Theory, computation and applications*, Lecture Notes in Computational Science and Engineering, vol. 11, Springer-Verlag, 2000.
10. B. Cockburn and C.-W. Shu, *TVB Runge-Kutta local projection discontinuous Galerkin finite element method for conservation laws II: General framework*, Math. Comput. **52** (1989), no. 186, 411–435.
11. _____, *The Runge-Kutta local projection P^1-discontinuous Galerkin method for scalar conservation laws*, M^2 AN **25** (1991), 337.
12. _____, *The local discontinuous Galerkin finite element method for convection-diffusion systems*, SIAM J. Numer. Anal. **35** (1998), 2440–2463.
13. _____, *The Runge-Kutta discontinuous Galerkin method for conservation laws V: Multidimensional systems*, J. Comput. Phys. **141** (1998), 199–224.
14. Glass,R.J., Steenhuis,T.S., Parlange,J-Y.: Wetting front instability 1.Theoritical discussion and dimensional analysis. Water.Resour.Res, **25**, 1187–1194 (1989)
15. Hairer, E. and G. Wanner: 1991, *Solving ordinary differential equations II*. Springer, Berlin.
16. Hoa,N.T.: A new method allowing the measurement of rapid variations of the water content in sandy porous media. Water.Resour.Res, **17**, 41–48 (1981)
17. E. E. Miller, R. D. Miller: Physical theory for capillary flow phenomena. JAP, **27**, 324-332, 1956.
18. Oden, J., I. Babuvska, and C. Baumann: 1998, 'A Discontinuous hp Finite Element Method for Diffusion Problems'. *Journal of Computational Physics* **146**, 491–519.
19. W. H. Reed and T. R. Hill, *Triangular mesh methods for the neutron transport equation*, Tech. report, Los Alamos Scientific Laboratory, 1973.
20. B. Rivière, M. F. Wheeler, and V. Girault, *Improved energy estimates for interior penalty, constrained and discontinuous Galerkin methods for elliptic problems I*, Comput. Geosci. **3** (1999), 337–360.
21. _____, *A priori error estimates for finite element emthods based on discontinuous approximation spaces for elliptic problem*, SIAM Journal on Numerical Analysis **39** (2001), no. 3, 902–931.
22. Shu, C.: 1988, 'Total-variation-diminishing time discretizations'. *SIAM J. Sci. Stat. Comput.* **9**(6), 1073–1084.

23. Tidwell,V.C., Glass,R.J.: X ray and visible light transmission for laboratory measurement of two-dimensional saturation fields in thin-slab system. Water.Resour.Res, **30**, 2873–2882 (1994)
24. M. T. van Genuchten, *A closed-form equation for predicting the hydraulic conductivity of unsaturated soils.* Soil Sci. Soc. Am. J., **44** (1980), 892–898.
25. M. F. Wheeler, *An elliptic collocation finite element method with interior penalties*, SIAM J. Numer. Anal. **15** (1978), no. 1, 152–161.

Lake Dynamics: Observation and High-Resolution Numerical Simulation*

C. von Rohden[2], A. Hauser[1], K. Wunderle[2], J. Ilmberger[2], G. Wittum[1], and K. Roth[2]

[1] Technische Simulation, Interdisziplinäres Zentrum für Wissenschaftliches Rechnen (IWR), Universität Heidelberg
[2] Institut für Umweltphysik, Universität Heidelberg

1 Objective

We aim at the quantitative understanding of the hydrodynamics of a small natural lake. Of particular interest here is the flow field and the vertical exchange during a typical summer stratification. The issue is an approach along two lines: (i) high-resolution measurements of temperature, electrical conductivity, and horizontal velocity and analyzed with a simplified effective model and (ii) a numerical solver is developed and demonstrated. The latter demanded a strongly anisotropic grid with a very high resolution and a massively parallel MIMD machine.

2 Measurements

2.1 Site

Lake Willersinnweiher is a 800 m long, 250 m wide and up to 20 m deep reservoir (mean depth ∼9 m) left after gravel mining in the 1960's and 70's. Due to its embedding in the aquifer system of the Rhinegraben it is evident that the water balance is predominantly determined by groundwater ([23]). The lake is completely mixed in late autumn and spring and shows a strong temperature dominated density stratification during summer. As a consequence of the urban surroundings and frequent public use the trophic state was often critical in the past ([18]) and its lower reaches, regularly becomes anoxic during summer. Therefore the extent of the redistribution of geochemically relevant species caused by the internal physical dynamics is of interest. This includes the occurrence and structure of currents and internal waves in respond to the wind forcing and the resulting effective vertical transport.

---

*This work has been supported by the German Research Foundation (DFG) through SFB 359 (Project D2) at the University of Heidelberg.

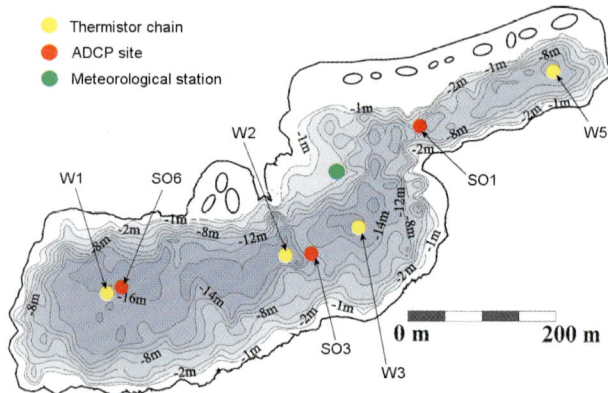

Fig. 1. Bathymetric map of Lake Willersinnweiher. Measurement sites are marked.

2.2 Data Acquisition

CTD-profiles (conductivity-temperature-depth) were taken weekly at five measurement sites with an automatic Idronaut Ocean Seven 319 probe. The vertical resolution is ∼2 m. The resolution of the temperature and conductivity sensor is 0.001°C and 0.1 μS/cm, respectively.

Temperature time series at four different sites (W_1, W_2, W_3, W_5, see figure 1) were acquired with bottom moored thermistor chains. The values were sampled every 5 minutes with a resolution between 0.012 and 0.035°C (figure 4).

Vertical profiles of the three water current velocity components were recorded with an 1200 kHz Acoustic Doppler Current Profiler (ADCP). The depth cell size usually was 40 cm with a temporal resolution of 5 minutes yielding a velocity resolution of 1–2 mm/s. The deployments were installed at different sites for typically one weeks duration. To minimize projection effects of horizontal velocity into vertical velocity when the ADCP is inclined ([12]), the ADCP was mounted in a cardanic suspension to keep pitch and roll angles below 2°.

Figure 2 shows a series of temperature profiles from the warm season. The epilimnion, i.e. the water layer below the surface which is mixed occasionally, descends from 2 m depth to some 7 m during the season (see figure 3). Below, a strongly stratified region follows, the metalimnion. Here the temperature changes slowly in time because of reduced exchange. The depth with the highest vertical density gradient, which is mainly caused by the temperature structure, is the thermocline. The hypolimnion — the deepest part of the water column (below ∼76 m asl) – differs from the metalimnion in cooler temperatures and weaker vertical gradients.

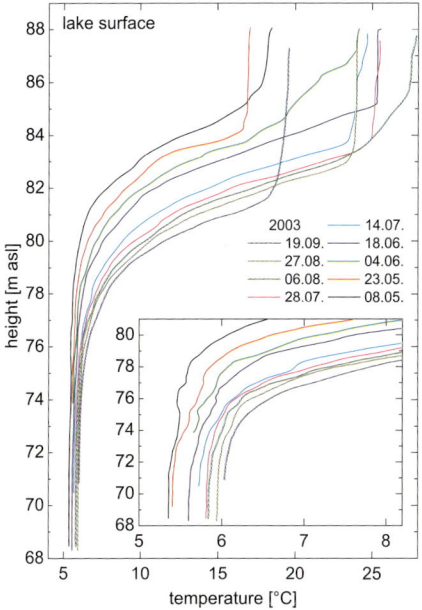

Fig. 2. Selection of temperature profiles during the stratified period in 2003.

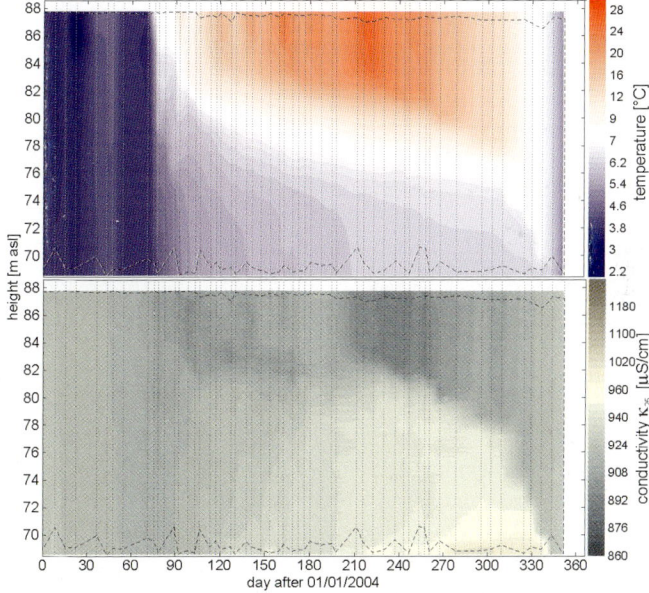

Fig. 3. Temperature and electrical conductivity (corrected to 25°C), exemplary for 2004. Vertical κ_{25}-gradients and temperature gradients are associated. Vertical dotted lines denote measured CTD-profiles. Dashed lines enclose the range of measured data points.

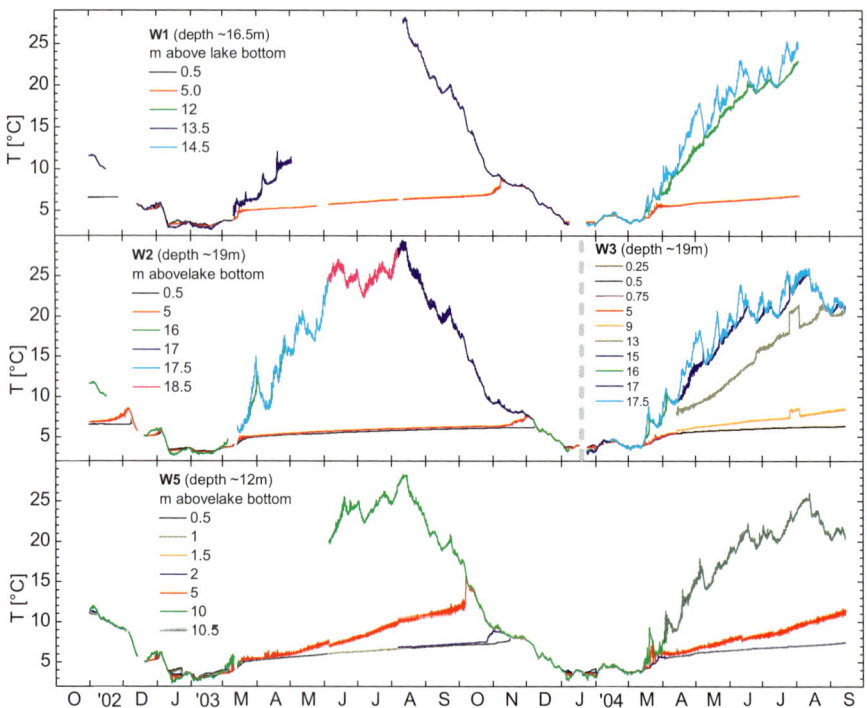

Fig. 4. Temperature time series from bottom mounted temperature sensor chains

2.3 Vertical Transport

Calculations

We aim at the quantitative description of the vertical exchange of heat during the stratified season between May and September of the years 2003 and 2004. We assume that convective exchange is negligible in permanently stratified depths and that the lake is horizontal homogeneous in this region for the observed timescales. That is, small scale turbulent motion and molecular diffusion are the only relevant vertical transport processes. Hence the vertical heat flux is $j_{h_z} \propto -K_z \cdot \frac{\partial T}{\partial z}$, where K_z is the effective thermal conductivity and accounts for molecular and eddy diffusion. In turbulent regions, the transport intensities are considered to be effective for all solutes governing the geochemistry of the lake.

Due to the strong variation of the density stratification (see right part of figure 5), we also expect a high variability of the vertical exchange. The stability is given by the square of the Brunt-Väisälä frequency

$$N^2 \equiv -\frac{g}{\rho_0}\frac{\partial \rho}{\partial z} \qquad [\text{s}^{-2}], \qquad (3)$$

Determination of vertical heat diffusion coefficients using the gradient-flux method

$$K_i = -\frac{\sum_{k=1}^{i}\left\{\frac{\Delta T_k}{\Delta t}\cdot V_k - \frac{Q_{Sed}(A_k - A_{k-1})}{c_p\,\rho}\right\} - (R_i - R_{i-1})\cdot\frac{A_i}{c_p\,\rho}}{\frac{\Delta T_i}{\Delta z}\cdot A_i} \quad (1)$$

i layer number, bottom-up
K effective vertical eddy diffusion coefficient including molecular diffusion D_T [m^2/s]
Δt selected time intervals for the heat balance and the calculation of mean K [s]
ΔT temperature change within Δt [°C]
V_i lake Volume in layer i [m^3]
A_i lake area on top of layer i [m^2]
Δz layer thickness [m]
ρ water density [kg/m^3]
c_p specific heat capacity of lake water [J/(kg·°C)]
Q_{Sed} heat exchange with the sediments [W/m^2]
$R_i - R_{i-1}$ radiation absorbed in layer i [W/m^2]

The temperature profiles are taken about weekly, interpolated to a 2 cm vertical grid and smoothed (figure 2). From this collection an initial and a final profile at a time are selected, each as an average over three different measuring sites to smooth out local horizontal inhomogeneities. The time span Δt of the profiles (budget period) is chosen to be long enough (e.g. 6 weeks) that the heat content in the water column represented by the profiles changed clearly. This time span is moved through the whole set of consecutive temperature profiles for calculation of successive mean transport coefficients K_z during the stratification periods in 2003 and 2004. For each selected depth the heat is balanced between the bottom and the current depth in 2-cm steps.

For each Δt we calculated a weighted mean of the global solar radiation based on data from the meteorological station at the Heidelberg University 30 km to the east of the lake. The heating in the water column is approximated by using Beer's law

$$R(z) = R_0 \cdot \exp(-\mu z). \quad (2)$$

R_0 is the global solar radiation multiplied by a factor of 0.435 to account for the 400–700 nm part of the radiation which penetrates into the water ([14]), μ is the rate of attenuation in m^{-1}. The dependence of μ on wavelength is neglected.

Time dependent values for the heat flux across the water-sediment boundary Q_{sed} are numerically estimated. We assumed that the flux is on the level of the molecular diffusion and driven by the depth and time dependent vertical temperature gradient in the sediment. As boundary conditions act the seasonally changing temperature signal in the hypolimnion and a constant (annual mean) temperature in a certain depth below the lake bottom. Again we used averaged values for Q_{sed} according to the budget periods. Because of higher temperatures in the lake than in the sediments during the summer, we found heat fluxes out of the lake all the time. The influence of this heat sink to the calculated K_z decreases rapidly within a few meter from the lake bottom into the lower hypolimnion. This is due to the X-Mozilla-Status: 8000 X-Mozilla-Status2: 00000000 decreasing ratio of the sediment area to the related lake volume in the upward direction.

where the density $\rho = \rho(T, \kappa_{25})$ of water is a function of conductivity (representing the ion concentration) and temperature.

The considerations are based on heat budget calculations. From regularly measured temperature profiles at different sites on the lake, the vertical eddy diffusion coefficients K_z were calculated from the thermal energy fluxes $H = -\rho \cdot c_p \cdot K_z \cdot dT/dz$ X-Mozilla-Status: 8000 X-Mozilla-Status2: 00000000 with dT/dz as the measured temperature gradients (flux-gradient method, e.g. [11, 15]). First, the vertical flux H through a horizontal cross section $A(z)$ at a certain depth z may be inferred from the temporal change of the heat content in the lake volume below $A(z)$. For the rather shallow and small lake investigated here, the temperature cannot be assumed as a conservative tracer. Two different sources are included into the heat balance in a simple way (details see box on page 603):

1) The upper part of the budget region (thermocline and upper metalimnion) is heated by absorption of solar radiation. The short wave part ($\sim$400-700 nm) of the global radiation generally penetrates a few meters deeper than the typical summer thermocline depth of 3–7 meter, with some variation depending on the lake's trophic state.

2) The exchange between the lake water and the sediments is an effective heat sink during the summer months. There is no indication for a geothermal heat flux from the deeper underground into the lake. However, we assume a heat flux out of the lake through the sediment boundary in summer, whereas during the convective mixing in winter the sediments would act as a source for the lake and release the heat stored during the summer. The flux is driven on the molecular level by the temperature gradient in the uppermost sediment layer. This is under the presumption of a constant mean annual temperature a few meters below the lake bottom.

Effective Transport Coefficients

Figure 5 shows the results for K_z and stability N^2. Averaged profiles are given for the early, mid and late summer periods in 2003 and 2004. The profiles extend from the thermocline to the lake bottom, the region where the vertical transport is supposed to be describable as Fickian diffusion process. The overall picture reveals a reduced transport in regions of strong stratification and an increase in K_z by about two orders of magnitude within a range of $\sim$8 m above the lake bottom (compare figures 5 and 2).

The flux-gradient calculations with eq. 1 include two parameters, the heat flux Q_{Sed} from the sediment and the light penetration depth $1/\mu$ (eq. 2) which are not measured directly. For details see the box on page 603.

We found K_z to be virtually constant in the metalimnion above $\sim$76 m asl. With the picture in mind that the stratification (N^2) and the extent of vertical transport act against each other, we conclude that K_z must be limited to the molecular level because N^2 changes by a factor of 100 in this depth section.

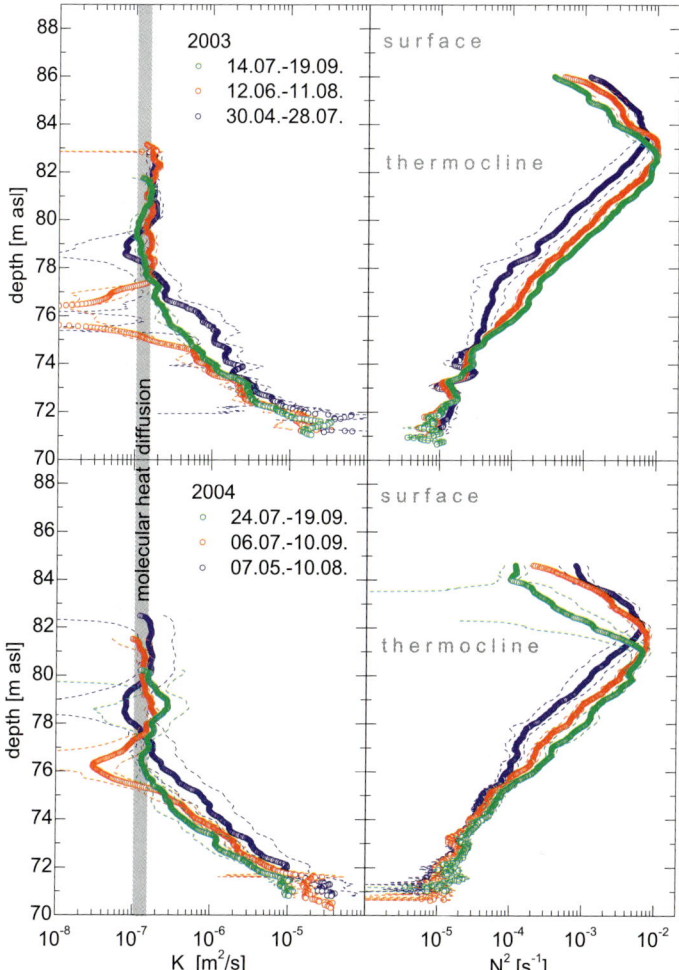

Fig. 5. Profiles of K_z and stability N^2 during the stratification period in 2003 and 2004. Each of the curves is averaged over several single profiles representing the early, mid and late summer time. The dashed lines denote the limits of variation (standard deviation). Regions with values of K_z significantly lower than heat conduction (red profiles around 76 m asl) are explained by the inflow of warm groundwater.

The transport intensity increases with decreasing stratification towards the bottom in the deep hypolimnion. This results from enhanced mixing due to bottom boundary effects and the impact of the sloping lake boundaries of the small Lake Willersinnweiher (e.g. [6, 7, 22]). In the lowest part of the hypolimnion (below ∼73 m asl), uncertainties in the heat balance and in the estimate of the external source Q_{Sed} cause errors of about one order of magnitude for the K_z.

The averaged K_z-profiles representing the mid summer (red symbols in figure 5) show conspicuous values around 76 m asl which significantly fall below D_T both in 2003 and 2004. We hypothesize that this results from the inflow of warmer groundwater, which was not included into the calculations.

With proceeding time, the thermocline and the temperature gradients move downwards, i.e. the stability N^2 increases in constant depths in the metalimnion down to ∼76 m asl. No systematic change of K_z is visible there beyond the K_z-minima. This corroborates the assumption of molecular heat transport.

2.4 Currents

Measurements

In general, currents are rather low (below 10 mm/s), reaching up to 50 mm/s in the hypolimnion after strong wind incidents. In figure 6 (upper panel) we present the north component of the measured flow velocities between August 3 and 10, 2004. This period is representative for a typical summer stratification and the velocity dynamics associated with it. In the lower panel, a section from August 7 to 9 from this record is set apart. The velocity profile extends over a 15 m water column. At the surface, the currents can reach several dm/s during strong winds. Because of the scaling this is not visible in figure 6.

In the hypolimnion (below 10 m) hardly any temporal or spatial structure can be resolved. Currents which can be distinguished from noise are detectable mainly in the epilimnion and metalimnion. The timescale for the vanishing of currents, even after relatively strong wind incidents, is on the order of 16 h. In the morning of August 7, a wind from north-east with an average speed of 4 m/s (figure 8) induced an increase in velocity in the epilimnion directly at the surface.

Fourier Analysis

In order to get information about temporal structure and evolution of internal oscillations the Fourier spectra (figure 7) of the east and north velocity component (figure 6) at selected metalimnion and hypolimnion depths and a spectrum of a temperature record at W3 in the metalimnion were calculated. We used the fast Fourier transform with a window length of 1024 data points,

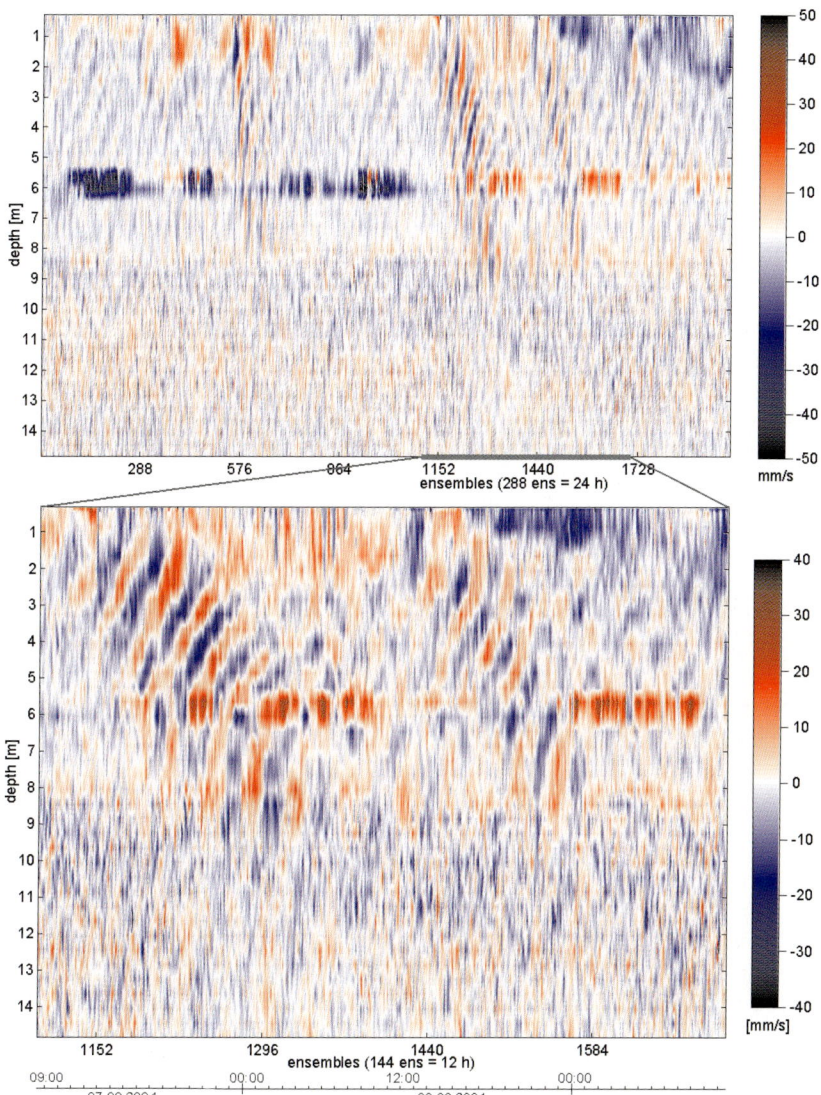

Fig. 6. Measured north component of the flow velocity at site SO6.

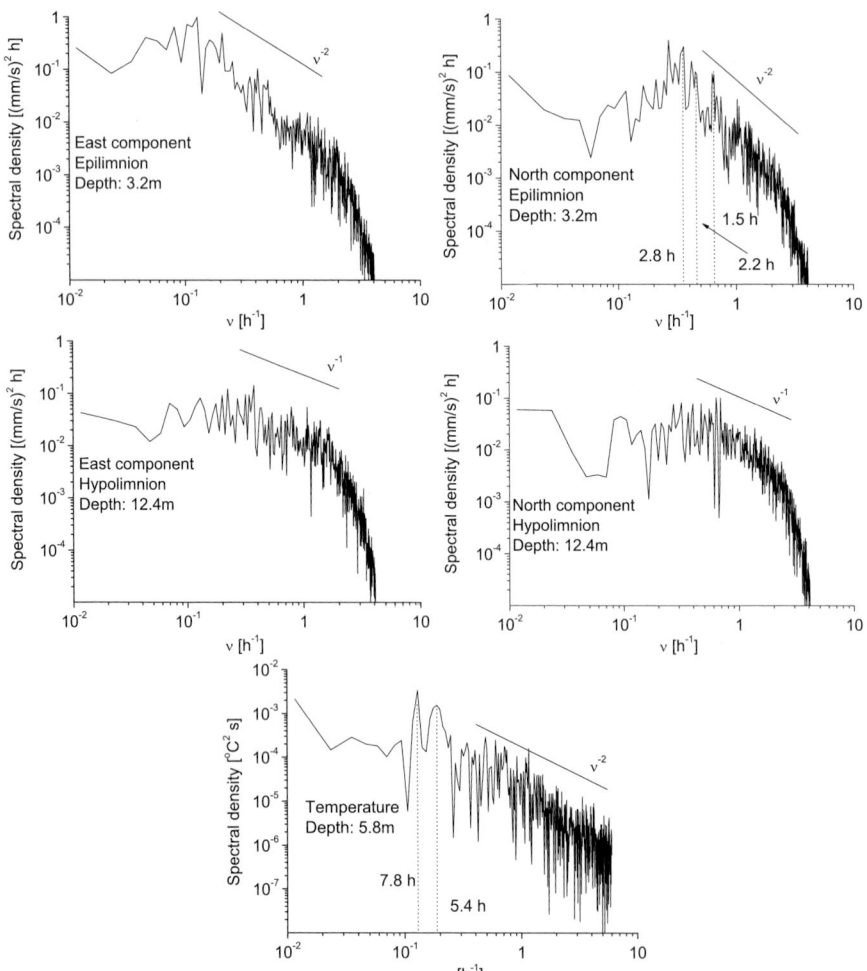

Fig. 7. Spectra of the east and north current velocities in the epilimnion (upper row) and hypolimnion (middle row) and spectrum of the temperature (see figure8). The Fourier analysis (FFT) was performed using a window of 1024 data points starting on 07 Aug. 2004. The time resolution of the measurements is 5 minutes.

where each point is a 5 minute mean. Before transformation, the mean of 3 adjacent depth cells and a truncated Gaussian filter with a total length of 35 minutes and a σ of 8.5 minutes was applied to reduce the noise.

The east component does not show any significant peaks neither in the epi- nor in the hypolimnion. The spectra are quasi-continuously. Thus the response of the lake to wind incidents is distributed over a broad band of possible eigenfrequencies. This was observed during the whole summer stratification period. The power law approximation ν^q yields $q \approx -2$ for epilimnion and

$q \approx -1$ for hypolimnion, respectively. This demonstrates that there is no strong dynamic coupling between the two.

In the spectrum of the north component of the velocity one can distinguish peaks at 2.8 h, 2.2 h, 1.5 h, with the same underlaying slope as the east component. Peaks can only be seen in the epilimnion further corroborating the near independence of hypolimnion currents from epilimnion currents (figure 7).

The decline of the frequencies higher then $2.5\,\mathrm{h}^{-1}$ in the velocity spectra is due to the smoothing with the Gaussian filter.

The temperature record (figure 8) from 5.8 m depth corresponds to the region of strong temperature stratification. With temperature gradients of 4.3 °C/m at this depth, the oscillation of 1.4 °C in temperature is linked to an elevation of the isotherms by 0.3 m, which illustrates the dynamics in Lake Willersinnweiher. A peak in the temperature spectrum at a period of 7.8 h and one at 5.4 h can be resolved, other peaks cannot be distinguished from the noise. Towards higher frequencies, the exponent of the power spectrum of temperature (figure 7, lowest part) is also close to -2, as it was observed by Garrett and Munk in the ocean ([5]). According to them, this can be interpreted in a way that internal waves in Lake Willersinnweiher are superpositions of linear waves. This adds further evidence to the existence of a universal wave spectrum, scaling from the oceans down to small lakes as reported in [10].

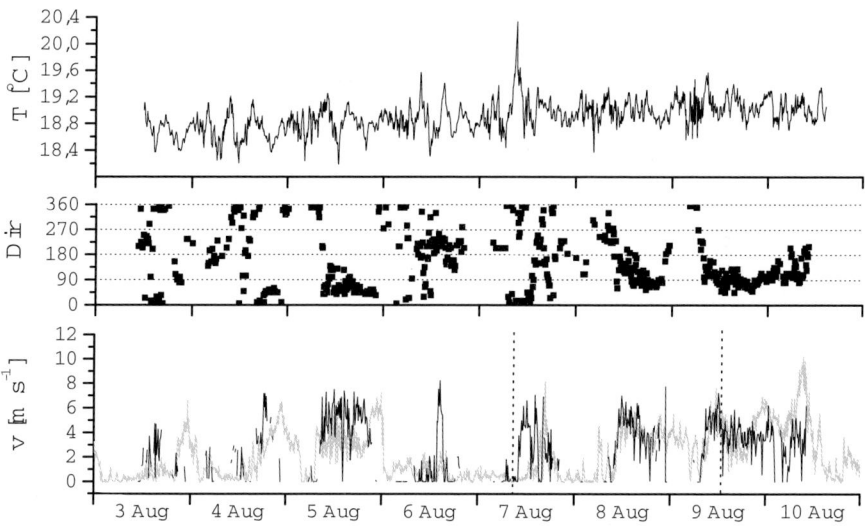

Fig. 8. Temperature time series at 5.8 m at site W3, wind speed and direction from the meteorological station on the lake. The wind data are not continuous. For a crude comparison, data from the station at the University of Heidelberg ∼30 km east of the lake are also plotted (gray). The vertical dotted lines define the time interval shown in figures 6 (lower panel) and 9.

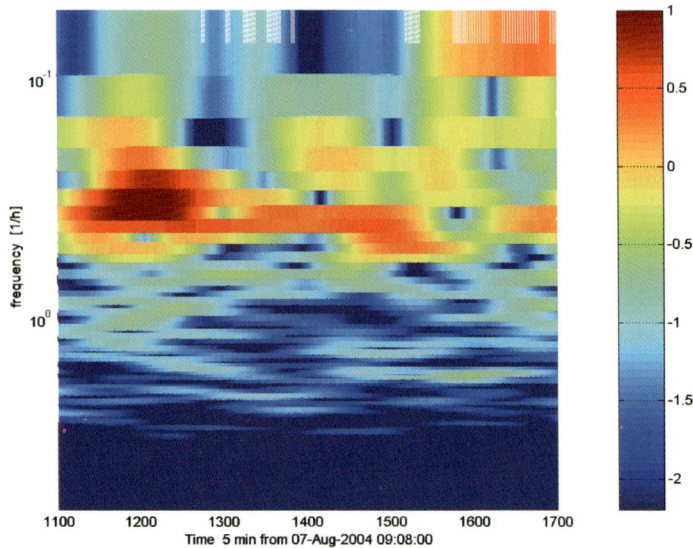

Fig. 9. Short Time Fourier Transform of the north component of the velocity at site SO6, 3.6 m. The color scale denotes the logarithm of the spectral intensity $[(\frac{mm}{s})^2/\frac{1}{h}]$. The time axis unit is 5 minutes and covers 50 hours.

Short Time Fourier Spectra

In order to get a better impression of the temporary frequency distributions we performed a short time Fourier transform (STFT). As example the results of the calculations on the time series at a depth of 3.6 m (figure 6, lower panel) are shown in figure 9. The window length is 256 data points, i.e. 20.3 h and a Gaussian filter with a σ of 5.3 h (64 data points) was used for anti-aliasing.

The spectra show a distinct peak at about $0.3\,h^{-1}$. This peak is highest about 5 h after the onset of the wind (see figure 8, lower panel). After a few hours, when the wind ceased, the peak decreases and is slightly shifted towards higher frequencies ($0.4\,h^{-1}$ or 2.5 h period). A new wind incident started at noon of August 8, 2004 maintained the intensity of the peak at about the reduced level with a broadening towards higher frequencies.

Multilayer System

The observed structure of the currents (figure 6) occurs frequently during summer in the metalimnion of the lake (depth range where strong density gradients exist) and can be interpreted as a multilayer system. The N^2–profile can be used to get the eigenfunction of the vertical displacement assuming a lake with flat bottom and straight walls (figure 10). While this is a crude

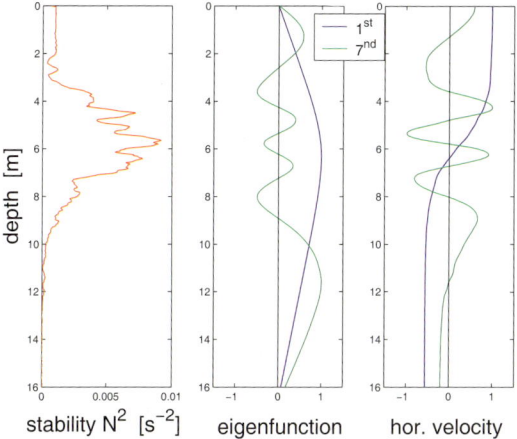

Fig. 10. Stability profile and calculated associated eigenfunctions of horizontal velocity.

approximation of the complicated lake topography, it still gives us valuable qualitative insight into the observed phenomena.

The first mode of vertical displacement has the highest elongation at about the depth of maximum stability, which is for example at 6 m water depth on October 3, 2004. The horizontal velocity is zero at that depth and shows opposite directions in the epi- and hypolimnion (figure 10).

One also can calculate the phase velocity $c_{ph} = (g \cdot H_{equ})^{\frac{1}{2}}$, with g as the acceleration of gravity and H_{equ} as the equivalent height. This gives a value of some $0.3\,\mathrm{ms}^{-1}$ for a summer stratification. Using 500 m as horizontal distance yields about 1 h for the resonant time period. This is not seen in the data, most probably because there is no forcing on this time scale. Moreover, the excitation due to wind forcing is only in the very top of the water column, while the first mode horizontal velocity includes the upper 6 m (figure 10). Calculating higher vertical modes leads to a vertical structure similar to the measurements (figure 6). For instance, a six layer system shows a vertical extension of the layers in the metalimnion of about 1 to 2 m (figure 10). The time scale of a standing wave across the lake is 3.5 h, using 180 m as length scale, which is exactly what we see in the spectra (north velocity component, figure 9).

These rather simple considerations can give only a rough idea and the system has to be modelled numerically to describe the structure properly. We can conclude that Lake Willersinnweiher responds to wind incidents mainly at the very top water layers, while the hypolimnion does hardly react. The reason for these only little dynamics might be an interaction of complex morphology and partial shielding of the wind due to lake surroundings.

3 Numerical Simulation

For the numerical simulation, a time period of two hours is considered. Within this time, steady wind conditions governed the surrounding area and almost constant velocities could be measured on the surface (measuring point S06). These velocities are assumed homogeneous over the whole surface and are used as boundary conditions. The temperature profiles, measured in vertical direction at several measuring points, are alike and justify a homogeneous distribution in horizontal direction serving as initial conditions.

Due to restricted computer resources, a period of 90 seconds real-time is computed.

3.1 Discretizing Lake Willersinnweiher

The geometry and corresponding mesh is generated by the software-tool *Complex Layered Domain Modeller* (CLDM) [3] from data in form of contour polygons. Figure 11 shows the geometry and the initial mesh consisting of 7179 hexaedra.

3.2 The Physical and Mathematical Model

The physical processes are described by the Navier-Stokes equations for an incompressible fluid extended with a thermal energy equation.

Under the assumptions that

1. density variations are neglected except for buoyant forces
2. all other fluid properties (ν, γ, κ) are constant
3. viscous dissipation is assumed negligible,

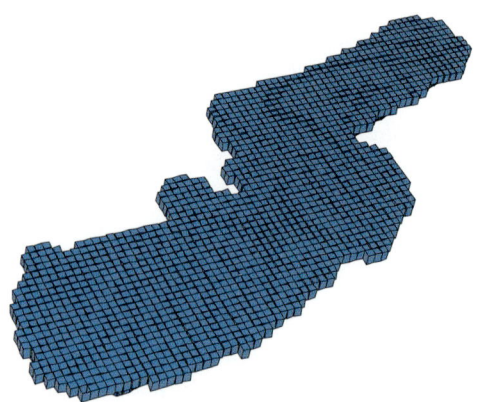

Fig. 11. Mesh of Lake Willersinnweiher

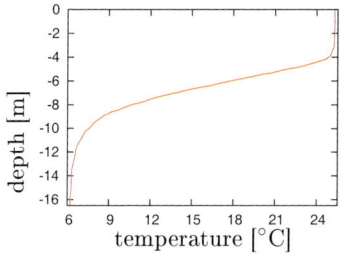

Fig. 12. Initial temperature distribution

the resulting system of coupled nonlinear partial differential equations (*Boussinesq Approximation*) reads:

$$\frac{\partial \mathbf{u}}{\partial t} + \mathbf{u} \cdot \nabla \mathbf{u} + \nabla p - \nu \Delta \mathbf{u} + \gamma \mathbf{g} T = 0 \tag{4a}$$

$$\nabla \cdot \mathbf{u} = 0 \tag{4b}$$

$$\frac{\partial T}{\partial t} + \mathbf{u} \cdot \nabla T - \kappa \Delta T = 0. \tag{4c}$$

The primitive variables $\mathbf{u}, p, T$ stand for velocity, hydrodynamic pressure, and temperature, respectively, t is time. The coefficients ν, γ, and κ denote the kinematic viscosity, volume expansion, and thermal diffusivity of water, respectively.

Boundary and Initial Conditions

On the surface (w), the measured constant velocities serve as boundary conditions, whereas at the waterside and on the ground (b) no-slip boundary conditions for the velocities are assumed:

$$\mathbf{u}_{\Gamma_w} = 0.24 \text{ m/s} \quad \mathbf{u}_{\Gamma_b} = 0 \text{ m/s}.$$

For the temperature, adiabatic conditions are assumed for all boundaries except on the surface, where the measured temperatures are prescribed (**n** denotes the normal):

$$T_{\Gamma_w} = 25.3°\text{C} \quad \mathbf{n} \cdot \nabla T_{\Gamma_b} = 0.$$

The computations start with zero initial conditions for the velocity and with the temperature profile shown in figure 12.

3.3 Spatial Discretization

The discretization in space is carried out with a vertex-centered finite volume scheme with colocated variables using continuous, piecewise trilinear

Table 1. The hierarchy of grids

level i	# of hexaedra	# of grid points	h_i[m]	anisotropy
0	7179	7722	3.00	2
1	14358	15444	1.50	3
2	28716	30888	0.75	7
3	57432	61776	0.38	13
4	114864	123552	0.19	27
5	229728	247104	0.10	53
6	459456	494208	0.05	106

trial functions. The LBB-stabilization is carried out using a local momentum balance equation on element level inducing a streamline-diffusion-type stabilization for the convective terms. This idea stems from [19] and was realized in *UG* [20] by [17, 13]. The discretization is locally mass-conserving and second-order consistent.

With the solution vector

$$\mathbf{x} = [\mathbf{u}, p, T]^T, \quad \mathbf{x} \in \mathbb{R}^n,$$

the continuous system (4) can be written in semi-discrete form

$$M\dot{\mathbf{x}} + kA(\mathbf{x})\mathbf{x} = kb,$$

with $b \in \mathbb{R}^n$, the scaled time step k, and with $A \in \mathbb{R}^{n \times n}$ being a nonlinear function of $\mathbf{x}$. The matrix $M \in \mathbb{R}^{n \times n}$ denotes the singular mass matrix.

The mesh generated by CLDM serves as the coarse grid of a hierarchy of anisotropically refined meshes. An anisotropic refinement strategy is applied, because the problem considered here requires a good resolution in vertical direction, in order to capture the boundary layer at the surface. Table 1 shows the number of hexaedra, the number of grid points n, the grid spacing in vertical direction h and the elements' anisotropy on level 0..6 used for the calculation presented in paragraph 3.6. As the number of unknowns per node is 5, the total number of unknowns on the finest level counts roughly $2.5 \cdot 10^6$.

3.4 Time Discretization

In order to avoid time step restrictions induced by the Courant-Friedrichs-Levy condition, a fully implicit time integration method is used. We apply the A-stable *Fractional-Step-θ* scheme [4], which is second order consistent and which uses three fractional steps ($t_{j-1} \to t_{j-1+\theta} \to t_{j-\theta} \to t_j$, $j = 1,..,N$).

With $\tilde{\mathbf{x}}^j$ denoting the approximate values of $\mathbf{x}$ at time step t_j, this scheme can be formally written as

$$F_\theta^{j,j-1}(\mathbf{x}^{j-1}, \mathbf{x}^{j-1+\theta}, \mathbf{x}^{j-\theta}, \mathbf{x}^j) = 0, \quad \theta \in \mathbb{R} \quad (5)$$

For the choice of θ and further details see for example [16].

3.5 Solving Strategy

For each time step the system of nonlinear algebraic equations (5) has to be solved three times. The nonlinear solution is found using the Newton method. The nonlinear defect of the iterate $\tilde{\mathbf{x}}$ is defined by

$$d_\theta^j(\tilde{\mathbf{x}}) := F_\theta^{j,j-1}(\tilde{\mathbf{x}}^{j-1}, \tilde{\mathbf{x}}^{j-1+\theta}, \tilde{\mathbf{x}}^{j-\theta}, \tilde{\mathbf{x}}^j).$$

The nonlinear iteration to solve (5) is considered to have converged if

$$||d_\theta^j(\tilde{\mathbf{x}})||_2 \leq 10^{-5}||d_{\theta_1}^{j-1}(\tilde{\mathbf{x}})||_2.$$

Each nonlinear iteration requires the solution of the linear equation derived by linearizing (5). The linear system is obtained by using a Quasi-Newton linearization

$$\mathbf{u}^n \cdot \nabla \mathbf{u}^n \approx \mathbf{u}^{n-1} \cdot \nabla \mathbf{u}^n \quad \text{and} \quad \mathbf{u}^n \cdot \nabla T^n \approx \mathbf{u}^{n-1} \cdot \nabla T^n \tag{6}$$

obtaining linear convergence [2].

On an average, about 5 Newton iterations are required to solve each nonlinear problem. The linear systems are solved using the geometric multigrid method with a V(2,2)-cycle applying the ILU as a robust smoother. To realize a robust behavior of the ILU-smoother, a special ordering of the unknowns is required. It is accelerated using a Bi-CGStab Krylov-space-method. For details, see [1, 9, 8, 21]

The computations using 128 nodes were performed on the HEidelberg LInux Cluster System (HELICS). On each node 2 Dual AMD Athlon 1,4 Ghz processors are installed with 1.5 GB RAM.

With these resources, the simulation of one time step solving three nonlinear problems took about 10 minutes on the finest grid level. The total execution time of the computation added up to 15 days, which corresponds to 2160 time steps.

3.6 Results

The computed results are evaluated at the location of measuring point S06. Figure 13 shows the velocity (left) and corresponding temperature (right) profiles on grid level 6 after $0, 30, 60$ and 90 seconds.

As expected from the convective character of the flow, a thin boundary layer at the surface with high velocities takes shape, whereas the flow in the rest of the lake is characterized by small velocities.

The changes in temperature are small and appear in a slight shift of the temperature profile in the direction of the driving velocity (figure 14). A balance between the convection of the temperature and the stable stratification prevents the temperature profile from being transported further.

After about 60 seconds a quasi-steady state is reached, where no noticeable changes in the global velocity and temperature field can be observed anymore.

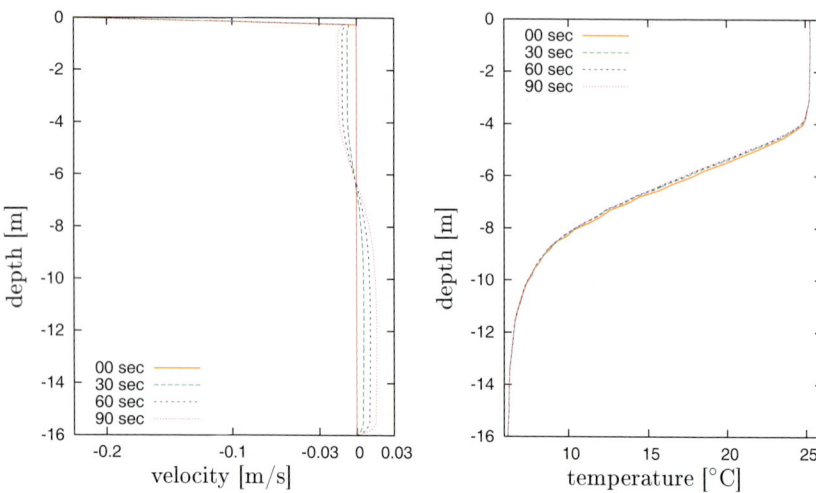

Fig. 13. Velocity and temperature profiles after 0, 30, 60 and 90 seconds

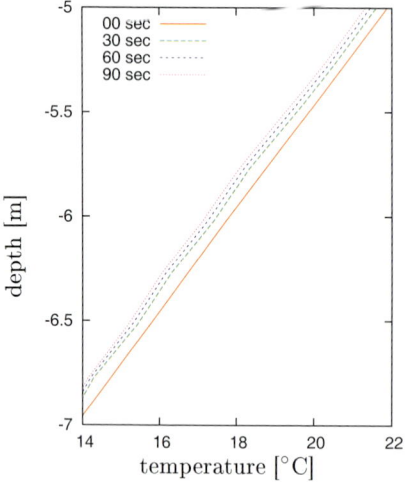

Fig. 14. Zoomed in temperature profiles after 0, 30, 60 and 90 seconds

It turns out that only above a certain grid resolution the physical processes can be reproduced properly. In particular, all computations below level 6 yield non-physical results. That is, after 30 seconds the solution on grid level 3, 4, 5 and 6 (figure 15) seem to converge to the same solution. However, after about 60 seconds this proves to be a false conclusion, especially with respect to the velocity field (figure 16). The velocity profiles on the coarser grids (level 3 to 5) show a strong global circulation, which after 90 seconds (figure 17 left) even starts to erode the stable stratification (figure 17 right). In contrast, the

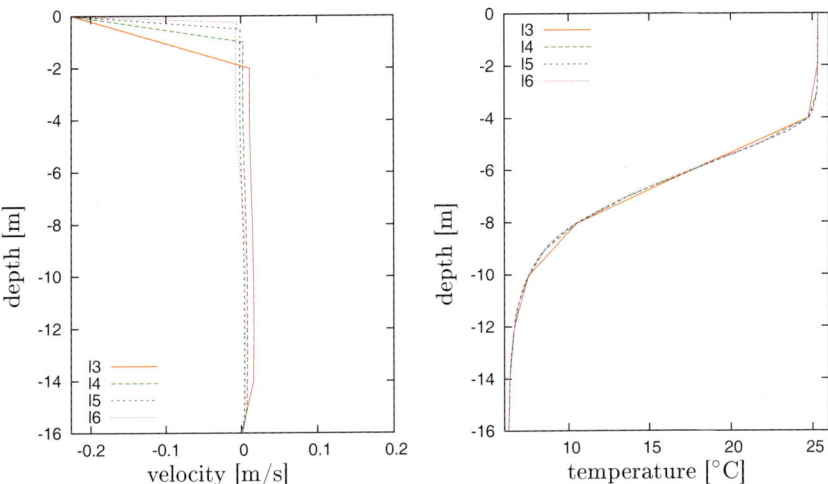

Fig. 15. Velocity and temperature profiles on level 3, 4, 5 and 6 after 30 seconds

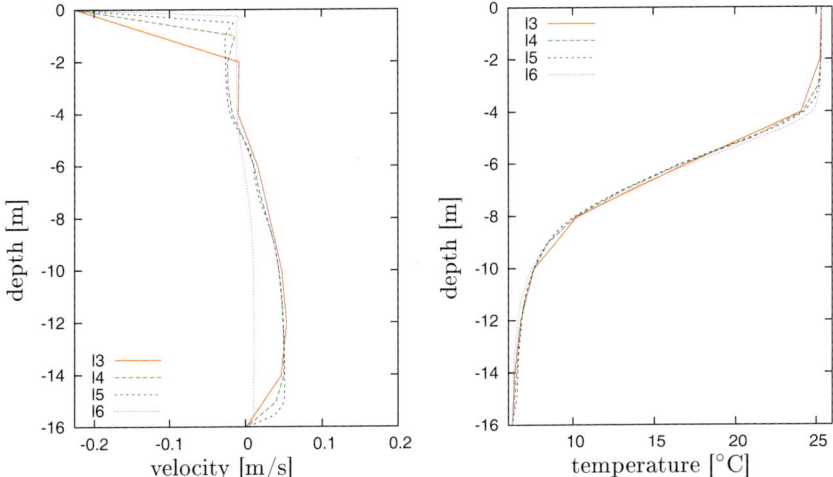

Fig. 16. Velocity and temperature profiles on level 3, 4, 5 and 6 after 60 seconds

velocity profile on level 6 shows no such global circulation and complies with measured data.

This behavior, that qualitatively different solutions are gained on different grids, underlines the necessity to compute solutions on grids with different resolution. Hence, in order to gain solutions with a good resolution in the context of complex problems as we face here, parallel codes using massive computer resources and fast methods like the applied multigrid are indispensable.

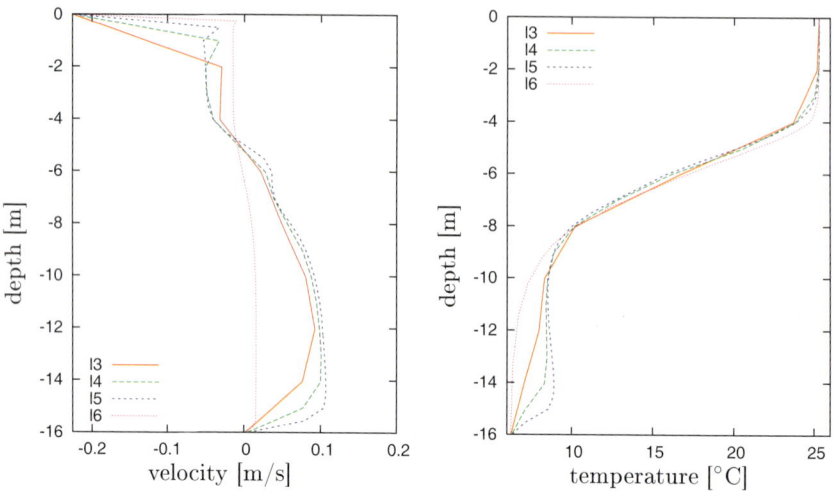

Fig. 17. Velocity and temperature profiles on level 3, 4, 5 and 6 after 90 seconds

4 Summary and Conclusions

We have demonstrated the rich set of hydrodynamic phenomena that operate in a natural lake. State-of-the-art experimental methods yield a wealth of data that are highly resolved in time and in one space axis. The orthogonal space is sampled only very sparsely, however. Still, such data allow the estimation of crucial effective parameters like vertical diffusion coefficients. Numerical simulations, on the other hand, yield high resolution results for the entire space. They are thus inherently more powerful than experimental observations and heuristic analysis whenever a complicated topography is crucial. However, even with the current most advanced solvers, such high-resolution studies of most phenomena on realistic time scales is beyond reach. Confining experimental and numerical approaches hence appears as the optimal way to reach a deeper understanding of the lake hydrodynamics.

References

1. Barrett, R, Berry, M., Chan, T., Donato, J., Dongarra, J: Templates for the solution of linear systems: building blocks for iterative methods. SIAM (1994)
2. Dennis, J., More, J.: Quasi-newton methods, motivation and theory, SIAM Review, **19**, 46–89 (1977)
3. Feuchter, D., Stemmermann, U., Wittum, G.: Description and generation of geometries and grids for layered domains. 17th GAMM Seminar Leipzig on Construction of Grid Generation Algorithms, 2001.
4. Glowinski, R., Periaux, J.: Numerical methods for nonlinear problems in fluid dynamics. Supercomputing. State-of-the-Art, 81–479 (1987)

5. Garrett, C., Munk, W.: Internal waves in the ocean. Annu. Rev. Fluid. Mech., **11**, 339–369 (1979)
6. Goudsmit, G.H., Peeters, F., Gloor, M., Wüest, A.: Boundary versus internal diapycnal mixing in stratified natural waters. J. Geoph. Res., **102**(13), 27903–27914 (1997)
7. Gloor, M., Wüest, A., Münnich, M.: Benthic boundary mixing and resuspension induced by internal seiches. Hydrobiologia, **284**, 59–68 (2000)
8. Hackbusch, W.: Multigrid Methods and Applications. Springer (1992)
9. Hackbusch, W.: Iterative Lösung großer schwachbesetzter Gleichungssysteme. Teubner (1992)
10. Hondzo, M., Haider, Z.: Boundary Mixing in a small stratified lake. Water Resour. Res., **40**, W03101, pp. 12, doi:10.1029/2002WR001851 (2004)
11. Jassby, A., Powell, T.: Vertical patterns of eddy diffusion during stratification in Castle lake, California. Limn. Ocean., **200**(4), 530–543 (1975)
12. Lu, Y., Lueck, R.B.: Using a broadband ADCP in a tidal channel. J. Atmos. Ocean. Tech., **16**, 1556–1567 (1999)
13. Nägele, S.: Mehrgitterverfahren für die inkompressiblen Navier-Stokes Gleichungen im laminaren und turbulenten Regime unter Berücksichtigung verschiedener Stabilisierungsmethoden, PhD thesis, IWR, University of Heidelberg (2004)
14. Ollinger, D.: Modellierung von Temperatur, Turbulenz und Algenwachstum mit einem gekoppelten physikalisch-biologischen Modell. PhD Thesis, University of Heidelberg (1999)
15. Quay, P.D., Broecker, W.S., Hesslein, R.H., Schindler, D.W.: Vertical diffusion rates determined by tritium tracer experiments in the thermocline and hypolimnion of two lakes. Limn. Ocean., **25**(2), 201–218 (1980)
16. Rannacher, R.: Numerische Methoden für Probleme der Ströomungsmechanik. Script, IWR-Numerical Methods, University of Heidelberg (2001)
17. Rentz-Reichert, H.: Robuste Mehrgitterverfahren zur Lösung der inkompressiblen Navier-Stokes Gleichung: ein Vergleich, PhD thesis, IWR, University of Heidelberg (1996)
18. Sandler, B.: Die Wirkung von Sanierungs- und Restaurierungsmaßnahmen auf die Nährstoffströme und die biotische Dynamik eines anthropogenen Gewässers, am Beispiel des Willersinnweihers/Ludwigshafen. PhD Thesis, ibidem-Verlag, Stuttgart (2004)
19. Schneider, G. Raw, M: Control volume finite-element method for heat transfer and fluid flow using colocated variables, Numerical Heat Transfer, **11**, 363–390 (1987)
20. Bastian, P., Birken, K., Johannsen, K., Lang, S., Eckstein, K., Neuss, N., Rentz-Reichert, H. Wieners, C., UG – a flexible software toolbox for solving partial differential equations. Computing and Visualization in Science, **1**, 27–40 (1997)
21. Van der Vorst, H.: Bi-cgstab: a fast and smoothly converging variant of Bi-CG for the solution of nonsymmetric linear systems. SIAM J Sci Stat Comp, **13**, 631–644 (1992)
22. Wüest, A., Lorke A.: Small-scale hydrodynamics in lakes. Annu. Rev. Fluid Mech., **35**, 373–412 (2003)
23. Wollschläger, U.: Kopplung zwischen Oberflächengewässer und Grundwasser: Modellierung und Analyse von Umwelttracern. PhD Thesis, University of Heidelberg (2003)

Part VIII

Computer Visualization

Preamble

This chapter contains three articles which describe software tools for the graphical visualization of the results of numerical simulations and measured data.

The first article *"Advanced flow visualization with HiVision"* describes the visualization software package HiVision which was primarily developed for the exploration of data in the area of Computational Fluid Dynamics (CFD). Its object-oriented design integrates several external toolkits such as VTK, Qt, and MPI.

The second article *"VisuSimple: An interactive visualization utility for scientific computing"* introduces an easy-to-use graphics software package which is based on the open-source visualization toolkit VTK. Its specialties are interactive visualization and graphics/movie-generation features.

The third article *"Volume rendering in scientific applications"* presents a software tool for visualizing 3D data sets by efficient volume rendering techniques. These methods allow for treating voxel-based data, for instance, by assigning appropriate opacities and surface properties.

Advanced Flow Visualization with HiVision*

S. Bönisch[1] and V. Heuveline[2]

[1] Institute of Applied Mathematics, University of Heidelberg
[2] Computing Center and Institute for Applied Mathematics, University of Karlsruhe

Summary. The features and the design concept of the visualization software **HiVision** are presented. The platform **HiVision** which was primarily developed for the exploration of data in the area of Computational Fluid Dynamics (CFD), is being used nowadays in numerous applications related to numerical simulation. The main strength of the **HiVision** package lies in its object-oriented design and in the integration of several powerful external toolkits (VTK,Qt,MPI). The features of **HiVision** are demonstrated by means of selected examples.

1 Introduction

Visualizing the results of flow computations is a challenging and demanding task. This is especially the case for time-dependent computations, where a huge amount of data has to be processed. In the past, a large number of techniques for analyzing and visualizing flow fields has been developed. Some of the most commonly used approaches are (see e.g. [8] or the survey article [7]):

- *Direct flow visualization,* like e.g. arrows ("glyph"), color coding or volume rendering;
- *2D visualization of slices,* like e.g. cutplanes;
- *Geometric and texture-based flow visualization,* like e.g. spot noise, texture advection, contouring (isolines,isosurfaces), streamlines, stream surfaces, particle tracing;
- *Feature extraction,* like e.g. extraction of vortices, shock waves or separation lines.

Many of these different techniques are included in the visualization software **HiVision** ([2]) developed by the authors during the last years. Although

---

*This work has been supported by the German Research Foundation (DFG) through SFB 359 (Project E) at the University of Heidelberg.

primarily developed for visualization in the area of Computational Fluid Dynamics (CFD), the platform **HiVision** is discipline independent and may be advantageously used in various areas such as structural mechanics and reactive flow simulations. The development of HiVision relies on the following model and design principles:

- **HiVision** is completely written in C++ and makes extensive use of the object-oriented concepts of this language (cf. Section 2). The use of C++ as a programming language makes it very easy to interface with other powerful packages written in C/C++.
- **HiVision** is based on the powerful Visualization Toolkit (VTK, [8, 6]) which provides many basic and advanced algorithms for the exploration of scalar, vector and tensor data (cf. Section 3).
- **HiVision** has a modern, flexible and ergonomic graphical user interface which is based on the Trolltech Qt library ([1]). Qt is a complete C++ application development framework which includes a class library and tools for cross-platform development and internationalization.
- **HiVision** is freely available under a GNU Open Source license and can be downloaded from http://www.hiflow.de/HiVision. As a consequence, interested users can not only get the software for free, but they are even allowed to modify and redistribute the code under some mild license restrictions.
- The visualization of large, instationary 3D data on a single processor machine is often intractable. The bottleneck using sequential platforms is not only due to the restriction on the IO-bandwidth but is also related to the high demand with regard to CPU-time and memory requirements. We account for this fact by providing a parallel version of **HiVision** which is able to exploit the resources of High Performance Computing environments. The parallel module of **HiVision** relies on the Message Passing Interface (MPI). An example of a parallel visualization algorithm implemented in **HiVision** is given in Section 3.2.

2 Software Design

As mentioned before, **HiVision** is written in C++ and takes great advantage of object oriented concepts, polymorphism as well as generic programming.

In **HiVision** the visualization network is encapsulated in a class called HvsPipeline, which is further subdivided into "HiVision Actor" classes, see Figure 1. The actors themselves are embedded in an inheritance hierarchy (cf. Figure 2): The general interface of all actors is encapsulated in a purely virtual class called HvsAbstractActor. This abstract interface is inherited by two other virtual classes (HvsScalarActor and HvsVectorActor) which contain the basic information needed in order to visualize scalar (resp. vector) data. All other actors contain the concrete implementations of the respective features

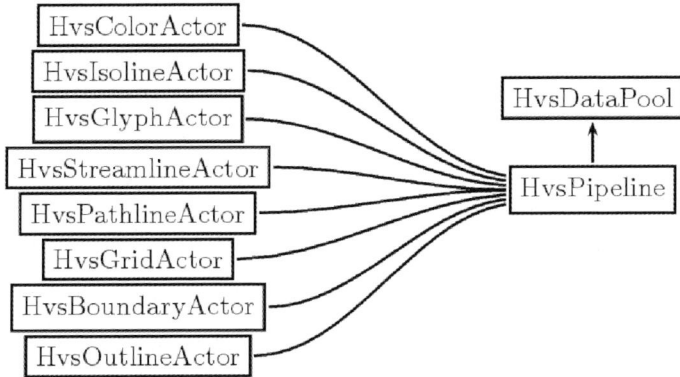

Fig. 1. The main classes forming the **HiVision** visualization network ("pipeline").

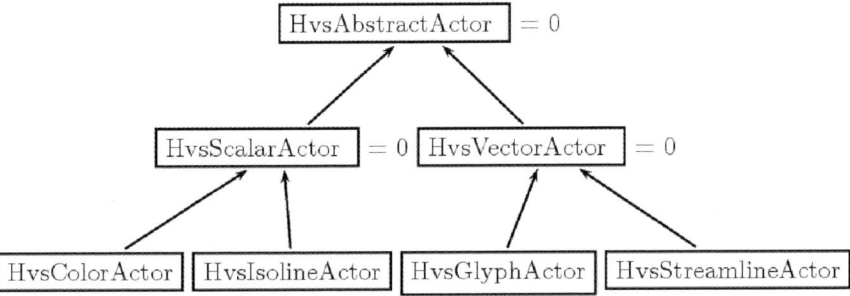

Fig. 2. Inheritance hierarchy of the **HiVision** "Actor" classes. Following the C++ syntax, the tag "=0" means that the corresponding class is purely virtual.

(e.g. HvsColorActor, HvsIsolineActor etc.) and are derived from one of the interface classes HvsScalarActor or HvsVectorActor.

The most important issue when designing an interactive visualization platform is the strict conceptual separation of the Graphical User Interface (GUI) and the underlying visualization pipeline. This paradigm is crucial in order to achieve transparency and extensibility of the software. The general idea of how this is achieved in **HiVision** is plotted in Figure 3, and an concrete example (visualization of isolines) is shown and explained in Figure 4.

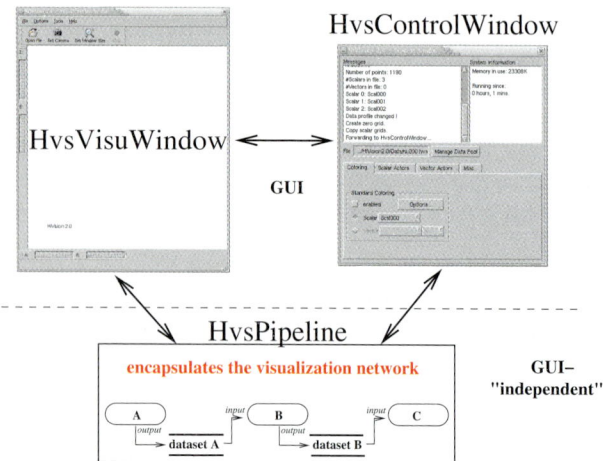

Fig. 3. Main classes of **HiVision**. The conceptual separation of the user interface (classes HvsVisuWindow and HvsControlWindow) and the visualization pipeline (class HvsPipeline) is an important design issue in **HiVision**.

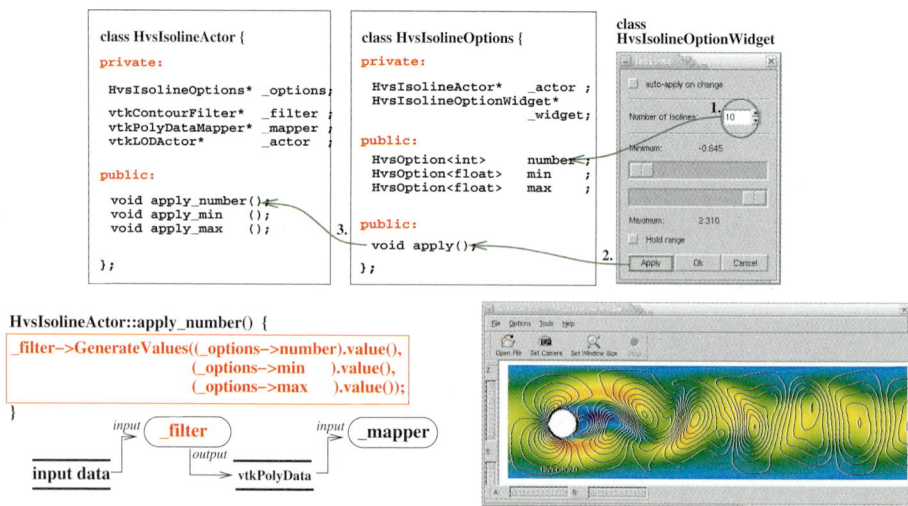

Fig. 4. Isoline creation as an example: Whenever the user changes a parameter in the user interface (class HvsIsolineOptionWidget), the parameter are *not* directly changed in the visualization pipeline, but rather stored in an intermediate object (class HvsIsolineOptions). After pressing "Apply", the changed parameters are passed to the respective part of the visualization pipeline (class HvsIsolineActor), which is itself *completely independent* of the GUI.

3 Features

In Section 3.1 we show examples of flow visualizations generated with **HiVision**. Section 3.2 is devoted to an advanced feature of **HiVision**, namely visualization of instationary 3D flow computations by means of parallel particle tracing.

3.1 Basic Features

HiVision offers extensive possibilities to explore interactively the results of flow computations. The following main features are available:

- 2D/3D unstructured grids;
- visualization of scalar data: texture mapping, colorbar, isolines, isosurfaces, cutplanes;
- visualization of vector data: glyph, hedgehogs, streamlines, pathlines, particle tracing;
- interactive creation of new data: vector composition, operators $\nabla\cdot$ and $\nabla\times$;
- animation: MPEG generation of instationary data;
- stereo rendering.

Some of these features are shown in Figures 5-6.

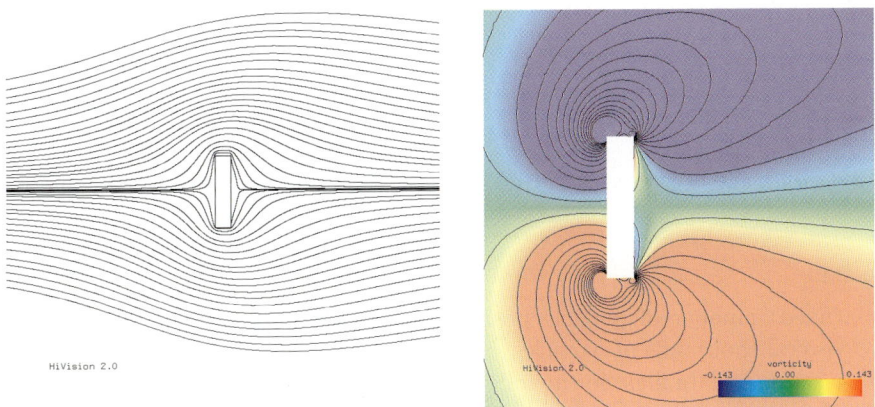

Fig. 5. Simulation of the flow around a rectangular obstacle (from [4]): Streamlines (left) and vorticity isolines (right).

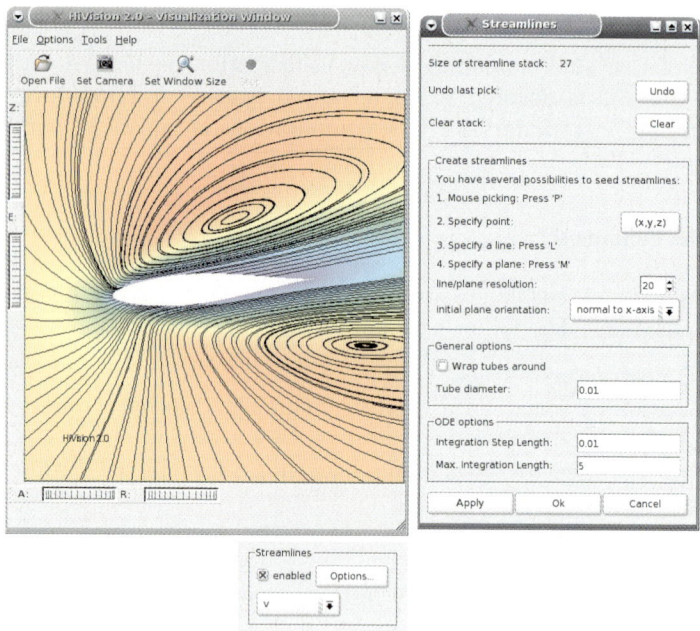

Fig. 6. Streamline visualization of the flow around a NACA airfoil (cf. [3]) with **HiVision**.

3.2 Parallel Particle Tracing

In this section we present the parallel approach implemented in **HiVision** for the visualization of instationary 3D flow computations by means of particle tracing.

First, we state precisely the problem of particle tracing from an algorithmic viewpoint.

<u>Given:</u>

- (fixed) number of particle positions, typically ranging from hundred to several thousands of particles
- a set of N data files (in the following identified with the numbers 0,1,...,N-1) containing the computed flow field v at N subsequent time steps t_n, $n = 0, ..., N-1$. Typically, such a data file contains at least the nodal values of v with respect to an underlying grid (e.g. FDM/FEM) and can be very large for 3D computations.

<u>Wanted:</u>

- A movie (e.g. MPEG) showing the temporal evolution of the particles.

Before we state the algorithm we formulate some **assumptions** on the underlying parallel computer architecture:

- We assume a computer cluster of P identical nodes.
- All nodes share a common file system and different nodes can access different files simultaneously (parallel input/output).
- On all nodes all needed graphic libraries are installed (e.g. VTK).

In our situation, all these preconditions are fulfilled, since we use the LINUX cluster HELICS ([5]). In the following we further assume $N = kP$ with $k \in \mathbb{N}$ in order to ease notations. The extension to the case $N \neq kP$ is straightforward.

Our proposed algorithm is depicted in Figure 7. The first step consists in a cyclic partitioning of the work: Process Π_k obtains the data files $k, P+k, 2P+k, \ldots$. Then each process performs a loop over "its" data files. For each data file/time step the following steps are performed:

- LOAD_FILE(work[i]): Load the data file containing the information of the velocity field v at time t_i. This step also includes the conversion of the raw data into an internal representation (see e.g. [8]).
- CREATE_START_PARTICLES: at the beginning ($i = 0$) the root process Π_0 is responsible for the definition of the initial positions of the particles to be traced. For example, this can be done interactively via the graphical user interface of **HiVision**.
- RECV_PARTICLES: In all other cases the process receives the actual positions of the particles from its (cyclic) predecessor.
- INTEGRATE_PARTICLES: Given the positions X_{n-1} of the particles at time t_{n-1}, solve for all $\xi \in X_{n-1}$ the IVP

$$\begin{cases} \dot{x}(t) = v(x(t)), & t \in \{t_{n-1}, t_n\}, \\ x(t_{n-1}) = \xi, \end{cases}$$

by means of an (explicit) standard ODE solver (e.g. Euler).
- SEND_PARTICLES: Send the new particle positions to the (cyclic) successor.
- RENDER_PARTICLES: Turn the particles at their new positions into graphic primitives and plot them, possibly together with other interesting features. Note that, unlike in the sequential case, the nodes do not have their own physical screens where they could plot the resulting picture. Instead they plot on a *virtual* screen exploiting the so-called "Offscreen Rendering" facilities of VTK.
- SAVE_PICTURE: Save the current picture in a format that is convenient for movie generation.

The final step (GENERATE_MPEG) consists in concatenating the single pictures to obtain a movie. Clearly, this step could be easily parallelized, too. But since the costs of the MPEG generation are negligible as compared with the rest of the work, we simply keep the movie generation in its sequential version.

In Figure 7 a schematic performance diagram for the main loop of this parallel algorithm is depicted. Note that the new particle positions are sent to the successor immediately after having integrated the ODE.

PARTICLE-TRACING-PARALLEL(N, P)

    // cyclic work partitioning
    int work$[N/P]$;

    process $\Pi[\text{int } p \in \{0, 1, \ldots, P-1\}]$
        for $i \leftarrow 0 \ldots N/P - 1$ **do**
            LOAD_FILE(work$[i]$);

            if $p = 0$ && $i = 0$ **then**
                CREATE_START_PARTICLES;
            else
                RECV_PARTICLES$((p-1)\%P)$;
            endif

            INTEGRATE_PARTICLES;
            SEND_PARTICLES$((p+1)\%P)$;
            RENDER_PARTICLES;
            SAVE_PICTURE;
        endfor

    end

    GENERATE_MPEG;

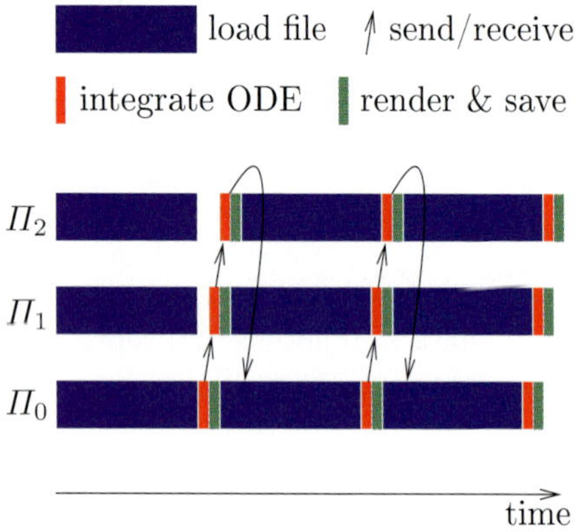

Fig. 7. Parallel algorithm for particle tracing. The problem parameters are the number of time steps N, and the number of processors P. Pseudo-code (left) and schematic performance diagram of the main loop for $N = 9, P = 3$ (right).

We applied our parallel particle tracing algorithm to the situation of a three- dimensional channel flow around a cuboid, and traced 1000 particles. Figure 8 shows some snapshots of the instationary fluid motion, visualized by means of particle tracing.

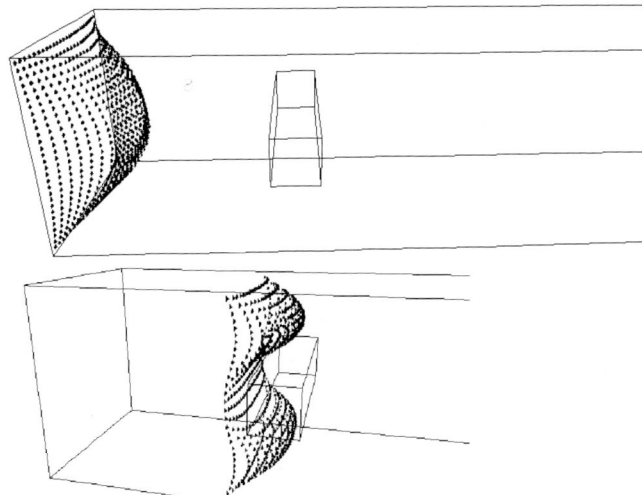

Fig. 8. Flow around a cuboid. 1000 massless particles are released from the left side of the channel and then traced through the fluid.

The performance of the algorithm for this configuration is depicted in Table 1, where T_{seq} (resp. $T_{par}(N,P)$) denotes the sequential (resp. parallel) runtime, and

$$S(N,P) := \frac{T_{seq}(N)}{T_{par}(N,P)},$$

$$E(N,P) := \frac{S(N,P)}{P},$$

are the "speedup" and the "efficiency".

Table 1. This table shows the performance of our parallel algorithm. The number of processors is P=7 (top) resp. P=15 (bottom).

P = 7	$T_{seq}(N)$	$T_{par}(N,P)$	$S(N,P)$	$E(N,P)$
N = 35	805 sec.	124 sec.	6.5	0.93
N = 70	1610 sec.	236 sec.	6.8	0.97
N = 100	2300 sec.	338 sec.	6.8	0.97

P = 15	$T_{seq}(N)$	$T_{par}(N,P)$	$S(N,P)$	$E(N,P)$
N = 35	805 sec.	72.7 sec.	11.0	0.74
N = 70	1610 sec.	120 sec.	13.4	0.89
N = 100	2300 sec.	166.8 sec.	13.8	0.92

References

1. J. Blanchette and M. Summerfield. *C++ GUI Programming with Qt 3*. Prentice Hall, 2004.
2. S. Bönisch and V. Heuveline. Hivision, an advanced framework for flow visualization. Preprint 2003-20, University of Heidelberg, 2003.
3. S. Bönisch, V. Heuveline, and P. Wittwer. Second order adaptive boundary conditions for exterior flow problems: non-symmetric stationary flows in two dimensions. Technical Report 2004-10, University Heidelberg, 2004.
4. S. Bönisch, V. Heuveline, and P. Wittwer. Adaptive boundary conditions for exterior flow problems. *J. Math. Fluid Mech.*, (7):85–107, 2005.
5. http://helics.iwr.uni heidelberg.de/. HEidelberg LInux Cluster System.
6. Inc. Kitware. *The VTK User's Guide*.
7. R. Laramee, H. Hauser, H. Doleisch, B. Vrolijk, F. Post, and D. Weiskopf. The State of the Art in Flow Visualization. *Computer Graphics Forum*, 23(2), 2004.
8. W. Schroeder, K. Martin, and B. Lorensen. *The Visualization Toolkit - an object-oriented approach to 3D graphics*. Kitware, Inc.

VisuSimple: An Interactive Visualization Utility for Scientific Computing*

Th. Dunne[1] and R. Becker[2]

[1] Institut für Angewandte Mathematik, Universität Heidelberg
[2] Laboratoire de Mathématiques Appliquées, Université de Pau et des Pays de l'Adour, France

Summary. In this work we describe an easy to use and easy to modify open-source program that was developed for the needs of several projects in the SFB 359. The utility is based on the powerful open-source visualization toolkit VTK, the *de facto* industry standard for visualization. It is an interactive visualization and graphics/movie-generation utility for multicomponent 2D- and 3D-data in the VTK-format. The data is the result of numerical computations such as e.g. computational fluid dynamics or computational structure mechanics. The main features of VisuSimple are shown based on examples.

1 Introduction

Motivation. Most software used in research is written for two simple reasons: to enable and facilitate research. Applications range from small systems of simulating and modeling fluids, structures, fluid-structure interaction, chemistry, reactive flows to large systems encountered in astrophysics, [1, 2, 3, 4]. All calculations create large amounts of visual data that needs to be visualized. Due to the wide range of topics in our group there is a long list of demands that the software has to satisfy. It should be capable or processing 2D- and 3D-data, generating output in several formats as an image (jpeg, ps, pdf) or a movie (mpeg, avi), capable of processing locally refined meshes with hanging nodes, capable of processing meshes based on different geometric types (triangles, quads, hexaeders, tetraeders), capable of extracting features (isolines, isosurfaces, cutplanes). In the course of time we have come to use multiple other tools for visualization. And we encountered the typical common problems with "outside" software. In some way or the other it does not do what one wants and it is not possible to modify it in the needed way; or it is possible but the needed investment is too large. The costs of either the software licenses or the required modifications were simply too high.

---

*This work has been supported by the German Research Foundation (DFG) through SFB 359 (Project E) at the University of Heidelberg.

Creation. With the introduction of the open-source library VTK ([5, 6]) for processing visual data the possibility of developing our own visualization software became feasible. VisuSimple was initially written 1999 by Roland Becker[3]. It is presently being maintained and further developed by Thomas Dunne[4] and Dominik Meidner[5]. It is written in the scripting language Tcl ([7, 8]), a programming language that can be learned quite quickly. Like VTK the authors chose to distribute the source code based on an open-source model. Based on this source code and knowledge of the VTK library it has become almost trivial to add new visualization methods or fix and extend present features.

Capabilities. VisuSimple is being used to visualize multicomponent 2D- or 3D- data by multiple projects in the SFB 359 [1, 2, 3, 4]. It aims to have a wide range of use and to be independent of discipline. It offers functions that are based on the standard methods of visualizing scalar and vector fields and subsets thereof such as cutplanes and isolines. Additionally it is capable of visualizing suites of data based on a series of results from stationary or instationary numerical computations. Presently one can (among other things) use VisuSimple to:

- read structured or unstructured, optionally compressed data files in VTK-format and the lesser known INP-format;
- visualize a 2D- or 3D- grid with optional local refinement based on hanging nodes; the grid may contain any mix of basic graphics elements (triangles, quads, hexaeders, tetraeders);
- process data files that contain an arbitrary amount of component data in scalar or vector form based on the nodes or cells of the mesh;
- visualize carpets of 2D scalar data, shading;
- visualize vector fields (as glyphs - multiple types pickable e.g. arrows, pins, lines... - normable, scalable, selectable visualization ranges);
- with all visualizations provide a selectable colormap;
- multiple camera positions can be changed, saved and loaded;
- interactively move and zoom around the visualization;
- extract values and grid-data off of the grid or mesh with the mouse;
- save the visualizations to common graphics formats: jpg, ps, pdf and tiff;
- make movie animations.

Code. Since visual data processing is not trivial, learning to use the VTK library (in Tcl or any other language) is not as easy as initially learning to use Tcl. Still, the VTK library facilitates this task considerably. The VTK library has practically become an industry standard as a visual data processing toolkit, so VisuSimple can also be seen as an excellent learning opportunity to using the VTK library.

---

[3] roland.becker@univ-pau.fr
[4] thomas.dunne@iwr.uni-heidelberg.de
[5] dominik.meidner@iwr.uni-heidelberg.de

We provide the code of VisuSimple free to the public under an MIT-like-license - anyone is free to download it and apply their own modifications and even redistribute these under basic restrictions.

Since VisuSimple is programmed using the scripting language Tcl and the excellent visual data processing toolkit VTK, modifications can be literally tested and integrated interactively into the source[6]. The language Tcl is easily learned and so modifications can even be made by Tcl-quickstarters.

VisuSimple is written in Tcl/Tk, using VTK's Tcl/Tk interface. Its version is at the moment 3.44 (as of 10.10.2005).

In the following sections we will briefly show the core functions of VisuSimple based on its individual actors. We refrain from explaining the basic inner workings of the code since essentially VisuSimple is based on the principles of VTK and VTK's actor based approach of processing data. Actors are responsible for the different aspects of visualizing data: displaying a mesh, displaying isolines, displaying hedgehogs, etc.. More details of the use of VisuSimple can be found in the user's guide.

2 The Actors

After having loaded a file and selected which scalar and/or vector datasets should be processed, one can specify in which ways they are exactly to be processed. All functions of visualization - be it the simple mesh, the carpet on the mesh, isolines, isosurfaces, vector-arrows, etc - are rendered into the view

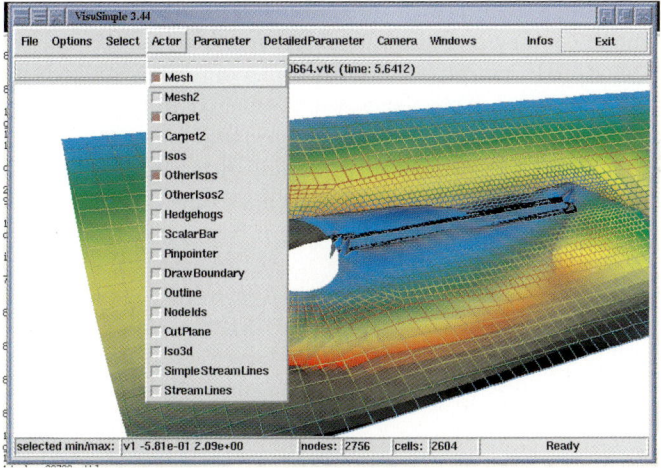

Fig.1. "Actors" menu opened.

---

[6]This is possible using the Interactor dialog box

area by **actors**. Actors (Fig. 1) can individually be turned on and off in the "Actors" menu. Presently VisuSimple offers the following functions as actors:

- **Mesh**: shows a wireframe of the mesh.
- **Mesh2**: same as Mesh, for mirroring purposes.
- **Carpet**: maps a texture onto the mesh, based on the selected scalar dataset and colormap.
- **Carpet2**: same as the Carpet actor, for mirroring purposes.
- **Isos**: displays isolines of the selected scalar field on the flat mesh or as an adjustable height field.
- **OtherIsos**: displays isolines of an *arbitrarily selected* scalar field on the flat mesh or as an adjustable height field.
- **OtherIsos2**: same as above for any other scalar field.
- **Hedgehogs**: displays the selected vector field as glpyhs e.g. in the form of arrows, pins or lines.
- **Scalarbar**: shows a legend of the selected dataset based on the colormap.
- **Pinpointer**: with this one can continuously pick and display datavalues out of the visualized grid with the mouse.
- **DrawBoundary**: puts a black line on the boundary of the mesh.
- **Outline**: draws a bounding box around the whole geometry.
- **NodeIds**: displays the indices of all nodes in the mesh.
- **CutPlane**: shows the intersection of a plane with the mesh.

All actors can be configured in two manners:

- The "**Parameter**" menu: in this menu (Fig. 2) many common and usually quick-to-do options are provided, such as changing the color or scaling of an actor. Each actor has its own submenu in the "Parameter" menu and in this menu each actor option itself has a submenu.
- The "**DetailedParameter**" menu: Each actor has its own item in the "DetailedParameter" menu which when selected will open a actor-specific dialog box. Each actor's dialog box will provide a means of changing all of the important configuration options. Some of the options in the "Parameter" menu will not be provided, such as mirroring or changing the actor's color.

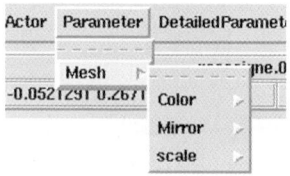

Fig. 2. "Parameter" menu.

In the following sections we will occasionally describe an actor's configuration possiblities, either just to give an initial description or due to the sheer volume of configuration options, but otherwise in general will we refrain from going into too great detail.

3 The Mesh Actor

The Mesh actor is responsible for displaying the wireframe grid that all datasets are based on. If the dataset contains a volume then only the bounding wireframe of the dataset will be shown.

The Mesh actor has options available in both the "Parameter" menu (Fig. 3) and the "DetailedParameter" menu. The options provided in the "Parameter" menu are setting the color, mirroring the actor and scaling a certain value associated with the actor.

For this actor the scale option specifically changes the opacity of the mesh. The dialog box (Fig. 4) that can be opened by selecting the "Mesh" item in the "DetailedParameter" menu. It offers many frequently occurring options: turning an actor on/off, setting the opacity, shifting the actor around in the viewing area and specifying that all changes are to be applied immediately.

If no other actors are present these two options will not be very useful. The opacity option is good if multiple actors are active e.g. the Mesh and the Carpet. With the help of this option one can *dim down* the intensity of the Mesh so that it will not distract too much while viewing the Carpet actor.

In Fig. 5 we show an example of a data file where the opacity of the Mesh actor has been reduced to 20%. (The border of the mesh is visible due to an additional actor: the DrawBoundary actor.)

With the shift-options (Fig. 6) one can move the Mesh actor up or down (e.g. by changing the z-value) and so away from the Carpet actor. This way the

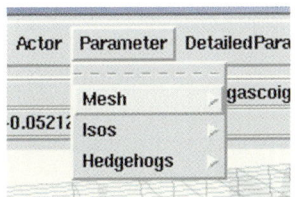

Fig.3. "Parameter" menu.

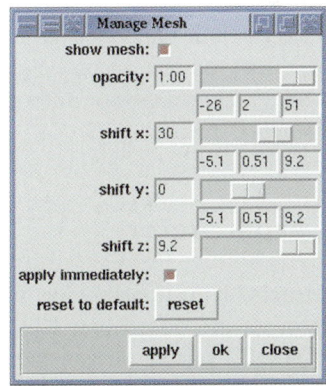

Fig.4. Mesh dialog box.

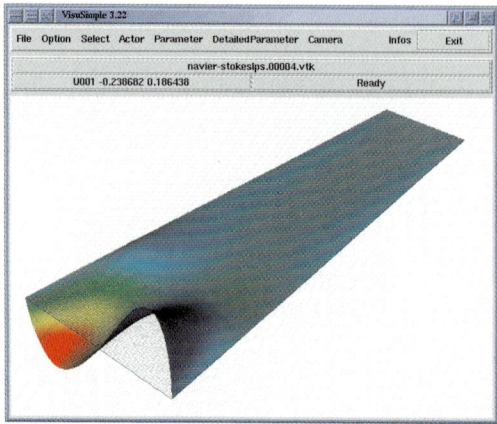

Fig.5. Mesh with reduced opacity.

Fig.6. Control for changing position.

mesh can still be provided in the viewing area but it will be less distracting. Almost all dialog boxes will have a grouping of graphical elements similar to the grouping of elements associated here with the "shift x" value (Fig. 6) . The setup of the elements is the following: the left element contains the value that is being setup. This element can be edited directly. The scale to the right of the element enables you to more easily change the value by sliding the scale button. More often then not though you will want to adjust the bounds or the step size of the scale. For this reason the editable elements above the scale are provided. They represent respectively the left bound, the resolution and the right bound of the scale. These values upon changing are applied immediately. You do not need to press enter oder wave the mouse around.

The Mesh2 Actor

The Mesh2 actor is in its function identical to the Mesh actor. It serves the purpose of acting as the second half of an incomplete axially symmetric geometry that can be flipped or mirrored with the help of the mirroring options in its "Parameter" menu.

4 The Carpet Actor

The Carpet actor is responsible for "putting a carpet" onto the wireframe grid that all datasets are based on. This is also known as texture mapping. An example of the Carpet actor was shown in the previous section.

Initially the carpet will be placed on top of the mesh. Alternatively one can scale the carpet to a certain height based on the selected scalar field. The

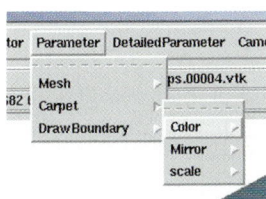 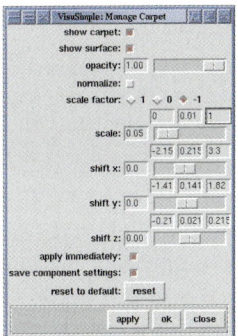

Fig. 7. "Parameter" menu.   **Fig. 8.** Carpet actor dialog box.

Carpet actor has options availible in both the "Parameter" menu (Fig. 7) and the "DetailedParameter" menu. The options provided in the "Parameter" menu are setting the color, mirroring the actor and scaling a certain value associated with the actor. The only difference for this actor is the scale option, here it specifically changes the height of the carpet.

The dialog box (Fig. 8) can be opened by selecting the "Mesh" item in the "DetailedParameter" menu. It offers options similar to those of the Mesh actor dialog box: turning an actor on/off, setting the opacity and shifting the actor around in the viewing area. Additionally these functions are provided.

- The texture on the Carpet actor can be turned off with the show surface option.
- The height of the Carpet actor can be normalized based on the minimum and maximum values of the selected scalar dataset.
- The scaling of the Carpet actor can be adjusted.
- The scaling of the Carpet actor can be quickly modified with factors -1,0 or 1. This way the Carpet actor can easily either be flipped around or made flat.

Normalizing the height of the carpet has the effect that VisuSimple will try to fit the height of the Carpet actor to the viewing area. It is still possible to scale the carpet though. Usually normalization will be turned on, but there are situations where one will want to see how the absolute values of the carpet diminish. In these cases simply turn normalization off.

The Carpet2 Actor

The Carpet2 actor is in its function identical to the Carpet actor. It serves the purpose of acting as the second halve of an incomplete axially symmetric geometry that can be flipped or mirrored with the help of the mirroring options in its "Parameter" menu.

5 The Isos Actor

The Isos actor will show the isolines of the selected scalar dataset. A wide range of options and values can be configured with this actor. Here are the options and the values they will initially be set to:

- **Position**: it will not be shifted in position.
- **Scaled height**: the height will not be scaled or normalized to any value (similar to a carpet). That means: it will be placed on top of the mesh and flat.
- **Number of lines**: it will consist of 10 lines.
- **Spread of lines**: the lines will be between the maximum and minimum values of the selected scalar dataset.
- **Opacity**: its opacity value will be set to 1.0.
- **Color**: it will be colored black.

As an example (Fig. 9) we show isolines with modified options. Most notably the height of the isolines have been set to indicate the scalar value. Similar to other actors the Isos actor has options available in both the "Parameter" menu and the "DetailedParameter" menu.

The actors dialog box (Fig. 10) offers many frequently occurring options: turning an actor on/off, setting the opacity and shifting the actor around in the viewing area. Additionally following functions are provided.

- The manner in which the isolines are to be shown.
- The number of isolines to be shown.
- The range of where isolines are to be shown.
- The Isos actor can be assigned a height to indicate the value of the selected scalar field. The height can be scaled and normalized based on the minimum and maximum values of the selected scalar dataset. The values

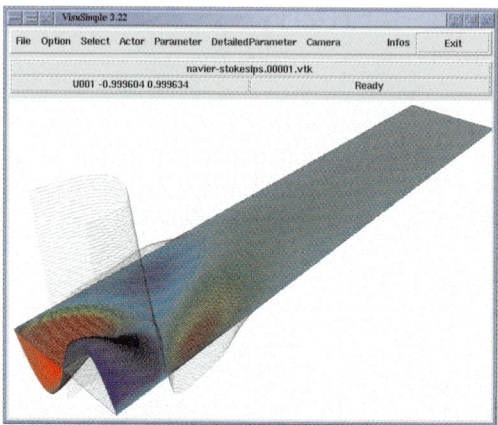

Fig.9. Example of Isos actor with height and reduced opacity.

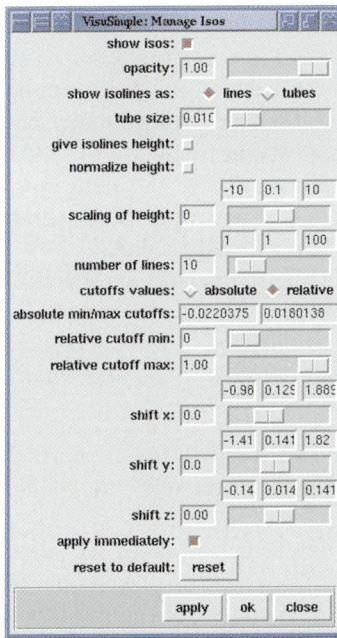

Fig.10. Isos actor dialog box.

for the range are relatively based values between 0 and 1. The relative numbers are internally scaled to the minimum and maximum values of the selected scalar dataset. This way one can for example specify that the isolines are to be shown between the 40% and 60% marks of the datasets extreme values.

The OtherIsos and OtherIsos2 Actors

These actors are in their function identical to the Isos actor. Additionally they let the user select different scalar fields to which the isolines are to be applied. This is useful when the isolines are to be shown for a fixed scalar component of the data and allows the user to change the selected component.

6 The Hedgehogs Actor

The Hedgehogs actor will show the vector field of the selected vector dataset. This is achieved by positioning glyphs - small arrows, pins or lines - at points on the mesh. Initially the glyphs are spread on all points, but it is also possible to only show a reduced number of glyphs spread out evenly on the mesh.

The size of the glyphs will also initially be set to indicate the length of the respective vector, but one can also set the actor to show all glyphs at a normalized length[7]. This is good when one for example only wants to see the general behavior of the vector field. Additionally one can specify that only vectors whose length is in a certain range are to be shown. This is very handy when for example visualizing an almost dormant flow in a crevice surrounded by otherwise quick flow. As an example (Fig. 11) we show the actor with normalized glyphs and a lower density value.

Similar to other actors the Hedgehogs actor has options available in both the "Parameter" menu and the "DetailedParameter" menu. A wide range of options and values can be configured in the dialog box (Fig. 12) for this actor. Here are some of the actor specific options and what they will initially be set to:

- **Normalized size**: the vectors will not be normalized to the length 1.
- **Density**: Set at 1-to-1. Alternatively one could for example only show one glyph for every 10 vectors.
- **Placement**: the arrows will be placed on all mesh points within the dataset. Alternatively they could be placed on a reduced number of points or spread out evenly independent of the mesh points.
- **Scaled size**: the size of vectors will be scaled to 1.0.
- **Cutoff**: all vectors between the minimum length and the maximum length will be shown.
- **Arrow type**: VisuSimple's own arrow type nr. 1 will be used.

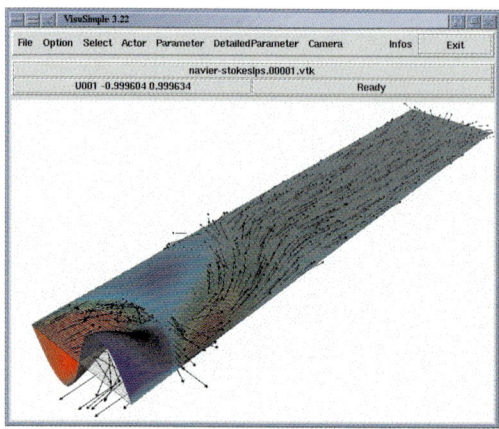

Fig.11. Example of Hedgehos actor with normalized glyphs at a lower density.

---

[7]The glyphs will still be scalable, albeit they will all have the same length.

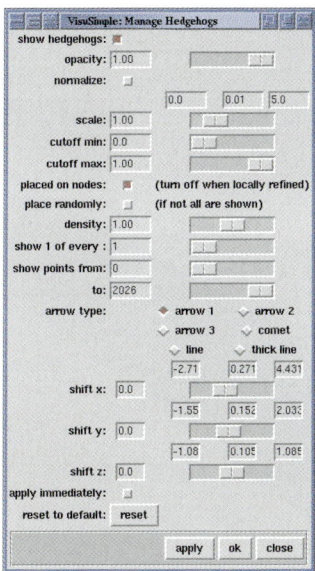

Fig.12. Hedgehogs actor dialog box.

7 The ScalarBar Actor

The ScalarBar actor will show a colorbar based on the value range of the selected scalar dataset. The actor can be set to change the title and show the scalar value range alongside the colorbar. As a graphic example (Fig. 13) we show the ScalarBar actor where some of the options have been changed.

Similar to other actors the ScalarBar actor has options available in both the "Parameter" menu and the "DetailedParameter" menu. The actor's dialog box (Fig. 14) offers many frequently occurring options: turning an actor on/off, setting the opacity and shifting the actor around in the viewing area. A wide range of options and values can be configured with this actor. Here are the options and what they will initially be set to:

- **Position**: it will be positioned on the right side of the viewing area. Its bottom left corner will be positioned at (82%,10%) of the width and height.
- **Scaled size**: the size of ScalarBar will be (20%,80%) of the width and height of the viewing area.
- **Shadowed**: the displayed text will have a shadow.
- **Number of ticks**: how many values should be shown for aligning colors to values.
- **Orientation**: the ScalarBar will be vertical.
- **Title**: the title will be that of the selected dataset.

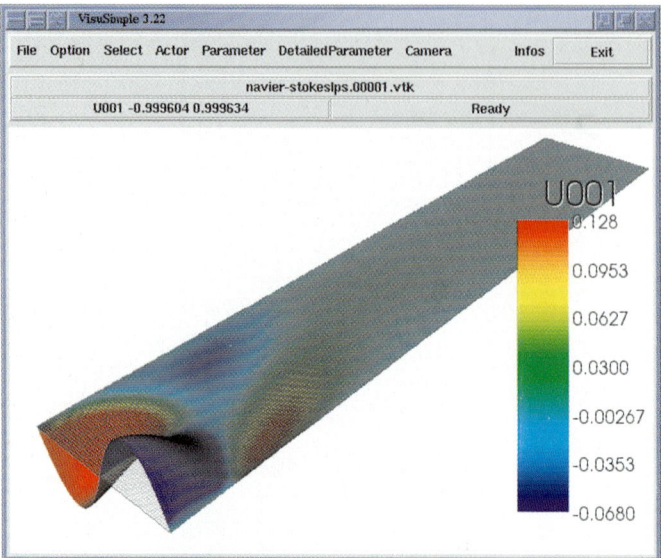

Fig. 13. Example of Scalabar actor.

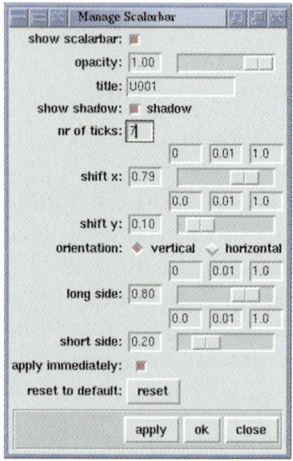

Fig. 14. Scalarbar actor dialog box.

8 The Pinpointer Actor

The Pinpointer actor is a tool for obtaining mesh information and scalar and vector data from the viewing area. The actor is represented by a small sphere that can either be moved around on the mesh or the carpet. Placement and movement of the actor is achieved by simply moving the mouse to the respective position on the mesh or carpet. The data and information to be

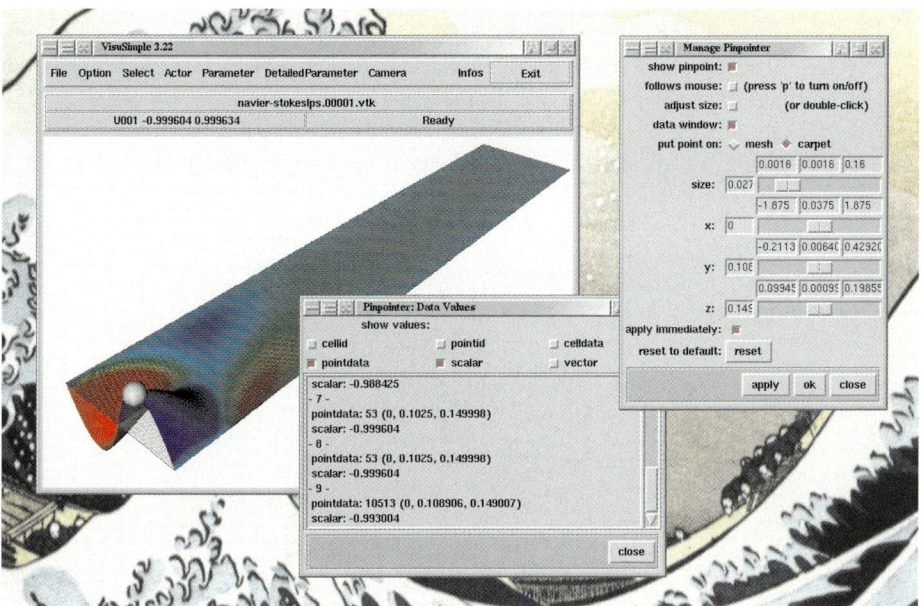

Fig.15. Pinpointer actor as a large sphere with dialog box and data window.

shown can be specified and viewed in the data window dialog box. This dialog box can be opened from the Pinpointer dialog box.

Here are the actor's options and the values they will initially be set to:

- **Position**: it will be positioned in the middle of the mesh.
- **Size**: its size will be adjusted to the size of the mesh.
- **Size adjustment**: its size will be adjusted to the size of the cells.
- **Attachement** : by default it is attached to the Mesh actor (and not the Carpet actor).
- **Data window**: the data window will not be shown and no information or datasets are selected.

As a graphic example (Fig. 15) we show the Pinpointer actor where some of the options have been changed. The size of the Pinpointer as purposely been enlarged to make it visible.

The Pinpointer actor can only be configured by means of the its dialog box. The dialog box (Fig. 15) offers the following options: turning the actor on/off, activating the Pinpointer actor to follow the mouse, opening the data window, attaching the actor to the mesh or carpet and manually positioning the actor. Additionally the dialog box offers an option of opening an additional dialog box: the data window. The data window (Fig. 16) is used to view requested information and data at the point of the Pinpointer actor.

The following information and data can be set to be viewed:

- The ids of the selected cell or node.
- The celldata of selected cell, i.e. the cell id and the ids of its points.
- The pointdata of the selected node, i.e. its id and coordinates.
- The scalar or vector values of the selected fields at the present point.

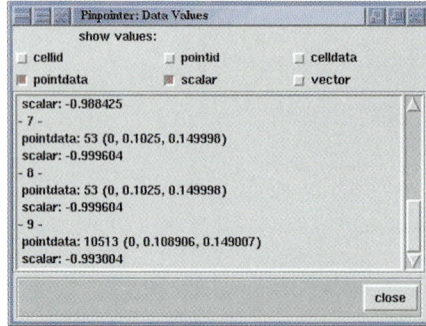

Fig.16. Pinpointer actors data window.

9 The NodeIds Actor

The NodeIds actor will display the id of each node at each node. This actor is only useful for situations where one needs to see the ids of many nodes simultaneously. If it is only necessary to see the ids at certain points, then using the Pinpointer actor would be a better approach.

That actor is faster and uses less memory in comparison to this actor. The NodeIds actor has two options available in the "Parameter" menu so that the displayed ids can be mirrored and the size of the ids can be scaled. As an example (Fig. 17) we show the actor when used with the file example.vtk from VisuSimple's examples directory.

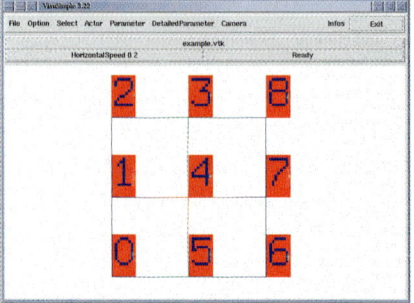

Fig.17. Example of NodeIds actor.

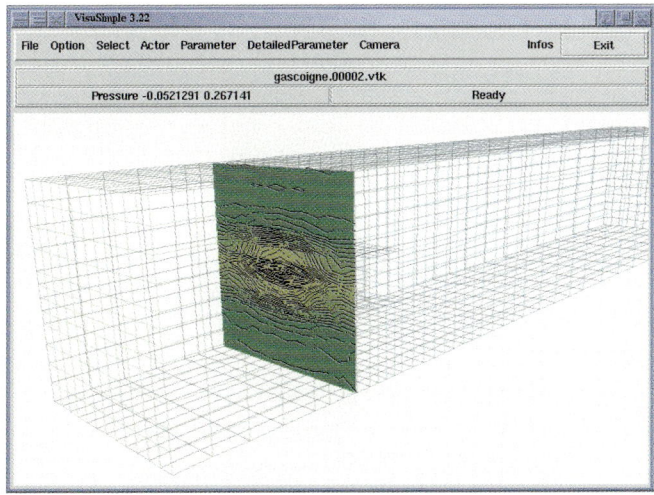

Fig.18. Example of Cutplane actor.

10 The CutPlane Actor

The CutPlane actor is meant to be used with 3D data. As the name suggests, it intersects the data with a plane. In this plane the selected scalar field will be shown. Technically it can also be used for 2D data. In this case the cut will result in a line. Some options that can be configured with this actor.

- **Plane**: which plane to be used can be chosen. Initially the xy-plane is set.
- **Position**: where within the bounding box the plane will be position. Initially it will be positioned at the center.
- **Isos**: if isolines are to be shown on the cutplane or not. Initially no.

As an example (Fig. 18) we show the CutPlane actor where some of the options have been changed.

11 The DrawBoundary Actor

The DrawBoundary actor will draw the exact boundary of the mesh. A similar actor is the Outline actor. The Outline actor though only draws a simple bounding box around the mesh. Only a minimum number options can be configured with this actor. Here are the options and the values they will initially be set to:

- **Position**: the boundary shown will be positioned where the boundary of the mesh is.
- **Opacity**: its opacity will be set to 0.5.

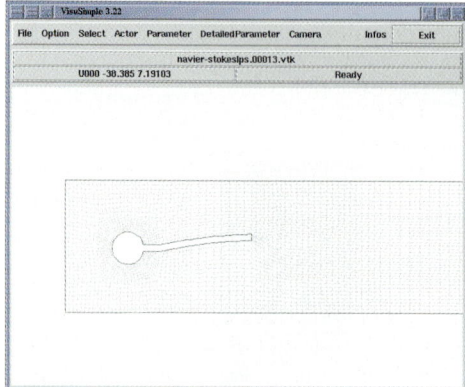

Fig. 19. Example of DrawBoundary actor.

As an example (Fig. 19) we show the actor in combination with Mesh actor. Take note that the inner boundary has also been recognized and "outlined".

12 The Outline Actor

The Outline actor will draw a bounding box around the mesh. In most cases its output will be equal to that of the DrawBoundary actor. The DrawBoundary actor differs from this actor by displaying the exact boundary of the mesh. No options can be configured with this actor. As an example (Fig. 20) we use the mesh from the previous section and show the actor in combination with Mesh actor. Take note that the inner boundary has *not* been recognized nor "outlined".

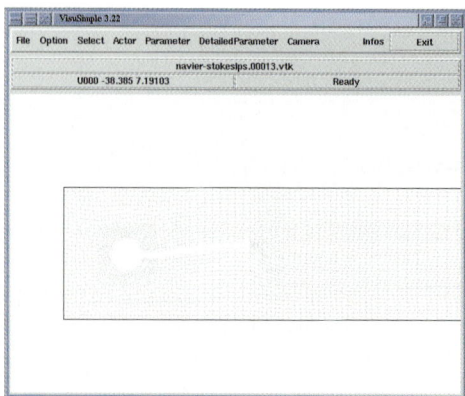

Fig. 20. Example of Outline actor.

References

1. R. Becker, M. Braack, R. Rannacher and T. Richter, "Mesh and model adaptivity for flow problems", in *Reactive Flows, Diffusion and Transport* (R. Rannacher, ed.), Springer, Berlin, 2005.
2. R. Becker, M. Braack, and T. Richter, "Parallel multigrid on locally refined meshes", in *Reactive Flows, Diffusion and Transport* (R. Rannacher, ed.), Springer, Berlin, 2005.
3. M. Braack and T. Richter, "Solving multidimensional reactive flow problems with adaptive finite elements", in *Reactive Flows, Diffusion and Transport* (R. Rannacher, ed.), Springer, Berlin, 2005.
4. R. Becker, M. Braack, D. Meidner, R. Rannacher and B. Vexler, "Adaptive finite element methods for PDE-constrained optimal control problems", in *Reactive Flows, Diffusion and Transport* (R. Rannacher, ed.), Springer, Berlin, 2005.
5. Will Schroeder, Ken Martin, Bill Lorensen "The Visualization Toolkit, An Object-Oriented Approach To 3D Graphics", Kitware, Inc. publishers, ISBN 1-930934-12-2
6. Kitware, Inc., "The Visualization Toolkit User's Guide", Kitware, Inc publishers. ISBN 1-930934-13-0
7. Paul Raines, Jeff Tranter, "Tcl/Tk in a Nutshell" O'Reilly, ISBN: 1-56592-433-9
8. Paul Raines, "Tcl/Tk Pocket Reference", O'Reilly, ISBN: 1-56592-498-3

Volume Rendering in Scientific Applications*

S. Krömker and J. Rings

Interdisziplinäres Zentrum für Wissenschaftliches Rechnen (IWR), Universität Heidelberg.

Summary. The increasing amount of 3D data in measurements and simulations needs a tool to visualize the interior in a convenient way. This can be done with so-called volume rendering techniques that allow for treating voxel-based data by assigning appropriate opacities to it.

1 Introduction to Volume Rendering

Volume rendering is based on the idea to cast rays through a 3D data set. The result of what encounters the rays is visualized in a 2D picture on the screen. Usually these data sets are scalar or vector valued data that have to be assigned colors and opacity values. These optical attributes enter a local light model that effects the casted rays.

The process of assigning such optical attributes is called classification. A segmentation according to the intensity histogram groups a certain range and assigns the same attributes or materials to this segment. This allows for pointing out characteristic features appearing as a peak in the histogram and having a certain meaning in the particular context. Large ranges of low densities or irrelevant data that obstruct the view can be turned totally transparent.

The skillful user is able to manipulate his 3D data in a way that the rendered image either looks quite realistic and/or reveals the important features in e.g. false color. Therefore a flexible user interface offers all these possibilities implemented in our program *Vrend*. The basic algorithm and first attempts with user interfaces were implemented during the PhD work done by Catalin Dartu [1].

---

*This work has been supported by the German Research Foundation (DFG) through SFB 359 (Project E) at the University of Heidelberg.

1.1 The Ray Casting Algorithm

A ray of light reflects from a surface or is transmitted through the surface due to the local optical properties of the specific material. Ray casting is a ray tracing algorithm but without scattering and iterative reflections.

The equation of transfer for luminesence (or radiance) L is a differential equation. It describes the attenuation along a ray $\mathbf{r}$ with respect to a certain angle ω as the sum of an extinction term as a sink, an emission function as a possible light source, and an integral consisting of a scattering kernel multiplied by the radiance.

$$\omega \cdot \nabla L(\mathbf{r},\omega) = \varphi(\mathbf{r})L(\mathbf{r},\omega) + \varepsilon(\mathbf{r},\omega) + \int_{S^2} k(\mathbf{r},\omega' \to \omega)L(\mathbf{r},\omega')d\omega' \qquad (1)$$

The volume rendering equation is derived from the integral form:

$$L(\mathbf{r},\omega) = e^{-\tau(\mathbf{r},\mathbf{r_B})}L_\mathbf{B}(\mathbf{r_B},\omega) + \int_{\Gamma(\mathbf{r},\mathbf{r_B})} e^{-\tau(\mathbf{r},\mathbf{r'})}Q(\mathbf{r'},\omega)d\mathbf{r'} \qquad (2)$$

$$\text{with} \qquad \tau(\mathbf{r},\mathbf{s}) = \int_{\Gamma(\mathbf{r},\mathbf{s})} \varphi(\mathbf{r'})d\mathbf{r'} \qquad (3)$$

The integrating factor $e^{-\tau(\mathbf{r},\mathbf{s})}$ simply is the attenuation along the ray with $\varphi(\mathbf{r})$ being the extinction coefficient, r_B denotes the boundary point on the ray. The term Q accounts for the sum of the source terms consisting of the emission function $\epsilon(\mathbf{r},\omega)$ and the integral with the scattering kernel $k(\mathbf{r},\omega' \to \omega)$. With the usual assumptions, which are first of all taking into account only a single scattering event, and then neglecting absorption between the light source and the voxel, this yields the volume rendering equation

$$L(x) = \int_x^{x_B} e^{-\int_x^{x'} \varphi(x'')dx''} \epsilon(x')dx' \qquad (4)$$

which can be evaluated for a sample spacing Δx with the rectangle rule along the viewing ray. It remains a summation of emission at the sample point x_i times the product of extinction happened so far:

$$L(x) = \sum_{i=0}^{n-1} e^{-\sum_{j=0}^{i-1} \varphi_j \Delta x} \cdot \epsilon_i \Delta x = \sum_{i=0}^{n-1} \epsilon_i \Delta x \cdot \prod_{j=0}^{i-1} e^{-\varphi_j \Delta x} \qquad (5)$$

where Δx is the spatial increment along the viewing ray and ϵ_i, φ_i are the emission and extinction functions at the ith sample point, respectively. The opacity α_i and the weighted color c_i of the sample x_i then are $\alpha_i \equiv 1 - e^{-\varphi_i \Delta x}$ and $c_i \equiv ((\epsilon_i/\alpha_i) \cdot \Delta x)\alpha_i$. This yields the standard volume rendering equation

$$L(x) = \sum_{i=0}^{n-1} c_i \cdot \prod_{j=0}^{i-1} 1 - \alpha_j. \qquad (6)$$

Shear-Warp Algorithm

The idea of shear-warp (see [4]) is an anamorphism that is a distorted projection: An image is distorted in such a way that it looks undistorted only when viewed from a specific angle. A shear in the stack of sliced images allows for the rays to go through the volume in parallel and to calculate an intermediate image by means of the volume rendering equation described in sec. 1 above. This image is warped to yield a final image of the cubic data on the 2D screen (see figure 1).

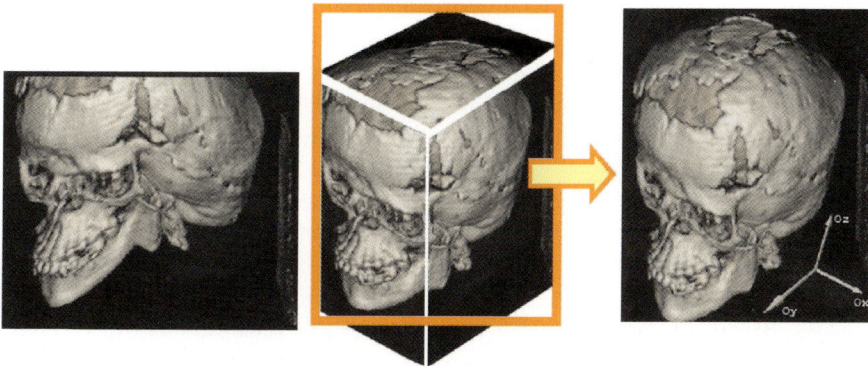

Fig. 1. Shear-warp illustration showing the intermediate image resulting from the ray casting through sheared slices (left) and the warp making the final image (right)

1.2 Perspective Shear-Warp

In virtual environments a perspective viewing transformation is indispensible. The stack of sliced images has to be not only sheared but also foreshortened according to the perspective diminuation. This yields an additional multiplication with a 4×4 perspective matrix in homogeneous coordinates. The local light models now give an appealing image in virtual environments. Unfortunately the algorithm is slow due to the computational work wich has to be done for each change of the viewing angle. Tracking the position of the observer forces a complete run of the algorithm for every movement of the observer's head.

A promising approach is to reduce the number of voxels and slices as long as the viewing angle changes rapidly. When it comes to rest, an image with best quality is rendered. Lots of intermediate images are rendered and mapped as semitransparent images onto slices of the cuboid perpendicular to the x-, y-, or z-direction, whichever is the best with regard to the viewing angle. For an implementation of such a texture based volume renderer into a geometry based renderer for stereographic vision see [6]. We did not follow this approach, first due to lack of demand for a combination of volume rendering in a virtual environment, and second due to a loss in quality of surface representations. Instead we concentrated on a practical user interface for a PC version of the program, keeping highest quality of the display.

2 Scanline Conversion

As shown in the middle part of figure 1, three images have to be computed due to the three faces of the cubic data set facing the viewer. In order to make the best use out of the shear part we arrange that the rays directly follow the scanline through the memory storage. We can improve the integration time by a better use of the cache functionality and by taking benefit of data coherence of the empty voxels inside the volume. By using object ordered ray casting we can change the ray integration order into a scanline ordered integration (see [5]).

After having done a segmentation due to the intensity histogram each voxel is assigned to its material properties by an entry in a lookup table. In case of total transparency this voxel does not contribute to the rendering equation. Therefore it makes sense to store the number of non-transparent voxels adjacent in memory that follow. We call this a run of length l.

2.1 Data Structure

We need two scan directions, e.g. x and y, since four of the six faces of a cuboid can be constructed from scanlines parallel to a given direction. The run length encoding is done only in the positive axis direction. For a given direction then each voxel has a value that represents the remaining length. This value together with the scan direction gives a complete definition of the current run. The access at any position inside the run is consistent. The amount of memory used to encode this type of scanline run-length is much lower than using an octree based run-length (see [1]).

3 Anisotropic Diffusion Filters

In many fields of applications we encounter noisy images due to bad recording. In order to make use of coherent data by means of run-length encoding the data itself are filtered before starting the segmentation and classification process. This comes along with a better image quality for noisy data with widely scattered segments.

Anisotropic diffusion filtering is able to enhances the structural features of the data. This denoising process can be described as a gradient flow which considers the structure tensor obtained by the gray value gradient of the unprocessed and pre-smoothed data. The task of finding the appropriate parameters can be done in a semi-automatic way with a user interface. The continuous filtering model, a coherence enhancing diffusion proposed by [7], is embedded into a more general class of filters which covers a lot of nonlinear diffusion models.

The general nonlinear diffusion model in image analysis is described by

$$\partial_t u(x,t) = div(D_u^p \nabla u(x,t)) + h(p, f(x), u(x,t))$$
$$on \ \Omega \times (0,T] \tag{7}$$

together with initial and boundary conditions

$$u(x,0) = f(x), \ (D_u^p \nabla u(x,t), n) = 0 \ on \ \partial\Omega \times (0,T].$$

$\Omega \subset \mathbf{R}^3$ is the area of the image, usually a cuboid with n being the outer normal to $\partial\Omega$. The unprocessed image $f = f(x)$ is the gray level function of a pixel x, and $u(x,T)$ is the gray level function of the processed image at time $T > 0$.

Adding a strictly monotone h, the so-called stopping function, in the right hand side of equation (7) prevents u from converging to the mean value for $t \to \infty$. We assume h to be in $\mathbf{C}^{0,1}$ with $h = 0$ for $u = f$ and monotonely increasing in $f - u$.

The symmetric and positive definite matrix $D_u^p \in \mathbf{R}^{3\times 3}$ depends on the derivatives of the gray level function u, and a set of parameters p to adapt the diffusion to the local structure. This happens by means of the so-called structure tensor.

Let $K_\sigma * u = u_\sigma$ be the convolution of u with a normalized Gaussian of standard deviation σ. The tensor product $\nabla u_\sigma \otimes \nabla u_\sigma = \nabla u_\sigma \nabla u_\sigma^T$ is a 3×3 matrix with eigenvalues $|\nabla u_\sigma|^2$ and 0 (twice). A diffusion correlated with this tensor would be able to smooth the data according to the local structures by enhancing the diffusion perpendicular to the eigendirection of $|\nabla u_\sigma|^2$.

Assuming the surface structure to have a preferred direction over a greater volume we look at the componentwise convolution $K_\rho * (\partial_{x_i} u_\sigma \partial_{x_j} u_\sigma), \ i, j \in \{1,2,3\}$. The structure tensor J then has the form

$$J := J_\rho(\nabla u_\sigma) = K_\rho * (\nabla u_\sigma \nabla u_\sigma^T)$$

$$= (\nabla u_\sigma \nabla u_\sigma^T)_\rho = \begin{pmatrix} a^2 & ab & ac \\ ab & b^2 & bc \\ ac & bc & c^2 \end{pmatrix}.$$

Let μ_1, μ_2, μ_3 be the characteristic values of J with $\mu_1 \geq \mu_2 \geq \mu_3 \geq 0$ and $\mathbf{v_1} = (v_1, v_2, v_3)$ be the normalized eigenvector to the eigenvalue μ_1. Since the eigenvectors of a symmetric matrix form an orthonormal basis, the eigenvectors $\{\mathbf{v_2}, \mathbf{v_3}\}$ of μ_2 and μ_3, respectively, span an orthogonal plane tangential to the direction of highest fluctuation of the gray value. This plane gives the direction of coherence.

Let the matrix $D_u^\rho = D_u^{\sigma,\rho} = D((\nabla u_\sigma \nabla u_\sigma^T)_\rho)$ be defined by

$$D_u^{\sigma,\rho} = \begin{bmatrix} \epsilon & 0 & 0 \\ 0 & \epsilon & 0 \\ 0 & 0 & \epsilon \end{bmatrix}, \quad if \ (\mu_1 - \mu_2)^2 = 0, \tag{8}$$

otherwise

$$D_u^{\sigma,\rho} = \epsilon v v^T + (\epsilon + (1-\epsilon)g((\mu_1-\mu_2)^2))\left[v^\perp\right]\left[v^\perp\right]^T, \tag{9}$$

with $g \in \mathbf{C}^\infty(\mathbf{R}^+)$ strictly positive and monotonely increasing. Furthermore it holds $g(0) = 0$ and $g(\infty) = 1$. For example the function can be of the form $g(d) = \frac{d}{1+d}$ or $g(d) = exp(\frac{-C}{d^m})$, $m \in \mathbf{N}$.

One filter step equals a time step in the diffusion equation. In each time step, the filter calculates a structure tensor. In order to enhance surfaces and to diffuse noisy spots or lines the anisotropic filter is designed with a big diffusion coefficient in directions of detected two dimensional coherent structures, and a small coefficient perpendicular to the detected surfaces. Thereby two dimensional structures are enhanced without blurring in direction perpendicular to the surface. Whereas noise is smoothed equally in all directions and thereby approaches the intensity of the neighboring voxels.

We implemented an interface with a slider for choosing the number of time steps to calculate. The filter may cause deformations to the surfaces since our algorithm does not contain a stopping function. For $t \to \infty$ the intensity tends to the mean value. The user should start with a small number of time steps in order not to lose important information.

The anisotropic diffusion filter can sometimes amplify random noise structures, so a median filter can be invoked as a prefilter in between filtering steps to reduce noise. The user can choose to have this prefiltering take place every 5 or 10 filter steps.

4 Application in Measurements

The technique of volume visualization is widespread in medicine but newly adapted to scientific imaging. In contrast to data from medical applications (CT-reconstruction or MR-imaging), the data is not a-priori voxel-based, i.e. there is no 3D-equidistant grid, and further data preparation (interpolation or filling of gaps via diffusion) is required.

Cross-plane laser induced fluorescence imaging is able to measure the distribution of specific chemicals in a turbulent flame (see figure 7 and for measurement and visualization techniques see [3] for details). The resulting images are slices through the flame front at very coarse intervals (see e.g. figure 3). The application of an anisotropic diffusion algorithm allows to fill areas of missing information by using the information in the source planes. In near proximity to the source planes and especially next to the cutting line volumetric data can be reconstructed with rather high resolution. 3D gradients of the measured quantity can then be extracted from the 3D data set for any point within the 3D volume with quality of the data increasing towards the cutting line.

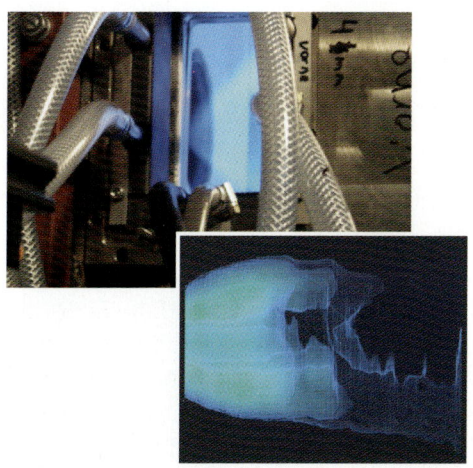

Fig. 2. Comparison of a picture of the real experiment with the volume visualization in similar colors

Based on this volumetric data set of intensities our volume rendering algorithm is used to visualize the measured flames with photorealistic virtual images. User supplied colors and material properties are given to those segments of the data which cause the main interest. Other intensities can be easily clipped by choosing them to be totally transparent. This is done interactively to explore the data by automatic segmentation based on the gray value histogram or by user defined segmentation.

The color comes into play via classification or user-supplied or automatic segmentation. In that process material properties like color and transparency are linked to ranges of the scalar data values. Multiple materials are generated and stored in lookup tables to be used by the ray tracing algorithm. Also the gradients of the intensities at the segment boundaries have to be stored, since these are the normals of the boundary used in the local light model. The user interface allows for defining these materials and mapping them to defined segments via graphical editing within a window showing the gray value histogram of the volumetric data set and the assigned material in a bar below. The surface model works with standard light models, therefore,

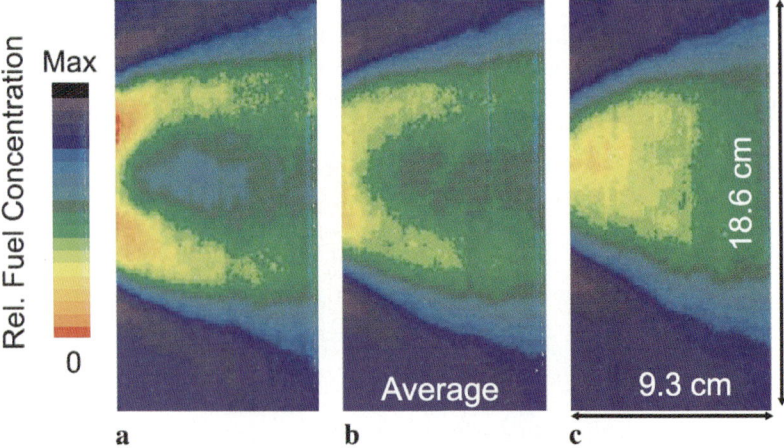

Fig. 3. Measurements of the average fuel distribution in a mixing chamber of a gas turbine in three parallel planes (A. Hoffmann, PCI, University of Heidelberg)

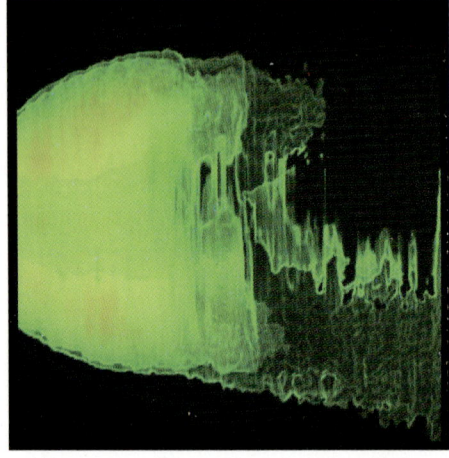

Fig. 4. Volume visualization of the averaged three-dimensional fuel distribution in the non-reactive swirl flow

specular or highlighted surfaces can be used to increase the realistic impression of the virtual image. Detection of these surface boundaries works by simply enhancing the topologic boundary of segmented objects (for details see [1]).

The more coherent the data, the longer are the lengthes of transparent segments, and the faster is the integration.

To illustrate the process see figure 2 for a comparison of the non-reactive swirl flow with the digital 3D object in similar colors. This object is based on laser induced fluorescence images as shown in 3. By the described interpolation a volumetric data set is produced (figure 4). The colorcode is chosen

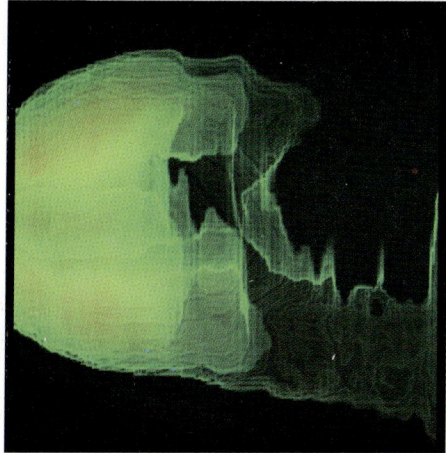

Fig. 5. Same volume visualization as in figure 4, after application of an anisotropic diffusion filtering

according to the false colors of the measurement. Data post-processing with coherence-enhancing anisotropic nonlinear diffusion filter was used in order to both improve the image quality and reduce the rendering time by making use of coherent data (see figures 5 and 6).

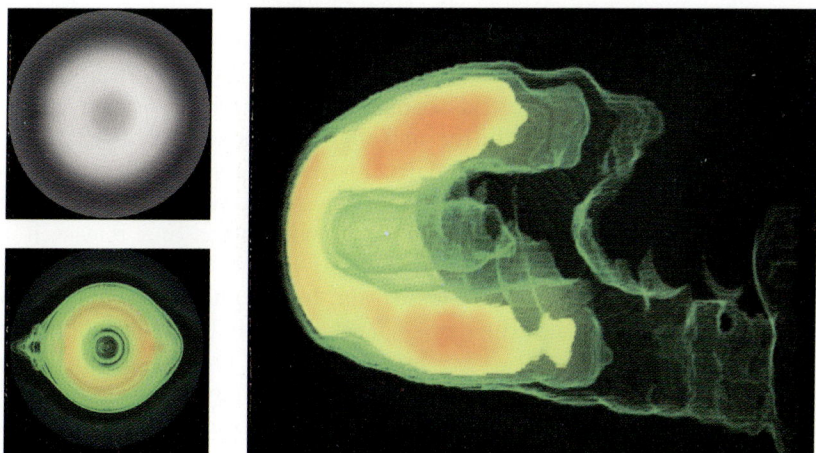

Fig. 6. Coherent 3D data can be transformed and moved efficiently. **Upper left:** cross section of interpolated intensities, **lower left:** segmented volume. **Right:** same data as in figure 5 but different viewing angle. Half of the data set is clipped and the color table and opacity is changed to point out areas of high temperature

5 User Interface

The handling of the program is described in detail in our tutorial (see [2]). An updated version can be found on our website where we also distribute the program. Here follows a summary of the main additions.

5.1 Segment Editor and Material Editor

By clicking the right mouse button within the window area, a local popup menu is prompted. Select the lowest item *Custom -Segments editor*, and a segments editor window is prompted.

In the bottom area, the segments editor has two options: to update surfaces and to update gradients of the volume data. Since you just loaded a new segmentation of the volume the two options of the segments editor have to be enabled by clicking with the left mouse button on the square buttons. Then click on the "Apply" button in order to update the volume parameters. You get prompted about the update in the window where you started the program. Any time you modify the view upon your volumetric object by using the rotation wheels in the left lower corner or the zooming wheel at the right edge of the window or by directly grabbing the object with the left mouse button, the draw style switches to a wire frame of the cube. As soon as you release the button, the program starts updating those sides of the volume facing the observer.

To modify the segmentation itself, move the mouse into the segments editor window and position it at a number shown in the area at the left lower corner. It corresponds to the white line in the histogram, showing the density on the x-axis and the frequency as height. Now press the middle mouse button. A green dot is added in the top area of the segments editor where the transparency is displayed as a bar diagram. In the bottom area you see a white dash and a default yellowish material assigned to this new segment. With a second mouseclick with the middle mouse button at the same position, you can delete it again. Pressing and holding the middle mouse button and release it at another position shifts the location. In this way a new segmentation can be positioned around the peaks of the histogram. Invoke the material editor by double clicking in the area of the new segment with the left mouse button.

Besides there have been two bigger additions to *Vrend 2.1*. One is the anisotropic diffusion filter introduced above that provides a filtering method which uses the structure tensor to ensure the filtering is mostly perpendicular to surfaces.

The second addition is auto segmentation which can be used to get a preliminary segmentation when exploring a new volume.

5.2 Auto Segmentation

Two different methods for automatical segmentation of the volume are implemented. The first method is the **Maximum method**. This method seeks

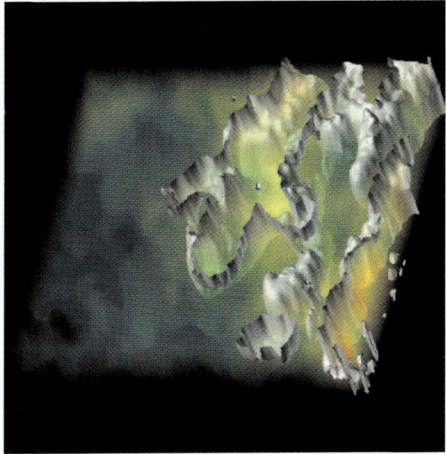

Fig. 7. Multiphase image showing a front of OH radicals (gray) within a temperature field (data courtesy of PCI, University of Heidelberg)

for peaks in the histogram. A peak marks the center of a segment under two conditions: The distance to a higher peak in the histogram must be larger than τ, where the *threshold* τ is a user defined input parameter. The second parameter, the *variance*, is a percentage value. It defines how much bigger the frequency of the tested peak must at least be with respect to the mean frequency in the histogram. This grasps only significant peaks for segmentation. A further user input parameter *offset* defines the width of a completely transparent segment at the beginning of the histogram. This offset is due to the fact that the low intensities often show some irregular noisy surrounding of the object of interest.

In the second method, the **Accumulation method**, the histogram is divided into n segments of equal area under the histogram curve. The algorithm starts at low densities and integrates the frequencies until a fraction $\frac{F}{n}$ has been added up, where F is the integrated frequency of the histogram. A segment boundary is set with respect to the user defined *minimum width* of a segment.

A menu in *Vrend* allows access to quick testing of these methods with varying parameters. Two different coloring schemes and an automatically set high transparency value allow fast exploration of a data set. Image features revealed this way then can be further explored manually.

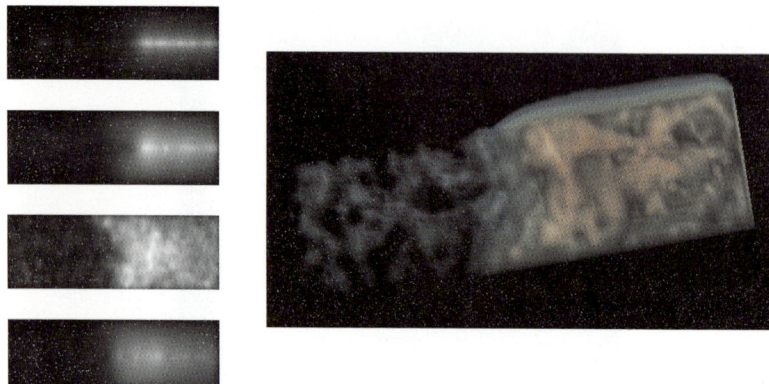

Fig. 8. Volume visualization in a turbulent mixing process from laser-induced fluorescence imaging. **Upper to lower left:** 2D sections as TIFF images (F. Zimmermann, PCI, University of Heidelberg), **right:** Volume rendered concentration due to laser image intensities

6 Data Format

The volume data format for *Vrend* consists of a header and a raw data part. The header contains six numbers. The first three hold the number of voxels in x, y and z-direction rerspectively. The other three parameters are the distances between two neigbouring voxels in x, y and z-direction. This allows the use not only of cubic grids but of a rectilinear grid.

The data part contains the density values for each voxel. It consecutively lists the densities for each voxel in each line of each slice of the volume. A density value d has the size of one byte ($0 \leq d \leq 255$).

6.1 TIFF Reader

A series of TIFF images can be read and transferred into a *Vrend* readable volume data file. The function to convert the images is invoked via a menu entry. It takes the image name prefix and the number of indices. The images are read by using methods of the standard libtiff libraries. In case that the TIFF images are colored, the red, green and blue value is averaged. This average, or respectively the gray value of a voxel is assigned as a voxel density to a slice in the volume data. The image stack is transferred into slices of the volume data set (see figure 8).

7 Distribution of the Program

The progam is actually distributed by direct request to the authors or as SGI IRIX 6.5 executable or PC Debian Linux 3.0 executable via the following website http://www.iwr.uni-heidelberg.de/groups/ngg/Vrend/.

Acknowledgment

We want to thank Catalin Dartu for implementation of the shear-warp and scanline conversion algorithm and for the first steps in direction of a user interface in the *vrend 1.0* version based on SGI openInventor libraries. We also thank Michael Winckler for discussion on auto segmentation and Alexander Dressel for helpful extensions due to filter analysis. And last not least we thank all previous users especially from medical imaging for valuable hints that improved the usability of the program.

References

1. Dartu, C.: Visualization of volumetric datasets. PhD Thesis, IWR, Universität Heidelberg (2001)
2. Dartu, C., Krömker, S.: VREND TUTORIAL, IWR Preprint 2001-18, Universität Heidelberg (2001)
3. Hoffmann, A., Zimmermann, F., Scharr, H., Krömker, S., Schulz, C.: Instantaneous three-dimensional visualization of concentration distribution in turbulent flows with crossed-plane laser induced fluorescence imaging. Appl.Phys. B, Lasers and Optics (2004)
4. Lacroute, P.G., Levoy, M.: Fast Volume Rendering using a Shear-Warp Factorization of the Viewing Transformation. ACM SIGGRAPH, Orlando, Florida (1994)
5. Lacroute, P.G.: Fast Volume Rendering using a Shear-Warp Factorization of the Viewing Transformation. Technical Report CSL-TR-95-678, Computer Systems Laboratory, Stanford University (1995)
6. Schulze, J.P., Niemeier, R., and Lang, U.: The Perspective Shear-Warp Algorithm in a Virtual Environment. IEEE Visualization '01 Proceedings, pp. 207-213 (2001)
7. Weickert, J.: Anisotropic Diffusion in Image Processing. Teubner Stuttgart (1998)

List of Contributors

Prof. Dr. Peter Bastian
AG Paralleles Rechnen
Interdisziplinäres Zentrum für Wissenschaftliches Rechnen (IWR)
Universität Heidelberg, Im Neuenheimer Feld 348
D-69120 Heidelberg, Germany
peter.bastian@iwr.uni-heidelberg.de

Prof. Dr. Roland Becker
Lab. de Mathématiques Appliquées
Université de Pau et des Pays de l'Adour
IPRA-Av. de l'Universite, BP 1155
F-64013 Pau Cedex, France
roland.becker@univ-pau.fr

Prof. Dr. Dr. h.c. Hans Georg Bock
AG Optimierung
Interdisziplinäres Zentrum für Wissenschaftliches Rechnen (IWR)
Universität Heidelberg, Im Neuenheimer Feld 368
D-69120 Heidelberg, Germany
bock@iwr.uni-heidelberg.de

Dr. Simina Bodea
AG Angewandte Analysis
Institut für Angewandte Mathematik
Universität Heidelberg, Im Neuenheimer Feld 294
D-69120 Heidelberg, Germany
simina.bodea@iwr.uni-heidelberg.de

Sebastian Bönisch
AG Numerische Mathematik

Institut für Angewandte Mathematik
Universität Heidelberg, Im Neuenheimer Feld 294
D-69120 Heidelberg, Germany
sebastian.boenisch@iwr.uni-heidelberg.de

Priv.-Doz. Dr. Malte Braack
AG Numerische Mathematik
Institut für Angewandte Mathematik
Universität Heidelberg, Im Neuenheimer Feld 294
D-69120 Heidelberg, Germany
malte.braack@iwr.uni-heidelberg.de

Dr. Thomas Carraro
Institut für Angewandte Mathematik
Universität Karlsruhe (TH)
D-76128 Karlsruhe, Germany
carraro@mathematik.uni-karlsruhe.de

Prof. Dr. Olaf Deutschmann
Institut für Technische Chemie und Polymerchemie
Universität Karlsruhe (TH)
Engesserstr. 20
D-76131 Karlsruhe, Germany,
deutschmann@ict.uni-karlsruhe.de

Thomas Dunne
AG Numerische Mathematik
Institut für Angewandte Mathematik
Universität Heidelberg, Im Neuenheimer Feld 294
D-69120 Heidelberg, Germany
dunne@iwr.uni-heidelberg.de

Dr. Ralf Markus Eiswirth
Physikalische Chemie
Fritz–Haber-Institut der Max–Planck-Gesellschaft
Faradayweg 4-6
D-14195 Berlin, Germany
eiswirth@fhi-berlin.mpg.de

Prof. Dr. Hans-Peter Gail
Institut für Theoretische Astrophysik
Universität Heidelberg
Albert–Ueberle-Straße 2
D-69120 Heidelberg, Germany
gail@ita.uni-heidelberg.de

Dr. Hai-Wen Ge
AG Mehrphasenströmungen und Verbrennung
Interdisziplinäres Zentrum für Wissenschaftliches Rechnen (IWR)
Universität Heidelberg
Im Neuenheimer Feld 294
D-69120 Heidelberg, Germany
heaveng@iwr.uni-heidelberg.de

Dr. Achim Gordner
AG Technische Simulation
Interdisziplinäres Zentrum für Wissenschaftliches Rechnen (IWR)
Universität Heidelberg, Im Neuenheimer Feld 368
D-69120 Heidelberg, Germany
achim.gordner@iwr.uni-heidelberg.de

Prof. Dr. Eva Gutheil
AG Mehrphasenströmungen und Verbrennung
Interdisziplinäres Zentrum für Wissenschaftliches Rechnen (IWR)
Universität Heidelberg
Im Neuenheimer Feld 294
D-69120 Heidelberg, Germany
gutheil@iwr.uni-heidelberg.de

Dr. Alexander Hanf
Physikalisch-Chemisches Institut
Universität Heidelberg, Im Neuenheimer Feld 253
D-69120 Heidelberg, Germany
ki7@ix.urz.uni-heidelberg.de

Andreas Hauser
AG Technische Simulation
Interdisziplinäres Zentrum für Wissenschaftliches Rechnen (IWR)
Universität Heidelberg, Im Neuenheimer Feld 368
D-69120 Heidelberg, Germany
andreas.hauser@iwr.uni-heidelberg.de

Prof. Dr. Vincent Heuveline
Rechenzentrum
Universität Karlsruhe (TH)
D-76128 Karlsruhe, Germany,
heuveline@rz.uni-karlsruhe.de

Johann Ilmberger
Institut für Umweltphysik
Im Neuenheimer Feld 229
D-69120 Heidelberg, Germany
johann.ilmberger@iup.uni-heidelberg.de

Dr. Oliver Inderwildi
University of Cambridge
Dept. of Chemistry
Lensfield Road, Cambridge CB2 1EW
United Kingdom
ori20@cam.ac.uk

Dr. Olaf Ippisch
AG Paralleles Rechnen
Interdisziplinäres Zentrum für Wissenschaftliches Rechnen (IWR)
Universität Heidelberg, Im Neuenheimer Feld 368
D-69120 Heidelberg, Germany
olaf.ippisch@iwr.uni-heidelberg.de

Prof. Dr. Dr. h.c. Willi Jäger
AG Angewandte Analysis
Institut für Angewandte Mathematik
Universität Heidelberg, Im Neuenheimer Feld 294
D-69120 Heidelberg, Germany
jaeger@iwr.uni-heidelberg.de

Priv.-Doz. Dr. Guido Kanschat
AG Numerische Mathematik
Institut für Angewandte Mathematik
Universität Heidelberg, Im Neuenheimer Feld 294
D-69120 Heidelberg, Germany
guido.kanschat@iwr.uni-heidelberg.de

Dr. Stefan Körkel
MATHEON
Institut für Mathematik
Humboldt-Universität zu Berlin
Unter den Linden 6
D-10099 Berlin, Germany
stefan@koerkel.de

Dr. Ekaterina Kostina
AG Optimierung
Interdisziplinäres Zentrum für Wissenschaftliches Rechnen (IWR)
Universität Heidelberg, Im Neuenheimer Feld 368
D-69120 Heidelberg, Germany
ekaterina.kostina@iwr.uni-heidelberg.de

Dr. Susanne Krömker
AG Visualisierung und Numerische Geometrie
Interdisziplinäres Zentrum für Wissenschaftliches Rechnen (IWR)
Universität Heidelberg, Im Neuenheimer Feld 368
D-69120 Heidelberg, Germany
kroemker@iwr.uni-heidelberg.de

Dr. Dirk Lebiedz
AG Reaktive Strömung
Interdisziplinäres Zentrum für Wissenschaftliches Rechnen (IWR)
Universität Heidelberg, Im Neuenheimer Feld 368
D-69120 Heidelberg, Germany
lebiedz@iwr.uni-heidelberg.de

Dr. Lubow Maier
AG Reaktive Strömung
Interdisziplinäres Zentrum für Wissenschaftliches Rechnen (IWR)
Universität Heidelberg, Im Neuenheimer Feld 368
D-69120 Heidelberg, Germany
luba.maier@iwr.uni-heidelberg.de

Dominik Meidner
AG Numerische Mathematik
Institut für Angewandte Mathematik
Universität Heidelberg, Im Neuenheimer Feld 294
D-69120 Heidelberg, Germany
dominik.meidner@iwr.uni-heidelberg.de

Dr. Erik Meinköhn
AG Numerische Mathematik
Institut für Angewandte Mathematik
Universität Heidelberg, Im Neuenheimer Feld 294
D-69120 Heidelberg, Germany
erik.meinkoehn@iwr.uni-heidelberg.de

Dr. Hoang Duc Minh
AG Optimierung
Interdisziplinäres Zentrum für Wissenschaftliches Rechnen (IWR)
Universität Heidelberg, Im Neuenheimer Feld 368
D-69120 Heidelberg, Germany

Dr. Sandra Nägele
AG Technische Simulation
Interdisziplinäres Zentrum für Wissenschaftliches Rechnen (IWR)
Universität Heidelberg, Im Neuenheimer Feld 368
D-69120 Heidelberg, Germany
sandra.naegele@iwr.uni-heidelberg.de

Priv.-Doz. Dr. Nicolas Neuss
AG Technische Simulation
Interdisziplinäres Zentrum für Wissenschaftliches Rechnen (IWR)
Universität Heidelberg, Im Neuenheimer Feld 368
D-69120 Heidelberg, Germany
nicolas.neuss@iwr.uni-heidelberg.de

Priv.-Doz. Dr. Karl Oelschläger
Institut für Angewandte Mathematik
Universität Heidelberg, Im Neuenheimer Feld 294
D-69120 Heidelberg, Germany
karl.oelschlaeger@urz.uni-heidelberg.de

Prof. Dr. Rolf Rannacher
AG Numerische Mathematik
Institut für Angewandte Mathematik
Universität Heidelberg, Im Neuenheimer Feld 294
D-69120 Heidelberg, Germany
rannacher@iwr.uni-heidelberg.de

Christian Reichert
Institut für Angewandte Mathematik
Universität Heidelberg, Im Neuenheimer Feld 294
D-69120 Heidelberg, Germany
christian.reichert@iwr.uni-heidelberg.de

Fereidoun Rezanezhad
Institut für Umweltphysik
Im Neuenheimer Feld 229
69120 Heidelberg, Germany
fereidoun.rezanezhad@iup.uni-heidelberg.de

Dr. Thomas Richter
AG Numerische Mathematik
Institut für Angewandte Mathematik
Universität Heidelberg, Im Neuenheimer Feld 294
D-69120 Heidelberg, Germany
thomas.richter@iwr.uni-heidelberg.de

Priv.-Doz. Dr. Uwe Riedel
AG Reaktive Strömung
Interdisziplinäres Zentrum für Wissenschaftliches Rechnen (IWR)
Universität Heidelberg, Im Neuenheimer Feld 368
D-69120 Heidelberg, Germany
riedel@iwr.uni-heidelberg.de

Jörg Rings
Institut für Meteorologie und Klimaforschung
Forschungszentrum Karlsruhe
Hermann-von-Helmholtz-Platz 1
76344 Eggenstein-Leopoldshafen, Germany
joerg.rings@imk.fzk.de

Christoph von Rohden
Institut für Umweltphysik
Im Neuenheimer Feld 229
D-69120 Heidelberg, Germany

Prof. Dr. Kurt Roth
Institut für Umweltphysik
Im Neuenheimer Feld 229
69120 Heidelberg, Germany
kurt.roth@iup.uni-heidelberg.de

Dr. Johann von Saldern
Physikalisch-Chemisches Institut
Universität Heidelberg, Im Neuenheimer Feld 253
D-69120 Heidelberg, Germany
johann.vonsaldern@web.de

Sven Sanwald
Interdisziplinäres Zentrum für Wissenschaftliches Rechnen (IWR)
Universität Heidelberg, Im Neuenheimer Feld 368
D-69120 Heidelberg, Germany
sanwald@iwr.uni-heidelberg.de

Dr. Andreas Schäfer
Panoratio Database Images GmbH
Theresienstr. 4
D-80333 München, Germany
andreas.schaefer@panoratio.com

Dr. Johannes Schlöder
AG Optimierung
Interdisziplinäres Zentrum für Wissenschaftliches Rechnen (IWR)
Universität Heidelberg, Im Neuenheimer Feld 368
D-69120 Heidelberg, Germany
j.schloeder@iwr.uni-heidelberg.de

Prof. Dr. Christof Schulz
Institut für Verbrennung und Gasdynamik
Universität Duisburg-Essen
Lotharstrasse 1
47057 Duisburg
christof.schulz@uni-duisburg.de

Prof. Dr. Jens Starke
Department of Mathematics
Matematiktorvet, Building 303
Technical University of Denmark
DK-2800 Kgs. Lyngby, Denmark
j.starke@mat.dtu.dk

Dzmitry Starukhin
Physikalisch-Chemisches Institut
Universität Heidelberg, Im Neuenheimer Feld 253
D-69120 Heidelberg, Germany
sdz@pci.uni-heidelberg.de

Dr. Christina Surulescu
Universität Stuttgart
Institut für Angewandte Analysis und numerische Simulation
Pfaffenwaldring 57
D-70569 Stuttgart, Germany

Dr. Irina Surovtsova
EML Research GmbH
Bioinformatics and Computational Biochemistry
Schloss-Wolfsbrunnenweg 33
D-69118 Heidelberg, Germany

Dr. Adela Tambulea
Universität des Saarlandes
FR 6.1 - Mathematik
Postfach 15 11 10
D-66041 Saarbrücken, Germany
tambulea@mx.uni-saarland.de

Dr. Steffen Tischer
Institut für Technische Chemie und Polymerchemie
Universität Karlsruhe (TH)
Engesserstr. 20
D-76131 Karlsruhe, Germany,
tischer@ict.uni-karlsruhe.de

Prof. Dr. Volker Schulz
FB 4 - Mathematik
Universität Trier
D-54286 Trier, Germany
volker.schulz@uni-trier.de

Prof. Dr. Ben Schweizer
Mathematisches Institut
Universität Basel
Rheinsprung 21
CH-4051 Basel, Schweiz
ben.schweizer@unibas.ch

Prof. Dr. Werner Tscharnuter
Institut für Theoretische Astrophysik
Albert–Ueberle-Straße 2
69120 Heidelberg, Germany
wmt@ita.uni-heidelberg.de

Daniela Urzica
AG Mehrphasenströmungen und Verbrennung
Interdisziplinäres Zentrum für Wissenschaftliches Rechnen (IWR)
Universität Heidelberg
Im Neuenheimer Feld 294
D-69120 Heidelberg, Germany
urzica@iwr.uni-heidelberg.de

Dr. Boris Vexler
Johann Radon Institute for Computational and Applied Mathematics
Austrian Academy of Sciences
Altenberger Strasse 69
A-4040 Linz, Austria
boris.vexler@oeaw.ac.at

Priv.-Doz. Dr. Hans-Jörg Vogel
Institut für Umweltphysik
Im Neuenheimer Feld 229
69120 Heidelberg, Germany
hjvogel@iup.uni-heidelberg.de

Dr. Markus Vogelgesang
AG Mehrphasenströmungen und Verbrennung
Interdisziplinäres Zentrum für Wissenschaftliches Rechnen (IWR)
Universität Heidelberg
Im Neuenheimer Feld 294
D-69120 Heidelberg, Germany

Priv.-Doz. Dr. Hans-Robert Volpp
Physikalisch-Chemisches Institut
Universität Heidelberg, Im Neuenheimer Feld 253
D-69120 Heidelberg, Germany
aw2@ix.urz.uni-heidelberg.de

Prof. Dr. Dr. h.c. Jürgen Warnatz
AG Reaktive Strömung
Interdisziplinäres Zentrum für Wissenschaftliches Rechnen (IWR)
Universität Heidelberg, Im Neuenheimer Feld 368
D-69120 Heidelberg, Germany
juergen@warnatz.de

Prof. Dr. Rainer Wehrse
Institut für Theoretische Astrophysik
Albert–Ueberle-Straße 2
69120 Heidelberg, Germany
wehrse@ita.uni-heidelberg.de

Prof. Dr. Gabriel Wittum
AG Technische Simulation
Interdisziplinäres Zentrum für Wissenschaftliches Rechnen (IWR)
Universität Heidelberg, Im Neuenheimer Feld 368
D-69120 Heidelberg, Germany
wittum@iwr.uni-heidelberg.de

Prof. Dr. Jürgen Wolfrum
Physikalisch-Chemisches Institut
Universität Heidelberg, Im Neuenheimer Feld 253
D-69120 Heidelberg, Germany
wolfrum@urz.uni-heidelberg.de

K. Wunderle
Institut für Umweltphysik
Im Neuenheimer Feld 229
69120 Heidelberg, Germany